* Israel claims Jerusalem to be its capital.
The United States recognizes Tel Aviv.

Human Geography

Human Geography

Cultures, Connections, and Landscapes

Edward F. Bergman

Lehman College of the City University of New York

Prentice Hall
Englewood Cliffs, New Jersey 07632

Library of Congress Cataloging-in-Publication Data
Bergman, Edward F.
    Human geography: cultures, connections, and landscapes/Edward F. Bergman
        p.   cm
    Includes bibliographical references and index.
    ISBN 0-13-121278-8
    1. Human geography.   I. Title
    GF41.B49   1995
    304.2–dc20   94-17511
                                                                            CIP

Acquisitions Editor: *Ray Henderson*
Editor in Chief: *Timothy Bozik*
Project Manager: *Joanne E. Jimenez*
Development Editor: *Elaine Silverstein*
Marketing Manager: *Leslie Cavaliere*
Editorial Assistant: *Pamela Holland-Moritz*
Copy Editor: *Karen Verde*
Interior Design: *Lorraine Mullaney*
Cover Design: *Amy Rosen*
Manufacturing Coordinator: *Trudy Pisciotti*
Page Layout: *Shari Toron*
Photo Research: *Teri Stratford*
Illustrators: *Maryland Cartographics Inc./Precision Graphics*
Cover Photo: *Pablo Bartholomew/Gamma Liaison*

Printed in the United States of America

10 9 8 7 6 5 4 3 2 1

ISBN 0-13-121278-8

Prentice-Hall International (UK) Limited, London
Prentice-Hall of Australia Pty. Limited, Sydney
Prentice-Hall Canada Inc., Toronto
Prentice-Hall Hispanoamericana, S.A., Mexico
Prentice-Hall of India Private Limited, New Delhi
Prentice-Hall of Japan, Inc., Tokyo
Simon & Schuster Asia Pte. Ltd., Singapore
Editora Prentice-Hall do Brasil, Ltda., Rio de Janeiro

*This book is dedicated
to the memory of my undergraduate geography advisor,
Professor Richard Hartshorne (1899–1992)
of the University of Wisconsin
and to my graduate geography advisor,
Professor W. A. Douglas Jackson
of the University of Washington*

▼
Contents

13

International Organization: Regional Organizations and Global Coordination 461

List of Maps

A Political Map of the World   *Front Endpaper*
A Physical Map of the World   *Back Endpaper*

WORLD MAPS

Preface

The cover of this book illustrates how different peoples originally developed their own local cultures in relative isolation from each other. They adapted their ways of life to the possibilities of the local environment, but, at the same time, they transformed the local environment as they exploited it to live. Humans are not passive, but interact with their environment. When peoples are introduced to new ideas and opportunities, they adopt some new customs and products, adapt others, and thereby create new cultural syntheses. Thus, cultures evolve and are modified as their members exchange goods and ideas with other cultures. In the process, people transform their own lives and landscape. Through history, new technologies of transportation and communication have accelerated these processes of cultural exchange, adaptation, and landscape transformation.

WHAT IS HUMAN GEOGRAPHY?

Many readers of this textbook may not have studied geography since grade school, where geography may have meant simply memorizing place names. Knowing where places are, however, is not all there is to human geography. Knowing place names is a tool for studying geography, just as counting is a tool for studying mathematics and reading is a tool for studying literature. In geography, as in any other field of study, we begin by gathering some basic information—in the case of geography, this is the *where*.

Once we know the names and locations of environmental features, people, and activities, then human geography can proceed to the challenging questions of significance: *Why* are people and activities located where they are? How do the features and activities at any one place *interact* to make that place unique, and what are the *relationships* among different places? What factors or forces *cause* these distributions of human populations and activities? And how and why are these distributions *changing*? Exploring these questions stretches your mind and imagination. Geography helps you understand current events, and it can provide you with information that is useful in deciding where to live, in seeking or building a career, and in forming your own position on political issues.

This book introduces the principal content of human and cultural geography, as well as the major

tools and techniques of the field. Human geography studies the geography of human groups and activities. Cultural geography focuses specifically on the evolution, interaction, and transformation of human cultures. Folk cultures, which are generally localized and preserve traditions, often contrast with popular culture, which generally embraces innovation and tends to diffuse rapidly. Geographers define principles and approaches for studying many aspects of each.

FOUR THEMES IN THIS TEXTBOOK

This textbook emphasizes four themes throughout the study of human geography.

- Geographers exercise both scientific and humanistic analytical skills;
- many basic principles of human geography can be studied and demonstrated in your own home town;
- human geography examines the interrelationships between humans and their natural environment; and,
- human geography is dynamic.

Geographers Exercise Both Scientific and Humanistic Analytical Skills

The scientific approach in geography employs rigorously logical tools and techniques to explain distributions. For example, if we want to study a phenomenon such as high agricultural productivity or national wealth, we first map it. Then we can compare that map with maps of other phenomena. If we find that the map of another phenomenon seems to coincide, we can investigate whether the factor on that other map causes high agricultural productivity or national wealth. For example, does the map of the world's most productive agricultural regions reflect the map of specific climates? Does the world pattern of manufacturing coincide with the location of raw materials? Throughout this book, the text and the captions to the maps often ask you to make such comparisons and investigate relationships. When you make these and other correlations, you will exercise analytical skills as you learn important information.

When making these correlations, you may be surprised to discover paradoxes. A paradox is a fact which is true, yet which seems opposed to "common sense" The small symbol calls your attention to paradoxes through the book, and you should consider this symbol a flashing yellow light. For example, you might assume that the most productive agriculture correlates with a certain climate and that the map of manufacturing reflects the locations of raw materials. In fact, neither correlation is absolutely clear. The text guides your investigation of why this is so.

This book contains many features to increase your awareness of *how* you are thinking and learning, as well as *what* you are thinking and learning. Awareness increases your command of analytical tools. A special essay discusses "The Scientific Method," and Focus Boxes analyze inductive and deductive methods of reasoning, statistical projections, and other intellectual tools. The text demonstrates the use of these tools in analyzing geographic material.

The use of scientific methods in geography is complemented by humanistic methods of analysis. Humanism emphasizes interpretation of subjective experience, including thoughts and feelings that may not be rational or logical. These thoughts and feelings influence human behavior, so the humanistic approach to geography enhances our understanding of why people organize territory, exploit the natural environment, and transform landscapes the ways they do. Humanistic geography incorporates insights drawn from psychology, economics, and other social sciences. This book quotes poems and novels, and it is illustrated with a diverse selection of photographs and even reproductions of paintings as "evidence" for your study of geography. This humanistic approach complements the scientific approach.

Human Geography is not Restricted to the Study of Exotic Lands and Peoples

The basic principles of human geography can be studied in your home town and even on campus. Where have new arrivals in your community come from, and why did they move? Where are local food crops and manufactured goods sold? How does distribution of the town newspaper diminish with distance from the town? Are the boundaries of city council districts manipulated to give one group an unfair advantage in elections? How many religions are represented by houses of worship in your town, and are these

centers of identifiable residential communities? Can you map the rents on commercial properties in your town, and how do these values reflect accessibility or, perhaps, perception of which neighborhoods are the most elegant?

Study of geography introduces a great range of careers, even though the professionals at work in many fields may not call themselves geographers. People studying or "doing" geography include planners designing new suburbs, transportation consultants routing new highways, advertisers targeting "junk mail" to zip codes where residents have specific income levels, diplomats negotiating treaties to regulate international fishing, and still more various occupations.

Human Geography Explores the Interrelationships Between Humankind and the Environment

The study of the Earth's climates, soils, vegetation, and physical features is called physical geography, and physical geography is itself a full course of study. Some knowledge of physical geography is essential to a human geography course, however, because physical geography sets the stage upon which humans act out their lives. A great deal of human effort is spent wresting a living from the environment, adjusting to it, or altering it.

Chapter 1 of this book, therefore, offers an overview of the Earth's physical environment. The discussion emphasizes how each element of the physical environment affects humankind—how the regularity of rainfall, for example, affects agriculture. The theme of human-environmental interaction then weaves through the book. Chapter 2 examines how people interpret and evaluate possibilities in their environment, and how they alter natural conditions. Any alteration in one of the Earth's natural physical and ecological systems, however, will trigger changes throughout the entire system. Because these systems are tremendously complex, many changes may be unexpected or even harmful. The information provided in Chapter 1 provides the foundation for appreciating these interelationships. For example, people "tame" rivers and build on their flood plains, but that construction can increase damages during occasional floods. People redistribute plants and animals around the Earth, but the relocated species can cause unexpected ecological damage. Chapter 9 discusses how cities change local climates; these changes affect human health. Pollution threatens the air we breathe, the water we drink, and the soil we till. Chapter 11 discusses urban waste management and recycling, as well as our growing demands for water. The epilogue, "Protecting the Global Environment," returns to this theme of human-environmental interaction. It emphasizes that environmental protection is one issue that today challenges all humankind, requiring international collaboration for a positive goal.

One of the unique features of this textbook is the number of its maps on which relief shading indicates the Earth's major mountain chains and plains. This shading consistently keeps before our eye the underlying variations in the configuration of the Earth's surface—the stage on which human activities take place. The shading is omitted from those maps where it might obscure the principle subject of the map.

Modern Geography is Dynamic

It is important to know the current distributions of people, languages, religions, cities, and economic activities, but none of these patterns is static. Transportation and communication ties among peoples and regions have multiplied, so social, political, and economic forces constantly redistribute activities. Maps of human activities reveal only temporary balances between forces for change and forces for stability. What happens *at* places depends more and more on what happens *among* places, and we can understand maps of economic or cultural activities only if we understand the patterns of movement that create them. Modern geography explores the forces at work behind the maps.

Each day's news reports events in which what happens is directly related to where it happens, and these events trigger changes in geography: A bountiful harvest in Argentina reduces food prices and thus improves the diet available to Africans. A new government in Africa redirects international alliances, economic links, and migration streams. U.S. movies and music diffuse U.S. language and culture around the world, while Americans themselves adopt new foods and words such as sushi. Protestant Christianity wins converts throughout Latin America; women are accepted into the priesthood, and governments open family planning clinics. Meanwhile, in some African and Asian countries, Islamic fundamentalists win political power and curb women's rights. These events remap world cultural, political, and economic landscapes. Today's dynamic geography doesn't just *exist*; it *happens*.

CONTEMPORARY ISSUES IN HUMAN GEOGRAPHY

▼▼▼▼▼▼▼▼▼▼▼▼▼▼▼▼▼▼▼▼▼▼▼▼▼

What you learn in your reading and in your geography classroom can help you better understand current events and form your own opinions on impor-

tant questions of the day. Each chapter of this book provides background material for understanding the news. Any number of topics in the book demonstrate this benefit, but here we might mention just two: the treatment of the topic of development, and treatment of the issue of gender justice.

Development

Each one of the Earth's billions of people aspires to superior material welfare, yet today many people live in conditions of deprivation. The world distribution of wealth and welfare does not coincide with the world distribution of raw material resources. If the possession of raw materials were the key to wealth, then Zaire and Mexico, for example, would count among the richest countries in the world, and Japan and Switzerland would count among the poorest. In fact, Zaire and Mexico are poor, and Japan and Switzerland are rich. An understanding of the resolution of this paradox is essential to understanding the world today.

Chapter 12 of this book, "National Paths to Economic Growth," goes beyond merely describing where there is wealth and where there is not. It weaves together a number of threads of understanding that have been introduced through earlier chapters. This makes clear the *why* of the *where*. Relevant economic principles are individually introduced in the text where appropriate, including adding value to raw materials, sectoral evolution, locational determinants for manufacturing, various nations' economic policies, and the patterns of world trade. Chapter 12 also suggests how and why the balance among the factors that support economic development is continuously shifting. New considerations arise virtually every day, such as the discovery of new resources, new technologies, and new governments with new policies. These changes redistribute advantages.

Economic development is only one aspect of development. Human development includes adequate nutrition, education, and political liberty. Chapter 4 analyzes the geography of health, and Chapter 6 explains the distribution of world food production and trade. Chapter 10 maps world education and freedom. In each case, the text not only describes the distribution, but explains it and suggests what factors might redistribute it in the future.

Gender Justice

Almost everywhere women are worse off than men are. They have less power, less autonomy, less money, more work, and more responsibility. The issue of discrimination against women—gender justice—is emerging as a major question of our time. This text does not separate this issue in an isolated chapter, but examines it as it plays a role in each topic through the book.

Chapter 4, on population, notes how discrimination against women can begin even against children in the womb. Some countries record higher rates of abortions of female fetuses than male fetuses. We see that in some countries females get less health care and nutrition. The result is unexpectedly low ratios of females to males in the overall national population. Discrimination against female farmers, discussed in Chapter 6, lowers agricultural productivity and the levels of nutrition in many countries. Chapter 8 examines the differing attitudes toward women taught by various religions. Ramifications of those teachings reach from the issue of priesthood for women into the treatment of women in national laws. Chapter 10 notes the variations in the percentages of males or females in school around the world, women's role in politics, and other issues of legal rights.

MAPS AND CARTOGRAMS

A great variety of maps illustrate this book. A special effort was made in this text to offer physical featuring on maps where these details help explain the geographic issue being illustrated. You will probably be familiar with traditional maps that illustrate distributions as mosaic patterns of color. On other maps, however, called *flow maps*, arrows represent movements of people or of goods, and on these the widths of the arrows often convey the quantity of each flow—the numbers of migrating Russians, for example (Figure 5-14, p. 174), or the quantities of oil moving in world trade (Figure 11-10, p. 397).

Many cartograms have also been specially designed for this book. A cartogram is a visual device much like a map, but on a cartogram physical distance is replaced with some other measure in order to convey a visual impression of the magnitude of something. Cartograms of such things as countries' populations, for example (Figure 4-2, pp. 126-127) or of countries' economic output (Figure 12-2, page 421) visually convey the relative population or wealth of different countries better than standard land area maps do. A population cartogram is especially useful when we want to illustrate human population dynamics or the conditions in which people live. The cartogram shows not only *where* people live in certain circumstances of health or wealth, but, at the same

time, it shows *what share* of the total world population lives in those circumstances.

A variety of other visual devices includes tables, bar graphs, and pie graphs, with which you are probably familiar, but also more complicated devices called histograms. A histogram shows the relationships among three values at once. Figure 12-6, for example (p. 425), shows what percentages of the labor forces of 20 countries are working in each of three sectors of the national economies. The captions for figures such as these have been carefully written to help you read these sophisticated images. Each contains a great deal of information.

Special care has been taken in preparing the captions for all illustrations. Each caption guides your eye through or into the figure. Many captions include questions to get you thinking about the information in the image. Many ask you to compare it with information shown on another image in the book in order to develop hypotheses of explanation.

FEATURES TO HELP YOU STUDY

Each chapter offers a number of features to help you study:

- *Key terms.* The key terms in each chapter are printed in **bold face** when each is first introduced. These terms are also listed at the end of each chapter along with the page on which each first appears. All key terms are defined in the Glossary at the end of the book. Some additional terms that may be unfamiliar to some readers but that are less central to the study of geography are printed in *italics*. These terms are defined carefully in the text.

- *Geographic Reasoning Boxes.* Geographic reasoning boxes discuss problems or pose questions for which there are no simple or clear answers. These require you to exercise your analytical skills to develop your own view points. Each geographic reasoning box includes questions to start you thinking about the material in the box.

- *Focus Boxes.* Focus boxes highlight individual problems or case studies of the topics in that chapter.

- *Summary.* Each chapter closes with a summary of its main points.

- *Questions for Investigation and Discussion.* At the end of each chapter you will find a list of questions about the subject of that chapter.

These are not review questions to test your reading. The answers to most of these questions cannot be found in the chapter. These questions encourage exploration beyond the text—into the library or out into the community. Some students may find these questions good suggestions for papers or research projects.

- *Additional Reading.* A short list of up-to-date readings provides sources for further information.

SUPPLEMENTS
▼▼▼▼▼▼▼▼▼▼▼▼▼▼▼▼▼▼▼▼▼▼▼▼▼▼▼

Visual Aids This text is supplemented by a package of figures selected from the text and made available in formats convenient for classroom use: color transparencies, black-and-white transparencies, and color slides. Two additional media supplements demonstrate how geography lends understanding to the daily news.

The New York Times Contemporary View Program A "custom edition" of *The New York Times* has been prepared especially for users of this book. It contains articles chosen to exemplify topics in the text.

ABC/ Prentice Hall Video Library This textbook is accompanied by a videocassette from ABC News. A variety of clips demonstrate again how many of the topics in the news reveal changing geography.

Instructor's Manual and Resource Guide The text is also accompanied by an *Instructor's Manual and Resource Guide.* This contains chapter-by-chapter suggestions for amplification of material in the text and for supplementary lecture topics, additional questions for investigation and discussion, additional bibliography, and a list of the relevant supplementary materials supplied.

Student Study Guide A *Study Guide* containing additional exercises, review, and a unique hypertext reference to key points in the text is available at a very low cost.

ACKNOWLEDGMENTS
▼▼▼▼▼▼▼▼▼▼▼▼▼▼▼▼▼▼▼▼▼▼▼▼▼▼▼

Countless colleagues, librarians, and generous individuals both in government and in the private sector helped with information for this text. Any errors in the text are the fault of the author, but I would like to thank the following people for their special help: Reider Brekke, Norwegian Trade Council; Amy Staples, National Museum of African Art; Laveta Emory, Smithsonian Institution; Pete Daniel, National Museum of American History; Priscilla Strain, National Air and Space Museum; Dan Beard, Commissioner of the U.S. Bureau of Reclamation; Pat O'Connell, Twin Cities Metropolitan Council; Clyde McNair, Agency for International Development; Dan Cintron, Japan External Trade Organization; Don McMinn, Covington and Burling; Greg Henniger, Hawkins, Delafield and Wood; Linda Carrico, U.S. Bureau of Mines; Barbara Mathe, American Museum of Natural History; Kerstin Erickson and Anne Stanley, European Union Information Service; William Usnik, American Express Travel Related Services; Leo Dillon, Office of the Geographer, U.S. Department of State; Jeffrey Hoover, Facts on File, Inc.; Roger Wieck and William Voelkle, the Morgan Library; John Rutter, U.S. Department of Commerce; Dr. Joseph E. Ryan, Freedom House; Alex de Sherbinin, Population Reference Bureau; Gary Cohen, *U.S. News & World Report*; Jerry Hagstrom, *The National Journal*; Robert Gaiser, Nassau County Planning Commission; Debbie Janes, U.S. Environmental Protection Administration; David Sackett, IBM; Adrian Smith, New York City Department of Parks; Anthony S. Hamer, Chase Manhattan Bank; Joseph W. Cherner, plus the many friends and colleagues whose photographs are used and credited in the text.

I also wish to thank my scholarly colleagues who provided thoughtful suggestions for improving the manuscript during its preparation. These include

Arthur Steele Becker, University of Nebraska at Kearney
Robert A. Bolding, Memphis State University
Fiona Davidson, University of Arkansas
James B. Kenyon, University of Georgia
James C. Hughes, Slippery Rock University
James Hathaway, Slippery Rock University
W.C. Jameson, University of Central Arkansas
Ann M. Legreid, Central Missouri State University
John C. Lowe, George Washington University
John Milbauer, Northeastern State University
Roger Miller, University of Minnesota Twin Cities
David J. Nemeth, University of Toledo
Thomas M. Orf, Prestonburg Community College
Madhu Rao, Bridgewater State College

Harry J. Schaleman, Jr., University of South Florida, St. Petersburg Campus
Paul Shott, Plymouth State College

I owe a debt of gratitude to many people. At Prentice Hall, Ray Henderson, Executive Editor, managed the project from its beginning stages through the journey to production. Ray Mullaney, Editor in Chief of Science and Math Development, with help from Elaine Silverstein, was responsible for the development of the book and contributed a very constructive review. Lorraine Mullaney, the designer, and Shari Toron, the page formatter, have given this new edition its fresh and appealing layout. The indefatigable Joanne Jimenez, Project Manager at Prentice Hall, brought all the pieces together into the completed volume you now hold. I have enjoyed working with all of these people, and I thank them.

Human
Geography

1 The Background to Human Geography

The grandeur of the Italian dolomite mountains overshadows this human settlement nestled in the Val de Funes.

Courtesy of Tony Stone Images.

When Russian astronaut Sergei Krikalev landed on Earth on March 25, 1992, after 313 days in space, he might have wondered if he had returned to the same planet. The physical landscape—the plains of Kazakhstan—had not changed. The political and cultural landscape, in contrast, had changed dramatically. Kazakhstan was no longer the Kazakh Republic of the Union of Soviet Socialist Republics. Instead, it had become an independent country and was busily developing its own economic and cultural life. The Soviet state, whose seal decorated Krikalev's spacesuit, was gone, and his home city of Leningrad had reclaimed its historic name of Saint Petersburg. Since Krikalev had rocketed into space, the Communist party had fallen from power in the USSR, the constituent republics had proclaimed their sovereignty, and the USSR had dissolved (Figure 1-1).

This episode demonstrates how rapidly geography can change. The USSR had exercised centralized political and economic power over one-sixth of the Earth's land surface, and its collapse is remapping cultural, economic, and political activities around the world. The disintegration of the USSR is one of the most spectacular events of recent years, but geography is never static. Distributions of people and of activities are continuously changing. This book will help you understand current distributions and also the forces that drive change.

Geography is one of the oldest fields of study. We may consider the first geographer to have been the first person who climbed a mountain or crossed a river to see what was on the other side, or who noticed that the environment or the people were different away from home and tried to understand why. He or she must have posed the basic questions still asked by geographers: Where is it, and what is it like there? After geographers have answered these two descriptive questions, they can go on to analyze questions of cause and meaning: Why is it there? How does that relate to other things? Geographers know that location matters, and that the events at different places relate to one another. Therefore, we can define **geography** as the study of the interaction of all physical and human phenomena at individual places and of how interactions among places form patterns and organize space.

THE DEVELOPMENT OF GEOGRAPHY

Curiosity about various lands and peoples was one of the forces that drove Alexander the Great (356–323 B.C.) on his career of military conquest, which took him from Greece all the way to India. Alexander's boyhood tutor was the philosopher Aristotle (384–322 B.C.), who emphasized that whenever possible a person should personally investigate to learn the truth. Alexander took a group of scholars along with him on his travels, and he is often credited with treating all the peoples he conquered as equals.

Every language has a word for the study that we call "geography," but the Greek word that we use, which means "Earth description," was the title of a book by the Greek scholar Eratosthenes (c. 275–195 B.C.). He was the director of the library in Alexandria, a city in Egypt founded by Alexander. The library was the greatest center of learning in the Mediterranean world for hundreds of years. No copies of Eratosthenes's book have survived, but we know from references by other authors that Eratosthenes accepted the idea that the Earth is round and that he even calculated its circumference with amazing precision. Eratosthenes also drew fine maps of the world as it was known to him. Hipparchus (180–127 B.C.), a later Greek scholar and director of the library, was the first person to draw a grid of imaginary lines on the Earth's surface to locate places precisely.

FIGURE 1-1

Russian astronaut Sergei Krikalev returned to Earth from a 10-month space voyage on March 25, 1992, to discover that the USSR, whose flag and symbol—CCCP—decorated his spacesuit, had dissolved. What other major changes occurred while he was in space? In 1994 Krikalev became the first Russian astronaut to fly aboard a U.S. space shuttle. (Courtesy of ITAR-TASS/Sovfoto/Eastfoto.)

2 *Chapter 1*

After the great age of Greek civilization gave way to the Roman era, the Roman Empire produced a number of geographical scholars and compilations of geographical learning. These culminated in the geography of Ptolemy, a Greek scholar who worked at the library in Alexandria between A.D. 127 and 150. Subsequently, in the long period between the fall of the Roman Empire and the European Renaissance of the fifteenth century, Western civilization accumulated little additional geographical knowledge. When Ptolemy's book reappeared in Latin in 1406, it was still taken as the most authoritative source of geographical knowledge.

Geography in the Non-European World

Outside Europe, however, geography made considerable advances. The expansion of the religion of Islam (discussed in Chapter 8) allowed travel and research across the wide region from Spain to India and beyond. Islam requires each believer to complete a pilgrimage to Mecca, in Arabia, at least once in his or her lifetime, if possible. This obligation encourages travel. Muslim scholars produced impressive texts describing and analyzing both physical environments and also the customs and lifestyles of different peoples.

China also developed an extensive geographical literature. The oldest known work dates from the fifth century B.C. *The Tribute of Yu* describes the physical geography and natural resources of the various provinces of the Chinese Empire. It interprets world geography as a nest of concentric squares of territory, ranging from the innermost zone of the imperial domain to an outermost zone inhabited by "barbarians." In A.D. 267, Phei Hsis made an elaborate map of the empire, and he is often called the father of Chinese cartography (map making). In the following centuries Chinese Buddhists wrote occasional accounts of travel to India to visit places sacred to the history of Buddhism, which had spread into China. Chinese maritime trade extended throughout Southeast Asia and the Indian Ocean, and it even reached East Africa, but the Chinese eventually withdrew from more extensive exploration, and few Chinese descriptive geography texts of this activity survive.

The Japanese and the Koreans also engaged in East Asian trade. In fact, the greatest world map of which we know that was produced before the European voyages of exploration was made in Korea in 1402. This map, the "Kangnido," was made by an unknown cartographer, but she or he drew on the combined knowledge of Korea, Japan, and China, including Islamic sources that were known to the Chinese. As a result, this map includes not only East Asia but also India, the countries of the Islamic world, Africa, and even Europe itself. It shows a far more extensive world knowledge than the Europeans had at the time.

The peoples of Africa, of the Western hemisphere, and of other areas may have had geographical accounts of the parts of the world known to them, but either these works have not survived or modern scholars have not yet fully inventoried and studied them.

The Revival of European Geography

Beginning in the fifteenth century, Europe's world exploration and conquest brought so much new geographical knowledge that Europeans were challenged to devise new methods of cataloguing or organizing all of it. Sir Walter Raleigh's book *The History of the World* (1614) is a milestone in European intellectual life because Raleigh for the first time assumed the uniformity of the human mind. The leading role in his book is taken by the collective person "mankind," with Raleigh as "his" biographer.

The Two Approaches to Geography Another key work of Western intellectual history was the book *General Geography* (1650) by Bernhard Varen ("Varenius" in Latin; 1622–1650), a German who taught at the University of Leiden, in Holland. Varen differentiated two approaches to geography. One he called "special geography." This was the description and analysis of any place in each of ten categories: (1) "the stature of the natives"; (2) employment; (3) virtues, vices, learning, and wit; (4) customs; (5) speech and language; (6) political government; (7) religion and church government; (8) cities and renowned places; (9) history; and (10) famous people and inventors. Today we call such an inventory analysis of any individual place a **regional geography**.

Varen defined a second approach to geography that he called "general geography." This was the discussion of universal laws or principles that apply to all places. Today we call this approach **topical** or **systematic** geography, and individual geographers concentrate on topics as diverse as the geography of soils (pedology), of life forms (biogeography), of politics (political geography), of economic activities (economic geography), and of cities (urban geography). The Association of American Geographers, a professional scholarly organization, today recognizes 51 topical specialties. Regional and topical geography are complementary. A regional study covers all topics in the region under review, whereas a topical study notes how a particular topic varies across regions.

Varen's book was a standard reference for over 100 years. Even Sir Isaac Newton (1642–1727) edited two editions in Latin, one of which was studied by students at Harvard in the eighteenth century.

The German philosopher Immanual Kant (1724–1804) taught that the study of geography was the natural complement to the study of history. History, he wrote, was the study of phenomena in their relations of time, or *chronology*, whereas geography was the study of phenomena in their relations of place, which he called *chorology*. Some scholars believe that Kant's 1784 essay "Ideas on a Universal History From the Point of View of a Citizen of the World" marks the beginning of the modern age of intellectual life.

The U.S. Congress created the title "Geographer to the United States" as early as 1781. Thomas Hutchins was the first person appointed to the position, and he carried out the first surveys of the national lands. Today the Office of the Geographer is within the Department of State, and the office publishes many useful and publicly available studies and maps.

We cannot fully understand the physical environment without understanding humankind's role in altering it, nor can we fully understand human life without understanding the physical environment in which we live. Therefore, modern geographic studies bridge the physical sciences and the social sciences.

The Two Subfields of Geography Either the regional approach or the topical approach in geography can be applied to either one of two subfields: **physical geography** or **human geography**. Physical geography studies the characteristics of the physical environment, and human geography studies human groups and activities. **Cultural geography**, a subfield of human geography, focuses specifically on the role of human cultures.

We cannot fully understand the physical environment without understanding humankind's role in altering it, nor can we fully understand human life without understanding the physical environment in which we live. Therefore, modern geographic studies bridge the physical sciences and the social sciences. This textbook of human geography, for example,

introduces aspects of physical geography, but in doing so, we always note how human life interacts with them. The book's approach to human geography is topical.

Geography through the Nineteenth Century

Through the nineteenth century scholarly geography developed as an ever more rigorous inquiry into the reasons for the distributions of phenomena. Geography courses were offered at leading American and European universities. Professor Karl Ritter (1779–1859) at the University of Berlin offered a program of study directed toward finding laws to describe how humans organize and use territory. Ritter is generally regarded as a founder of modern human geography.

The German explorer and naturalist Alexander von Humboldt (1769–1859) wove together virtually all knowledge of the Earth sciences and anthropology in his great multivolume book *Cosmos*. Von Humboldt is usually regarded as the co-founder, with Ritter, of modern geography. Von Humboldt's brilliant interrelating of the phenomena of physical nature and the world of humankind had an enormous intellectual impact in the United States (Figure 1-2). This was partly because European Americans had long thought of America as the "New World," a veritable "Garden of Eden" over which they had been given dominion.

George Perkins Marsh (1801–1882), an American scholar and diplomat, expanded on von Humboldt's theme of the interconnections between humankind and the physical environment. While Marsh served as U.S. ambassador to several Mediterranean countries, he was impressed by humankind's destruction of an environment that ancient authors had described as lush and rich. Marsh's book *Man and Nature, Or Physical Geography as Modified by Human Action* (1864) was one of the earliest key works in what would become today's environmental movement.

FIGURE 1-2

This portrait bust of the explorer and naturalist Alexander von Humboldt was the first statue of a non-U.S. citizen to be erected in New York City's Central Park. It went up in 1869, the one hundredth anniversary of von Humboldt's birth. Today he gazes across the street at the American Museum of Natural History, where statues of the explorers Lewis and Clark gaze back.
(EFB photo)

Why Study Geography?

Who is a geographer today? And why should anyone study geography? Today almost all of us are geographers, whether or not we call ourselves that. Geographers include people who plan the need for future schools, who develop real estate, who choose locations of new stores for retail chains, or who plan road tours for rock groups; journalists, diplomats, executives of multinational corporations, and farmers, who choose which crop to plant on the basis of world markets.

The information that any citizen needs to make an informed decision on an important question of the day is largely geographic. And as world events force rapid change upon us daily, the geography of our world changes too. Almost any topic of the day's news will bring about redistributions—new geographies—of people and activities. Studying geography will not always help you predict the future, but it will help you understand the forces behind these changes. Therefore, when changes come, you may be less surprised. Sometimes you will be able to see changes coming, make changes work for you, or even make the changes yourself.

Geography is one of the most diverse fields of academic study, so it offers an introduction to a unique breadth of eventual careers. A great variety of careers involve studying or "doing" geography, even though the professionals at work may not call themselves geographers. As you read through the topics in this text, ranging from newspaper distribution to agricultural supplies, from urban planning to the delineation of voting districts, and from multinational marketing of consumer goods to international treaties on environmental preservation, think always of the many ways you might develop a career analyzing or managing any of these activities.

Geographers' studies of where? what? when? why? and why there? necessarily prompt thoughts about "how ought?". Geographers' expertise can inform public debates on major issues of the day—not just to help us understand the world, but to help improve it.

BASIC CONCEPTS GEOGRAPHERS USE

In Western civilization formal geographic studies arose as a search for meaningful ways of organizing and finding relationships among the tremendous number of facts gathered to describe the globe. Some of the terms that geographers use in organizing these facts will be familiar to you, but others may carry specific technical meanings or be entirely new. The distribution of a phenomenon means its position, placement, or arrangement throughout space, and three terms can be used to define the distribution of a phenomenon under study: density, pattern, and dispersion (Figure 1-3). In addition, the terms **location** and **region** merit individual discussion.

Location

Location is a unique attribute of any place. Geographers distinguish between **absolute location** and **relative location**. The absolute location of a place, or its **site**, pinpoints a spot in terms of the global geographic grid. The relative location of a place, by contrast, called its **situation**, tells us where it is relative to other places.

The Geographic Grid The Earth is a sphere with a diameter of about 8,000 miles (12,875 km) and a circumference of about 25,000 miles (40,000 km). It rotates continuously about an axis which penetrates the Earth's surface at the North Pole and the South

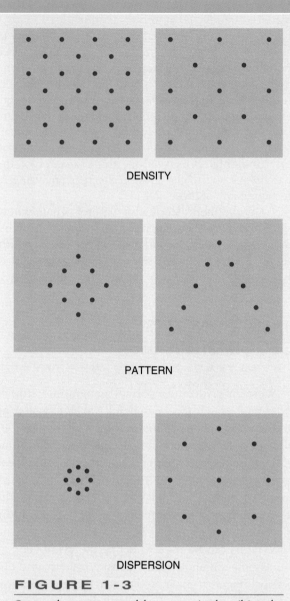

DENSITY

PATTERN

DISPERSION

FIGURE 1-3

Geographers use several key terms in describing the extent or the spread of observed features. The three elements common to all distributions are density, pattern, and dispersion.

Density is overall frequency of occurrence of a phenomenon, the number of items per unit of area.

Pattern is the geometric arrangement of the distribution of a phenomenon. Typical words used to describe patterns are *linear, centralized, uniform,* and *random.* Geographers note when the occurrence of a phenomenon forms a pattern, or when two different phenomena occur in the same pattern. The ability to recognize patterns is a learned skill that geographers exercise in common with art historians and others in the visual arts.

Dispersion refers to the extent of the spread of the feature relative to the size of the area being studied.

Pole. An imaginary plane perpendicular to the axis of rotation which passes through the Earth halfway between the poles is called the plane of the equator, and this plane intersects the Earth's surface at the Earth's imaginary midline, or "waist," called the **equator**. We can use the North Pole, the South Pole, and the equator as reference points for defining positions on the Earth's surface.

Latitude is angular distance measured north and south of the equator. As shown in Figure 1-4a, we can project a line from any given point on the Earth's surface to the center of the Earth. The angle between this line and the equatorial plane is the latitude of that point. Latitude is expressed in degrees, minutes, and seconds. There are 360 degrees (°) in a circle; 60 minutes (') in a degree; and 60 seconds (") in a minute. Latitude varies from 0° at the equator to 90° at the North and South Poles.

A line connecting all points of the same latitude is called a parallel because it is parallel to all other lines of latitude. Parallels are imaginary lines, and there can be an infinite number of them—one for each degree of latitude, or for every minute, or for any fraction of a second (See Figure 1-4b).

Longitude is angular distance measured east and west on the Earth's surface. Longitude is also measured in degrees, minutes, and seconds. It is represented by imaginary lines, called meridians, extending from pole to pole and crossing all parallels at right angles (See Figure 1-4c). Meridians are not parallel to one another, except where they cross the equator. They are farthest apart at the equator, become increasingly close northward and southward, and finally converge at the poles.

The **prime meridian**, the meridian from which longitude is measured, was chosen by an international conference in 1884. It is the meridian passing through the Royal Observatory in Greenwich, England, just east of London. Longitude is measured both east and west of the prime meridian to a maximum of 180° in each direction. Thus, the location of any spot on the Earth's surface can be described with great precision by reference to its latitude and longitude. If we say that the dome of the U.S. Capitol in Washington is located at 38°53'23" N. Lat. and 77°00'33" W. Long., we have described its position within about ten steps (Figure 1-5).

The same international conference which agreed on the prime meridian also agreed to divide the world into 24 standard time zones, each extending over 15 degrees of longitude. Today's world time zone map does not consist of exact vertical stripes; it bends for the convenience of individual countries (Figure 1-6).

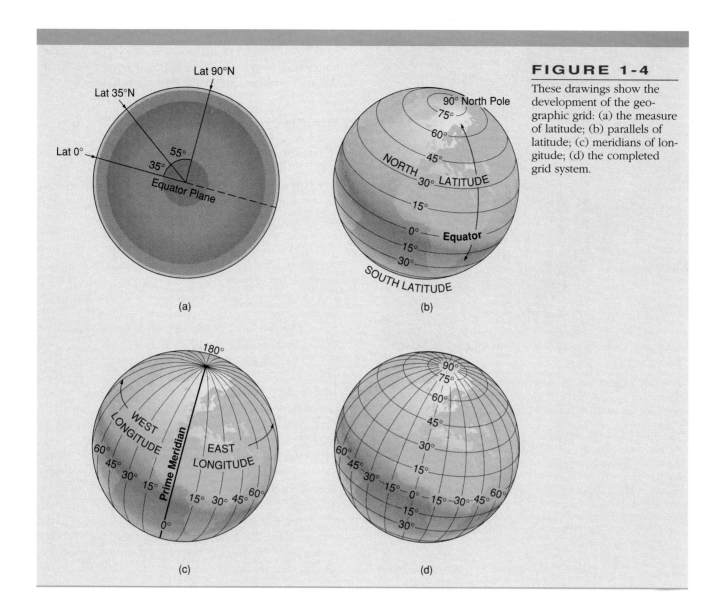

FIGURE 1-4

These drawings show the development of the geographic grid: (a) the measure of latitude; (b) parallels of latitude; (c) meridians of longitude; (d) the completed grid system.

Relative Location The relative location of a place, or its situation, describes its location relative to other places. Relative location influences accessibility, indicated by such terms as nearer and farther, easier or more difficult to reach, between, and on the way or out of the way. Knowledge of the relative situation of a place helps us understand how it interacts with the rest of the world.

The study of relative location suggests that distance can be measured absolutely, in terms of miles or kilometers, but it can also be measured in other ways. It can, for instance, be measured in time. Most people use *time distance* measurements in their daily lives without realizing it ("The store is about 20 minutes from here."). Sometimes it can be quicker to reach a place that is far away than to reach another place that is, in absolute distance, closer. This may be the result of rugged **topography** (surface relief) in one direction, direct or winding routes, or different methods of transportation (Figure 1-7).

Geographers also frequently measure *cost distance* between two places. Transportation over land is usually more expensive than transportation over water, so two seaports that are far apart in absolute distance may be close in cost distance. They may even be closer to each other, in cost distance, than either is to inland cities. That explains why, in the past, many political units were built up around the shores of bodies of water. Classical Greece, for example, consisted of cities surrounding the Aegean Sea, and the Roman Empire grew around the Mediterranean Sea (the word *mediterranean* means "middle of the world").

The Background to Human Geography   **7**

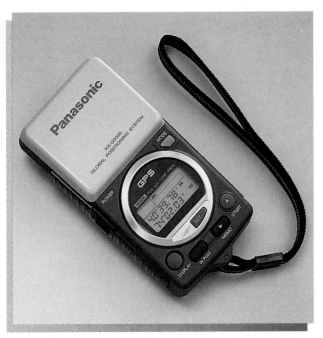

FIGURE 1-5

This tiny hand-held Global Positioning System (GPS) device is a spinoff of the satellite navigation system developed for the military. Twenty-four orbiting satellites are constantly transmitting signals that indicate their positions and the time of the transmission. At the press of a button, the receiver reads three satellites' positions and determines the time elapsed between transmission and reception of the three signals to figure its distance from each of the satellites. From those readings, it plots exactly where you are. These devices are installed in aircraft, yachts, and even cars equipped with computer-generated maps, so travelers never need be lost. Where is the GPS in this photograph? (Courtesy of Panasonic Inc.)

(The Statue of Liberty in New York Harbor)

No matter how we choose to measure distance, distance forces us to make some effort to overcome it by paying some cost to move or transport items. We call that effort or cost the **friction of distance**.

The Concept of a Region

A **region** is a territory that exhibits a certain uniformity. Of all the features that can be found in any landscape, a geographer chooses only a few specific criteria and then maps those criteria to define a region. The criteria chosen may be either physical or cultural. Sometimes the criteria chosen are strictly descriptive. We call those *formal regions*: a region of mountains, for instance. Other regions can be defined on the basis of how they are organized. We call these *functional regions*. Examples are the suburban Denver region, which supplies commuters into downtown Denver, or any governmental jurisdiction, such as a county or state (see Figure 1-8).

The choice of criteria to define a region takes place in the geographer's mind. Therefore a region is a concept, an abstract idea. The geographer's concept of a region is similar to the historian's concept of a *period*. Both time and space are continuous. How you divide them up into units of analysis depends on your purpose. Art historians define *styles*, sociologists define *societies*, and many other scholarly fields define their conceptual units of analysis.

Some regions are demarcated clearly. Among regions based on natural phenomena, for instance, watersheds can usually be mapped distinctly, as can a region defined as "all territory over 1,000 feet (305

The World Time Zones

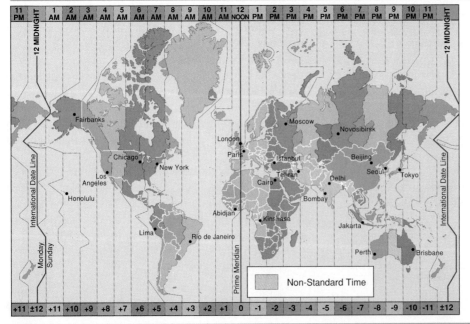

FIGURE 1-6

Each of the Earth's 24 time zones represents 15° of longitude or one hour of time. The number in each time zone on this map represents the number of hours by which that zone's standard time differs from Greenwich mean time, plus or minus. The colored stripes indicate the standard time zones; gray areas have a fractional deviation. These irregularities in the zones are dictated by political and economic factors.

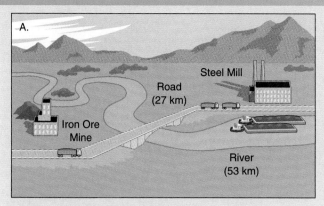

A.

Iron Ore Mine

Road (27 km)

Steel Mill

River (53 km)

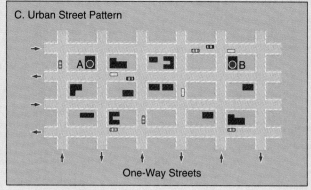

C. Urban Street Pattern

A○ ○B

One-Way Streets

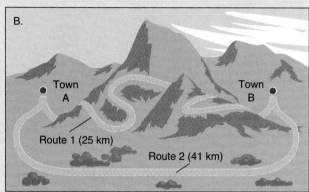

B.

Town A

Route 1 (25 km)

Route 2 (41 km)

Town B

D. Different Elevations

B

A

FIGURE 1-7

Drawings A and B illustrate how the shortest distance between two points may not be the cheapest or fastest distance. Transportation by water is almost always cheaper than transportation over land. Drawings C and D illustrate how the time or cost distance from any point A to a point B may not be the same as from point B to point A. In ways like these, cost distance and time distance seem to warp absolute distance.

Formal and Functional Regions

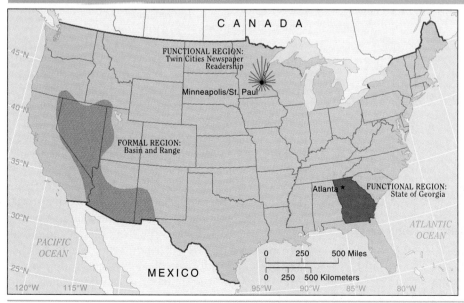

FIGURE 1-8

A geographer could define almost any number of formal or functional regions within the borders of the United States. Formal regions can be described as homogeneous in some way. Functional regions are defined by some organizing activity.

meters) above sea level." Among regions defined on the basis of human activity, national borders are usually clearly marked.

In other cases, however, the regions that geographers define are less distinct in the landscape. Later in this chapter, for example, we will map regions of different climates. On Earth those climatic regions are not sharply defined; they imperceptibly merge into one another. The geographer's definition and trained discrimination draw the line between one climatic region and its neighbor.

Regions defined by cultural phenomena often merge or overlap as well. The people who live between two big cities, for example, may listen to the radio station broadcasting from City A or City B or to both. If a geographer wants to define the market regions of the two radio stations, he or she might draw a line dividing households that listen to the station broadcasting from City A more than 50 percent of the time from those that listen to the station broadcasting from City B more frequently.

Sometimes people confuse regions based on natural phenomena with regions based on cultural criteria. For example, many people refer to Africa as a region, but what element of uniformity does it exhibit? Africa is a continent, and some people unconsciously accept that as sufficient to be considered a region. To the ancient Romans, however, northern Africa and southern Europe formed a single region focused on the Mediterranean Sea, and that region was distinct from sub-Saharan Africa, which was another region (see Figure 1-9). The African continent is not in fact homogeneous by any criteria of physical environment or by the history or cultures of its many peoples. Whenever anyone refers to any territory as a region, we must be sure that we know and agree on exactly what territory the region includes and what the criteria of homogeneity are.

PORTRAYING THE EARTH

▼▼▼▼▼▼▼▼▼▼▼▼▼▼▼▼▼▼▼▼▼▼▼▼▼▼

Maps are probably the most common and useful tools used in geographic studies. A map is a graphic representation of any area on which only selected data are shown. Maps can show distance, direction, size, and shape, as well as a great variety of other information, such as road patterns, the distribution of population, and any of an infinite number of other facts or combinations of facts. Mapping the phenomena under study is essential to understanding distributions and spatial relationships.

FIGURE 1-9

Roman ruins haunt the North African desert. These, in Morocco, demonstrate that to the ancient Romans the African continent was not a region. The lands surrounding the Mediterranean Sea formed one region, and that region was bounded to the south by the Sahara Desert. (Courtesy of the Moroccan National Tourist Office.)

Map Scale

Map makers must reduce the world to manageable proportions. The relationship between length measured on the map and the corresponding distance on the ground is called the scale of the map. Knowing the scale of a map makes it possible to measure distance, determine area, and make comparisons of size. Scale is usually shown on a map in one of three ways: graphic, fractional, or verbal (see Figure 1-10). A *graphic scale* allows us easily to determine approximate distances on the ground by measuring them directly on the map. A *fractional scale* compares map distance with ground distance by proportional numbers expressed as a fraction or ratio. A *verbal scale* will tell us, for example, that "one inch equals six miles."

The adjectives "large" and "small" are often used to describe map scales. A large-scale map has a relatively large fraction, which means that its denominator is small: 1/10,000, for example. A large-scale map portrays only a small portion of the Earth's surface but in considerable detail. A small-scale map, by contrast, has a large number in the denominator: for example, 1/10,000,000. A small-scale map can portray a large area, but in limited detail.

Map Projections

It is impossible to render the spherical Earth on a flat sheet of paper without some distortion, just as it is impossible to peel an orange and flatten the peel without tearing or wrinkling it. A **map projection** is a method of portraying the Earth or any portion of it on a flat map. Hundreds of projections have been devised, but cartographers must always choose between making a map on which shapes are accurate (called a *conformal map*), or one on which sizes are accurate (called an *equal-area*, or *equivalent map*; see Figure 1-11). No map can be both conformal and equivalent. Conformal maps are good for finding direction, but poor for conveying scale, while equal-

area maps are good for showing the relative sizes of countries, but poor for showing the actual shapes of countries. Most maps represent some compromise between conformality and equivalence.

Appendix 1 describes the methods for making different projections, but here we will examine briefly the advantages and disadvantages of a few popular projections.

Advantages and Disadvantages of Various Map Projections

The advantages and disadvantages of any projection can be exemplified by one popular example, the Mercator projection, devised in 1569 by the Flemish geographer and cartographer Gerardus Mercator (A on Figure 1-11). Mercator produced some of the best

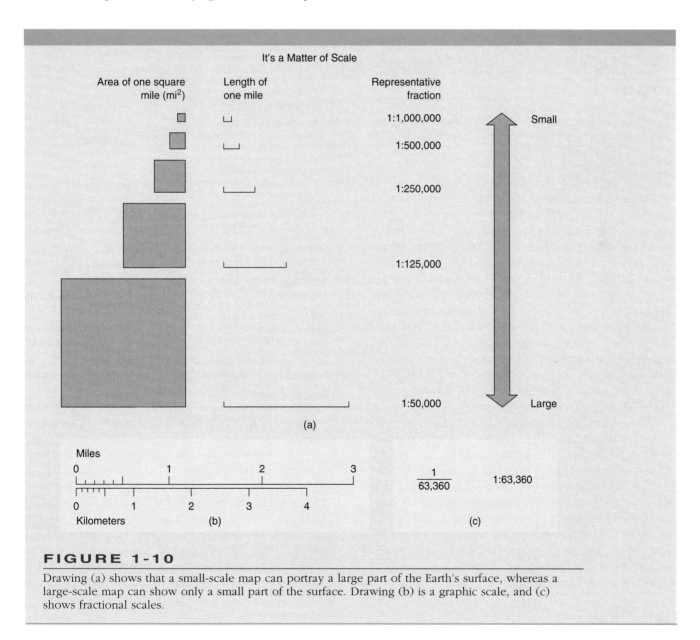

FIGURE 1-10

Drawing (a) shows that a small-scale map can portray a large part of the Earth's surface, whereas a large-scale map can show only a small part of the surface. Drawing (b) is a graphic scale, and (c) shows fractional scales.

Conformal and Equivalent Projections

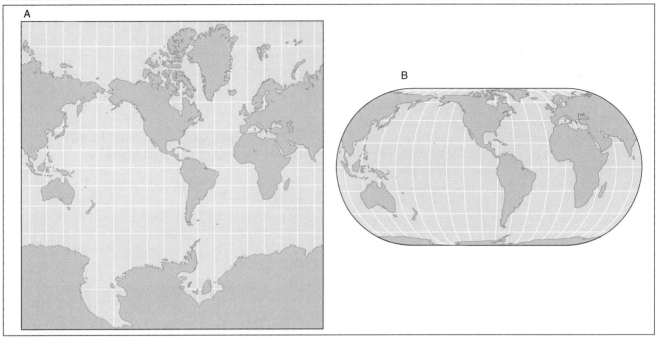

FIGURE 1-11

Flat maps necessarily distort the Earth. A *conformal* projection (A) displays the correct shapes but exaggerates sizes in high-latitude areas. An *equivalent* projection (B) can portray accurate sizes, but shapes are extremely distorted around the edges of the map.

maps and globes of his time. He decorated one book of maps with an illustration of the mythological figure Atlas holding the world on his shoulders, and books of maps have been called atlases ever since. Mercator also developed a special-purpose projection that became inordinately popular for general-purpose use.

On the Mercator map parallels and meridians form a perfect rectangular grid. The map is mathematically adjusted to achieve conformality. This projection is accurate at the equator and relatively undistorted at the low latitudes, but distortion of size increases rapidly in the middle and high latitudes. Area is distorted by 4 times at the 60th parallel of latitude and by 36 times at the 80th parallel.

The Mercator projection was designed to facilitate oceanic navigation. It shows lines of constant compass direction as straight lines. Mercator wrote, "If you wish to sail from one port to another, here is a chart, and a straight line on it. If you follow this line carefully you will certainly arrive at your destination." Mercator was well aware of the limitations of his projection, however, and he warned, "The length of the line may not be correct. You may get there sooner or you may not get there as soon as you expected, but you will certainly get there."

The Mercator projection serves its original purpose excellently, but it has serious limitations for most other uses. Its popularity for use in U.S. classrooms and atlases has created many misconceptions, one of which is a confusion about the relative sizes of land masses. For example, on a Mercator projection the island of Greenland appears to be as large or larger than the continents of Africa, Australia, or South America. In actuality, Africa is 14 times larger than Greenland, South America is 9 times larger, and even Australia is 3.5 times larger. Some scholars argue that this inflation of mid- and high-latitude countries' sizes leads to a corresponding inflation in the students' minds of the relative "importance" of those countries in the world, and to a relative diminution of the importance of the tropical countries.

Furthermore, the Mercator map distorts direction and, therefore, situation. As an experiment, draw a straight line between two places on a Mercator map, and then compare that route with the route between those same two points traced by a string stretched between the two points held taut against a globe. On a Mercator map, for example, Hawaii seems to be on the way between California and Japan, but in fact it is not. In 1976 President Gerald

Ford, while flying from Washington, D.C., to Tokyo, stopped in Alaska to give a speech. The political opposition argued that this was a tax-supported campaign trip crudely disguised as a diplomatic trip. To argue their point, they printed a Mercator map showing that the president's visit to Alaska had taken him far out of the way. In reality, a direct route from Washington to Tokyo crosses over Alaska.

Several scholars have argued that common use of the Mercator projection in the United States has also bolstered American political isolationism. The United States looks safe and isolated by broad ocean "moats," so some Americans came to feel that they need not concern themselves with international affairs.

Other Rectangular Projections In January 1990, the American Cartographic Association urged book and map publishers, the media, and government agencies to abandon all square-gridded, rectangular maps. The four common rectangular maps that came under fire from the A.C.A. were Mercator's, Gall's (drawn in 1855), Miller's (1942), and Gall-Peters's (1967). All four distort shapes and land masses and lead to misunderstandings about world relationships (Figure 1-12).

Insets and Other Conventions

Other conventions of map projections or printing can lead to misunderstanding. For example, in order to save space, many maps of the United States tuck a separate map of Alaska into the corner below the U.S. Southwest. This is called an *inset* map. This common practice lead many U.S. schoolchildren to think of Alaska as an island off the coast of California. If the mapmaker also prints the inset map at a smaller scale than that of the other states, as is common, the confusion about Alaska's location may be aggravated by confusion about its size. In fact, Alaska is one-fifth the size of the lower 48 states combined. In January 1990, in order to counter this misrepresentation and the resulting confusion in some students' minds about exactly where Alaska is, the Alaska state legislature passed a resolution "respectfully requesting that all major United States magazines, newspapers, textbook publishers, and map publishers...place Alaska in its correct geographical position...in the northwest corner."

Several world maps in this book cut out great expanses of the world's oceans so that we can map human activities on the land in greater detail. This shifts the relative positions of the continents and tucks Australia and New Zealand south of Asia—to the west of where they really are. Clear divisions on the maps mark the lines along which the map has been cut, and each portion has markings of latitude and longitude for proper positioning. The first such map is Figure 2-20 on page 74. If you want to recheck the continents' relative postions shown more accurately, you can always refer to the front or back endpaper maps—or better still—a globe.

Miller's and Gall-Peter's Projections

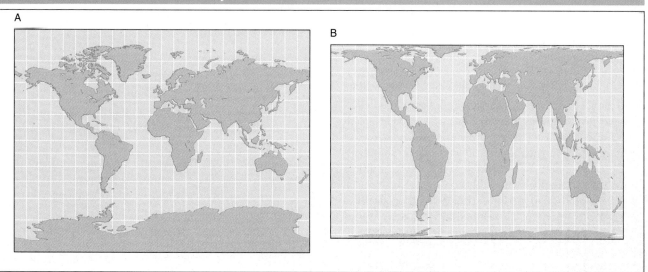

FIGURE 1-12

Many scholars have criticized both Miller's cylindrical projection (A) and (B) Gall-Peters's equal-area projection for unacceptably distorting the Earth's shapes and landmasses.

It is conventional to put North at the top of a map, but that is not real. In fact, any conventional view "conceals" a great deal of information simply because it is conventional—we are not really looking at the map, but only being reminded of something we have "seen" often before.

The inability to understand mapmakers' conventions troubled Mark Twain's fictional character Huckleberry Finn when he and Tom Sawyer were blown east in a balloon. Huck insisted that they were not over Indiana yet because the color of the land below had not changed from the green of Illinois to the pink of Indiana, as shown on his school atlas. "What's a map for?" he demanded. "Ain't it to learn you facts?" Our answer is "yes," but a map "works" for you only if you understand its projection and the explanation of its colors and symbols found in its *key*, usually at the bottom of the map.

REMOTE SENSING

The term remote sensing refers to any method of measuring or acquiring information by use of a recording device that is not in physical contact with the object being studied. Aerial photography was first tried in 1858 but it was not used widely until three-quarters of a century later. Color infrared film, which senses, in part, beyond the visible portion of the electromagnetic spectrum, was developed during World War II. Today thermal infrared, microwave, radar, and sonar sensing systems provide new ways to examine the Earth.

The United States and several other countries have launched numerous satellites that monitor the Earth with a variety of photographic and electronic techniques, and some of these images are used as illustrations in this book (see, for example, Figure 4-3). TIROS, the first weather satellite, launched in 1960, started the era of daily weather and storm prediction that we now take for granted. Prior to satellite monitoring, the only way coastal areas received warning of an impending hurricane was if a plane or ship happened to encounter the storm at sea. In 1900, a fierce storm that suddenly struck the coast near Galveston, Texas, killed 12,000 people. Today weather satellites continuously monitor the formation of severe storms and then track them, helping to predict their paths.

In the early 1960s the United States also sent aloft the first photo reconnaissance satellites, known as "spies in the skies." The satellites' images, along with information from electro-optical and radar reconnaissance, gave U.S. policymakers insights into activities in other countries.

The early Gemini missions of the same period produced the first successful photographic observations of large-scale geologic structures from space. This success led to the development of the LANDSAT (land satellite) program. Information provided by LANDSAT data aids in crop yield estimates, forest inventory, and rangeland management. LANDSAT also surveys soil for erosion and moisture content, monitors droughts, maps flood plains, explores for oil and minerals, provides data for land use planning decisions, discovers ocean circulation patterns, and assists environmental scientists studying wildlife habitats and pollution problems. Today British, French, German, Russian, Chinese, Indian, Japanese, and other nations' satellites have joined America's in the skies.

In the early 1990s, the U.S. space shuttle *Columbia* deployed two Laser Geodynamics Satellites, LAGEOS I and II. These measure movements of the Earth's crust (discussed at the end of this chapter) and anomalies and changes in the Earth's gravity. A worldwide cooperative ground network of 29 stations communicates with these satellites.

Satellites and planes equipped with specialized sensing devices can even read history as it is recorded below the surface. Information available includes changes in plant growth or drainage patterns brought about by human activity, and densely compacted soils and erosion created by traffic on long-buried footpaths. Archeologists have used this information to find long-lost cities.

GEOGRAPHIC INFORMATION SYSTEMS

One of the latest developments in geographic techniques is the **Geographic Information System (GIS)**. GIS is the use of computer systems to organize, store, update, and map information, and to investigate that data. Most of the information is the sort that geographers have traditionally mapped, but computers' capacities to store and to correlate information constitute a technical revolution. GIS analytic capabilities also suggest new applications for the information traditionally collected.

A GIS system begins with the accumulation of a geographic database, or geographic information prepared in a way that a computer can store it and then print it. Traditional maps are composed of three basic features—points, lines, and areas. For example, a typical map identifies a spring as a dot, a stream as a squiggle, and a lake as a blue patch.

A GIS can reproduce such a graphic representation, but it stores the data differently. In one way of storing data, called the *vector* method, map features are usually represented by X, Y coordinates. Points are identified as a single coordinate pair. Lines are identified as a connected set of points (like "connect-the-dots" pictures), and areas, such as lakes, are identified by the coordinates defining their borders. This is not unlike the geographic grid system, with each point having a value for its latitude and longitude. A computer can store vast amounts of information organized according to coordinates of such a grid.

Another format for storing information is called *raster*. It uses an imaginary grid of cells to represent the landscape. Points are stored as individual Column, Row entries. Lines are stored as a set of connected cells, and areas are identified as all of the cells within the interior of a feature. This is comparable to drawing a map by filling in the squares on a sheet of graph paper. The degree of precision of a raster data structure depends on the size of the cells used. For example, if each cell represents one acre, and if a stream passes through one cell, the entire acre/cell will be identified as a stream, and a researcher could not be sure whether the stream is located at the northern edge of the acre/cell, the southern edge, or whether it winds around within it.

A map "works" for you only if you understand its projection and the explanation of its colors and symbols found in its key.

Therefore, in this case, the geographer would have to be sure to use cells that represent areas smaller than one acre. Raster-based systems greatly simplify data storage and computation. Therefore, a geographer storing any information or studying any problem must choose whether a vector or a raster system is more appropriate for his or her research.

One of geographers' tools of logical technique is the hypothesizing of cause-and-effect relationships between two phenomena that have been found to have comparable distributions. For example, geographers study the relationships between the distributions of certain types of climate and certain soil types. Traditionally, in order to compare information on two maps, geographers have had to glance back and forth between the two (as you will be asked to do repeatedly in this textbook), or else to make one of the maps on some transparent material and place it on top of the other. Neither method is always easy.

It is easy, however, for a GIS to make such a correlation, because the information that would be illustrated on each map is already linked electronically by its identification number on the grid system of the database. The computer can search through all information stored about any given point (or cell) almost instantly. The geographer can seek correlations among vast amounts of data across great distances at the press of a button. This is the equivalent of being able to pile up every existing map, one on top of the other and all of them transparent, in order to see how patterns and distributions correlate (see Figure 1-13).

A geographer might ask, for example, what places are characterized by the following four variables: Hispanic population, most adults having more than 12 years of education, median household income over $50,000, and most households containing children under 18 years of age. The computer could in a moment produce a map showing the places having all four variables. A magazine distributor might then try to sell subscriptions to a new magazine directed to those people in those neighborhoods. This is called *niche* or *target marketing*.

More and more government agencies and private firms are making data available in GIS form, and a great variety of computer software packages have made GIS manipulation of data possible in the classroom or office. A GIS can even exhibit correlations among data that the geographer may not, originally, have thought to seek. What hypotheses, for example, might we formulate if a GIS shows a correlation between those neighborhoods in a metropolitan area housing the greatest concentrations of Asian Americans and those that have the most (or fewest) swimming pools? Or a correlation between high-income neighborhoods and those where most people are brown-eyed? Sometimes we face an embarrassment of data that gives us answers before we know what questions to ask. Whether or not these correlations *mean* anything must still be determined by the geographer's ability to formulate and investigate explanatory hypotheses.

THE PHYSICAL ENVIRONMENT OF THE EARTH

The Earth is so far the only home of humankind. This chapter will introduce an understanding of some basic physical attributes of this home. The Earth's

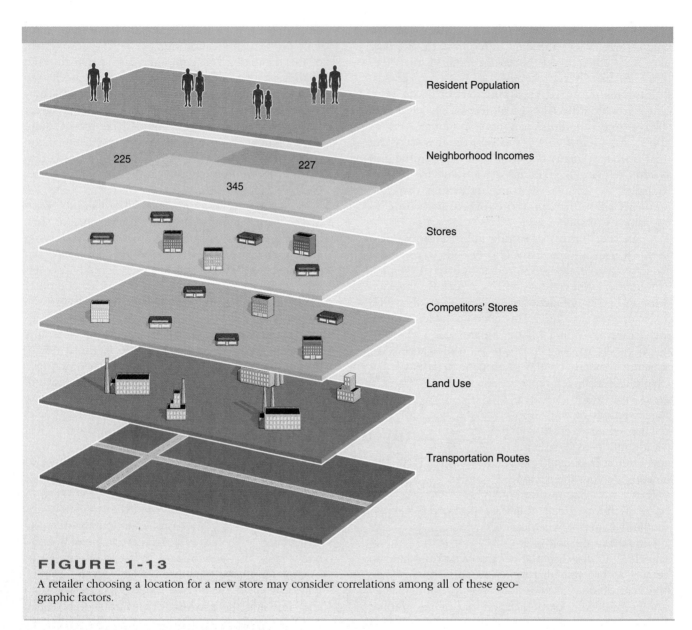

Resident Population

Neighborhood Incomes

225 227 345

Stores

Competitors' Stores

Land Use

Transportation Routes

FIGURE 1-13

A retailer choosing a location for a new store may consider correlations among all of these geographic factors.

physical environment is extraordinarily complex and diverse, yet it is ultimately fragile. Therefore, Chapter 2 will examine in detail the many ways humankind redesigns and transforms this physical environment.

Earth-Sun Relations

Life on Earth is dependent on solar energy, so the Earth's relationship to the sun is vitally important. The perpetual motions of the Earth continually change the distance between the Earth and the sun and the portion of the Earth that faces the sun.

The Earth *rotates* toward the east on its axis, making one complete rotation in 24 hours. It also *revolves* around the sun in an elliptical orbit, complet-

ing the full orbit in slightly more than 365 days. An imaginary plane that passes through the sun and through the Earth at every position in its orbit is called the plane of the ecliptic. The Earth's axis is tilted at an angle of 23.5° from perpendicular to the plane of the ecliptic, and it is always parallel to itself as it revolves around the sun, pointing toward Polaris, the North Star. Therefore, the angle at which the sun's rays strike different parts of the Earth's surface changes throughout the year (see Figure 1-14). This causes the changes of the seasons.

The Tropic of Cancer at 23.5° N. Lat. and the Tropic of Capricorn at 23.5° S. Lat. define the region within which the sun is ever directly overhead at noon. This region is called the *tropics*. The sun appears to migrate from north to south and back

In 1993 a bitter feud developed between, on the one hand, biologists who wanted to stop the harvest of forests in the U.S. Pacific Northwest in order to protect the environment of the rare spotted owl, and, on the other, timber interests and loggers who fought to protect their industry and jobs. Secretary of the Interior Bruce Babbitt compared this collision of interests to a "train wreck," and he pointed out that while the arguments proceeded in court, the spotted owl slipped to the brink of extinction.

Secretary Babbitt argued that these confrontations occur partly because scientific knowledge of the United States's natural resources is inadequate, and to remedy that inadequacy he proposed the creation of a National Biological Survey (NBS). This new agency, which came into existence by presidential action on November 12, 1993, is to be equivalent to the U.S. Geological Survey that was established in 1879. The NBS will consolidate federal biological research, and it will inventory every animal and plant species in the United States

and its habitat. It is hoped that this new treasure trove of data will enable scientists and regulators to see correlations never observed before and to devise plans to protect fragile species before they become threatened with extinction.

NBS is still in its formative stages, and the techniques and standards to be used for data collection and storage are not yet defined. There can be no doubt, however, that GIS will play a key role in organizing the vast amounts of information.

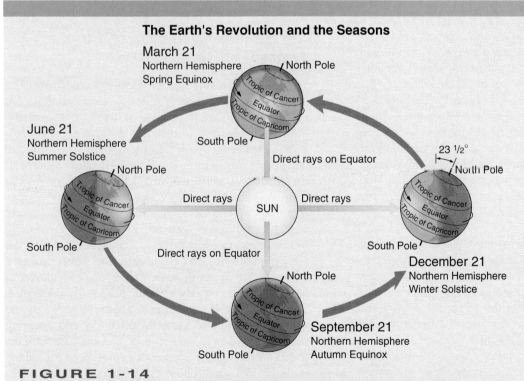

The Earth's Revolution and the Seasons

FIGURE 1-14

The Earth makes one revolution around the sun in 365 1/4 days, and during the year different places receive different intensities of sunlight as well as different total amounts. The June solstice is a time when sunlight is concentrated in the Northern Hemisphere; at the time of the December solstice, sunlight is concentrated in the Southern Hemisphere. At the times of the two annual equinoxes, the sun's noon rays strike directly upon the equator. Only on those two days is the length of the daylight and darkness equal all over the Earth. The exact dates of all four days varies by a day or two on our calendar.

again within the tropics. The sun's direct rays reach the Tropic of Cancer each year at the time of the June solstice, cross the equator on their journey south at the September equinox, reach the Tropic of Capricorn at the time of the December solstice, and recross the equator on their journey northward again at the March equinox. Thus, daylight lasts longest in the Northern Hemisphere at the time of the June solstice (which is summer season in the Northern Hemisphere but is winter in the Southern Hemisphere), and the reverse is true south of the equator.

Weather and Climate

The Earth is enveloped by an atmosphere that supplies most of the oxygen that animals must have in order to survive, as well as the carbon dioxide plants need. The atmosphere helps maintain a water supply, which is essential to all living things. It serves as an insulating blanket to moderate temperature extremes, and it shields the Earth from much of the sun's ultraviolet radiation, which otherwise would be fatal to most life.

The atmosphere consists of a mixture of gases, along with small amounts of impurities, and it is held to the Earth by gravity. The atmosphere's attachment to the Earth, however, is loose, and the atmosphere has actions of its own. The atmospheric envelope is energized by solar radiation, stimulated by earthly motions, and affected by contact with the Earth's surface. The atmosphere reacts to these influences by producing an infinite variety of conditions known as weather. Climate is the aggregate of weather conditions over a long period of time. Climate has direct and obvious influences on agriculture, transportation, and human life in general, as well as on all major aspects of the physical landscape—soils, vegetation, animal life, hydrography (water features), and topography.

Insolation and Temperature Incoming sunlight, called insolation, is the only important source of energy for the Earth's atmosphere, and the atmosphere as a whole maintains a radiation balance. That is, as much energy leaves the atmosphere for space as enters it from the sun. This heat balance, however, does not exist within individual parts of the globe. Tropical areas receive a large energy surplus, while high latitudes experience an energy deficit. There are two reasons for this. In the low latitudes the sun is more directly overhead through more of the year than it is in the higher latitudes. Therefore, its rays are more intense, just as a flashlight beam is more intense shining directly onto a surface than when the beam is cast across a surface at an angle. In addition, through the course of a year the tropical latitudes

receive a greater total period of light than do the higher latitudes. Therefore, latitude is the dominant influence on temperature through the year.

This influence is moderated by the fact that oceans respond very differently than continents do to insolation. In general, land heats and cools faster and to a greater degree than water does, and, therefore, the oceans act as great reservoirs of heat. They moderate temperature extremes. In summer they absorb heat and store it; in winter they give off heat and warm up the air. Therefore, both the hottest and the coldest areas of the Earth are found in the interiors of continents, distant from the moderating effects of the oceans. A continental climate experiences greater seasonal extremes of temperature than does a maritime climate. In Portland, Oregon, on the west coast of the United States, for example, the average daily temperature is 39°F in January and 66°F in July, a difference of 27 degrees. By contrast, at about the same latitude deep in the continental interior at Minneapolis, Minnesota, the average daily temperature is 13°F in January and 73°F in July, a difference through the year of 60 degrees. People in Minneapolis need more varied wardrobes than do people in Portland.

The General Circulation of the Atmosphere and Oceans

The tropics experience an energy surplus, and the higher latitudes experience an energy deficit, but there is a persistent shifting of warmth toward the high latitudes and of coldness toward the low latitudes. This is accomplished by movements of air and water. In general, hot air rises in the equatorial regions, cools as it rises, and falls again in the middle latitudes (see Figure 1-15). As a result of the Earth's rotation, any free-moving object in the atmosphere will drift sideways from the direction of movement (to the right in the Northern Hemisphere and to the left in the Southern Hemisphere). Therefore, air falling in the northern hemisphere tends to fall clockwise, and in the southern hemisphere it falls counterclockwise. These movements set up the major world wind belts. The trial-and-error discovery and mapping of these wind belts was one of the greatest achievements of the age of European exploration, for, in the days of sail, the discovery of this global pattern allowed exploration to proceed. The pattern of world circulation of air is replicated in the world circulation of ocean currents.

Vertical Temperature Patterns The atmosphere generally allows short-wave radiation from the sun to pass through to the Earth's surface but inhibits the

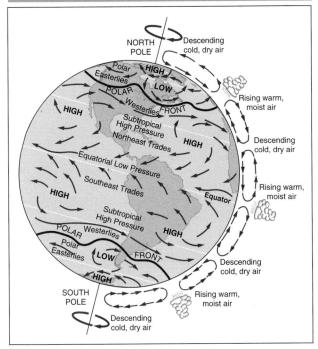

Atmospheric Circulation

FIGURE 1-15

The foci of subtropical high pressure cells in both the Northern and Southern Hemispheres act as giant pinwheels spinning clockwise in the Northern Hemisphere and counterclockwise in the Southern Hemisphere. These create the trade winds and the westerlies. These subtropical highs follow the sun northward in the Northern Hemisphere summer and southward in the Northern Hemisphere winter, shifting the wind belts north and south along with them through the course of the year.

escape of long-wave radiation from the Earth into space. This heats the lower atmosphere. This same trapping and heating function is served by the glass in a commercial greenhouse, so it is usually referred to as the **greenhouse effect**. You have experienced the greenhouse effect if you have ever left a car with the windows rolled up sitting in the sun.

 Light converts to heat at the ground, and the air warms from the ground up. Also, the atmosphere is heavier and denser nearer the surface of the Earth. Therefore, temperatures usually decrease with altitude in the air or with elevation on Earth. It seems paradoxical that the closer one rises to the sun, the cooler the air temperature is, but this explains why summer vacationers may seek relief from lowland heat by traveling up into the mountains. Countries that have hot lowland areas often have highland summer resorts. Before the invention of air conditioning, many countries, including Spain and India, had two capital cities, a highland summer capital and a lowland winter capital (see Figure 1-16).

The most prominent exception to a normal condition is a **temperature inversion**, a situation in which temperature increases at higher altitudes. Inversions are relatively common, but they are usually of brief duration. Inversions decrease the possibility of precipitation, and they can also significantly increase air pollution. The stagnant air limits the natural upward dispersal of pollutants from motor vehicles, industry, and other sources. If a temperature inversion remains over a major city for a few days, poor air quality can become a health hazard.

The Global Distribution of Precipitation

Warm air can hold more moisture than cool air can, so when moist warm air rises or cools, it drops its moisture as precipitation (see Figure 1-17). Precipitation occurs unevenly over the surface of the Earth (Figure 1-18, p. 22). It is largely based on latitude, but many other factors are involved, and the overall pattern is very complex.

The most conspicuous feature of the worldwide precipitation pattern is that most of the wettest areas lie in the tropics. The only other areas of conspicuously high precipitation are narrow zones along the west coasts of North and South America between 40 and 60 degrees of latitude. Orographic precipitation is triggered in these areas by a combination of frequent onshore westerly air flow and local conditions.

Monsoons

Monsoons are a variation in the generalized pattern of global wind and pressure systems. A **monsoon** is a seasonal reversal of winds, accompanied by distinct precipitation patterns. During the summer, moist winds flow from the ocean toward the shore, bringing heavy rainfall. In contrast, during the winter the winds blow out to sea from the continents, which results in a pronounced dry season.

More than half of the world's population inhabits regions subject to monsoons (Figure 1-19, p. 22). Monsoonal rains are essential for the production of food and cash crops. The failure, or even late arrival, of monsoonal moisture can cause economic disaster and starvation.

The Quality of Rainfall

Precipitation varies in quantity and seasonality in different parts of the world, but it may also vary in its quality, which has major effects on human activity. The concept of rainfall quality includes the characteristics of intensity and regularity.

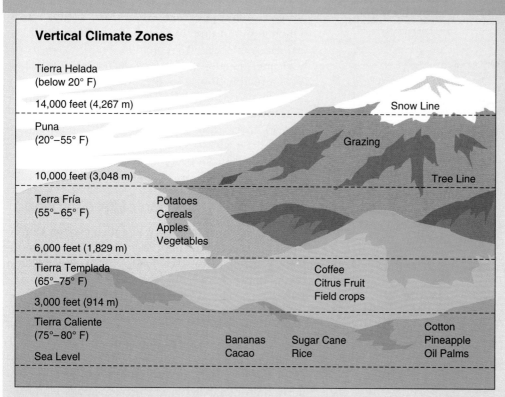

Vertical Climate Zones

Tierra Helada
(below 20° F)

14,000 feet (4,267 m) — Snow Line

Puna
(20°–55° F)

Grazing

10,000 feet (3,048 m) — Tree Line

Terra Fría
(55°–65° F)

Potatoes
Cereals
Apples
Vegetables

6,000 feet (1,829 m)

Tierra Templada
(65°–75° F)

Coffee
Citrus Fruit
Field crops

3,000 feet (914 m)

Tierra Caliente
(75°–80° F)

Bananas   Sugar Cane   Cotton
Cacao   Rice   Pineapple
                         Oil Palms

Sea Level

FIGURE 1-16

Throughout much of Latin America, elevation defines these vertical climate zones. Distinct crops characterize each zone.

Intensity of Rainfall In temperate areas precipitation frequently takes the form of gentle rainfall, whereas in tropical regions it may come in torrential and destructive downpours. In northern Nigeria, for example, 90 percent of all rain falls in storms of more than one inch (25 mm) per hour. That is, by way of comparison, half the average monthly rainfall of London. In Ghana, cloudbursts regularly occur at a rate of 8 inches (200 mm) per hour, four times London's monthly total. In Java, an Indonesian island, a quarter of the annual rainfall comes in showers of 2.4 inches (60 mm) per hour, more than Berlin gets in an average month.

Sudden storms are the worst possible way to get water. Much of the water gets lost, carrying away great quantities of precious topsoil. The first heavy drops in a downpour clog the pores of the soil with fine particles washed from the surface. After only a few minutes of such a deluge, the soil cannot absorb more than a fraction of the rain. More than two-thirds of the water may then run off in sheets and rivulets, which leads to tremendous erosion. Studies in the West African nation of Burkina Faso found that in one year nearly 90 percent of all erosion took place in just 6 hours.

Regularity of Rainfall Another component of the quality of rainfall is its regularity. In Western Europe and eastern North America, rainfall does not vary much from month to month. New York's rainiest month receives only one and one-half times as much rain as its driest; London's two times, Berlin's two and one-half. Delhi receives about the same total annual rainfall as London, but Delhi, by contrast, receives only 0.2 inch (10 mm) of rain in November, and more than 7 inches (175 mm) each in July and August. Zungeru, in central Nigeria, gets 54 times as much rain in the wettest month as in the driest.

Soil can absorb and hold only a limited amount of water for crops to use. Think of this as a bank account deposited in the wet season and drawn on in the dry season. Although many tropical countries get what seems to be adequate annual rainfall, they get it in a lump sum, like a huge win at a casino. Most is squandered, and the little that is left in the soil bank is used up within a few months. For example, in Agra, India, where the Taj Mahal is located, rainfall exceeds the current needs of vegetation only in July and August. As the rains decrease in September, the water stored in the soil is used up in just 3 weeks, leaving 9 months of the year in which water supplies are inadequate.

Human societies can adjust to variations in rainfall if these variations are regular or predictable. Irrigation can help even out the supply of water over the growing season. In the Sahel, south of the Sahara, farmers grow fast-maturing crops such as sorghum and millet that shoot from seed to ripened ear in 3 or 4 months.

It is much more difficult, however, to cope with rains that come irregularly from year to year. In Europe and much of North America, the amount of annual rainfall varies by less than 15 percent per year on average. In many of the tropical and subtropical regions, by contrast, it fluctuates from 15 to 20 percent, and in the semiarid and arid lands, by up to 40 percent or more. In the lean years, crops fail. If the rains are late, the growing season is cut short, and yields are greatly diminished. Local famine can follow. That is why, on the Indian subcontinent, the monsoon is awaited with such hope and trepidation, and a delay of a week or two causes panic.

CLASSIFYING CLIMATES
▼▼▼▼▼▼▼▼▼▼▼▼▼▼▼▼▼▼▼▼▼▼▼▼▼

Climate is the result of the interaction of a number of different elements, but climates are classified on the basis of temperature and precipitation. The classification system that is the most widely used today was developed by Wladimir Köppen (1846–1940), a Russian-born German climatologist who was also an amateur botanist. Köppen identified five climate groups on the basis of the mean annual and monthly values of temperature and precipitation, combined and compared in a variety of ways. Köppen used vegetation as the indicator of these parameters. Each of Köppen's five major climatic groups is then subdivided according to various temperature and precipitation relations. The system presented here, which includes six major categories, is modified a bit from Köppen's final version (Figure 1-20).

The relationship between rainfall and temperature at a place determine its humidity. Absolute humidity is the actual amount of water vapor in the air, while relative humidity is the amount of vapor in the air relative to what the air could hold if it were saturated. Warm air can hold more vapor than cool air; thus, an increase in temperature results in a decrease in relative humidity because the air's capacity for holding vapor is increased. As a result of these factors, two places may receive the same amount of

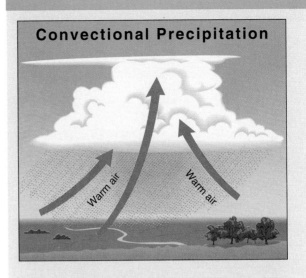

Convectional Precipitation

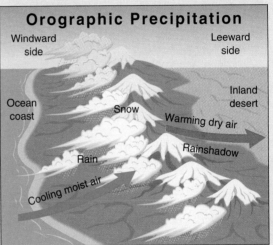

Orographic Precipitation

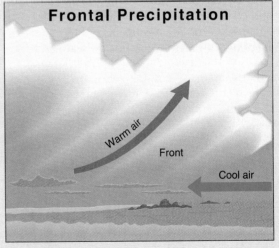

Frontal Precipitation

FIGURE 1-17

These are the three main circumstances under which moist air rises, cools, and drops its moisture as precipitation.

The Background to Human Geography **21**

Average Annual Precipitation

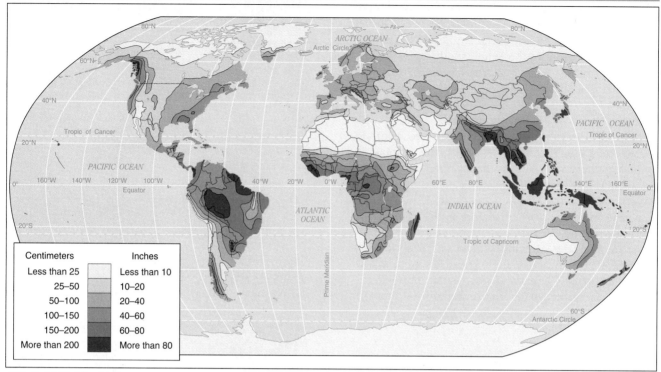

FIGURE 1-18

Average annual precipitation ranges from the 460 inches received on Mount Waialeale, Hawaii, to several desert locations where no precipitation has ever been recorded.

precipitation, but the warmer of the two may be arid, while the cooler may be humid. This affects the distribution of all types of vegetation, including crops. Cool areas can be productive with less precipitation than can hotter areas. For example, crops that thrive in parts of Canada that have little rainfall cannot be grown in parts of the United States that have the same amount of precipitation but are warmer.

Monsoonal Regions

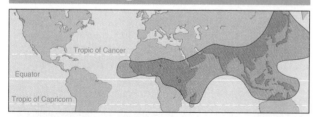

FIGURE 1-19

Along South Asia the onshore wind in summer is stronger than the offshore winds in winter, but in East Asia the out-blowing winter monsoon is stronger than the onshore summer flow.

THE GLOBAL PATTERN IDEALIZED

From the data on the characteristics and distribution of the basic climate types, we can construct a simplified representation of the distribution of Köppen climatic types as they would appear on a hypothetical continent of low and uniform elevation (see Figure 1-21, p. 26). Its shape, with its great bulk in the Northern Hemisphere, corresponds roughly to that of the actual landmasses on Earth. This hypothetical continent illustrates how the climatic types would be distributed without the modifications and complications caused by the varying shapes, sizes, positions, and elevations of the Earth's actual landmasses.

The figure portrays the generalized distribution of all but the highland (F) climate. Four of the climatic groups (A,C,D, and E) are defined principally by temperature, which means that they usually stretch east-west. The dry climatic group (B) is defined by moisture conditions, and its distribution cuts across those defined by temperature.

Any such idealized, simplified representation is called a **model**. A good model simplifies reality

22 *Chapter 1*

enough to help us understand what our assumptions should lead us to expect. Then we can compare our model with reality, and if the real world deviates from the model in only minor ways or in only a few places, we may be able to explain those local deviations due to local variations from our assumptions. This is an example of *deductive reasoning*. If, however, the real world bears little resemblance to our model, then we have to start over with new assumptions.

A good model serves as a predictive tool. For example, Köppen's model allows us to state, with some degree of assurance, that at a particular latitude and a general position (eastern, central, western) on a continent, a certain climatic type is likely to be found. Moreover, it helps us understand the locations of the climatic types relative to one another. The "real world" holds many refinements and modifications to this global pattern, but the schematic model shows the general alignments with considerable validity.

CLIMATIC CHANGE THROUGH TIME

▼▼▼▼▼▼▼▼▼▼▼▼▼▼▼▼▼▼▼▼▼▼▼▼

The Earth's climate has not always been as it is today. Our understanding of how the Earth's climate has changed is increased by the existence of great ice caps tens of thousands of years old. Much greater areas of the Earth were covered with ice in the past.

Glaciers and Ice Ages

Today about 10 percent of the Earth's land surface, amounting to some 6 million square miles (15 million km^2) is covered with ice. More than 96 percent of that area comprises the Antarctic and Greenland ice caps. In addition, many of the high mountains of the world contain glaciers, which are large natural accumulations of land ice that flow downslope or outward from their centers of accumulation (Figure 1-22, p. 26).

Today, most glaciers are small, but over the last few million years, periods of relatively low temperature and large accumulation of ice ("Ice Ages") have been interspersed with periods of melting. Several Ice Ages occurred during the epoch known as the Pleistocene, which began at least 1.5 million years ago and ended less than 10,000 years ago. At the maximum Pleistocene extent of ice, ice covered about one-third of the total land area of the Earth—nearly 19 million square miles (47 million km).

One way of reading the history of the Earth's climate is to drill holes in the deep ice caps and examine various properties of the ice layers laid down each year by snowfall. The past climate can be inferred from the relative amounts of light and heavy oxygen in the layers of ice; the relationship indicates the temperature at the time each layer of snow fell. In 1992, the average global temperature was 59.2°F (15.13°C), but analysis of the oxygen in each layer of ice suggests that the climate was different in the past.

The analysis of ice extracted from depths of more than 9,000 feet in Greenland suggests that the climate has changed frequently and abruptly a number of times over the past 250,000 years. Previous studies had shown that there were abrupt changes in climate during glacial periods, but the new results suggest that the same was true in the periods when the glaciers had retreated. The depths of ice may have been subject to bending and warping that would distort the historic record they store, but early results of analysis suggest that the average global temperature may have changed as much as 18°F in a couple of decades during interglacial periods. At one time during the last interglacial period, the average temperature plunged 25°F to ice-age levels for about 70 years.

Scientists now believe that the Earth has experienced a period of relatively stable temperatures for the 8,000 to 10,000 years since the last glacial epoch. This stability, however, seems to have been exceptional in the long history of the Earth, and the reason for this has lately become an important puzzle for scientists to solve. It is during this period of stable temperatures that human civilization has risen and flourished. If this stability is truly unusual, then the current climate may get either warmer or colder much more quickly than had been believed—in spans of decades or even less. This would happen for reasons we do not understand and probably could not control.

Gradual climatic changes spread over centuries would allow farmers to adjust to altered growing conditions and coastal cities to deal with rising sea levels, but humanity might have difficulty adjusting to abrupt changes. For example, at one relatively warm period between the last two glacial epochs the climate melted enough polar ice to raise sea levels some 30 feet. If, on the other hand, the climate became colder, the northern and southern regions would be covered with snow much longer each winter. Glaciers would advance, reflecting more of the sun's energy back into space, and chilling the climate even more.

The degree to which the atmosphere demonstrates the greenhouse effect depends on its relative content of various gases. Some gases intensify the greenhouse effect. These are called *greenhouse gases*, and they include water vapor, carbon dioxide, methane, nitrous oxide, and ozone. Therefore, the higher the atmosphere's content of these gases, the warmer the Earth's atmosphere can be expected to

FIGURE 1-20

In areas of tropical climate (A), the average temperature of even the coldest month exceeds 64°F (18°C). Tropical areas are subdivided into those that receive precipitation throughout the year and those that experience seasonal variations. The areas of dry climate (B), where evaporation equals or exceeds precipitation, are subdivided between true deserts and steppes, which are slightly more moist. In mild midlatitude climate regions (C), the average temperature exceeds 50°F (10°C) at least eight months of the year, and the average temperature of the coldest month is between 64°F (18°C) and 27°F (-3°C). In severe midlatitude climates (D), the coldest month averages below 32°F (0°C), while four months record average temperatures above 50°F (10°C). Polar regions (E), have no months with an average temperature over 50°F (10°C). In the areas labeled highland climate, (F) on this map, altitudinal variations cause significant climate changes within short distances.

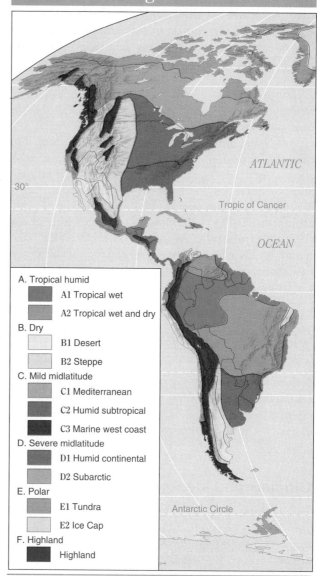

The Climatic Regions

A. Tropical humid
   A1 Tropical wet
   A2 Tropical wet and dry
B. Dry
   B1 Desert
   B2 Steppe
C. Mild midlatitude
   C1 Mediterranean
   C2 Humid subtropical
   C3 Marine west coast
D. Severe midlatitude
   D1 Humid continental
   D2 Subarctic
E. Polar
   E1 Tundra
   E2 Ice Cap
F. Highland
   Highland

become. Examination of old ice provides hints about the effects of rising levels of greenhouse gases. In one study, analysis of microscopic air bubbles in Greenland ice showed a remarkable correlation between a rise in atmospheric carbon dioxide and an increase in temperature.

Volcanic activity is one important factor in determining the chemical composition of the atmosphere. For example, Mount Pinatubo in the Philippines erupted on June 16, 1991, spewing upward some 20 megatons of sulfur dioxide. This created a layer of sulfuric and sulfurous acid droplets about 15 miles up in the atmosphere, and these droplets reflect insolation. Scientists predicted that this single volcanic explosion would significantly alter the Earth's temperatures.

Measurements made in the summer of 1992 confirmed that the amount of insolation reaching the Earth had indeed dropped nearly 4 percent. Satellite readings indicated that air temperatures over the globe had dropped by 1°F and were continuing to drop; the cooling may eventually reach 2°F. The effect is expected to last for 3 to 5 years. Mount Erebus in Antarctica regularly spews thousands of tons of chlorine vapor into the atmosphere, and we do not know the ramifications of this gas. We need to know a lot more about volcanic activity, how it affected the atmosphere in the past, and what its impact is today before we can understand changes in the Earth's climate.

Humankind's Impact on Global Warming

Many scientists hypothesize that human activities since 1850 have increased the release of greenhouse gases into the atmosphere so greatly that future global warming may be **anthropogenic** (human-caused). The greenhouse gases increased by human activity include synthetic chemicals called *chlorofluorocarbons*, or CFCs. *Nitrous oxide* comes from chemical fertilizers and emissions from internal combustion engines. *Methane* is produced by grazing livestock, rice paddies, burning wood, and termites that multiply where humans disturb the forests. The combustion of carbon-containing, or fossil, fuels, mainly coal and oil, yields carbon dioxide. The reduction of the world's forests also contributes to the amount of carbon dioxide in the atmosphere because trees naturally absorb great amounts of it. Annual increases in the atmospheric concentrations of carbon dioxide have been measured since 1958; by 1992, concentrations were 13 percent over 1958 levels.

If overall global temperatures rise, the patterns of temperature and precipitation will change. Heat

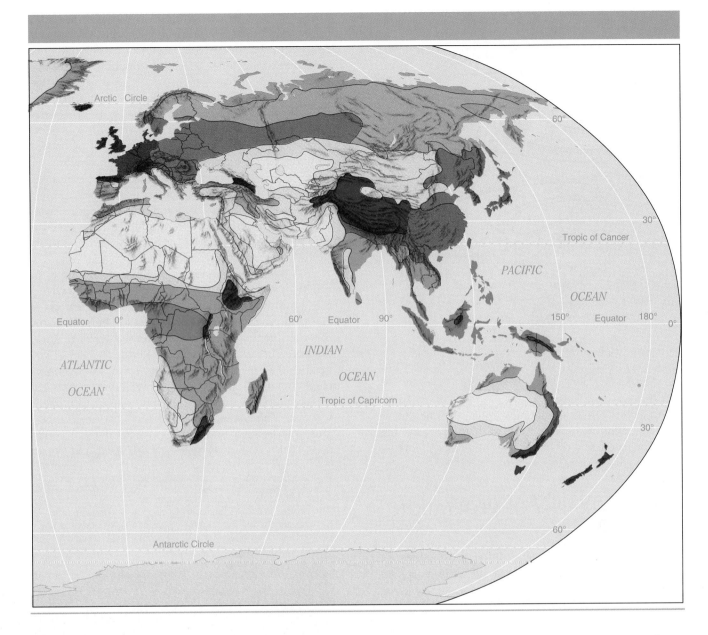

and drought will become more prevalent in the middle latitudes, and milder temperatures will prevail in higher latitudes. Some arid lands might receive more rainfall. Ice caps will melt, and global sea levels will rise, flooding many populous areas. Current living patterns over much of the Earth will be affected.

Many scientists believe that we must reduce emissions, particularly from smokestacks and internal combustion engines, in order to prevent these effects. Some estimate that the world's population will have to decrease its consumption of fossil fuels by at least 50 percent. Per capita emissions of carbon dioxide from the United States are nearly 8 times those from China and 25 times those from India, but the emissions from these countries will soar as their economies grow. Burning natural gas produces fewer

emissions than burning coal or oil does, and solar, wind, and nuclear energy produce no carbon dioxide emissions. Greater use of these energy sources could be encouraged. The Climatic Change Convention adopted at the United Nations Conference for Environment and Development in Rio de Janeiro, Brazil, in June 1992 targeted a goal for the industrial nations to stabilize their emissions of carbon dioxide at 1990 levels by the year 2000. Environmental monitors insist, however, that stabilization is not enough; emissions must be cut back.

No scientific data foretell the future, but they underlie growing concern about the stability of the Earth's present climatic pattern and humankind's possible role in disrupting natural processes. Dr. J.W.C. White of the University of Colorado wrote, "If the

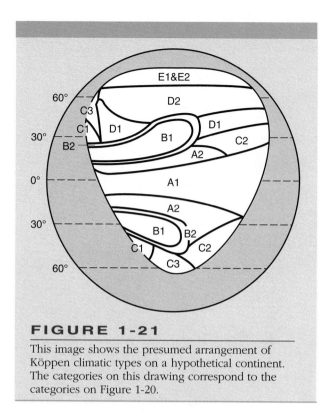

FIGURE 1-21

This image shows the presumed arrangement of Köppen climatic types on a hypothetical continent. The categories on this drawing correspond to the categories on Figure 1-20.

Earth came with an operating manual, the chapter on climate might begin with a caveat that the system has been adjusted at the factory for optimum comfort, so don't touch the dials."

THE GEOGRAPHY OF VEGETATION

▼▼▼▼▼▼▼▼▼▼▼▼▼▼▼▼▼▼▼▼▼▼▼▼▼▼▼▼

Plants (flora) belong to many different categories, or *taxa*. One important biological distinction can be made between lower plants, which reproduce without seeds, and higher plants, which reproduce by means of seeds. Lower plants include *bryophytes* (mosses and liverworts) and *pteridophytes* (ferns and club mosses); higher plants include *gymnosperms* (cone-bearing evergreens) and *angiosperms* (flowering plants). Most trees, shrubs, grasses, crops, weeds, and garden flowers are angiosperms. Angiosperms are often subdivided into the *woody plants*, which have stems composed of hard fibrous material, and *herbaceous plants*, which have soft stems.

Climatic Climax Vegetation

One of the difficulties in mapping world vegetation is that it is often changing. Even the most casual observer cannot help but notice such change in the local environment. Weeds invade the garden; grass

pops up through cracks in the sidewalk; untended public gardens grow wild; forests reclaim land that has been abandoned by farmers. Occasionally we read that an ancient complex of temples abandoned and completely overgrown by jungle vegetation has been rediscovered somewhere in the tropics.

A full-grown jungle rich in varieties of species does not appear overnight, but develops in slow stages. At any fresh spot the community of plants will naturally evolve through a series of increasingly complex stages of types of vegetation. The natural progress of this evolution will be from lower to higher species. The series that evolves at any spot differs, depending on the environment, but in each case the natural succession of plant types is called a primary series, or prisere, for that location.

A prisere has not developed to its final form if the variety of species that dominate in a place is still changing and if these plants are modifying certain aspects of the local environment, such as the soil beneath them, the conditions of light and shade, and the local drainage. In such cases, the plant species are in fact making the site more hospitable for a succession of plant communities, each of which will outcompete and thereby replace the existing community. Ultimately, however, one community will evolve that can compete most successfully under the existing physical conditions. This community will establish itself and persist

FIGURE 1-22

The Mount Robson Glacier in British Columbia, Canada, illustrates how glacial "rivers" of ice can flow outward and downward from large natural accumulations. (Courtesy of John Daniels.)

Deductive Geography

These pages have mapped climates around the world and then simplified the map into a generalized model. We can now use that model to guess what the climate might be at any place. This example of reasoning with a model introduces the differences between an **inductive** approach in geography and a **deductive** approach.

One reason we travel is that we think that other places are different from the place where we are. Travel agents entice us to visit far-off places. "Visit exotic Marataria" trumpet their brochures, "and see things you've never seen before!" If you've already been to Marataria, how about a trip to lovely Jusopia, with its unique wonders and attractions? Travel agents emphasize the uniqueness of each place because they want us to buy tickets to each destination one after the other.

Geographers often take a similar view of the world, analyzing the ways in which specific elements of a place combine to give that place an individual character. Regional studies analyze and synthesize the unique attributes of different places. The popular *National Geographic* magazine similarly focuses on the individuality of each place. Geographic studies that gather information about place after place and describe each as unique are called *idiographic* (idio = distinct; graphic = description). Geography has a long history of such accumulation of knowledge about the world through exploration and analysis of discovery. Aristotle is said to have answered his pupil Alexander the Great's questions about world places with the quintessentially inductive answer, "Go and see." Most geographers still do travel and explore whenever they can.

Even the travel agent, however, would have to admit that places do have similarities. If you want to vacation on a tropical beach, you could choose any one of several in the Caribbean. They are similar enough so that you have an idea of what you're choosing when you choose that category of vacation spot. So it is with big cities. Each city does have distinguishing characteristics, but London, Paris, New York, and Caracas also have much in common. If one has traveled a lot or studied many places, one soon begins to see many ways in which places are similar. A traveler begins to form categories, and each place the traveler visits loses just a bit of its individuality in his or her mind in order to become an example of a category. The traveler sees each place's individuality, but also its similarity with other places. An approach to knowledge that seeks general rules is called *nomothetic* (law-establishing). Urban planners look for similarities from city to city in the hope of using one city's experiences to help solve similar problems in another.

The two ways of looking at the world represented by the travel agent and the urban planner represent, respectively, inductive and deductive ways of viewing reality. *Inductive thinking* begins with specific cases and leads up to general conclusions. It proceeds step by step, building up a picture of the world from fact after fact. *Deductive thinking*, on the other hand, starts from general assumptions about reality. It creates models and then compares real-world specimens with those models. If we find that any individual real-world specimen deviates from the attributes of our model, we try to explain that

deviation from the model. For example, we can design in our minds a model "big city," and the model big city has certain attributes. One of those attributes is probably traffic jams. If, perhaps, the big city of Melbourne, Australia, does not have traffic jams, we try to explain why. Deduction is a way of describing places' uniqueness by explaining why they are not what you expect them to be.

In some examples of deductive research, geographers imagine the existence of a perfectly flat surface with absolutely no variations on it. This is called an **isotropic plain**. Geographers can then explore through logic alone how things would be distributed on such a surface. Even an isotropic plain would demonstrate the effects of the abstract qualities of distance, direction, and relative position.

To give just one example of deductive thinking, consider the concept of distance. Distance is a powerful factor in our lives. We try to overcome the cost and inconvenience of it. The distribution (the geography) of items in your home kitchen cabinets demonstrates your attempt to overcome the cost of distance. You probably put the items you use most often on the closest, most convenient shelves, and the items you use less often farther away toward the top. Most people do, and I deduce that you do. If your cabinets demonstrate a significant deviation from this expected behavior, I would seek an explanation.

Köppen's model continent is an example of deductive geography. Later in this book we will see how the geographers Johann von Thünen and Walter Christaller used simple deductive notions to build theories of the use of land and of the distribution of cities across a landscape.

indefinitely, as long as the underlying physical conditions are relatively balanced. This stable association of vegetation in equilibrium with local soil and climate conditions is called the **climatic climax vegetation**.

The concept of climax vegetation enables us to think of the Earth's vegetation not as a static patchwork, but as a dynamic system continually accommodating environmental changes and human interference. It is more useful than the term "natural vegetation," which is actually hard to define. What seems to be "natural" vegetation at any place and time may be evolving.

Many places do not exhibit climatic climax vegetation because the flora is not yet fully mature. All land on Earth was once fresh and uninhabited by life forms—lava flows, recently uplifted sea floors, and so forth. Although vegetation covered some places millions of years ago, it has penetrated other places only recently. Areas of tropical rainforest in Africa and South America, for example, have been evolving without interruption for millions of years, but northwest Europe and much of North America were scoured by glaciers just a few thousand years ago. The glaciers left a bare, inorganic surface, and the soils and vegetation located there today have had only a relatively short time to develop. This vegetation is still changing, and for many places we do not know what the local climatic climax vegetation will eventually prove to be. In addition, cataclysmic events such as floods, mudslides, or wildfires episodically clear substantial areas of their plant cover, but after each such event the vegetation resumes its slow evolution toward climax.

Furthermore, the plant cover over much of the Earth today is the result of human interference. Climatic climax vegetation has been destroyed, and in many places humans have interfered for so long that we do not know what plants grew there before human occupancy, or whether those plants could re-establish themselves if the area were left alone for a long period. In China today, for example, there are grasslands that are spotted with towns named "Wang," which means "oak tree." We assume that oaks must have grown there in the past, but we cannot be sure whether the climate

has changed so that oaks no longer grow there or whether human interference has destroyed the oaks.

Humans have devoted much of the Earth's surface to pastures or crops, and people prevent these areas from evolving to any other state. An environment thus artificially maintained or arrested is called a plagioclimax, or false climax. If the farmer abandons the field, however, priseral evolution recommences.

Diversity Indices

Wherever the local environment remains stable enough to support evolution over long periods of time, fragile superstructures of species develop. *Mutation* within species seems to be random; that is, each generation of any given species of plant or animal will have a proportion of significantly new forms. If any new form survives and multiplies, the species has become modified, or a new species has evolved. Many new forms do not survive to multiply and establish themselves; others persist for long periods and then die out. Each species is a sort of natural "experiment." Climate and other aspects of the local natural environment can affect the process of evolution of new species, but generally the longer the period during which evolution has continued uninterruptedly at a place, the more complex the community of species there will be.

The concept of climax vegetation enables us to think of the Earth's vegetation not as a static patchwork, but as a dynamic system continually accommodating environmental changes and human interference. It is more useful than the term "natural vegetation," which is actually hard to define. What seems to be "natural" vegetation at any place and time may be evolving.

This explains why the oldest vegetative complexes on Earth, the great tropical rainforests, are so rich in species. They have uniquely high **diversity indices,** or numbers of different species per unit of area. A 2.5 acre (1 hectare) plot in Brazil has been discovered to support 450 species of trees, in comparison with ten species in a typical New England forest. The entire United States only counts some 650 tree species.

Species of plants and animals may migrate. However, because of the seeming randomness of evolution, many of the species at any place may be unique to that place, and certainly the mix of species at any place is always unique. Each spot on Earth has a unique inventory of biological resources.

ECOLOGICAL SYSTEMS

The term **ecological system**, or **ecosystem**, refers to all the organisms in a given area and the totality of interactions among the organisms and between the organisms and the nonliving portion of the environment. Thus, an ecosystem is a biological community or an association of plants and animals. In an ecosystem, all of the parts are interrelated, and therefore a change in any component will change other parts. Ecosystems are so complex that the full ramifications of any change anywhere in the system can seldom be predicted. For example, the most powerful computers cannot model the interrelationships among the soil, the insects, the wildlife, and the vegetation in a meadow.

Energy, water, and nutrients flow through ecosystems largely by direct passage from one organism to another in linkages called **food chains**. In a food chain one organism eats another, thereby absorbing energy and nutrients; the second organism is consumed by a third, and so on. The matter of who eats whom may be extraordinarily complex.

Scientists have become concerned about contamination of human foods as they have begun to trace pollution and poison through natural food chains. For example, recent studies prove that polychlorinated biphenyls—PCBs, a class of industrial pollutants—are absorbed by algae in water. The algae are in turn eaten by small plankton, which are eaten by larger plankton, which are eaten by cod. The cod are eaten by seals which are eaten by polar bears. At each step, the concentration of PCBs increases, so they become very concentrated in the bears' fatty tissue. When Inuits (Eskimos) eat the polar bears, they are contaminated in turn. For another example, ciguatera fish poisoning is an increasingly common ailment among humans. This disease is caused by the ciguatera toxin, for which there is neither a diagnostic test nor a cure. Ciguatera is produced by a single-celled protozoan that attaches itself to marine algae growing on coral reefs. Small herbivorous reef fish ingest the protozoan as they eat the algae. These fish are then eaten by larger carnivorous fish such as tropical and semitropical grouper, red snapper, and barracuda, and the fat-soluble toxin becomes concentrated. When people consume these fish, they ingest the highly concentrated toxin. These examples of contamination concentrating upward in food chains may provide clues about what happens when we humans use chemicals in the production of foods.

Biomes

An ecosystem can range in magnitude from a planetary system that encompasses all the life on the planet to a single fallen log or even a drop of water. The most useful scale of ecosystem for geographic study is that of the biome. A **biome** is a large recognizable assemblage of plants and animals in functional interaction with its environment. It is usually identified and named by its dominant vegetation association. Animals (fauna) occur in much greater variety than do plants, but they are generally less prominent in the landscape, and they do not provide the clear evidence of environmental interrelationships that plants do. Animals are mobile and therefore more able to adjust to changes in the environment than plants are.

Humankind domesticated a great variety of plants and animals that are native to each different biome. **Domestication** is the process of adapting plants and animals to intimate association with humankind, to the advantage of humankind. Once species were domesticated, many were redistributed around the globe, but usually to environments similar to the species' native or original environments. Bananas, for example, were first domesticated in Southeast Asia, but humans transplanted them to similar environments in Africa and Central and South America, where the greatest production is today. Different natural environments still offer different advantages in the production of various agricultural goods. These differences affect local diets and international trade.

Scholars commonly accept the definitions of ten major biomes (Figure 1-23). The interrelationships of climate, flora, and fauna are exceedingly complex, but the global distribution of biomes shows a striking correspondence with that of climatic types (compare Figure 1-23 with Figure 1-20).

1. Tropical Rainforest The tropical rainforest, or selva, occurs principally in tropical wet climatic regions. The rainforest contains a variety of tall trees that never lose their leaves. Only where there are gaps in the canopy, as along riverbanks, does light reach the ground, resulting in a dense "jungle" undergrowth (Figure 1-24).

 The soil under this dense vegetation is, paradoxically, surprisingly infertile. This is because the interior of the selva is characterized by heavy shade, high humidity, windless air, and continual warmth. Litter accumulating on the forest floor is rapidly acted upon by plant and animal decomposers. The upper layers, or *stories*, of the forest are areas of high productivity, and the nutrients in this ecosystem are concentrated in the vegetation, not in the soil.

Rainforest animals are largely *arboreal* (tree-dwelling), although the interior includes vast num-

FIGURE 1-23

The infinite variety of the world's physical environment is best comprehended when generalized into these ten biomes. Compare the biomes shown on this map with the climate zones shown on Figure 1-20.

The Major Biomes of the World

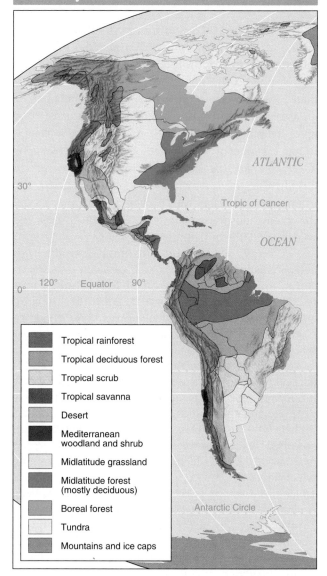

Tropical rainforest
Tropical deciduous forest
Tropical scrub
Tropical savanna
Desert
Mediterranean woodland and shrub
Midlatitude grassland
Midlatitude forest (mostly deciduous)
Boreal forest
Tundra
Mountains and ice caps

bers of *invertebrates* (animals without backbones), including insects.

2. Tropical Deciduous Forest The canopy of the tropical deciduous forest is less dense than that of the rainforest, and trees are shorter. There is a dry season that lasts for weeks or months, so many of the trees shed their leaves. This allows sunlight to penetrate to the forest floor and a dense understory to thrive. This biome does not correspond directly with any specific climatic type, indicating great complexity of environmental relationships.

Animal life includes more ground-level vertebrates than exist in the rainforest.

3. Tropical Scrub Scrub vegetation consists of low, scraggly trees and tall bushes, usually with an extensive understory of grasses. Trees are often openly spaced (Figure 1-25). Far fewer species are found in the scrub than in the tropical forests. Scrub areas are extensive in the drier portions of the A climatic regions.

The fauna consists of a varied mix of ground-dwelling mammals and reptiles and of birds and insects.

4. Tropical Savanna The plant cover of savanna lands consists primarily of tall grasses with a mix of trees and shrubs (Figure 1-26). Most savannas experience wildfires, almost annually. The fires burn away the nonfertile portion of the grass, but they do not cause any harm to the shrubs and trees. When the next season's rains arrive, the grasses spring into growth with renewed vigor.

Animal life in the savanna varies among continents. African savannas are rich in large animals such as elephants, zebras, and giraffes. The Latin American savannas, by contrast, have only sparse populations of large animals. The Asian and Australian savannas are intermediate.

This biome correlates with the tropical wet and dry climatic type, but imperfectly. The correlation is best in the parts of South America and Africa where contrasts in seasonal rainfall are greatest.

5. Desert Deserts often exhibit surprisingly rich vegetation and animal life (Figure 1-27). Many plants are drought-resistant; they conserve what moisture they receive and reproduce hastily during brief rainy periods.

Animal life is inconspicuous but diverse. There are few large animals, but small mammals and other species rest in burrows and crevices during the day and prowl at night.

6. Mediterranean Woodland and Scrub This biome, dominated by a dense growth of woody shrubs corresponds with the Mediterranean mild midlatitude climate region (Figure 1-28). The trees are primarily broadleaf evergreens. Their small, waxy leaves retain water during the dry season, and most of the plants have deep roots. Late summer fires can be followed by winter floods, and if the winter rains arrive before the vegetation resprouts, erosion can be severe. Seed-eating burrowing rodents are common, as are some bird and reptiles.

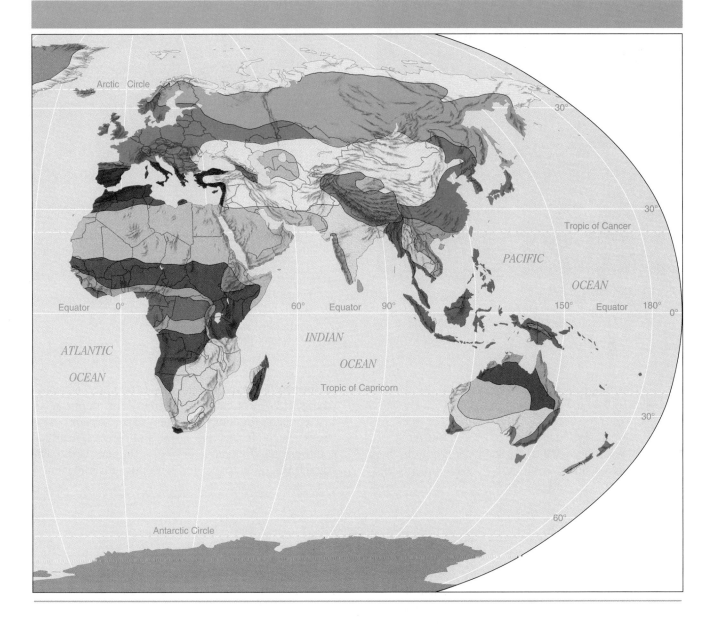

Southern California exemplifies a Mediterranean environment, and the region periodically suffers late summer fires, followed by mudslides when the fall rains come (Figure 1-29). This biome is particularly susceptible to damage, and, as noted above, the impoverishment of the Mediterranean basin prompted George Perkins Marsh's early writings about humankind's role in environmental degradation.

7. Midlatitude Grassland The midlatitudes of all continents contain vast areas of relatively continuous grassland (Figure 1-30). The insufficiency of water plus the frequency of fires prevents the growth of trees or shrubs except along streams. The wetter areas that support tall grasses are called prairies; in the drier areas, called steppes, short grasses predominate. Areas of midlatitude grassland where humankind cultivates domesticated grasses are among the world's great "breadbaskets." Other grasslands provide pasturage for grazing animals, often domesticated ones.

In the Northern Hemisphere the grassland biome occurs largely in regions with the steppe climatic type, but in the Southern Hemisphere the correspondence is less exact.

8. Midlatitude Deciduous Forest Extensive areas on all Northern Hemisphere continents and more limited tracts in the Southern Hemisphere were originally covered with forests of largely broad-leafed deciduous trees. Very little of this vegetation remains, however, because a large portion of it has been cleared for human use.

FIGURE 1-24

Rainforest vegetation is generally dense and multistoried, as exemplified by this scene from the tropical island of Borneo. (Courtesy of Julia Nicole Paley.)

FIGURE 1-25

This lion stalks his prey in a thorn scrub environment in northern Namibia. (Courtesy of Tom L. McKnight.)

In Northern Hemisphere regions the deciduous forest gives way toward the north to needle-leaf evergreens. The pine forests of the southeastern United States seem to be "natural vegetation," but most of this land was only recently (in geologic terms) uplifted from the ocean floor, and this vegetation is probably not climax.

9. Boreal Forest Boreal forest, or taiga, is one of the world's most extensive biomes. This northern forest contains the simplest assemblage of plants. Most trees are coniferous pines, firs, and spruces, nearly all of them needle-leaf evergreens (Figure 1-31). This biome is closely associated with the subarctic climatic type.

FIGURE 1-26

These topi (a kind of antelope) graze the savanna lands of the northern Serengeti Plain in Kenya. Natural wildlife attracts tourists from around the world and makes tourism one of Kenya's principal money-earning industries. (Courtesy of Tom L. McKnight.)

FIGURE 1-27

California's Mojave Desert is not without vegetation, but agriculture can be practiced only near infrequent springs. (Courtesy of John Daniels.)

FIGURE 1-28

This Moroccan scene typifies a Mediterranean environment. The vegetation reflects long-term human intervention.
(Courtesy of the Moroccan National Tourist Office.)

The harsh climate, floristic homogeneity, and slow plant growth yield only limited supplies of food for animals. Mammals are represented by fur bearers such as bears and wolves and *ungulates* (hoofed animals) such as caribou and moose. Birds are numerous and diverse in summer, but in the winter they depart for warmer regions. Insects are abundant in summer but absent in winter.

10. Tundra The tundra is essentially a cold desert in which moisture is scarce and summers are so short and cool that trees are unable to survive. The dwarf plant forms consist of a mixture of species, including grasses, mosses, lichens, flowering herbs, and shrubs (Figure 1-32).

Animal life is dominated by birds and insects during the summer. There may be a few species of mammals, but almost no reptiles or amphibians.

The Earth's taiga and tundra regions hold considerable concentrations of mineral resources—oil and gas as well as mineral ores—but the difficulties of working in these environments increases the costs of exploiting the resources. It is difficult to build roads, railroads, or pipelines because the ground is frozen solid in the winter, and during the short sum-

FIGURE 1-29

These terrible fires that swept the hills above Malibu, California, in November 1993, may have been set by an arsonist, but fires can be started by lightning.
(Courtesy of Theodora Litsios/Gamma-Liaison, Inc.)

mer the surface becomes deep mud and the constructed facilities break up.

Exploitation also presents dangers to these environments themselves. Chemical decomposition is slow in the cold weather, so refuse and oil spills degrade slowly and may threaten animal life. These considerations have raised great controversy over the exploitation of Alaskan resources.

FIGURE 1-30

This midlatitude grassland in Kansas has been planted in domesticated grasses for human consumption.
(Courtesy of the Kansas Wheat Connection.)

FIGURE 1-31

The town of Kennicott in southern Alaska's boreal forest was built when copper was discovered there in 1898, but the town was abandoned after the mine closed in 1938. Decay and chemical decomposition occur very slowly in this cold climate. Therefore, any environmental pollution lingers.
(Courtesy of Julia Nicole Paley.)

SOILS

▼▼▼▼▼▼▼▼▼▼▼▼▼▼▼▼▼▼▼▼▼▼▼▼

Soil is an infinitely complex and varied mixture of disintegrating mineral particles, decaying organic matter, living organisms, gases, and liquid solutions. This precious medium is spread thinly across most continental surfaces: its average worldwide depth is only about 6 inches (15 cm).

The development of soil is initiated by the physical and chemical disintegration of rock that is exposed to the atmosphere and by the action of water percolating down from the surface. The disintegration is called *weathering*, and the dissolving away of minerals by percolation of groundwater is called *leaching*. Four factors then interact in soil formation: (1) the parent rock; (2) the environmental conditions, including temperature, moisture, and topography; (3) time; and (4) the biological factor of the living and dead plants and animals incorporated into the soil (Figure 1-33).

Scientists first assumed that the parent rock was the most important of these factors. Studies of soil types around the world, however, have taught us that

their distribution is closely related to the distribution of world climate types. Climate, rather than parent material, is the principle determinant of soil type.

Soil Fertility and Productivity

The innate fertility of a soil is due largely to the chemical composition of the parent rock, but other factors, such as climate, affect soil fertility. We earlier noted the paradox that the soils under lush tropical rainforests are often surprisingly infertile. This is largely explained by the rapid recycling of the ecosystem's nutrients by living things. Another reason is that the tropical soils are excessively leached. When farm settlers clear the vegetation, the infertility of the soil disappoints their hopes.

 In drier grassland areas, by contrast, vegetation is sparser, so more nutrients are left in the soil. These factors explain the double paradox that the luxuriant rainforest vegetation can be found growing on poor soil, while some of the richest soils in the world are found under humble grasses in arid regions.

Fertility refers only to the availability of nutrients for plant growth, but human geographers are

FIGURE 1-32

Tundra can be found at high latitudes or high in mountains where trees can no longer survive. Trees just barely hang on at the bottom of this Alaskan scene, but the higher valleys sustain only rudimentary species of vegetation.
(Courtesy of Julia Nicole Paley.)

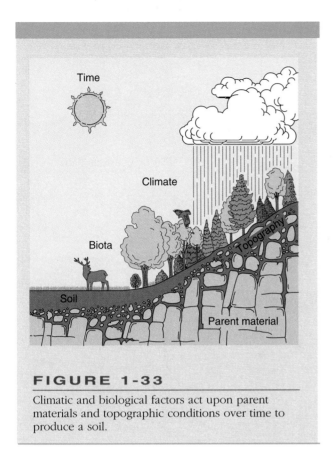

FIGURE 1-33

Climatic and biological factors act upon parent materials and topographic conditions over time to produce a soil.

interested in soil's usefulness to humankind. A soil's productivity is influenced by considerations of its structure (including physical characteristics such as the size of particles), the needs of the particular crop being planted, details of the local climate, local agricultural practices, and many other variables. Fertile soils are not always productive, and the most productive soils are not always the most fertile.

Different soils must be managed differently. Through the thousands of years of the evolution of human agriculture, humankind has devised many techniques of plowing, planting, cultivating, fertilizing, draining, and irrigating. Humankind has learned which crops will grow in soils that are either acidic or alkaline. Excess acidity normally can be reduced by the addition of an alkaline substance, such as pulverized limestone. An excess alkaline or saline content in arid soils usually can be removed by abundant irrigation and good drainage. Desert soils may be extremely fertile and productive if irrigated.

Some soils yield abundantly if they are simply cleared, plowed, and seeded. Examples of such productive soils include those derived from limestone, soils deposited on *flood plains* (the lowlands along a riverbank, which may be subject to periodic flooding), fine-textured wind-deposited soils called loess, and certain volcanic soils.

Less fertile soils may respond to fertilization or crop rotation. For example, the soils of the eastern United States developed under an original forest cover in a humid climate. They have been thoroughly leached, their nutrient content is relatively low, and their natural fertility is limited. However, they generally have good structure, and local rainfall usually is dependable. These soils respond well to careful management, so with proper fertilization, drainage, and crop rotation, they can—and have— become productive.

TOPOGRAPHY

The Earth's topography varies from flat plains to rugged high mountains (see the endpaper map at the back of the book). Even the land surface underneath the world's oceans is variegated, although that is of less immediate interest to humankind. These variations in surface configuration affect almost all other aspects of the Earth's physical geography (climate, vegetation, and soils), as well as almost every aspect of human life.

The Earth's topography results from the effects of long-term processes on the geologic foundation. We refer to these processes as geomorphic—"geo" meaning earth and "morph" meaning form.

Massive Crustal Rearrangements

Prior to the twentieth century, most Earth scientists assumed that the planet's crust was rigid, with continents and oceans fixed in position and significantly modified only by changes in sea level and periods of mountain building. Today, however, the theory of **plate tectonics** is widely accepted. This theory holds that the upper portion of the Earth's crust consists of a mosaic of about 20 rigid plates embedded in an underlying, somewhat plastic layer.

Plate tectonics provides a framework for understanding crustal rearrangements. About 200 million years ago the present continents were connected as a single landmass, called Pangaea, but Pangaea broke apart, and the present continents moved away from one another. This process is called *continental drift*. Still today these plates continue to move apart, come closer together, or slide past one another at the rate of a few millimeters to a few centimeters per year (Figure 1-34).

The rate of movement of the plates at their contact points with other plates is important, because these contact points are the locations of most mountain

The Background to Human Geography **35**

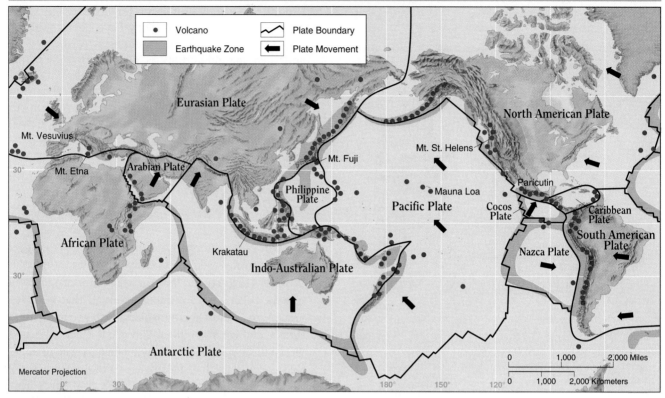

FIGURE 1-34

Volcano and earthquake zones characterize the boundaries among the Earth's tectonic plates.

building, earthquakes, and volcanic activity, sometimes with devastating effects to human life and property.

Earthquakes and Volcanoes

Deformation of the Earth's crust continues in a great variety of processes that are together labeled *diastrophism*. These include *warping* (the uplifting or depression of broad surfaces), *folding* (bending), and *faulting* (breaking apart).

Zones of weakness in the Earth's crust are called *fault zones* or planes, and the vibrations in the crust produced by shock waves from sudden displacements along faults are called *earthquakes*. These fault movements result from an abrupt release of energy from a long, slow accumulation of stress, which is an ongoing process in crustal deformation. The center of motion is called the *focus*, which is underground, and the spot directly above the focus is called the quake's *epicenter*.

As noted above, crustal activity is concentrated along the contact points between tectonic plates. For example, California is tectonically active because two

plates meet along several fault lines. In the Mediterranean region, the African, Eurasian, and Arabian plates collide, causing severe earthquakes in southeast Europe and southwest Asia. The Himalayas stand where South Asia is converging with Asia, and Mount Everest is still being pushed up at a rate of about 0.394 inches (1 cm) per year. The so-called "Ring of Fire" around the Pacific Ocean is always rumbling with earthquakes or volcanic eruptions.

The LAGEOS satellites mentioned earlier in this chapter can detect crustal movements to within one centimeter. The more we know about the rate, intensity, and duration of change of the contacts at tectonic plate interfaces, the better we will be able to predict earthquakes and volcanic activity.

Earthquakes are usually measured on the Richter scale. The scale is logarithmic, so each successively higher number represents a quake ten times stronger than the preceding number. The scale numbers range from 0 to 9 but theoretically have no upper limit. Any earthquake registering 8 or above is considered to be catastrophic, but even lesser shocks can cause great damage to human settlements. The most violent known earthquakes have recorded 8.9

on the Richter scale. San Francisco suffered earthquakes rated 8.3 in 1906 and 6.9 in 1989, and the earthquake that struck Los Angeles in January 1994 rated 6.7.

Underwater earthquakes may trigger great sea waves, called tsunamis. These are unrelated to tides, but they are often improperly called "tidal waves." A tsunami may be a wall of water as much as 50 feet high traveling at a speed of up to 450 miles (720 km) per hour. When a tsunami strikes a coast, it may sweep away entire communities. On July 12, 1993, an earthquake recorded at 7.8 on the Richter scale occurred off the southeast coast of the Japanese island of Hokkaido. It triggered a tsunami at least 30 feet high that obliterated a district on Okushiri, a small island known for its fishing and resorts. Nearly 300 people died.

Volcanoes are eruptions onto the Earth's surface of molten material from the interior of the earth, called magma (Figure 1-35). Magma which has cooled and solidified is called lava, and some soils that develop from lava are among the world's richest. Farmers will often cling to the slopes of volcanoes that are inactive (called *dormant*, sleeping) but that may erupt again. These farmers risk the loss of everything, even their lives, in order to enjoy the rich harvest.

The example of a farm family planting a crop on the edge of a volcano, earning there its livelihood from that natural richness, raising children, and even singing through the workday may be an appropriate image of humankind struggling and thriving in the context of the Earth's great natural forces. The next

FIGURE 1-35

An erupting volcano can put on a spectacular show. This eruption of the volcano Arenal surprised Costa Ricans in 1991; the volcano had long been described as "extinct." (Courtesy of Gregory G. Dimijian/Photo Researchers, Inc.)

chapter will examine in greater detail the many ways in which humankind manipulates and transforms the Earth.

SUMMARY

Geography is the study of the interaction of all physical and human phenomena at individual places and of how interactions among places form patterns and organize space.

Classical Greek and Roman geography culminated in the geography of Ptolemy, and when Ptolemy's book reappeared in Latin in 1406, it was still taken as the most authoritative source of geographical knowledge. Outside Europe, the Islamic world, the Chinese, the Japanese, and the Koreans developed extensive geographical literatures. Other peoples may have, but either these works have not survived or modern scholars have not yet fully inventoried and studied them.

Europe's world exploration and conquest challenged scholars to catalogue and organize new knowledge. Bernhard Varen differentiated special geography—today called regional geography—and general geography—today called topical or systematic geography. Either approach can be applied to either physical geography or human geography. Cultural geography focuses specifically on the role of human cultures.

Scholarly geography developed as an ever more rigorous inquiry into the reasons for the distributions of phenomena. Professor Karl Ritter and Alexander von Humboldt are usually regarded as co-founders of modern geography. George Perkins Marsh's writings were key to what would become today's environmental movement. Today almost all of us are geographers, and the field offers an introduction to a unique breadth of eventual careers.

The distribution of a phenomenon means its position, placement, or arrangement throughout space. The absolute location of a place, or its site, pinpoints a spot in terms of the global geographic grid. The relative location of a place, by contrast, called its situation, tells us where it is relative to other places. A region is a territory that exhibits a certain uniformity.

A map is a graphic representation of any area on which only selected data are shown. Cartographers must always choose between making a conformal map or an equivalent map. Techniques of remote sensing include aerial photography, aerial heat-detecting sensors, weather satellites, photo reconnaissance satellites, and specialized sensing devices. Geographic Information Systems are computer systems that organize, store, update, and map information. Some use a vector method, others a raster method. Computers can seek correlations among vast amounts of data across great distances at the press of a button. Whether or not these correlations mean anything must still be determined by the geographer's ability to formulate and investigate explanatory hypotheses.

The perpetual motions of the Earth continually change the distance between the Earth and the sun and the portion of the Earth that faces the sun. This causes the changes of the seasons. The Earth's atmospheric envelope is energized by solar radiation, stimulated by earthly motions, and affected by contact with the Earth's surface. It produces an infinite variety of conditions known as weather. Climate is the aggregate of weather conditions over a long period of time. Tropical areas receive a large energy surplus, while high latitudes experience an energy deficit, but air and water currents shift warmth toward the high latitudes and of coldness toward the low latitudes. The atmosphere generally allows short-wave radiation from the sun to pass through to the Earth's surface but inhibits the escape of long-wave radiation from the Earth into space in a phenomenon called the greenhouse effect.

Climates are classified on the basis of temperature and precipitation. We can construct a model of the distribution of climatic types, thereby demonstrating deductive thinking. Deductive studies often use an isotropic plain.

Through the Earth's history, periods of relatively low temperature and large accumulations of ice ("Ice Ages") have been interspersed with periods of melting. Volcanic activity is one important factor in determining the chemical composition of the atmosphere, but future global warming may be anthropogenic.

The concept of climax vegetation enables us to think of the Earth's vegetation not as a static patchwork, but as a dynamic system continually accommodating environmental changes and human interference. Each spot on Earth has a unique inventory of biological resources. An ecosystem is a biological community, and a biome is a large recognizable ecosystem. Scholars commonly accept the definitions of ten major biomes.

Soil is a complex and varied mixture of disintegrating mineral particles, decaying organic matter, living organisms, gases, and liquid solutions. Climate is the principle determinant of soil type. Fertile soils are not always productive, and the most productive soils are not always the most fertile.

The upper portion of the Earth's crust consists of a mosaic of about 20 rigid tectonic plates embedded in an underlying, somewhat plastic layer. Deformation of the Earth's crust continues through diastrophism. Zones of weakness in the Earth's crust are called fault zones or planes, and the vibrations in the crust produced by shock waves from sudden displacements along faults are called earthquakes.

Volcanoes are eruptions onto the Earth's surface of molten material from the interior of the earth.

KEY TERMS

geography (p. 2)
regional geography (p. 3)
topical geography (p. 3)
systematic geography (p. 3)
physical geography (p. 4)
human geography (p. 4)
cultural geography (p. 4)
absolute location (p. 5)
relative location (p. 5)
site (p. 5)
situation (p. 5)
equator (p. 6)
latitude (p. 6)
longitude (p. 6)
prime meridian (p. 6)
topography (p. 7)
friction of distance (p. 8)
region (p. 8)
map projection (p. 11)

QUESTIONS FOR INVESTIGATION AND DISCUSSION

1. What would be the effect on the annual march of the seasons if the Earth's axis did not maintain parallelism during its revolutions?

2. How does weathering play a role in the survival of life on Earth?

3. How could an increase in the greenhouse effect change the way of life on Earth?

4. Nations try to preserve their national cultural heritage and their national resources. Governments generally seem less concerned, however, about preserving their natural biological diversity, the uniqueness of each nation's floral and faunal heritage? Can you explain why?

5. How does the geographers' notion of a region compare with historians' "period," psychologists' "personality," art historians' "style?"

6. Why are there so many different kinds of map projections?

7. What is the climatic climax vegetation in your area?

8. In which biome do you live? If you live in a farming region, what crops are most common?

9. What is the latitude and longitude of your town?

10. What is the importance of the greenhouse effect to the heating of the Earth's surface?

11. Plot the average monthly rainfall and temperature where you live. Which are the hottest and coldest, wettest and driest months? In which climate type do you live?

38   *Chapter 1*

Geographic
  Information
  System (GIS) (p. 14)
greenhouse effect (p. 19)
temperature inversion (p. 19)
monsoon (p. 19)
model (p. 22)
anthropogenic (p. 24)
inductive and deductive
  geography (p. 27)
isotropic plain (p. 27)
climatic climax
  vegetation (p. 28)
diversity index (p. 28)
ecosystem (p. 29)
food chain (p. 29)
biome (p. 29)
domestication (p. 29)
soil (p. 34)
plate tectonics (p. 35)

ADDITIONAL READING

BARRY, ROGER G. and RICHARD J. CHORLEY. *Atmosphere, Weather and Climate*, 6th ed. New York: Routledge, Chapman and Hall, 1993.

COX, C. BARRY and PETER D. MOORE. *Biogeography: An Ecological and Evolutionary Approach*, 5th ed. Cambridge, MA: Blackwell Scientific, 1993.

GAILE, GARY L. and CORT J. WILLMOTT, eds. *Geography in America*. Columbus, Ohio: Merrill Publishing, 1989.

HAINES-YOUNG, ROY, DAVID R. GREEN and STEVEN COUSINS, eds. *Landscape Ecology and Geographic Information Systems*. Philadelphia, PA: Taylor & Francis, 1993.

JAMES, PRESTON E. and GEOFFREY J. MARTIN. *All Possible Worlds: A History of Geographical Ideas*, 3rd ed.. New York: John Wiley & Sons, 1993.

JOHNSTON, R. J., ed. *The Challenge for Geography: A Changing World, A Changing Discipline*. Cambridge, MA: Blackwell, 1993.

MAGUIRE, DAVID, MICHAEL GOODCHILD and DAVID RHIND, eds. *Geographical Information Systems: Principles and Applications*. New York: John Wiley & Sons, 1991.

McKNIGHT, TOM L. *Physical Geography*, 4th ed.. Englewood Cliffs, NJ: Prentice Hall, 1993.

MOORES, ELDRIDGE, M, ed. *Shaping the Earth: Tectonics of Continents and Oceans*. New York: W. H. Freeman and Co., 1990.

The Scientific Method

In the Middle Ages people thought that they had to prove things in order to know them. There are logical reasons, however, why it is impossible to prove almost anything. For instance: Is there a hippopotamus in the room in which you are sitting right now? You probably cannot see one, hear one, smell one, or feel one, but how can you prove that one isn't there? How can you be absolutely sure that one isn't hiding? Perhaps the sandwich that you had for lunch has fermented into a drug that blinds you to hippos. Or maybe the Frisbee that knocked you on the head last week has affected your senses. Or maybe some other reason—no matter how improbable—prevents you from perceiving the hippopotamus. How can you know for certain? If you think about it long enough, you could go crazy, but you will not be able to prove that you are in a hippopotamus-less room. Early in the Renaissance people decided that you really do not have to prove things. You can assume something is true until you are proven wrong. This way of proceeding can help you accumulate knowledge faster. One of the first people to articulate this idea was William of Ockham (or Occam), who lived from about 1285 until 1349. He cut away all the complications from medieval thinking with a much simpler idea, appropriately called *Ockham's razor*. "Entities must not unnecessarily be multiplied," he said. In other words, the simplest explanation that fits the facts is most probably true. If you are given a choice among possible explanations for any observation or phenomenon, you should prefer the simplest. If you cannot see, hear, smell, or touch a hippopotamus in your room, there probably isn't one there.

THE SCIENTIFIC METHOD DEFINED

Ockham's articulation of his razor eventually led to the step-by-step technique of investigating phenomena that we call the **scientific method**. This is a set of procedures used in the systematic pursuit of knowledge. If you want to learn about the world, follow these three steps: (1) Observe the world, (2) explain what you see as simply as possible, and (3) hold on to your tentative explanation until it is disproved.

OBSERVE THE WORLD

At the time of the Renaissance many people believed that the statements given in ancient Greek or Latin texts or in the Bible should override their own observations. People were not confident that they were as smart as the ancients had been, and, in addition, they believed that much of what we see in the world is the work of the devil. Therefore, if you believe your own senses you may be fooled and suffer eternal damnation.

The Bible records that Joshua said "Sun, stand thou still," and "the sun stood still" (Joshua 10.10). This passage was taken as proof that the sun goes around the Earth. Anyone who believed that the Earth goes around the sun was wrong and was considered heretical. Thus, when the Italian astronomer Galileo (1564-1642) wrote that the Earth goes around the sun, he was tried, convicted, and punished for heresy. "I think," he wrote in his defense, "that in discussions of physical problems we ought to begin not from the authority of scriptural passages, but from sense experiences and necessary demonstrations. God is no less excellently revealed in Nature's actions than in the sacred statements of the Bible." A Church court nevertheless forced him to apologize for teaching that the Earth goes around the sun.

EXPLAIN WHAT YOU SEE AS SIMPLY AS POSSIBLE

Your explanation of what you see is called a hypothesis. A **hypothesis** is a proposition tentatively assumed in order to draw out all of its consequences and so test its accord with all facts an investigator can gather. You must remember to cut away (with Ockham's razor) all unnecessary complications and explain what you see as simply as possible. If you cannot see, hear, smell, or touch a hippopotamus in your room, you should not hypothesize that one is hiding or that you have been drugged. Hypothesize that no hippopotamus is there.

It is not always easy to suggest a hypothesis. Sometimes it is hard to think of any possible explanation for an observed phenomenon, and it takes a brilliant insight to see a link between two facts that had never occurred to anyone else before. Biologist Sir Peter Medawar wrote: "Hypotheses appear in scientists' minds along uncharted byways of thought...they are imaginative and inspirational in character...they are adventures of the mind."

HOLD ON TO YOUR HYPOTHESIS UNTIL IT IS DISPROVED

If your hypothesis withstands test after test, then you may become confident of it. If I hold out my hand and let go of my pencil, I do not know for certain that it will fall to the floor. I know that that has happened before so regularly that people have gained considerable confidence in the notion of gravity. The force of gravity is a **theory**, which is a hypothesis that has been given probability by experimental evidence. I cannot, however, be absolutely certain that my pencil will fall if I let go. If I let go and my pencil hovers in the air, then the theory of gravity will have to be discarded, and we will have to hypothesize a new theory. Gravity is only a theory.

It should be noted that because the sun rises in the East and sets in the West, a geocentric (Earth-centered) theory that the sun goes around the Earth is actually the simplest possible explanation of the observed phenomena, and it was held by

thoughtful people, including Aristotle, for a long time. The geocentric theory cannot, however, be reconciled with additional observations of celestial phenomena. The Polish astronomer Copernicus (1473-1543) first proposed the heliocentric (sun-centered) theory to replace the geocentric theory. The weight of observation supports the heliocentric theory. It explains all observed celestial phenomena better.

A theory that has withstood testing over a long time is often called a **law**, but this term is confusing and should be avoided. What does it mean to say that there is a "law of gravity"? Who passed it? Who enforces it?

The English philosopher and statesman Sir Francis Bacon (1561-1626) said, "We cannot command Nature except by obeying Her." In other words, we do not pass laws for Nature to obey. We study Nature's principles and try to understand them. Once we do understand them, however, we can harness them to work for us. Medieval magicians waved magic wands, recited magic spells, and commanded Nature to change one thing into another, but their efforts were unsuccessful. Modern chemists, however, can break up materials and reconstitute them into new products, because the chemists understand and obey the principles of molecular bonding.

We do not always understand Nature's principles as completely as we think we do. Theories that have been held for a long time

and that seem "obvious" can be overturned overnight. For example, Sir Isaac Newton's theory that time is an absolute value was overturned by Albert Einstein's theory of relativity. Therefore, a true scientist will never say that he or she is certain of any theory. Scientists proceed on the basis of hypotheses. I only hypothesize that I am sitting on a chair, that I am wearing a blue shirt, and that I hear cars honking in the street outside. Who knows, maybe there was something about that sandwich I had for lunch.

GEOGRAPHIC COINCIDENCE AND CAUSE-AND-EFFECT

In geography, one use of the scientific method is to try to explain patterns. Geographers note when the occurrence of a phenomenon forms a pattern, or when two different phenomena occur in the same pattern. In Chapter 1, for example, we learned that certain climates often appear at corresponding latitudinal and continental locations around the world. This is a pattern. We also learned that the patterns of certain types of climates and of certain biomes are the same. This ability to recognize patterns may seem simple, but actually it is a learned skill which geographers exercise in common with art historians and others in the visual arts.

Once a pattern is discovered, then geographers hypothesize

about what might be the cause of the pattern. When two phenomena exhibit the same pattern, they are called **coincidental**, and geographers hypothesize whether one of those phenomena might cause the other. There may be a cause-and-effect relationship between them, but we cannot immediately assume that one of the phenomena causes the other. We have to hypothesize a cause-and-effect relationship and then investigate it.

When scientists test a hypothesis, they experiment to see whether it is repeatable. Some scientists argue that the repeatability of results is actually what defines a science. If a chemist in Los Angeles performs an experiment and reports its results, a chemist in Paris must be able to repeat the experiment with exactly the same results.

If we geographers want to test a hypothesis about a cause-and-effect relationship between two phenomena, however, we cannot isolate the phenomena we want to test as easily as a chemist can isolate two chemicals in a test tube. The world is too complicated. An economic development policy that works in Thailand may not work in Argentina. This may be because of any one of a great number of factors that differentiate Thailand from Argentina, and none of these other factors may be under our control. In order to test geographical cause-and-effect hypotheses, we must compare as many different places as possible, and we must search for exceptions.

When beginning geographers make a few observations of a coincidental occurrence of two phenomena, they may hastily hypothesize that some factor A always causes some factor B: "Natural resources cause national wealth"; "A certain climate causes civilization"; "A certain religion causes

social peace." The assumption that one factor alone is sufficient to explain any event is called a *single-factor hypothesis*. In many cases, however, single-factor hypotheses are too simplistic, and we need to consider several factors.

Further research and study often reveal exceptions to our first observations. The world is rich in examples of different environments, of different ways in which human cultures have evolved and adapted to those environments, and of different ways in which various aspects of cultures have interacted. Therefore, we can test our hypothesis by hunting for exceptions to our first observations. If we cannot find exceptions or examples in the world today, we must explore the other dimension: time. We may find an exception or example at some place at some time in the past. The study of geography and the study of history are often complementary.

Some people tease geographers for being obsessed with finding unique cases of phenomena, places where things are different, even "odd." They say that geography is just a sophisticated game of *Trivial Pursuit*. The reasons that geographers collect examples of phenomena and exceptional cases, however, are to test cause-and-effect hypotheses and also to catalog the tremendously rich variety of phenomena on Earth. If a geographer finds just one example of a way of life, for example, on a tiny Pacific island or in a mountain valley in Switzerland, then our catalog of possibilities has been enlarged. The options among which we may choose to organize our own lives have increased.

Because geographers cannot always disentangle the interrelationships among the phenomena

we study, we must research cause-and-effect links by stating and restating certain questions very carefully. If any factor "A" (a certain climate, natural resources, or a certain religion) is coincidental with phenomenon "B" (a certain soil, national wealth, or environmental degradation) in one place, then we might hypothesize that phenomenon A is causing B. If we later find B someplace else where there is no A, then we might restate our hypothesis. We might hypothesize that A sometimes causes B, but that B can also be caused by some other factor C. Then we have to guess what that factor C might be. Maybe B can be caused by a factor D, an E, or an F, as well. Sometimes many factors combine to explain the phenomenon under investigation. Similarly, if we hypothesize that A is causing B at one place, and if we then observe A without B in another place, we might hypothesize that A has not yet had time to trigger B in that place.

We must carefully differentiate between factors that are *necessary* to cause an event and factors that are *sufficient* to cause an event. If A is necessary to cause B, then we cannot find B without A. If A is necessary but not sufficient to cause B, then A plus some other necessary factor must appear. We must investigate what that other factor is. If A is both necessary and sufficient to cause B, then we will find B wherever there is A. For example, observation of the world today suggests that domestic tranquillity is necessary for a country to achieve economic prosperity. Many peaceful countries, however, are poor. Therefore, we may hypothesize that domestic tranquillity is necessary but not sufficient to bring prosperity. Additional necessary factors must be identified.

Maybe A and B are always found together, but there is actually no relationship between them. In common speech this is usually called "*mere* coincidence." Maybe A and B are always found together, but both are actually caused by some third factor Q. Chapter 1 gave an example of such a case: Certain kinds of vegetation and certain soils are found together, but one does not cause the other. Rather, both are caused by climate. Still another hypothetical relationship is that B causes A.

There are, then, at least four possible hypotheses if any two phenomena A and B are found together: (1) A may cause B; (2) B may cause A; (3) both may be caused by some factor C that is yet to be identified; (4) or the coincidence of A and B may be *mere* coincidence, and there is no causal relationship between A and B at all.

PARTIAL EXPLANATION

The problem of national economic prosperity illustrates that in geography any one thing seldom explains another thing completely. Usually, one factor explains only part of another factor. It contributes to an explanation.

The science of statistics breaks down the degree of variation in one factor that can be attributed to any one of several other factors. When statisticians describe the variation of two measures together, called *covariation*, they use the word *explain* in a very special sense. To statisticians, an explanation does not really answer the question of why something happens when something else happens. To them, an *explanation* means only that one factor varies with the other with a definable regularity. When a sta-

tistician says that "variation in factor X explains 80 percent of the variation in factor Y," the statistician is saying that they vary together 80 percent of the time. The statistician is not explaining anything in the normal sense of the word *explain*.

In geography, as in everyday life, we can often list probable causes, and sometimes we can even rank those causes in order of their probable explanatory force, but we cannot always assign exact relative weights to the individual causes. We can say, for example, that the United States is a rich country because it has many natural resources, it has a history of relative political stability, and its economic system has encouraged people by allowing them to enjoy the rewards of their enterprise. These, however, are only a few of the relevant factors, and we cannot measure statistically the exact contribution of each of them. That is, we cannot say that natural resources account for 25 percent of national wealth, domestic tranquillity 35 percent, and the economic system 40 percent. There may be many other factors involved, and we do not know how to allocate explanatory power among them.

In some cases geographers record that two things are found at the same place at the same time, but the degree of cause-and-effect relationship between them—if any—simply is not known. Any one or combination of a tremendous number of other factors may be relevant. These are challenging questions for continuing research, and we will note a great number of such questions in this textbook.

Throughout the book you will be asked to compare two or more maps and to hypothesize from them what cause-and-effect relationships might exist among the

factors mapped. Does the most productive agriculture coincide with certain climates? Do the most generous national food supplies coincide with the highest percentages of the national populations working in agriculture? What factors coincide with the ratios of females to males in national populations? Do the countries that have the largest fuel reserves enjoy the highest per capita use of energy? Does the world pattern of manufacturing coincide with the location of raw materials? When you make these and other correlations you may be surprised at what you find. Form your own hypotheses of explanation.

Key Terms
Scientific method (p. 40)
Hypothesis (p. 40)
Theory (p. 41)
Law (p. 41)
Coincidental (p. 42)

Additional Reading
AMEDEO, DOUGLAS AND REGINALD G. GOLLEDGE. *An Introduction to Scientific Reasoning in Geography.* New York: John Wiley & Sons, 1975.

BRONOWSKI, J. AND BRUCE MAZLISH. *The Western Intellectual Tradition.* New York: Harper & Row, 1960.

HARVEY, DAVID. *Explanation in Geography.* London: Edward Arnold, 1969.

2 | Cultures and the Transformation of Landscapes

Maintenance of these rice terraces on the Indian Ocean island of Bali transforms the natural landscape into a product of human engineering.

Courtesy of Jim Steinberg/Photo Researchers, Inc.

The Earth's natural environment discussed in Chapter 1 is the result of complex interactions of physical and ecological systems. This chapter will examine in greater detail the relationships between human beings and the natural environment.

First we will examine the ways in which the physical environment affects human activities. Human societies must adapt to local climatic conditions, soils, hydrography, vegetation, and other attributes of a local environment as they learn to exploit the local resources. Human societies are not, however, passive. As any human society adapts, it actively and deliberately alters natural conditions. The plagioclimax (false climax; refer back to Chapter 1) vegetation that covers much of the Earth, for example, is maintained for human benefit. Thus, interaction between the environment and humankind is reciprocal—the environment affects human life and cultures, and humans alter and transform the environment.

THE DEVELOPMENT OF CULTURES IN ISOLATION

▼▼▼▼▼▼▼▼▼▼▼▼▼▼▼▼▼▼▼▼▼▼▼▼▼▼▼▼▼

Imagine that deep in the Brazilian rainforest there is a human community that is completely isolated. The people carry on their lives in total ignorance of the rest of us here on the planet, knowing nothing of international trade, of international political affairs, or even of the government of Brazil, which exercises no effective jurisdiction over them. Certainly they are affected by the complicated affairs of the rest of us. They see our airplanes in the sky, and their local environment is changing because the rest of us are polluting the atmosphere, but they explain these things either as natural phenomena or as the work of spirits. These isolated people—let's call them the "Lonely" people—remain convinced that they are the only humans on the Earth (see Figure 2-1).

An isolated society such as this must depend entirely on its local environment for all of its needs, including food, clothing, and shelter. Thus, for example, the Lonely people would have learned over time which local plants and animals can be eaten and which local animals can be trained to work for humans. If these people receive no imports from the outside world, neither do they produce anything for others. Everything produced locally is consumed locally, and any surplus food or goods are stored for emergencies or else just wastes away.

As the Lonely people make use of their local environment, they also transform it. They can clear forests or plant new ones, drain swamps or dig irrigation systems, fertilize the soil or deplete and waste it through poor agricultural practices. If they burn local vegetation they make their own contribution to world atmospheric pollution. They can even redesign local landforms, as many societies have reduced steep mountainsides to stepped terraces for agriculture. People are never entirely passive; rather, they interact with their local environment.

The Concept of Culture

The Lonely people will have developed in isolation their own distinctive **culture**. This word is often used to mean only fine paintings or symphonic music, but to social scientists it means everything about the way a people live: what sort of clothes they wear (if any); how they gather or raise food; whether they recognize marriages, and, if they do, whom (and how many spouses) they think it proper to marry and how they celebrate marriage ceremonies; what sorts of shelters they build for themselves; which languages they speak; which religions they practice; and whether they keep any animals for pets. Human cultures vary significantly in all of these aspects. Of all the ways in which the behavior and beliefs of individuals differ, a social scientist chooses a few specific criteria and then defines that bundle of attributes of shared behavior or belief as a culture.

The definition of a culture is a concept, just as the definition of a region is a concept. Therefore, when we define cultures, we must observe the same precautions that we observed in defining regions. Back in Chapter 1 we wrote: "Whenever anyone refers to any territory as a region, we must be sure that we know and agree on exactly what territory the region includes and what the criteria of homogeneity are." Similarly, whenever anyone writes of a cultural group, or a culture, we must be sure that we know and agree on exactly whom the culture includes and what the criteria of inclusion are.

The definition of a culture may include a great number of characteristics or just a few. For example, all social scientists agree that language is an important attribute of human behavior. Two people who share a language share something very important. If, however, those two people hold different religious beliefs, feel patriotism for two different countries, and eat different diets, social scientists may insist that although those two people share one *attribute* of culture, they do not share *one culture*. The great number of English speakers around the world today share few attributes other than language, so we would not say that they all share one common culture. If, however, two people do share a language, religious beliefs, political affiliation, and dietary preferences, then social scientists would agree that those two people share a culture.

FIGURE 2-1

This haunting photograph shows the first contact between these formerly isolated people and the rest of humankind. On August 4, 1938, a U.S. scientific expedition entered the supposedly uninhabited Grand Valley of the Balim River in western New Guinea. They found over 50,000 people living in Stone Age circumstances totally unaware of the existence of anyone else on Earth. We cannot say for certain whether any totally isolated groups live on Earth today, but it is less probable each year.

(Courtesy of the Department of Library Services, American Museum of Natural History.)

A group's culture includes its patterns of behavior as well as its possessions. An object of material culture is called an **artifact**, which means, literally, "a thing made by skill." The ways groups do things—interpersonal arrangements, the family, education, and so forth—are called **sociofacts**, and the group's poetry, music, language, and religion are called **mentifacts**. For example, religion is a mentifact, a religious congregation is a sociofact, and a church building is an artifact.

A **subculture** is a smaller bundle of attributes shared among a smaller group within a larger, more generalized culture group. For example, Italian Americans, Chinese Americans, and African Americans share subsets of cultural attributes within the larger American culture. Sometimes even one single attribute—shared loyalty for a sports team, for example—can bind individuals so strongly that that single attribute is termed a subculture.

Ultimately, cultural affiliation may be a matter of the feelings or the preferences of the individuals. Two people may share a great number of cultural attributes—enough for us "objectively" to insist that they share a culture—yet they may hate each other and even kill each other because they differ in one cultural attribute that an "objective" observer would argue is trivial. Each day's newspaper brings proof of this statement. Warring gangs form in urban neighbor-

hoods that outsiders see as culturally homogeneous. In some cases people who share religious beliefs kill each other over the results of soccer games. Our study of political geography in Chapter 10 will cite examples of countries that outsiders regarded as homogeneous that have nevertheless broken out in civil war. Outsiders did not see the cultural fault lines within the countries. This textbook will always make it clear whether we are investigating any particular attribute of culture, or whether we are speaking of a particular bundle of attributes defined as a culture.

Cultural Landscapes

As people modify their local landscape, we say that they transform the landscape from a **natural landscape**, one without evidence of human activity, into a **cultural landscape**, one which reveals the many ways people modify their local environment. Aspects of cultural landscapes include both the treatment of the natural environment and also houses and other parts of the **built environment** (see Figure 2-2).

The evidence of human activity is so ubiquitous on Earth that we might be justified in saying that the entire Earth is today a cultural landscape. Travelers quickly notice, however, that the cultural landscape displays a tremendous number of variations. Rural China does not look like rural France or rural Nigeria. One factor in explaining these variations is that the natural environments in these places were different even before humankind set to work on them. The "raw material" that human societies transformed into cultural landscapes affected what humans would do in each. Some writers argue that variations in the physical environment explain even the variations among human cultures. Theories that emphasize the role of the environment in human life are called **environmentalist theories**.

Another factor in explaining variations in cultural landscapes, however, is that each cultural group creates a distinct cultural landscape; a cultural landscape is an artifact. Therefore, a landscape can be thought of as a language, and the better we can learn to "read" a landscape, the more it will reveal to us about the culture that produced it.

In the past, human societies developed in greater isolation from one another than today, and the extraordinary diversity of human cultures and cultural landscapes testifies to human ingenuity. Different peoples who live in very similar environments but isolated from one another developed astonishingly different lifestyles. Conversely, some aspects of cultures that have developed in different physical environments are startlingly similar. No direct cause-and-effect relation-

FIGURE 2-2

This aerial photograph of the Grand Valley of the Balim was taken a few weeks before the expedition actually contacted the valley's inhabitants. It must have been a surprise to look down into a supposedly uninhabited valley and to see these neat clearings and settlements outlined by irrigation ditches. This was clearly a cultural landscape, one modified by considerable human effort.
(Courtesy of the Department of Library Services, American Museum of Natural History.)

ship between a physical environment and any aspect of culture can be assumed.

CULTURAL ECOLOGY AND ENVIRONMENTAL DETERMINISM

▼▼▼▼▼▼▼▼▼▼▼▼▼▼▼▼▼▼▼▼▼▼▼▼▼▼

The study of the ways societies adapt to environments is called **cultural ecology**. In contrast, the simplistic belief that human events can be explained entirely as the result of the effects of the physical environment is called **environmental determinism**.

In the past, when the link between the local environment and local culture was very strong everywhere, many writers directed their thoughts to this influence. Many ancient Greeks believed that the principal determinant of a people's culture was the way they made their living and that this, in turn, was determined largely by their physical environment.

Certainly distinctive lifestyles have evolved as ways of exploiting diverse natural environments—fishermen by the seas live differently from cowboys on the grassy plains—but today a smaller percentage of people are tied to occupations that exploit the local physical environment, and even their lifestyles have been transformed by imported goods, techniques, and cultural products and ideas.

Many writers have emphasized climate as a major control on human affairs. The Greek physician Hippocrates (460?-377? B.C.) argued in his book *Airs, Waters, Places* that civilization flourishes only under certain climatic conditions. Those just happened to be the conditions in which Hippocrates lived. Most people are prejudiced in favor of their home environments.

Historian Arnold Toynbee (1889–1975) turned Hippocrates' idea that a hospitable climate nourishes civilization upside-down with his own **challenge-response theory**. Toynbee argued that people need the challenge of a difficult environment to put forth their best effort and to build a civilization. A rich environment encourages only sluggishness. Today whenever anyone expresses a preference for an environment of seasonal change, as in the northern United States, over the almost tediously fine weather of southern California, that person is suggesting the challenge-response theory.

Aristotle identified climate as the principal influence on a people's political spirit, and this theory was later elaborated by the French political philosopher Jean Bodin (1530–1596). Bodin attributed differences among peoples to three climatic belts. The northern cold zone, he argued, produced a physically vigorous but mentally slow type of person, tending toward democratic government. The hot south, by contrast, produced lazy people, intelligent but politically passive, and thus satisfied to live under despotism. In between, in the temperate zone, wrote Bodin, the optimal mixture of intelligence and industry promoted the best government—the monarchy of Bodin's own place and time.

Later writers nominated topography as the principal determinant of a people's political life. The French theorist Montesquieu (1689–1755) argued that rugged landscapes, for example, produce individualistic peoples capable of developing democracies. We hear an echo of this today in the state motto of West Virginia: "Montani Semper Liberi," "Mountain men are always free."

Still other writers have emphasized the influence that the local rainfall regime can exercise over a people's political life. The political scientist Karl Wittfogel (1896–1988) argued that in environments that are either particularly wet or particularly dry the need for organization and care of waterworks

One goal of studying cultural geography is to understand why other people act the way they do. **Ethnocentrism** is a tendency to judge foreign cultures by the standards and practices of one's own, and usually to judge them unfavorably. Practices in other cultures which may at first seem strange to us, however, may in fact be sensible and rational. Conversely, some aspects of our own culture may seem strange or even offensive to others. Only by studying the extraordinary diversity of cultures can we truly understand even our own. Most Americans, for example, assume that a man should have one wife and a woman one husband at a time, but almost any number of spouses are allowed in a series. This shocks some people, admittedly some Americans as well as some people of other cultures.

Tibetans assume that a woman should marry all sibling brothers. This is called *fraternal polyandry*. Through thousands of years, fraternal polyandry has prevented unsustainable population increases and the fragmentation of landholdings in poor mountain valleys. The social ramifications of fraternal polyandry boggle an American mind. An American might ask how you identify the father of any given child, but to a Tibetan it makes no difference, and a Tibetan would in fact consider the question prurient.

Any geography book will contain many examples of ways of life that contrast with your own. None is necessarily "right" or the best for everybody. All people have to overcome the initial assumption that "different from" the way we do things ourselves is "worse than" our way,

and people everywhere can learn to appreciate and respect the integrity of other peoples' behaviors. Cable television entrepreneur Ted Turner once issued a dictum forbidding writers and newscasters at his Cable Network News to use the word "foreign." In Turner's view, the world is pejorative and implies peculiarity or oddness. It has been replaced by "international" or other alternatives.

Learning that other peoples do things differently from the way we do, and coming to accept that ("Live and let live") is a minimum requirement for getting along in a diverse world, but shrewd people have always taken a step beyond that. They consider whether somebody else's way might be superior to their own, and, if it is, they adopt it.

encourages despotism. He called these political conditions "hydraulic societies."

Many environmentally deterministic theories are intriguing, but exceptions can always be found. For example, the "hydraulic" Dutch have managed their waterworks democratically for hundreds of years. The ability to find exceptions to any theory should warn us that no simple theory completely explains the relationship between the environment and technological development or politics. Any theory that promises to explain human affairs so simply is attractive, but it is also false. Human affairs are not simple, and when we examine any world situation or historic event carefully, environmental determinism, or any other single-factor explanation, proves insufficient.

A milder form of environmental rhetoric personalizes nature by giving it the ability to think or to act. John Ruskin (1819–1900) labeled this style the **pathetic fallacy**. In typical examples, nature "offers" certain resources to humankind, or certain environments "discourage" settlement. In the above discussion of the ideas of Hippocrates and Toynbee, we wrote that the former argued that "a hospitable climate *nourishes*

civilization" while Toynbee wrote that "a rich environment *encourages* only sluggishness." Both of these are examples of pathetic fallacies. In other cases of the pathetic fallacy, nature has feelings or encourages specific feelings in people (see Figure 2-3).

Nineteenth-century French geographers proposed the theory of **possibilism** as an antidote to environmental determinism. Possibilism insists that the physical environment itself will neither suggest nor determine what people will attempt, but it may limit what people can profitably achieve. The choices and constraints involved in utilizing the natural environment are often as much cultural, economic, political, and social as they are technological. The reason that we do not grow bananas in Canada is that it is easier and cheaper to grow bananas in Costa Rica.

Environmentalism Today

Assumptions about the physical environment's influence in human affairs still affect the economic and political debates of our time. For example, today

FIGURE 2-3

Rembrandt's *The Mill* (c. 1650) is described in the official guide to the U.S. National Gallery of Art as "melancholy, conveying a mood of sublime sadness…through the stark simplicity of a windmill silhouetted in the fading light against the mist-filled sky…incredibly moving." Can a landscape have feelings? Does it convey these feelings to you? Does any landscape necessarily evoke the same emotional response from all viewers, no matter how different the viewers' backgrounds or experiences?
(Courtesy of the National Gallery of Art.)

many of the Earth's tropical regions are poor. Chapter 1 explained that the tropical environment presents disadvantages to human societies. Tropical heat multiplies the activity of organisms that are inimical to people and to agriculture, and it reduces human work efficiency. Tropical rainfall regimes may be unfavorable in quality even if sufficient in quantity, and storms frequently are violent. Tropical soils are poor, and the tropics are also cursed with a disproportionate share of natural disasters. These facts have often been cited as sufficient explanation for the relative poverty of the tropical regions.

Another school of thought, however, explains the relative poverty of these regions differently. Its defenders point out that many tropical areas were long held as colonies by countries in the temperate latitudes. This colonial experience left physical, cul-

tural, and economic legacies (discussed in Chapter 3) which, combined with the continuing patterns and terms of world trade (discussed in Chapter 12), best explain tropical regions' current poverty. Each side in this debate is partly correct. Which side you take will determine how you allocate the "blame" for the current poverty—whether on uncontrollable environmental factors, on the rich, formerly colonial powers, or on the peoples and governments of the tropics themselves. Furthermore, your assumptions will determine what solutions you suggest to the problem of poverty in the tropics.

Another example of an environmentalist assumption is the belief that a simple cause-and-effect relationship exists between the possession of natural resources and national wealth. In Chapters 11 and 12, however, we will examine this relationship carefully.

We will learn that in fact the possession of natural resources and national wealth are not clearly related, and then we will explain this paradox.

All forms of environmentalist assumptions—determinism, the challenge-response theory, and pathetic fallacies—remain common in popular discussion and debate. Often they seem to be harmless figures of speech, but they are intellectually dangerous. They predispose conclusions and close the mind to alternative explanations. Watch carefully for these simplistic traps.

PERCEPTION AND BEHAVIORAL GEOGRAPHY

▼▼▼▼▼▼▼▼▼▼▼▼▼▼▼▼▼▼▼▼▼▼▼▼▼▼▼▼

The political commentator Walter Lippmann (1889–1974) differentiated "the world outside" from "the pictures in our heads." "The world outside" is the way things really are, but "the pictures in our heads" may be based on preconceptions, misperceptions, or incomplete understanding. Geographers call these pictures in our heads **mental maps**.

The psychological theory of **cognitive behavioralism** argues that people react to their environment as they perceive it. In other words, people make decisions on the basis of "the pictures in their heads." They must act, however, in "the world outside." If there is a difference between the way a situation is and the way that we think it is, then we may get into difficulty by trying to do something impossible or by failing to take advantage of some opportunity. Cognitive behavioralism follows the philosophy of *phenomenology*, which states that our knowledge derives from our personal world experience.

We cannot understand why people make the decisions they do or act as they do by studying only the real environment in which they acted. We must discover what was in their heads when they made their decisions. In 1898, for example, Senator George Hoar of Massachusetts cast a key vote for the annexation of Hawaii because, according to his personal papers, he drew a line on a map from Alaska to southern California, saw that Hawaii fell inside that line, and concluded that its possession was necessary for the country's defense. No one has been able to discover what sort of map he was looking at (perhaps one with an inset map of Hawaii?), but his misperception affected history.

Therefore, environmentalistic theories cannot explain human cultures and how they transform landscapes. Study of environmental influences on human life must be supplemented by study of how people perceive the environment and how these perceptions and thoughts affect their behavior. If we combine the theory of cognitive behavioralism, which examines perception and behavior, with the theory of possibilism, which examines the results of human behavior in the environment, then we have a more precise and careful approach to examining human–environmental relationships. The study of human behavior, then, must be our next topic.

We cannot understand why people make the decisions they do or act as they do by studying only the real environment in which they acted. We must discover what was in their heads when they made their decisions.

Behavioral Geography

Behavioral geography is the study of how we perceive the environment and of how our thoughts and perception influence our behavior. Where we choose to live, shop, or visit depends on our feelings about places—whether we think certain places are good or bad, beautiful or ugly, safe or dangerous (Figures 2-4 and 2-5). Our behavior, including the way we transform a natural landscape into a cultural landscape, reveals our thoughts, feelings, attitudes, and preferences. Therefore, behavioral geography draws upon a variety of theories about human behavior formulated in psychology, economics, anthropology, and other disciplines.

The essay on *The Scientific Method* noted that one definition of a true science is that it reaches conclusions that are predictable and regular. H_2O (water) combined with SO_2 (sulfur dioxide) will yield H_2SO_3 (sulfurous acid) with predictable regularity, and, therefore, chemistry is a science. Whenever we study human behavior, however, we confront the question of just how predictable and regular, or even logical or rational, human beings really are. Our answer to this question will determine whether we feel confident in calling geography, economics, and sociology, for example, "social sciences," or whether we should call them "social studies." Chapter 1 noted that Karl Ritter, one of the founders of human geography, tried to find laws to describe how humans organize and use territory. Attempts to find laws exemplify the nomothetic approach to phenomena. The philosophy that laws and regularity can be found in data is called *logical positivism*. Geographers who support logical positivism rely on quantification, statistical description of patterns, and statistical manipulation and testing of hypotheses.

Other geographers insist that logical positivism dehumanizes geography by ignoring real human experience, which includes fear, imagination, sentiment, and other feelings that may not be rational or logical. Subjective experience cannot be scientific, but it must be accepted as evidence. Human mental processes can interact with "facts" in surprising ways. For example, today the economies of many East Asian countries are growing rapidly, and scholars are seeking the causes for this rapid growth. Several scholars have noted that

Some Students' Mental Maps

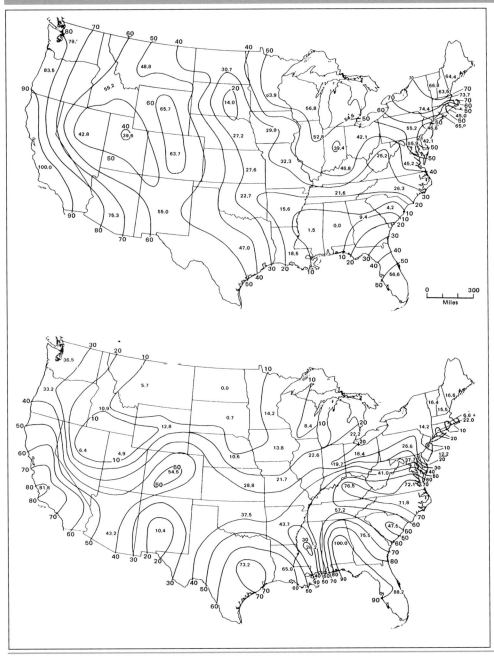

FIGURE 2-4

These mental maps reflect the states' residential desirability, according to the opinions of students at the University of California at Berkeley (above) and the University of Alabama (below) in 1966. High numbers indicate a positive image in the students' minds. Students at both places seemed to like where they were, but the Alabamans had a better opinion of California than the Californians did of Alabama. Both seemed to have positive images of Colorado, and low images of the Dakotas. These preferences are not necessarily based on personal experience. We all have opinions of places we have never visited, and a visit might surprise us. (From Peter R. Gould, "On Mental Maps," in *Michigan Interuniversity Community of Mathematical Geographers, Discussion Paper #9*, 1966. Reprinted with permission.)

A Mental Map of Boston

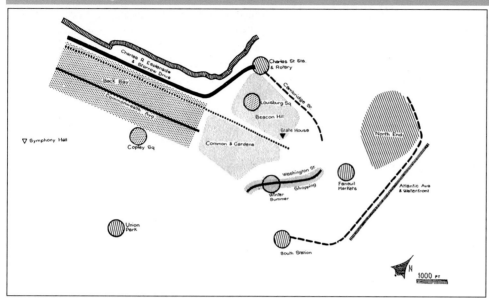

FIGURE 2-5

This composite mental map shows Boston's distinctive places according to a scheme developed by urban planner Kevin Lynch. Lynch discovered that people's mental maps seldom cover territories as regular grids do. People map space in terms of paths or routes; edges, such as the banks of rivers; nodes; districts and landmarks. (Reproduced from Kevin Lynch, *The Image of the City.* Cambridge, MA: MIT Press, 1960, p. 147. Copyright 1960 by the Massachusetts Institute of Technology and the President and the Fellows of Harvard College.)

these East Asian societies have cultural traditions rooted in Confucianism, a philosophy discussed in Chapter 8. These scholars hypothesize that Confucianist traditions may be a factor conducive to economic growth. No cause-and-effect link between Confucianism and economic growth has ever been proven, yet many Western business executives have come to believe that a link does exist. Government officials in the East Asian societies that have Confucianist traditions may or may not believe that a link exists; we cannot know their thoughts for certain. It is certain, however, that they know that Western executives believe in the link. East Asian governments of countries with Confucianist traditions are advertising that fact in Western business periodicals in order to lure investment by Western executives who believe in the link. If the Western executives believe in the link (whether it "really" exists or not), they will invest in those countries, and *make the link true.* In this example, the interplay between "reality" and "subjective evidence" is too complicated and illogical for "scientific" analysis.

Therefore, many geographers argue, a scientific geography cannot deal with the whole range of human experience. These geographers examine literary texts, art works, and other subjective evidence. They practice *humanistic geography*, taking its name from the Renaissance "humanistic" ideas about the central importance of human beings, their thoughts, and their ideas. Humanism places the emphasis on interpretive understanding. Whereas positivists may find human behavior maddening or frustrating when it is irrational, humanists find it fascinating and even endearing. Unpredictability is a sort of freedom. The

areas of study described below illustrate both positivist and humanist approaches in geography.

The Theory of Economic Man Geography has borrowed some ideas about human behavior from economics. For example, one school of English and Scottish economists devised a theoretical individual called **economic man**. Economic man is an actor who, in any situation, behaves according to three assumptions: (1) he has all the information necessary to make decisions; (2) he is completely rational; and (3) he tries to minimize the effort he expends to achieve any goal (psychologists call this "the principle of least effort") and to maximize the profitability of any action he undertakes. Because of this consistent drive to maximize his return from any activity or investment, economic man is often called *man the optimizer* or *man the maximizer.* Beginning with these assumptions, economists have devised economic models based on the conclusions that naturally follow. This example of deductive reasoning from a model exemplifies logical positivism.

The contemporary American geographer Richard Morrill has translated this model into geographic terms. In one of his books, he explores the conclusions about geography that result logically if you assume that "spatial organization is the outcome of man's attempt to use his territory efficiently."

- To maximize the net utility, or productivity, of areas and places at minimum effort.
- To maximize spatial interaction at minimum effort (or cost).

52 *Chapter 2*

- To bring related economic activities as close together as competing aims permit.

Professor Morrill acknowledges that "Perhaps, after all, local physical and human variation will prove more significant than the ordering principles I stress. Even so, I feel that it is valuable to seek out the commonality in seemingly disparate landscapes." The book exemplifies the deductive positivist approach.

In this *Human Geography* textbook, we already used the model of economic man in Chapter 1. The discussion of inductive and deductive geography assumed that you would try to minimize the effort of reaching for items in your kitchen cabinets. That assumption about your behavior led to us to deduce a description of the geography of items in your kitchen cabinets. Later in this book we will present other deductive geographic models by Alfred Weber, Heinrich von Thünen, and Walter Christaller that employ the model economic man.

The idea of economic man has proven useful in many social sciences, but we all know in truth that none of its three assumptions are ever completely fulfilled. For example, people seldom have all the information relevant to the problem they face. In addition, they do not always seek to maximize their profit or return; sometimes people work just hard enough to gain satisfaction, or a satisfactory return. We may, for example, choose to invest our money in financial institutions that help finance projects we support, even though those institutions earn a rate of return that is less than maximum. Therefore, alternative models of economics and geography have been devised in which scholars have replaced the theoretical man the maximizer with a theoretical **man the satisficer**.

Furthermore, humanist scholars remind us that behavior is determined by feelings, sentiments, and preferences that are not suited to neat mathematical or economic modeling. Psychologists who study performance in the workplace, for example, know that people are motivated by many things other than money, including interest in their job, status, curiosity, self-esteem, and peer acceptance. These motivations may be difficult to quantify. In Chapter 6 we will see that the European peoples subsidize agriculture partly because they *like* to have agricultural countrysides surrounding their crowded cities. Behavioral geography tries to define and analyze how such feelings affect geography.

Theories about the Perception of Hazards One group of geographic studies has focused on how we perceive hazards, either natural (such as earthquakes) or human-made (such as chemical plants).

How we perceive these affects our actions or reactions toward them.

People often flock to settle volcanic rims and floodplains. Why do people continue to build where disaster will inevitably strike? Or even to rebuild where disaster has struck and will inevitably strike again? The San Francisco region suffered a terrible earthquake in 1906 and another in 1989. Seismologists predict that the area will suffer still another—and probably massive one—before 2010. Why do people continue to live there?

Hurricanes are intense destructive storms that form in the warm tropical oceans. Winds may reach 150 miles (240 km) per hour, and the destructive power with which a hurricane strikes land can do considerable damage. Chances are one in four that a hurricane will hit southeast Florida in any given year. Why do people live in southeast Florida?

There are many reasons why people settle in these environments, ranging from simple ignorance of danger, to sentiment ("that's our home"), to the allure of beautiful spots, to financial necessity ("that's our property" or "that's where we can make a good living"). The occasional financial losses from volcanic eruption, fire, or flooding in some dangerous environments may in fact be less than the return from rich harvests won through most years. Furthermore, from a practical point of view, insurance policies are often written so that people are virtually required to rebuild after a disaster.

Most important, however, is probably what psychologists call *denial* ("It won't happen."). People generally do not like to think about painful subjects, and, in fact, a certain amount of denial of fear is necessary for any of us to go about our daily life.

Theories of Aesthetic Reactions to Landscapes
Our individual notions of usefulness or of aesthetic value affect our assessment of landscapes. Many urbanites, for example, agree that deserts may be beautiful, and some deserts, such as Arizona's Painted Desert, are major tourist attractions. A farmer, however, might be appalled at such a landscape (see Figure 2-6). We evaluate landscapes according to our individual criteria. Tourism is at least partly inspired by attitudes toward environments.

In Western civilization, the romantic movement forever changed the way we view nature. The notion that nature is sublime and divine was invented in the eighteenth century. Weather, for example, does affect our moods, and yet it received almost no notice in prose fiction until the late eighteenth century. Nineteenth-century novelists, however, noted it frequently. They used weather to reveal characters' feelings; for example, a passionate or angry scene often

FIGURE 2-6

This black soil photographed near Martinsburg, Iowa, is so rich that a farmer might find this a beautiful view. What is your aesthetic reaction?
(Courtesy of John Daniels.)

took place against the backdrop of a violent storm. The way we regard mountains also exemplifies this new view of nature. For over a thousand years of Western civilization they were simply considered a nuisance—unproductive, obstacles to communication, and a refuge of bandits. By 1739, however, the English poet Thomas Gray wrote of his trip over the Alps, "Not a precipice, not a torrent, not a cliff, but is pregnant with religion and poetry!" The interest in nature paralleled new interest in the individual self, in states of feeling that affect and are affected by our perception of the external world.

The contemporary American geographer Yi-fu Tuan has defined the study of how people evaluate landscapes according to personal emotional or sentimental criteria **topophilia** ("love of place"). Topophilia investigates, for example, the pathetic fallacies found throughout art and literature. Works of fiction become documents for scholarly analysis, providing rich descriptions of landscapes and individuals' attitudes toward them. For example, John Steinbeck's great novel about the American drought of the 1930s, *The Grapes of Wrath*, provides material for the study of landscape degradation. Students of topophilia suggest that one reason why the American landscape bears many allegorical place names is that many westward-moving pioneers brought John Bunyan's

allegory *The Pilgrim's Progress* (1678) with them. People who read about the tribulations of the fictional character Christian as he passes the Slough of Despond on the way to Celestial City naturally dubbed forbidding terrain Desolation Peak (Wyoming) or Hell Creek (Colorado).

The different cultural landscapes created by different cultures provide evidence of aesthetic tastes. A classic investigation of English landscape tastes, for example, by contemporary geographer David Lowenthal, concluded that predominant English landscape values include preferences for the bucolic, the deciduous, the tidy, and the antiquarian. The British antilittering slogan "Keep Britain Tidy" contrasts with its U.S. counterpart "Keep America Beautiful." What are the subtle differences in attitude between the words "tidy" and "beautiful"?

One consistent aspect of Euro-American cultural identity through history has been a sentimental interpretation of the American landscape as a "blank slate" upon which Europeans were given an opportunity to start afresh, as if in a new Garden of Eden. The words of two Euro-American artists illustrate this (Figure 2-7). When the American landscape painter Thomas Cole went to visit Europe in 1828, his friend the poet William Cullen Bryant wrote:

> ....thy heart shall bear to Europe's strand
> A living image of our own bright land....
> Fair scenes shall greet thee where thou goest— fair
> But different—everywhere the trace of men....
> Gaze on them, till the tears shall dim thy sight,
> But keep that earlier, wilder image bright.

Exactly 140 years later the American landscape photographer Ansel Adams refused to visit Europe. He wrote, "I am afraid to visit Europe, to see all [its] ancient towns, all [its] fairy-tale castles because, as I understand, all the landscape in Europe is converted into overcultured scenery. I'll never be the same after such a trip. I might lose my identity." Both Adams and Bryant believed that it is essential to American cultural identity to be surrounded by a certain amount of "wilder," "natural," "uncultured" landscape.

Aesthetic perception influences both our feelings about environmental preservation and also our attitudes toward our built environment—our homes and cities. We know that people's attitudes toward environments were different in the past, and our own attitudes toward the past influence our feelings about touring historic sites and preserving landmarks. In the United States, for example, preservationists regret that

FIGURE 2-7

This view of New York's Catskill Mountains without any evidence of human intervention dates from just before the painter Thomas Cole's trip to Europe. It exemplifies what Bryant called the "wilder image" of America. *Sunrise in the Catskills*, by Thomas Cole, 1826.
(Courtesy of the National Gallery of Art.)

old buildings may be torn down and the land dedicated to new uses, but there is nothing deliberate about this destruction of the visible past. In some countries that have undergone political or cultural revolutions, by contrast, such as China, old buildings have been deliberately destroyed as symbols of old political and social systems (see Figure 2-8).

Consideration of aesthetics leads to recognition of deeper philosophical and religious influences in environmental transformation. For example, the design and layout of a society's entire built environment may reflect that society's ideas about harmony with nature. Many cultures in the past and present, including most Asian and native American cultures today, believe in aligning the elements of the built environment with the cardinal points of the compass, with the elements of the natural environment (such as winds and water bodies), or with celestial bodies (see Figure 2-9). Sometimes these massive constructions serve as calendars, but they may also serve to align human activities with certain "forces" in the universe.

The placement and design of temples, gravesites, homes, business establishments, and even whole cities is called in Chinese **feng-shui**. Feng-shui originated in early philosophical questioning of humankind's place in nature. It was practiced in China by the third century, and from China it spread into Vietnam, Korea, and Japan. Migrants from East Asia have more recently spread interest in feng-shui around the world, and today it exercises considerable influence, for example, on the real estate market, architecture, and interior design in the western United States and Canada. Feng-shui is sometimes confused with **geomancy**, but geomancy is more strictly the attempt to foretell the future by means of studying configurations of landmarks.

Cultures and the Transformation of Landscapes   **55**

FIGURE 2-8

This ruin of a once-proud temple standing isolated in rural Vietnam reminds local farmers that their land was once inhabited by a different people, the Cham people, of a different culture praying to long-forgotten gods in a long-forgotten language.
(Courtesy of Julia Nicole Paley.)

Some people insist today that maintaining harmony with nature has assumed a fully religious quality, and they speak of the need for "environmental ethics." The use of this phrase deliberately elevates environmental preservation to an ethical matter—one above considerations of economics or mere convenience. In Chapter 8 we will question whether the followers of different religions treat the environment differently.

Other Influences on the Perception and Organization of Territory
Studies which have grouped men, women, children, the elderly, the disabled, and individuals categorized in many other ways have taught us that each of these groups often perceives the environment in significantly different ways. Therefore, each group may almost be said to

live in a different environment. This knowledge has proven useful in the design of new environments, such as housing. The layout of housing for hearing-impaired people, for example, might provide for visual communication between rooms.

Proxemics is the study of how people perceive and use space. Research in proxemics has discovered that each of us lives in concentric "bubbles" of space. We allow people to get closer to us and to invade each of these bubbles under different circumstances, or else we bridle at invasion. We maintain *public distance* from strangers in large open areas, but we accept the closer stance called *social distance* in formal meetings, and we can be comfortable if we stand within still closer *personal distance* with our friends. We let very few people, however, penetrate the innermost zone, that of *intimate distance*.

Students of proxemics have learned that the measurement of each of these distances differs among individuals and, strikingly, among cultural groups. One study, for example, found a significant proxemic difference between black Americans and white Americans. The study found 22 inches to be the most comfortable distance between two standing black Americans conversing, whereas for white Americans that figure was 27 inches. Black Americans

FIGURE 2-9

Stonehenge, on Salisbury Plain in Wiltshire, England, was built about 1500 B.C. It seems to have served both as a place of worship and as an observatory. The sun and the moon probably rose and set over certain stones at certain times of the year, although it is hard for us today to reconstruct those astronomical sightings precisely. So-called "primitive" peoples all over the Earth constructed astonishingly accurate observatories as places of worship.
(Courtesy of the British Tourist Authority.)

and white Americans, therefore, might not agree on what constitutes "crowding" in a room. Even greater differences have been measured between citizens of different countries.

The lessons to be learned from proxemics range from how best to behave at a cocktail party to how best to design buildings or even whole cities. We do not know for certain the degree to which these proxemic differences among groups are cultural or biological. Human behavior is rooted in biology and physiology, but it is filtered through culture. People of different cultural backgrounds actually rely on their eyes, their ears, and their noses differently, even if they have the same physical sensory capacity.

The more we learn about the body's sense organs, the more we suspect that proxemics may have an important biological aspect. Most modern cultures de-emphasize the importance of our sense of smell, for instance, and so we gradually lose awareness of how ambient smells affect our thoughts. Nevertheless, they may do so.

Furthermore, scientists are just now discovering that humans, like most animals and even plants, give off clouds of odorless molecules called *pheromones*. These pheromones are detected by other peoples' vomeronasal organ, a small organ inside the nose which was once thought to be a useless vestige. The vomeronasal nerves are connected to the brain's center for basic drives and emotions, the hypothalamus. Therefore, we may react to other people partly as a result of a cryptic sensory system that exists without conscious awareness. Our reactions to other individuals may be much more biological and visceral than scientists have traditionally thought.

Some scientists believe that humans exhibit **territoriality**. Many animals lay claim to territory and defend it against members of their own species. Most aspects of territorial behavior among animals have to do with spacing, protecting against overexploitation of that part of the local environment on which the species depends for its living. Applying these studies to human behavior is complicated by the interplay between human biology and culture. People defend their standing space in a crowd (they bump back), urban gangs defend their "turf," and nations defend their land. Some scientists studying pheromones have suggested that pheromones may play a role as a physiological mechanism to control human population. In studies of territoriality, however, as in studies of proxemics, it remains unclear to what degree the observed behaviors are biologically determined or culturally determined.

BIOLOGICAL EVOLUTION AND TECHNOLOGICAL EVOLUTION

▼▼▼▼▼▼▼▼▼▼▼▼▼▼▼▼▼▼▼▼▼▼▼▼

As plants, insects, fish, birds, and other animals developed and spread around the globe, they all divided into a great number of species. Each species evolved distinctive biological capabilities and so survived to fill an **ecological niche**, which is a particular combination of physical, chemical, and biological factors that a species needs in order to thrive.

Humankind, by contrast, has for the most part adapted itself to widely varying conditions through cultural and technological evolution, rather than through physical differentiation. Humans survive Minnesota winters, for example, by making warm clothes, not by evolving furry bodies. Humans have developed cultural or technological tools adequate to survive anywhere. Scientific research stations can even be maintained in the frozen wilderness of Antarctica, although many basic supplies must be brought to the local residents. Thus, humans have been able to establish a presence in every biome.

The science of genetic engineering offers the possibility that we might assume responsibility for our own biological evolution, as we have for that of domesticated animals. Some people fear that this capability might be misused.

Two Theories of the Evolution of Societies

Two theories emphasize the role that technological and economic evolution plays in the interaction between societies and the environment. Both theories have been used to suggest that certain developments in the past were inevitable or even that certain events are inevitable in the future.

The Theory of Human Stages One theory of human cultural evolution argues that all human cultures pass through certain stages of development, and it defines these stages by the way in which the culture exploits the environment. The Roman general Marcus Terentius Varro (116-27 B.C.) first articulated this theory in a compendium of geography, and the theory remained almost unchallenged until the nineteenth century.

Varro argued that humankind originally derived its food from things that people hunted or harvested naturally from the abundant Earth. He referred to humans in this stage of evolution as **hunter-gatherers**. Varro stated that humans then domesticated animals and moved into the subsequent evolutionary stage of **pastoral nomadism**. Pastoral nomads have no fixed residences, but drive their flocks from one place to another to find grazing lands and water. If their movements are regular and seasonal—between mountain and lowland pastures, for example—their movements are called *transhumance*. The evolutionary stage of pastoral nomadism was in turn followed by settled agriculture. Agriculture was at first **subsistence**, which means that people raised food only for themselves, not to sell to others, but subsistence agriculture slowly evolved into **commercial agriculture**, which is raising crops to sell. The final stages of social and economic evolution were urbanization and "industry."

Varro's theory was generally accepted until Alexander von Humboldt pointed out that the theory of stages could not be applied everywhere. The native peoples of the Western Hemisphere, for example, had never experienced a pastoral stage.

There are still human groups living at each "stage" in this theory of evolution. Very few people today are hunter-gatherers, but in poor societies pastoral nomadism may still be a successful low-technology adaptation to arid or semiarid environments. Pastoral nomadism may be found in parts of Africa and in the dry interior of Asia, but today most pastoral nomads are settling down. Many governments are forcing them to settle in order to control them. The government of Kenya, for example, is forcibly settling the Masai people. Transhumance among fixed dwellings is still widely practiced. In advanced societies, however, as in the U.S. West, modern methods such as "riding herd" in a helicopter on animals that are wearing individual electronic identification tags is hardly a "primitive" activity. Subsistence agriculture is almost everywhere being commercialized. These developments result from the increasing interconnectedness of the Earth discussed in Chapter 3.

Historical Materialism **Historical materialism** is a school of thought that tries to write a plot for human history based on the idea that human technology has increased humankind's control over the environment. This understanding might then be used to predict the future. Karl Marx (1818–1895) was the founder of historical materialism. His friend and collaborator Friedrich Engels (1820–1895) said that Marx did for history what Darwin did for biology, that is, identified the underlying progressive force.

Historical materialists ask: What is the "story" of human history? What thread or theme can we trace in it through time? If we focus on politics, it is apparent that empires and entire civilizations rise and fall. Some periods are peaceful; others plagued by wars. Some rulers have been good, others bad all through recorded history with no apparent pattern of change.

In what aspect of human affairs can we demonstrate evolution?

Marx said that evolution is to be found in humankind's conquest of the physical environment in order to meet material needs. A contemporary ruler may or may not be wiser than one in the past, but a modern tractor can do more work than a horse dragging a wooden plow. Therefore, historical materialists insist, history's plot is the advance of technology over the environment, and this continuous improvement of productive technology encourages social evolution. The degree of control a society has over its environment (the "stage" in its development) will determine any society's economic system. The society's legal, political, and social life will be built on that (see Figure 2-10). Theories such as Marx's which suggest that social, economic, psychological, and other structures lie beneath and control individual and group action are called *structuralist theories*. Structuralism often leads to a commitment to change these structures to improve society.

Marx's historical materialism differs from environmental determinism. Marx said that the degree of control a society has over its environment is the key to understanding that society. Thus, control over the environment is more important than the nature of the environment itself.

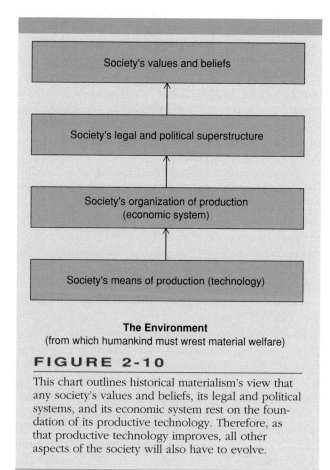

The Environment
(from which humankind must wrest material welfare)

FIGURE 2-10

This chart outlines historical materialism's view that any society's values and beliefs, its legal and political systems, and its economic system rest on the foundation of its productive technology. Therefore, as that productive technology improves, all other aspects of the society will also have to evolve.

Marx predicted that, eventually, technology would produce such material abundance that economists and politicians would no longer study and dispute the allocation of scarce goods, but, rather, the distribution of abundant goods—"From each according to his abilities; to each according to his needs."

Some of what Marx wrote has turned out to be misguided guesses, and some of it is simply false, but much of what Marx wrote is generally accepted. Today no one doubts that science progresses almost inexorably, and that scientific progress often upsets society's economic, political, and even philosophical assumptions. The invention of new birth control methods, for example, and of prenatal testing, have challenged our laws and even our ideas of life. Many of the political movements that have used Marx's name, however, disgrace his memory. The Communist parties that seized power in Russia in 1917 and that seized power or were imposed on other peoples in the twentieth century bear no relationship to the humanitarian society that Marx envisioned material abundance would bring. Long after the collapse of the monstrous tyrannies imposed around the world in the name of communism, the nations which have achieved material abundance will still be confronted with the question of why there is any poverty or deprivation among their citizens.

Criticisms of Historical Materialism Materialism assumes that methods of production are responsible for all other aspects of human societies. The factors on the bottom of Figure 2-10 always cause those toward the top.

The philosopher Bertrand Russell (1872–1970), however, argued that the factors at the top of the figure can change the factors at the bottom. He wrote: "Methods of production change...owing to intellectual causes,...that is to say, to scientific discoveries and inventions....This intellectual causation of economic processes is not adequately realized by Marx." Thinkers like Marx who emphasize the power that economic forces have to change or create new ideas are called *materialists*. In contrast, thinkers who emphasize the power that ideas have to cause economic and social changes are called *idealists*. A variety of both materialistic and idealist ideas can be found in this book. Later in Chapter 8, for example, we will examine the idealist argument that a society's predominant religion can affect that society's economic development.

Most recently some economists and ecologists have attacked historical materialism from a new direction. Historical materialism notes that in the past, technological progress offered ever-higher material standards of living to human populations, and historical materialists assume that this rate of technological advance will continue. This optimism places some of our leading scientists and industrialists among "Marxist thinkers," a fact that might surprise them.

New critics, however, argue that past experience should not encourage confidence that technology will continue to provide rising standards of living for the proliferating human population. Marx himself never imagined the explosive growth of the human population that has occurred since his death. At the time of his death, the human population was only about 1.5 billion, and it was growing at an annual rate of less than 1 percent. Today the human population is about 5.7 billion, and it is rising at an annual rate of over 1.7 percent. Given this increasing population, historical materialists' optimistic confidence that technology will always solve problems of scarcity might lead humankind to disaster.

The rate of human population increase today is discussed in greater detail in Chapter 4, but concerned scholars and politicians around the world are asking whether the human population is, in fact, outpacing technological development, or whether it might do so in the future. Some scholars point out that technology and welfare are not equally distributed on Earth today, and they insist that we should examine the reasons for that distribution. We could either redistribute world welfare (through foreign aid or investment, for example), or else we could encourage more places to generate higher technology and welfare locally (by changing local government policies, for example). People of different political and economic philosophies offer different answers to these arguments. By the time you finish reading this textbook, you will have more information on which to form clearer ideas of your own.

CULTURE AND ENVIRONMENT IN U.S. HISTORY

▼▼▼▼▼▼▼▼▼▼▼▼▼▼▼▼▼▼▼▼▼▼▼▼▼

Each of the theories of the influence of the environment on human affairs contains some truth, but none of them can explain all events in history or in the present. The history of the United States demonstrates this. The history of Europeans in the Western Hemisphere began with a misconception: Columbus planned his voyage to Asia assuming the Earth to be much smaller than it is and having no knowledge of the American landmass.

U.S. history often used to be told as the story of Europeans pressing westward across the continent and adapting their culture to new environmental circumstances. This approach neglected the contributions of native Americans, who taught the Europeans how to adapt to the various North American environments, as well as the many contributions of African Americans and various Asian peoples. Even in analyzing the adaptations of European culture, however, U.S. history teaches that culture can be very conservative and that preconceptions and misperceptions can blind settlers to opportunities and difficulties in new environments and slow the process of cultural adaptation.

The English thought that the environment of eastern North America was similar to that of England, so they anticipated no problems in the transfer of grains or livestock. They often failed to utilize superior native plants and animals. In fact, in eastern North America the summers are hotter, the winters colder, the sun brighter and hotter, the rainfall harder, and the environment generally suited to a much greater variety of vegetation than in England. The colonists nevertheless attempted to reproduce their lifestyle from home.

The settlers learned farming and fishing techniques from friendly native Americans, and they only slowly evolved a material culture adapted to the eastern seaboard environment. Wood, for instance, was more available than it was in the deforested British Isles, so Americans came to rely on wood for a fuel, building material, and raw material for making tools and machinery. They also arduously learned and devised new techniques for utilizing the forest soils. When U.S. expansion reached the semiarid and treeless Great Plains, early observers assumed that they faced a wasteland (Figure 2-11). Major Stephen Long, the leader of an 1820 exploratory expedition, described the Plains as "wholly unfit for cultivation, and of course uninhabitable by a people depending upon agriculture." He labeled the Plains "The Great American Desert," and settlers avoided the region for the next 50 years.

Pioneers eventually devised new agricultural techniques, however. Because the settlers had not seen such soil in Western Europe and because there were no trees, the settlers at first assumed that the soil must be poor. New techniques of turning and plowing the rich prairie soils evolved through laborious experimentation, and the most suitable seed grains were eventually brought over by Eastern European immigrants, particularly Ukrainians, who came from similar grasslands. Barbed wire substituted for wooden fencing, and windmills drew up precious water. Most national lawmakers came from the relatively well-watered East, and they failed to understand that the settlement of a drier area would necessitate different laws regarding homestead size and water rights. Thus, environmental misperceptions hindered even the formulation of a suitable framework of law.

FIGURE 2-11

This rich harvest feeds people around the world. In 1820, however, when explorer Stephen H. Long looked across these treeless Nebraska plains, he thought them unfit for agriculture, and labeled them "The Great American Desert." This misconception delayed their settlement for decades.
(Courtesy of the Nebraska Department of Agriculture.)

At the end of the nineteenth century historian Frederick Jackson Turner propounded a "Frontier Theory" of U.S. history. Turner argued that as the front of advancing European settlement rolled from East to West across the continent, any given spot experienced successive waves of fur traders, cattlemen, farmers, and then urban development. In this way U.S. history exhibited a "continuous rebirth of society." Turner's theory was rooted in Varro's theory of stages of human evolution. For Turner, the native Americans and early fur traders represented the primitive stage of hunter-gatherers, and it was natural and inevitable that they should be replaced.

By the time Turner spoke, native Americans were virtually in danger of extinction, and some European Americans were beginning to feel guilty about their responsibility for that. Others, however, seized upon Turner's frontier thesis to rationalize the replacement of the natives by the coming of the white Europeans' civilization to North America. They argued that the replacement was "scientifically inevitable" and, therefore, "morally neutral." Sociologist Lester Ward criticized that interpretation of history. He answered vehemently in his book *Dynamic Sociology* that precisely what differentiates humankind from the brute animals is that human behavior is based on conscious choices, and therefore the result of human choices can never be "scientifically inevitable" and "morally neutral."

Questions

1. Does the fact that we confusedly call our reserves "parks" reflect confusion about what we really want?

2. Should we let our natural reserves become human-made parks?

3. How can we enjoy our reserves and yet save them?

In this photograph, which is artificial: the soil, drainage, topography, and selection of vegetation in the foreground, or the skyscraper in the background? In fact, all are. New York City's 840-acre Central Park was constructed over the 20 years after 1858. The land was still beyond the fringe of urban development, but a design was approved, the land was blasted and sculpted, and vegetation was planted to achieve an ideal. (EFB photo)

Gardens and parks are landscapes that humans have molded according to aesthetic notions. These are based on the assumption that in some ways human designs are superior to nature.

Tastes in garden design vary among cultures, and within any culture they change through time. In all cases, however, gardens and parks reflect the prevailing human notions of how best to "improve" nature. The eighteenth-century park designer Lancelot Brown earned his nickname, "Capability Brown," because whenever he saw a natural landscape, he commented that it had great "capabilities," upon which he would improve. In the nineteenth century Calvert Vaux and Frederick Law Olmsted "built" New York City's Central Park, including its topography and watercourses, and they chose and planted imported species of vegetation. Today an army of engineers and landscape architects is required to maintain the park, yet naive visitors marvel that the city has "preserved" this "natural environment" in the midst of sky-scrapers. In truth, the park is scarcely less artificial than the surrounding skyscrapers.

Minnesota's huge Mall of America encloses a seven-acre interior landscape which, according to official Mall publications "bring[s] the outdoors indoors… inspired by Minnesota's natural habitat—forests and meadows, river banks and marshes." Minnesota's natural conifers, however, cannot survive in a constant temperature of 70°F (21°C). Therefore, the Mall's "indoor natural Minnesota forest" is populated by tropical orange jasmines, black olives, oleander, hibiscus, and other species flown in from the South. Few visitors seem to notice, and none complain. The Mall is, in fact, a terrarium of exotic species.

Parks and gardens are different from **reserves** or **preserves**. These are areas that are set aside to save natural environments, including, usually, the natural wildlife. Most of the properties in the U.S. National Park System, such as Yellowstone National Park, the world's first national park (1872), were originally designated as reserves. These enclaves of nature attract rapidly growing numbers of visitors who demand facilities ranging from hotels to hiking trails to designated scenic viewing spots. The press of their numbers endangers the pristine character of the property. In other words, the people destroy what they come to see. In other cases, properties that were originally reserves are increasingly engineered in order to save them. If, for example, water were allowed to fall over Niagara Falls 24 hours per day, and if the Falls were not shored up with concrete and steel, the attraction would erode away.

THE IMPACT OF HUMAN ACTIVITIES ON THE ENVIRONMENT

So far in this chapter we have examined many theories about the ways in which the natural environment affects or conditions human life. The exact extent of this conditioning is controversial; human perceptions and actions must always be studied in light of the complex mixture of biological, psychological, and cultural factors that affect our behavior. We have learned some important ideas geography has traded with psychology, economics, history, sociology, and studies of art and literature.

Now we will examine the many ways in which human activity alters the natural environment. Humankind has altered or manipulated some environments so much that it is impossible to differentiate the "natural" from the anthropogenic, and little "natural environment" is left to identify. George Perkins Marsh actually argued that "whereas [others] think the Earth made man, man in fact made the Earth." No place on Earth has fully escaped human influence. Human beings are engineers, and even satellite photographs taken from space show the imprint of human habitation on the land. Land reclamation, drainage, irrigation, cultivation, livestock grazing, mining, vegetation removal, cities, and transport facilities all transform the Earth's surface. These and still more transformations were made to benefit humankind—to direct more energy or food to human uses, for example.

Any change in one of the Earth's natural physical and ecological systems, however, will trigger changes throughout the entire system. Because these systems are tremendously complex, many changes may be unexpected or even harmful. After humans have altered a natural system, we may wish to restore it. We may even make efforts to do so, but no natural system is ever quite the same as it was before human intervention.

The next few pages discuss a few ways in which human manipulation of the environment may already have gone beyond its original purposes and caused changes which necessitate still further anthropogenic alteration. Humankind has assumed responsibility for adjusting Earth systems for our benefit, and therefore we can question our stewardship, or at least the success rate of the alterations we have made. These pages represent, to some degree, a "report card" on human maintenance of the planet.

Different cultures treat the natural environment differently, and any society's ability to alter the environment depends on its technological level and its wealth. Both rich and poor societies are capable of doing enormous damage to natural systems; rich societies, however, are better equipped to repair damage—to reforest denuded landscapes, for example.

The following pages focus on interventions that have severely disrupted natural processes with negative repercussions to humankind. The principal interventions discussed will be in weather and climate, in hydrography, in the distributions of plants and wildlife, and in soils.

Climate Modification

People have probably always wanted to do something about the weather or the climate, but no signifi-

cant steps were taken until the invention of techniques for *seeding* clouds with dry ice and silver iodide in the 1940s. Snowfall and rainfall can thus be artificially induced. The grand hopes for these techniques were never fully realized, but clouds are still seeded in some countries, and the techniques are also used to clear clouds over some airports.

Perhaps the boldest experiment in altering the Earth's environment today is the current effort by Russian scientists to increase the amount of solar radiation reaching the northern latitudes. In February 1993, a Russian satellite released a drum out of which a mirror of eight thin aluminum and plastic segments unfurled like a fan to reach a diameter of 20 meters (65 feet). This device reflected a beam of sunlight 2.5 miles wide across part of the Atlantic Ocean, and then across Europe, including Russia. This feeble light was too faint and traveled too fast (17,000 miles—10,625 km—per hour) to illuminate the ground, but it was visible from Earth and astronauts tracked it from space. Several such devices in permanent orbit could turn the long Russian winter nights into daylight. Savings in municipal lighting costs alone could pay for the devices within a few months, but the ambitious Russians foresee the eventual illumination of areas tens of miles across—even the extension of planting and harvesting seasons in the far North. The potential ramifications of this activity are staggering.

Ozone: Where We Need It and Where We Do Not Want It

Chapter 1 noted that human activities may be adding enough greenhouse gases to the atmosphere to contribute to global warming. Human activities may also be altering the atmosphere by depleting the layer of **ozone** in the upper atmosphere. Molecules of ozone (O_3) are created in the upper atmosphere, where they form a layer that protects the Earth from the sun's ultraviolet rays. Ultraviolet rays cause sunburn, and they have been linked to the development of cataracts and of weakened immune systems in humans and other animals. They are known to damage crops, marine organisms, and DNA. Exposure to ultraviolet rays is thought to be the cause of some skin cancers.

In 1974, scientists first theorized that the ozone could be depleted by chlorofluorocarbons (**CFCs**), synthetic chemical compounds containing chlorine, fluorine, and carbon atoms. CFCs were invented in the 1930s and found widespread use in air conditioning and refrigeration units and in the manufacture of plastics. When CFCs are released, they drift into the upper atmosphere, where the chlorine atoms break down ozone molecules into standard oxygen molecules (O_2).

GEOGRAPHIC REASONING
Modern Alienation from Nature

Questions

1. What was the weather last Tuesday? Was that unusual for the time of year?

2. Can you name the hottest, coldest, driest, or wettest month in the area where you live?

3. Does a river flow near you? Where does it come from, and where does it go?

4. How do you note the change in the seasons?

5. Are you conscious of changing your diet through the year?

6. Our ability to buy and enjoy frozen food has diminished our recognition of the changing seasons. Can you name some vegetables available fresh each month, or, perhaps, a flower that blooms each month in the area where you live?

Many modern people are quite unaware of major characteristics of their physical environment. Unless they were raised on farms or in families in the construction industry or some other outdoor business, most are surprisingly unaware even of weather and climate.

The 31 species of flowers in this vase could never naturally bloom at the same time of the year. This painting is, therefore, self-consciously unnatural—an intellectual in-joke. In seventeenth-century Europe, exotic flowers and paintings like this signified intellectual conquest of nature and political conquest of the world. *Vase of Flowers*, by Jan Davidsz de Heem, 1606–1683. (Courtesy of the National Gallery of Art.)

In 1985, satellite images first revealed a hole in the atmosphere's ozone layer over Antarctica. This hole varies with the seasons and weather patterns, but at times the depletion has been as much as 97 percent. Similar but less severe depletion of the ozone layer and concomitant intensification of ultraviolet radiation has also been discovered over northern latitudes.

In 1987, the leading industrial nations agreed that they would reduce production of CFCs by 50 percent by 1998, but as concern increased, they agreed to phase out production entirely by January 1, 1996. They have also agreed to share replacement technologies with the developing countries, thereby setting an important precedent for financial assistance to the developing countries for environmental preservation. Researchers developed substitutes in record time, and already by mid-1993 reports indicated that the buildup of CFCs was slowing. The worst ozone destruction should occur around the year 2000, and then the ozone layer is expected to begin a recovery taking 50 to 100 years.

Although the upper-atmospheric layer of ozone is beneficial, ozone in the lower atmosphere, in contrast, is highly dangerous to delicate lung tissue. Unfortunately, increasing amounts of it are being produced. The processes that produce ozone in the lower atmosphere depend on a supply of fuel, such as hydrocarbons, a source of energy, such as the sun, and a catalyst, such as reactive nitrogen compounds. In cities, nitrogen compounds and hydrocarbons are given off by industrial smokestacks and by motor vehicles. Therefore, it is on the sunniest days that these elements combine to produce the most dangerous photochemical air pollution. The concentrations of ozone that build up in the air of large cities present a worldwide health crisis. Still more ozone is being produced in the southern tropics from burning of fields and grasslands in Brazil and savannas in southern Africa. Intense tropical sunshine seems to have reacted with chemicals from this burning to produce an immense cloud of ozone over the South Atlantic Ocean. This cloud was discovered only in 1992.

Ironically, studies of world urban areas suggest that the relatively high ozone content of the lower atmosphere, which is unhealthy for us to breathe, may offer some protection from the sun's ultraviolet radiation.

Acid Rain

Acid rain is another one of the most perplexing anthropogenic environmental problems of recent decades. This term refers to the deposition of either wet or dry acidic materials from the atmosphere on to the Earth's surface. Acid rain is most conspicuously associated with rainfall, but the pollutants may fall to Earth with snow, sleet, hail, or fog, or in the dry form of gases or particulate matter.

There is no universal agreement on the origin of acid rain, but evidence indicates that the principal human-induced sources are sulfur dioxide emissions from smokestacks and nitrogen oxide exhaust from motor vehicles. These and other sulfur and nitrogen compounds are expelled into the air, where they may be carried hundreds or even thousands of miles by winds. In the air they combine with moisture to form sulfurous and nitric acids that sooner or later fall to Earth.

Acidity is measured on the pH scale, which is based on the relative concentration of active hydrogen ions. The scale is logarithmic and ranges from 0 to 14, where the lower end represents extreme acidity (battery acid has a value of 1) and the upper end extreme alkalinity (lye has a value of 14). Rainfall in clean, dustfree air is slightly acidic, due to the reaction of water with carbon dioxide to form a weak carbonic acid. Today, however, precipitation with a pH of less than 4.5 (the level below which most fish perish) is being recorded, and an acid fog with a record low of 1.7 (8,000 times more acidic than normal rainfall) was experienced in California in 1982.

Several hundred lakes in the eastern United States and Canada have become largely devoid of life in the last quarter century, primarily due to deposition from acid rain (Figure 2-12). The precise effects of acid rain on forests are not clearly understood, but evidence increasingly indicates that it is primarily responsible for killing trees around the world. Crops and human health are also at risk, and even buildings and monuments are being destroyed. For example, acid deposition has caused more erosion to the Parthenon—an ancient marble temple in Athens, Greece—in the most recent quarter century than took place in the previous 24 centuries. China is the world's biggest user of sulfur-rich coal, and the Chinese government has only recently become

alarmed at the billions of dollars of damage done to forests, farm crops, and buildings by acid rain.

It is difficult to coordinate action on this problem because the pollution is experienced at great distance from its source. Downwind locations receive unwanted acid deposition from upwind locations. Thus, within the United States, New Englanders accuse the Midwest, but people in Ohio are reluctant to finance expensive cleanups that will benefit forests in Maine. The problem also crosses international borders. Japan has some of the strictest emissions laws in the world, yet rain in Japan is often as sour as grapefruit juice. Japanese scientists say that China is responsible for half of the problem, and South Korea for another 15 percent. Japan has donated to China devices called *scrubbers* that clean smokestack emissions. In 1979, 35 countries signed the Convention of Long-Range Transboundary Air Pollution, and in 1985, European countries promised further reductions in sulphur dioxide output. Eastern Europe suffers from the pollution blowing from Western Europe.

Scientists believe that about half the acid rain that falls on Canada comes from U.S. sources, and Canada has expressed dissatisfaction with U.S. delay in mitigating the problem. The U.S. Federal Clean Air Act of 1990 set limits on pollutants which utilities could discharge in terms of pounds of pollutant per unit of heat produced in utility boilers. One limit is set for the period 1995–99, and a stricter level begins in 2000. Since Congress passed this act, U.S. coal-fired plants have invested billions of dollars in mitigation efforts. As a

FIGURE 2-12

Scientists from the U.S. Environmental Protection Agency (EPA) regularly check readings from this sampling buoy floating in Chesapeake Bay. Instruments in the buoy record the acidity of the water, as well as its temperature, salinity, and other chemical characteristics.
(Courtesy of S.C. Delaney/U.S. Environmental Protection Agency.)

Externalities

It is generally cheaper to prevent pollution at its source than it is to clean up pollution. The principal question, however, is: Cheaper for whom? If I own a factory that adopts an inexpensive industrial process that heavily pollutes the local air and local waterways, I am saving myself money, I am saving my customers money, and I am able to stay competitive and pay my workers' wages. I am, however, imposing a burden on the local taxpayers—many of whom may be my own employees—who suffer ill health and who will pay to clean up the local waterway.

Any ramifications of an activity that do not rebound directly to the responsible party and are not, therefore, accounted for in that party's costs or profit are called **externalities**. Externalities may be either positive or negative. In the example above, my factory is imposing negative externalities (costs) on others. An example of a positive externality is education. It is usually provided by the government, and its benefits are assumed to diffuse throughout the local economy. One of the principle purposes of government is to assume responsibility for those activities which profit everybody but whose costs

Deforestation in the foothills of the Himalayas in Nepal (left) inevitably causes death and destruction 500 miles (800 km) downstream in Bangladesh (right) where workers are frantically building dikes against the flooding rivers. Should Nepal pay for damage in Bangladesh? Should Bangladesh pay Nepal not to deforest the mountainsides? (Courtesy of Agency for International Development/Delores Weiss.)

cannot generally be allocated by market forces.

Within any one country, national laws govern the allocation of costs of environmental cleanup between the public and the private sectors. In the democratic capitalist system in the United States, the allocation of costs among private industries, governments, and other interested groups is subject to the give-and-take of the political process. The cost of lowering air pollution from automobiles, for example, is being shared by the fuel companies (cleaner-burning fuels), the car manufacturers (cleaner-burning engines), and the car-driving public (more inspections and tighter emission controls).

At the international level, however, among independent countries, the allocation of cleanup costs is a subject of considerable dispute. Environmental disruption in one country may export negative externalities to another. Downwind countries feel that upwind countries seldom do enough to reduce the emissions that cause acid rain. Similarly, downriver countries complain about water pollution or flooding from countries upriver. Deforestation in Nepal, for example, increases the runoff from the monsoon rains, and that causes floods that drown people downriver in Bangladesh.

result, sulfur emissions have declined by almost 35 percent, despite the fact that the use of coal has almost doubled. Still, the problem persists, and heroic efforts and staggering costs will be required to achieve cleaner emissions. Meanwhile, the Canadian federal and provincial governments have passed laws requiring strict reductions in Canada's own acid rain emissions.

The Human Impact on Hydrography

For thousands of years, humans have altered the flow of streams and rivers. Humans have rerouted streams to irrigate agricultural fields and supply urban needs, and they have engineered great dams and levees to achieve security for human settlements against floods. Humankind's continuing efforts to manage water, however, illustrate that sometimes we must choose between trying to dominate nature and learning how better to accommodate human activities to it. One of the greatest civilian losses of life in U.S. history followed the collapse of California's Saint Francis Dam in 1928. About 450 people died, and although the precise cause of the collapse was never officially determined, the dam had probably been ill-constructed.

Flooding is usually referred to as a "natural disaster." In some cases, however, humankind has so altered natural drainage conditions that the disaster must be recognized as, in part at least, human-induced. When people build levees along rivers and settle the surrounding floodplains, they are gambling that floods will never breech the levees, and yet the construction of levees in itself creates conditions which make eventual flooding more probable. Floodplains naturally reduce the force and height of floodwaters by spreading, slowing, and storing them (Figure 2-13). Once a river is constricted by levees, however, it moves faster. Bends and islands that normally slow the water are erased. The river's sediments are no longer deposited to fertilize the floodplain, but instead they may be dropped within the narrow channel just inside the levee. In time, this raises the height of the river above the surrounding occupied land, thus increasing the risk of disaster in the event of failure. Reservoirs, designed to allay flooding, are often kept high to allow for water skiing, boating, and other recreational opportunities.

By 1991, according to a U.S. federal task force on floodplains, floodplain land in 17,000 U.S. communities occupied more than 145,000 acres and included nearly 10 million households and $390 billion in property. This development, according to the report, has come at a "high price extracted annually in deaths, personal injury and suffering, economic loss and damage to or destruction of natural and cultural resources." Despite extensive and expensive efforts through the years to control floods (the U.S. Army Corps of Engineers alone had built 10,500 miles of levees and flood walls), inflation-adjusted flood damages per capita were almost 2.5 times as great in the period from 1951 to 1985 as from 1916 through 1950.

The upper Mississippi and Missouri River valleys provide an excellent case study. After a disastrous flood in 1927, Congress passed the 1928 Mississippi River Flood Control Law, mandating the Corps of Engineers to confine the Mississippi River to a prescribed channel. Over the following 65 years, the Corps spent billions of dollars to build over 300 dams and reservoirs and thousands of miles of levees and floodwalls and to operate pumping stations to regulate water on the Mississippi and its tributaries. Elaborate computer systems measured rainfall and tracked the flow to the Gulf of Mexico.

In 1993, however, unusually heavy rains fell during the spring and summer in the upper Mississippi and Missouri River valleys. The rivers bounded beyond their banks, flooding millions of acres of prime farmland in nine states, forcing tens of thousands of people to evacuate their homes, halting river traffic, and causing billions of dollars in damages (see Figure 2-14). More than 50 people died. These floods forced Americans to choose between building still more elaborate flood control works, at

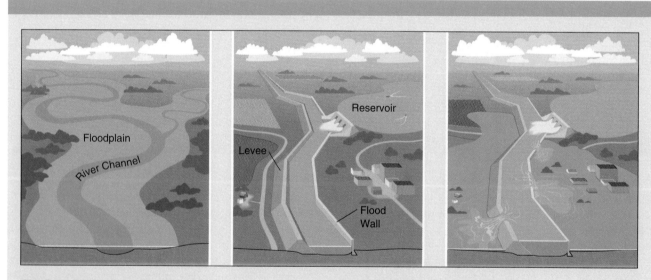

FIGURE 2-13

Rivers naturally flood periodically. When they are artificially narrowed and straightened by levees and floodwalls, however, they flow faster, and they can rise above the level of the surrounding land. Then, in flood years, floodwaters may overtop or breach the levees, crashing over the human developments. Levees prevent floods in most years, but when they fail, the losses are greater.

Midwest Flood Damage in 1993

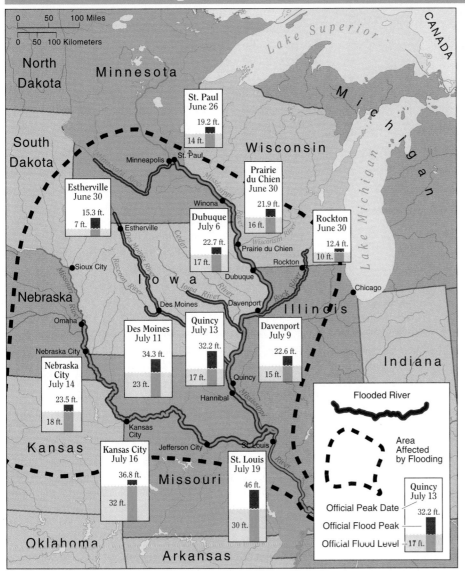

FIGURE 2-14

The floods of 1993 caused extensive damage throughout the upper Midwest.

enormous expense, or, alternatively, dismantling some of the floodworks and trying to re-create more natural river systems.

Tulsa, Oklahoma, earlier had already chosen the latter alternative. After a disastrous 1984 flash flood, the city and the Corps of Engineers scooped out a series of lakes in a greenway corridor along Tulsa's Mingo Creek flood plain (Figure 2-15). Most of the lakes are dry, and the lake bottoms are usually occupied by soccer and baseball fields and tennis courts. When there is a flood, however, the lakes store and slow down the water. They are connected by a network of "trickle trails," the edges of which serve as jogging paths in dry times and low-flow flood channels in wet ones. At the same time, development in the flood plain is strictly regulated. Developers must build their own detention ponds to make up for the increased flow that paving

causes. The Tulsa system has worked perfectly during storms, and it may become a national model.

Whether a community decides to build more waterworks or to design a more natural drainage system requires balanced consideration of several factors: the cost of building the protection facilities; the value of the uses to which the land is put if it is protected (agricultural, commercial, residential); the value of the alternative uses, if unprotected (parks, recreation, or wildlife refuge); and the cost of occasional flood damage in either case.

The Human Impact on Vegetation

The overall human impact on the Earth's vegetation has been to diminish natural variety, degrade natural complexity, and redistribute plant species around the

Cultures and the Transformation of Landscapes **67**

FIGURE 2-15

The city of Tulsa, Oklahoma, reduces flood damage by dedicating floodprone lands to recreation.
(Courtesy of Valerie McCaw, Department of Public Works, Tulsa, Oklahoma.)

globe. The first of these points is well illustrated in the tropical rainforests, the clearing of which is today a major international concern.

Rainforest Removal As recently as 1950, tropical rainforests covered about 15 percent of the Earth's surface. That percentage declined to 12 percent by 1975, and destruction of these forests is continuing today at so great a pace that they might cover only 7 percent of the Earth's land surface by the year 2000. The area of temperate forest, by contrast, has remained at about 20 percent of the Earth's surface, due largely to reforestation.

The tropical forests are home to perhaps half the world's biotic species, about five-sixths of which have not yet even been described and named, and clearance of the tropical forests endangers many species with extinction. When any species is gone, it takes with it hard-won lessons of survival encoded in its genes over millions of years. The destruction of species is like burning vast libraries of unique books before we check to see what knowledge they contain. The loss of biotic diversity means that valuable potential resources—pharmaceutical products, new food crops, natural insecticides, industrial materials—may disappear before they are even discovered.

The gases released from burning of the forest contribute to the thinning of the Earth's ozone layer and add to the atmosphere's carbon dioxide content. The burning of the trees also ends their helpful function of removing carbon dioxide from the atmosphere.

Tropical deforestation imposes incalculable costs on societies. These include loss of fisheries, loss of wood for fuel, soil erosion and silting of rivers, drought, flooding, declining agricultural productivity, and the impoverishment of the local inhabitants. Indigenous peoples, thinly dispersed through the forests, are ill-prepared to defend themselves, and introduced diseases have wiped out whole peoples before anybody from outside has even had a chance to talk to them.

The tragic irony is that the economic benefits anticipated are largely illusory. Much of the forest clearing, especially in Latin America, happens in response to the social pressures generated by over-crowding and poverty in societies where most of the people are landless. The government throws open "new lands" for settlement, and the settlers clear the land for crop growing or livestock raising. The result almost always is an initial increase in productivity, followed in only two or three years by a pronounced decline in fertility due to nutrient loss, erosion, and infestation by weeds. Agriculture can generally be sustained only through expensive fertilization. Substantially greater long-term income could be achieved through harvest of the diverse riches of the rainforest. For example, harvesting the edible fruits, rubber, oils, and cocoa from 2.5 acres (1 hectare) of Brazilian forest yields more than double the value of the forest cut over for timber or used for grazing.

Commercial logging accounts for about a quarter of the world's total annual rainforest loss, but in West Africa and Southeast Asia logging is already necessarily coming to an end, and the timber trade is moving to the forests of central Africa and the Amazon basin. In 1989, the World Bank estimated that fewer than 10 of the 33 countries then exporting tropical forest products would still be able to do so by the year 2000.

Recent discoveries in some of the world's densest rainforests testify to human intervention in the past. Discoveries of pot shards and the remnants of ancient fields testify that forests previously thought to be virgin have been harvested, or even burned and cultivated. The distribution of the Iriartea palm in Central America provides another sort of evidence. Harvesting the delicious heart of this palm kills the tree. Studies have found that those areas most accessible to humans have been cleared of the palm, but the palm still flourishes in less accessible areas. This distribution cannot be explained by any other factor of local topography or soils, so it suggests that humans harvested this tree in the past.

A discovery like this is tremendously important for ecological and geographical studies. It means that widespread patterns of plant and animal distributions that were previously thought "natural," but that were inexplicable in terms of the local natural environ-

The Three Gorges Dam

Just as the U.S. Midwest was suffering disastrous flooding in 1993, the Chinese began work on a 607 foot (186 m) high dam stretching 1.2 miles (1.9 km) along the third of the Yangtze River's three gorges. The dam is intended to create a lake 350 miles (220 km) long, provide power for new industry, and save millions of people downriver from the threat of flooding.

Three Gorges will not be the highest dam in the world, nor will its reservoir be the biggest, but its planned output of electricity will be the greatest: 17,680 megawatts, compared to the 10,300-megawatt capacity of the Guru-Raúl Leoni Dam in Venezuela; the 7,400-megawatt capacity of the Itaipu Dam on the Brazilian-Paraguayan border; and the 6,495-megawatt capacity of the Grand Coulee Dam in the United States.

The main argument for the Three Gorges Dam, however, is not for the electricity, but for flood control. Downriver Chinese have built levees for many years; today the lower Yangtze flows so far above the level of the surrounding towns that it is known as the "hanging river." During the worst recorded break in the downriver levees, in 1870, millions of people perished.

Critics argue that the Three Gorges Dam will force 1.2 million people to move and that it will partly submerge the spectacular natural scenery, which is as famous as the Grand Canyon. The major technical question is the possibility of silting. The river carries 523 million tons of sediment each year. If the dam slows the river, the silt might drop in the reservoir. If it does, then the dam will only create the world's largest and most expensive mudpie.

The Yangtze River and Its Gorges

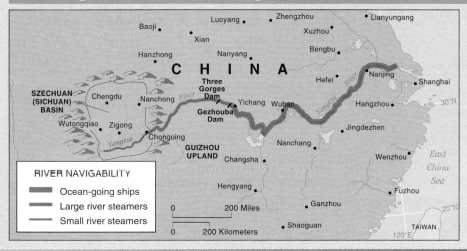

Communications between Sichuan and the rest of China depended on precarious roads along the Yangtze River's gorges until the Tang dynasty (618–906) cleared the gorges for ship and barge traffic.

ment, may in fact be the historical result of societies that have since disappeared or moved on.

Also, studies suggest that the ability of the rainforest to regenerate itself naturally is much greater than previously thought. It is difficult to ascertain the age of rainforest trees because they do not exhibit annual growth rings, but researchers estimate that the vegetation on an average rainforest plot will turn over within just 100 years.

These discoveries encourage optimism that the tropical rainforests are more resilient than once thought, and that they can even recover from some human disturbance. In the past, however, less extensive patches of forest were cleared, and they were given ample time to recover. Today's much larger scale human intervention may be devastating.

The Redistribution of Plant and Animal Life

Both plants and animals can migrate great distances without human intervention. Intercontinental winds can carry seeds, or birds can carry them undigested

Diffusion of African Honeybees

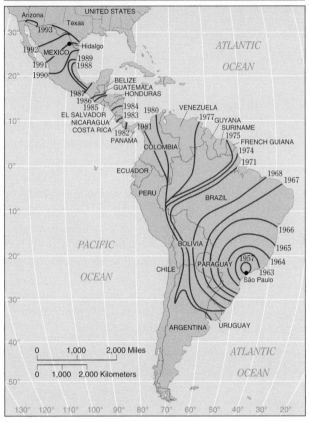

FIGURE 2-16

African honeybees were inadvertently released from a laboratory outside São Paulo, Brazil, in 1957. They diffused throughout lowland eastern South America and northward through Central America before they first reached the United States in 1993. (Data from "Africanized Bees in the U.S.," by Thomas E. Rinderer, Benjamin P. Oldroyd and Walter S. Sheppard, © December 1993 by *Scientific American, Inc.*)

in their stomachs. The discovery of birds far from their expected habitats lends excitement to the scientific hobby of bird-watching. In 1993, for example, ornithologists from throughout North America raced to New Jersey to confirm the sighting there of a whiskered tern (*Chlidonias hybridus*). This bird is common in Europe, Africa, and Asia, but it had never before been seen in the Western Hemisphere. The lone male probably flew from the British Isles.

Also in 1993, 36 years after African "killer bees" were inadvertently let loose from a laboratory in Brazil, these bees caused their first documented death of a human being in the United States (see Figure 2-16). The bees will probably spread throughout the warm southern third of the United States, but we cannot predict how they will adjust to the climate, the resident population of European bees, North American bee diseases, or whatever natural enemies they may find in the United States.

Human activities have transferred plants and animals from one region to another since ancient times, but the global transplantation of organisms accelerated with the European exploration of the sixteenth and seventeenth centuries and again with the speedier transportation and expanded trade of modern times. Today flora and fauna in virtually every part of the planet are a mixture of natives and exotics. Some exotics have settled benignly into their new habitats, but others have aggressively eaten or crowded out natives to the point where some natives have been extinguished and others threatened with extinction.

Humans have introduced alien species to new environments both accidentally and deliberately. Successful redistributions of domesticated crops or animals have allowed regional specializations and multiplied world food supplies. Some deliberate redistributions, however, have turned out for the worse. For example, Australians imported the tropical toad (*Bufo marinis*) from South America in 1935 to control sugarcane beetles. Unfortunately, the toads are nocturnal feeders and the beetles are abroad by day. By night the toads flourish, reproduce phenomenally well, and eat up everything they can find. Today the toads are thriving. Some are the size of dinner plates, and nobody knows what to do.

The nutria, a South American rodent, provides another example (Figure 2-17). It was introduced into Louisiana in 1937 to be raised commercially for its fur.

FIGURE 2-17

Some people may think that the nutria (*Myocastor coypus*) is a cute little animal, yet this South American native has wrought tremendous environmental damage in Louisiana. (Courtesy of Michael Carloss, Louisiana Department of Wildlife and Fisheries.)

70 *Chapter 2*

GEOGRAPHIC REASONING
An Eleventh Biome

Questions

1. If you live in an urban biome, list the ways in which your environment differs from the natural biome surrounding your city: climate, animals, plants, and other factors.
2. About how many days per year, on average, do you leave your urban biome to visit a more "natural" biome? Do you enjoy those visits?
3. Would you be happy if it never rained where you live?
4. What survival skills would you teach a young scout in the urban biome?
5. Can humans create this urban biome anywhere on Earth? At what cost?

Chapter 1 defined ten Earth biomes. About one-half of the human species today, however, lives in cities, and we might consider "the urban environment" as an eleventh biome. The description of this "urban" biome exercises our deductive logic.

Within modern cities, let us say, heating and cooling keep the temperature range between 60°F and 80°F year round. Precipitation is an inconvenience. The inhabitants have little idea how much of it falls, but generally, the less of it that falls, the happier they are. There is little surface erosion. Precipitation runs off paved surfaces and goes somewhere (where?). The depth or quality of the bedrock is of concern only to architects, tunnelers, and others in the construction industry. Similarly, only

golf course and park maintenance workers know or care much about the fertility of whatever soil is exposed. The simplest forms of vegetation (mosses, lichens, and grasses) try to establish themselves on urban surfaces, but almost all vegetation is selected, planted, and tended by humans.

The wildlife consists of human beings, cats, dogs, pet birds, pigeons and other birds that scavenge on human refuse, and great numbers of rats and insects. Considerable numbers of other animals may be kept in confinement awaiting slaughter for food, or in wholly artificial pseudonatural environments called zoos.

This biome covers rapidly increasing areas of the Earth.

A few escaped into the marshes, and their multiplication there has devastated natural ecosystems. They consume prodigious amounts of the fragile swampgrass vegetation, and they chop down more than ten times as much as they consume. They can grow as long as 3 feet (1 m) long and weigh as much as 15 pounds (7 kg). They threaten to eliminate fish and shellfish breeding grounds. The state government has sponsored nutria hunting contests, subsidized and promoted the nutria fur industry, and even underwritten cooking contests to devise tasty dishes based on nutria meat in order to reduce the numbers of the proliferating pests. Additional examples of unforeseen negative results from the deliberate redistribution of animal or plant species around the world would fill an encyclopedia. A U.S. government report estimated that in the single year of 1991, 15 of the most harmful exotic species of plants and animals brought to the United States in the past 100 years cost the United States $134 billion, most of it in crop damage.

Recent studies have emphasized the extent to which international shipping has inadvertently introduced alien life into distant ecosystems. The ballast tanks of ocean-going ships transport veritable soups of

marine organisms. A 1992 sampling of the ballast waste of 159 Japanese cargo ships that arrived at Coos Bay, Oregon, found 367 different types of plants and animals, mostly in larval form, representing all the major marine habitat groups: plants, plant-eaters, meat-eaters, scavengers, and parasites. In any bay or estuary, ecological factors such as water quality, salinity, time of year, temperature, and number of predators and competitors vary from time to time and place to place. If the combination is not exactly favorable when an invader arrives, the invader may not become established. But if all circumstances are favorable, a window of opportunity is open. The researchers estimated that about 5 to 10 percent of the introduced species became established in Oregon waters. Some scientists wonder whether all bays and estuaries in the world will eventually have the same life forms, those that can compete most successfully against all other species.

Soil Erosion

Some ecologists have referred to the global loss of potentially productive topsoil as a "quiet crisis." The United Nations Global Assessment of Soil Degradation focuses attention on four processes of human-induced soil degradation.

1. *Water erosion*, which occurs in all countries.
2. *Wind erosion*, which is widespread in both arid and semiarid regions, and which is nearly always caused by a decrease in the vegetative cover through overgrazing, poor agricultural practices, or deforestation.
3. The many types of *chemical deterioration*, which include repeatedly cropping soils without adding fertilizers; salinization, usually caused by poorly drained irrigation systems; acidification caused by overapplication of acidifying fertilizers; and pollution by industrial or urban wastes.
4. The types of *physical deterioration*, which include compaction by machinery or trampling by livestock; and waterlogging caused by human intervention in drainage systems.

Figure 2-18 shows the ratio of soil degradation by type for the whole world and for each world region. The processes responsible vary greatly from region to region. In Europe, for example, 17 percent of soil degradation is caused by physical degradation, whereas in North America only 1 percent is caused by physical degradation.

Figure 2-19 shows the most common causes of soil degradation. The causes also vary greatly from continent to continent. In North America, for example, 66 percent of soil degradation is attributed to poor agricultural practices, whereas in Africa 49 percent is attributed to overgrazing.

Figure 2-20 maps the areas of greatest concern. Soil can regenerate under proper care, but in many areas, economic or population pressures drive local people to overexploit the environment. Attempts to farm inappropriate areas can even lead to virtual desertification.

Desertification

The term **desertification** has come into use since the 1960s to refer to the expansion of desert conditions into areas that previously were not deserts. The

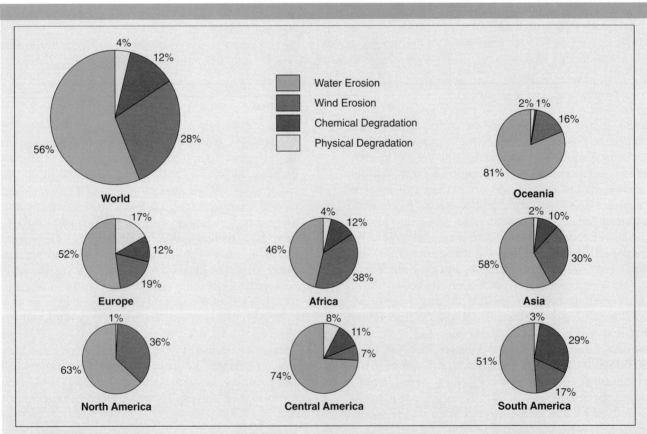

FIGURE 2-18

These graphs of the types of soil degradation reveal that, worldwide, water erosion is most common, but that the percentage of degradation for which it is responsible varies greatly among continents. Chemical degradation, for example, is important in Latin America, Africa, and Europe, but negligible in North America.

(Data from World Resources Institute, *World Resources 1992–93*. New York: Oxford University Press, 1992, p. 114.)

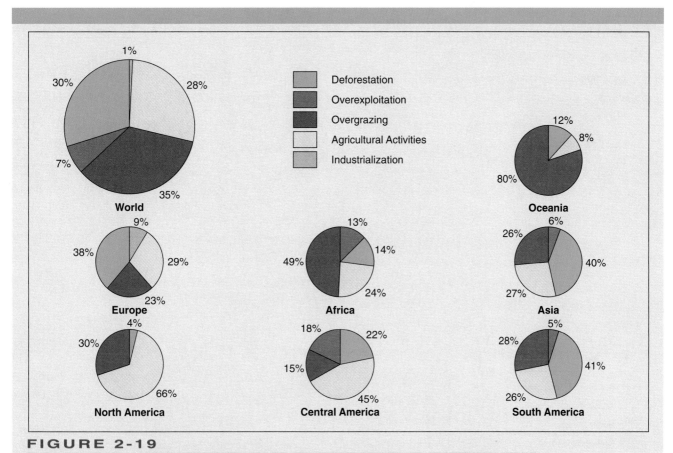

FIGURE 2-19

The causes of soil degradation also vary greatly from continent to continent. Industrialization is important in Europe, but not so important anywhere else.
(Data from World Resources Institute, *World Resources 1992–93*. New York: Oxford University Press, 1992, p. 114.)

enlargement of deserts can come about through causes that are entirely natural, the most important being recurrent drought. It can also be caused by the activities of humans in degrading the land through overgrazing of livestock, imprudent agriculture, deforestation, and improvident use of water resources. The process of desertification is now generally accepted as a product of human-induced environmental degradation superimposed on a natural drought situation.

Desertification normally is associated with the margins of existing deserts and implies an expansion of the present desert into a spreading peripheral zone. However, this is not always the case. For example, the "Dust Bowl" conditions in the North American Great Plains in the 1930s were not peripheral to any existing desert, yet they represented a clear and spectacular example of desertification long before the term was in common use.

The specter of desertification was brought to worldwide attention in the late 1960s by the onset of a 6-year drought in the African Sahel (Figure 2-21).

The Sahel is a subhumid to arid region on the southern margin of the Sahara Desert that occupies parts of 11 countries.

As the Sahelian drought intensified, from 1968 to 1974, vegetation disappeared, millions of livestock perished, and thousands of survivors migrated to an uncertain future in already overcrowded lands, particularly cities to the south of the drought-stricken region. Rainfall increased through the 1980s, but has decreased again in the 1990s. Every year has been "deficient" in rainfall, and the cumulative effect has been overwhelming.

Traditional life in the Sahel was dominated by nomadic herding of goats, cattle, and camels. Crop growing was mostly limited to favored river valleys. In general, human use of resources was kept in balance with environmental conditions. During the 1950s and 1960s, however, human and livestock populations soared, due to improved medical conditions and above-average rainfall. Dry-land farming and some irrigation agriculture spread into traditional nomadic

World Soil Degradation

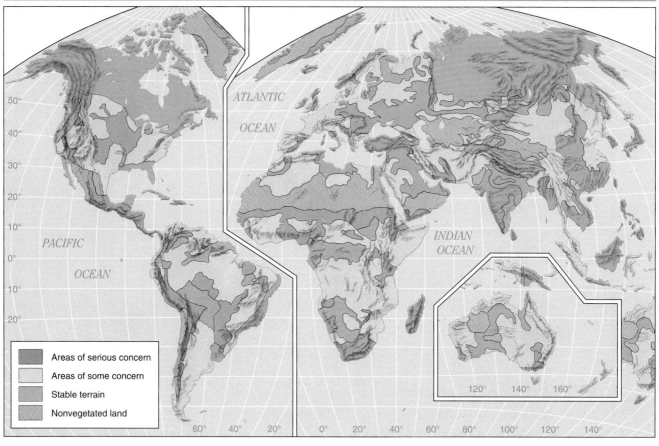

Areas of serious concern
Areas of some concern
Stable terrain
Nonvegetated land

FIGURE 2-20

Soil degradation is a matter of at least some concern almost anywhere people live.
(Data from World Resources Institute, *World Resources 1992–93*. New York: Oxford University Press, 1992, p. 117.)

lands, tilling soil that never should have been broken. In addition, colonial rulers and then later newly independent governments wanted to increase tax collections, and to achieve this they forced the people to shift from subsistence agriculture to cash crops. This changed the age-old mutually supportive relationship between the pastoral nomads and the settled agriculturalists. At the same time the enormous demand for firewood (the cooking and heating fuel for 80 percent of the Sahelian population) resulted in the elimination

Rainfall in the Sahel

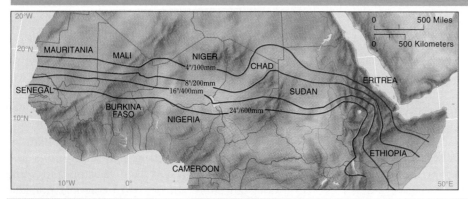

FIGURE 2-21

The isohyets (lines of equal rainfall) show the amount of annual rainfall increasing steadily southward from the Sahara Desert, but agriculture and even life is still precarious throughout this Sahelian region.

of most trees and large shrubs. Overgrazing removed most grasses and small shrubs, and newly independent countries closed their borders to traditional pastoral migration routes. Then came year after year of searing drought. As a consequence, land that formerly supported vegetation became desert.

Desertification is not limited to the Sahel. It is also being experienced in southern Africa, in the Middle East, in India, in western China, in Chile and Peru, in northeastern Brazil, in the southwestern United States, in Mexico, and in other places. The United Nations Environment Program estimates that 8,100 square miles (21 million ha) of land is being desertified in the world each year.

The semiarid environment is fragile, but it is also resilient. The North American Dust Bowl recovered from its devastation of the 1930s. If human and livestock pressure on the Sahelian land could be relieved, and if a few years of "normal" rainfall would occur, we could expect striking improvement in the landscape.

Environmental Reclamation

The successful restoration of Western Australia demonstrates that over-exploited and polluted landscapes can be restored to productivity. Since white settlers arrived in Western Australia in 1829, they have chopped down more than half of the forests that once covered one-third of the continent. They replaced native plants with clover or wheat. Today two-thirds of Australia is farmland or grazing land, but this area has begun to shrink. Nearly 1.25 million acres (500,000 ha) were used by farmers in 1980, but by 1990 that area had shrunk by 75,000 acres (30,000 ha). Part of the explanation is the impoverishment of the soils, which were either washed or blown away or poisoned by salt.

Soil salinity is widespread. Trees once sheltered the soil from heavy rain and soaked up water, but when the trees are gone, the water table rises, and mineral salts are rinsed out of the soil. As the water evaporates, a crust of salt bleaches the surface of the soil, killing plants. In 1982, Western Australia was thought to have about 110,000 saline acres (44,000 ha); by 1991, that number had risen to 180,000 saline acres (72,000 ha).

Western Australia was the last part of the country to be cleared for farming. Government support for land clearance continued into the 1980s, but the government changed that policy after 2 years of drought in 1981–82. New land is still being cleared, but degraded land is being rescued at almost the same pace. Rescuing land is difficult and expensive. Planting trees lowers the water table and helps the soil recover, but trees flourish only if they are protected by native shrubs. Shelter belts of vegetation that protect the soil

from the winds can raise crop yields by 25 percent. The importation of Mediterranean and South African grasses that are suited to arid soils allows the land to support twice as many grazing animals.

What was once a source of pollution from mining is today helping to restore agriculture. Aluminum companies tore up forests around Perth in order to strip mine the bauxite below (Figure 2-22). Once they had extracted the ore, they left mounds of red mud. Experiments have proven that when this mud is mixed with the thin and sandy soils of the coastal region, the soils retain moisture. Therefore, spreading this mud on local fields stabilizes applications of fertilizer, and farm yields rise dramatically. In addition, the rivers stay clean because the fertilizer is not flushed into local rivers.

In conclusion, interaction between the environment and humankind is reciprocal—the environment affects human life and cultures, and humans alter and transform the environment. The Earth's delicately bal-

FIGURE 2-22

Spreading the residue from bauxite mining on fields in Western Australia reduces the loss of phosphorous from local sandy soils. This helps the soils retain moisture, thus increasing their productivity. Furthermore, it reduces the flow of nutrients from fertilizers and animal stockyards into local water bodies. The nutrients had been encouraging the multiplication of algae. Algae had been accumulating and rotting on the shores, stinking, polluting the shoreline, and killing wildlife and fish. Alcoa of Australia has won awards from the United Nations Environment Program for this work and for its reforestation after bauxite mining. (Courtesy of Alcoa of Australia Limited.)

Cultures and the Transformation of Landscapes **75**

anced physical and ecological systems provide the stage upon which humans live. The stage establishes certain conditions of human life, yet the evidence of human intervention in natural systems is virtually ubiquitous. The physical environment provides a stage, but humankind sets the scenery.

When humans alter or manipulate complex, interrelated natural systems, this activity necessarily triggers changes throughout the systems. Humans cannot foresee the repercussions of their acts, but humans must hope to be able to deal with them. One intervention leads to another, and the results of each succeeding intervention can only be solved by still another intervention. The process is endless. To assume the responsibility for this is, perhaps, to be human.

SUMMARY

Interaction between the environment and humankind is reciprocal—the environment affects human life and cultures, and humans alter and transform the environment.

People in isolation develop their own distinctive culture. A group's culture includes artifacts, sociofacts, and mentifacts. People transform their local landscape from a natural landscape into a cultural landscape. Each cultural group creates a distinct cultural landscape. The extraordinary diversity of human cultures and cultural landscapes testifies to human ingenuity. No direct cause-and-effect relationship between a physical environment and any aspect of culture can be assumed.

Cultural ecology studies the ways societies adapt to environments. The simplistic belief that human events can be explained entirely as the result of the effects of the physical environment is called environmental determinism. Many environmentally deterministic theories are intriguing, but when we examine any world situation or historic event carefully, environmental determinism, or any other single-factor explanation, proves insufficient.

Possibilism insists that the physical environment itself will neither suggest nor determine what people will attempt, but it may limit what people can profitably achieve.

People react to their environment as they perceive it. Therefore, study of environmental influences on human life must be supplemented by study of how people perceive the environment and how these perceptions and thoughts affect their behavior.

Behavioral geography draws upon a variety of theories about human behavior formulated in other disciplines. Rigidly scientific approaches in geography must be supplemented by humanistic geography, which emphasizes interpretive understanding. Theories in geography about human behavior include the applications of the theory of economic man, theories about the perception of hazards, theories of aesthetic reactions to landscapes, and studies of proxemics and territoriality.

Humankind has for the most part adapted itself to widely varying conditions through cultural and technological evolution, rather than through physical differentiation. Varro's theory of human stages and Marx's theory of historical materialism emphasize the role that technological and economic evolution plays in the interaction between societies and the environment.

Humankind has altered or manipulated some environments so much that it is impossible to differentiate the "natural" from the anthropogenic, and little "natural environment" is left to identify. Any change in one of the Earth's natural physical and ecological systems, however, will trigger changes throughout the entire system. After humans have altered a natural system, the natural system is never quite the same.

Humans have tried to alter the climate by seeding clouds for rain and even increasing the amount of solar radiation reaching the northern latitudes. Human activities add greenhouse gases to the atmosphere and contribute to global warming. Human actions are also depleting the protective layer of ozone in the upper atmosphere, while at the same time creating increasing amounts of ozone in the lower atmosphere, where it is highly dangerous. Human burning of fossil fuels causes acid rain, which can cause damage to wildlife and human works. Acid rain is a negative externality.

Flooding is usually referred to as a "natural disaster." In some cases, however, the disaster must be recognized as, in part at least, human-induced. The overall human impact on the Earth's plant and animal life has been to diminish natural variety, degrade natural complexity, and redistribute species around the globe.

Some ecologists have referred to the global loss of topsoil as a "quiet crisis." Soil can regenerate under proper care, but in many areas, economic or population pressures drive local people to overexploit the environment. Overexploited and polluted landscapes can be restored to productivity.

KEY TERMS

culture (p. 45)
artifact (p. 46)
sociofact (p. 46)
mentifact (p. 46)
subculture (p. 46)
natural landscape (p. 46)
cultural landscape (p. 46)
built environment (p. 46)
environmentalist
  theories (p. 46)
cultural ecology (p. 47)
environmental
  determinism (p. 47)
challenge-response
  theory (p. 47)
ethnocentrism (p. 48)
pathetic fallacy (p. 48)
possibilism (p. 48)
mental map (p. 50)
cognitive
  behavioralism (p. 50)
behavioral
  geography (p. 50)
economic man (p. 52)
man the
  satisficer (p. 53)
topophilia (p. 54)
feng-shui (p. 55)
geomancy (p. 55)
proxemics (p. 56)
territoriality (p. 57)
ecological
  niche (p. 57)
hunter-
  gatherers (p. 57)
pastoral
  nomadism (p. 57)
subsistence
  agriculture (p. 58)
commercial
  agriculture (p. 58)
historical
  materialism (p. 58)
garden or park (p. 61)
reserve or preserve (p. 61)
ozone (p. 62)
CFCs (p. 62)
acid rain (p. 64)
externalities (p. 65)
desertification (p. 72)

QUESTIONS FOR INVESTIGATION AND DISCUSSION

1. Make your own mental maps of residential desirability of your town, of the United States, of the world.

2. What are some ways that technology has increased the amount of habitable land? Heating, cooling, irrigation, transportation, etc.

3. Go to a local art museum and find examples of "pathetic fallacies."

4. Evaluate your own local environment as it would be evaluated by an industrialist, by a farmer. What about a tourist? Try to find a selection of tourist guides to your community. Is it recognizable to you as described?

5. How would a historical materialist explain the paradox that in a country as rich as America, some people remain poor and hungry?

6. What are the principal harmful effects of acid rain? Does your area suffer from acid rain or export pollutants to other regions?

7. What effect could continued industrialization have on acid rain around the world?

8. To some degree "natural" events are in fact caused by humankind, and humankind can ameliorate the impact of "natural disasters." List the ways in which the desertification of the Sahel, for example, was worsened by government action, and ways the governments might have ameliorated its effects.

9. Search the newspapers for an example of a "natural" disaster. To what degree did human action create the circumstances for the disaster to occur?

ADDITIONAL READINGS

BOAL, FREDERICK W. and DAVID N. LIVINGSTONE, eds. *The Behavioral Environment*. New York: Routledge, 1989.

DOWNS, ROGER and DAVID STEA, eds. *Image and Environment: Cognitive Mapping and Spatial Behavior*. Chicago: Aldine, 1973.

GOLLEDGE, REGINALD and ROBERT STIMSON. *Analytical Behavioural Geography*. New York: Routledge, 1987.

GOULD, PETER and RODNEW WHITE. *Mental Maps*. Boston: Allen & Unwin, 1986.

HALL, EDWARD T. *The Hidden Dimension*. Garden City, NY: Doubleday, 1966.

Harmful Non-Indigenous Species in the United States. Washington, DC: U.S. Office of Technology Assessment, 1993.

HOSKINS, W. G. *The Making of the English Landscape*. London: Hodder & Stoughton, 1955.

MAHAR, DENNIS J. *Government Policies and Deforestation in Brazil's Amazon Region*. Washington, DC: The World Bank, 1992.

MORRILL, RICHARD. *The Spatial Organization of Society*. North Scituate, MA: Duxbury Press, 1974.

PIMENTEL, DAVID, ed. *World Soil Erosion and Conservation*. New York: Cambridge University Press, 1993.

READER, JOHN. *Man on Earth: A Celebration of Mankind: Portraits of Human Culture in a Multitude of Environments*. Austin: University of Texas Press, 1988. Reprint. New York: Harper & Row, 1991.

SPROUT, HAROLD and MARGARET SPROUT. *The Ecological Perspective on Human Affairs*. Princeton, NJ: Princeton University Press, 1965.

THOMAS, WILLIAM, ed. *Man's Role in Changing the Face of the Earth*. Chicago, IL: University of Chicago Press, 1956.

TURNER, B. L. II, W. C. CLARK, R. W. KATES, JOHN F. RICHARDS, JESSICA T. MATHEWS and WILLIAM B. MEYER, eds. *The Earth as Transformed by Human Action: Global and Regional Changes in the Biosphere over the Past 300 Years*. New York: Cambridge University Press, 1990.

3 Cultural Diffusion

Scotland's familiar White Horse joins other international brand name symbols in this lively neon display. Consumer products salespeople strive to achieve global recognition of such symbols—called logos. Thus, Coca-Cola becomes a sign and word in a new international language. The fact that most signs here are in Japanese reveals that this photograph was taken in Osaka, Japan.

Courtesy of Jean Kugler, 1991/FPG International.

*C*hapter 2 introduced the imaginary Lonely people. They lived deep in the Brazilian rainforest completely unaware of the existence of anyone else on Earth. They had developed their own distinctive culture, and, because they did not engage in any trade or exchange with other peoples, their lifestyle was dependent on the materials provided by the local environment, and their cultural development was entirely independent and separate from that of other peoples.

The evolution of individual human cultures in isolation has, through history, yielded to the increasing interconnectedness of human societies. This chapter analyzes how the exchange of goods, ideas, and techniques among peoples has increased, and how the original individual characteristics of world cultures have come to be overlain or commingled with new shared characteristics. This mingling has created new and original balances and combinations.

The second section of this chapter analyzes a process that largely defined the past 500 years of human history: the worldwide diffusion of European culture with European global conquest and economic dominance. European political dominance has retreated, but it left behind cultural, political, and economic legacies. Just now, at the end of the twentieth century, we are reaching the end of this era and entering a new age of human history.

EARLY IDEAS ABOUT CULTURAL EVOLUTION AND CULTURAL DIFFUSION
▼▼▼▼▼▼▼▼▼▼▼▼▼▼▼▼▼▼▼▼▼▼▼▼▼▼▼

Human cultures are never static. They evolve and develop as peoples devise new techniques. In the nineteenth century many anthropologists upheld theories that cultures generally evolve in a uniform manner. That is, most societies pass through the same series of stages and arrive ultimately at a common end. These scholars held that sources of cultural change were embedded in the culture from the beginning, so the course of development was internally determined. This theory was called **evolutionism**. Varro's theory of uniform stages in the development of human societies discussed in Chapter 2 was an early example of evolutionism.

Early in the twentieth century, however, evolutionism was attacked by a new school of thought called **diffusionism**. Diffusionists held that aspects of civilization were actually developed in very few places. Then various aspects of cultures, such as clothing styles, diet, language, music, and architecture spread out from the place they originated and were adopted by other peoples. This process is called **cul-**

tural diffusion, and the process of adopting some aspect of another culture is called **acculturation**. Cultural diffusion can be actively imposed, as when an outside power conquers a region and tries to impose its way of life. It can also be freely chosen, as when one group discovers and adopts some aspect of a different culture that it considers superior to its own. Many diffusionists argued, for example, that most aspects of modern Western civilization were developed in ancient Egypt and diffused out from there.

The American anthropologist Clark Wissler (1870–1947) called the centers of cultural development *geographical culture centers*. He stated a principle, called the *age-area principle*, that if traits diffuse outward from a single geographical culture center, the traits found most widely distributed around that center must be the oldest traits. This description of cultural diffusion suggests concentric waves spreading out from a stone dropped into a pool. Carl O. Sauer (1889–1975), who taught geography at the University of California at Berkeley, elaborated geographical studies of cultural diffusion.

GEOGRAPHIC FACTORS IN CULTURAL DEVELOPMENT
▼▼▼▼▼▼▼▼▼▼▼▼▼▼▼▼▼▼▼▼▼▼▼▼▼▼▼

Scholars who prefer the evolutionist point of view of culture emphasize how cultural changes result from local initiatives and developments. Scholars who prefer the diffusionist point of view emphasize how changes result from influences from other places. Geographers and other social scientists must carefully examine the balance between evolutionist and diffusionist explanations of cultural phenomena.

Endogenous Factors

Endogenous factors or *site factors* of cultural development are elements of the specific local environment or of local cultural history and evolution. A study of endogenous factors can be summarized by the statement that "X" is there or is happening there because "Y" is also there. A map of the global distribution of pigs (Figure 3-1), for example, reveals that few are found in the Near East. Why? What other local attributes of the Near East might explain this lack of pigs? It cannot be any attribute of the local physical environment, because the local physical environment could support pigs. Attributes of local culture can explain the absence of pigs. The local population is overwhelmingly Jewish and Muslim, and both Judaism and Islam forbid the consumption of pork.

Geographers traditionally begin inventories of specific places with the local physical geography: climate, landforms, soil, vegetation and so forth. Each of these attributes is itself a product of many factors, and together they create a unique local physical environment. Human activities are carried out on this stage, and human activities interact with the local physical environment to create a cultural landscape. For example, the climate of a place affects its soil, the soil affects local agriculture, and the nature of the agricultural system affects local political life. In turn, however, political life will affect agriculture, which in turn will affect the soil. Cause and effect may be circular and almost infinitely complicated.

Exogenous Factors

In contrast to endogenous factors, *exogenous factors* (generated outside) or *situation factors* refer to the way a particular place interacts with other places. The map of the world distribution of pigs reveals a significant concentration in Iowa. What explains that concentration? Iowans could not possibly eat so much pork. The answer is that Iowans raise pigs and ship pork for consumption elsewhere. In turn they bring in other foods. The Iowan economy is not isolated and self-sufficient; it depends on exchange.

This principle of local specialization and exchange explains why there might be a concentration of pigs somewhere on Earth, but still we must answer the specific question "Why is this concentration found in Iowa?" Pigs were not first domesticated there; they are not even native to Iowa. The answer is that pig production is concentrated there today because a great supply of corn is there to feed the swine, markets are convenient, and the industry settled there years ago. Endogenous factors

World Distribution of Pigs

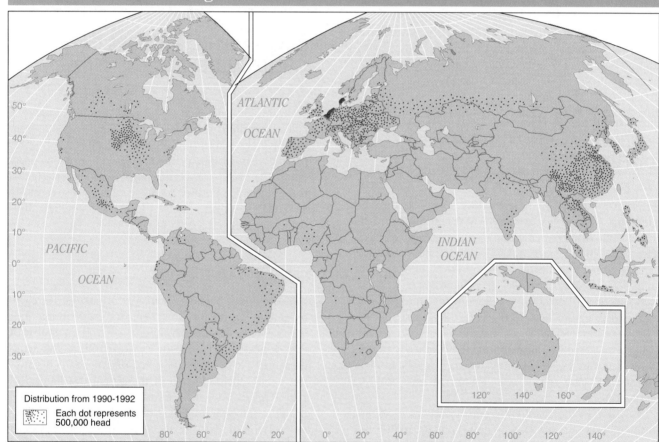

Distribution from 1990-1992
Each dot represents 500,000 head

FIGURE 3-1

Today about 43 percent of the world's pigs are raised in China, 6 percent in the United States, 4 percent each in Russia and Brazil, and 2 percent in the Ukraine. Some countries raise none, including countries that would be better off if they could increase their protein intake. What factors account for this distribution? (Data from World Resources Institute, *World Resources 1994–95*. New York: Oxford University Press, 1994, pp. 296–97.)

of explanation and exogenous factors of explanation are not contradictory. They complement each other.

The Great Shift in Cultural Geography

Through history, global communication and transportation have increased, and trade and other cultural exchanges have multiplied. As this has occurred, the balance of factors that explains the local activities and culture at any place has tipped steadily away from endogenous factors and toward exogenous factors.

Trade and exchange release people from dependence on their local environment and allow them to draw resources from around the world. Fewer and fewer people anywhere rely on their local environment for all of their needs, so the link between the resource endowment of any local environment and the well-being of the people who live there has been weakened. The Japanese, for example, live in an environment that is poor in resources, yet the Japanese have grown rich through trade.

Nor do peoples today exhibit local cultures which are devoid of imported cultural influences. More people eat imported foods and combine them in imported recipes. They wear imported clothes in imported styles, and they fashion their built environments out of imported materials in styles which originated among distant peoples. Find a picture of the most isolated people you can, and chances are high that they will be wearing or carrying something that they did not make themselves (Figure 3-2). This interconnectedness among distant peoples and places is a principal theme of this book. You undoubtedly exemplify it repeatedly in actions through your daily life. Many articles that you use and other aspects of your daily life draw on materials and ideas from around the world.

Diffusionist arguments have been used to minimize or underestimate the ingenuity of various peoples, but exogenous factors do not everywhere or always out-balance endogenous factors, nor does cultural diffusion explain the distribution of all cultural phenomena. Sometimes the same phenomenon occurs spontaneously and independently at two or more places. In the history of mathematics, for example, the idea of the zero and its use as the basis of a numerical place system was conceived in two places: among the Maya, an Indian people of Central America, and among the ancient Hindus or perhaps the Babylonians before them. Diffusionists once argued that the Maya must have learned how to use the zero from Phoenicians who sailed across the Atlantic, but there is no evidence for this. We should

FIGURE 3-2

It took 5 days of hiking through jungle to reach the village in which this young girl lives—a member of one of the tribal hill peoples of Thailand—yet much of the fabric and decoration of her exquisite costume of traditional design is of modern machine manufacture. The village is not entirely isolated but does exchange goods with the rest of the world. (Courtesy of Julia Nicole Paley.)

never assume that any aspect of culture has diffused just because similar cultural artifacts are found in two places. To demonstrate diffusion, we must be able to illustrate its path from one culture to another.

Some places enjoy great accessibility, while others are more isolated. Some engage in more trade and exchange than others. These conditions and activities determine whether exogenous or endogenous factors predominate at any given place. The conditions at any place at any time represent a temporary balance between the two sets of factors. It is clear, however, that interconnectedness and exchange among the world's peoples and places is growing. What happens *at* places depends more and more on what happens *among* places, and the patterns on maps of economic or cultural factors can be understood only in terms of the patterns of movement that create them.

Cultural Diffusion   **81**

Geographers must study not just *stasis*, which is motionlessness, but also *kinesis*, movement. Today geography doesn't just *exist*; it *happens*.

CULTURE REALMS AND CULTURAL DIFFUSION

▼▼▼▼▼▼▼▼▼▼▼▼▼▼▼▼▼▼▼▼▼▼▼▼▼▼▼▼

Today we call the place where a distinctive culture originates the **hearth area** of that culture, or **cultural hearth.** Aspects of material culture, sociofacts, and mentifacts all have culture hearths, and all diffuse. Therefore, the variety of cultural attributes open to geographic investigation is limited only by the imagination and curiosity of the geographer.

The entire region throughout which a culture prevails is called a **culture realm**. The use of the phrase "the Christian World," for instance, implies that the prevalance of Christianity across a large region unites the peoples of that region in ways that might significantly be contrasted with, for instance, "the Islamic World." Each is a great culture realm, and we will see in Chapter 8 how the prevalence of each of those religions encourages other similarities across those realms.

Smaller subcultural realms can be differentiated even within individual countries. In the United States, a number of regions can be identified and mapped (Figure 3-3). Regions such as these, defined by widespread popular perception of their existence by people inside them or outside them, are called *vernacular regions.*

The presence or impact of any cultural attribute may diminish away from its hearth area, just as the volume of a sound diminishes with distance. This phenomenon is called the **distance decay** concept. As we travel away from any city, for example, the percentage of the local population that reads that city's newspapers, tunes in to its radio and television stations, or relies on its other services shrinks. We might try to define and map the territories of influence of the cities of Los Angeles and of San Diego, for instance, as culture regions. To the immediate south of Los Angeles

people probably subscribe to Los Angeles newspapers, watch Los Angeles television stations, and rely on service from the Los Angeles airport. As we move southward, however, farther away from Los Angeles and closer to San Diego, the influence of Los Angeles diminishes and yields to that of San Diego.

Geographers try to achieve precise measurements of phenomena such as distance decay. A geographer might, for example, try to write a mathematical equation to define what percentage of households at a certain distance from downtown Los Angeles will subscribe to Los Angeles newspapers. That equation could then serve as a model to compare distribution of Los Angeles newspapers with patterns in New York, St. Louis, and other metropolitan regions.

Physicists have a precise mathematical formula to express the gravitational force that one object exerts on another, and social scientists have tried to find analogous formulas in their study of human activities. As early as 1868 the philosopher Henry C. Carey suggested that "Gravitation is here [among human activities], as everywhere else in the material world, in the direct ratio of the mass, as in the inverse one of the distance." This is called the **gravity model**, and scholars have investigated whether it describes other phenomena. In 1929, for example, the marketing consultant William J. Reilly postulated that the movement of persons between two urban centers is proportional to the product of their populations and inversely proportional to the square of the distance between them. Modern geographers have experimented with other mathmatical formulas to describe interaction across distance.

What happens *at* places depends more and more on what happens *among* places, and the patterns on maps of economic or cultural factors can be understood only in terms of the patterns of movement that create them. Geographers must study not just *stasis*, which is motionlessness, but also *kinesis*, movement. Today geography doesn't just exist; it *happens*.

THE IDENTIFICATION OF CULTURAL REALMS

▼▼▼▼▼▼▼▼▼▼▼▼▼▼▼▼▼▼▼▼▼▼▼▼▼▼

In the built environment, architecture and settlement patterns are among the most immediately evident clues to the extent of culture realms.

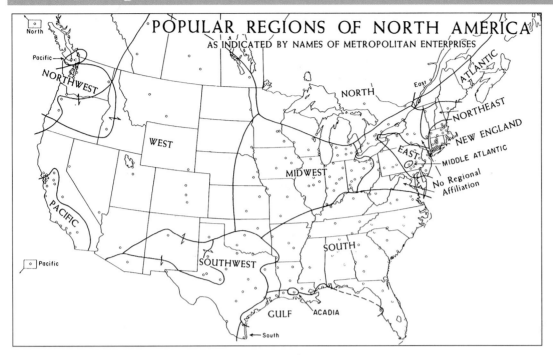

FIGURE 3-3

The United States is significantly homogeneous as a culture realm, but local culture regions can nevertheless be mapped. The regions on this map are based on one single criterion: telephone directory listings of local metropolitan enterprises. Thus, they reflect regions as their inhabitants define them self-consciously. Outsiders to each of these areas, however, might have mental maps that differ from those of the local citizens. An Easterner, for instance, might consider Kansas City to be in the West, but to a Kansas Citian, home is in the Midwest. Does this map concur with your mental map of the United States? (After Wilbur Zelinsky, "North America's Vernacular Regions," *Annals*, Association of American Geographers, vol. 70, March 1980, p. 14. By permission of the Association of American Geographers/Blackwell Publishers.)

Buildings and Monuments

People usually rely on local materials for building, so in one place stone may be the traditional building material, in another place brick, and in another wood (see Figures 3-4 and 3-5). Innovations in the uses of these materials may diffuse across cultures.

Individual styles of architecture often represent adaptations to climatic conditions, so a particular style may be adopted in different regions with similar climates (Figure 3-6). Cultural preferences are sometimes so powerful, however, that an architectural style may diffuse beyond the limits of where its building materials can be found and even beyond the architecture's physical comfort range. The style of the Italian Renaissance architect Andrea Palladio (1508–1580), for example, spread to England because of aesthetic preferences, despite the fact that Palladian buildings are uncomfortable in England's damp, cool climate. From England the preference diffused to America, even to areas that lacked both the

appropriate climate and the necessary building materials. As Europe and the United States came to dominate other regions, Palladian architecture diffused throughout the world (see Figure 3-7, p. 86 and look ahead to Figure 3-23, p. 105).

Public statuary and monuments proclaim local cultural values. Many cities in the southern United States erected monuments commemorating the Confederate cause, but as African Americans gained political power throughout the South, they insisted that these be taken down. In the Eastern European countries that were formerly Communist, cities have also redecorated their public places. In Budapest, Hungary, for example, statues of heroic workers and Soviet liberators are being replaced by statues of ancient Hungarian royalty and Christian saints.

Settlement Patterns

The designs of settlements reflect cultural differences, so the look and layout of whole towns and cities

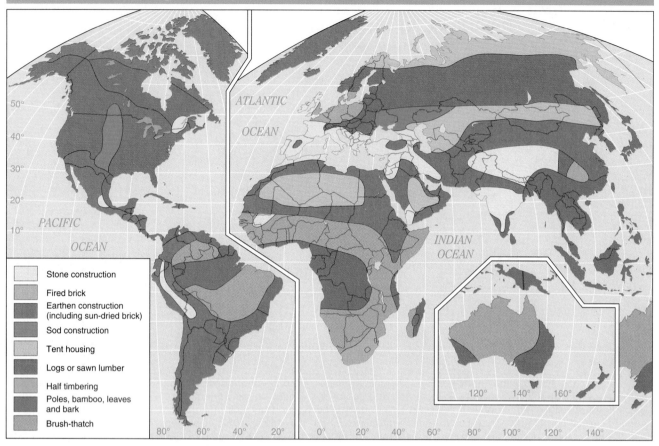

FIGURE 3-4

Mud is the most popular building material in the world, but locally available materials—stone, brick, bamboo, or wood—are everywhere used in traditional or "vernacular" architecture. (Data courtesy of Goode's World Atlas, 18th ed., copyright © Houghton Mifflin.)

reveal the cultural backgrounds of their builders to a trained observer. Among rural societies, for example, some cluster housing settlements, and others isolate settlements in individual farmsteads.

Clustered farm villages range from a few dozen houses up to thousands of houses in large agro-towns (Figure 3-8). There are no dwellings in the surrounding farmland, but the farmers journey out to work in the fields each day. The farm buildings are usually concentrated together with the human settlement. Such compact villages may be found in many forms: irregular; wandering along a principal street, river, or canal; clustered about a village common; or checkerboard. Clustering may reveal the need for common security against bandits or invaders, family or religious bonds, or communal land ownership. It may be the deliberate action of the government to supervise the population or to provide education or health care. Clustering may also have environmental reasons—people may cluster at water sources, for example, or

on dry places when the surrounding land is swampy. Clustering is more common among farmers than among livestock ranchers, except in Africa, but it generally characterizes settlement across much of Europe, Latin America, Asia, Africa, and the Middle East.

Isolated farmsteads, by contrast, characterize those areas of Anglo-America, Australia, New Zealand, and South Africa that were settled by emigrating Europeans, and also some parts of Japan and India. The conditions for isolated farmstead settlement usually include peace and security in the countryside; initial agricultural colonization of the region by individual pioneer families rather than by socially cohesive groups; agricultural private enterprise, as opposed to communalism; unit-block farms in which all of the land belonging to any one farmer is in a block rather than in scattered parcels; and well-drained land where water is readily available.

The history of the United States provides examples of both types of settlement. In the southern

FIGURE 3-5

The mosque at San, in the West African country of Mali, demonstrates that when people have mud and very little wood, they can still build extraordinary structures. This building is in the distinctive Dyula architectural style, named for a trading people who diffused the style as they moved around West Africa. It is quite different from the Middle Eastern architectural styles that many Americans expect mosques to reflect.
(Courtesy of Professor J. Markusse, University of Amsterdam.)

colonies in the seventeenth and eighteenth centuries the settlement pattern was characterized by widely dispersed, relatively self-sufficient plantations. In New England, by contrast, the settlers advanced westward in tiers of adjacent, well-planned towns. This reflected the tightly bound religious character of the New England communities.

PATHS AND SPEEDS OF CULTURAL DIFFUSION

Tracing a course of diffusion may teach us a great deal about how peoples and cultures interact and influence one another. The paths of cultural diffusion are seldom so simple as the image of concentric rings spreading out from a stone dropped into a pool presented earlier in this chapter. The complex patterns and speed of diffusion provide measures of cultural interactions. Three basic types of diffusion are relocation diffusion, contiguous or contagious diffusion, and hierarchical diffusion.

Relocation diffusion occurs from widely separated point to point. A nomadic tribe, for example,

might wander until it finds a physical environment similar to the one that it left, and then the tribe might settle down. Chapter 2 noted the example of relocation diffusion when Ukrainians brought grain seeds from their homeland to the similar environment of the U.S. Great Plains.

Contiguous diffusion occurs from one place directly to a neighboring place. Contiguous diffusion that occurred in the past is often revealed by the dispersion of artistic styles. For example, during the fifteenth and sixteenth centuries Italian Renaissance art spread to the Low Countries of Northern Europe (today's Belgium and the Netherlands), along a path that reflected trading and financial links. As long ago as the fourth century B.C., Alexander the Great's conquests carried Hellenistic Greek art far into Asia. Central Asian statues of Buddha sculpted during the centuries immediately following these conquests

FIGURE 3-6

The building in the left foreground is a bungalow, which is a Hindi word for a low-sweeping, single-story house with a roof extending out over a veranda. Such houses were first built in the mountain foothills of northern India, but the style was copied throughout the British Empire and eventually the United States. Today in English the word *bungalow* means almost any sort of small house.

This photograph shows Simla in the 1880s. Simla was the summer capital of the British Indian Empire between 1864 and 1947. It was cooler up at Simla (7,100 feet—notice the vegetation) than down at New Delhi in the plains about 170 miles (275 km) to the south. The pseudomedieval English cathedral in the background seems an odd presence in northern India, but wherever the British went, they took their culture and traditions, including architecture.
(Courtesy of the Library of Congress.)

FIGURE 3-7

This house, called the Morris-Jumel Mansion, was built in 1765 in Manhattan 10 miles north of Wall Street. It is in the Palladian style, which is appropriate for the Italian climate, but in New York it would have been impossible to keep warm. In Italy it would have been built of stone. Here, virtually on the frontier of Western civilization, it was built entirely of wood masquerading as stone. The choice of style demonstrates the power of prevailing aesthetic taste, even over comfort.
(Courtesy of Washington Headquarters Association, New York City.)

resemble Greek gods (see Figure 3-9). (The fact that we use "B.C." here reveals that we live in the Christian realm; if we lived in the Islamic realm we would have used the Muslim calendar.)

Sometimes cultural diffusion does not occur contiguously across space, but downward or upward in a hierarchy of organization. When such **hierarchical diffusion** is mapped, it shows up as a network of spots rather than as an ink blot spreading across a map. The Roman Catholic church illustrates both an organizational hierarchy and a geographical hierarchy. Each parish priest is answerable to his bishop, who from his cathedral church ("cathedra" is Latin for "throne") presides over a diocese, which includes many parishes. The bishops, in turn, look up to the Vatican (Figure 3-10). An announcement from the Vatican diffuses to the cathedrals and then down to the parishes around the world. In hierarchies more information generally travels up and down the hierarchy than across any level of the hierarchy. Therefore,

news from a parish about an innovation in community service or in a church ceremony might reach the cathedral and even the Vatican before it reaches neighboring parishes.

Advanced societies usually exhibit well-developed hierarchies of cities, and we can study how the hierarchy evolved through time. Once a hierarchy is developed, the diffusion of any specific phenomenon might be traced from the few biggest cities "down" to the many more smaller cities and then further "down" to the even more numerous small towns and villages until the landscape is covered (Figure 3-11).

One attempt to map the hierarchical diffusion of information in the United States focused on the flow of Federal Express Corporation overnight parcels in 1990. New York City was found to account for nearly 17 percent of all parcels shipped among major cities. New York's most important links were with four cities that serve as regional information capitals: Los Angeles, Dallas-Fort Worth, Chicago, and Atlanta. The higher a city was found to be in the hierarchy, the more parcels it sent, and the greater the distances it sent them. As we might expect, the flow of information was chiefly up and down the hierarchy, not among the cities at any given level. The map of America's most heavily traveled domestic airline routes (Figure 3-12) also reflects the national hierarchy of cities.

FIGURE 3-8

An expert on African cultures would immediately recognize the pattern and spacing of the buildings in this view as typical of the Zaria people of Nigeria.
(Courtesy of National Museum of African Art, Eliot Elisofon Archives, Smithsonian Institution.)

FIGURE 3-9

This richly dressed young man is the Indian Prince Siddhartha before he became the Enlightened One ("Buddha") and gave up his worldly wealth. The statue exhibits a mix of styles: the curly hair, stocky physique, relaxed stance, and deeply carved drapery are Greek characteristics, but the facial expression is typical of Indian mysticism. The nimbus of light around the prince's head originated with Iranian statues of sun gods.

Where could all these cultural influences have come together and this extraordinary figure have been carved? The answer is the region of Peshawar in today's Pakistan, the crossroads of Eurasian travel routes, where the Khyber Pass opens from Central Asia into the plains of the Indian subcontinent. This statue dates from the late second or early third century A.D.
(Courtesy of the Seattle Art Museum.)

Many phenomena diffuse hierarchically around the world today. The spread of a disease or of a cultural innovation such as a clothing fashion or a music hit can often be traced among the world's principal metropolises before it reaches "down" into the smaller cities in each country. This path of diffusion reveals the world's interconnectedness.

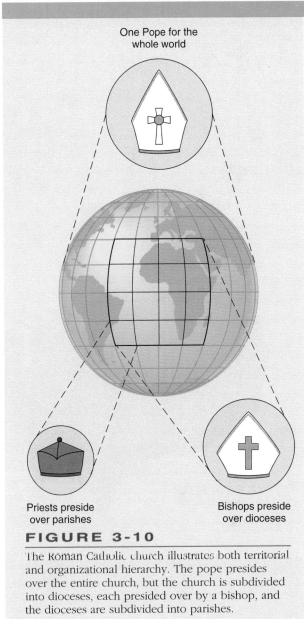

One Pope for the whole world

Priests preside over parishes

Bishops preside over dioceses

FIGURE 3-10

The Roman Catholic church illustrates both territorial and organizational hierarchy. The pope presides over the entire church, but the church is subdivided into dioceses, each presided over by a bishop, and the dioceses are subdivided into parishes.

Barriers to Diffusion

There are many barriers to cultural diffusion. Oceans and deserts are examples. They increase the cost distance or time distance between places. Other topographical features, such as mountains and valleys, historically have blocked human communication and contributed to cultural isolation. In Asia, the deeply cut valleys of the Tongtian, Mekong, Nu, and Brahmaputra rivers make overland travel between the Chinese cultural realm and the Indian subcontinent extremely difficult, so these two realms and great concentrations of population are surprisingly isolated from one another. Look ahead to compare world

Market Penetration by Television

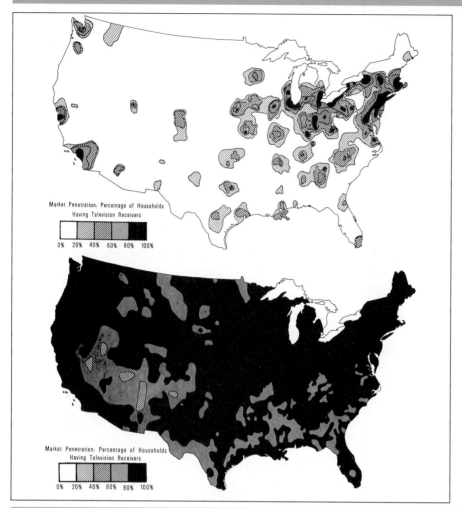

Market Penetration: Percentage of Households Having Television Receivers

0% 20% 40% 60% 80% 100%

Market Penetration: Percentage of Households Having Television Receivers

0% 20% 40% 60% 80% 100%

FIGURE 3-11

These maps show the increasing *market penetration*, or percentage acceptance of television sets in U.S. households from 1953 to 1965. This was before cable television or satellite dishes, so viewers had to receive transmissions from metropolitan centers. How would the pattern of diffusion differ with today's technology? [Reprinted with permission from Brian J.L. Berry, "The Geography of the United States in the Year 2000," *Transactions of the Institute of British Geographers*, LI (November 1970), 21–53.]

Major Air Passenger Routes in the United States

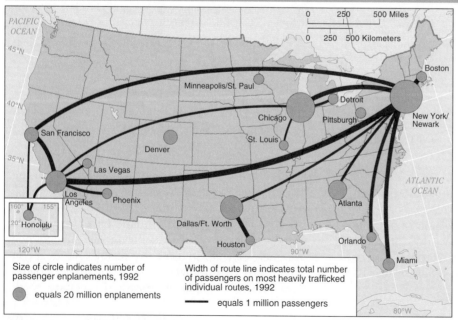

Size of circle indicates number of passenger enplanements, 1992

⬤ equals 20 million enplanements

Width of route line indicates total number of passengers on most heavily trafficked individual routes, 1992

▬ equals 1 million passengers

FIGURE 3-12

This map is a good indicator of the urban hierarchy of the United States. Any airport that has a large number of total enplanements, but few or no routes with over two million passengers, such as Minneapolis/St. Paul or Denver, must be serving as a hub of many local regional routes with small numbers of fliers on each.

population distribution (Figure 4-1) with the back endpaper map of the world's topography. In Europe, the Alps have impeded the flow of human populations, thereby contributing to cultural differentiation.

Other barriers to cultural diffusion include political boundaries and even the boundaries between two culture realms. Hostile misunderstanding, distrust, and competition between two culture groups can hinder any communication and exchange between them.

DIFFUSION CHALLENGES STABILTY
▼▼▼▼▼▼▼▼▼▼▼▼▼▼▼▼▼▼▼▼▼▼▼▼▼

The isolation in which we described the imaginary Lonely people living is very different from the way most of us live today. Modern people no longer exploit their own local environments and develop their cultures in isolation. They are interconnected by transportation and communication of goods, people, ideas, and capital. To describe all of this movement, we use the term *circulation.*

In the past, travel and transportation were more difficult and expensive than they are today, whether we measure the cost in time, money, or in any other unit. The friction of distance was so high that only a few things moved far, and those things moved slowly. The place of origin and the path of diffusion of any innovation could be fairly well traced. Over the past 200 years, however, technology has reduced the fric-

tion of distance and accelerated the diffusion of cultural elements. We often say that these developments in transportation and communication have "shrunk" the world. Electronics, for example, disengaged communication from transportation. Originally information could move only as fast as a person could carry it, but in 1844, Samuel Morse demonstrated the first intercity telegraph line. This passed information almost instantaneously between Baltimore and Washington, D.C. Baltimore newspapers had previously carried news of events which had occurred in Washington two days before; the time distance between the two cities had been two days. After Morse demonstrated the telegraph, however, a Baltimore evening paper printed Washington political news of that same day. The paper concluded, "This is indeed the annihilation of space." Telephones have been said to annihilate space by allowing you, figuratively, to be in two places at the same time.

Today the global news industry exists specifically to ferret out and diffuse innovation. New ideas, products, and processes spread around the globe overnight. A new American hit song is playing on Brazilian, Australian, and Kenyan radio stations within a few weeks. What we do not adopt, we at least know about.

Although we have access to news and cultural elements from all parts of the world, we continue to be selective in what we pay attention to (Figure 3-13). Our own backgrounds, our education, perceptions and prejudices, and the media to which we are

One Newspaper's View of the World

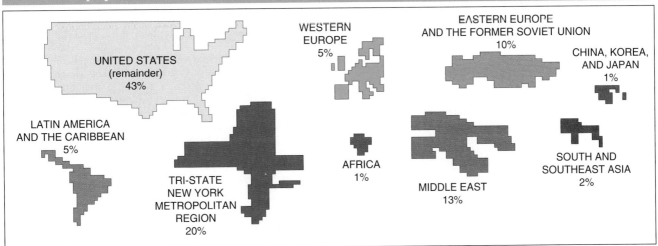

FIGURE 3-13

The relative sizes of world regions on this illustration correspond to the coverage devoted to each on the front page of *The New York Times* for two months in 1993. New York has a large and influential Jewish population, and this is reflected in the large amount of coverage of news from the Middle East. Miami has a large and influential Latin American population, and Miami has important trade links with Latin America. Considering those facts, how do you think "the view from Miami" would differ? How about Los Angeles? How about the city in which you live? Measure your local newspaper and see.

exposed affect our understanding of other peoples and places and how we rank their relative importance. For a great variety of reasons we are more knowledgeable about or more interested in some places than others. These reasons are not necessarily related to the size or population of a particular region.

Today no pattern of culture realms is stable. This textbook will detail how some religions are winning new converts, expanding geographically and assuming new influence in world affairs, whereas others are withering away. Some languages are demonstrating the flexibility to adapt to new communication technology and are therefore gaining users at the expense of others. The global popularity of American popular culture, for example, diffuses American language. The relative rise and fall of nations' influence in world political or economic affairs wins or loses new adherents to those nations' products, cultural artifacts, or lifestyles. If the Lonely people were to be discovered today, people in other nations would most likely adopt some aspect of Lonely culture—perhaps Lonely music, or a food or medicinal crop (Figure 3-14). At the same time, the Lonely way of life would be transformed by the infusion of products and ideas from the outside world.

Folk Culture

Today's rapid pace of cultural innovation and diffusion requires us to make a distinction between folk culture and popular culture. The term **folk culture** refers to a culture that preserves traditions. Folk groups are often bound by a distinctive religion, national background, or language, and folk cultures are conservative and resistant to change. Most folk culture groups are rural, and relative isolation helps these groups maintain their integrity. Folk culture groups, however, also include urban neighborhoods of immigrants struggling to preserve their native cultures in their new homes. The term folk culture suggests that any culture identified by the term is a lingering remnant of something, that it is embattled by the tide of modern change.

Cultural geographers have identified culture hearths and routes of diffusion of a surprising number of folk cultures across the United States. Geographers who have investigated house types, for example, have concluded that diffusion paths conformed to the migrations of various immigrant groups (see Figure 3-15). Barns and other structures also frequently exhibit distinct architectural styles that reveal the origins of their builders. Folk geographic studies in the United States range from studies of folk songs, folk foods, folk medicine, and folklore to objects of folk material culture as diverse

FIGURE 3-14

Dr. Michael Balick of the New York Botanical Garden learns plant uses from this traditional healer of the Kekchi Maya people in Belize. The knowledge that "primitive" cultures have accumulated of local plant uses, called *ethnobotany*, may hold keys to the cures of many diseases. Dr. Balick is researching potential anti-AIDS and anticancer agents for the National Cancer Institute. Take a good look at the healer's clothing; it reflects how fast these people are being acculturated and their knowledge is being lost. (Courtesy of Dr. Michael Balick.)

as locally produced pottery, clothing, tombstones, housing styles, farm fencing, and even knives and guns. Not everyone all across the United States even today buys the same nationally advertised and distributed goods.

In the United States, the Amish provide an example of a folk culture. The Amish are a sect within the Mennonite Protestant religious denomination formed in Switzerland in the sixteenth century. Early Amish immigrants to the United States settled in Pennsylvania and spread across the Midwest. The

90 *Chapter 3*

Paths of Diffusion of House Types

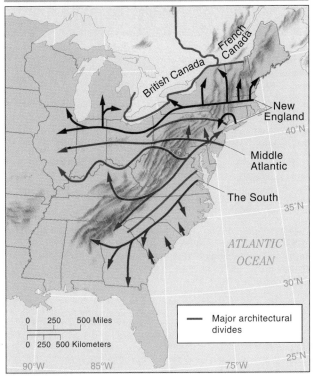

FIGURE 3-15

Geographers have traced the diffusion of three distinct types of houses from the origins of each along the eastern seaboard toward the West. (From Rooney, Zelinsky, and Louder, eds., *This Remarkable Continent.* Texas A&M Press, 1982.)

Amish stand out in the United States because they wear plain clothing and shun modern education and technology. They prosper by specializing their farm production and marketing their produce, but they severely curtail the choice of goods that they will accept in return.

Many African Americans have long felt that their traditional folk cultures were taken from them during the years they suffered in slavery, yet careful investigation has yielded an astonishingly rich variety of cultural elements in some African-American communities today that are traceable to specific African origins. These range from *geophagy*, the eating of dirt in order to supplement certain vitamin deficiencies and to provide kaolin (an antidiarrhea and anticramps agent); to the meticulous sweeping of front yards (a tradition traced from villages of coastal Nigeria to rural Georgia); to patterns of speech discussed in Chapter 7.

The United States is relatively homogeneous in culture, modern, and technologically advanced. Cultural diffusion and change churn across the country rapidly and repeatedly. In such a country the identification and study of folk cultures helps us appreciate the inherent richness of some of our remaining folk traditions. Furthermore, it contributes to our understanding of the histories of the many threads that weave together in our cultural tapestry and might remind us of the value of preserving some of these threads.

In countries that are less modern and less technologically advanced than the United States, however, the traditional culture of a place, its native culture, is its folk culture. Some newly independent countries are today emphasizing or even resurrecting aspects of traditional folk cultures as a way of enhancing national identity and community. One of the first acts of the government of newly independent Eritrea in 1993, for example, was to open a national university mandated to research and define folk literature, music, and other distinct Eritrean traditions.

Popular Culture

In contrast to folk culture, **popular culture** is the culture of people who embrace innovation and conform to changing norms. Popular culture may originate anywhere, and it tends to diffuse rapidly, especially wherever people have time, money, and inclination to indulge in it. Here again, the number of aspects of this culture that geographers can study is virtually limitless.

Popular material culture usually means mass culture—that is, items such as clothing, processed food, books, tapes, and household goods that are mass produced for mass distribution. Mass manufacture lowers the cost of items, but it relies on homogeneous consumer taste. The United States has long been the world's largest relatively homogeneous consumer market, but the worldwide expansion of huge corporations and the multiplication of franchises have meant that today we can find identical shops selling identical goods around the world.

Material popular culture is largely defined by consumption, so it is usually more closely related to social class, as defined by income and education, than folk cultures are. The consumer items people buy are largely determined by what they can afford. In other words, General Motors mass produces both Cadillacs and Chevrolets, but the company manufactures these brands for different potential customers.

The popular culture of the United States exhibits a great degree of national homogeneity, but individual consumer goods do win varying degrees of market penetration in different regions of the country. The soft drink Doctor Pepper, for example, originated in the U.S. South, and it still enjoys its greatest market

The Diffusion of Anglo-American Religious Folk Songs

The diffusion of white spiritual songs from a culture hearth in eighteenth-century New England exemplifies folk cultural geographic studies. These songs spread in a southwesterly direction into the Upland South and then into the Lowland South, where many of them are still popular. Some individual migrants carried the songs westward with them, but, for the most part, the acceptance of these songs diffused from community to community.

Barriers existed against the flow of these songs in other directions. Roman Catholicism hindered the diffusion of these Protestant songs into Louisiana, and Roman Catholicism plus political-cultural barriers and language prevented the diffusion of these simple songs to the North into Quebec.

At the same time as these songs were spreading to the Southwest, religious folk songs largely disappeared from their culture hearth in the North. This may be because of the rapid urbanization and popularization of culture in that region.

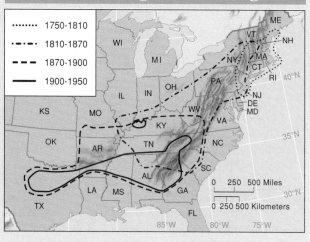

Diffusion of Religious Folksongs

Legend:
- ········· 1750-1810
- ·-·-·- 1810-1870
- - - - 1870-1900
- —— 1900-1950

Protestant religious folksongs diffused from New England toward the southwest. (Reproduced by permission of the American Folklore Society from *Journal of American Folklore* 65:258, Oct.–Dec. 1952. Not for further reproduction.)

penetration there. The U.S. South is today less culturally distinct than it was decades ago, but a cultural heritage, much of which lingers from earlier folk traditions, still exists.

Geographers investigate the origins and diffusion paths not only of popular material culture, but of popular sociofacts and mentifacts as well. New ways of living, working, and playing, or innovations in education and in employer-employee relations, for example, diffuse. Sport is an important part of popular culture, and geographers have identified popular culture regions defined by sporting preferences, either in terms of sports people like to watch (spectator preferences) or play (participant preferences). Regions differ in their popular entertainments—movies that are big hits in Seattle, for example, may play to empty houses in Houston. The radio stations in different regions across the United States play varying mixes of country, gospel, rock, classical, and other popular kinds of music. Geographers have investigated these and many more attributes of popular culture.

FORCES THAT FIX THE PATTERN OF CULTURES

Despite the force of diffusion, a number of factors tend to fix the geography of culture realms. Culture leaves its mark on the landscape. The fixed pattern of activities, land uses, transport routes, and even individual buildings guides, restricts, or predisposes future patterns and activities. The construction of a factory, for example, represents a great investment of money, and once the factory is operating it relies on a local work force and develops ties to local suppliers. An industrial complex such as this cannot easily be picked up and moved. **Inertia** is the term for the force that keeps things stable. All of a people's fixed assets in place—railroads, pipelines, highways, airports, housing, and more—are called the **infrastructure**.

Historical geography is the subfield within geography that studies the geography of the past and how geographic distributions have changed. Historical geographers can sometimes read landscapes as if, through time, the landscape has been overlain with layer after layer of peoples using the land in different ways and organizing it for different purposes. The landscape is like an old manuscript on which a reader can discern earlier erased writings (see Figure 3-16). Scholarly recreations of past land uses were defined as *sequent occupance* studies by the geographer Derwent Whittlesey in 1929. In recent decades historical geographers have also developed an alternative approach that focuses on how the transformation, use, and organization of the landscape is continuously changing.

Culture includes a set of values and ways of doing things, and culture groups seldom get displaced or eliminated entirely. Culture is learned behavior, and cultural norms are handed down through generations

FIGURE 3-16

An abandoned city prompts us to ask, "Who was here before us? What did they do? What happened to them?" This ruin is in Morocco.
(Courtesy of the Moroccan National Tourist Office.)

surprisingly unchanged. Groups may be quick to take up new techniques or products, but tradition is a powerful force in human life, and imported ideas or lifestyles only slowly transform a people's entire cultural inheritance. This inertia is reinforced by the existence of territorially sovereign states and the enormous power that governments exert over their citizens to teach and to enforce norms of behavior. This power includes, as we have seen, the power to resurrect and promote folk traditions. Lee Kuan Yew, prime minister of Singapore from 1965 through 1990, emphasized the role of culture in various Asian countries: "Culture is very deep-rooted; it's not tangible, but it's very real: the values and perceptions, attitudes, reference points, a map up here, in the mind."

Each local culture is today a unique mix of what originated locally and what has been imported, and each unique culture is a local resource, just as surely as the minerals under the soil and the crops in the fields.

Most peoples value their culture, and they usually try to preserve key aspects of it. This influences the way they interact with other peoples. A people's self-consciousness as a culture may be codified in their religion, as it is most notably with the Jews, or in their sense of their own history. This is called **historical consciousness**. Many Americans have difficulty understanding other people's historical consciousness, because one aspect of American culture is an optimistic denial that history can shackle future opportunity. Other peoples, however, nurture their traditional cultures and their historical consciousnesses. Many people act the way they do because that is the way their ancestors acted, or in order to right wrongs or to account for deeds or misdeeds that took place long ago. Several peoples fighting in the Balkans in the early 1990s, for example, feel that they are avenging medieval battles.

Furthermore, some peoples take a long-term view of historical events which is different from popular American attitudes. Traditional Chinese, for example, who are educated to appreciate the antiquity of their civilization, may unself-consciously speak of units of time of hundreds of years even in casual conversation. To such people, Westerners are tremendously impatient, while to many Westerners, people who dwell on historical facts are stubborn, living life while looking in the rearview mirror.

Various peoples' historical consciousness strongly influences current international affairs. For example, in 1879 500 Chilean troops overpowered 130 Bolivian troops who were celebrating Carnival and seized 160 miles of Bolivian coastline on the Pacific Ocean plus great deposits of copper and saltpeter. To this day Bolivia refuses to establish diplomatic relations with Chile. Each year on March 23,

Bolivians celebrate "The Day of the Sea." The Bolivian Coastal Museum in downtown La Paz features maps identifying Bolivia's "rightful" land claims. Countries often go to war to avenge old hatreds that they have harbored and nurtured. We will note examples of such wars in Chapter 10.

One major theme in geography is the tension between forces of change and forces for stability. Cultures diffuse, cultures change, and peoples can transform themselves and their behavior, but cultures and culture realms also have elements of stability. Any cultural pattern or distribution maps the current balance between those forces.

AUTARKY AND TRADE

The study of how various peoples make their living and what they trade with other peoples is **economic geography**, which is an important component of human geography. When the Lonely people were isolated, they neither imported nor exported any goods. They were totally dependent upon the resources of the local environment, and they were self-sufficient. Such economic independence is called **autarky**. As peoples and regions come into contact with others, however, they begin to exchange goods. At first people export only whatever they have in surplus and have no use for, and they view imports as luxuries unnecessary to their way of life. Trade, however, entangles peoples. The opening of trade connections with other peoples triggers far-reaching economic and cultural changes.

For example, imagine three isolated communities in three different environments (Figure 3-17). Village A is located along a river plain. The people there catch fish, which they fry, they grow rice, which they eat as rice cakes, and they distill rice into sake (rice wine). Village B lies on the slope of nearby hills. The people there have domesticated grapes, which they have learned to distill into wine, and oats, which they make into oat bread. They have also domesticated goats, which clamber about the rugged hillsides. They milk the goats for dairy products, and they roast goat meat. The people in village C, up on a nearby plateau, grow corn for cornbread and to be distilled into whiskey, and have domesticated cattle, which they eat as broiled steaks.

The construction of a road and the commencement of trade among the three villages will trigger at least four tremendously significant changes. First, each village will have access to the products of each of the other villages. The people of village A, for example, will taste broiled steaks and the people of

village B will be introduced to fried fish. We might call this the simple *addition* of cultural possibilities.

Second, wholly new cultural combinations, or *permutations,* will appear: fried steaks, broiled fish, roast beef, and more. A good example of cultural permutation is the appearance of pita fajitas on menus in southern California. A pita fajita is a piece of pita bread, an item brought to southern California by immigrants from the Near East, stuffed with fajita ingredients, usually beef or chicken, which were introduced to southern California by Mexican immigrants. Nobody knows what genius first devised this permutation.

Third, village B will probably grow larger than village A or village C. This is because of its relative position. It is between the other two villages, and so, as the most convenient spot, the village will probably develop the largest market.

Fourth, the residents of all three villages will begin to enjoy the new cultural possibilities. People see how imports can raise their standard of living, and in order to pay for the imports they want, they dedicate more local effort to producing items for trade. In addition, the existence of trade and markets gives each village, for the first time, an incentive to produce surpluses of its local goods. Agronomists (economists who specialize in agriculture) say that, in general, "The market produces the surplus." In other words, if farmers have a market where they can sell their surplus for a profit, they are likely to produce a surplus. Furthermore, when people specialize in producing an item, they become more efficient, and the quantities they produce increase rapidly. Therefore, the total amount of food produced in all three villages may well rise significantly.

All of these developments in the economic life in the three villages will be paralleled by evolutions in cultural life in the three villages. Residents of one village may convert to the religion of another, or, perhaps, the religions will blend. The languages, styles of architecture, music, games, and other customs of all three villages may add up and also permutate. Intermarriage may also blend and mix the three peoples' physical attributes.

Trade leads to specialization of production and greater production, but it does not benefit all regions equally. Why international trade causes some regions to prosper and others to fall behind is a principal subject of research in economic geography. Many different reasons will be suggested throughout this book.

As trade multiplies, more of what people produce anywhere is consumed elsewhere. Conversely, a growing percentage of the things people use are produced elsewhere. Eventually, if the terms of trade are favorable, people dedicate most of their efforts to

FIGURE 3-17

The inhabitants of three isolated villages in three different environments learn to use local resources. When those villages are connected by a road, however, the lifestyles of all three villages may be transformed, and their economic and cultural possibilities will multiply.

producing their export products. They rely on imports not only for luxuries but even for their necessities. They have surrendered their self-sufficiency and become dependent on trade. This evolution from autarky to dependence upon trade is accelerating for almost all peoples today.

How Geographers Study Trade

Geographers ask three questions about trading systems. First, how do the people at any specific place make their living? Second, where do specific economic activities locate? All activities have **locational determinants**, which are the specific factors considered when people choose where to locate that activity. Geographers identify the locational determinants for each activity. Finally, how do specialization and trade affect other aspects of human geography? The first two questions are complementary—almost the same question asked from two perspectives.

Geographers often approach the first two questions—how people at any place make their living and where specific activities locate—by mapping the production of various items. These maps indicate the quantity of the item produced in each country and may also indicate each country's share of world production. In 1992, for example, about 721 million metric tons of crude steel were produced in the whole world. Japan produced 14 percent of that, the United States 12 percent, China 11 percent, Russia 9 percent, Ukraine 6 percent, and so on.

Production figures alone, however, do not reveal why any place is producing what it is producing, nor do they reveal where any country's production is being used. Some countries import steel, whereas others export steel. Furthermore, some countries, such as the United States, import one type of steel and export another type, or even import steel at one end of the country and export steel from another region. Therefore, after we have gathered production figures of any item, we also need to examine the balance of imports and exports in order to understand the distribution of the final use of any product. Geographers' statistics and maps of production must always be supplemented with statistics about trade and maps of trade. Trade reveals the items set in motion.

Some places produce materials primarily or solely for export. For example, Brazil is the world's second largest producer of iron ore, but it uses only one-third of what it produces. The remaining two-thirds are exported in exchange for machinery, fuels, and other goods. Each year the Ivory Coast produces about one-third of the world's cocoa beans. Why? The people there do not eat the beans. Rather, the terms of trade make it profitable for the people of the Ivory Coast to export them in exchange for other goods, including necessities such as food.

What explains this extraordinary concentration of cocoa beans in the Ivory Coast? What are the locational determinants for cocoa bean production? Did the cultivation of cocoa beans originate there? No, it originated in Mexico many centuries ago. Could the concentration of cocoa bean production in the Ivory Coast result from some unique local environmental advantage? Could the water be special? Obviously not. The climate and the soils of the Ivory Coast are fine for the cultivation of cocoa beans, but so are the climates and soils of many other places. The concentration of cocoa bean production in the Ivory Coast results from political and economic decisions that were made within a context of global production, marketing, and consumption trading patterns. Those decisions and choices may be called the economic, political, and diplomatic environments of cocoa bean production, and they rival or even supersede the physical environment as locational determinants.

Regions or peoples have not always freely chosen to enter into the system of production and exchange. Colonialism forced specialized production on vast areas. In some areas the peoples were forced to buy goods, and they had to develop export specialties in order to pay for these goods. The British actually went to war to force the Chinese to buy opium from British India (1839–1842). Opium is an addicting drug, but all kinds of desirable goods— electronic equipment, leisure activities, even new

styles of clothing—stimulate peoples' wishes to enjoy them. This has been called the creation of new *felt needs*. People begin to depend on the availability of desirable goods, and in that way they become more deeply involved in the web of trade and circulation.

The share of any country's territory that is devoted to export production may be small, but the shares of the national population and the national income that are involved are rising everywhere. Even for a country as large and as rich as the United States, the share of the nation's economic life which was accounted for by foreign trade tripled between 1955 and 1990, from 10 percent to over 30 percent of total national output. Furthermore, even within individual countries, growing cities create markets for food from surrounding rural regions. This draws the rural population out of subsistence autarky and into production for national urban markets.

Geographers' statistics and maps of production must always be supplemented with statistics about trade and maps of trade. Trade reveals the items set in motion.

The Impact of Trade on Culture

A local culture will influence the degree to which people participate in trade. A culture can either limit circulation or encourage it. Some peoples, such as the Dutch, traditionally have embraced circulation and exchange (Figure 3-18). Others, such as the Albanians from 1945 to 1990 or the Myanmarese still today, prefer isolation. We earlier noted the Amish compromise: They market their goods but limit which products they will accept in return.

When a local economy is transformed from self-sufficiency to specialized production for export, and when people become increasingly dependent on imports, every aspect of local culture is affected. The way people make their living is in itself a key aspect of their culture, and all aspects of culture interact. It often happens, for example, that those who profit from a new trading system and exchange with other cultures challenge and overturn local politics and traditions.

When an economy is tied into trade, even the selection of items that are produced for local use— the local folk culture—can become dependent on whether those items are exportable. Local artifacts that cannot be exported may no longer be produced even for local consumption. The production of items of local culture that win broader markets, on the other hand, may be increased. These, however, may be altered in order to increase the exports of the

FIGURE 3-18

This advertisement, which was widely placed in international business magazines, demonstrates how aggressively the Dutch seek interaction and trade with other peoples. (Courtesy of Netherlands Foreign Investment Agency.)

item. For example, many native American tribes have altered their traditional craft items such as baskets, jewelry, and clothing in order to win broader markets for their production. In these ways people slowly surrender their own traditional material culture, and they may not even be aware of it.

Trade as Diffusion

Economic exchange is a form of cultural diffusion. Every object in trade is an artifact of the culture that originates it, so the introduction of a new item into a different culture—a sale—reduces the buyer's cultural uniqueness. Salespeople are, in effect, cultural missionaries.

Both production and consumption are more than simply material aspects of existence. Goods are building blocks of lifestyles, of social and cultural identity, and of self-consciousness. Housing, clothing, personal transportation, and other choices of posses-

sion or use all constitute a language. The study of the meanings of such symbols and signs is called *semiotics*. Every morning when you choose what clothes to wear, you are choosing to say something. Choosing goods is an act of self-definition. Changes in the goods produced or consumed change cultures.

Distribution of popular material culture can overwhelm folk culture. In 1991, the Disney Company chose Singapore as its Southeast Asian distribution center. The countries of the region are being flooded by over 16,000 Disney products. Will the children of Southeast Asia be able to resist Goofy and Donald Duck T-shirts? If not, will they also remember their culture's traditional folk fables and mythical characters? How many of the characters and tales of traditional U.S. folk culture—or even of genuine historical heroes and heroines—have survived the onslaught of mass-produced and mass-marketed merchandise related to comic strip characters? Who is most famous in the United States today: Paul Bunyan, John Chapman ("Johnny Appleseed;" 1774–1845), or Garfield the cat?

Today cultural diffusion through the exchange of consumer goods is swamping the world. The cheapest way to manufacture and distribute consumer goods is in large quantities. This might mean blanketing several different culture realms with one product. The integrity of each culture realm, however, resists homogenization. Therefore, cross-cultural advertising and marketing of consumer goods present fascinating cultural-geographic questions. Why do some products have worldwide appeal, whereas others are successful only in geographically restricted markets? What patterns of culture realms and markets does this reveal? Can salespeople deliberately break down cultural differences? A few consumer products have already achieved almost global diffusion: Coca-Cola, Sony, Kodak, and Levi's. Why do people around the world dress differently, eat different foods, and speak different languages, but almost everywhere emulate the Marlboro Man? Is there any place where people refuse cola drinks (Figure 3-19)? Yes, Utah. Why there? Mormons do not drink caffeine.

All peoples, to some degree, defend their traditional cultures. In Iran in the early 1990s, for example, some 3 million men known as *Bassijis*, "those who are mobilized," patroled the streets battling prostitution, drugs, alcohol abuse, and atheism, but also objects and values which they interpreted as imported from the West, including stereos, pop music, videos, lipstick, and indecorous dress among women. Their commander, Ali Reza Afshar, said, "This war goes to the root of our existence. While physically there is no loss of life, our young people are being felled by cultural bullets, and this cultural corruption makes our young impotent to rebuild the nation."

Cultural Diffusion **97**

FIGURE 3-19

Some consumer goods practically blanket the Earth.
(Courtesy of Julia Nicole Paley.)

Mass manufacturing and marketing can lower the cost of goods and raise standards of living, but at some sacrifice of cultural identity. Consumer goods are arguably the most significant means of cultural diffusion today. We must remember, however, that material consumer goods are only one aspect of culture. In some places people wear blue jeans to political demonstrations against the United States. They like blue jeans, but that does not mean that they like everything about U.S. culture.

THE CHALLENGE OF CHANGE

▼▼▼▼▼▼▼▼▼▼▼▼▼▼▼▼▼▼▼▼▼▼▼▼▼▼

Maps of culture realms, of regions of economic specialization, and of political jurisdictions reveal current distributions of specific human activities. Human activities, however, are not static. They are dynamic, continuously organizing and reorganizing, forming and reforming. Therefore, no pattern of the organization and use of territory is stable. Distributions and patterns are disrupted and reshaped repeatedly, just as the patterns in a kaleidoscope are. A map of human activities at any time is like a snapshot of the dancers in an intricate minuet. We cannot understand the dance—or the activity—if we do not know the principles of its choreography or become familiar with the music. For geographers to answer the question "Why there?" requires the study of processes—the music and the choreography.

Geographers always need to understand what makes things move and redistribute themselves. It requires some force to overcome inertia. In nature, vari-

ations in air pressure cause the winds to blow from points of high pressure to low pressure. Gravity causes water to flow to the sea. In studies of human activities, variations in local job opportunities or political liberties cause people to migrate. Variations in interest rates or in profitability cause investment capital to migrate, and the places that attract capital investment enjoy economic development. People go on pilgrimages to holy sites by the force of their religious beliefs. Ideas may flow under the power of conversion. Variations in places' natural endowments trigger trade flows. Produce flows to markets, raw materials to manufacturing sites, tourists to attractions or to beaches. Variations in transport capability or freight rates direct flows of traffic. All of these forces overcome geographic inertia.

Furthermore, geographic inertia, the friction of distance, has been reduced. The cost and time of moving almost anything—people, food, energy, raw materials, finished goods, capital, information—has steadily fallen, so many activities have been significantly released from the constraints of any given location (see Table 3-1). Few things are quite so fixed in place as they were in the past. If an activity can move or relocate freely, we call it **footloose**. As French president François Mitterand likes to say, "History has accelerated." Geographic redistributions have too.

The geography of global manufacturing, which we will examine in detail in Chapter 12, provides an example of continuous redistribution. Factors that redistribute manufacturing include new products, new technologies, new raw materials (the replacement of metals with plastics, for example), new sources of traditional raw materials, new technologies of manufacture, new governments with new policies, growing and shrinking labor supplies, and the opening and development of new markets. When the executives of huge global corporations decide where in the world to invest in new factories, they must balance these and still more factors, and each of these factors changes every day.

Connectivity Expands Market Areas

Improvements in connectivity unify previously isolated markets. This means that if any item is widely traded and widely available, and if local circumstances force up its price at one place, buyers can import it from elsewhere. World supplies and information flows become more important than local conditions anywhere. Already by the late nineteenth century, for example, world markets existed for wheat, so wheat cost nearly the same in major markets around the globe. Today prices for commodi-

FOCUS
Economic Determinism

Economic determinism is the belief that economics is the primary force in social life. This belief received a powerful impetus in the late eighteenth century with the development of the theories of "economic man" discussed in Chapter 2 and with the expansion of world trade and investment. By 1848, Karl Marx and Friedrich Engels would insist that economic forces were overcoming all other cultural aspects of human life. They wrote in the *Communist Manifesto:*

The bourgeoisie has through its exploitation of the world market given a cosmopolitan character to production and consumption in every country.... All old-established national industries ... are dislodged

by new industries ... that no longer work up indigenous raw material, but raw material drawn from the remotest zones; industries whose products are consumed, not only at home, but in every quarter of the globe. In place of the old wants, satisfied by the productions of the country, we find new wants, requiring for their satisfaction the products of distant lands and climes. In place of the old local and national seclusion and self-sufficiency, we have intercourse in every direction, universal interdependence of nations....

Marx and Engels believed that the spread of capitalism created

an international economy that would put an end to nationalism. This sounds like much of what we read in the press today, yet it was written almost 150 years ago.

Since these words were written nationalism, which we will examine in Chapter 10, and religion, which we will examine in Chapter 8, have repeatedly proven to be forces that are at least as powerful as economics. Each day's newspaper still reports ways in which these other two forces confound economic determinism. Poor nations, for example, waste money on wars, and people make economic sacrifices for their religious beliefs. As we have seen before and will emphasize repeatedly, human behavior can seldom be explained by any single factor.

TABLE 3-1
Transport and Communication Costs Expressed in 1990 Dollars

Year	Average ocean freight and port charges per short ton of import and export cargo	Average air transport revenue per passenger mile	Cost of a 3-minute telephone call New York to London
1920	$95	na	na
1930	$60	$0.68	$244.65
1940	$63	$0.46	$188.51
1950	$34	$0.30	$ 53.20
1960	$27	$0.24	$ 45.86
1970	$27	$0.16	$ 31.58
1980	$24	$0.10	$ 4.80
1990	$29	$0.11	$ 3.32

Courtesy Institute for International Economics.
Sources: *Historical Statistics of the United States; Statistical Abstract of the United States.*

ties, manufactured goods, and even human labor are seeking common world levels. An architect, for example, can easily fly anywhere on Earth. Therefore, any modern architect competes with every other one on Earth. In a small world, each person must ask himself or herself: "Are you as good as the best in the world? Is your product or service as inexpensive as the cheapest competition?" If a particular task, for example, such as the assembly of computers or even the writing of software, can be done almost anywhere on Earth, will everyone on Earth who can perform that task eventually be getting the same pay?

Connectivity Compresses Space and Time

The telegraph and telephone began the systematic annihilation of space that has continued with electronic mail, facsimile machines, computer modems, and other electronic devices collectively known as the **electronic highway**. More people with personal electronic devices are plugged in everywhere all the time (Figure 3-20). School children with personal computers can tap into networks of information that were unavailable to the world's leading scholars 50 years ago.

The activities of communicating with other people through an electronic network or even of playing a game alone with a computer create a new mental world—a new "place" where that activity is occurring, called *cyberspace*. That extension of reality is called **virtual reality**. The term may be applied widely. Many office workers, for example, collaborate through electronic networks without sitting down together in one office. One may be at home tending children, a second in a car on the road, and a third in an airport waiting room. The electronic network connects them in a *virtual office*. Videoconferencing creates life-sized holographic images of individuals who are scattered around the world; the participants see each other as if they were all sitting together in one virtual room.

The compression of space compresses time. People scattered around the world can work on the same project simultaneously, and by relaying the project around the world, work can proceed 24 hours per day. In 1992, for example, a U.S. financial exchange asked Texas Instruments Corporation to design a hand-held electronic bidding device for traders. The company's design department began working on the problem in Dallas. At quitting time the Dallas workers electronically sent what they had done to designers in Tokyo, who, when their work-

FIGURE 3-20

European film producer John Halmi relied on his MLink to stay in touch with his office in 1993 while he scouted locations in Papua New Guinea for his upcoming cinematic remake of *Tarzan*. This briefcase-sized satellite terminal can turn any spot on Earth into a phone booth immune to the forces of nature, civil unrest, or power or service failures. All you need to do is unfold the lid (which doubles as a phased-array antenna), aim it at any of four Immarsat satellites that hover 22,500 miles (36,000 km) over the Earth, and dial your call. The unit can also receive calls, transmit faxes, and allow personal computers to tap into remote data banks.
(Courtesy NEC America, Inc.)

day ended, passed the project on to designers in Nice, France. By the next day in Dallas, Texas Instruments could show the potential customer a computer-generated photo of what the product would look like and offer a fairly accurate estimate of the cost of producing it.

Whenever improvements in transportation and communication bring people together, ideas and additions and permutations among cultures multiply.

The Principle of Time–Space Convergence
New transport or communication facilities trigger a greater increase in interaction (called the *convergence rate*) between more distant places than between two closer or intervening places. Geographers call this the *time–space convergence* effect. For example, when a new road passes through several towns between two large cities, the interaction between the two large cities grows faster than does interaction among the

smaller towns or between either large city and any of the towns. Thus, big cities usually grow faster than the smaller ones; the new route augments their dominance. Whenever new means of transportation or communication are constructed, this principle of time-space convergence redistributes relative locations and locational advantages.

Situations Change

Accessibility can be a resource as valuable as mineral deposits or fertile soil, but any place's relative location and accessibility can change (Figure 3-21). The opening of a new highway stimulates the development of new housing and new roadside services, such as shopping malls. At the same time it can choke older shopping malls by rerouting traffic past them. Similarly, at a larger scale, great cities rise, and countries prosper if they are advantageously situated within national or international patterns of trade. Isolated areas usually fall behind and stagnate. When the Europeans learned how to sail around Africa in order to reach Asia, they brought about one of history's greatest redefinitions of relative locations. Europeans drew to the African and Asian seacoasts much of the commerce that had previously moved

only by long and difficult overland caravan journeys. New seacoast cities sprang up, such as Bombay in India, Rangoon in Burma (today known as Yangon in Myanmar), and Hong Kong on the coast of China. Major cities in the interior of the Eurasian and African continents were suddenly "on the back road." Some of these, such as Kashgar and Samarkand, have never recovered their earlier importance.

The opening of new transport routes always redistributes accessibility. The tunnel under the English Channel between England and France (the *Chunnel*) will redistribute land values and many activities in southern England and northern France. Lille, in northern France, will become more convenient to London, Paris, and Brussels, and it will probably grow (Figure 3-22, p. 104).

Changes in transportation technology also redistribute accessibility. In the 1950s, the Suez Canal was the principal route of oil shipments to Western Europe from the Middle East. New oil tankers, however, grew so large that they were incapable of squeezing through the Canal and had to sail around Africa. This rerouting placed Capetown "between" the Persian Gulf and Rotterdam, Europe's chief port for importing oil. Capetown's ship supply and repair facilities won new business.

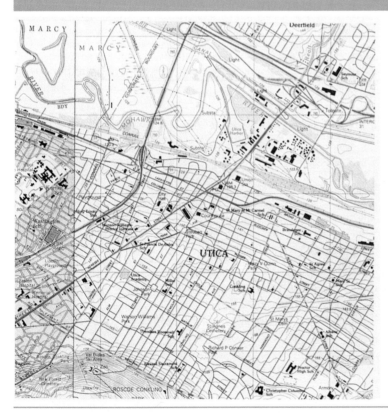

FIGURE 3-21

This map shows how changing means of transportation changed the relative location of Utica, New York, within the state, and, at a smaller local scale of analysis, they also changed the relative location of places within the city. Utica arose where the Mohawk River could be forded and was later bridged, and the city prospered as a commercial outpost on the frontier. The city industrialized when coal could be brought up from Pennsylvania by the 1837 Erie and Chenango Canal, whose route was later paralleled by the tracks of the New York, Ontario & Western Railroad, seen entering the map from the lower left side. When the Erie Canal and later the New York Central Railroad (the tracks labeled Conrail cutting across the map) were built, new industry multiplied around Utica Harbor and in the big buildings downtown along that railroad. In the nineteenth century workers lived close to these factories, and downtown Utica was a busy place. Since World War II modern interstate and local highways provided accessibility to the suburbs, and virtually all new industrial development has occurred there. New buildings can be seen to the north across the river from the old downtown. Housing has also spread into the suburbs, and shopping and office malls have pulled more activities out of the downtown area.

GEOGRAPHIC REASONING
Tombouctoo

Questions

1. Compare Tombouctoo's location with that of Trabzon on the Black Sea in modern-day Turkey. Among which great empires was Trabzon once a major contact and trading point? How important is it today?

2. Do you know of any other once-great cities that have declined as trade routes bypassed them?

3. Conversely, which great ports developed on the west coast of North America as trade with Asia multiplied in the twentieth century?

4. The geographer Ellen Churchill Semple opened her book *American History and Its Geographic Conditions* with the sentence: "The most important geographical fact in the past history of the United States has been their location on the Atlantic opposite Europe; and the most important geographical fact in lending a distinctive character to their future history will probably be their location on the Pacific opposite Asia." This was written in 1903. Has the twentieth century proved her correct?

In modern slang "Timbuktoo" is a synonym for "nowhere," and few Americans even know that there really is such a place. In fact, Tombouctoo is a city in Mali at absolute location 16°46' N. Lat.; 3° W. Long. It plays no major role in today's world, but in past centuries it was a major urban center. Tombouctoo is situated on the Niger River where the river bends farthest north into the Sahara Desert. This puts Tombouctoo on a cultural border between nomadic Arab peoples to its north and the settled black African peoples to its south. The city was settled by the Taureg people from the north in 1087, and, as a major contact point between north and south, it prospered as a market for slaves, gold, and salt and as a point of departure for trans-Saharan caravans to North African coastal cities. From these cities Tombouctoo's fame as a center of almost mythical riches spread around the Mediterranean. It was a center of Islam in West Africa, and it boasted a great university.

In the fifteenth century, however, Portuguese expeditions sailing around West Africa outflanked the trade routes on which Tombouctoo depended. The city declined rapidly. It was captured by the Moroccans in 1591, and by the time the French captured it in 1893, it was little more than an extensive ruin. Today it is a small city of 10,000 people who live surrounded by few remnants of the city's former greatness.

The fabulously rich medieval King of Mali is shown here enthroned by the city of Tombouctoo, spelled Tombuch, just to the left of his throne. This map is from the Catalan Atlas (c. 1375), which was owned by the king of France. A search for the gold of Mali was one impetus for the ensuing European exploration and colonization of Africa.
(Courtesy of the Bibliothèque Nationale, Paris.)

Internal Asian Trade Routes

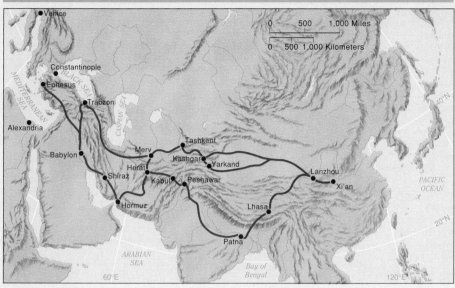

Venice, Trabzon, Kashgar, and Merv were once great cities on major trade routes, but when sea routes opened around Africa and Asia, these and many other cities were suddenly "on the back road." Few ever recovered their former economic strength.

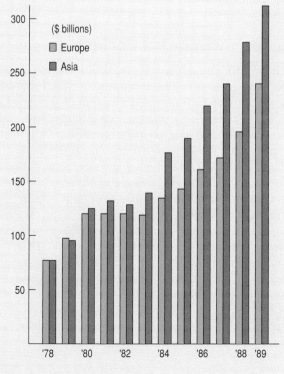

This chart shows the slow but steady rerouting of U.S. trade from Europe to Asia. It is just one piece of evidence confirming Ellen Churchill Semple's prediction of 1903. (Courtesy U.S. Dept. of Defense, as reprinted in *Forbes*, October 1, 1990.)

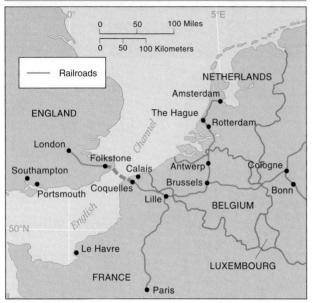

Route of the English "Chunnel"

FIGURE 3-22

The idea of a tunnel under the English Channel has long been alternatively a dream and a nightmare to the French and the English, depending upon whether they were at peace or at war with each other. In these times of peace, the Chunnel has become a reality, and new flows of traffic will relocate activities across Northern Europe.

A place's relative location can also change if the *territorial scale of organization* of an activity—that is, the extent of territory within which that activity occurs—changes. For example, when trade barriers between countries fall, activities redistribute themselves. Significant redistributions will occur throughout North America as the economies of the United States, Canada, and Mexico merge through the new trade pacts discussed in Chapter 13. All activities in Europe are readjusting from a national scale of organization to a supranational scale represented by the European Union.

Conversely, when a large territory is broken up into smaller individually organized territories, a place's accessibility can be reduced. Calcutta was a prosperous city when Bangladesh and India were united as one colonial region. Since Calcutta's service area—the state of Bengal—was split between two countries, however, Calcutta has experienced a decline in access that has resulted in devastating poverty. In 1991, the Union of Soviet Socialist Republics—15 percent of the Earth's land surface—broke up into 15 independent countries. That political change required reorganizing the scale of all political, economic, and cultural activities throughout the vast territory.

Being familiar with the distributions of human activities is important, but that is not sufficient to understand contemporary geography. These distributions are continuously changing. If you learn the reasons why things get distributed the way they do, however, then you will have learned something that will be useful for the rest of your life.

THE GLOBAL DIFFUSION OF EUROPEAN CULTURE

Despite the rich variety of indigenous local cultures around the globe, the world is increasingly coming to look like one place. In consumer goods, architecture, industrial technology, education, and housing, the European model is pervasive. To ethnocentric Westerners who presume the superiority of their own culture, or to westernized locals, this may seem natural. To them, this may be "modern" life, or "progress," or "development."

A dispassionate observer, however, might expect more diversity, more styles and models of development. Why are so many people around the world imitating Western examples and adopting aspects of Western culture? In much of the world, acculturation to the Western way of life is rapidly replacing both the positive and negative features of other cultures. It is the most pervasive example of cultural diffusion in world history (Figure 3-23). An understanding of this process of European cultural diffusion over the past 500 years is essential to understanding the world we live in today.

Europe's Voyages of Contact

Europeans came to play a central role in world history and world geography because it was they who paved the way for the modern system of global interconnectedness. In the fifteenth century the great cultural centers of the world—the Inka and Aztec empires in the Western Hemisphere, the Mali and Songhai in Africa, the Mughal in India, Safavid Persia, the Ottoman Turks, the Chinese, and all of the lesser empires and culture realms—were still largely isolated from one another (Figure 3-24).

The European voyages of exploration and conquest connected the world (Figure 3-25, p.108). It was not inevitable that the Europeans would be the ones to do this. The Chinese were actually richer and more powerful than the Europeans, and earlier in the fifteenth century they had launched great fleets of exploration that had reached the east coast of Africa.

FIGURE 3-23

By studying this picture from a news magazine, can we tell who or where this man is? He is dressed as a modern, westernized man anywhere in the world today. The photo suggests that he knows something about modern computers; the one beside him is a U.S. brand, although many of its key parts were manufactured in Singapore. The buildings behind him are in a Roman classical or Palladian style. This picture seems entirely "Western."

In fact he is Huzur Saran, an Indian computer scientist. He studied in the United States but works today at the Indian Institute of Technology. The photo was taken in New Delhi, the British-built former imperial capital of India. (Courtesy of Dilip Mehta/Contact Press.)

We can hardly imagine how different world history and geography would be if the Chinese had continued their initiatives and gone on to explore and conquer Africa and the Western Hemisphere. But they did not. Instead, for internal political reasons, the Chinese withdrew into themselves, and the Europeans were the ones to draw the world together.

The first European initiative was Prince Henry of Portugal's conquest of the city of Ceuta on the north coast of Africa in 1415. He learned about trans-Saharan caravan routes and the riches of West Africa, which inspired him to finance studies of naval skills. Improvements in sailing technology allowed the Portuguese to sail down the west coast of Africa to reach beyond the Sahara Desert. Prince Henry, known as "The Navigator," launched the era of European seaborne colonial empires. Soon Spain, Holland, France, and England were racing to secure colonies. In contrast to these seaborne empires, Russia at the same time forged eastward overland from the Russian homeland in Europe west of the Ural Mountains. The Russian Empire pushed across

Siberia to the Bering Strait, crossed over into Alaska, and eventually established colonies down the North American west coast as far south as today's California.

This European outreach triggered the **Commercial Revolution**, between about 1650 and 1750. A tremendous expansion of trade followed the development of the first ocean-going freighters that could carry heavy payloads over long distances. The evolution of superior ships was paralleled by the evolution of superior naval gunnery and of additional useful technologies such as clocks, which were perfected in order to determine longitude at sea. This first era of outreach ended with the death of the British explorer Captain James Cook in 1779. By then, for the first time in world history, Europeans could draw a fairly accurate outline map of all of the world's continents and islands.

Expansion and Cultural Diffusion

From that time on, Europe did not actually originate every "modern" idea and then impose it on the rest of the world. Europeans learned from others, too, and then transplanted around the world the ideas that they had adopted from elsewhere. "Knowledge is power," said Sir Francis Bacon, and because it was the Europeans, and not the Chinese or the Inka or any other group, who had contacted all other civilizations, Europe became the clearinghouse of world information and products. Global diffusion was fixed hierarchically, with Europe as the apex. Europe became cosmopolitan, that is, familiar with many parts of the world. Other peoples, no matter how great their native civilizations were, remained more localized in the world Europe was creating. In a hierarchical system of diffusion, as we have seen, more information flows up and down the hierarchy than among the places at lower levels.

Consider, for example, what Europeans did with agricultural products. They took sugarcane from Asia and planted it in the Caribbean region. They replanted bananas from Southeast Asia to South America, cocoa from Mexico to Africa, and coffee from Arabia to Central America. All of these foods remain major products of international trade and important factors in the export economies of many countries.

Europeans not only relocated the production of many goods around the world, they also introduced many products into world trade, both for the European home market and for other overseas markets that they pioneered. Europeans introduced many Indian goods, for instance, into China, and South American products into Africa and Asia. An Englishman stole tea plants from China and cultivated them in Calcutta, leading to the development of a tea

The World's Major Culture Regions in the Year 1500

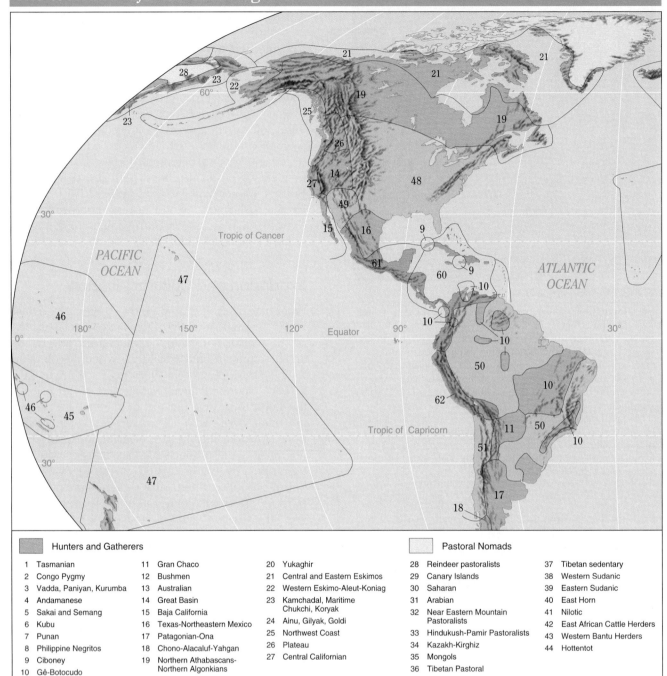

	Hunters and Gatherers						Pastoral Nomads		

1	Tasmanian	11	Gran Chaco	20	Yukaghir	28	Reindeer pastoralists	37	Tibetan sedentary
2	Congo Pygmy	12	Bushmen	21	Central and Eastern Eskimos	29	Canary Islands	38	Western Sudanic
3	Vadda, Paniyan, Kurumba	13	Australian	22	Western Eskimo-Aleut-Koniag	30	Saharan	39	Eastern Sudanic
4	Andamanese	14	Great Basin	23	Kamchadal, Maritime Chukchi, Koryak	31	Arabian	40	East Horn
5	Sakai and Semang	15	Baja California			32	Near Eastern Mountain Pastoralists	41	Nilotic
6	Kubu	16	Texas-Northeastern Mexico	24	Ainu, Gilyak, Goldi			42	East African Cattle Herders
7	Punan	17	Patagonian-Ona	25	Northwest Coast	33	Hindukush-Pamir Pastoralists	43	Western Bantu Herders
8	Philippine Negritos	18	Chono-Alacaluf-Yahgan	26	Plateau	34	Kazakh-Kirghiz	44	Hottentot
9	Ciboney	19	Northern Athabascans-Northern Algonkians	27	Central Californian	35	Mongols		
10	Gê-Botocudo					36	Tibetan Pastoral		

FIGURE 3-24

These major culture regions interacted somewhat, but their individual isolation was significant enough to allow each a distinct cultural individuality. The Europeans' voyages of exploration and conquest would soon scramble this map. (Adapted from G. W. Hewes, "A Conspectus of the World's Cultures in 1500 A.D.," *University of Colorado Studies*, no. 4, 1954.)

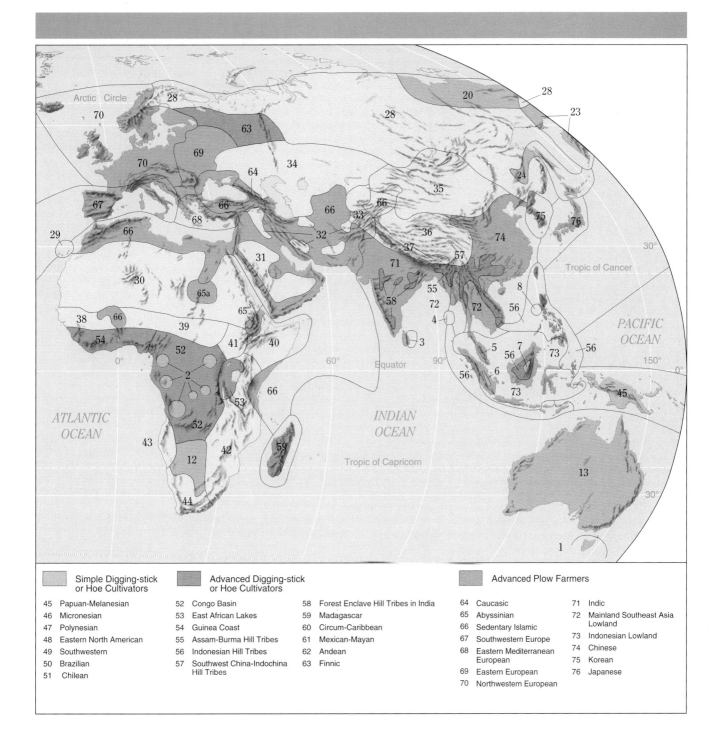

	Simple Digging-stick or Hoe Cultivators		Advanced Digging-stick or Hoe Cultivators				Advanced Plow Farmers	

45	Papuan-Melanesian	52	Congo Basin	58	Forest Enclave Hill Tribes in India	64	Caucasic	71	Indic
46	Micronesian	53	East African Lakes	59	Madagascar	65	Abyssinian	72	Mainland Southeast Asia Lowland
47	Polynesian	54	Guinea Coast	60	Circum-Caribbean	66	Sedentary Islamic		
48	Eastern North American	55	Assam-Burma Hill Tribes	61	Mexican-Mayan	67	Southwestern Europe	73	Indonesian Lowland
49	Southwestern	56	Indonesian Hill Tribes	62	Andean	68	Eastern Mediterranean European	74	Chinese
50	Brazilian	57	Southwest China-Indochina Hill Tribes	63	Finnic	69	Eastern European	75	Korean
51	Chilean					70	Northwestern European	76	Japanese

Cultural Diffusion   **107**

Major European Voyages of Exploration

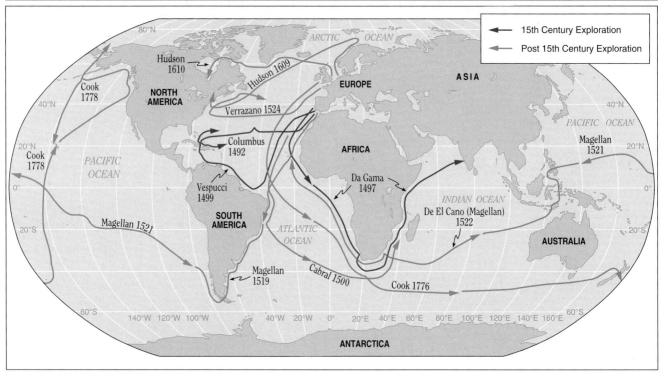

FIGURE 3-25

Each of the European voyages of exploration was a daring enterprise. The Portuguese first established a string of bases around the coasts of Africa and Asia all across to Japan, where they established a trading post in 1543. Other European nations followed, and the race for discovery of new riches, of new converts to Christianity, and for help against Christianity's powerful Islamic foe was launched.

industry in India and Ceylon (today known as Sri Lanka). Europeans created world markets, and they profited by controlling every stage: production, transportation, and marketing.

Europeans also redistributed people and ideas. The culture of North America today is considered "Western," yet it draws on a diverse blend of contributions. The native American cultures were overlain by several European and African traditions, with additional significant inputs by various Asian peoples. The world migrations of peoples triggered by European expansion will be examined in detail in Chapter 5.

The story of Coca-Cola, one of the world's most familiar consumer products, exemplifies cultural blending. The popular drink was formulated in Atlanta, Georgia, and today symbolizes westernization so powerfully that the large-scale infusion of Western products into the non-Western world is often referred to as the "Coca-colonization of the world." The two original ingredients from which Coca-Cola takes its name, however, are coca, a Quechua native American word for the South American tree whose leaves supply a stimulating drug (used today to make cocaine and crack), and cola, a Mandingo word for the West African nut

that supplied the other original stimulating ingredient. Westerners borrowed the knowledge of both of these ingredients from their far-flung hearth areas, combined them, and marketed the drink worldwide.

 Westernization has come to be synonymous with modernization, but modern world culture is by no means exclusively a Western product. For Westerners to think it is so is presumptuous. For non-Westerners to think that it is, is to miss the many contributions of non-Western peoples. In a system of hierarchical diffusion, however, the direct source of much information for most non-Western places is the West, and this causes resentment.

The Industrial Revolution

Europe pulled ahead of the rest of the world economically as it underwent the tremendous transformation of the **Industrial Revolution**. Between about 1750 and 1850, Europe evolved from an agricultural and commercial society to an industrial society relying on inanimate power and complex machinery. Again, we cannot fully explain why Europe experienced this transformation before any other part of the world, but we can list several factors that enabled Europe to industrialize. The voyages of discovery and conquest resulted in an influx of precious metals and other sources of wealth, stimulating industry and a money economy. The expansion of trade encouraged the rise of new institutions of finance and credit. In the mid-sixteenth century the joint stock company was developed. This was a way many investors could share both potential profit and risk in new enterprises. The creation of markets where stocks could readily be bought and sold granted capital new **liquidity**, or easy conversion from one form of asset to another. This created, in the words of English writer Daniel Defoe, "strange unheard-of Engines of Interest, Discounts, Transfers, Tallies, Debentures, Shares, Projects."

In 1709, Abraham Darby first smelted iron ore using coke instead of charcoal, and in 1769, James Watt designed the steam engine, which multiplied the energy available to do work. Subsequent inventions and technical innovations in manufacturing, applied first to textiles and then across a broad spectrum of goods, dramatically increased productivity. Factories and industrial towns sprang up, canals and roads were built, and later the railway and the steamship expanded the capacity both to transport raw materials and to send manufactured goods to markets (Figure 3-26). New methods of manufacturing steel, chemicals, and machines played important parts in the vast changes. These innovations occurred first in Great Britain and subsequently spread to the continent of Europe and to the United States.

Beginning in the eighteenth century, Europe also first experienced an **Agricultural Revolution**. This development, to be examined in detail in Chapter 6, both increased food production and released agricultural workers from the land, thereby creating a supply of industrial labor.

As a result of the Industrial and Agricultural Revolutions, Europe and European settlements around the world drew far ahead of any other places and peoples in productive capacity. As recently as 1800, the per capita incomes of the various regions of the world were close. If we index the Western European per capita income in the year 1800 as 100 units of wealth, then estimated per capita incomes in North America were 125, in China 107, and in the rest of the non-European world 94. By 1900, however, European and North American incomes were several times those of non-Western peoples.

Commercial Contacts and Economies

At the beginning of the age of the European voyages, European demand for foreign products such as

FIGURE 3-26

This is the world's first iron bridge, built over England's Severn River in 1779. It demonstrated iron's strength, and the bridge's bold design invited other uses for the new material. Its builder, John Wilkinson, launched the first iron boat in 1787, and when he died he had himself buried in an iron coffin.
(Courtesy of the British Tourist Authority.)

spices, sugar, fruits, and North American furs grew rapidly. Soon the Europeans were no longer content to trade with native peoples for these goods, and the Europeans themselves established overseas estates and plantations and applied large-scale techniques to specialized production.

European commercial plantations at first concentrated along the coasts, but in the nineteenth century the railroad allowed penetration of the continental interiors to superior agricultural lands or, later, as Europe industrialized, to mineral deposits. The world's railway network expanded from 125,000 miles (200,000 km) in 1870 to over 625,000 miles (1 million km) by 1900. What had been European treaty ports and coastal enclaves became vast inland empires (Figure 3-27). The steamship also greatly increased the possibility of transporting African minerals, and new quantities of minerals supplied Europe's multiplying factories. Between 1840 and 1870, the world's merchant shipping rose from 10 million tons to 16 million tons, and then it doubled in the next 40 years.

New cities emerged in the non-European world as coordinating centers for these commercial activities; new ports sprang up along the seacoasts. The major seaport cities from the Straits of Gibraltar around Africa, across the Indian Ocean, and all throughout South Asia are the products of European contact (Figure 3-28). The same is true for most of the port cities of the Western Hemisphere. The railroads

Cultural Diffusion   **109**

European Intervention in Africa

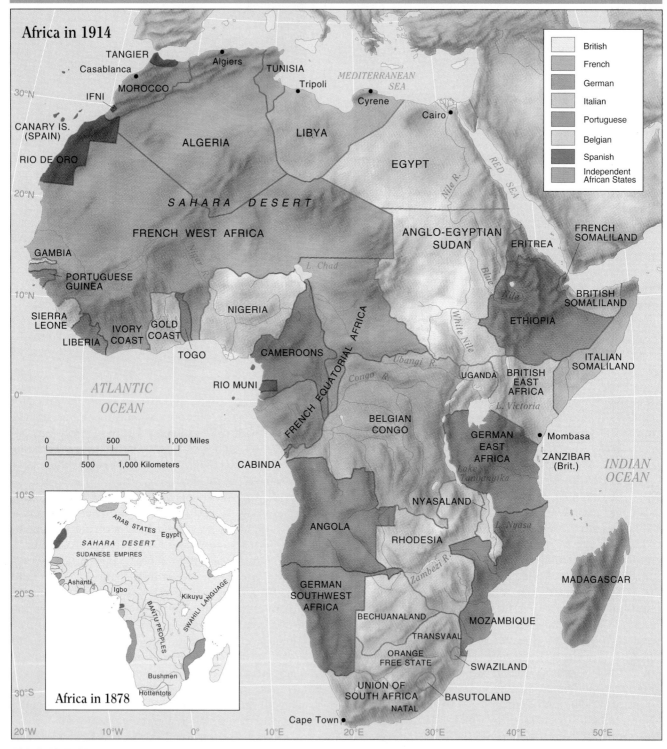

Africa in 1914

British
French
German
Italian
Portuguese
Belgian
Spanish
Independent African States

FIGURE 3-27

Europeans' treaty ports and trading stations along the African coast were transformed into inland empires when the Europeans decided to parcel out the continent in order to prevent competitive war among themselves. Industrial Europe demanded African raw materials, and railroads allowed the Europeans to draw them out of the African interior. The native African peoples were not consulted in the political reapportioning, which was completed at a conference in Berlin in 1884–85.

European Intervention in Asia

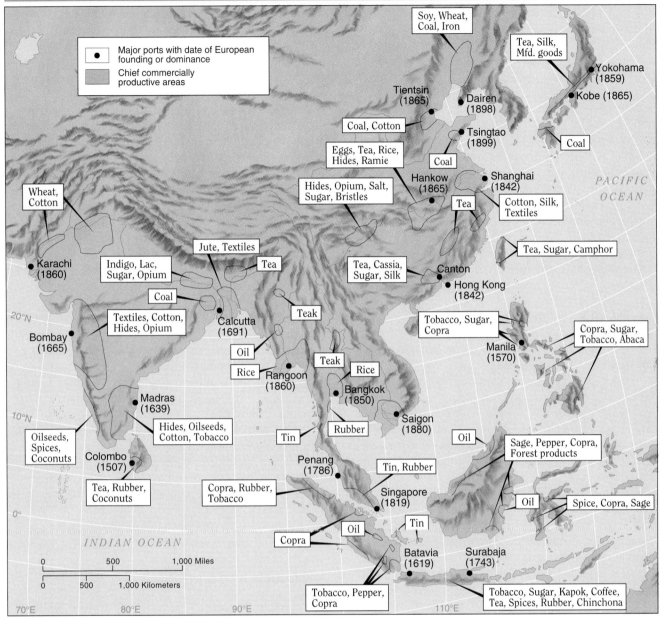

FIGURE 3-28

Many of Asia's port cities were at first spigots that Europeans tapped into the continent's flank to draw off Asia's wealth. The ports expanded into plantations for the production of goods valuable in international trade, and then the plantation regions expanded into political colonies, as in Africa.

and associated commercial economies at first affected only a small percentage of the population, but over time an increasing share of the population and territory were drawn into the emerging global economy. In some countries, however, such as India, the modern commercial economy still overlies a traditional subsistence economy. This economic dichotomy is called *dualism*. There may be little exchange between the two economies unless refugees from the collapse of the traditional economy flee to the slums of the modern cities.

Political Conquest

In two waves of exploration and conquest—the first extending from 1415 to 1779 and the second occurring at the end of the nineteenth century—Europe (and, in the second period, the United States) conquered most of the rest of the world. One of the original reasons for this conquest was the European nations' wish to protect their investments in foreign lands and to control these lands as markets for themselves. European countries divided up the rest of the world in order to keep their own rivalry under control. Superior military power guaranteed their ascendancy over the indigenous population. The Europeans' confidence in their military superiority was expressed by British satirist Hilaire Belloc, who wrote in honor of the newly invented Maxim machine gun: "Whatever happens we have got/ The Maxim gun, and they have not."

The United States and most of Latin America won independence between 1775 and 1825, but between 1875 and 1915 about one-quarter of the Earth's land surface was distributed or redistributed as colonies among a half-dozen imperialist states. Of all the countries in the world today, the only ones never ruled by Europeans or by the United States are Turkey; Japan (although it was occupied by the United States from 1945–1952 and its constitution was imposed on it); Korea (which was ruled by Japan from 1910 to 1945 and remains split today); Thailand (left as a *buffer state* between the French and English empires); Afghanistan (a buffer between the English and Russian empires); China (which was nevertheless divided into foreign "spheres of influence"); and Mongolia (ruled by China). The peoples of every other country on today's world map experienced European or American imperial rule.

Therefore, European cultural attributes linger as a legacy of European rule in most of the countries of the world today. They still predominate or overlay native pre-European traditions, and few people anywhere in the world remember what may have locally preceded the overlay of European culture. European concepts of law, for example, drastically changed native societies, especially European ideas of property rights and land ownership. Before the Europeans came, land was generally considered a good held in common for all members of the community. Local political leaders apportioned and adjudicated land use and tenure by customs which brought the community together. Neither the leaders nor anyone else *owned* land. The idea that any individual could own land and determine alone how to use it was largely unknown. Europeans introduced their idea of "ownership" of an "estate" which could be bought, sold, or mortgaged by individual contract. European-style ownership introduces liquidity, but it dissolves social cohesion. Land was no longer a common good, and the regulation of its use was no longer a shared community affair.

Native American chiefs, for instance, did not by their own customs have the right to transfer land out of tribal control, and they frequently did not actually understand what Europeans meant when they "bought" native American lands. In the history of the United States, innumerable wars were sparked when native Americans returned to hunt or harvest unoccupied land which the Europeans insisted the natives tribes had

sold. Throughout Africa and Asia, the Europeans often simply assigned ownership to the local political leaders. This ended traditional egalitarian systems and created new classes of rich and poor. The descendants of many of these newly enriched leaders remain great landholders today throughout the Near East, for example. Many traditional societies crumbled under the transformation of land into a private commodity.

European law tended to transform labor into a commodity as well. Sociologists differentiate a group of people held together by traditional networks of rights and responsibilities, called a *community*, from a group interacting as more self-interested individuals, called a *society*. Traditional communities were not idyllic; they restricted individual liberties, and slavery and serfdom were not unknown. These constraints, however, were often balanced by strong webs of responsibilities and rights which usually kept anyone from being entirely outcast and starving. The European idea of a self-regulating market for individual labor is more abstract and impersonal than traditional community values. It cut traditional ties of both rights and responsibilities. In addition, Europeans required the use of money as a universal measure of value. Natives were forced to work for wages or to sell goods for money in order to pay taxes. Chapter 2 noted how this contributed to the degradation of the Sahel environment.

Europeans also brought their forms of administration, government, centralized state authority, written arrangements, uniformity, secularization, economic planning, public accounting and treasury control, central administration, and decision making. In many cases the civil services that the Europeans left behind remain the pride of new nations, as in India.

Cultural Imperialism

European rule was marked by **cultural imperialism**, which is the substitution of one set of cultural traditions for another, either by force or by degrading those who fail to acculturate and rewarding those who do. Europeans seldom doubted that native cultures were inferior and that native peoples needed "enlightenment." Therefore, the Europeans destroyed other ways of life, including religious and political traditions as well as physical artifacts such as art and architecture (Figure 3-29).

One reason for this is the nature of Christianity. It is a proselytizing religion, which means that its adherents try to convert others to their faith. The natives in many areas accepted that there is truth in all religions, and their tolerance helps explain why they were open to acculturation.

112 *Chapter 3*

FIGURE 3-29

This page is from a book called the *Grolier Codex*, one of only four books known to survive from the considerable libraries of Central America's Mayan civilization. The rest were burned by the Spanish. This book contains astronomical records regarding the recurrence of the planet Venus. The vertical columns of markings along the left are countings of days. The warrior, perhaps mythological, stands before an incense burner.
(Courtesy of Justin Kerr.)

Additionally, Europeans believed that their military and technological superiority presumed European superiority in all other aspects of life. The conquerors often failed to appreciate the value of the civilizations which they conquered, or even their science and technology, so the Europeans did not learn as much from others as they might have. For example, they might have learned agricultural techniques from native Americans.

European cultural imperialism began with the systematic training of local elites. The missionary schools produced converts who proselytized among their own people, helping to eradicate the local culture. Later the government schools turned out bureaucrats and military officers who helped govern their own people. A second channel of transmission was *reference-group behavior*. People who wish to belong to, or be identified with, a dominant group often abandon their traditions in order to adopt those of the dominant group.

The returned slaves who carried the first wave of westernization to West Africa wore black woolen suits and starched collars in the tropical heat. In India the native officer corps imitated English officers, complete with waxed mustaches. Premier Lee of Singapore, quoted above, a Chinese, even personally boasted of being "the last true Englishman east of Suez."

Local elites also adopted Western ways because they were made to feel ashamed of their color and their culture. Given the racism and cultural imperialism of the rulers, natives could succeed only by succeeding on the whites' terms, by adopting their ways. Colonial rule was an experience in racial humiliation, as the autobiographies of the new national leaders of every African and Asian country recount (Figure 3-30). The colonial school systems implanted in children's minds an image of the power and beneficence of the "Mother Country." Children in the Congo, for instance, knew more about Belgium than about their own land and peoples. History texts used in French African colonies opened with the words, "Our ancestors the Gauls…." Schooling focused on each colony's ruler, so the native peoples scarcely knew that other countries existed. Residents of the Congo, for example, referred to all whites as Belgians.

Colonial rule was also based on intellectual institutions and the power to define tradition. For example, from the founding in 1784 by Sir William Jones and others of the Asiatic Society of Bengal, the accumulation of knowledge about India was spearheaded by Europeans. The definition of what was Indian culture came to depend on European study of Indian languages and texts. This standardized and made authoritative what Europeans thought to be "classics" of Indian thought and literature, not only for Europeans, but for Indians themselves. Archeological surveys led by Europeans determined what monuments were fit for description and preservation as part of "the Indian heritage." Census operations in the 1860s and the establishment of an ethnographic survey described the peoples and cultures of India. Indian arts had to be defined, collected, and preserved in museums. Art schools were founded. Educated Indians increasingly came to learn about their own culture through the mediation of European ideas and scholarship.

Self-Westernization By the end of the nineteenth century, the elites of the entire non-Western world were taking the Europeans as their reference group. Only three major nations were never colonized—the Turks, the Chinese, and the Japanese—but these were all militarily humiliated, and this experience was traumatic for them. All three had had empires of their own, and their defeats forced them to reconsider all of the assumptions on which their institutions and daily lives were based. All three preempted or coopted Western civilization, and to some degree this saved them from Western political rule.

Mustafa Kemal (1881–1938) seized power in Turkey and forced the country to undergo self-westernization (Figure 3-31). In China, the Boxer Rebellion of 1900 challenged Western influence but failed to overthrow it. The Republican Revolution of 1911 attempted to modernize the country, but China's subjection to the West continued until another leader, Mao Zedong, applied an alternative brand of westernization, communism, although in a uniquely Chinese

FIGURE 3-30

In the 1935 film *Saunders of the River*, Commissioner Saunders supervises and keeps the peace among tribes depicted as dangerous children in a British African territory. The movie is racist and illiberal, and in this key scene Saunders prods a submissive native in the chest with his cane while lecturing on the superiority of British civilization. That young African was portrayed by the actor Jomo Kenyatta, who would later serve as president of Kenya. What could he have been thinking when this scene was filmed? What could he have thought years later while viewing the film in the presidential palace of independent Kenya?
(Courtesy of the Kobal Collection.)

FIGURE 3-31

Mustafa Kemal Ataturk, whose surname means "father of the Turks," recognized the technological superiority of the West, and he consciously turned his people toward westernization. One of his changes was the transformation of the writing of Turkish from the Arabic into the Roman alphabet. Here he teaches a class in an Istanbul public park.
(Courtesy of the Turkish Government Tourist Office.)

adaptation. The Japanese were forced to open their society to Western trade in 1853, after centuries of near-total isolation. They decided to become thoroughly Western but to retain control of the process. The Japanese escaped the domination of the West by embracing the West. "Western science with Eastern ethics" became their slogan. The Meiji Restoration in 1867 launched the westernization of Japan with astonishing speed. Japan sent students to the United States and Europe and adopted Western science, technology, and even many cultural traits so successfully that by 1904 the country was able to defeat a Western power—Russia—in war (Figure 3-32).

Although in World War II Japan was defeated in its attempt to extend its empire, its early military successes discredited the Western powers and helped encourage non-Western peoples, particularly in Asia but later in Africa too, to pursue national independence after 1945. Despite Japan's military defeat, it later developed into one of the world's dominant economic powers.

Regional Self-Consciousness in the World Today World War II created in the minds of non-European peoples a new definition of themselves and of their relationship to Europe. Japanese propaganda proclaimed that Japan's only purpose in the war was to liberate Asia from the whites. The Japanese devised the slogan "Asia for the Asians" to mobilize all Asian peoples against Europeans. Most peoples of Asia, however, had never before thought of themselves as "Asians," a distinct group in oppo-

FIGURE 3-32

This is a Japanese woodblock cutaway view of a German battleship that visited Yokohama in 1873. This woodcut would later provide a virtual blueprint for the technologically backward Japanese Navy to copy.
(Daval Foundation: from the Collection of Ambassador and Mrs. William Leonhart, Courtesy of Arthur M. Sackler Gallery, Smithsonian Institution, Washington, DC)

sition to Europeans. "Asian" is in fact a European geographical description that was previously unknown to the people it categorizes, and the people thus categorized never saw any logical basis for solidarity simply in being "Asian." Asia had never had the cultural cohesion which Christianity gave Europe. Koreans and Cambodians, Japanese and Malays, Burmese and Manchu are culturally more different from one another than Italians are from Germans, Spaniards from Scots, or Portuguese from Poles. "Africa" is also a European geographical term. All of the many peoples who today call themselves "African" share first and foremost only a conviction that they are *not European.* Only recently have some peoples of that continent begun to think of themselves as having common "African" interests, and that common interest still is largely common opposition to outsiders.

Thus, many Asian and African peoples have come to accept and even favor the geographic labels that were originally put on them by European outsiders. As they do this, they are actually repeating the earlier experience of the many native peoples of the Western Hemisphere, who had been indiscriminately categorized by the European outsiders as "Indians." In this way, Eurocentric terms have taken on powerful symbolic political values.

The period of European relations with Africa and Asia that began 500 years ago is ending. Only two tiny Spanish holdings survive on the African continent—Ceuta and Melilla—and the Asian empires of the Dutch, English, Portuguese, and French have been surrendered except for a few islands. Great Britain is scheduled to yield Hong Kong to China in 1997, and in 1999 Portugal will return the Chinese territory of Macao, which it has held since 1557.

The West and Non-West Since Independence

The fixation with the West among the elite in the non-European world did not end when these countries won political independence. Those who assumed power in Africa and Asia were descendants of the Western-educated class who had become disappointed with their rulers, and their demands for independence included quotations from Western political writers whom they had read at Western universities.

Few of these Western-educated elites developed indigenous models of development because they were not themselves members of the traditional elites, such as chiefs or religious precolonial leaders.

Cultural Diffusion **115**

They were themselves westernized, and they continued the diffusion of Western models. They started building in the middles of their commercial cities and went on building outward. The technical term for the favored cities is **growth poles**. The leaders hoped to convert their whole national territories to modern Western societies, but they did not realize how long it would take the majority of their people to benefit.

The departure of the colonial rulers caused a local status vacuum, because the traditional sources of status, such as religion, family, and customs, had faded. The new power groups—politicians, bureaucrats, and businessmen—defined their status in the only way they knew, which was to exhibit Western material goods. Their housing, clothing, and means of transportation all mimicked Western status symbols.

In many countries the new rulers have practiced a sort of internal colonialism on their own people. For example, the rulers who assumed power in Latin America following the independence movements of the early nineteenth century were of European stock, and their descendants still dominate this region. Peoples of native American or African backgrounds have been treated as subject groups, forced to adopt European culture, religion, and language, and subjected to discrimination. The situation is not very different in Africa and Asia. The "colonizers" are generally the westernized elite; the "colonized" are all those who do not belong to this group, which is often the majority of the population.

Government by an elite privileged clique is called an *oligarchy*. Those of us living in democratic countries unconsciously assume that most governments in the world represent their people and work for their welfare. This is dubious. A great many governments do not represent the people. The governments of many countries have even been called *kleptocracies*—government by theft. In these cases foreign aid, or any system of funneling assistance to the poor through their governments, cannot guarantee that the needy will receive the assistance. It might end up in secret Swiss bank accounts. The presidents or rulers of many of the poorest countries in Latin America, Africa, and Asia are regularly counted among the world's richest individuals.

Westernization Today

The diffusion of Western culture continues today. Western culture diffuses from the top of societies, from the examples and activities of the local elites. Young people also diffuse it by adopting Western dress and lifestyles as status symbols. International studies reveal that these two groups—the rich and the young—are everywhere the most cosmopolitan consumers, and most consumer items in international trade are artifacts

of Western popular culture. Even the schools have become instruments of westernization. Their syllabuses emphasize modern, urban activities and values. The young sometimes emerge oblivious to their traditional culture, or even despising it.

The adult media reinforce the message. Western television programs, movies, advertisements, and videos penetrate millions of homes and implant Western values. Night after night, on television screens around the world, the images of the good life are images of the life among the wealthy in the United States. This imagery has dramatically changed behavior. In traditional societies, for instance, sexuality is associated with childbearing, but in modern lifestyles it is not. Some governments have hypothesized that imported Western TV programs are responsible for lowering their national birth rates. Consumption patterns also change because people want to copy what they see. This occurs regardless of any specific advertising (Figure 3-33).

The media in developing countries often produce their own coverage of domestic news and local events, but they tend to buy coverage of foreign events from Western news agencies. Few developing countries report on events in other developing countries. Thus, even when developing countries' views of themselves are their own, their perspectives on the rest of the world are at least partly those of the West. Table 3-2 shows the amount of news coverage specif-

FIGURE 3-33

Television, particularly U.S. television programming, penetrates practically every region. How do these people interpret what they see? How does it transform their cultural inheritance? The truth is that we do not know the answers to these questions.
(Courtesy of Raghu Rai/Magnum Photos, Inc.)

ic world regions devote to news from each of the other world regions.

Tourism provides still another channel of westernization. Westerners are attracted to "different" and "unspoiled" places, but the presence of tourists changes the places they visit. National cultures can degenerate into commercialized spectacles and shoddy souvenirs. Many local people abandon their own material culture and adopt that of the visitors.

Global flows of professionals and of professional education are another powerful force for the diffusion of westernization. The rich Western countries export professional services, and people from around the world attend Western schools for professional education. The elite and professionals of most countries have been educated in Europe or the United States. These graduates, who hold influential roles in their societies, have been acculturated to Western ways.

The diffusion of Western standards seems to carry an especially powerful impact in design professions. Western architects, civil engineers, and urban planners—or non-Western individuals educated in Western schools—are transforming the built environment around the world. The poor countries' cultural landscapes are increasingly "modern," and there is only one global cultural landscape of modernity. There is no prototype other than the Western. The cultural landscape, in turn, influences and conditions the lives and even the thinking of the area's inhabitants. Increasingly, peoples' cultural landscapes everywhere are built by individuals who are either Westerners or else educated to Western standards and ideals. Local cultures are lost.

World flows of capital overwhelm the poor countries. Countries are being transformed by capital investment, and capital is invested according to the standards of the societies which export the capital. Under this barrage of westernization, many traditional cultures and social structures are being radically transformed, and these transformations are not always in accord with the people's conscious wishes. In some cases whole cultures disappear, and their disappearance reduces cultural diversity and impoverishes all of us.

TABLE 3-2
Location of News Stories, November 19th 1991 (percentage of broadcast time)

	BROADCASTING REGIONS									
	Southern Africa	Latin America	North America	West Europe	E. Europe & USSR	Japan	China	Middle East	Other Asia	Australia/ New Zealand
Aust & NZ	0	0	0	0	0	4	0	0	0	56
China	0	0	0	0	1	5	64	2	3	0
W. Europe	19	4	20	63	21	0	12	43	12	20
E. Europe & USSR	5	8	10	23	80	6	14	28	12	4
Japan	0	0	0	0	3	78	10	0	2	2
Latin America	0	92	7	1	1	0	7	0	2	0
Middle East	8	2	16	19	8	0	19	85	5	19
North Africa	5	0	7	1	1	0	6	17	0	0
North America	6	7	80	11	12	1	17	13	5	14
Other Asia	2	1	1	9	0	8	21	8	80	10
Southern Africa	80	0	9	3	0	0	7	1	1	0

Source: Calculated from News Agency Survey IIC. Data from the *Economist*, Feb. 8, 1992.

Contemporary Reevaluations

In 1898 Winston Churchill, future prime minister of Great Britain, was a 25-year-old journalist. After witnessing the Battle of Omdurman, which eliminated the last Sudanese resistance to British imperialism, Churchill wrote: "These extraordinary foreign figures … march up one by one from the darkness of barbarism to the footlights of civilization … and their conquerors, taking their possessions, forget even their names. Nor will history record such trash."

History is written by the winners, and if the winners devalue all cultures but their own as "barbarism" and "trash," then much of value is destroyed, and the losers, in effect, have their history stolen from them. After three-quarters of a century of British rule and cultural imperialism, the descendants of the Sudanese warriors would in fact have little knowledge of their ancestors' preconquest culture and civilization. That is why to many people the study of their history is a fierce reclaiming of their history, of their independent identity, of their self-respect (Figure 3-34).

The second generation after independence of many non-European peoples has attempted to revive its own cultural history and values. European intervention in colonial societies, however, was so profound that it is virtually impossible to reconstruct what was there before. Much of what is today called "traditional" is in fact the result of European codification or impressions of rules. All modernizers around the world today face a problem. Many might like to recreate some idealized model of what their society was like in the past, but that is impossible. Modernizers also reject total conversion to Western ideas and standards. Therefore, they must produce some synthesis of civilizations. This requires difficult decisions and compromises. The Japanese, for example, adopted much of Western science and technology, but they have protected and maintained other aspects of traditional Japanese culture, such as an especially high regard for group effort and teamwork.

Many of the problems that countries face today—problems such as urbanization, pollution, and cultural confusion—are problems that Europe faced first but that European culture has been unable to solve. Some observers have spoken of the "exhaustion" of Western modernity, and the opening of the postmodern world. Europe may no longer be so central as it once was (Figure 3-35). Today more and more people everywhere are involved with more than one culture. Latin America is reaching out directly to Africa, Asia to Latin America. Perhaps world cultural diffusion will be less hierarchical in the future. More world societies might stop being passive receptacles of Western influence and become active shoppers in a global cultural

FIGURE 3-34

The extraordinary 1981 international hit movie *Lion of the Desert* was financed by the government of Libya in order to teach Libya's history from Libya's point of view, rather than Hollywood's. Anthony Quinn starred as Omar Mukhtar, who defeated Italy's attempted conquest of Libya from 1911 to 1931. Not all countries have been able to tell their own versions of history this way, either to international audiences or even to their own people.
(Courtesy of Film Stills Archive, Museum of Modern Art.)

bazaar, picking and choosing what they want, and then turning it into something of their own.

In our search for solutions to world problems we might look to the other world cultures. The worldwide revival of non-Western cultures is both a search for new cultural solutions to world problems and a revival of self-respect on the part of peoples whose cultures were overlain by European culture. All peoples will contribute.

SUMMARY

An isolated society depends entirely on its local environment for all of its needs, but the evolution of individual human cultures in isolation has, through

FIGURE 3-35

The strong, bold forms of this sculpture by Jacques Lipchitz (1891–1973) demonstrate that African art was one of the main influences on the twentieth century Western art movement called cubism. Lipchitz was one of the founders of cubism, along with Pablo Picasso and Georges Braques, all of whom acknowledged their artistic debt.
(Courtesy of Hirshhorn Museum and Sculpture Garden, Smithsonian Institution, Gift of Joseph H. Hirshhorn, 1966.)

history, yielded to the increasing interconnectedness of human societies.

The place where a culture originates is called the culture hearth. Various aspects of cultures may spread out and be adopted by other peoples in a process called cultural diffusion. The impact or frequency of any cultural attribute may diminish away from its hearth area; this phenomenon is called distance decay. Some aspects of culture may also develop variations as people who carry that culture wander outward and away from one another. Tracing a course of diffusion may teach us a great deal about how peoples and cultures interact. Diffusion may be by relocation, or it may be contiguous or hierarchical.

Barriers to cultural diffusion may be topographic, political, or even cultural. Diffusion does not explain the distribution of all cultural phenomena. Sometimes the same phenomenon occurs spontaneously and independently at two or more places.

Cultures diffuse, cultures change, and people can transform themselves and their behavior, but cultures and culture realms also have elements of stability. Any cultural pattern or distribution maps a current balance between the force of change and of stability. Cultures are constantly changing and evolving through time, and these changes may result either from local initiatives and developments or else as the result of influences from other places. Each local culture is a mix between what originated locally and what has been imported. Endogenous factors, or site factors, are elements of the specific local environment or of local cultural history and development. Exogenous factors or situation factors refer to the way a particular place interacts with other places. Endogenous factors of explanation and exogenous factors of explanation complement each other. As global communication and transportation have increased, the balance of factors that explain the local activities and culture at any place has tipped steadily away from endogenous factors and toward exogenous ones. What happens at places depends more and more on what happens among places.

Today's rapid pace of cultural innovation and diffusion requires us to make a distinction between folk culture and popular culture. Folk culture preserves traditions. Most folk culture groups are rural, and relative isolation helps these groups maintain their integrity, but folk culture groups also include urban neighborhoods of immigrants struggling to preserve their native cultures in their new homes. Popular culture, by contrast, is the culture of people who embrace innovation and conform to changing norms. Popular culture may originate anywhere, and it tends to diffuse rapidly, especially wherever people have time, money, and inclination to indulge in it. Here again, the number of aspects of this culture that geographers can study is virtually limitless.

In studying economies and trade, geographers note how the people at any place make their living, where specific economic activities locate, and how economic activities and trade affect other aspects of cultural geography. The role any place plays in systems of trade may depend on its situation or relative location. Accessibility can be a resource, and accessibility can change with changes in transport routes, technology, or the territorial scale of organization of any activity.

All human activities find territiorial expression, and maps of culture realms, regions of economic specialization, or political jurisdictions reveal current distributions of human activities. These activities, however, are dynamic, continuously organizing and reorganizing, forming and reforming. Therefore, their distributions and patterns are disrupted and reshaped repeatedly. Geographers need to understand what forces make things move and redistribute themselves.

Despite the rich variety of indigenous local cultures around the globe, the European cultural model is widespread. Europeans came to play a central role in world history and geography because they paved the way for the modern system of global interconnectedness. Global diffusion was fixed hierarchically, with Europe as the apex. Europe conquered most of the rest of the world, and European political domination imposed European concepts of government, law, property, and other aspects of culture. The spread of Western culture continues today, and many traditional cultures and social structures are being radically transformed. In some cases whole cultures disappear.

The second generation after independence of many non-European peoples has attempted to revive their own cultural history and values. Perhaps world cultural diffusion will be less hierarchical in the future.

KEY TERMS

evolutionism (p. 79)

diffusionism (p. 79)

cultural
  diffusion (p. 79)

acculturation (p. 79)

cultural hearth (p. 82)

culture realm (p. 82)

distance
  decay (p. 82)

gravity model (p. 82)

relocation
  diffusion (p. 85)

contiguous
  diffusion (p. 85)

hierarchical
  diffusion (p. 86)

folk culture (p. 90)

popular
  culture (p. 91)

inertia (p. 92)

infrastructure (p. 93)

historical
  geography (p. 93)

historical
  consciousness (p. 93)

economic
  geography (p. 94)

autarky (p. 94)

locational
  determinants (p. 95)

footloose activities (p. 98)

electronic highway (p. 100)

virtual reality (p. 100)

Commercial
  Revolution (p. 105)

Industrial Revolution (p. 108)

liquidity (p. 109)

Agricultural Revolution (p. 109)

cultural imperialism (p. 112)

growth poles (p. 116)

QUESTIONS FOR INVESTIGATION AND DISCUSSION

1. What were a few distinctive products of your region 50 years ago? Religious observances? Food products? Clothing or costumes? Architectural styles? Games? Have they been exported from the region? Are they still typically produced? Attend a local street fair or celebration. What aspects of the festival are different from such a festival 100 miles away?

2. For examples of distance decay, map the homes of faculty or classmates at your college or in your class or of customers at a local store.

3. Take a photo of a local landscape, and then analyze how different aspects of it reveal where it is—the dress of the inhabitants, vegetation, languages on signs, and so forth.

4. How many cities are served by nonstop flights from your town?

ADDITIONAL READING

BLAUT, J. M. *The Colonizer's Model of the World.* New York: The Guilford Press, 1993.

BRAUDEL, FERNAND. Civilization and Capitalism: 15th–18th Century. English translations published by Harper & Row, NY: Volume I: *The Structures of Everyday Life* (1981); Volume II: *The Wheels of Commerce* (1982); Volume III: *The Perspective of the World* (1984).

CURTIN, PHILIP D. *Cross-Cultural Trade in World History.* Cambridge: Cambridge University Press, 1984.

FEATHERSTONE, MIKE, ed. *Global Culture: Nationalism, Globalization and Modernity.* New York: Sage Publications, 1990.

FERNANDEZ-ARMESTO, FELIPE, ed. *The Times Atlas of World Exploration: 3000 Years of Exploring, Explorers, and Mapmaking.* New York: HarperCollins, 1991.

HUGILL, PETER J. *World Trade Since 1431: Geography, Technology, and Capitalism.* Baltimore, MD: Johns Hopkins University Press, 1993.

NOBLE, ALLEN G., ed. *To Build in a New Land: Ethnic Landscapes in North America.* Baltimore, MD: Johns Hopkins University Press, 1992.

UPTON, DELL, ed. *America's Architectural Roots: Ethnic Groups That Built America.* Washington, D.C.: National Trust for Historic Preservation Press, 1986.

WOLF, ERIC R. *Europe and the People Without History.* Berkeley: University of California Press, 1982.

ZELINSKY, WILBUR. *A Cultural Geography of the United States.* Englewood Cliffs, NJ: Prentice Hall, 1992.

4 Human Population: The Distribution and the Pattern of Increase

China is home to almost one quarter of all people on Earth, so the family planning policy it started in 1979 has significantly affected the overall rate of human population increase.

Courtesy of Doria Steedman/The Stock Market.

This chapter introduces the world's **population geography**, that is, the distribution of humankind across the Earth. Some areas are densely populated, but much of the Earth is virtually uninhabited. The relative distribution of humankind is constantly shifting as the result of two factors. One is the fact that different countries have different internal population dynamics, such as the numbers of births and deaths, and the consequent rates of increase or decrease of the total population. The second factor is human **emigration** out of some places and **immigration** into others. This chapter will examine human distribution and human population dynamics, and the following chapter will examine how migration has affected the distribution of peoples in the past and how continuing migration affects world affairs today.

In studying population geography we will occasionally include ideas and information about **demography**, which is the study of individual populations in terms of specific group characteristics. These characteristics may include the distribution of ages within the group, the ratio between the sexes, income levels, or any other characteristic. Demography means "describing people."

THE DISTRIBUTION OF THE HUMAN POPULATION

▼▼▼▼▼▼▼▼▼▼▼▼▼▼▼▼▼▼▼▼▼▼▼▼▼▼

In 1994, the population of the Earth was about 5.7 billion, but people are not evenly distributed across the landscape (see Figure 4-1). About 90 percent of the population is concentrated on less than 20 percent of the land area.

Figure 4-1 reveals three major concentrations of population: (1) East Asia, where the combined populations of eastern China, the Koreas, and Japan total almost 1.5 billion people; (2) South Asia, where India, Pakistan, and Bangladesh together account for another billion; and (3) Europe from the Atlantic to the Urals, with almost another billion. Southeast Asia forms a secondary concentration, with just under 500 million people; the eastern United States and Canada are home to another 175 million.

These five population concentrations, plus parts of West Africa, Mexico, and areas along the eastern and western coasts of South America, are densely populated, with more than 60 people per square mile (25 per square kilometer). More than half of the Earth's land area, by contrast, has fewer than 2 people per square mile (1 per square kilometer). A sur-

prising percentage of the Earth's land surface is virtually uninhabited: central and northern Asia, northern and western North America, and the vast interiors of South America, Africa, and Australia. The countries that occupy these spaces are enormous in area, but they contain relatively few people. Mongolia, for instance, is almost 4 times as big as the state of California, but Mongolia's 2.3 million people would fit into Los Angeles. The three African countries of Chad, Niger, and Mali together cover 39 times the area of South Korea, but their combined population is about half that of South Korea.

National Censuses

A national **census** is an official enumeration of the population, and national censuses usually include additional information relating to social and economic conditions. All of the census data presented in this textbook are the official reports of national and international agencies. Figure 4-2 on page 126 shows the official population of each country. There are weaknesses in some of these data, however, and a reader should be aware of them. Counting a country's population might seem to be a fairly easy task, but in fact throughout history—and in many countries today—it has been difficult or impossible. Most states counted their people only to record how many were on the tax roles—or ought to have been. The Canadian census of the late seventeenth century was probably the world's first census undertaken solely to count the population. The U.S. Constitution requires a decennial census (every 10 years) for the purpose of reapportioning seats in the House of Representatives. This was a revolutionary new idea in 1789.

The United Nations annually publishes the *Demographic Yearbook*, but the figures in it are, in many cases, only estimates. Some countries lack the administrative apparatus necessary to carry out a full census. Portions of the country may be inaccessible, and the population may be mobile. Portions of the population may avoid being counted out of superstition, distrust of the central government, or the wish to avoid taxation. In April 1990, one-third of American households failed to return their census forms.

Nigeria exemplifies a country in which politics militates against an accurate census. The census of 1991 counted 89 million Nigerians, but the count is politicized because the balance of population among the country's 30 states determines how much money each state gets from the federal government. The numbers dictate the political, ethnic, and religious balance of the country, and no one wants to be part of a well-documented minority. Nigeria's population is dominated by three broad ethnic groups: the Hausa-Fulani in the north, Yorubas to the west, and Igbos in the east. The last fairly reliable census, in 1952-53, gave the largest share of the population to the Hausa-Fulani. Consequently, when Nigeria received its independence in 1960, northerners dominated its first government. Subsequent censuses were fiascoes. The count of 1962 suggested that in the intervening decade the population of the east had increased by 71 percent, that of the west by 70 percent, and that of the north by only 30 percent. The government ordered a recount, and the north's increase magically rose to 67 percent. Most northern and western politicians agreed to accept that and to assume that the national population was 55.6 million, with 30 million in the north. A 1973 census demonstrated similar creative counting techniques, and thus the north has been dominant in every government. Southerners have accepted that, provided always that each region has a share of people in the nation's top jobs.

The 1991 census was the first deliberately to avoid questions of religious or tribal affiliation, and it arrived at a total of about 25 million fewer people than most international agencies had estimated. The government released the national totals in 1992, but it withheld regional data, and when they were released, they were challenged.

Even the U.S. census is politically contested. Each decade individual states and cities argue that their populations were undercounted in the census. They claim greater political representation and larger shares of federal funding.

The world's ten most populous countries contain almost two-thirds of all humankind: China (1,238 million people), India (931), the United States (263), Indonesia (201), Brazil (161), Russia (150; 266 million together with the other states of the Commonwealth of Independent States), Japan (126), Pakistan (135), Bangladesh (128), and Nigeria (89). Figure 4-2 is a world map drawn so that the size of each country reflects its population, not its land surface area. The most populous countries are shown as the biggest, and the least populous as the smallest, no matter how big or small their land surface areas are. Maps like this are called **cartograms**.

EXPLAINING THE PATTERN OF POPULATION DISTRIBUTION

Several factors combine to explain the distribution of the human population. These include the physical environment, history, and local differences in rates of population increase.

The Environmental Factor

It is often thought that a place's population density reflects the productivity of the local environment. Some scholars differentiate simple **arithmetic density**—the number of people per unit of area—from **physiological density**—the density of population per unit of arable land. Furthermore, we might assume that a fertile region could support a higher population density than a less fertile land; in other words, the fertile area has a high **carrying capacity**. In an infertile region the same arithmetic density could impose a greater strain on local resources.

In autarkic societies entirely dependent on local agriculture, physiological density may in the past have suggested national well-being. Today, neither arithmetic density nor physiological density fully explains the local standard of living anywhere or allows us to judge whether any region is overpopulated. Trade and circulation free societies from the constraints of their local environments and allow them to draw resources from around the world. Local carrying capacity has become less important in determining how many people can live in a given area or how well they can live. The physiological density of a highly industrialized country is irrelevant to its welfare. Some of the most densely populated areas in the world support some of the richest populations, while in others the people are poor. Conversely, people in some of the sparsely populated regions are rich, and in other sparsely populated regions the people are poor. No clear correlation between density and welfare can be drawn. Environmental factors that help explain population patterns are climate, topography, and soil.

Climate The clearest relationship between population distribution and environmental factors is that population densities are low in most of the world's cold areas and dry areas (compare Figure 4-1 with Figure 1-18 and Figure 1-20). The major exceptions to this rule are places where rivers flowing through dry areas provide water for irrigation, such as India, Pakistan, and Egypt (see Figure 4-3 on p. 129).

The Earth's coldest areas are inhabited only where mines or other special resources make it worthwhile to support populations at work there, as in parts of central Asia or high in the mountains of South America. People will settle in harsh areas if it is profitable to do so; the settlement in those areas demonstrates the points emphasized in the discussion of possibilism in Chapter 2.

Some warm and wet equatorial regions are also sparsely populated, such as the Amazon Basin in South America, the Congo Basin in Africa, and the

FIGURE 4-1

This map shows that large areas of the Earth's surface are surprisingly uninhabited. About 75 percent of the total human population lives in the Northern Hemisphere between 20 degrees and 60 degrees North Latitude, and even great expanses of that area are sparsely populated.

island of New Guinea. Despite the high biological productivity of these regions, conventional agriculture has not proved successful. This failure is attributed to nutrient-poor soil or to competition from other plants. Furthermore, heat multiplies biological activity, and this provokes another hazard—the flourishing of life forms hostile to humans and their agriculture. Insects such as mosquitoes, tsetse flies, blackflies, and sandflies thrive, as do the diseases they carry. In addition, these environments support countless forms of parasites, microbes, and fungi that weaken and kill humans, wilt and blight their plants, eat crops alive in the fields, or quietly feast on them in granaries or storerooms.

Heat also affects human productivity. Work generates heat, and the body has to lose excess heat to work efficiently. The productivity of manual workers decreases by as much as one-half when the temperature is raised to about 95°F (35°C), as is quite common in the tropics. The invention of the air conditioner, on the other hand, has made it more comfortable for people to settle in the warm regions of the United States, and this has transformed the country's population geography. Virtually all of the country's population growth since 1970 has taken place in the southern "Sunbelt." In 1992, an economist estimated that U.S. citizens had spent to that date a total of roughly one trillion dollars on air conditioners and an additional $25 billion each year on electricity to keep cool. This illustrates again how cultural and technical innovation can overcome the constraints that the theory of environmental determinism would suggest.

Tropical Asia is home to several great concentrations of people, but these concentrations are mostly in seasonal regions that are not wet year-round. Some of these, such as the island of Java or the Deccan Plateau of India, offer rich volcanic soils, and others are well-adapted to the cultivation of rice paddies (Figure 4-4, p. 129).

Rice terraces are extremely productive, yielding the highest number of calories per acre of any known crop. Paddy flooding plus intensive supervision and care limit the growth of competing vegetation, and algae in the water supply nitrogen as a plant nutrient. Furthermore, it seems almost always possible to increase the yield of a rice terrace by working it just a little bit harder, so rice paddies can

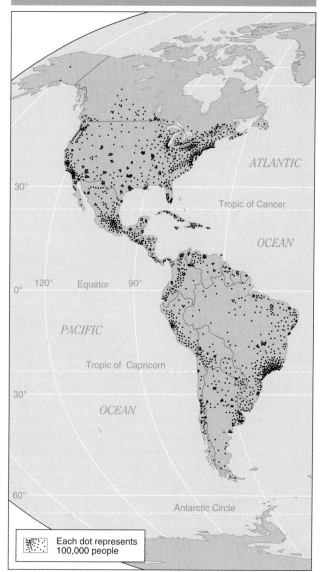

The Distribution of the Human

Each dot represents 100,000 people

support increasing numbers of cultivators on a given unit of cultivated land. In addition, rice terraces create a stable ecosystem. They do not break down the physical system.

In conclusion, most of the world's population is concentrated in areas of seasonal environments that are not too wet, too hot, too dry, or too cold. In fact, the plurality of the Earth's population lives in areas that are designated "C" climate on Figure 1-20.

Topography Topography often affects population distribution, although its effect is less prominent than that of climate. People tend to settle on flat lands because of the ease of cultivation, construction, and transportation. Thus, most of the densest concentrations of population in the world

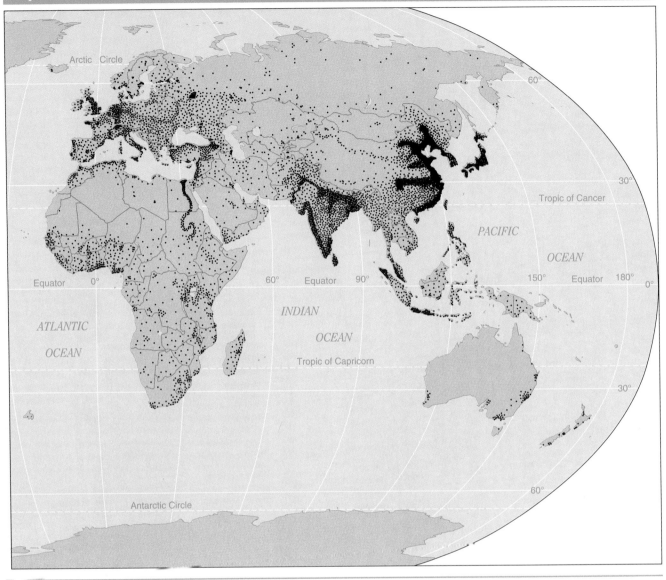

are found on level terrain (compare Figure 4-1 with the back endpaper map).

Flatness alone, however, is insufficient to attract a large population. Usually, flat lands that are densely populated are also characterized by other favorable environmental factors, such as fertile soil, available water supply, and moderate climate. By contrast, flat lands in central Siberia, western Australia, and central Brazil are relatively empty of population because other factors, such as the nature of the climate or soils, diminish the area's productivity.

Conversely, sloping land (hills and mountains) usually has a sparser population density. Steep slopes have thin, infertile soils, and agricultural opportunity is limited to the narrow valleys or the grazing of the slopes by livestock. Nevertheless, high density is found in many mountainous portions of South America, Japan, New Guinea, Southeast Asia, Central Europe, and elsewhere. Specialized attractions, such as valuable forest or mineral resources or recreational opportunities, often help explain such concentrations. Moreover, in most mountainous areas the settlements actually are concentrated in valley bottoms, even though they may be small and constricted. Thus, the connection between topography and population density is too inconsistent to validate a direct relationship.

Soils People usually settle areas of fertile and potentially productive soil unless there are powerful negative factors. Thus, most river floodplains are densely populated unless they are located in extremely cold or dry areas. Floodplain soils are typi-

Human Population: The Distribution and the Pattern of Increase **125**

A Cartogram of the World Population

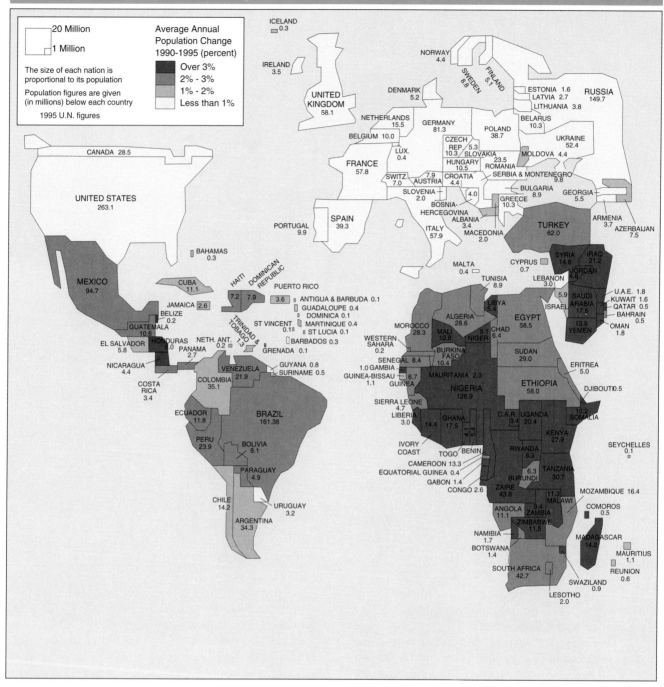

FIGURE 4-2

On this population cartogram the size of each country reflects its population, not its actual land area. On a population cartogram the whole African continent is smaller than India, because it is less populous. South America is also very small. The color coding conveys additional information; it indicates rates of annual population increase. Almost equal quarters of the human population live in countries in which the annual rate of increase is less than 1 percent; between 1 and 1.4 percent; between 1.5 and 2.1 percent; and over 2.2 percent. (Data from the World Resources Institute, *World Resources 1994-95.* New York: Oxford University Press, 1994, pp. 268-69.)

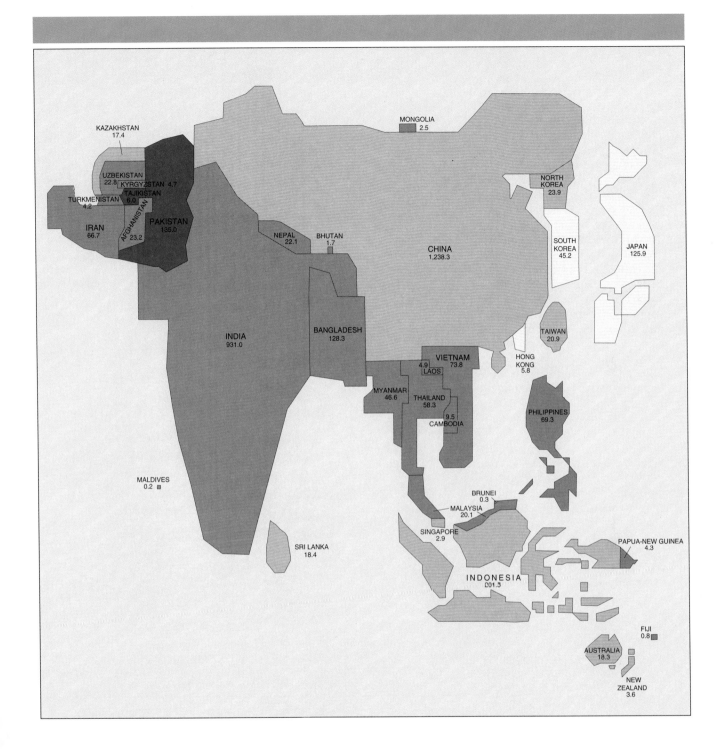

KAZAKHSTAN 17.4
UZBEKISTAN 22.8
KYRGYZSTAN 4.7
TAJIKISTAN 6.0
TURKMENISTAN 4.2
AFGHANISTAN 23.2
IRAN 66.7
PAKISTAN 135.0
NEPAL 22.1
BHUTAN 1.7
MONGOLIA 2.5
NORTH KOREA 23.9
SOUTH KOREA 45.2
JAPAN 125.9
CHINA 1,238.3
INDIA 931.0
BANGLADESH 128.3
TAIWAN 20.9
HONG KONG 5.8
VIETNAM 73.8
LAOS 4.9
MYANMAR 46.6
THAILAND 58.3
CAMBODIA 9.5
PHILIPPINES 69.3
MALDIVES 0.2
SRI LANKA 18.4
BRUNEI 0.3
MALAYSIA 20.1
SINGAPORE 2.9
PAPUA-NEW GUINEA 4.3
INDONESIA 201.5
FIJI 0.8
AUSTRALIA 18.3
NEW ZEALAND 3.6

FOCUS
Cartograms

Standard maps try to depict the Earth's features in their correct locational and size relationships. Sometimes, however, it may be more important to convey a visual impression of the magnitude of something than to convey a visual impression of exact spatial locations. To do this cartographers design special maps called cartograms. All cartograms replace physical distance with some other measure. The two main types of cartograms are *area cartograms* and *linear cartograms*. On an area cartogram, a region's area is drawn relative to some value other than its land surface area. Figure 3-13 was a cartogram on which world regions were drawn according to the amount of press coverage they receive in a major U.S. newspaper. In Figure 4-2, the countries of the world are drawn according to the relative sizes of their populations. The bigger a country is drawn on the cartogram, the greater is its population.

When a cartographer makes a conversion from physical space to something else, he or she tries to retain as many spatial attributes of the conventional map as possible. If recognizable shape, proximity, and continuity are preserved, it is easier for the viewer to compare the cartogram with a standard map. Each of these attributes of space can be retained, however, only by distorting one or more of the others.

Linear cartograms draw our attention to distance. The transit map, for example, is an example of a time-distance map. The subway map of Manhattan distorts the island's true shape in order to clarify the subway lines and stops. Subway riders think in terms of stops, not distances. Therefore, on many transit maps the distance between stops is shown as a uniform distance.

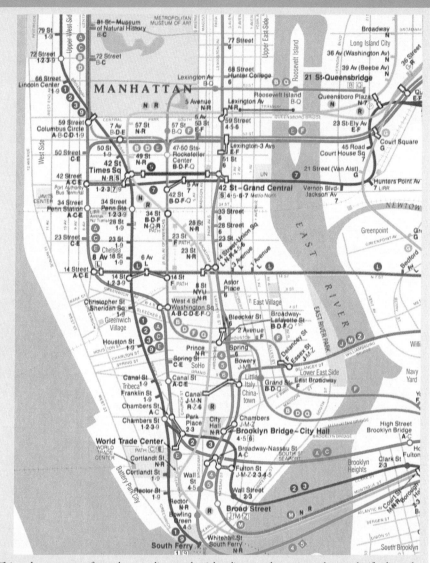

This subway map of manhattan distorts the island's true shape in order to clarify the subway lines and stops. Subway riders think in terms of stops, not distances. Therefore, on many transit maps the distance between stops is shown as a uniform distance. (Reprinted from the New York City Subway Map. © 1992 New York City Transit Authority.)

Cartograms may look strange to us because the scale depends on something other than the physical units we expect. Cartograms are not, however, lies or tricks. Their purpose is to portray some other important aspect of reality.

Area cartograms of such things as countries' populations (Figure 4-2) or economic output (Figure 12-2) visually convey the relative population or wealth of different countries better than standard land area maps do. The viewer must be certain to note when a figure is a cartogram. Most maps in this book are standard land-area maps, but the book also contains cartograms. These are especially useful when we want to illustrate human population dynamics or the conditions in which people live. A cartogram shows not only *where* people live in certain circumstances of health or wealth, but, at the same time, it shows *what share* of the total world population lives in those circumstances.

cally fine-textured and full of nutrients, so they can be used intensively for agriculture. The soils under the midlatitude grasslands are also usually rich, but these areas—including the U.S. Great Plains and stretches of Ukraine and Argentina—are often given over to large-scale grain farming, and population densities are not high. People do not necessarily avoid areas of poor soils, however, because some poor soils can be enriched with fertilizers.

History as an Explanatory Factor

History helps explain patterns of human settlement. The populations of China and the Indian subcontinent achieved productive agriculture and relative political stability thousands of years ago. These peoples domesticated many plants and animals very early. Intensive cropping and irrigation yielded food supplies adequate to support rising populations. Western Europe's population multiplied when the Europeans gained material wealth and improved their food supplies as the result of world exploration and conquest. Later, during the Industrial and Agricultural

FIGURE 4-3

This photograph illustrates why the ancient Greek historian Herodotus called Egypt "the Gift of the Nile." A thin strip of cultivable land winds through uninhabitable desert and then fans into a delta where it meets the Mediterranean Sea. Each year the Nile River floods and recedes, leaving a narrow plain of rich black mud. The photograph was taken by a camera held by an astronaut. (Courtesy of NASA.)

FIGURE 4-4

The construction and intensive cultivation of rice terraces such as these on the Indian Ocean island of Bali supports a very high human population density.
(Courtesy of Professor J. Markusse, University of Amsterdam.)

Revolutions, European productivity multiplied again. Migrants from Europe settled and brought European technological sophistication to more sparsely populated areas throughout the world. Thus, some of today's secondary population concentrations grew as extensions of Europe. These include eastern North America, coastal South America, South Africa, and Australia and New Zealand. It has been estimated that Europeans and their descendants increased from 22 percent of the world's population to 35 percent in the period from 1800 to 1930.

When we look at the maps of current world population distribution, we should remember that considerable population reductions have occurred in some areas. The arrival of Europeans triggered wholesale depopulation of some areas; for example, large proportions of the native American population

Human Population: The Distribution and the Pattern of Increase **129**

succumbed to disease and mistreatment after their first contact with Europeans. We cannot calculate what the total population of the Western Hemisphere, Australia, New Zealand, or the South Sea islands would be today, or its distribution, if the native peoples had not suffered these losses. Similarly, we cannot compute what the population of sub-Saharan Africa would be today if millions of Africans had not been drawn out as slaves.

Human population distribution can also be affected by human demarcation of territories, particularly political territories. Chapter 2 noted that new national governments restricted nomadic migration in the Sahel, and that action intensified the problem of desertification there. The governments on different sides of an international border may have different environmental policies, and these can drastically alter the carrying capacity of the environment for low-technology societies. Furthermore, a national government may manipulate the distribution of its population. It might subsidize or command the settlement of harsh territories in order to occupy them effectively, or it might establish settlements along its borders for defense. In Chapters 9 and 12 we will see how some countries, such as Brazil and Russia, plan and organize the settlement and economic exploitation of their vast sparsely-populated frontier regions.

WORLD POPULATION INCREASE

World population density is increasing. Five hundred years ago the Earth's population was around 300 million. About 1800, the Earth's population reached 1 billion; by 1930, it reached 2 billion; and by 1975, 4 billion. Today there are about 5.7 billion people, and their numbers are increasing at a rate of about 93 million per year (see Figure 4-5).

It is doubtful, however, whether the present rate of increase will continue into the future. In the second half of the 1960s the rate of growth of the world's population began to slow down. The average annual population increase fell from 2.06 percent in the period 1965-1970 to 1.73 percent for the period 1985-1990, and it has remained at about 1.7 percent per year since then. Because of past growth, however, the number of people added each year is still rising. In 1975 the annual addition was about 72 million. In 1992, it was 93 million. It is projected to peak between 1995 and 2000, at about 98 million annually.

Rapid population growth will continue to be the dominant feature of global demographics for at least the next 30 years. A **population projection** is a forecast of the future population, assuming that current

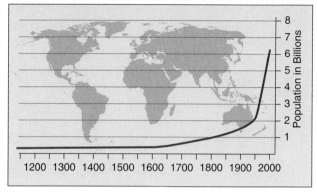

The Increase of Human Population

FIGURE 4-5

The rate of increase of the human population accelerated from about 1550 until about 1950, and then accelerated to a still higher rate of increase after 1950. The rate of increase may now be slowing.

trends remain the same or else change in defined ways. The smallest rise or fall in the percentage of increase today would increase or decrease the total population in the next century by hundreds of millions of people, but present projections made by the United Nations expect the global population to increase to 6.25 billion in 2000, 8.5 billion in 2025, and 10 billion in 2050. Significant growth will probably continue until the total human population reaches about 11.6 billion around the year 2150. Many people wonder whether it might be difficult to sustain the world's economy and political order with twice as many people as today.

Figure 4-2 shows how greatly the rate of population increase varies among countries. In each case, the population growth rate represents a balance among several demographic statistics. The **crude birth rate** is the annual number of live births per 1,000 people. The **crude death rate** is the annual number of deaths per 1,000 people. The difference between the number of births and the number of deaths is the **natural population increase or decrease**. For individual countries or regions, the natural increase or decrease can then be modified by subtracting the number of people who emigrated out of that area and adding the number of people who immigrated into that area. The result is the overall population growth or population decrease, and that figure can be expressed as a percentage of the total population. For example, the U.S. population increased by some 22 million people between 1980 and 1990, and that was an increase of 9.8 percent over the decade. Geographers also investigate each country's **fertility rate**, which is the number of children born per year per 1,000 females in a population, and the **total fertility rate**, which is the number of

130 Chapter 4

FOCUS
Statistical Projections

Statistical projections are statements of future possibilities based on current trends. Projections are not estimates, because the person who makes a projection states clearly that this is what *will* happen if what is happening now continues to happen at exactly the same rate. Nor is the person who makes a projection necessarily making a prediction, because he or she is fully aware and states clearly that any of the relevant variables may change.

For example, most people grow a few inches taller between the ages of 10 and 18. If we were to project that growth rate to middle age, we would project most people to grow considerably taller than they usually actually do grow. Nevertheless, it is accurate to project that if our friend George grew 5 inches taller between the ages of 10 and 18, to a height of 5 feet 9 inches, *and if he kept on growing at that pace*, George definitely would reach 7 feet tall by the age of 42. We know, however, that people generally stop growing by their early twenties. This example should be a warning that, in retrospect, some projections can seem downright silly. For example, one famous nineteenth-century prediction warned that, given the rising number of horses, Manhattan Island would be buried under several feet of horse manure by the year 1950. In other cases, publicizing a projection can change people's behavior, and in that way the projection can sabotage its own probability. Warning a spendthrift of how soon his money will run out, for example, may inspire him to slow his rate of spending.

No one can predict the future, and changes in the relevant variables may make the numbers of anything projected actually turn out much higher or much lower than projected. Projections assume that the future will bring no changes, and this assumption is the basic weakness of all projections; how can we assume that nothing will change in the future? History teaches that change is actually more normal than continuity. In the case of population studies, projections are based on current values for a great number of variables: birth rates, fertility rates, total fertility rates, death rates, and more. Any one of these variables may change, however, and changing interactions among them would cause the future to turn out very different from a projection. A sharp drop in total fertility rates would lower the total future population below projections; a sharp rise would greatly increase the future human population.

Some scholars project the number of years it would take any country's population to double at present rates of increase. That figure is called the country's **doubling time**. The annual rate of population growth for Uganda, for example, is 3 percent. Any number compounding at 3 percent annually doubles in just 23 years. A number compounding at only 1 percent annually, by contrast, which is the rate of population growth for the United States, takes 70 years to double.

children an average woman in a given society would have over her lifetime. A total fertility rate of 2.1 stabilizes a population, so this total fertility rate of 2.1 is called the **replacement rate**. If a country's total fertility rate falls below the replacement rate and the country does not experience immigration, its population will gradually fall. Each of these statistics varies considerably for different countries.

The Distribution of the Population Increase

The overall population projection conceals wide variations in the rate of population increase from region to region and country to country. For example, in 1993 the United Nations estimated annual growth for the period 1990-1995 to be 2.9 percent for Africa, 1.8 percent for Latin America and Asia, but only 1.1 percent for North America and 0.3 percent for Europe. If today's different regional rates of population increase continue into the future, the population variations between regions will widen. For example, in 1985 the populations of Western and Central Europe and of sub-Saharan Africa were about equal at 480 million, but by 2025 the population of Europe will be about 540 million, whereas that of sub-Saharan Africa will be 1,600 million. Today Africa has about 12 percent of the world population, but its share is projected to increase to 19 percent by 2025.

A significant variation in overall rates of increase can be noted between the rich countries and the poor countries. Ninety-five percent of the population increase projected to the year 2025 will be in countries that are today poor, those least able to support it.

Human Population: The Distribution and the Pattern of Increase　　**131**

If those countries hope to raise their incomes per capita (per person), they must achieve a rate of economic growth greater than their rate of population growth. Some 70 percent of the total projected population increase from 1989 to 2025 will occur in just the 20 countries listed in Table 4-1. In the rich countries of Europe and North America, by contrast, natural population growth has slowed or stopped altogether, and total fertility is at or below the replacement level.

The countries that have the highest rates of population increase have the lowest median ages. The median is the number above which half the examples in a set lie and below which half the examples lie; in other words, if a country's median age is 15, half the population is older than 15 and half is younger. Today about one-third of the world's population is under 15 years of age, but that proportion is as low as 15 percent in Italy and as high as 42 percent in Bangladesh (Figure 4-6). The median ages of the populations in many countries in Latin America, Africa, and the Near East are between 15 and 18 years. The relative percentages of a nation's population in each age group are often represented by a graphic device called a **population pyramid** (Figure 4-7).

TABLE 4—1
Projected Population Increases, 1989–2025 (in millions)

Country	Increase
India	592.2
China	357.1
Nigeria	188.3
Pakistan	144.4
Bangladesh	119.4
Brazil	95.4
Indonesia	82.7
Ethiopia (with Entrea)	65.6
Iran	65.6
Zaire	63.5
Mexico	61.5
Tanzania	57.5
Kenya	52.5
Vietnam	50.8
Phillipines	49.0
Egypt	39.9
Uganda	36.8
Sudan	34.4
Turkey	34.0
South Africa	28.0

In a country whose population has a low median age, a high percentage of the total population is drawing on national resources but has not yet reached its most productive working years. The national budgets of these countries are consumed by the challenges of feeding, clothing, and schooling youngsters. As the youngsters reach their mature working age, the national economy must be able to provide productive jobs for them, or else their frustration may break out in civil disorder. This underscores the need for economic development. In Mexico, for example, a million people enter the labor force annually. In the United States, with a population almost three times as great and an economy almost 30 times larger, the annual addition is only 2 million.

In contrast, the median ages of the populations of wealthy countries are higher. In Western Europe, for example, the median age is 34; in the United States it is nearly 33. These countries face a different challenge from that faced by the poor countries. Increasing percentages of their populations are either approaching retirement or are already retired. In these countries national economies with shrinking labor forces must be able to support increasing numbers of pensioners. The **dependency ratio** of a country suggests how many of its people are in their most productive years. It is defined as the ratio of the combined child population (less than 15 years old) and adult population over 65 years old to the population of those between 15 and 64 years of age. In the poor countries, this ratio may be as high as 80, while in the rich countries it is usually about 50.

Total Projected Population Increases

Even if worldwide crude birth rates and total fertility rates were to fall tomorrow, the total number of people on Earth would continue to grow. This is because the number of young women reaching childbearing age is larger than ever before. In Brazil, for example, the fertility rate has dropped 30 percent since 1965, yet the birth rate has dropped by only 19 percent, and the total number of births each year has risen from 3 million in the late 1950s to almost 4 million today.

The Earth's population will go on rising, but the sooner the world's total fertility rate falls to the replacement rate, the lower the total population will be at which the numbers eventually level off. Some people believe that the sooner the Earth reaches a constant population, **zero population growth** overall, the better. When will this happen? At what figure?

The United Nations offers a range of answers to this question. In 1993 the total fertility rate for the world was 3.3. If that drops to the replacement rate as soon as 2015, then the world's population will sta-

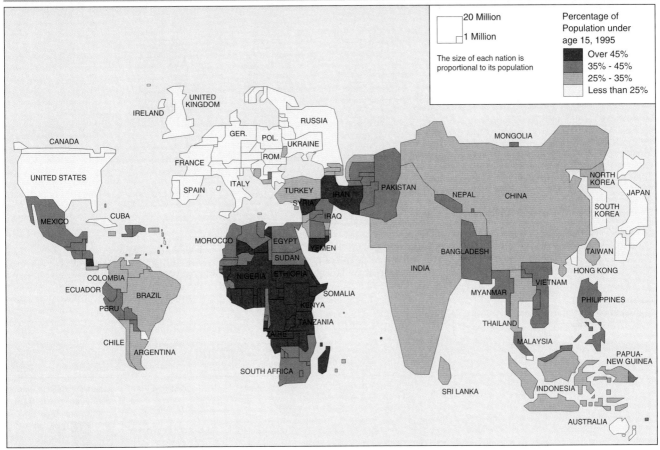

FIGURE 4-6

This cartogram, drawn on the cartographic outline of Figure 4-2, maps where the national populations are young. It also conveys a visual idea of what proportion of the world population is young. Comparison with Figure 4-2 suggests that populations are young where rates of population increase are high. (Data from the World Resources Institute, *World Resources 1994-95*. New York: Oxford University Press, 1994, pp. 270-71.)

bilize at about 8 billion. If the world reaches replacement total fertility by the year 2035, then its population will stabilize toward the end of the 21st century at 10.2 billion. If, however, the total fertility rate does not fall to replacement level until 2055, then the population will not stabilize until it has reached 13 billion. The difference between the two extreme estimates, 5 billion people, is almost the Earth's present population.

THE DEMOGRAPHIC TRANSITION MODEL

▼▼▼▼▼▼▼▼▼▼▼▼▼▼▼▼▼▼▼▼▼▼▼

Scholars who have studied population growth in the countries that are today rich have defined a model to describe these countries' historical experi-

ence. That model is called the **demographic transition**. The demographic transition was a pattern of growth that exhibited three distinct stages. In the first stage, both the crude birth rate and the crude death rate were high. Therefore, the population did not increase rapidly. All areas of the world experienced this stage in the past, when the human population was a fragile number in continual danger of actual extinction, locally if not globally, by periodic epidemics.

As medical science developed, however, these countries entered the second stage of the demographic transition. During this stage, crude death rates dropped dramatically (Figure 4-8). **Infant mortality**, which is the number of infants per thousand who die before reaching 1 year of age, usually fell almost immediately (Figure 4-9). Another factor that lowered death rates was the improvement in the

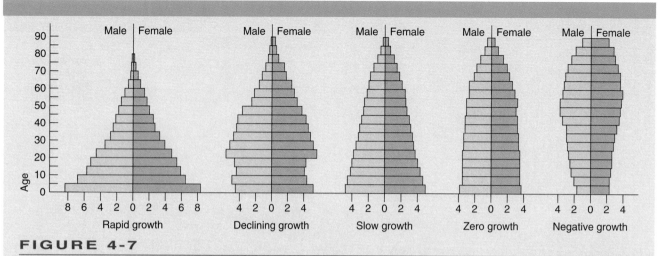

FIGURE 4-7

These are examples of a graphic device called a population pyramid, which shows the age and sex structure of a country's population. As any particular group of individuals of the same age (called a cohort) grows older together, the horizontal bar that represents that age group moves toward the top of the pyramid. The population pyramid suggesting rapid growth is typical of the poor countries, where birth rates are high and life expectancies are limited. Broad bases indicating many children taper to narrow tops of fewer older people. The population pyramids of the rich countries, by contrast, resemble the example marked "Negative growth." In them, both crude birth rates and crude death rates are low, and life expectancies are long. The image more closely resembles a column than a pyramid.

Crude Death Rates

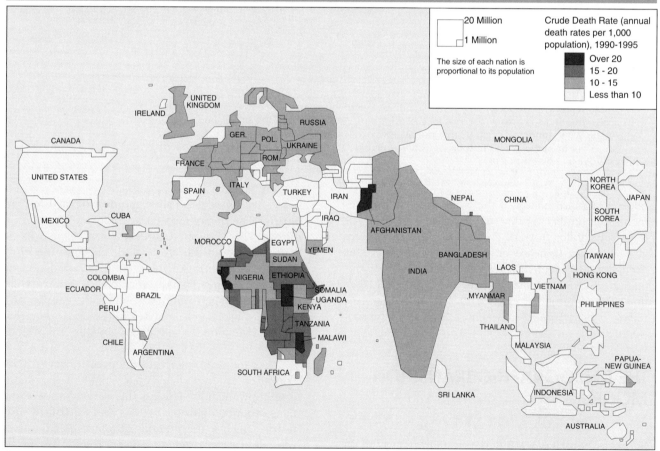

FIGURE 4-8

Crude death rates are falling almost everywhere, but this cartogram reveals that they remain high in some countries that are home to large numbers of people. Environmental disasters, local famine, epidemics, or war can still substantially increase a country's crude death rate. (Data from World Resources Institute, *World Resources 1994-95*. New York: Oxford University Press, 1994, pp. 272-73.)

Infant Mortality Rates

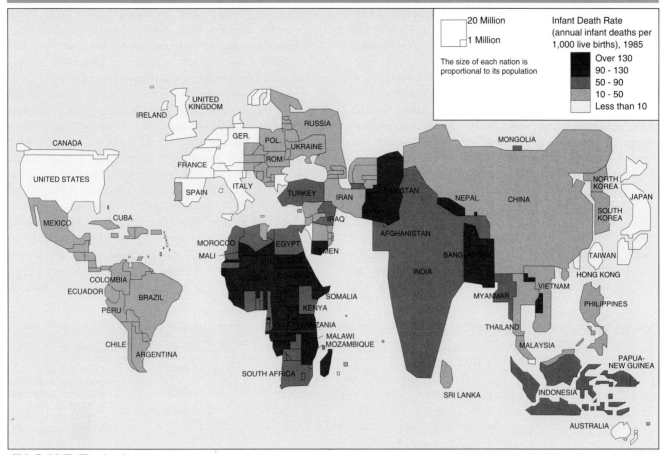

FIGURE 4-9

This cartogram shows that tragically high infant mortalities still plague many populous countries. (Data from World Resources Institute, *World Resources 1994-95*. New York: Oxford University Press, 1994, pp. 272-73.)

quantity and quality of food that resulted from the Agricultural Revolution (discussed in Chapter 6). As crude death rates dropped, however, crude birth rates remained high. Thus, in this second stage of the demographic transition, the rates of natural increase were very high.

Several theories were offered to explain the persistence of high birth rates during the second stage of the demographic transition. Children are economic assets in traditional, largely rural societies; they provide more hands to help in the fields. Also, people traditionally expected their children to look after them in their own old age. When infant mortality rates were high, parents had many offspring to ensure that some would survive into adulthood. When infant mortality rates fell, the next one or two adult generations still did not believe that their children would survive, so they continued to have many children.

The rich countries today demonstrate a third stage of the demographic transition. Their crude

death rates remain low, but their crude birth rates have also dropped. Total fertility rates are at or below the replacement level. This decline in childbearing occurred along with economic growth, urbanization, and rising standards of living and education. Today few rich countries have rates of natural population increase higher than 1 percent.

Once the current generation has passed in many rich countries, those countries' populations might actually fall in each succeeding generation. Italy, for example, has the lowest total fertility rate in the world (1.3), and at current rates its population will begin to decline early in the 21st century. Reasons cited by Italians for having fewer children typify the social changes in the rich countries: Women are pursuing careers, day-care services are limited, urban housing is short, and abortion has been legalized. As this demographic transition model has been described in the rich countries, the transition took place over many decades.

Is the Model Relevant Today?

In the period from about 1950 to 1980, scholars studying world population believed that the demographic transition model would eventually apply to all countries of the world. They noted that birth rates and total fertility rates were still high in the poor countries of the world. They described these countries as being in the second stage of the demographic transition, and they concluded that when these countries achieved economic development, they, too, would enter the third stage. Therefore, the 1974 World Population Conference in Bucharest, Romania, concluded that "development is the best contraceptive."

Several characteristics of the demographic transition model do apply in today's poor countries. Crude birth rates and total fertility rates remain relatively high (Figure 4-10). The average total fertility rate for the countries of West Africa, for example, is 6.5. In many of these countries it remains economi-

cally rational for the poor to have more children—for all the same reasons as it was in the past for the populations of today's rich countries (Figure 4-11). It can even be argued that some African countries face labor shortages in agriculture, and because they lack capital for agricultural machinery, increases in rural labor forces would increase food output.

Furthermore, in even the world's poorest countries, infant mortality rates are dropping because antibiotics and immunization today help many babies to survive who would have died in past years (Figure 4-12). Modern medicine also keeps people alive longer, and the combined effect of lowering infant mortalities and keeping people alive longer expands **life expectancies**, the average number of years that a newborn baby within a given population can expect to live (Figure 4-13). Today life expectancies in even the poor countries are lengthening fast. For example, today life expectancy in Mexico, a poor country, is already higher than life

Crude Birth Rates

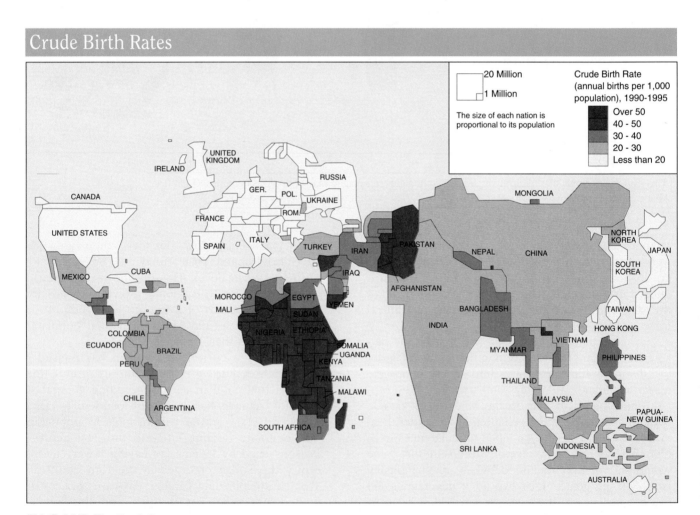

FIGURE 4-10

This cartogram illustrates that birth rates remain high among a large share of the Earth's population. Today's rich countries generally have low rates, but the poor countries still show high rates. Compare this figure with Figure 4-7. Countries with high birth rates have young populations. (Data from World Resources Institute, *World Resources 1994-95*. New York: Oxford University Press, 1994, pp. 270-71.)

FIGURE 4-11

Children can be an economic asset if they are put to back-breaking labor in the fields or rice paddies at young ages, as in this photograph taken on the Indonesian island of Java. (Courtesy of the Agency for International Development.)

expectancy was in France as recently as 1950, and France is a rich country clearly in the third stage of the demographic transition. The combined effect of lowering infant mortalities and lengthening life expectancies is to increase total populations.

Some countries have at times viewed population as a potential source of strength, and they have tried to increase the growth rates of their populations. In the 1970s, for example, the government of Iran encouraged Iranians to have more children. Each child was called "an arrow in America's heart." The population did grow rapidly, but soon the government realized that a rising population would inevitably be a problem for the future, and by the early 1990s it was trying to lower the country's population growth rate.

Therefore, some circumstances in today's poor countries are comparable to the second stage of the demographic transition as it was experienced in the countries that are today rich.

NEW DEVELOPMENTS ARE CHANGING DEMOGRAPHY

▼▼▼▼▼▼▼▼▼▼▼▼▼▼▼▼▼▼▼▼▼▼▼▼▼

By the time of the 1994 opening of the Third UN Conference on Population, in Cairo, Egypt, most experts recognized that two conditions were forcing them to abandon the demographic transition model to describe what is happening in the poor countries

today. First, in today's poor countries total fertility rates are dropping much faster than they did in the countries that are today rich. In the 20 years from 1970-75 to 1990-95, the world total fertility rate fell from 4.5 to 3.3. In Asia the decline was from 5.1 to 3.3; in Africa the decline was from 6.6 to 6.0, and in South America the decline was from 4.6 to 3.2. Demographers watch total fertility rates closely because that statistic is a good indicator of what the future will bring. Second, today falling total fertility rates are not necessarily associated with improving incomes and standards of living. They are falling even in countries where the economy is stagnating or even declining. Bangladesh, for example, is one of the world's poorest and most traditional agrarian countries. Infant mortality is high, women have low social status, and most families depend on children for social security. Nevertheless, the total fertility rate dropped from 7 to 5.5 between 1970 and 1991. Per capita incomes in many African economies actually fell during the 1980s, but so did total fertility rates.

Today's decreases in total fertility rates in the poor countries are fundamentally different from those that occurred generations ago in today's rich countries, and new causes must be found to explain them. Theories of explanation focus on the growing role of family planning programs, new contraceptive technologies, and the educational power of mass media.

Family Planning Programs

During the demographic transition, the concept of family planning was not quickly accepted. Early lead-

FIGURE 4-12

These Haitian children are getting protection against measles. Such health measures greatly reduce the mortality rates among children and can trigger high population growth. (Courtesy of the Agency for International Development/ John Metelsky.)

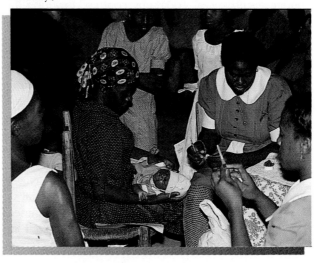

Human Population: The Distribution and the Pattern of Increase　　**137**

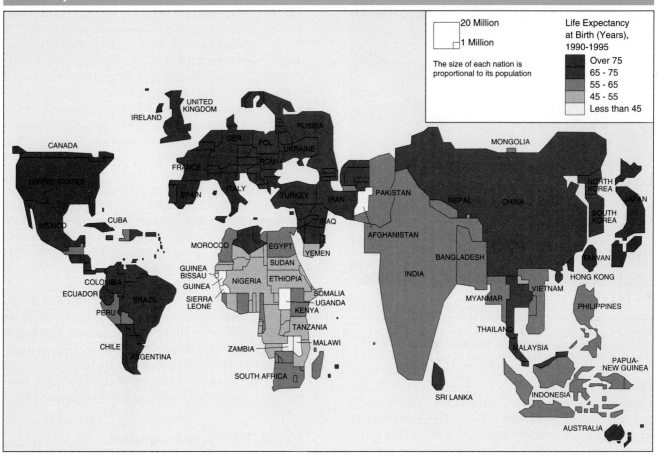

FIGURE 4-13

Life expectancies are lengthening everywhere, but much of the human population still cannot be expected to reach what we would call "old age." Compare this cartogram with Figures 4-8 and 4-9. Countries with high death rates have low life expectancies. (Data from World Resources Institute, *World Resources 1994-95*. New York: Oxford University Press, 1994, pp. 270-71.)

ers in the population control movement in Europe and the United States, such as Margaret Sanger (1883-1966), were even arrested for immorality or creating public disturbances. Today, by contrast, lowering the crude birth rate has become the focus of most countries' efforts to reduce overall population increases. According to the United Nations Sixth Population Inquiry, conducted in 1990, some 90 percent of the peoples of developing countries live under governments committed to reducing the rate of population growth (Figure 4-14). Today national public services seek clients and have removed or lowered many of the barriers to the availability of contraception.

A traditional cultural preference for large families is strong in some areas, and, as emphasized above, having children may even be the economically rational choice for some families in poor countries. Surveys have shown, however, that sizable portions of the populations of most poor countries want small-

FIGURE 4-14

This Indonesian government billboard tries to reduce traditionally high family sizes by suggesting that smaller families can be healthier, happier, and better off.
(Courtesy of the Agency for International Development.)

138   *Chapter 4*

er families than have traditionally been normal, and demand for family planning services has risen rapidly. The use of contraception among women in the poor countries rose from 10 percent in 1960 to over 50 percent in 1990. Some low-income areas like Sri Lanka and the populous Indian states of Kerala (population about 35 million) and Tamil Nadu (population about 58 million) have lowered fertility and at the same time have improved noneconomic indicators of quality of life such as literacy, infant mortality, and life expectancy. These societies' common approach to development includes universal access to education, health, and family planning, combined with relative equality in status, employment opportunities, and full political rights for women.

Contraceptive Technology

During the demographic transition, modern contraceptive methods were not yet invented. Today, however, modern technology has provided new means of birth control. As access to newer forms of contraception has spread, couples have become less likely to depend on traditional methods such as periodic abstinence and withdrawal.

Sterilization is the most widespread form of birth control in the poor countries. Some 98 million women and 35 million men around the world had chosen that solution by 1990. Several drugs and devices offer new, easier, and more certain strategies. The drug RU486, for example, induces a miscarriage if used in the early stages of pregnancy. The French government is part owner of the patent, and it has referred to RU486 as "the moral property of women." Still another recent invention is a steroid-filled capsule embedded in a woman's arm. It provides contraceptive protection for 5 years. The injectable contraceptive Depo-Provera prevents pregnancy for 3 months. More than 30 million women have been using it worldwide since 1969, although the U.S. government approved it for use only in 1992. Female condoms, which have been available since the mid-1980s, fail to prevent pregnancy at a higher rate than male condoms do, but they give women control over exposure to sexually transmitted diseases. Use of each of these drugs and devices is spreading, but substantial unmet demand for each has been shown to exist in many developing countries.

These new techniques are lowering reliance on abortion. About 28 million abortions are performed annually in the poor countries, another 26 million in the industrial world. The World Health Organization (WHO, a United Nations agency) estimates that improperly performed abortions kill about 300,000 women around the world each year.

The Role of the Mass Media

During the demographic transition, information about contraception diffused slowly because educational levels were low and mass communication was limited. Today, widespread communication and the influence of mass media in developing countries have accelerated the diffusion of novel ideas about family planning in both urban and rural environments. Chapter 3 already noted that modern television programs, for example, detach sexuality from childbearing, and their diffusion has been credited with lowering fertility rates in several countries.

In many countries, mass media campaigns have been shown to shape family planning decisions. The governments of Ghana, Nigeria, Gambia, Zimbabwe, and Kenya, for example, all broadcast family-planning messages on both radio and television.

Enhanced education increases the probability of the use of contraceptives, but it is not a necessary condition. In most of the developing world over the past two decades, the use of contraception has risen among all educational groups, and the gap in the prevalence of family planning according to educational level has narrowed.

Enhanced information has almost everywhere undermined traditional preferences for large families. This has not necessarily come about in response to improved living circumstances, as the demographic transition model suggests. In some cases, in contrast, this has come about in response to economic contractions. In most African countries, for example, incomes actually fell through the 1980s. For the first time, many families found that their standards of living fell, and, in response, they decided to limit their families.

Obstacles and Opposition to Population Control

Despite technological advances, family-planning programs face obstacles. Manufacturing, distributing, and teaching people how to use contraceptive devices is expensive. The World Bank noted in 1993 that worldwide prices of condoms ranged from $2-3 per 100 in China, Egypt, and Tunisia to over $70 per 100 in Brazil, Venezuela, and Burundi. To some degree, the high prices were caused by nations' import restrictions or inefficient marketing systems. Two-thirds of all condoms, diaphragms, and sponges are still used in the industrialized world.

Organized religion can block birth control programs. Speaking in Mexico in May 1990, Pope John Paul II reiterated the Roman Catholic church's position: "If the possibility of conceiving a child is artificially eliminated in the conjugal act, couples shut themselves off from God and oppose His will." The Roman Catholic church also adamantly opposes abortion. Some other religions also preach against specific forms of birth control.

Another obstacle to birth control is the low status of women. In numerous societies women lack political and economic rights, have limited access to education, and in general exercise little control over their own lives. The only way for them to gain status is to have children—especially male children. The United Nations estimates that fully half of the married women in the poor countries do not want any more children, yet many have little or no access to methods of birth control. For example, in Pakistan, Bangladesh, and Arab countries where women suffer low status, crude birth rates remain high. In contrast, on the Indonesian island of Java, where women enjoy greater rights and better education than in many other Muslim regions, crude birth rates are falling.

In culture realms where male children are preferred to females, birth rates frequently remain high because if a couple's first child is female, they will continue having children until a male child is born. In these traditional societies, a girl is often regarded as just another mouth to feed, a temporary family member who will leave to serve her husband's kin. Also, in many societies a woman's parents must provide her with a dowry, which is a substantial financial settlement at the time of her marriage. A son, on the other hand, means more muscle for the farm work, someone to care for aged parents, and someone who, in several Asian and traditional African religions, can burn the necessary ritual offerings to ancestors. A son carries on the family name. The financial preference for male children may be reduced in societies in which women are winning the right to work outside the home.

In societies in which men believe that the number of their children is a measure of their masculinity, family planning relies on contraceptive techniques that do not involve men's participation or even knowledge. In Mexico, for instance, an estimated 60 percent of women who receive government-sponsored birth control do so secretly.

In cities, large families offer the advantage that someone in the family might find employment in the daily or casual labor market discussed in Chapter 9. Overall, however, fertility rates are generally lower in urban areas than in rural areas. Moreover, the larger the city, the more likely women are to use contraception. Family-planning services may be easier to reach. Children cost more to raise in big cities, and space is at a premium. Roughly half of the world's people will live in cities by 2000, and this will probably lower total fertility rates.

Birth Control Programs: Some Examples The most dramatic and significant change in fertility has occurred in China, where the annual rate of natural population increase dropped from 2.9 percent to 1.4 percent between 1965 and 1990. Because China is home to about 22 percent of the human population, that decline alone accounts for much of the change in world trends. In 1979, China launched a one-family/one-child policy with the aim of limiting the population to 1.2 billion by the year 2000. The government offered incentives to reduce childbearing, such as financial rewards and special privileges for small families, and penalties for exceeding the targets. By 1992, the government announced that the national total fertility rate had dropped to 1.73. Unfortunately, China's policies have perpetuated traditional prejudice in favor of males. Women are allowed to have a second child only if the first is a girl, but sterilization is obligatory after the second child. Women have been coerced into having abortions, and there have been reports of *female infanticide*, the killing of female children. China is determined to continue to reduce the crude birth rate, but the total population will nevertheless reach 1.27 billion by 2000.

India's population has grown from 342 million at independence in 1947 to 880 million by 1992. If India maintains its growth rate, it will pass 1 billion by the end of the century and overtake China as the world's most populous country by the middle of the next century. Variations in crude birth rates within India reflect those around the world. Family sizes are smallest in regions where education and social status of women are high, and population growth is concentrated in the areas with the lowest levels of education, especially the northern states where literacy among women is as low as 5 percent.

In the mid-1970s, a government-sponsored family planning drive was given coercive powers, but, after widespread reports of forced sterilizations, the government backed down. Today radio and television advertising and information have brought to all Indians the message that small families are healthier and happier. Sterilization accounts for about 90 percent of India's program. The Indian social security system guarantees financial security to people in their old age, and it has been found that when people are convinced that they will be taken care of, they stop having many children.

Recent trends in Zimbabwe, Kenya, and Botswana illustrate why fertility is falling in tropical Africa. Those three countries have increased contraceptive use dramatically. In Botswana few couples used contraception 10 years ago; today around 27 percent do so. In Zimbabwe the figure is 40 percent.

140 *Chapter 4*

Kenya has long had one of the highest rates of natural population increase in the world, estimated in the 1980s at 4.1 percent per year, but even Kenya has recently reduced its crude birth rate. Key factors in Kenya seem to have been the government's promotion of family planning, a well-developed rural health system that can deliver contraceptive services, and the widening realization that families are financially better off with fewer children.

Brazil's total fertility rate has been cut in half in one generation. By the year 2000, Brazil will probably have 170 million people, 50 million fewer than demographers predicted in the 1970s. The falling fertility rate has been attributed to the spread of contraceptive devices and to access to television. More Brazilian households have television sets than refrigerators (a situation actually quite common around the world), and television transmits images, attitudes, and values of modern, urban, middle-class existence in which families are small, affluent, and consumer-oriented.

Will the Decline in Total Fertility Rates Continue?

Lack of women's rights and the preference for sons throughout Asia and Africa still prevent total fertility rates from declining to replacement levels. Even in the rich countries, the projections that declining total fertility rates will lead to overall population decreases may be premature or exaggerated. In several advanced countries, rates have recently risen. For example, total fertility rates in Sweden sank to a low of 1.6 in 1978, then rose to 2 in 1989 and 2.1 in 1990. Similar though smaller rises have been noted in Denmark, Norway, and Iceland. Even in Germany the rate climbed from its record low of 1.28 in 1985 to 1.5 in 1990. In the United States too, the rate has been creeping upward, to 2.

The explanation for these trends could lie in the growing ability of women to earn their own living outside the home. Sweden offers the most extensive day-care and parental leave benefits in the world. In the United States some employers are offering day-care programs, and women's salaries are rising. When child-care costs drop as a percentage of women's wages, the fertility rate rises.

The population dynamics in the United States make it an anomaly among rich countries. It still welcomes more immigrants each year than do the rest of the industrialized countries combined (a topic discussed in the next chapter). That raises the country's rate of population increase. In addition to the sheer number of immigrants, another important factor is that fertility is about 25 percent higher for foreign-born than for native-born women. Therefore, in 1992 the U.S. Census Bureau projected that the United States would be the only major industrial country whose population will continue to rise—to at least 383 million by 2050.

SEX RATIOS IN NATIONAL POPULATIONS

It is biologically natural for about 105 or 106 male children to be born for every 100 females. This ratio occurs everywhere in the world except in countries where amniocentesis and abortion are available and where there is a strong preference for male children. The demographic impact is dramatic. In South Korea, for example, where fetal testing to determine sex is common, male births exceed female births by 14 percent. It must be assumed that female fetuses are aborted at a higher rate than male fetuses.

In China, the male-female birth ratio has changed dramatically as sound scanners have become available to check the sex of a fetus. In the censuses of 1954 and 1962 the ratio was the expected 104 and 105; by 1982, however, it rose to 107.5, by 1990 it was 113.8, and by 1992 it had reached 118.5. These ratios suggest that some 12 percent of all female fetuses, or about 1.7 million female fetuses, were aborted or are unaccounted for. In Taiwan it is illegal to use amniocentesis to check the sex of a fetus, but from 1985 to 1990 the male-female birth ratio changed from 106.7 to 110.3. Some states of India have limited or banned the use of sound scanners and amniocentesis.

Although it is natural for more males to be born than females, women, given similar care, tend to live longer than men. Therefore, the natural ratio between females and males in a human population should be 1 to 1, or slightly higher. The United Nations estimated in 1990, however, that, worldwide, men outnumbered women. Figure 4-15 shows how the national ratios of women to men vary around the world.

What Determines National Sex Ratios?

 Some scholars who have looked at Figure 4-15 have noticed that the ratios of females to males are high in the rich countries of North America, in Western Europe, and in Japan. They have hypothesized that sex ratios vary with national wealth. Almost all countries with deficits of women are more or less poor, and few rich countries have such a deficit. Rising national wealth might

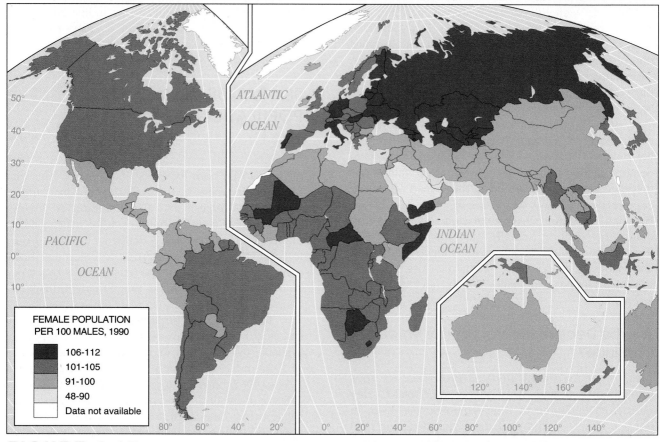

Female to Male Population Ratios

FEMALE POPULATION
PER 100 MALES, 1990

- 106-112
- 101-105
- 91-100
- 48-90
- Data not available

FIGURE 4-15

The ratios of females to males in populations vary greatly around the world. What factors could explain these great variations? (Data from *Handbook of International Trade and Development Statistics*. New York: United Nations Publications. Reprinted with permission.)

be expected to reduce female mortality for two reasons, both of which follow from the fact that women's mortality at childbirth is potentially high everywhere. First, birth rates are generally lower in richer countries, thus reducing maternal mortality. Second, improved medical care is generally available in richer countries during that critical period.

This hypothesis fails, however, when one examines the ratios in the poor countries. These countries do not show a consistent pattern. Some poor countries do not have deficits of women. Tropical Africa, for example, ravaged by poverty, hunger, and disease, has a substantial excess rather than a deficit of women.

In fact, variations in sex ratios can be explained only by a combination of national economic and cultural factors. In the countries of North America and Europe and in Japan, women may suffer many kinds of discrimination, but they do not seem to suffer in access to medical care. Moreover,

in these areas social and environmental differences tend to increase mortality among men. In most of Asia and Africa, however, women do not get the medical care, food, or social services that men get. As a result, fewer women survive than would be the case if they had equal care. In India, for example, the death rate is consistently higher for women at almost all ages except immediately after birth. A 1990 UNICEF study found that Indian girls are not fed as well as boys and get less medical attention.

Great variations exist within each world region. In Asia, Japan is at European status, and in some countries of Southeast Asia women outnumber men. In South Asia, however, the ratios of women to men are low. India and Pakistan exhibit two of the lowest ratios in the world, except for those Middle Eastern countries with large numbers of immigrant single males.

Sharp contrasts exist even within individual countries. In the northern Indian states of Punjab and

142 *Chapter 4*

Haryana, for instance, which are among the country's richest, the ratio of women to men is a remarkably low 0.86, whereas the state of Kerala in southwestern India has a ratio higher than 1.03, similar to that in Europe, North America, and Japan.

In China, the status of women rose after the Communists came to power in 1949, but most recently it seems to be declining. This may be the result of the birth control drive and the one-child policy. The preference for boys may be unfavorably affecting the survival prospects of female children, possibly explaining a rise in the female infant mortality rate. China's 1990 census revealed a female/male ratio of 0.938, but for children under 1 year of age the ratio was 0.898.

It seems almost universally true that women's positions are generally more favorable if they can earn an income outside the home. That reduces boys' advantage as potential supporters in the parents' old age, and females suffer less relative deprivation from birth. Later in life working women can rely on their own resources. A lot more needs to be learned about attitudes toward female employment outside the home.

Some demographers hypothesize that the societies that limit women's rights do not count all women in the census. If this undercounting of females actually occurs, then the resulting statistics would magnify the sex ratio imbalances or even report gaps where they do not in fact exist. Certainly women are better off in countries that acknowledge female deprivation and seek to remedy it. Efforts to do this can be influenced by education and suffrage.

The sharp contrasts in female/male ratios among national populations examined here testifies to widespread restriction of women's rights. National birth rates, as we have seen, may be affected by favoritism of male children and by the degree of education that women receive. Throughout this text we will return repeatedly to examples of discrimination against women, and we will see how this discrimination affects almost every aspect of human geography. In Chapter 6, for example, we will see how discrimination against women can reduce food supplies in Africa and other regions in which women are responsible for raising most food. In Chapter 8 we will see the differing attitudes toward women taught by various religions. Ramifications of those teachings reach from the issue of priesthood for women into the treatment of women in national laws. Chapter 10 notes the variations in the percentages of males or females in school around the world, women's role in politics, and other issues of legal rights. The question of gender justice is winning new attention around the world, and variations in the geography of this issue will be noted in several instances.

THE GEOGRAPHY OF HEALTH

Worldwide, birth rates and total fertility rates are falling. The long-term effect of those developments tends toward falling rates of population increase. We cannot conclude our discussion of world rates of natural population increase, however, until we have examined other factors in the equation, that is, global health and global death rates. Global death rates are currently falling, and they will probably continue to fall. More people staying alive longer will put upward pressure on the rate of natural population increase. The changes in world rates of death and disease are discussed below.

Changes in World Death Rates

Changes in the death rates mapped in Figure 4-8 could drastically alter the rate of human population increase. Natural disasters, wars, malnourishment, and epidemics all affect world death rates (Figure 4-16). Natural disasters claim tens of thousands of lives each year. A tropical storm in Bangladesh in May 1991 took as many as 200,000 lives. War also tragically increases the death rate. Iraq, a nation of some 19 million people, suffered the loss of an estimated 120,000 young male soldiers in its war with Iran between 1980 and 1988, and tens of thousands more in its 1991 war with United Nations forces. Those losses will depress the nation's birth rate and affect its demographic balance

FIGURE 4-16

The little girls playing by the coffin store on the Indonesian island of Java seem to be unbothered by the reminder that local death rates are high.
(Courtesy of the Agency for International Development.)

Human Population: The Distribution and the Pattern of Increase **143**

for decades. The victims of war also include noncombatant civilians.

World death rates would fall if the people of many countries could be better nourished. The geography of world food production and of nutrition will be examined in Chapter 6. In fact very few people anywhere starve to death, but millions become weak and succumb to diseases that a well-nourished individual would survive. The most significant alterations in the world's death rates, however, are expected to continue to result from changes in the world geography of disease.

The Epidemiological Transition

As today's rich countries evolved from the second stage of the demographic transition to the third stage, their populations grew older and richer, and the main causes of ill health and death changed. Today more people in them die from degenerative diseases such as heart disease, stroke, and cancer, and fewer die from infectious and parasitic diseases. The study of the incidence, distribution, and control of disease is called **epidemiology**, so this transition in the leading causes of death in a society is called the **epidemiological transition**.

Diseases may be grouped into three broad categories:

1. *Infectious diseases*, caused by the entrance, growth, and multiplication of foreign organisms within the body.
2. *Chronic* or *degenerative diseases*, which result from the long-term degeneration of the body.
3. *Genetic* or *inherited diseases*, caused by characteristics of the chromosomes and genes inherited from our parents.

As a result of the epidemiological transition, the principal types of disease, and hence the principal causes of death, are very different in the rich countries and the poor countries (Figure 4-17). In the developing countries as a whole, infectious and parasitic diseases account for almost half of all deaths, nearly all of which occur among children under 5 years of age. Modern medical science could lower the death rates in these countries significantly. In the rich countries, by contrast, circulatory and degenerative diseases are the main killers. Lowering death rates in these countries becomes increasingly difficult, because combating these diseases usually requires changes in the behavior of adults, who are the main sufferers.

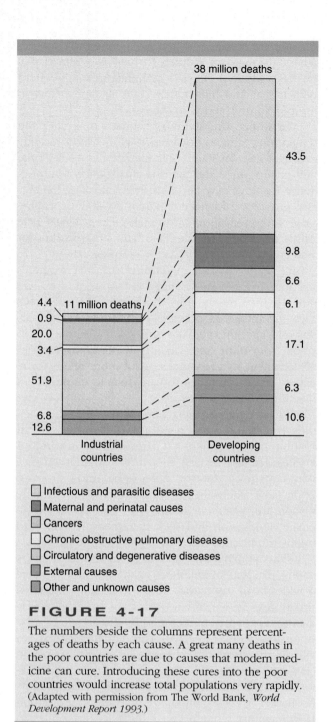

FIGURE 4-17

The numbers beside the columns represent percentages of deaths by each cause. A great many deaths in the poor countries are due to causes that modern medicine can cure. Introducing these cures into the poor countries would increase total populations very rapidly. (Adapted with permission from The World Bank, *World Development Report 1993.*)

Infectious Diseases

Humans eat plants and animals, and humans in turn provide a home for invading microorganisms. Some are harmless, but others cause severe damage. Today the epidemiology of infectious diseases is complicated by the multiplication of worldwide transportation and communication linkages. One traveler can introduce a dangerous disease across the globe within a few hours. In addition, medical science is learning that the disruption of the Earth's natural habitats can trigger the evolution of new **pathogens**, or disease-causing organisms.

Some infectious diseases are caused by bacteria (microscopic one-celled organisms), including tuberculosis (TB), cholera, syphilis, typhoid fever, and tetanus. Other infectious diseases are caused by viruses, another form of microscopic organisms smaller than bacteria. Viruses are known to cause cancer in animals and are suspected of causing cancer in humans; they also cause measles, mumps, yellow fever, poliomyelitis, influenza, and AIDS. Protozoa are still another kind of one-celled animals that can cause disease, including sleeping sickness and several kinds of malaria.

Pathogens often travel from one human or animal to another by way of intermediate hosts, called **vectors**. The most common vectors are insects (including mosquitoes, flies, and ticks), but worms, snails, and even house pets can also function as vectors. Some diseases, called *vectored diseases*, can be transmitted only by means of a vector because the pathogen itself changes in the body of the vector, usually as a part of its regular life cycle. Malaria, for example, is a vectored disease, and a person catches malaria only by being bitten by an infected mosquito (the vector), not by touching a person who is already suffering from the disease.

Disease pathogens and vectors each have their own life requirements, and therefore the spread of a vectored disease can be fought by attacking either the pathogens or the vectors. Vectored diseases will not spread beyond the vector's habitat. African sleeping sickness, which affects both people and their cattle, is an example. It is caused by small parasites called trypanosomes which are passed around by tsetse flies, so it is restricted to the range of these vectors. The tsetse fly is restricted to the area within approximately 14° of the equator in Africa and to elevations below about 4,000 feet (1250 m). Sleeping sickness has plagued Africa for centuries, but today it is largely under control.

Malaria, by contrast, occurs widely because approximately 35 species of anopheles mosquitoes can act as vectors of plasmodia (the disease-causing protozoa), and these species live in a wide variety of climates. DDT and other insecticides developed shortly before and during World War II allowed greater control of infectious diseases because they can be applied over wide areas without the active participation of the local population. Mosquitoes, however, develop resistance, so every year 100 million people worldwide still suffer from malaria. About 80 percent of the estimated 1.4 to 2.8 million malaria-related deaths each year occur in sub-Saharan Africa. Scientists fear that as new resistances develop, the number of deaths from malaria could rise in coming years.

In contrast to the vectored diseases, the pathogens of nonvectored diseases may travel directly from one person to another without any intermediate host. Some pathogens may be transmitted by a handshake, a kiss, any other form of touching, or even standing close enough to be reached by tiny particles of moisture from exhaled air. Others can be transmitted only through prolonged contact or through contact with infected blood or other body fluids.

In some parts of the world, diseases caused by parasitic worms (hookworm, trichina worm, flukes, and others) have been virtually eradicated, while elsewhere they still debilitate and kill large numbers of people.

We do not understand the causes and methods of transmission of all diseases, and in some cases the categories used here are not exclusive. Some diseases, for example, can be transmitted either directly or indirectly through a vector. Sometimes humans can act as vectors. Cancer is classified as a degenerative disease, yet some forms of cancer may be viral diseases. A great deal of research remains to be done in epidemiology.

Many diseases are **endemic**. That is, a state of equilibrium exists between humans and microorganisms whereby the latter is widespread in a community without causing extensive death. The carriers of the organism may nevertheless lack energy and suffer long-term deterioration. Tuberculosis and syphilis may be regarded as endemic in the United States and Western Europe. An **epidemic** is a sudden and virulent outbreak, and a worldwide epidemic is called a **pandemic**. This century's most terrible influenza pandemic was that of Spanish Flu in 1918, during which some 22 million people died, but many people also died during the pandemics of Asian Flu (1957) and Hong Kong Flu (1968).

The Battle against Infectious Diseases   The introductions of sulfa drugs in the 1930s and antibiotics in the 1940s greatly diminished the fatal effects of many bacterial diseases, including TB, pneumonia, typhoid fever, bacterial dysentery, rheumatic fever, syphilis, yaws, and cholera. Most viral infections are not treatable with drugs, but vaccination has proven useful in preventing outbreaks of viral diseases, such as polio. Others, such as yellow fever, are controlled by eradication of mosquito vectors as well as by vaccination.

As recently as the 1970s scientists tended to see infectious diseases as a series of problems to conquer, ticking off victories like notches on a gun. Nobel Prize winning Australian virologist Sir F. Macfarlane Burnet wrote in 1962 that the late twentieth century would be witness to "the virtual elimination of infectious disease as a significant factor in social life." To write about infectious disease, he wrote, "is almost to write of something that has passed into history." In 1969, U.S. Surgeon General William H. Stewart declared, "The war against diseases has been won."

Human Population: The Distribution and the Pattern of Increase   **145**

Humankind hoped to achieve over all diseases the triumph that had been achieved over smallpox, the biggest cause of death by infection in human history. The last known smallpox case was cured in Somalia in 1977, and the disease's last victim—a British medical photographer—died in 1978. Variola, the virus that causes smallpox, survives only in a few scientific laboratories, and many scientists have argued that these last few surviving pathogens should be killed. Variola has the distinction of being not only the first disease pathogen to be so completely conquered, but also the first to have its entire genetic makeup, its genome, recorded. The complete DNA sequence of variola, containing about 186,000 letters of genetic code, has been written. Today variola could be reproduced in a laboratory.

Humankind is on the verge of achieving a similar victory over leprosy, a disease caused by a bacterium. The disease deadens nerves, and it has crippled and blinded millions of people since biblical times. From 1983 to 1993 the treatment—a combination of antibiotics—reduced the number of cases worldwide from 12 million to 3.1 million. Some 81 percent of all cases can be found today in India (a full 64 percent), Indonesia, Myanmar, Nigeria, and Brazil, and WHO hopes to eradicate the disease by the year 2000.

TB is currently the leading pathogenic cause of adult deaths (almost 3 million annually), and new drug-resistant strains have recently developed. Today approximately one-third of the human population carries TB bacilli in their bodies, although in the vast majority of cases the disease is latent and cannot be transmitted.

Cholera also remains worrisome. This disease is caused by a bacterium, *Vibrio cholerae*, that produces a toxin that causes severe diarrhea. New strains of cholera appear regularly (139 are known), and they are spread through contaminated water and poor sanitation. Antibiotics are not essential to recovery, but body fluids must rapidly be replaced. The world's first pandemic of cholera surged out of India in 1817, and the seventh began in Asia in 1960. That strain, number 138, has caused tens of thousands of deaths worldwide, and it still afflicts large areas. In 1992, a new 139th strain, however, was identified in the Indian subcontinent, threatening an eighth pandemic.

New Viral Strains Can Appear Victory over infectious disease, however, was declared prematurely. The more we learn about how, why, and where viruses emerge, the more we understand that human actions can create breeding grounds for new viruses and augment the number of pathways viruses can take to new populations. Alterations of the environment can trigger the emergence of new viral strains and alter the distributions of pathogens, possibly quickly. These include deforestation and urbanization, which bring virus-carrying rodents and people together; international trade, which unwittingly transports virus-bearing insects from country to country; medical practices (organ transplants); sexual practices; and farming methods that allow viruses from birds and mammals to mingle.

The more we learn about how, why, and where viruses emerge, the more we understand that human actions can create breeding grounds for new viruses and augment the number of pathways viruses can take to new populations.

Many influenza strains, for example, originate in Asia, because there the most common animal hosts for the influenza virus—pigs, ducks, and chickens—live in close proximity to each other and to human beings. Inside these animal's intestines, viruses mix and recombine in new combinations in a process called *antigenic shift*. Ducks are the most common reservoirs of new viral strains, and China arguably has more ducks than people. In the early 1990s, in the U.S. Southwest a wholly new virus emerged among deer mice whose environment had been disrupted. That virus spread to humans, and by the time it was isolated and grown in a laboratory in November 1993, it had spread from California to Louisiana, and 27 people had died.

Because diseases can now easily spread around the globe, the U.S. National Academy of Sciences declared in 1992 that global public health surveillance systems are inadequate and warned of new epidemics of infectious diseases. The Academy's report stated that "there has been no effort to develop and implement a global program of surveillance for emerging diseases or disease agents." The Academy feared that even the well-known insect-borne infections that were virtually eliminated from the United States and Europe by the mid-1970s, including malaria, could return.

The Effect of Aids Acquired immune-deficiency syndrome (AIDS) is a viral disease that destroys the body's ability to fight infections. It is believed that separate strains of the AIDS virus, called the human immunodeficiency virus (HIV), emerged in East Africa and in West Africa sometime in the 1950s. From those hearth regions, the virus has spread worldwide. People do not actually die of AIDS but of the opportunistic infections that follow it. In July 1994, WHO estimated that 16 million people worldwide had contracted the virus and that over 4 million had developed the actual disease. The greatest concentrations were in Africa, which had over 8 million infections, the Americas (over 2 million), South and Southeast Asia (over 2.5 million), and Western Europe (about 0.5 million).

By the year 2000, about 6 million people will have died from AIDS-related illnesses, another 15 million people will be sick, and the total number infected may approach 40 million. The death toll could rise to 2 million per year. In the United States over 200,000 people had died by 1994, over 250,000 more had been diagnosed with the disease, and an estimated 1 million were infected.

The disease is now spreading fastest in the poor countries, and by 2000, 90 percent of all infections will probably occur there. Already by 1990 the virus had infected one-third of the population in some parts of Africa, and in several African countries AIDS-related illnesses were the leading cause of adult death (Figure 4-18).

AIDS has disproportionately heavy effects on already weakened economies and weakened family structures in developing countries. It typically strikes young adults who are just entering their most productive years. Already in some African countries, the cost of orphanages is one of the fastest rising items in the national budget. Sub-Saharan African populations will continue to rise, but recent studies there and in Brazil, Thailand, and Haiti do suggest that AIDS will significantly raise infant mortality rates and lower life expectancies into the 21st century. Fear of AIDS, however, may increase the use of condoms, and this may lower fertility.

Tobacco-Related Deaths

In 1993, the World Bank officially listed tobacco products among the world's leading causes of preventable death. Epidemiologists have forecast that one-fifth of the people now living in the developed world—more than 250 million people—may die prematurely of smoking-related illnesses. In the United States the rate of smoking is decreasing, but the Centers for Disease Control and Prevention estimated that in 1990 418,690 people died prematurely from causes related to "tobacco addiction."

In many other countries, particularly developing countries, the rate of smoking and the numbers of

FIGURE 4-18

This young mother in Kenya stops to read the warnings about AIDS. The spread of this disease across Africa casts doubt on some projections of population increase there. (Courtesy of Wendy Stone/Gamma Liaison.)

tobacco-related deaths are rising rapidly. In some countries, including France, Thailand, and Japan, the government itself operates a tobacco monopoly or has a financial stake in the promotion of tobacco products, and in these countries warnings against the dangers of smoking are not strong. One-third of the world's cigarettes are smoked in China (some 60 percent of men smoke there), but we do not have reliable data on related deaths. WHO estimates that if one-third of the world's young adults become cigarette smokers, and one-third of them are eventually killed by the habit, then tobacco, which now accounts for about 3 million annual deaths worldwide, will account for over 10 million deaths a year—10 percent of all deaths—by 2030. International tobacco corporations employ the most sophisticated techniques to diffuse and increase the use of their product.

Disability-Adjusted Life Years

The world's poor countries suffer not only high mortality rates but also a substantial burden of disability, ranging from polio-related paralysis to blindness. To measure the overall burden of disease, the World Bank has developed a new measure. This statistic, called the *disability-adjusted life year* (DALY), combines healthy life years lost because of premature mortality with those lost as a result of disability. The total loss of DALYs is referred to as the global burden of disease.

Human Population: The Distribution and the Pattern of Increase   **147**

Questions

1. Atlases of disease occurrence spark hypotheses and investigation of the causes of many diseases. The map shows the incidence of heart disease throughout the 48 contiguous United States. What factors may explain the different rates in different regions?

2. Do the principal diseases and causes of death in your community differ from national norms? Inquire at your local health department.

Localized outbreaks of disease invite geographic study. We might exercise the scientific method and find the cause of an outbreak by studying the phenomena that occur in place together.

One classic epidemiological/geographic study was a correlation between cholera deaths and the use of a neighborhood water pump in London. In the 1850s, different water companies supplied water to London neighborhoods, and each neighborhood had a pump. In 1854, cholera struck one neighborhood of the city. Doctor John Snow was a local physician who had long suspected that cholera was a water-borne infection, although bacteria were unknown at the time. Dr. Snow plotted cholera deaths with pins on a map showing the location of each pump. The concentration of pins around one pump lead him to hypothesize that the pump carried the infection and to request authorities to remove its handle. The number of new cholera cases dropped almost to zero.

Observers still today correlate disease with local geographic attributes in a search for the causes for disease. In the early 1990s, for example, female residents of Long Island, New York, became alarmed that local rates for breast cancer were 13 times national rates. The women feared that the cause was something in the environment, and their hypotheses ranged from toxic wastes to pesticides, fertilizers, water contamination, chemical factories, termite poisons, and electromagnetic

Death's Dispensary, by George J. Pinwell, circa 1866.
(Courtesy of the Philadelphia Museum of Art, Art Media Collection.)

fields (EMFs) from the many wires criss-crossing the island.

The National Centers for Disease Control, however, has suggested that the cause does not have to be sought in the environment. The characteristics of the women themselves sufficiently explain the high rate of breast cancer. To statistically unusual degrees the local population consists of Jewish women of Eastern European origin, affluent, well-educated, late to bear their first children, tending toward small families, having family histories of breast cancer, and tending not to breast feed. Each of these eight factors is known to raise the risk of breast cancer. The high incidence of breast cancer on Long Island can be statistically explained by the coincident occurrence of several of these factors within the population itself. (You might want to review the meaning of *statistical*

The Mortality Rate from Heart Disease

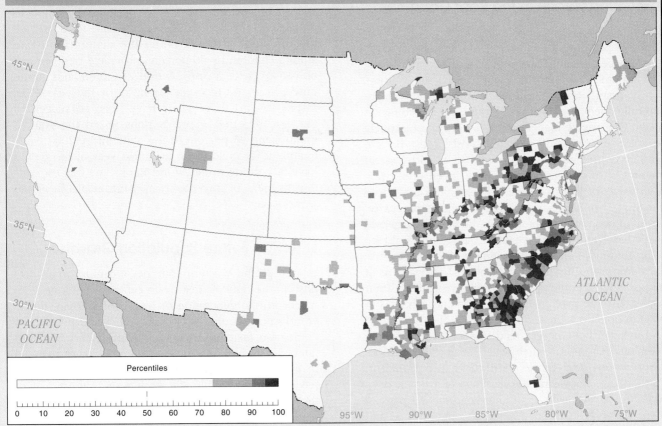

This figure maps the total mortality rate (deaths per 100,000 people) from all heart diseases in the 48 contiguous states. The South has distinctly high rates, and hypotheses to explain this have included Southern cooking, water softness, trace elements in Southern soils, altitude, and still other factors. Known risk factors for heart disease include smoking, cholesterol, lack of exercise, tense personality types, and blood sugar. It is known that blood pressure is significantly higher in the South than in other areas among men, women, and even children of all races, but research on causes remains inconclusive. (From B.A. Goldman, *The Truth About Where You Live*. New York: Times Books/Random House, 1991.)

explanation in the essay on the scientific method.)

Not only on Long Island, but worldwide, concern has been mounting over whether EMFs—from fuse boxes and computers and electric blankets as well as power substations and power-lines—may be a cause of cancer, particularly among children. The U.S. Environmental Protection Agency in 1990 classified EMFs as a "probable" carcinogen, but then changed the classification to "possible." That triggered a popular movie on the subject—Eddie Murphy's 1992 *The Distinguished Gentleman.* In 1992, the Swedish government found that children who lived close to high-voltage lines definitely had a greater chance of developing leukemia, and a 1994 study of Canadian and French electric company employees concurred that EMFs and leukemia are linked. Research continues.

Figure 4-19 shows the variations in per person loss of DALYs across regions. Regional differences in loss of DALYs as a result of disability are surprisingly small. The greatest variations in the loss of DALYs arise from differences in premature mortality.

Improving Health in the Poor Countries

At present, spending on health care accounts for almost 8 percent of total world production of goods and services. Some 87 percent of that is spent in the rich countries—41 percent in the United States alone. World health could improve if more money were carefully spent in the poorest countries. Spending on health is a private matter, and it is difficult objectively to balance lives, suffering, and costs. Most people who have the money to do so will spend a great amount on an operation to grant a loved one a few more years of life; that same amount of money could vaccinate ten thousand young strangers far away. From an economic point of view, however, the World Bank has suggested that current spending could be redirected more effectively in almost all countries.

Improving health in developing countries starts at home. What people do with their lives affects their health far more than anything governments can do, but what people do is greatly influenced by their education and income. Poor people tend to spend extra income on things that improve their health, such as better food, safer water, and better sanitation and housing. Furthermore, improving the earnings of women does more for family health than improving the earnings of men. Women usually spend more income on food and less on alcohol and cigarettes. Better education, especially for girls, is clearly related to better health. Better-educated women are more likely to get prenatal care, resulting in healthier babies. Furthermore, around the world, it is largely women who buy and prepare food, keep homes clean, care for children and the elderly, and initiate contacts with the health system.

Preventive medicine is cheaper than curative medicine, but the introduction of preventive medicine often requires changes in customs. The simplest but most important innovation is acceptance of public and private cleanliness. Even sanitation, however, may require money, even if only for soap and sandals, and people may hesitate to sacrifice food or other immediate necessities for an uncertain future benefit. Drugs are essentially curative, but a sick person is more willing to submit to an injection or to take a dose than one who is told that a shot will prevent future sickness. If a patient is cured, news of the cure spreads quickly, and people come to the clinics. Their new confidence in the clinic may extend to preventive inoculations, which further reduce death from infectious diseases.

In September 1990, the United Nations sponsored a World Summit for Children, attended by representatives of 149 countries, 71 of them by their heads of state or government. That meeting set a goal of immunizing 80 percent of the children in the developing countries against diphtheria, whooping cough, measles, polio, tuberculosis, and tetanus by 1995. The leaders adopted a program intended to reduce the infant mortality rate by one-third, reduce the maternal death rate and the malnutrition rate among children by 50 percent, provide a basic education to 80 percent of the world's children, and assure universal access to clean water, all by the year 2000. Efforts toward those goals have been so successful that by 1993 the United Nations raised the goal to immunizing 95 percent of the children by 2000. Such treatment for children in their first year of life would have, according to the World Bank, "the highest cost-effectiveness of any health measure available in the world today."

Effect on Future Population Growth

What took the Western countries a century to accomplish in the fight against death can now be done elsewhere in one generation or less. If health care in the world's poor countries is improved, world population could rise faster than presently projected. If death rates among children in poor countries were reduced to those prevailing in the rich countries, 11 million fewer children would die each year. Almost half of these preventable deaths are a result of diarrhea or respiratory illness, exacerbated by malnutrition. In addition, every year 7 million adults die of conditions that could be inexpensively prevented or cured. About 400,000 women die from the direct complications of pregnancy and childbirth. Maternal mortality rates are, on average, 30 times higher in developing countries than in high-income countries. The UN population projections rest on the assumption that life expectancy in the poor countries in 2020 will be no higher than it was in North America in 1950. In fact, life expectancies in many of these countries are already rising.

The world death rate is expected to fall, and that will tend to increase the rate of natural population increase, at least temporarily. Most world agencies assume, however, that that increase will quickly be offset by the rapid fall in birth and fertility rates that the world is already exhibiting.

Is the Earth Overpopulated?

Many people fear that the Earth may become overpopulated, or that it is already. People disagree, however, on

The Global Burden of Disease

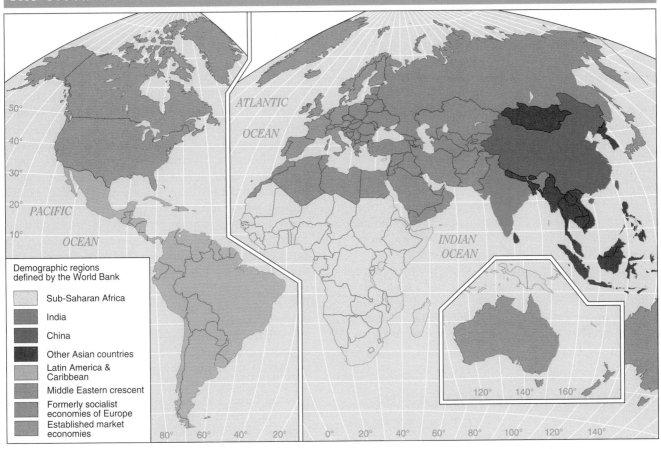

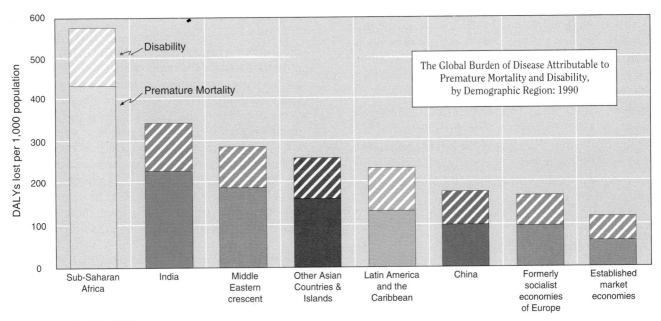

Demographic regions defined by the World Bank: Sub-Saharan Africa; India; China; Other Asian countries; Latin America & Caribbean; Middle Eastern crescent; Formerly socialist economies of Europe; Established market economies

The Global Burden of Disease Attributable to Premature Mortality and Disability, by Demographic Region: 1990

FIGURE 4-19

Both the burden of disability and the burden of premature mortality are highest in sub-Saharan Africa, but the figures for the burden of disability are high in all world regions. (Data courtesy of World Development Bank, *World Development Report 1993: Investing in Health*.)

the meaning of the word overpopulation. Some people just don't like people, and they complain that their neighborhood is overcrowded, or that new construction and crowding are everywhere lowering the quality of life. Studies of rats do demonstrate that crowded rats suffer mental and physical breakdown. Are we like rats? Is human territoriality equivalent? We don't know.

At a global scale, are many people hungry today? Yes. Are many people on Earth today without clothing and shelter? Yes.

This deprivation does not, however, result from insufficient food on Earth to feed everyone; we will examine this in detail in Chapter 6. Nor does it result from insufficient raw materials, as we shall see in Chapter 11. The problems are problems of organization and distribution. These are political problems, not technological problems. Will these political problems ever be solved? Again, we don't know.

Could mounting numbers of people ever place demands on materials so great that per capita supplies of goods will necessarily fall? That question will be answered by future developments in technology, and we cannot possibly foresee what those can be.

THE AGING HUMAN POPULATION

Humankind as a whole is growing older for the first time in world history. The aging began early in this century with improvements in health and hygiene, and it is projected to end in the middle of the next century. By then the age composition of the population is projected to stabilize, and the practical limits to human longevity will be approached. In 1900, there were 10 to 17 million people aged 65 or older, making up less than 1 percent of the total population. By 1992, the 342 million people in that age group made up 6.2 percent of the human population, and by 2050 the number of people in that age group may expand to one-fifth of the total projected population (Figure 4-20).

Aging in the Rich Countries

For most rich countries, substantial natural population increases seem to be over. As people have fewer children and live longer, ratios of the old to the young rise rapidly (Figure 4-21). In those countries in which fertility was already low by 1950 (the countries in Europe, the United States, Canada, Japan, Australia, and New Zealand), the proportion of the population over 65 years of age will jump from 12 percent in 1990 to 16 percent in 2010 and 19 percent in 2025. By that time, 3.2 percent of the population in these countries will be over 75.

This presents both unprecedented opportunities and problems. Questions arise about the possibility of sustaining economic growth and about the equitable distribution of wealth among different generations. Increasing shares of national wealth and expenditures must be directed to health care and other problems of the elderly. New social services, housing alternatives, transportation, and health care programs will have to be devised. How long, for example, can elderly people live alone? Almost every service provided by government at every level will require rethinking to cope with an elderly clientele. In 1993, 13 million people aged 65 or over were living alone in Western Europe. The European Union declared 1993 "The European Year of Older People," and international committees began planning for the necessary targeting of government services.

Taxes may rise on the working population to support pensioners. People may have to work longer to earn a pension. This would have the double advantage of increasing the number of contributors to pension funds and decreasing the number of recipients. The United States began making changes in the social security program in 1983 by increasing taxes, curbing benefits, and raising the age for retirement with full entitlement. Britain, France, Italy, Germany, and other European countries are following the U.S. example.

The settlement patterns of retired people can bring about a national or even international geographic redistribution of wealth. Within the United States, for example, pensioners in Florida and Arizona are today spending money earned in Illinois and New York. In addition, the United States is the only country that makes social security payments to noncitizens who have retired after working in the country. Because many of these people retire to their native countries, these payments drain capital that otherwise would be used for investment in the United States. For some poor countries, however, these same payments represent a significant contribution to the balance of trade and payments. People in the Dominican Republic, for example, receive more than $1.5 million each month in U.S. Social Security benefits. If citizens of one country retire to another country, the economic future of the country the pensioners are leaving may be hobbled by the drain on its capital sent to pensioners abroad. The Northern European countries, for example, fear that elderly Northern Europeans may all choose to retire in the milder South.

The World's Aging Population

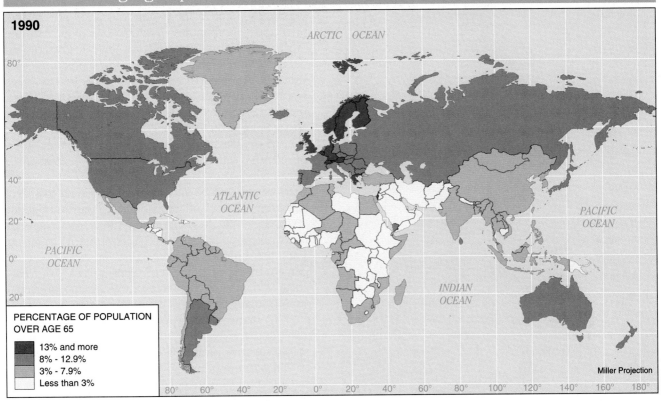

1990

PERCENTAGE OF POPULATION OVER AGE 65
- 13% and more
- 8% - 12.9%
- 3% - 7.9%
- Less than 3%

Miller Projection

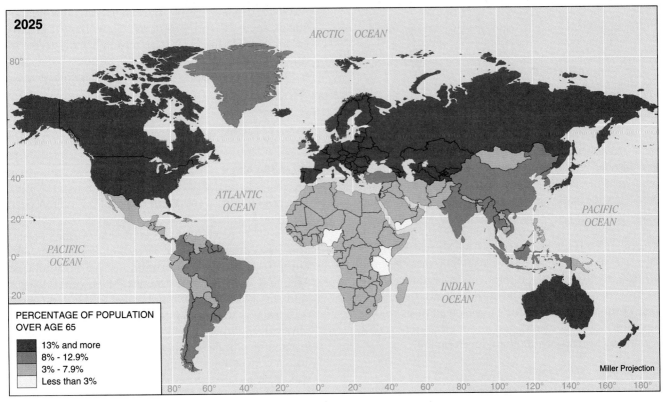

2025

PERCENTAGE OF POPULATION OVER AGE 65
- 13% and more
- 8% - 12.9%
- 3% - 7.9%
- Less than 3%

Miller Projection

FIGURE 4-20

This projection of the world's aging population shows widespread percentages of the population 60 or older by 2025. How old will you be in that year? (Data courtesy of the U.S. Census Bureau.)

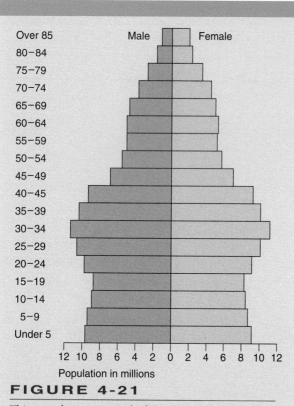

Over 85, 80–84, 75–79, 70–74, 65–69, 60–64, 55–59, 50–54, 45–49, 40–45, 35–39, 30–34, 25–29, 20–24, 15–19, 10–14, 5–9, Under 5

Male / Female

12 10 8 6 4 2 0 2 4 6 8 10 12
Population in millions

FIGURE 4-21

This population pyramid of the United States in the year 1991 shows a bulge of people between 20 and 45 years of age--the *baby boom* of 1945-1965, although slowly rising fertility does give a pyramidal shape to the bottom cohorts of people under 20. The United States continues to welcome immigrants, and fertility rates are high among them, but the nation's median age will continue to rise. (Data courtesy of the U.S. Census Bureau.)

The leveling-off and aging of a country's population also affects the long-term use of the land. Land use decisions in the United States have always been based on the assumption that there would be future growth and population increase, but that may no longer be the case. Decisions that are being made in America today, for instance, and which will be made through your lifetime, may leave their mark on the landscape for many generations to come. These include decisions about the use of land, about the care of the environment, about preservation of wilderness areas or of farmland, and about urban renewal or continuing spread of suburban development.

Aging in the Poor Countries

The problems associated with rapidly aging populations are today restricted to the rich countries, but quite early in the next century countries now labeled young and poor will themselves be aging fast. Today's enormous numbers of young and poor people will grow older in countries in which the crude birth rate is dropping and fewer young people are coming along behind them. Therefore, the median ages of those countries will rise sharply. Those countries where fertility fell rapidly between 1950 and 1990, including particularly China, Colombia, Indonesia, Mexico, and Thailand, will see rapid rises in their over-65 populations. In these countries, the proportion of over-65s will double from 5 percent in 1990 to 10 percent in 2025. The proportion of citizens over 65 years of age will be accelerating, suggesting that they will grow further and probably still faster after about 2025. By 2020, for example, the median age of Mexico's population, which was about 17 years in 1990, will probably be 33.4 years. By 2050, about 20 percent of the population of China will be over 65—some 270 million people. Many poor countries will soon be aging faster than the rich countries are now.

In many of those countries today, the elderly are cared for by their families. National welfare or social security programs do not exist. As the ratios of old to young rise in those countries, they will have to build up their national incomes in order to be able to look after their elderly. They will also have to devise appropriate national systems of social security and pensions. They have perhaps two generations' time in which to do this. In Chapter 12 we will examine the various paths that national governments are taking to increase their national wealth.

SUMMARY

The population of the Earth is about 5.7 billion people, but 90 percent of the population is concentrated on about 20 percent of the land area. Major population concentrations are in East Asia, South Asia, and Europe. Secondary concentrations are in Southeast Asia, and eastern North America. Much of the Earth's land surface is virtually uninhabited.

Factors that explain this distribution include the physical environment, history, and local differences in rates of population increases. Most of the population is concentrated in areas of C climate, but people will settle in harsh areas if it is profitable to do so. In some places local population densities may be higher than local environmental resources might support, but trade and circulation free societies from the constraints of their local environments.

The populations of China and the Indian subcontinent long ago achieved productive agriculture. Europeans multiplied when they gained material wealth and improved their food supplies as a result of world expansion and conquest, and later again during the industrial and agricultural revolutions. European migrants took technology to more sparsely populated areas. Population reductions have occurred in the Western Hemisphere and sub-Saharan Africa.

World population is increasing, but the rate of growth is slowing. The rate of increase in each country represents a balance among the crude birth rate, the crude death rate, the fertility rate, the total fertility rate, and migration. Most of the increase is occurring in the poor countries. The history of population growth in today's rich countries traces a demographic transition. In the first stage the crude birth rate and the crude death rate were high. In the second stage material well-being and medical science improved, so crude death rates dropped. Therefore, rates of population increase rose. In the third stage crude birth rates have dropped, so rates of natural population increase are low.

Experience in today's poor countries does not exactly follow this pattern. Crude birth rates are dropping fast, even without significant increases in material well-being. Theories of explanation focus on the growing role of family planning programs, new contraceptive technologies, and the educational power of mass media. Family planning efforts face obstacles in traditional cultures, religions, and in some cases, the cost of devices.

Even if worldwide crude birth rates and total fertility rate were to fall tomorrow, however, the total population will continue to grow because the number of young women reaching their childbearing years is larger than ever before. The sooner the world's total fertility rate falls to the replacement rate, the lower the total population will be at which the projected numbers level off. Current projections range from a maximum of 8 to 13 billion people some time in the 21st century.

It is natural for more males to be born than females, but women, given similar care, tend to live longer than men. Therefore, the natural ratio between females and males in a human population should be 1 to 1, or slightly higher. Worldwide, however, men outnumber women, and the sex ratio varies greatly in different countries. This may be due to a combination of economic and cultural factors.

Global death rates are currently falling, and they will probably continue to fall. More people staying alive longer will put upward pressure on the rate of natural population increase. As national populations grow older and richer, the main causes of ill health and death change. This transition in the leading causes of death in a society is called the epidemiological transition. Human actions can create breeding grounds for new viruses and augment the number of pathways viruses can take to new populations. Alterations of the environment can trigger the emergence of new viral strains, such as AIDS, and alter the distributions of pathogens, possibly quickly. Preventive medicine is relatively inexpensive, and it could save millions of lives.

Humankind as a whole is aging. The aging began early in this century and is projected to end in the middle of the next century. In the rich countries, questions arise about the possibility of sustaining economic growth and about the equitable distribution of wealth among different generations. Many poor countries must quickly build up their national incomes and devise national welfare or social security programs.

Many people fear that the Earth may become overpopulated, or that it is already. People disagree, however, on the meaning of the word overpopulation. Many people on Earth today lack sufficient food, clothing, and shelter, but this is caused by political problems, not technological problems. We do not know whether mounting numbers of people ever place demands on materials so great that per capita supplies of goods will necessarily fall.

KEY TERMS

population geography (p.122)
emigration (p. 122)
immigration (p. 122)
demography (p. 122)
census (p. 122)
cartogram (p. 123)
arithmetic density (p. 123)
physiological density (p. 123)
carrying capacity (p. 123)

QUESTIONS FOR INVESTIGATION AND DISCUSSION

1. If you can, investigate the numbers of children each generation in your family has had.
2. Investigate the various churches' positions on family planning and birth control.
3. Investigate the family planning efforts and experience in any one country.
4. If your local region were entirely dependent on local agriculture, what would be the carrying capacity of your local environment? Does your region now export or import food?
5. In what year do you intend to retire? What percentage of the national population will be working to support you?

population
  projection (p. 130)
crude birth rate (p. 130)
crude death rate (p. 130)
natural population
  increase (decrease) (p. 130)
fertility rate (p. 130)
total fertility rate (p. 130)
replacement rate (p. 131)
doubling time (p. 131)
population pyramid (p. 132)
dependency ratio (p. 132)
zero population
  growth (p. 132)
demographic
  transition model (p. 133)
infant mortality rate (p. 133)
life expectancy (p. 136)
epidemiology (p. 144)
epidemiological
  transition (p. 144)
pathogen (p. 144)
vector (p. 145)
endemic (p. 145)
epidemic (p. 145)
pandemic (p. 145)

6. What factors might intervene to change current population projections? How could either death rates or birth rates be either raised or lowered?

7. Compare the maps of the world's physical environment and the world population map. What correspondences do you expect to find? What deviations do you find from what you expected to find? Can you explain the deviations? How has physical geography affected the distribution of people in any given continent or region?

8. How might continued population growth affect the poor countries of the world? How could a decrease in their rates of growth affect them?

ADDITIONAL READING

BRODEUR, PAUL. *The Great Power-Line Cover-Up: How the Utilities and the Government are Trying to Hide the Cancer Hazard Posed by Electromagnetic Fields.* Boston, MA: Little, Brown, 1993.

GESLER, WILBERT M. *The Cultural Geography of Health Care.* Pittsburgh, PA: University of Pittsburgh Press, 1991.

International Bank for Reconstruction and Development. *World Development Report 1993: Investing in Health.* New York: Oxford University Press for the World Bank, 1993.

JONES, HUW. *Population Geography.* London: Paul Chapman, 1990.

LIVI-BACCI, MASSIMO. *A Concise History of World Population.* London: Blackwell, 1992.

MANN, JONATHAN, WITH DANIEL TARANTOLA AND THOMAS W. NETTER, EDS. *AIDS in the World 1992.* Cambridge, MA: Harvard University Press, 1992.

MORSE, STEPHEN S., ED. *Emerging Viruses.* New York: Oxford University Press, 1993.

OLSHANSKY, S. JAY, BRUCE A. CARNES AND CHRISTINE K. CASSEL. "The Aging of the Human Species." *Scientific American*, April 1993.

Population Reference Bureau. *Population Bulletin.* Washington, D.C., quarterly.

The State of World Population 1993. New York: United Nations Population Fund, 1993.

TEITELBAUM, MICHAEL S. AND JAY M. WINTERS, EDS. *Population and Resources in Western Intellectual Traditions.* Cambridge, England: Cambridge University Press, 1989.

World Population Prospects. New York: United Nations Population Division, periodic.

5 Human Migration

Haitians pile into rickety boats and set out for the United States, hoping to build new and better lives.

Courtesy of Randy Taylor/Sygma.

Now that we have recorded the distribution of humans across the globe, examined reasons for that distribution, and investigated population dynamics, we can investigate another factor of human distribution—the process of human *redistribution*. Human beings do not stay put; they never have. Wanderings and migrations of peoples have distributed and redistributed populations throughout human history, and significant redistributions are still going on today.

REASONS FOR MIGRATION

The principal causes for human migration may be divided into push factors and pull factors. **Push factors** include anything that makes a person want to leave and seek a better life elsewhere: starvation, political or religious persecution, or any other disagreeable circumstances. **Pull factors** attract people to new destinations. Economic opportunity and the promise of religious and political liberty are powerful pull factors. Push and pull factors are often attributed to physical geography. For example, we can say that migrants from northern cold regions have found the warmth of the U.S. Sunbelt attractive and that environmental disruption in the Sahel, by contrast, has been found intolerable by emigrants from that region. Notice how careful wording avoids the pathetic fallacy discussed in Chapter 2. The environment never actually pushes or pulls, but people may *decide* to migrate after consideration of environmental factors.

 Cognitive behavioralism emphasizes that people choose a particular destination because of what they think they will find there, not because of the reality of conditions at that place. Therefore, migrants can be surprised by the reality of what they find at their new homes, and sometimes they are disappointed (see Figure 5-1). An old tombstone inscription in Waldoboro, Maine, records the experience of one disappointed migrant group: "This town was settled in 1748 by Germans who migrated to this place with the promise and expectation of finding a prosperous city, instead of which they found nothing but a wilderness." Many other migrants to the United States, by contrast, have found success beyond their original hopes.

Migration has not always been voluntary. Some migrations have been forced, and millions of people suffered the tragedy of migrating in slavery. Others have fled as refugees. People who seek refuge in foreign countries are called **external refugees**, whereas those who are forced to flee from one region of a country to another region of that country are called

internal refugees. For example, the civil war that has raged in Angola since the 1970s has created millions of internal refugees within that country, as well as many external refugees.

Many migrants complete whatever documentation is required for them formally to settle in their host countries, but today great numbers of people are crossing borders without completing legal papers. These migrants are called **undocumented**, **irregular**, or **illegal immigrants**.

ROUTES OF MIGRATION

In 1888, the British geographer E.G. Ravenstein proposed a *law of migration* that was a variation on the gravity model discussed in Chapter 3. Ravenstein suggested an inverse relationship between the number of migrants and the distance between their homeland and their target destination. In other words, most people migrate to nearby locations; fewer migrate greater distances.

Migration flows today, however, are more complicated than this. Movements of people can better be explained by studying flows of trade or by mapping the most convenient transport routes away from those places people want to leave. The flow of information to potential migrants is another factor that plays a role in directing them. People who migrate and find success in their new homes write or telephone those who stayed behind, and their stories encourage others to follow. In addition, successful immigrants pave the way for new arrivals by providing them with employment and financial assistance. In this way, linkages called **migration chains** are forged.

Another frequent flow of information can often be traced between formerly imperial powers and their former colonies. Examples of this type of information and migration flow are between Great Britain and South Asia, between France's former colonies in Africa and France itself, and between the Philippines and the United States. In these cases potential migrants are partially acculturated to their target before they leave home.

Information flows can be astonishingly place-specific. In 1990, an estimated 80 percent of the Hispanics living in metropolitan Washington, D.C., were from the tiny Central American country of El Salvador. Many of them were undocumented. Nobody knows exactly how word first spread throughout El Salvador to trigger this migration.

Immigrants from one region may form distinct concentrations in their new homelands. We already noted in Chapter 2, for example, that immigrants from

The thriving City of Eden as it is appeared on Paper | *The thriving City of Eden as it appeared in Fact*

FIGURE 5-1

Charles Dickens's comic novel *Martin Chuzzlewit* (1844) recounts these gentlemen's dismay when they migrate to the American frontier and find that it is much more primitive than land developers had suggested. Many of Dickens's American readers took offense, but in fact such misrepresentations were common.

Eastern European grasslands settled on America's plains, and Chapter 3 noted that immigrant populations may settle in distinct urban neighborhoods.

Many migrants intend to stay in their new location only until they can save enough capital to return home and enjoy a higher standard of living there. These people are called **sojourners**, and they are usually men who are either unmarried or have left their families in their home country. Many sojourners eventually decide to stay, and then they send for their wives and families. The United States has always attracted great numbers of sojourners. In various years between 1890 and 1910 the number of Italians leaving for Italy equaled 75 percent of the number of Italians arriving in the United States. Even today many Greeks, Irish, Caribbean peoples, and Africans shuttle back and forth between their homelands and the United States.

Some migrations have numbered in the millions of people. Some small migrations, however, have been disproportionately significant because the migrants played key roles in transforming the government, language, economics, or social customs in their new homelands. As we shall see later in this chapter, the impact of the international migrations of Indians and Chinese, for example, or of Caribbean blacks into the United States, are not fully realized by their numbers alone.

Large-scale migration continues around the world today. Several continuing migration flows are intranational, that is, within one country. Chapter 9 will examine the migration of rural peasants everywhere into the cities, and Chapter 10 will examine countries that deliberately redistribute populations to sparsely-populated national frontier regions.

This chapter will review the prehistoric human migration out of Africa that first distributed humans around the Earth. Then we will move ahead to the migrations and redistributions of peoples that resulted from the global diffusion of European political power and culture after 1500. Only if we understand the major migrations since 1500 can we understand the mixed populations of many countries today and, in some cases, the civil strife which afflicts many individual countries. These migration streams still affect contemporary routes of cultural diffusion, distributions and flows of capital and commodities, and even continuing migration routes. Contemporary international migration includes the movement of people from poor countries into the rich countries, such as the flow of Mexicans into the United States and of Arabs and Africans into Europe, and the streams of political refugees fleeing tyranny and persecution in their homelands. Pressures for international migration challenge many nations' political and economic stability.

THE ORIGIN AND SPREAD OF HUMAN BEINGS

Humankind evolved as a distinct species in specific natural conditions. Many humanlike species have evolved over the past 10 million years, always, it seems, in Africa. Several of these, such as the "ape-humans" Dryopithecus and Ramapithecus, ventured out of Africa as much as 10 million years ago, and evidence of them has been found as widely dispersed as Britain, China, and India. Recent dating of skull fragments found on the Indonesian island of Java suggests that they may be up to 1.8 million years old. This may mean that two or more separate species of Homo erectus evolved independently from a common ancestor, but scholars are presently uncertain.

We cannot be certain why Africa should have been the crucible of human development, but the answer may lie in its interlocking habitats of rainforest, wooded savanna, and open grassland. Most of the physical attributes that link humans most closely with apes and monkeys developed in a forest environment, but our distinctive human qualities developed in a series of species (including Homo habilis and Homo erectus) over a period of several million years in a more open environment.

Through the long period during which humans were evolving in the forests, ape-humans may have developed color vision, as well as acute senses of both smell and taste. These qualities must have been of great survival value in the forest. Other sensory and physical capabilities that may date from this forest-dwelling period include forward-looking eyes that allow stereoscopic vision and our versatile shoulder joint, which allows us to throw objects either underhand or overhand.

Several million years ago, the Earth went through a cool climate phase, and eastern and central Africa became drier. The forests in which monkeys and apes had flourished shrank and were partly replaced by grasslands. As a result of this change, our ancestors came down from the trees. The continuing evolution on the grasslands helps explain humankind's subsequent phenomenal geo-

graphical success story. We achieved a fully erect stance, and our gait became two-legged. Our hands were freed to carry things, and they became superlative instruments of manipulation. At the same time, human brains enlarged and became more complex.

Some scientists argue that these many years of evolution during which humans constantly and intimately interacted with nature imbued us with a deep, genetically-based emotional need to affiliate with the rest of the living world. This argument is called the **biophilia hypothesis**. According to this hypothesis, urbanization alienates us from nature and detracts from our psychological well-being. It may be a long time before a truly urban human species evolves.

The first humans that could leave their environment of origin are called Australopithecines, "southern apes," and they evolved on the savanna about 4 million years ago. Large animals were abundant in that environment, but they were probably hard to catch, so the diet of the Australopithecines must have been largely made up of the seeds, roots, and tubers which still compose most of our own diet. These human adaptations to life on the grassland seem to have prepared us for life in other environments.

Over five billion human beings inhabit the Earth today, yet every modern human shares a common ancestral group that evolved between 100,000 and 200,000 years ago. Several groups of humans or near-humans have become extinct, but all surviving humans are in one single subspecies within the genus Homo—"human"—called Homo sapiens sapiens, which means wise, wise human.

Prehistoric Human Migrations

Modern Homo sapiens probably spread from Africa to southern Europe and central, southern, and southeastern Asia quite quickly from about 100,000 years ago onward (see Figure 5-2). This was a glacial period. Much water was held in the polar ice caps, so sea levels were low. Land bridges and coastal corridors existed where they do not exist today. For example, around 30,000 years ago Japan could have been reached from Korea.

The peopling of the Americas is still obscure. Isolated sites indicate that the Americas were probably reached at least 30,000 years ago by people who had traveled across a land bridge that joined Siberia to Alaska. Some of their descendants may have reached southern Argentina at least 12,000 years ago. Humankind's most recent colonization seems to have been of the Pacific Islands by people who came from the western Pacific.

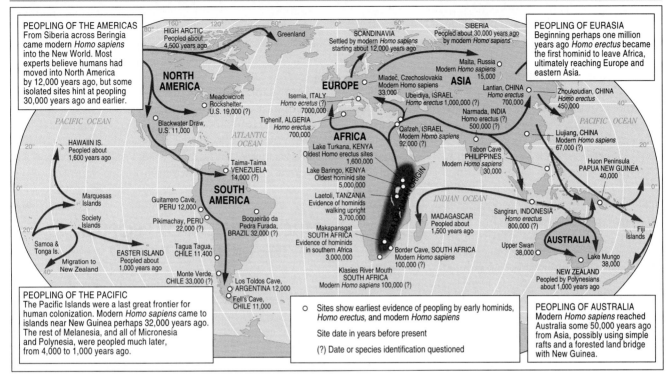

PEOPLING OF THE AMERICAS
From Siberia across Beringia came modern *Homo sapiens* into the New World. Most experts believe humans had moved into North America by 12,000 years ago, but some isolated sites hint at peopling 30,000 years ago and earlier.

PEOPLING OF EURASIA
Beginning perhaps one million years ago *Homo erectus* became the first hominid to leave Africa, ultimately reaching Europe and eastern Asia.

PEOPLING OF THE PACIFIC
The Pacific Islands were a last great frontier for human colonization. Modern *Homo sapiens* came to islands near New Guinea perhaps 32,000 years ago. The rest of Melanesia, and all of Micronesia and Polynesia, were peopled much later, from 4,000 to 1,000 years ago.

PEOPLING OF AUSTRALIA
Modern *Homo sapiens* reached Australia some 50,000 years ago from Asia, possibly using simple rafts and a forested land bridge with New Guinea.

○ Sites show earliest evidence of peopling by early hominids, *Homo erectus*, and modern *Homo sapiens*

Site date in years before present

(?) Date or species identification questioned

FIGURE 5-2

This map shows the latest information about the migrations of humans out of Africa to settle around the Earth.

Human Races

According to the theory of evolution, if one species disperses, its different dispersing groups encounter different environmental circumstances. If, through generations, different specific mutations thrive and multiply in each of the new environments, different **races** of the original species can be identified. If the races are different enough so that they can no longer interbreed, then they must be recognized as distinct species. This has not happened in the case of humankind. Human beings are one single species, and all humans can successfully interbreed. Compared to this biological community of humankind, subdivisions are insignificant.

Human beings have, nevertheless, evolved certain secondary physical characteristics as groups have moved and settled in different environments around the globe. These secondary physical characteristics include body bulk and basal metabolism, nose shape and skin color, lung capacity, and the ratio of red to white corpuscles in the blood. These and many more physiological responses to the physical environment may influence human performance and behavior. For example, peoples who have for many generations lived in countries at high elevations, such as Bolivia

and Nepal, have developed lung capacities that allow them to work at those elevations better than can peoples whose ancestors have always lived in lowland areas. Therefore, the national teams from highland countries did notably well competing at the 1968 Olympic Games in Mexico City at 7,350 feet (2,200 m) above sea level. These influences and differences among peoples should not be exaggerated, but neither should they be denied.

Anthropologists have traditionally used secondary characteristics to divide all humankind into races. The traditional criteria for these subdivisions have been external features such as eyefolds or skin color. A number of different classification systems have been based on these criteria, the most common of which divides humankind into three races: Caucasoid ("white"), Mongoloid ("yellow"), and Negroid ("black").

A second classificatory system is based on an internal criterion, the analysis of blood types. Studies and classifications of blood types have helped explain the distribution of certain diseases, and in that way they have helped direct researchers toward cures. The sickle-cell trait, for example, a chemical abnormality in the red blood cells, can cause chronic anemia, but it may play a role in protecting its carrier against malaria. It is very common in Equatorial Africa, and it is carried by 9 percent of African Americans. Research has shown, however, that the trait crosses "racial" lines. It is widespread in India, Greece, and Turkey. Classifications of people by blood type yields a different set of categories ("races") from classification of people by skin color or other external features. Individuals cross over categories, and this casts doubt on the true significance of either system of categories.

Geneticists—scientists who study the passage of traits from parents to offspring—approach the question of human variation from a cellular or molecular level. They are currently working on a third system of racial classification: global surveys of *genetic markers*, which are variations in proteins and enzymes that reflect differences in people's genetic makeup. The inheritance of genetic traits is controlled by the arrangement of molecules within an individual's deoxyribonucleic acid (DNA), which is contained within the person's chromosomes. Researchers who study genetic markers reach conclusions that do not always agree with those of any other classification system. Genetic studies place the Japanese, Koreans, and Inuits, for instance, close to the Caucasoids, although those groups have traditionally been included among Mongoloids.

Some geneticists believe that they have sketched a family tree of all humankind, depicting the divergence of the world's peoples from a single stock. From this one stock a three-branch family tree evolved: North Eurasians, Southeast Asians, and Black Africans. The Black Africans stayed "home," and the others migrated away (see Figure 5-3).

All of these events occurred deep in prehistory. Some hypotheses are controversial, and all are open to new testing and the discovery of new information. Research and hypothesis formulation about the origins and diffusion of humankind continue in geography, biology, anthropology, archeology, and, as we shall see, linguistics. This is an exciting frontier of human self-knowledge.

In fact, it is impossible to make unambiguous distinctions among races. How many races there are, which groups of people are in each, and even the very meaning of the concept of race are all open to question.

Nevertheless, attempts at racial classification continue. Although they frequently are based on scientific curiosity or methodologies, they have also been used to buttress racist ideologies. **Racism** is a belief in the inherent superiority of one race over another and the linking of human ability, potential, and behavior to racial inheritance. Racism is wrong, immoral, and inju-

rious to both those who discriminate and those discriminated against. Racists' search for "racial purity" is nonsense in a world in which people have intermingled as they have throughout human history.

Two of the leading researchers on an ongoing study of human genes, the Human Genome Diversity Project, Dr. Luigi Cavalli-Sforza and Dr. Mary-Claire King, submitted joint testimony to the U.S. Senate Committee on Government Affairs in 1993. They called racism "an ancient scourge of humanity" and expressed the hope that their work would "undercut conventional notions of race and underscore the common bonds between all humans."

THE DIFFUSION OF PEOPLES IN THE MODERN WORLD

For purposes of understanding contemporary geography, we will begin our study of major migrations 500 years ago. That date was long after humans first migrated out of Africa and just before the European voyages of exploration and conquest. In 1500, the many peoples and culture regions shown back on Figure 3-24 were relatively isolated from one another.

The great migrations that followed are sometimes described as migrations of **ethnic groups**. The word *ethnic* comes from the Greek for "people," and it suggests that the people described are "different from *us*." It conveys the idea that the group is in the minority or "not normal" for a particular place and time. We might use the term "ethnic" casually to refer to an immigrant settlement in a big city, a restaurant serving cuisine of another country, or a form of costume. The word *ethnic* refers to a cultural group; the word *race*, in contrast, refers to a group defined by biological distinctions.

We have already seen that the word "race" is dubious; "ethnic group" is equally dubious. Many people consider the term condescending and, therefore, perhaps best avoided. In this book it will be used only to refer to groups that are self-consciously minorities wherever they are. The definition of an ethnic group and the specific identification of who is in it and who is out of it may depend upon almost any attribute of culture, allegiance, or historic background. Ultimately, as with culture groups, one's ethnic background may even be largely a matter of self-identification. For example, new migrants to the United States often struggle to retain their inherited culture for a generation or two, but the generations born in the new homeland may quickly acculturate and even disdain their parents' distinct self-identification. Sometimes

Genetic Similarity Among Humans

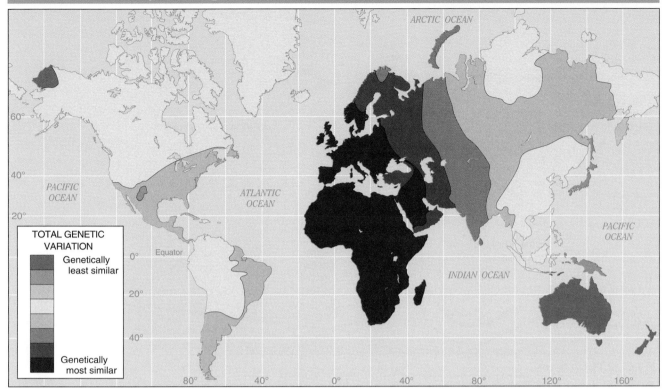

TOTAL GENETIC VARIATION

Genetically least similar

Genetically most similar

FIGURE 5-3

The patterns on this map show the peoples who are genetically most similar and genetically least similar. The greatest extremes of differentiation are between Africa and Australia, which suggests that either Africa or Australia was the origin of the dispersion and the other point was the farthest extension. Archeological evidence found so far indicates that Africa was the origin. (Cavalli-Sforza, Luigi, Luca and Paolo Menozzi, Alberto Piazza, *The History and Geography of Human Genes.* Reprinted with permission from Princeton University Press.)

people are aware of their ethnicity only when someone of their group wins celebrity and they want to bask in reflected glory (an Olympic athlete, for example) or hide from the shame (a gangster); when they are discriminated against; or when politicians appeal to ethnicity. In New York state, for example, the plurality of the population was of Italian background in 1982, but that fact may not have been important to the people until Mario Cuomo emphasized that he was the first Italian American to run for the governorship and appealed for Italian Americans' votes.

The most significant population transfers since 1500 have been migrations of Europeans to the Americas, Africa, Australia, and Asia; migrations of blacks out of Africa; and migrations of Indians and Chinese that were largely instigated or made possible by European expansion.

European Migration to the Americas

During the past 500 years, the largest substitution of population occurred in the Western Hemisphere

(Figure 5-4). The native American population fell victim to slavery, warfare, and, most important, to European diseases, including smallpox, influenza, measles, and typhus. These diseases do not seem to have occurred before in the Western Hemisphere, so the native Americans had no immunity to them. One possible reason for the virulence of Eastern Hemisphere diseases in the Western Hemisphere is that people in the Eastern Hemisphere lived with many domesticated animals, and to some degree the history of human disease was shared with domesticated animals, as we noted in Chapter 4. Because people in the Western Hemisphere had few domesticated animals, there was only a small reservoir for developing disease pathogens. The only Western Hemisphere disease that may have migrated to the Eastern Hemisphere was syphilis, but this is not certain.

Estimates of the native American populations at the time of the European conquests range from 30 million to 100 million. The numbers of deaths were appalling. There were between 12 and 25 million native Americans in today's Mexico before the arrival

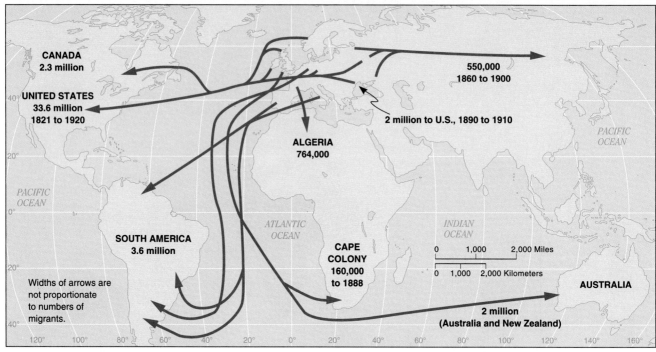

FIGURE 5-4

Europeans regarded the Western Hemisphere as a "New World" open to European settlement, and their massive migrations forever changed hemispheric demographics.

of the Spaniards; by 1600 that number had declined to about 1.2 million. The population of the Inka Empire in South America plummeted from about 13 million in 1492 to 2 million by 1600. The estimated 5 million native Americans in Brazil shrank to only 220,000 today. In today's United States and Canada the native American population was about 2 million in 1492, but 500 years later there were only 530,000.

Some 40 million Europeans migrated to North America. Today they and their descendants make up the majority of the population. The number of self-identified native Americans in the United States tripled to 1.8 million between 1960 and 1990, but this increase probably reflected growing cultural pride and, therefore, self-identification, rather than natural increase among the population. In 1990, native Americans still constituted less than 1 percent of the national population. Canada's 500,000 legally acknowledged "status Indians" constitute only 2 percent of that country's population.

Both the United States and Canada have mixed policies of both assimilating native Americans and at the same time setting aside lands for the preservation of native American cultures. In both countries native Americans retain preferred hunting and fishing rights over substantial areas in addition to their tribal lands,

or reservations, over which they exercise more substantial, but never complete, control (Figure 5-5). Canada has moved to grant the Inuit people political domain over roughly one-fifth of the country, the

FIGURE 5-5

This is a meeting of the tribal council of the Cherokee nation. U.S. law allows such councils considerable power on native American lands.
(Courtesy of Nancy J. Pierce/Photo Researchers.)

FOCUS
How Did Latin America Get Its Name?

There are some justifications for labeling Central and South America "Latin America." "Latin" suggests that certain important aspects of the culture spring from the cultural tradition of ancient Rome, and, in fact, most of "Latin America" was long ruled by either Spain or Portugal, both of which are "Latin" countries. Most "Latin Americans" today speak Spanish or Portuguese, both languages descended from ancient Latin, and most of the peoples there belong to the Roman Catholic church. Italy itself, however, played no historic role in "Latin America." Why, then, don't we call "Latin America" "Luso-Hispanic America?" ("Luso" meaning Portuguese, because today's Portugal was the ancient Lusitania) or "Ibero-America?" The more one thinks about the term "Latin America," the more mysterious its origin becomes.

The truth is that the term "Latin America" originated as political propaganda. During the U.S. Civil War, France's Emperor Napoleon III thought he saw an opportunity to take over Mexico. France had never had any colonial interests there, but Napoleon invaded on the pretext that France, a Latin nation, was avenging military humiliations suffered by the Spanish and was therefore "defending Latin honor." On July 3, 1862, Napoleon published a letter in the French newspaper *Le Moniteur* introducing the term "Latin America" to justify what was really French imperialist aggression. Eventually the French were defeated and retreated, but Napoleon's sly propagandistic term long survived him.

---

eastern half of the former Northwest Territories, now the territory of Nunavut. Continuing litigation of other extensive native claims embroils national politics.

In the United States native American lands represent 2.2 percent of the total land area, and the extent of native American rights—even on their own lands—continues to generate political controversy. For example, the U.S. Indian Gaming Act of 1988 was designed to generate wealth on impoverished reservations by allowing the tribes to operate casinos. Over 70 tribes in 19 states have seized this opportunity, and many are profiting handsomely. The Foxwoods Casino in Connecticut, owned by the 260 Mashanatucket Pequots, netted almost $500 million in 1993, and the Pequots are subsidizing a revival of native American crafts and traditions across the United States. Several state governments and private casino interests, however, are litigating the constitutionality of the 1988 law and trying to close the casinos on tribal lands.

About 4.5 million Europeans, mostly from the Mediterranean area, migrated to "Latin America," and that region developed a complex racial structure. Ibero-American (descended from Spain and Portugal) societies were composed as pyramids of varying proportions. All such societies had a large base of native Americans and blacks, a lesser number of people of mixed race, and a minority of whites on top. (In such societies, a person of mixed white-black ancestry is usually called a **mulatto**; one of mixed black-native American a **zambo**; and one of mixed white-native American a **mestizo**.) Several Latin American colonies and later the successor independent states institutionalized racial status rankings. Native Americans lost their rights in their own homelands, even though they may have constituted the majority of the population. Only since the mid-1980s have some South American countries, including Peru, Venezuela, Colombia, and Brazil, set aside lands for native Americans. In Brazil, the natives are not citizens, but "wards of the State," and they ostensibly enjoy protected status. Miners, loggers, and ranchers, however, have invaded lands set aside for natives, as occurred in the U.S. West in the nineteenth century.

Native Americans still make up one-third to one-half or more of the populations of Mexico, Guatemala, Ecuador, Peru, and Bolivia. In Mexico the native Americans have played a conspicuous political role since the election of the native American Benito Juárez as president in 1857, but native Americans may be waking to power in other countries. A Peruvian of Japanese ancestry, Alberto Fujimori, won that country's presidency in 1990 partly by explicitly allying himself with the nonwhite peoples. In 1993, a native American, Victor Hugo Cárdenas, won the vice-presidency of Bolivia. Among the southernmost midlatitude states Chile claims to be 95 percent mestizo, but the native Americans were practically exterminated in Uruguay and Argentina. In Central America most people are mestizos. Costa Rica is the only country in which whites form the majority.

The African Diaspora

Black peoples have migrated out of Africa, either freely or in slavery, for centuries. These migrations are known, collectively, as the **African diaspora**, from the Greek word for "scattering." This word was first used by the Jews to describe their worldwide migrations following the destruction of the Temple in Jerusalem in 70 A.D.

Racial slavery, as the modern world has come to know it, originated in medieval Islamic societies. Muhammad himself owned slaves, and Arabs and Persians invented the long-distance slave trade that transported millions of black captives from sub-Saharan Africa out to a realm stretching from Islamic Spain across to India (Figure 5-6). By the fifteenth century, when Christian Europe launched its geographic expansion, a slave system based on African labor, including sugar plantations and a fully developed slave trade from Africa, was already in place across the Muslim world.

The Islamic slave system lasted over a period of 12 centuries, and in total numbers it probably surpassed the African slave trade to the Western Hemisphere. The trade diminished slowly in Islamic societies, lingering longest in the Arab countries. It was finally banned in Saudi Arabia in 1962 and in Mauritania in 1980. The absence of a large population of black survivors across this enormous stretch today can be explained by the high mortality rate, by assimilation, and by the practice of castrating slaves. In central India today, however, there are communities of black descendants of African slaves.

The U.S. Department of State, the International Labor Organization (a United Nations agency), and other international organizations insist that Arabs from northern Sudan continue to enslave the black peoples of the southern Sudan today. War has raged there between the Arabs and the black peoples for decades.

The European Slave Trade European transportation of black slaves to the Western Hemisphere had a dramatic impact on population geography. In the fifteenth century Europeans took a few hundred slaves from West Africa to Europe and the Atlantic Islands, and in the 1520s Europeans began transporting slaves to the Western Hemisphere (Figure 5-7). Black Africans were proposed as an alternative labor supply to the dying native Americans. Some Europeans viewed the enslavement of blacks as a twofold humanitarian gesture: It would save native American

FIGURE 5-6

Malik Ambar was a great Indian military leader in the seventeenth century. A number of black Africans rose from slavery to prominence across the Islamic world, either through service as court officials or bureaucrats or in military service. (Ross-Coomaraswamy Collection, Museum of Fine Arts, Boston.)

FIGURE 5-7

Tens of thousands of Africans passed through this old slave market on Gorée Island off the coast of today's Senegal on their way to the Western Hemisphere. In 1992, Pope John Paul II stood in the doorway in the background, known as the "Door of Tears," through which tens of thousands of Africans were taken on the way to the Western Hemisphere, and blessed them and their descendants. Today the Senegalese government is restoring the island's historic fortications as a museum and raising money to build an international conference center. (Courtesy of Brian Seed/ TSW Chicago.)

lives and at the same time convert the Africans to Christianity. Fray Bartolomé de las Casas (1474–1566), whose writings contain the strongest denunciations of brutality toward the native Americans and who won the title "Protector of the Indians," was one of the earliest advocates of black slavery as an alternative to forced native American labor. He lived to abhor the black slave trade.

From 1526 until the trans-Atlantic slave trade came to an end in 1870, about 10 to 12 million slaves reached the Western Hemisphere from Africa (Figure 5-8). The sources of the slaves reflected patterns of European colonialism in Africa. The Portuguese African colonies of Angola and Mozambique provided the slaves of Portuguese-ruled Brazil, for instance, whereas West Africa supplied most of the slaves for North America. Migration of blacks to the Western Hemisphere far exceeded migration of whites until the middle of the nineteenth century.

The pattern of resettlement in the Western Hemisphere reflected the uses to which slave labor was put. Blacks were, for the most part, brought to work on tropical and semitropical plantations in the U.S. Southeast, to the islands of the West Indies, and, throughout Latin America, to the *tierra caliente* shown on Figure 1-16. Mexico was an exception because there the slaves were used in the cities and mines. Mexico never had a significant rural black population, and eventually Mexico's black population died out or mixed with the other groups almost entirely. Mexico abolished slavery at independence in 1821. Slavery was ended in the United States in 1865 and in Brazil in 1888.

Whereas black Africans were settled in the *tierra caliente*, Europeans established Latin America's seaports, but they found the *tierra templada* more comfortable. Many of Latin America's great cities are in this elevational zone, including Caracas, Venezuela; São Paulo, Brazil; and Medellín, Colombia. The *tierra fría* was left to the native Americans, but if rich mineral deposits were found there, the deposits were worked by native Americans under European supervision. Those native American groups that had traditionally

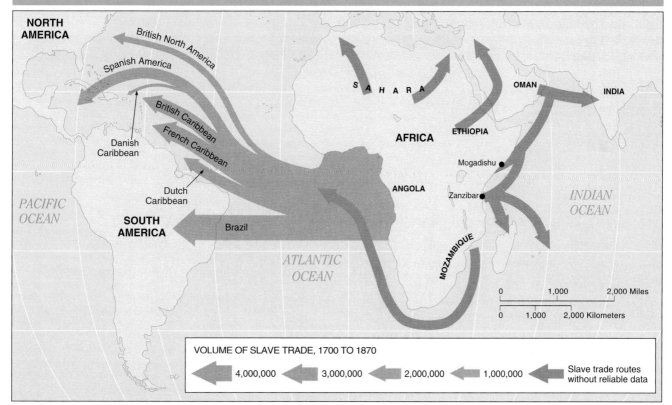

The Movement of Slaves Out of Africa

VOLUME OF SLAVE TRADE, 1700 TO 1870

4,000,000 3,000,000 2,000,000 1,000,000 Slave trade routes without reliable data

FIGURE 5-8

Arab slave dealers first transported black slaves North and East, throughout the entire region from Spain to India. Later, European colonial outposts in Africa determined the sources and destinations of slaves for the Western Hemisphere. The horrors of the African slave traffic were such that about four times as many were captured in the interior of Africa as reached the coast. Notice how relatively few were brought to today's United States. The plurality went to Brazil.

lived at the high elevations could in fact work best there, where lack of oxygen distressed both whites and blacks.

Blacks in the Americas Today Today descendants of slaves constitute the majority of the population on most West Indian islands and in Belize. In South America, blacks still occupy, for the most part, the lowland areas. After centuries of intermarriage, tropical South American countries still have significant minority black, zambo, and mulatto populations. The population of Brazil is mixed, and Brazil probably has the second largest combined black and mulatto population of any country in the world, after Nigeria. Brazil's 1988 Constitution even specifies that the state recognizes land ownership by *quilombo* communities—communities of runaway slaves. Several hundred *quilombo* communities have maintained their lands for as long as 200 years, and their populations may number up to 500,000.

In the United States, about 12 percent of the population is African American—27 million people—

giving the United States another one of the world's largest black national populations. The U.S. black population was originally concentrated in the Southeast (the historic and vernacular "South" shown on Figure 3-3), but blacks began to migrate to the cities of the North just before the time of World War I (see Table 5-1). Between 1910 and 1920 job opportunities in war-related industries pulled rural blacks north at the same time as a boll weevil infestation destroyed southern cotton fields and threw many black laborers out of work. These pulls and pushes accounted for most of the 1.5 million black migrants to the North between 1910 and 1945.

After the 1940s, the main push factor was the revolution in southern agriculture. The mechanical cotton picker, introduced in 1944, made many agricultural workers obsolete. As the blacks were pushed off the farms they continued to be pulled into the cities, and particularly the cities of the North, by the promise of economic opportunity and enhanced civil liberties (Figure 5-9). Between 1945 and 1970, 5 million more blacks followed the 1.5 million who had

TABLE 5-1
Estimated Net Black Migration by Region, 1871–1980

Intercensal period	South	Northeast	North central	North total	West
1871–80	–60	+24	+36	+60	(na)
1881–90	–70	+46	+24	+70	(na)
1891–1900	–168	+105	+63	+168	(na)
1901–10	–170	+95	+56	+151	+20
1911–20	–454	+182	+244	+426	+28
1921–30	–749	+349	+364	+713	+36
1931–40	–347	+171	+128	+299	+49
1941–50	–1,599	+463	+618	+1,081	+339
1951–60	–1,473	+496	+541	+1,037	+293
1961–70	–1,380	+612	+382	+994	+301
1971–75	+14	–64	–52	–116	+102
1976–80	+195	–175	–51	–226	+30

Note: Numbers in thousands. Plus sign (+) denotes net in-migration; minus sign (–) denotes out-migration.
Source: U.S. Bureau of the Census 1981.

168 *Chapter 5*

migrated North since 1910. This migration of southern rural blacks to the cities of the North was one of the greatest migrations in human history—many times the total number brought to the United States forcibly through more than 2 centuries of slavery. Between 1910 and 1970, the percentage of African Americans living in the South fell from 70 percent to 50 percent.

The flow patterns of this migration are clear. The states of the lower Mississippi Valley region, particularly Arkansas, Louisiana, Mississippi, and Alabama, supplied migrants to Milwaukee, Chicago, and Detroit. The black migrants to the eastern cities Washington, D.C., Philadelphia, and New York, by contrast, had headed north from coastal Virginia, the Carolinas, and Georgia. The population of New York City, which was only 2 percent black in 1900, rose to 11 percent black by 1950 and 21 percent black by 1970. These migrants' subsequent experiences will be discussed in Chapter 9.

Since the mid-1970s, African Americans have been migrating from the North to the South. Some of these migrants are actually the same people who migrated north 20 or 30 years ago, whereas in other cases the migrants are the children or grandchildren of those who migrated north. The reasons for this new migration are not completely understood, but new opportunities have opened for blacks in the South, and conditions in many northern cities have deteriorated. Some 52 percent of African Americans lived in the South in 1980, but 56 percent by 1990. Whereas the original migration to the North was a rural-urban migration, this new migration is from northern cities to southern cities. Therefore, the pattern of black settlement in the United States today shows urban concentrations in the North and West, but little rural black settlement in those regions. African Americans still form high percentages of the local populations throughout the South.

The twentieth century has also witnessed a steady migration of West Indian and Caribbean blacks into the United States. Between 1960 and 1990 alone, about 2 million Caribbean blacks migrated to the United States. A high percentage of these people brought capital or job skills, and they and their descendants have played a role in twentieth-century U.S. politics and culture disproportionate to their numbers. Statistical descriptions of America's foreign-born blacks and of their children reveal high educational attainment, high income levels, substantial representation in professional and managerial occupations, and low levels of unemployment. These statistics more closely match the statistical descriptions of America's highest-achievement whites—the Jews and the Irish—than they match the description of America's often-underprivileged native blacks. The first African-American Chairman of the Joint Chiefs of Staff, General Colin Powell, was the son of immigrants from Jamaica. This flow of migrants and sojourners to the United States, combined with the cultural and economic impact of U.S. tourists and investment in the Caribbean, has caused a degree of acculturation referred to as the "Americanization of the Caribbean."

The cultural and economic might of black Americans overall exerts a tremendous attraction among blacks in Africa. Black America's literature and consumer products blanket the continent, and black American civil rights leaders are heroes, a tradition begun by W.E.B. DuBois (1868–1963), who settled in Ghana (Figure 5-10, p. 172). Senegalese poet-president Léopold Senghor (1906–) specified poet Countee Cullen (1903–1946) among African Americans whose works sustained racial pride throughout Africa under colonialism. African American leaders in politics, the arts, and business have instituted periodic African/African-American summit meetings. The first was held in Abidjan, Ivory Coast, in 1991, and the second, held in Libreville, Gabon, in 1993, attracted over 1,000 Americans and 3,500 Africans, including some 20 heads of state or government. These meetings have strengthened the cultural and economic links between the United States and Africa.

Because of America's black population and its many achievements, American culture has diffused throughout Africa to a surprising degree. It is displacing the cultural influence of Africa's European former rulers, despite the fact that the United States never had any African colonies.

European Migration to Africa

Europeans migrated to Africa and settled in substantial numbers only where environmental conditions made possible a Mediterranean or European style of life and commercial agriculture.

East Africa, Northwest Africa, and West Africa

The East African highlands attracted white settlement after the opening of the Suez Canal in 1867. Europeans claimed the fertile lands in today's Kenya, Tanzania, and Uganda, establishing there what was for them a pleasant and prosperous lifestyle. The natives were crowded onto less fertile lands, and capital investment improved the economic situation of the whites, but not of the native peoples. Much of the black population ended up as workers on the white-owned farms or in white enterprises. Tens of thousands of whites remain scattered throughout East Africa, but their economically-privileged position is now under pressure from the black-majority governments.

Northwest Africa is called in Arabic the *Maghrib*, "the island," because the Mediterranean to

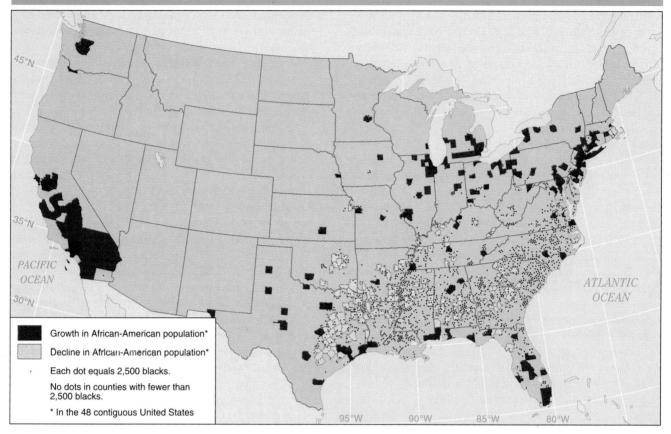

Distributions and Migrations of African Americans, 1950

Growth in African-American population*

Decline in African-American population*

· Each dot equals 2,500 blacks.

No dots in counties with fewer than 2,500 blacks.

* In the 48 contiguous United States

FIGURE 5-9

Each dot on these maps represents 2,500 African Americans, and the colors indicate whether each county gained or lost black population in the previous decade. The counties that contain fewer than 2,500 African Americans have no dots, and in some large cities the density of black residents becomes so great that it is impossible to distinguish the number of dots. The map for 1950 (a) reveals that during the decade of World War II, the black populations of many northern cities increased rapidly. Also, many blacks moved to California and Washington, where airplane manufacturing and other industries grew. Meanwhile, black populations declined in many southern rural counties. Map (b) indicates that between 1980 and 1990 the migrations continued. The rural South continued to lose black residents, while the black populations of cities in almost all regions increased. The black populations of many western states remain negligible. (Data courtesy of U.S. Bureau of the Census.)

the north and the vast Sahara to the south isolate this region, which itself supports settlement and agriculture. Here too European colonial governments reserved the best lands for European settlers. During French rule of Algeria from 1830 to 1962, some 1.2 million Europeans settled there—about as many as in all 26 countries of tropical Africa combined. When Algeria, Morocco, Tunisia, and Libya gained their independence, most white-owned farms were confiscated or bought by Arabs or Arab governments and the land was often redistributed to the peasantry. Few Europeans remain in these countries as farmers, although French ties remain strong with Algeria and Morocco, and tens of thousands of French citizens live in those countries engaged in business or retired.

West Africa never experienced substantial white settlement. The heat and humidity discouraged a European lifestyle, so the region's population never exceeded 1 or 2 percent white, most of whom were colonial administrators and entrepreneurs. The incidence of fatal disease among whites caused the area to be called "The White Man's Graveyard." Today whites still come or serve as business executives, technicians, or even government advisors, but they still make up only insignificant percentages of West African populations.

South-Central and South Africa Southern Africa includes large areas that Europeans found well-suited to settlement and creation of a European lifestyle.

Distributions and Migrations of African Americans, 1990

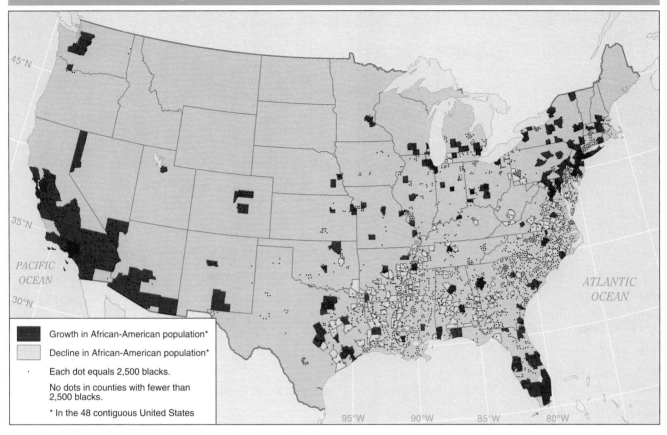

Growth in African-American population*

Decline in African-American population*

· Each dot equals 2,500 blacks.

No dots in counties with fewer than 2,500 blacks.

* In the 48 contiguous United States

Several colonies did achieve considerable prosperity—for the whites—but always based on repression and exploitation of the native black majority. As these colonies won independence, white racist regimes yielded to black-dominated governments (Figure 5-11).

The Republic of South Africa. In 1671, the Dutch established Capetown colony as a staging point for voyages to the East, and numerous Dutch farmers, called **Boers**, came to settle. Britain eventually conquered and annexed all of southern Africa, and in 1910 it formed the independent Union of South Africa. This state became a republic in 1961 and drew into increasing international seclusion in the face of opposition to its policy of racial segrega-

tion, called **apartheid**. The white minority perpetuated a strictly segregated society, with each ethnic group holding a fixed status. Laws enacted by the white minority government divided the population into four groups: whites (British and Boers combined, totaling 13 percent of the total population in 1992), Asians (2 percent), people of mixed race (or "colored," 9 percent) and blacks (75 percent).

The colored, concentrated in Cape Province, are mulatto products of early mixing between Boers and Hottentots. Most Asians came to Natal to work on the sugar plantations, but eventually many rose to middle-class status or even achieved considerable wealth in the capital, Durban. The South African government stopped immigration of Asians in 1913, and it main-

Human Migration **171**

FIGURE 5-10

U.S. civil rights leader W.E.B. Du Bois migrated to Ghana after that African country achieved independence in 1957. Ghana's president Kwame Nkrumah invited Du Bois to come over to edit a series of publications on black history. As a result, Du Bois earned fame throughout Africa. (Courtesy of AP/Wide World.)

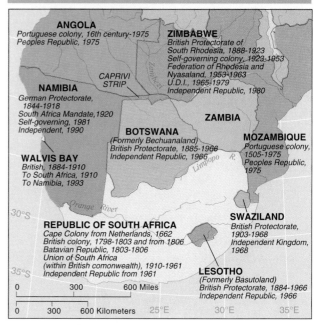

FIGURE 5-11

Substantial numbers of Europeans settled in the midlatitude lands of southern Africa when these were colonies, and today's independent countries are still working to achieve ways for whites and majority blacks to live together.

tained for 50 years the legal fiction that they would someday return to Asia. It did this in order to restrict Asian political rights. Only in 1963 was the Asians' residency recognized as permanent.

The government planned to move the blacks into ten segregated **Bantustans**, on the basis of tribal affiliations (Figure 5-12). The Bantustans covered only 13 percent of the territory of South Africa, rural areas of low agricultural potential and without other resources. The blacks could never support themselves in these lands, so they had to sell their labor within South Africa, where they were immigrant workers without political rights. By 1990, the per capita incomes of each of the groups were $4,720 for whites; $1,655 for Asians; $1,174 for people of mixed race; and $560 for blacks.

By the late 1980s, the system of apartheid and Bantustans had begun to collapse under the pressures of international sanctions and internal rebellion. A new governmental system was devised and free elections open to people of all races were held in April 1994. The Bantustans were merged into South Africa, and Nelson Mandela, a black leader who had been imprisoned for 27 years during the struggle for civil rights, won the new presidency.

The new democratic government will face tremendous pressures. Blacks understandably want the life of conspicuous affluence that South African whites have enjoyed, but South Africa does not have the wealth to provide that for all its people. The new government will also try to ease black intertribal rivalries that triggered massacres during the civil rights struggle.

Other Southern African States. The Portuguese colonies of Angola and Mozambique did not have many white settlers when they became independent in 1975, and most of those settlers fled, often destroying what they could not take with them. Botswana also had few white inhabitants when it gained independence in 1966, and when Namibia gained independence from South Africa in 1990, whites made up only 5 percent of its population. Zambia won independence in 1964 with a black-dominated government which, in 1975, nationalized all private landholdings and other enterprises. Today few whites remain.

The white population of Zimbabwe (previously known as the colony of Southern Rhodesia) unilaterally declared independence from Britain in 1965 in order to forestall the granting of rights to the majority black population (Figure 5-13). Fifteen years of international boycotts and internal civil warfare followed until a new

The New Map of the Republic of South Africa

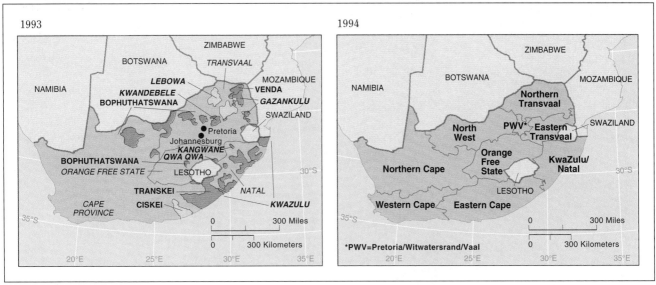

FIGURE 5-12

South Africa's new 1994 constitution abolished the Bantustans and established one central government and nine provincial legislatures.

constitution was internationally accepted. It guaranteed the whites, who constituted 6 percent of the popula-

FIGURE 5-13

The modern country Zimbabwe takes its name from this city ruin. Zimbabwe (which means "stone house" in the Bantu language) was first occupied by Iron Age people in the third century, abandoned at some time after that, and re-occupied in the late ninth century. The buildings were once richly decorated with stone carvings and gold and copper ornaments. There is evidence that fifteenth-century kings here ate off Chinese porcelain.
(Photo by Eliot Elisofon, National Museum of African Art, Eliot Elisofon Archives, Smithsonian Institution.)

tion, political protection in a system of racial representation. This system lasted until 1987, when the whites' privileged status was ended. By 1990, due to white emigration and a high black birth rate, whites made up less than 1 percent of the population.

European Migration to Asia and Australasia

Although the Spanish, French, Dutch, Germans, Portuguese, British, and Americans all created empires in Asia, few people from these countries settled permanently in the Asian colonies. In no case did any of these peoples come to make up significant proportions of mainland Asian populations. The most numerous European settlers in Asia can conveniently be divided into those who migrated by land—the Russians—and those who migrated by sea—settlers from the British Isles who went to Australia and New Zealand, which together are often referred to as Australasia.

Russian Expansion Ever since the Russians rose up against their Mongol rulers in the sixteenth century, they have sent a steady stream of settlers to the East, over the Ural Mountains into Asia. Between 1860 and 1914, almost 1 million Russians resettled in the East, and this flow continued within the political framework of the Union of Soviet Socialist Republics (see Figure 5-14). In 1990, the population of the cen-

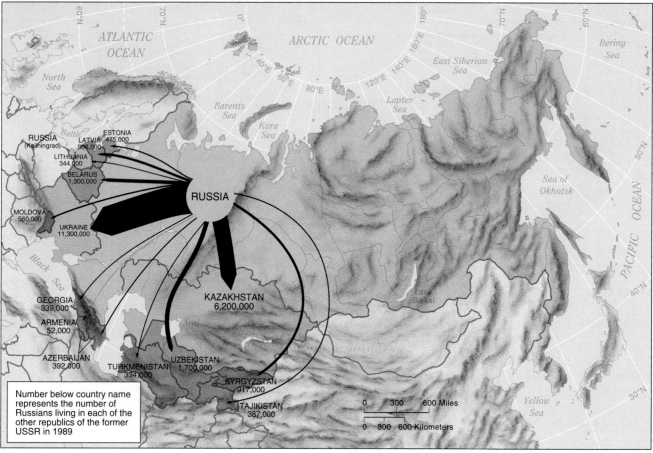

FIGURE 5-14

Russians migrated from their homeland around Moscow both westward toward the Baltic Sea and eastward deep into Asia. Recent political developments have caused a reverse flow.

tral Asian Republic of Kazakhstan, which was ostensibly a Republic of the Kazakh people, was 38 percent Russian. Russians migrated particularly into the cities in the non-Russian republics. In 1990 Tashkent, for example, the capital of the Uzbek Republic and the third largest city in the USSR, was well over 50 percent Russian.

The dissolution of the USSR and the achievement of independence by the central Asian Republics in 1991 triggered substantial migration of ethnic Russians into Russia, although many of these have not lived in Russia for generations. Many other Russians, however, have stayed in the newly independent republics. The Russians' economic, political, and cultural role as important minorities in these republics will be defined through the 1990s.

Australia and New Zealand Over 2 million Europeans, mostly from the British Isles, moved to Australia and New Zealand in the nineteenth century,

and another wave of 3 million Europeans (half of them British) arrived between 1945 and 1973. As in the Americas, the natives fell victim to disease and abuse. Latest censuses count only 300,000 **Aborigines** in Australia (of a total population of 17 million) and 295,000 **Maoris** among New Zealand's 3.3 million people. The Maoris may regain control over parts of the islands of New Zealand, pending outcome of judicial review of the 1840 Treaty of Waitangi. Legal actions can be expected to last until the end of this century.

In both Australia and New Zealand, commercial farming and ranching thrived from the beginning of white settlement. Wool, wheat, and later frozen meat were sent to markets within the British Empire. In Australia, white settlement in the dry areas was hazardous and only intermittent until spectacular mineral discoveries attracted some people inland. In general, however, the white populations of both Australia and New Zealand have always been highly urbanized and concentrated along or near the coasts (see Figure 5-15).

Population Distribution in Australia and New Zealand

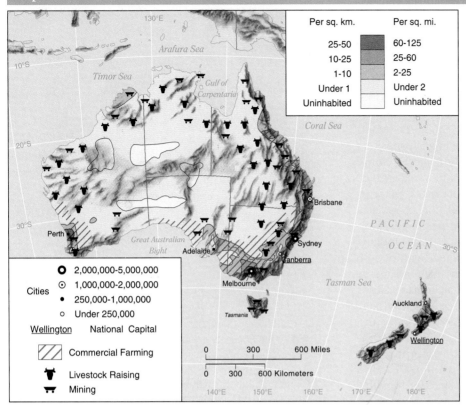

FIGURE 5-15

Vast stretches of inner Australia are virtually uninhabited.

The Migration of Indians

When the British ruled the Indian subcontinent they encouraged migration to other parts of their empire. Sometimes these migrants have been undifferentiated as "East Indians," whether they migrated from what is today India, Pakistan, or Bangladesh, but most did emigrate from today's India. They have been more important than their numbers alone might suggest, because wherever they went they formed a bourgeois infrastructure that ranked both socially and economically higher than the natives, as noted already in the South African state of Natal.

Indians settled in both Myanmar and Singapore, but since Myanmar achieved independence in 1948 the Indians have been persecuted there. Over 200,000 left in the 1980s, but several hundred thousand remain. Singapore is about 6 percent Indian. Indians went also to the Pacific Ocean island of Fiji, but continuing friction between Indians and Fijians has triggered Indian flight to Australia and New Zealand.

Indians still form key elements of the populations in several West Indian countries, even where their percentages are small. Belize is about 2 percent Indian and Jamaica about 3 percent, but Trinidad and Tobago is about 40 percent Indian, and Guyana on the South American mainland is about 50 percent Indian.

Indian migration to British East Africa surpassed white migration there. Indians generally made up less than 5 percent of the populations, but they dominated the commercial infrastructures of the new countries that emerged in the 1960s (Uganda, Kenya, Malawi, and Tanzania), and their high visibility made them targets of the majority blacks' ethnic hostility. Uganda even expelled over 80,000 Indians, who emigrated to Great Britain, Canada, and the United States. Several African countries have recently tried to entice the Indians to return. Over 70 percent of the economy of Kenya was said to be in Asian hands in 1990, and Uganda is offering economic privileges to Indians who return or their descendants.

Today more than 10 million overseas Indians represent one of the world's most diverse diasporas, and they have enjoyed widespread success. The several million Indians in Britain own about 60 percent of all independent retail outlets there. Indian immigration to the United States rose 126 percent during the 1980s, faster than any group except the Vietnamese, and their success in the United States is exceptional. In 1990, first-generation Indians enjoyed a median family annual income of $52,908, the highest of any foreign-born group. Nearly half of the 1 million Indians in the United States are professionals (including about 4 percent of U.S. doctors), and almost half of all U.S. motels are owned by Indians.

Several reasons have been cited for Indians' notable international success. One, as suggested above, is that they have been able to take advantage of the network of commercial outposts established by the former British Empire. A second reason is their facility with English, which is increasingly the world language of business. India has more English speakers than any country except the United States. Some Indians have themselves cited the cosmopolitan character of Indian culture as a third explanatory factor in their international success. The population of India is so diverse—in terms of skin colors, languages, religions, and other cultural attributes—that many Indians learn during childhood how to cooperate with people of all backgrounds. A fourth factor is that Indians in many countries help one another. If an Indian cannot borrow money from a local commercial bank to start a new business, local fellow Indians often raise the necessary capital. Hard work and frugality are additional factors often cited by Indian entrepreneurs themselves, but these characteristics often describe groups that still do not enjoy the success that the Indians have. The Indians demonstrate that entrepreneurial skill is one of the most valuable resources in the world.

The Overseas Chinese

From early in the nineteenth century until about 1930, waves of South Chinese left their homeland to work in the British, French, and Dutch Asian empires (see Figure 5-16). Their success mirrored that of the Indians. At the time the nations of Southeast Asia gained their independence, their economies were firmly in the hands of these small Chinese minorities, who were often linked across international borders by bonds of family, clan, and common home province.

The Chinese commercial supremacy and their international links still survive, so that today these 44 million **Overseas Chinese** constitute a formidable economic power throughout Southeast Asia. Furthermore, the Japanese have made considerable investments in these countries, and the Japanese have often preferred Chinese as business partners to native peoples. This partnership between Japanese and Overseas Chinese has everywhere bolstered the economic supremacy of the Overseas Chinese.

The Overseas Chinese have often faced native ethnic hostility, just as the migrant Indians have. For example, ethnic resentments dominate the history of

The Migrations of Overseas Chinese

FIGURE 5-16

The number after each country is the percentage of its population that is Chinese. Even where the Overseas Chinese make up small minorities of the population, they exercise financial influence far beyond their numbers.

Malaysia. When the British ruled Malaysia as a colony, they founded the cities of Singapore and Kuala Lumpur. The populations of both cities came to be overwhelmingly Chinese, although the majority of the population of the entire colony was Malay. Malaysia was granted independence in 1957, but animosity was so great between Chinese and Malays that in 1965 Singapore broke off as an independent, Chinese-dominated city-state. Even in Malaysia, however, the Chinese minority still controls the economy, although the Malay-dominated government blocks Chinese and Indians from advancement in government positions and encourages Malay entrepreneurship. The population of Malaysia today is about 59 percent Malay, 32 percent Chinese, and 9 percent Indian.

In Indonesia the 5 million Chinese control an estimated 80 percent of the country's private industry. Indonesian corporations that are owned by ethnic Chinese count among the world's largest, and the Indonesian government requires these Chinese-owned corporations to invest in businesses owned by other ethnic groups. A 1965 Communist coup attempt, supposedly backed by China, ignited anti-Chinese sentiment, and during the next year (*The Year of Living Dangerously* in the title of a Hollywood film about the events) as many as 500,000 ethnic Chinese were slaughtered. Violence against the Chinese has flared

FOCUS
Ethnic Mixes in the Caribbean

As a result of migration during the colonial period, the population mixes on the islands of the Caribbean and the lands around its shores demonstrate an astonishing variety of peoples from around the world. The native American populations were conquered and ruled by several different European countries: Spain, France, Britain, Denmark, Holland, and the United States. Each of these imperialist powers imported slaves or brought settlers not only from its homeland, but also from other places in its global empire. Blacks were usually brought to work the plantations, and Asians from India, China, Malaysia, and Indonesia were brought to mind the shops and government offices. European powers periodically traded Caribbean colonies, superimposing each time new settlements of people of still more diverse backgrounds.

When the Caribbean countries received independence, each tried to weld a nation out of its diverse peoples. The coat of arms of Jamaica reads: "Out of Many, One People." This is the same motto as that of the United States ("E Pluribus Unum"). Not all Caribbean countries, however, have enjoyed racial harmony. Surinam received its independence from the Dutch in 1975, and its last census before independence, in 1972, demonstrates the degree of ethnic mix there. It was then 35 percent Indian, 30 percent African, 12 percent Indonesian, 8 percent Chinese, 8 percent Bush Negro (descendants of runaway slaves who fled into the jungle), 3 percent native American, and 4 percent European.

Many people feared that independence would trigger ethnic violence, and in the 18 months before it, one-quarter of the total population fled. From 1980 to 1990 the country was ruled by a military dictator and disturbed by civil insurrection. Its economy collapsed, despite rich deposits of bauxite and Dutch and American foreign aid. Democracy may be achieved in the 1990s. The estimated 1990 population mix was 37 percent Indian, 31 percent African, 15 percent Indonesian, 10 percent Bush Negro, 2.6 percent native American, 1.7 percent Chinese, and 1 percent European.

periodically through the early 1990s, and the future status of Overseas Chinese in Indonesia remains unclear.

Overseas Chinese play a similar elite role in the Philippines, one-third of whose economy is estimated to be in the hands of Chinese, and Thailand's commercial infrastructure is almost entirely Chinese. Thailand's largest corporation, Charoen Pokphand, was founded by two brothers from South China in 1921, and today it is one of the world's largest producers and exporters of foods. It owns production and distribution facilities from Argentina to China. The "Bumble Bee" brand tuna popular in America is one tiny part of another Chinese-Thai multinational conglomerate. Thailand is a major producer and exporter of food and fish products.

Chinese migration was not restricted to Southeast Asia. Ambitious Chinese also established themselves along the west coast of the Western Hemisphere from British Columbia to Ecuador, and in small but significant numbers in the Caribbean. Tens of thousands of Chinese migrated throughout the U.S. West in the mid-nineteenth century, and a new wave of Chinese migration to the United States that began in 1965 is discussed later in this chapter.

Today tens of thousands of Chinese are emigrating from Hong Kong before that colony reverts to Chinese rule in 1997. Many of them have headed for Australia or for Canada, because both of those countries extend special welcome to Chinese who bring money to invest. Wealthy Chinese immigrants have made Vancouver, Canada, which was 15 percent Chinese in 1991, a major financial center. Other Hong Kong Chinese purchased East German passports just before German unification in 1990, and over 50,000 more were attracted to Panama in the 1980s by the sale of naturalization papers by that government.

The Overseas Chinese maintain international business contacts. One economist has estimated that although the Overseas Chinese population is only 4 percent of that of China, their total income is about two-thirds as big as China's. The overseas financial network is linked with the strong economies of Hong Kong and Taiwan, and it has already proven to be a valuable resource for China itself. Overseas Chinese have raised investment capital, invested in mainland industries, introduced the latest technologies to China, and distributed Chinese products worldwide. Thus, the Overseas Chinese help account for the rapid growth of the Chinese economy.

FOCUS
Indigenous Peoples

In several cases the impact of international migrants in their new homes has been deleterious to the welfare of the natives, or **indigenous peoples** of that region. The year 1992, the quincentenary of Columbus's voyage across the Atlantic, became, to the surprise of many people, an occasion not only to praise what Columbus had achieved, but to reappraise the history of European impact on the world for the past 500 years.

For several years a United Nations Committee has been composing a "Universal Declaration of the Rights of Indigenous Peoples," which will eventually be presented to the General Assembly for ratification. The committee estimates that there are today 300 million indigenous people in 70 countries, defined as "descendants of the original inhabitants of a land who were subjugated by another people coming after them." The United Nations hopes to ensure some degree of self-determination for these peoples. Self-determination does not necessarily mean political separatism, but the right to pursue and protect their own culture. The United Nations General Assembly designated an "International Year for the World's Indigenous Peoples," beginning December 10, 1992, which was International Human Rights Day.

The struggles of indigenous peoples won further world attention when the 1992 Nobel Peace Prize was awarded to Rigoberta Menchú, a Quiché Indian from Guatemala. She has written and spoken out to defend civil rights in her country, and the prize was given "in recognition of her work for social justice and ethno-cultural reconciliation based on respect for the rights of indigenous peoples."

MIGRATION TODAY

Human migration has by no means come to an end. In 1992, the World Bank estimated international migrants of all kinds at 100 million. Of these some 35 million were in sub-Saharan Africa, and 13 to 15 million were in each Western Europe and North America. An additional 15 million were in Asia and the Middle East, where a few countries had particularly heavy immigrant concentrations. Large-scale migrations still make daily news. The United Nations *Universal Declaration of Human Rights* affirms anyone's right to leave his or her homeland to seek a better life elsewhere, but it cannot guarantee that there will be any place willing to take anyone.

The greatest migration flow today is from the world's rural areas into the world's cities, a topic discussed in Chapter 9. Regarding international migration flows, people are trying to move from the poor countries to the rich countries and from politically repressive countries to more democratic countries. In addition, civil and international warfare inevitably create large numbers of refugees.

Pressures for migration are growing, and in coming years they may constitute the world's greatest political and economic problem. Smuggling gangs fill boatloads of Chinese destined for California, Haitians take to the seas in rickety boats to reach Florida, and thousands of Africans clamber aboard planes destined for Europe. In 1993, 156 Iraqis chartered an Egyptian airliner and flew to London to request asylum. "There are no distances any longer in this world," said Austria's Director of Immigration, Dr. Manfred Matzka. "There are no islands."

Immigrants may be poor, uneducated, or unskilled, but usually they are also enterprising. They seek opportunity, and it takes courage, stamina, and determination to pull up roots, say good-bye to all that is dear and familiar, and head away to a foreign and perhaps hostile land. Illegal migration, with all its hazards, is truly daring. Many Latino men who wait on Los Angeles street corners, hoping for work, have faced more risk than most Americans will ever know. In 1993, Mexico's President Carlos Salinas de Gortari said, "I don't want any more migration of Mexicans to the U.S., especially those migrants who are very courageous and take the impressive risk of going to the U.S. looking for jobs. That is precisely the kind of person I want here."

Political Refugees

Many migrants claim to be political refugees, but it is often difficult to discern who is a political refugee and who is only an "economic refugee" seeking a higher material standard of living. The 1951 Geneva Convention defines an international **refugee** as someone with "a well-founded fear of being persecuted in his country of origin for reasons of race, religion, nationality, membership of a particular social group or political opinion."

Some countries define "social group" more broadly than others. For example, several European countries and Canada recognize persecution due to sexual orientation as a claim to refugee status, and U.S. immigration courts have extended asylum to gay people who can prove discrimination in specific countries. In 1984, the European Parliament accepted the legal principle that in some countries women constitute a persecuted "social group" because of the way they are customarily treated. The United Nations Commission for Refugees accepted this idea in 1985, and Canada has extended refugee status to women from Saudi Arabia, recognizing that customs there place women "in a more vulnerable position than men."

Worldwide, international wars and civil wars boosted the numbers of political refugees in the late 1980s. The United Nations High Commissioner for Refugees estimated that on December 31, 1992, 19.7 million people had been driven across international borders, and an additional 24 million people had been forced into exile within their own countries (Figure 5-17). Asia had the largest number of refugees with some 7.2 million, although that figure was down from 8.6 million in 1991. Africa followed with 5.4 million refugees, while Europe was in third place with 3.6 million. Iran was then granting asylum to the largest number of refugees in the world, a total of 4.1 million, including 2.9 million Afghans and 1.2 million dissident Shiite Muslims. Iran was followed by Pakistan with 1.6 million Afghans. Malawi ranked third, with about 1 million refugees fleeing civil war in neighboring Mozambique. Political upheaval anywhere multiplies refugees, and in the event of such mass flight from civil wars, it seems futile legalistically to assign one cause or another to the migration.

International Refugees

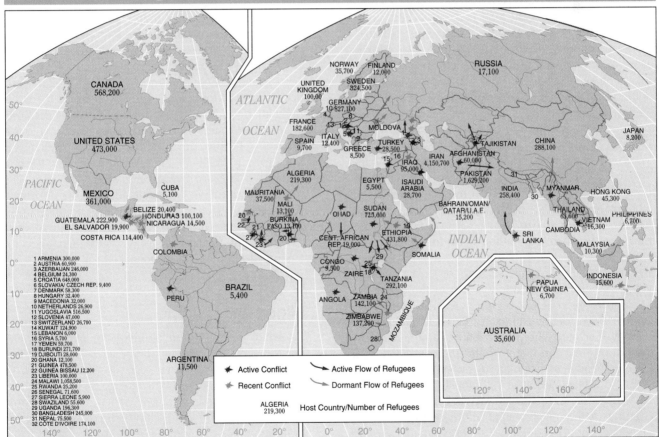

FIGURE 5-17

This map shows the world international refugee situation on December 31, 1992. It does not include the millions of people in countries around the world who were "internal refugees," that is, within their own home countries but not at their own homes because of civil wars or persecution. (Recreated with permission from the United Nations High Commission for Refugees, from *The State of the World's Refugees: The Challenge of Protection*. Published by Penguin Books, The Penguin Group, copyright © 1993 by the United States High Commissioner for Refugees.)

THE IMPACT OF INTERNATIONAL MIGRATION

The impact of international migration is even greater than the numbers might suggest. Immigrants are often greeted with apprehension, and in many receiving countries immigration has become an explosive political issue. This is for several reasons. First, official statistics substantially underestimate actual numbers; undocumented migration is rising everywhere. Second, migrants are usually in the peak years of fertility. Therefore, they are playing an increasing role in the total population growth in the rich target countries. Chapter 4 noted that in the United States, for example, total fertility among foreign-born women is about 25 percent higher than among native-born women. Third, migrant settlements are generally concentrated in a few places within a country. This usually adds to the immigrants' visibility and increases the perception of cultural differences.

The peoples of the rich and democratic countries are beginning to exhibit "compassion fatigue," and doors are closing all around the world. Moreover, as several of the world's major economies, including the United States and European countries, suffered recession in the early 1990s, native hostility against immigrants increased. Immigrants who were once grudgingly tolerated because they were willing to work in low-wage or dirty jobs suddenly came to be seen as competitors and became targets of wrath. Today no country in the world allows free immigration. Democratic governments, however, have generally proven incapable of coping with illegal immigration; once immigrants are in, democratic countries find it difficult to muster the brutality necessary to evict them.

Every demographic impetus toward migration will be multiplied over the next two or three decades as populations rise in the poor countries. Either substantial flows of investment capital and technology will be directed to the poor countries and create new job opportunities in them, or else the combination of large increases in population, excess labor supply, rising social and political turbulence, and persistent or worsening inequalities between rich and poor countries will inevitably stimulate migration in search of better living conditions. Furthermore, the spreading of images and information about the rich countries through global communications media and social networks is strengthening their pull.

Some poor nations may be relieved to see their unemployable or troublesome people emigrate, especially if those people send money to their families back home. Several countries have come to depend on **workers' remittances**, that is, money sent home from abroad, for a substantial share of their income. In 1991, the estimated total world value of international remittances was about $75 billion. This was second only to oil in its value in international trade, and larger than all international development assistance ($46 billion in that year). Workers' remittances from abroad constituted 10 percent of the gross national product (GNP) of Egypt, 11 percent of Morocco's, 18 percent of Jordan's, and 5 percent of Pakistan's. These countries' economies become dangerously dependent upon the status of their citizens in foreign countries. Studies have shown that remittance income may result in little overall productive investment in the receiving country. It is used for current consumption, health, and education of the recipient households.

The social impact of out-migration can be severe. If males migrate away alone, as is often the case, they leave wives and children at home. This transforms social patterns and sex-based economic and social roles in the home community. In some cases, new opportunities open for women, but in most others, women's autonomy is curtailed by social expectations and by an increased burden of work.

Many rich countries today accept foreign workers, but they do not grant them citizenship or the rights that citizenship would bring. The host countries accept them only as sojourners, and it may be unclear whether the migrant workers themselves hope to stay. Many migrants hesitate to apply for new citizenship because by doing so they lose citizenship rights in their native lands, including rights of property ownership and inheritance. The conditions under which the host countries grant citizenship, or even guarantee full civil and legal rights to aliens, become matters of international concern.

Migration to Europe

An estimated 15 million people poured into Western Europe between 1980 and 1992, and a million or two more are still arriving each year. Most have come from North Africa or Turkey, but the numbers of migrants from Asia and sub-Saharan Africa are increasing. These numbers do not include people who migrated from one European country to another.

The number of people arriving in Europe who claim political refugee status has mushroomed. Up until the mid-1970s, only about 30,000 per year claimed refugee status, but as civil turbulence wracked many Asian and African states, and as a claim of refugee status has come to be seen as a side door for people fleeing economic deprivation, asylum requests rose dramatically. They reached 700,000 in 1992 and cost host governments billions of dollars. The majority of applicants are denied refugee status, but many manage to stay on anyway after appeals that last for years.

Large numbers of immigrants have settled in Great Britain (in 1990 almost 9 percent foreign-born, mostly West and East Indian), in France (about 11 percent foreign-born, mostly North African Arabs and African blacks), in Belgium (8 to 9 percent foreign-born, mostly Turkish and North African), in Spain (about 2 percent North African), and in the Netherlands (about 4 percent foreign-born). In the former West Germany in 1990, the resident foreign population was 8.4 percent of the total, mostly Turks, Poles, and peoples from what was then Yugoslavia.

The European countries have not been notably successful, economically or socially, at absorbing the immigrants. Already in 1982, a United Nations report on international migration noted that "while Western Europe's foreign workers have now become entrenched as a structural element of labour markets in most of the receiving countries, their economic integration has not been paralleled by their social integration in the host countries." Many immigrants from the Near East and North Africa are Muslims, and, as we shall discuss in Chapter 8, their religion sometimes clashes with traditional European Christian culture. These immigrants do not always want to be acculturated to European traditional culture, so misunderstanding and mistrust arise both among the immigrants and among the residents of the host countries. It is one thing to deal with immigrants, even great masses of them, who see the new host society as admirable and the place of their future prosperity, as many millions of immigrants have viewed the United States for 200 years; it is quite another problem to deal with newcomers who reject many of the basic beliefs of the society to which they migrate.

The immigrant peoples have higher birth rates than the Europeans do, and compared to the Europeans around them, these groups achieve lower levels of education, occupy substandard housing, and receive inferior services. These conditions have triggered riots in immigrant ghettoes in several European capitals (Figure 5-18).

The collapse of Communist party governments in Eastern Europe multiplied migration into Western Europe from that direction. The crumbling of Eastern Europe's Communist bloc has enabled many of its citizens to seek their fortunes in the West, and the joy with which Western Europe greeted the political development was quickly followed by alarm. Roughly

FIGURE 5-18

North African immigrants in Brussels offer a variety of services and goods to their compatriots.
(Courtesy Robert Mordant.)

2 million people left Eastern Europe in 1990, of whom 400,000 were from the Soviet Union. If the economies of Eastern European countries deteriorate in the 1990s, these numbers could rise (Figure 5-19). Western European troops are now patrolling borders to keep people out that were previously guarded on the eastern, Communist side to keep people in. War in the successor states of the former Yugoslavia also sent tens of thousands of refugees fleeing westward at a rate, from 1990 to 1993, of about 200,000 per year.

A few Eastern European expatriates have migrated back to Eastern Europe, bringing specialized education, job skills, and much-needed capital. These people have assumed leading positions in several Eastern European countries. In the early 1990s, citizens of Canada, Britain, Australia, and the United States were serving as government ministers in Ukraine, Hungary, Latvia, and several other Eastern European countries. In August 1993, the U.S. Department of State warned 12 U.S. citizens serving in the Latvian Parliament, including three in the cabinet, that they were in danger of losing their U.S. citizenships. A 1951 U.S. law strips of U.S. citizenship anyone serving in an official capacity in a foreign government.

Western Europe's New Exclusionary Policies

Many Western Europeans increasingly look wistfully back to a bygone era when their countries were sup-

FIGURE 5-19

This may be the future appearance of the international borders between the rich countries and the poor countries—a line of soldiers pushing back a desperate crowd. In this photo, taken in 1991, the soldiers are Italian, the refugees Albanian.

(Courtesy of G. Giansanti/Sygma.)

posedly homogeneous, orderly, and virtually all white. The changes in the makeup of the European population, however, are irrevocable. For generations Pakistanis have lived in Britain, Algerians in France, and Turks in Germany, and, even though they are still regarded by some people as alien, they are going to stay. Most European states are nevertheless now restricting new entry.

Many of the new arrivals in Western Europe had passed through one or more countries on their way from their homelands. The European governments argue that even if these people were refugees when they fled their homelands, they ceased being refugees once they had escaped. Therefore, they are only "country-shopping" for the most favorable new site to settle. In the early 1990s, European governments decreed that no refugee can win asylum by coming overland through a third country. They also lengthened the lists of countries from which visitors' visas were necessary and began fining airlines for bringing people without visas.

Germany is the richest and most populous country in Europe. It has taken in considerably more immigrants and refugees than any other European country, yet Germany is a special case for several reasons. For one, Germany is one of several countries in which the definition of an individual's citizenship is based on the citizenship of his or her ancestors. This concept is called **jus sanguinis** (law of blood). It contrasts to the concept used in the United States that an individual's

citizenship is based on his or her place of birth, a concept called **jus soli** (law of soil). Germany was defeated in World War II and was divided into zones of occupation—East and West—by the victorious powers. These zones became East and West Germany. Legally, however, these were "provisional" countries, so the status of immigration and naturalization law in both was confused and could not be changed until East and West were unified. This did not take place until 1990. Up until then—and even now—only people of German ancestry could immigrate freely.

Starting in 1955, prosperous West Germany signed "employment contracts" with many individual countries that dictated the terms under which foreign citizens could come to work in West Germany. Many thousands of foreigners did come. In addition, West Germany's 1949 "Basic Law" (which served in place of a Constitution) made it possible for any foreigner who wanted asylum to apply for it and stay in the country for years. Under this lenient policy, in 1992 alone some 448,000 asylum seekers, plus 250,000 war refugees from the Balkans and 100,000 more illegal immigrants, entered Germany.

Overwhelmed by these numbers, united Germany changed its Constitution in 1993. It restricted entrance to new asylum seekers and immediately began sending people back to the country they were last in. At the same time, however, conditions for citizenship were extended to anyone who had lived legally in Germany for 15 years. In addition, people between the ages of 17 and 23 were made eligible if they had lived in Germany for 8 years and attended school for 6 years.

France has received about 100,000 immigrants annually since 1989, and tensions have been increasing between France's Muslim immigrants and government policy. Government policy, for example, allows no religious expression of any kind in French schools, and when girls of Turkish and Moroccan background have shown up at school wearing headscarfs, they have been suspended. In 1993, France deported a local Muslim imam (religious leader) who had backed the girls by declaring that "Allah's law takes precedence over French law." At the same time France made it more difficult for children born in France of foreign parents to become French citizens. France abandoned *jus soli*. Previously, residents born in France became French automatically at age 18. Under the new law, they must apply in person between the ages of 16 and 21, and citizenship can be refused. The government also restricted potential immigrants' access to France and enhanced authorities' powers to expel illegal aliens. In 1993, the Minister of the Interior, Charles Pasqua, called for a policy of "zero immigration."

By the early 1990s, many Europeans felt that they had lost control of their borders, and as Europe pulls closer together in the European Union, it may build higher walls against outsiders.

Migration to Australia

About 90 percent of Australia's 17 million people are of British and European ancestry, and only 4 percent are Asian. Australia has indicated, however, that it might absorb more settlers. The country abandoned an explicit "White Australia" policy in 1973, but it still has to decide on which basis it will open its borders: refugee status, merit, investment, family ties, or other criteria. Today Australia accepts about 100,000 immigrants per year, roughly one-third of whom are Asian, and the Asian proportion is rising.

Migrations of Asian Workers

An estimated 4 to 5 million Asians work abroad, and in many cases it is doubtful whether many will ever return to their homelands. Meanwhile, their remittances to their homelands are an important element in the economies of their home countries.

The absence of these workers lowers the local unemployment rates in their homelands, but it also constitutes a loss of important skills needed back home. Although the home country might not have the resources to reward these workers adequately, it frequently needs their contributions. A young Filipino woman, for example, may be needed in her home town as a school teacher, but staying there she would earn $40 per month. If she is earning $150 per month as a maid in Singapore or $384 per month as a maid in Hong Kong, she is better off, but the Philippines suffers. In 1990, the Filipino consulate in Hong Kong estimated that 52,000 Filipino women were working as maids in Hong Kong. Many were college graduates.

In the 1970s and the 1980s, the oil wealth of the Arab states drew into them many Asian workers who seem little disposed ever to go home. A full 70 percent of the work force in the Arab Gulf states was foreign in 1985, 63 percent Asian. The Arab states grant these people few rights, and their children born in Arab states are not citizens. The 1991 Mideast War unsettled the status of millions of people in Iraq and Kuwait, and it caused all the oil-rich countries to reconsider their immigration policies. As long as the Middle East remains an international hotspot, and as long as Arab oil supplies are important to world prosperity, any destabilizing migration into Arab countries is of world concern.

About 1.5 million Vietnamese fled their homeland after the 1975 military victory of communist North Vietnam and the merger of the South with North Vietnam. The United States and several other countries have accepted thousands of these people, but in 1994 the United Nations ruled that Vietnamese fleeing their country would no longer automatically be considered political refugees. Many Vietnamese are today returning to their homeland.

Migration to Japan

The topic of migration to Japan can be dealt with quickly. There is virtually none. Of all of the world's rich countries, Japan limits immigration the most strictly. Between 1945 and 1993, only 222,000 foreigners acquired Japanese citizenship, including those who married Japanese. The number of permanent foreign residents has remained remarkably stable at about 900,000, significantly less than 1 percent of the national population. Japan's illegal population is reportedly increasing, but it is still minute at perhaps 300,000 total.

The most significant minority in Japan is the Japanese of Korean ancestry—about 0.5 percent of the total population. These people are a reminder that Japan ruled Korea from 1910 until 1945, and they suffer discrimination.

MIGRATION TO THE UNITED STATES

The single largest migration flow for the past 150 years has been migration to the United States. In 1992, the United States welcomed 974,000 new immigrants—more than all other industrialized nations combined. These numbers do not include illegal immigrants, who could number an additional half million each year. Immigration contributes to the national population growth in two ways. One is simply in the numbers of immigrants. The second factor is that immigrants have high birth rates. Therefore, immigrants are responsible for about one-third of U.S. population growth. Only two other rich countries are projected to experience significant continuing population increase in coming decades—Canada and Australia—and both of these countries are also targets for migration.

The United States was dubbed **the melting pot** in a 1914 novel of that title, but various groups have to some degree at least maintained their cultural identities. Perhaps it would be better to call the United States a **cultural mosaic**, which is what Canada proudly calls itself. In the 1990 census, the plurality of Americans (23.3 percent) claimed German ancestry. The next largest self-identified groups were Irish (15.6 percent); English (13.1 percent); African American (9.6 percent); Italian (5.9 percent); American (5 percent—it is unclear what that may have meant to these people); Mexican (4.7 percent); French (4.1 percent); Polish (3.8 percent); and American Indian (3.5 percent).

The History of U.S. Immigration Policy

Immigration into the United States was totally unrestricted until the late nineteenth century, when the government enacted rules explicitly designed to keep out the Chinese and other Asians (Figure 5-20). Two events in 1890 triggered the growth of **nativism**, which is an attitude or policy favoring the native inhabitants of a country as against immigrants. First, the director of the Census Bureau announced that all lands thought suitable for farming (basically, those east of the 100 degree line of longitude, to the west of which annual rainfall is usually less than 20 inches per year) were all occupied to a minimum of 25 people per square mile. In other words, there was no longer a frontier of unsettled farmlands. Second, stud-

FIGURE 5-20

This cap pistol reflects the virulency of U.S. anti-Asian animosity in 1876.
(Courtesy of the Smithsonian Institution.)

ies first revealed that the immigrants had higher total fertility rates than native-born Americans. Many nativists took this to mean that people defined as "them" were threatening to overwhelm people defined as "us." At the same time, the source area of most immigrants shifted suddenly from Northwest Europe to South and Eastern Europe (Figure 5-21). Unlike earlier immigrants, these newcomers, including large numbers of Italians, Slavic peoples, and Jews, were seen as "different." Therefore, in the nativists' minds, immigration had to be stopped. In U.S. history most nativists were white people and, therefore, themselves the descendants of immigrants; generally they were not advocates of rights for native Americans, that is, the Indians

No significant steps to limit immigration, however, were taken until the 1920s. Then the government issued immigration quotas for various nationality groups. The first quotas favored British, Irish, and Germans, who had been the dominant segments of the population back in 1890. By 1920, however, substantial numbers of Southern and Eastern Europeans had already immigrated, and the ethnic makeup of the country had already changed. After the 1920s, a maximum quota was set on immigrants from the Eastern Hemisphere (in practice, Europe), but there was in theory no ceiling on immigrants from the Western Hemisphere. The overall effect of the legislation still favored Europeans.

The Immigration Reform Law of 1965 changed the rules again in a way that eased immigration for Asians and Latin Americans. Supporters of the bill argued that it would cause no transformation in U.S. ethnic makeup. Attorney General Robert Kennedy, testifying in 1964, asserted confidently that the number of immigrants "to be expected from the Asia-Pacific triangle would be approximately 5,000." In 1965, Nicholas Katzenbach, Kennedy's successor as Attorney General, said, "If you look at the present immigration figures for the Western Hemisphere countries, there is not much pressure to come to the United States from these countries. There are in a relative sense not many people who want to come." The bill passed, and its consequences astonished its supporters. The number of Asians and Latin Americans migrating to the United States exploded. Observers had simply underestimated the pressures in Asia and Latin America for emigration, as well as the attraction of the United States.

The United States has continued to change immigration rules. The Immigration Control and Reform Act of 1986 granted amnesty to illegal residents who had been in the United States since 1982, and at the same time it imposed penalties on employers who hire illegal residents. Lax enforcement, however, and easily obtained false documents have defeated this barrier.

The 1990 census counted 19.8 million foreign-born U.S. residents, the highest number ever. More people came to the United States in the 1980s than in any previous decade in national history. Furthermore, the source areas of U.S. immigrants had changed significantly, and this will bring long-term consequences (Table 5-2 and Figure 5-22). In 1990, three-quarters of all Americans were non-Hispanic whites; by 2050 that percentage may fall below half. People who are 30 years old or older and who live in the United States today might have the experience of having grown up in one country but growing old in another—without moving.

In 1990, the United States rewrote the national immigration legislation still again. The new legislation granted preference to immigrants bringing either skills or money. Such a policy is new in U.S. history although it has been Canadian and Australian policy for years. The nation's new rules favor immigrants who bring job skills or knowledge in mathematics, engineering, and the sciences, thus inviting the world's skilled workers to come, and foreign students already here to stay. The legislation set aside 120,000 visas annually for skilled workers and their families and 10,000 for unskilled workers and their families. It lifted the overall ceiling from 530,000 immigrants per year to 700,000 in 1992–1994 (some 500,000 of whom will be members of families being reunited) and 675,000 thereafter.

Sociologist Nathan Glazer has pointed out that an increasing number of migrants to the United States are not fleeing desperate poverty. "The first impact of prosperity," he argues, "will be to increase [migration]. Look at China. These people don't come from the backward areas; they come from the progressive parts. As they learn how to run a business, they say to themselves, 'Why not go to the United States and do even better?'" The U.S.

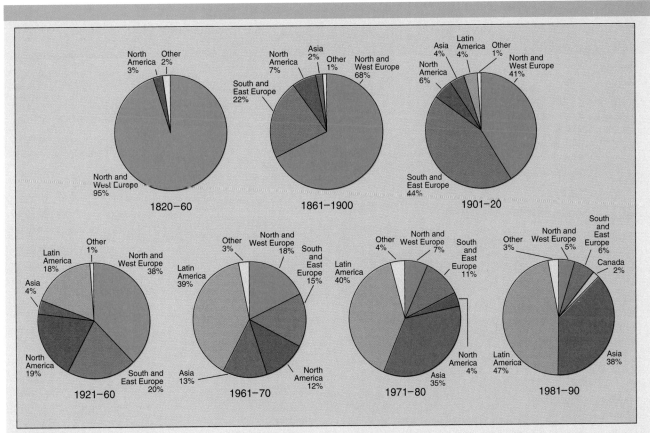

FIGURE 5-21

These graphs show how the source areas of immigrants to the United States have changed through history. The percentages of immigrants from Northern and Western Europe have dropped steadily, as the percentages from Latin America and Asia have risen. These trends have affected the composition of the national population. (Based on data from U.S. Dept. of Justice, Immigration and Naturalization Service.)

1990 legislation did in fact set aside a special quota for 10,000 millionaires able to invest in the U.S. economy. Millionaires do not have to demonstrate special skills or education, but 30 percent of the millionaires must invest either in rural areas or wherever the unemployment rate is 1.5 times the national average.

The Number and Distribution of the Foreign-Born in 1990

In 1990, the 19.8 million foreign-born people in the United States represented 7.9 percent of the total population, up from 6.2 percent in 1980. Of these people, the largest number came from Mexico (21.7 percent of the total), the Philippines (4.6 percent), Canada (3.8 percent), Cuba (3.7 percent), Germany (3.6 percent), Britain (3.2 percent), Italy (2.9 percent),

South Korea (2.9 percent), Vietnam (2.7 percent), and China (2.7 percent).

Some 60 percent of the total were not U.S. citizens, but that percentage varied greatly by source area. As a rule, the percentages of foreign-born from the European countries who were not citizens were low (28.1 percent for the Germans and 24.2 percent for the Italians, for example), but the percentages of the Latin Americans who were not citizens were high (84.6 percent of the people from El Salvador; 84.9 percent of those from Nicaragua; and 83.1 percent of those from Guatemala). This may be explained by the fact that the Europeans had individually been here longer and perhaps numbered fewer sojourners among them.

California is home to almost one-third of the foreign-born today (6.5 million people, 21.7 percent of the state's population), and the next ranking concentrations are in the states of New York (2.8 million foreign-born, or 15.9 percent of the state total);

TABLE 5-2
The Racial Composition of the U.S. Population in 1990 and 1980

Race and Hispanic origin	1990 Census		1980 Census		Number change	Percent change
	Number	Percent	Number	Percent		
RACE						
All persons	248,709,873	100.0	226,545,805	100.0	22,164,068	9.8
White	199,686,070	80.3	188,371,622	83.1	11,314,448	6.0
Black	29,986,060	12.1	26,495,025	11.7	3,491,035	13.2
American Indian, Eskimo, or Aleut	1,959,234	0.8	1,420,400	0.6	538,834	37.9
American Indian	1,878,285	0.8	1,364,033	0.6	514,252	37.7
Eskimo	57,152	0.0	42,162	0.0	14,990	35.6
Aleut	23,797	0.0	14,205	0.0	9,592	67.5
Asian or Pacific Islander	7,273,662	2.9	3,500,439	1.5	3,773,223	107.8
Chinese	1,645,472	0.7	806,040	0.4	839,432	104.1
Filipino	1,406,770	0.6	774,652	0.3	632,118	81.6
Japanese	847,562	0.3	700,974	0.3	146,588	20.9
Asian Indian	815,447	0.3	361,531	0.2	453,916	125.6
Korean	798,849	0.3	354,593	0.2	444,256	125.3
Vietnamese	614,547	0.2	261,729	0.1	352,818	134.8
Hawaiian	211,014	0.1	166,814	0.1	44,200	26.5
Samoan	62,964	0.0	41,948	0.0	21,016	50.1
Guamanian	49,345	0.0	32,158	0.0	17,187	53.4
Other Asian or Pacific Islander	821,692	0.3	NA	NA	NA	NA
Other race	9,804,847	3.9	6,758,319	3.0	3,046,528	45.1
HISPANIC ORIGIN						
All persons	248,709,873	100.0	226,545,805	100.0	22,164,068	9.8
Hispanic origin*	22,354,059	9.0	14,608,673	6.4	7,745,386	53.0
Mexican	13,495,938	5.4	8,740,439	3.9	4,755,499	54.4
Puerto Rican	2,727,754	1.1	2,013,945	0.9	713,809	35.4
Cuban	1,043,932	0.4	803,226	0.4	240,706	30.0
Other Hispanic	5,086,435	2.0	3,051,063	1.3	2,035,372	66.7
Not of Hispanic origin	226,355,814	91.0	211,937,132	93.6	14,418,682	6.8

*Persons of Hispanic origin may be of any race.

Florida (1.7 million; 12.9 percent); and Texas (1.5 million; 9 percent). New York City had the largest foreign-born population (2 million people), but metropolitan Miami had the highest percentage of foreign-born (45 percent). The 1.3 million foreign-born in Los Angeles made up 38.4 percent of that city's population, ranking it just behind Miami. The state of Wyoming had the smallest number of foreign-born (7,647), but Mississippi had the lowest percentage (0.8 percent).

New Arguments Over Immigration

In the United States, as in Europe, the terms of immigration are being questioned. The early 1990s saw many U.S. citizens out of work or pessimistic about national job growth, and combined legal and illegal immigration accounts for about 40 percent of the annual growth of the U.S. labor force. In 1992, 1.3 million foreign workers were granted temporary or permanent authorization to work in the United States; this was more than the net number of jobs the economy created in that year and six times the number to be created by a program that President Clinton was recommending to Congress.

Many people argue that immigrants cost Americans money for public services, while others argue that immigrants' labor contributes to American economic growth. There is truth in both arguments, but any full and complete accounting is impossible. The political arguments hinge on the basis of percep-

Concentrations of Hispanic Americans

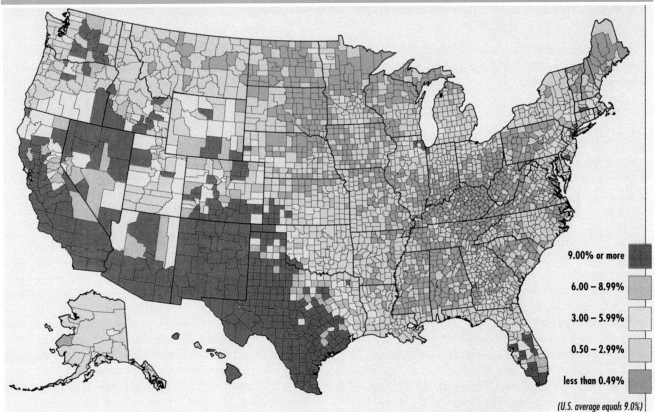

9.00% or more
6.00 – 8.99%
3.00 – 5.99%
0.50 – 2.99%
less than 0.49%
(U.S. average equals 9.0%)

FIGURE 5-22

This map shows the percentage of the population in each county that was Hispanic in 1990, whether or not the individuals were citizens of the United States. Most of the territory colored red on the map, counties in which Hispanic persons make up more than their 9 percent share of the overall national population, was taken from Mexico in war. Today, Mexicans boast that Los Angeles is the world's second largest "Mexican" city.

In 1990, four states were home to 71 percent of U.S. Hispanics: California, Texas, Florida, and New York. In California and Texas Mexicans make up over 75 percent of the Hispanics. New York's Hispanic population is evenly split between Puerto Ricans and other Hispanics, and Florida has the most diverse Hispanic population. [Reprinted with permssion of *American Demographics* (July 1991): 14.]

FOCUS
Asian Americans

Asian Americans are the fastest-growing segment of the U.S. population, making up 3 percent of the total in 1990.

Asian Americans were once highly concentrated in a few states (California and Hawaii) or clustered in urban centers, but today they are dispersing more widely throughout the country. Almost 40 percent still live in California, but from 1980 to 1990 their numbers rose 124 percent in Alabama, 162 percent in New Jersey, 166 percent in Texas, 245 percent in Rhode Island, 76 percent in Mississippi, and 73 percent in Louisiana.

The mix of the Asian-American population also changed dramatically. Until 1970, about two-thirds were Japanese or Chinese. In 1970, the Japanese were the largest group, but in 1980 the Chinese surpassed them. Today the mix is extremely varied.

Many Asian Americans have done well financially. In 1990, the median household income of Asian Americans, whose education levels are well above the general population's, was 18 percent higher than that of non-Hispanic whites.

Chapter 3 quoted Ellen Churchill Semple that the future of the United States will be determined by "their location on the Pacific opposite Asia." Is today's immigration pattern evidence of her perceptiveness?

---

tion, and today Americans are hearing the same anti-immigrant arguments that were voiced in 1890: the country is full, the newcomers are different, and they will never assimilate.

Many Americans have come to accept at least a case for doing something about claims to asylum. As late as the mid-1980s, asylum requests were running at about 20,000 per year, but by 1992 they had risen to more than 103,000. Many of these may be abuses of the claim. Therefore, early in 1994, the United States began imposing a $130 fee on applicants who can pay and doubled the applicants' wait for a work permit to 6 months.

On August 10, 1993, Governor Pete Wilson of California took the unprecedented step of sending a letter about immigration to President Clinton and, at the same time, printing that letter as a full-page announcement in newspapers from coast to coast. Governor Wilson recommended that the Constitution be amended to deny citizenship to American-born children of illegal immigrants (who account for two-thirds of 1992 births in Los Angeles County Hospital), that a tamper-proof identification card be created for U.S. citizens and legal residents, and that Congress repeal federal mandates that make illegal immigrants eligible for health, education, and other benefits. "It's time to amend the Constitution so that American citizenship belongs only to those children of legal residents of the United States," he wrote, "not to every child whose mother can make it to an American hospital."

One may or may not agree with Governor Wilson's suggestions, but a broadside as powerful as this from the governor of America's most populous state by itself changes the nature of public debate. It places specific items on the national agenda. President Clinton responded that the idea of a national identity card "ought to be examined." United States immigration policy will continue to be a political issue in the 1990s, and perhaps the nation will follow France and move away from its historic *jus soli* definition of citizenship.

THE BRAIN DRAIN

Many nations are losing their best-educated people and most skilled workers through emigration. This migration from less-developed to more-developed countries has been called the **brain drain.** In some cases, a country's schools qualify students for professional careers faster than the economy can absorb them. In other cases, the skilled or educated seek political freedom, and in still other cases the country's best students go abroad for training to help their own people, but they stay abroad and never return home. Africa, in particular, has lost an estimated one-third of its highly educated manpower in recent decades, but even a country as advanced as Ireland can suffer. Between 1982 and 1990, nearly one in twenty of the total national population emigrated, and the emigrants took skills with them that Ireland could use to build its own future. Among those who took their degrees in 1987, half the engineers had departed by 1990, along with one-third of the medical graduates and three-quarters of the architects. No

GEOGRAPHIC REASONING
Is Immigration a Substitute for Education?

Questions

1. Which is cheaper, importing a trained labor force or educating native-born workers? In the short run? In the long run?

2. What are the costs of inadequately training native-born Americans?

3. Does immigration of skilled workers discourage any country from educating its own labor force?

Many critics of immigration have long argued that the labor force provided by immigration has allowed the United States to slight the education and development of the nation's native work force. They have suggested that the country invest in educating and training its own workers rather than rely on immigrant labor. In his 1895 speech "Let Down Your Bucket," the black educator Booker T. Washington (1856–1915) encouraged Southern industrialists to seek their labor force around them in the willing blacks who had so long "tilled your fields, cleared your forests, builded your railroads and cities," rather than "to look to the incoming of those of foreign birth and strange tongue and habits...to buy your surplus land, make blossom the waste places in your fields, and run your factories." This would require "helping and encouraging them ... and education."

Today, 100 years later, some U.S. industrialists insist that skilled immigrant labor is needed because the national labor force still has not been properly trained. A spokesperson for the National Association of Manufacturers defended the 1990 legislation that welcomed skilled immigrants, saying of the U.S. labor market: "The bottom line is that the talent isn't there. It certainly might be there in 20 years. But we are where we are today because the American educational system has failed us." An editorial in the *Wall Street Journal* on February 2, 1990, insisted, "Our own view remains that the problem is not too many immigrants, but too few....As long as we don't train enough [Americans] ourselves, immigration is a saving grace." The 1990 Economic Report of the President agreed: "With projections of a rising demand for skilled workers in coming years, the nation can achieve even greater benefits from immigration."

country can afford a drain of its most highly trained people. Several countries have levied taxes on educated or skilled emigrants in order to prevent them from leaving, although the policy's ostensible goal is to recapture the government's investment in their education. The former Soviet Union, under pressure from the United States, dropped emigration restrictions against Jews, and today Jewish immigrants from Russia and Ukraine sustain Israel's world-class computer software and other high-technology industries.

The United States profits most from the drain of skilled labor from other countries, and, in fact, the nation's scientific establishment would suffer without them. About one-third of the nation's Nobel Prize winners in the sciences have been foreign-born. Thousands of senior scientists, engineers, and other professionals migrate to the United States each year, and about half of the young foreigners who come for education never return home. Each year from 1960 to 1990, the percentage of U.S. Ph.D.s in science and engineering granted to foreign students rose, while the number of Americans getting such degrees barely held steady. The percentage of noncitizens earning advanced degrees in all fields in the United States grew from 15 percent in 1972 to 26 percent in 1991, and in technology and the sciences foreign nationals earn more than half of all degrees granted.

In November 1993, the United States granted asylum to all Chinese students who were studying in the United States at the time of the Chinese crackdown on antigovernment demonstrators in 1989. Thus, over 49,000 Chinese who had applied, almost all of whom were students, will stay. Had these people returned to their homeland with their new education, they might have been able to make significant contributions. Instead, the United States will benefit from their skills.

Data suggest that a higher proportion of immigrants to the United States are either earning degrees here or else bringing skills learned elsewhere. In 1993, 24 percent of people who had migrated to the United States between 1980 and 1993 had a bachelor's degree or higher, compared with 19 percent of those who had migrated before 1980 and 20 percent of U.S.-born adults.

The drain of scientists, engineers, and physicians from poor areas undermines the ability of these areas to develop. Any advanced country is reluctant to expel these skilled immigrants, who would probably just go to another advanced country anyway. How these skilled professionals are to be kept in their homelands, or attracted back, is a problem that must be solved if the gap between the world's rich and poor countries is not to widen further.

SUMMARY

Human migrations have redistributed people throughout history, and significant migrations continue. Push factors drive people away from wherever they are, and pull factors attract them to new destinations. Some migrations have been forced.

Most people migrate to nearby locations; fewer migrate greater distances, but migration flows can better be explained by studying flows of trade, transport routes, and connections between formerly imperial powers and their former colonies. Immigrants from one region may form distinct concentrations in their new homelands. Sojourners intend to stay in their new location only until they can save enough capital to return home.

Many humanlike species have evolved over the past 10 million years, always, it seems, in Africa. Most of the physical attributes that link humans most close-

ly with apes and monkeys developed in a forest environment, but our distinctive human qualities developed over a period of several million years in a more open environment. Modern Homo sapiens probably spread from Africa to southern Europe and central, southern, and southeastern Asia quite quickly from about 100,000 years ago onward. Isolated sites indicate that the Americas were probably reached at least 30,000 years ago, and humankind's most recent colonization seems to have been of the Pacific Islands. Human beings are one single species, and all humans can successfully interbreed. Compared to this biological community of humankind, subdivisions are insignificant.

In 1500, the Earth's many peoples and culture regions were relatively isolated from one another. The most significant population transfers since 1500 have been migrations of Europeans to the Americas, Africa, Australia, and Asia; migrations of blacks out of Africa; and migrations of Indians and Chinese that were largely instigated or made possible by European expansion.

Today people are trying to move from the poor countries to the rich countries and from politically repressive countries to more democratic countries. Many migrants claim to be political refugees, but it is often difficult to discern who is a political refugee and who is only an "economic refugee" seeking a higher material standard of living. The impact of international migration is even greater than the numbers might suggest. Immigrants are often greeted with apprehension, and in many receiving countries immigration has become an explosive political issue. The peoples of the rich and democratic countries are beginning to exhibit "compassion fatigue," and doors are closing all around the world. The single largest migration flow for the past 150 years has been migration to the United States, and U.S. immigration policy will continue to be a political issue in the 1990s. The brain drain of educated people and skilled workers from poor countries undermines their ability to develop.

KEY TERMS

push factors (p. 158)
pull factors (p. 158)
external refugees (p. 158)
internal refugees (p. 158)
undocumented,
    irregular, or illegal
    immigrants (p. 158)
migration chain (p. 158)
sojourner (p. 159)
biophilia
    hypothesis (p. 160)
race (p. 161)
racism (p. 162)
ethnic group (p. 162)
mulatto (p. 165)
zambo (p. 165)
mestizo (p. 165)
African diaspora (p. 166)
Boers (p. 171)
apartheid (p. 171)
Bantustan (p. 172)
Aborigine (p. 174)
Maori (p. 174)
Overseas Chinese (p. 176)
refugee (p. 178)
indigenous
    peoples (p. 178)
workers'
    remittances (p. 180)
jus sanguinis (p. 182)
jus soli (p. 182)
melting pot (p. 183)
cultural mosaic (p. 183)
nativism (p. 184)
brain drain (p. 188)

QUESTIONS FOR INVESTIGATION AND DISCUSSION

1.   Do you know how many generations of your family have moved either to this country or within this country? What push and pull factors motivated them?

2.   Has your local community seen net in-migration or net out-migration of its native black population in recent years? If in, where from? If out, where to? Are first- or second-generation Americans a significant share of your local black community? If so, where did they come from? Africa? The Caribbean?

3.   How many different national backgrounds did the 1990 census reveal in your community? Have newcomers settled in distinct neighborhoods? Which push and pull factors motivated them?

4.   Are any major refugee streams reported in this week's headlines? If so, what are the push factors?

5.   What definition of *refugee* do you think should be accepted internationally?

6.   What do you think should be U.S. immigration policy?

ADDITIONAL READING

ASANTE, MOLEFI and MARK T. MATTSON. *Historical and Cultural Atlas of African Americans.* New York: Macmillan, 1991.

GIOSEFFI, DANIELA, ed. *On Prejudice: A Global Perspective.* New York: Anchor Books, Doubleday, 1993.

KOTKIN, JOEL. *Tribes: How Race, Religion, and Identity Determine Success in the New Global Economy.* New York: Random House, 1993.

LAYARD, RICHARD, and OLIVIER BLANCHARD, RUDIGER DORNBUSCH and PAUL KRUGMAN. *East-West Migration: The Alternatives.* Cambridge, MA: MIT Press, 1992.

PAN, LYNN. *Sons of the Yellow Emperor: A History of the Chinese Diaspora.* Boston: Little, Brown, 1991.

RISCHIN, MOSES, ed. *Immigration and the American Tradition.* Indianapolis: Bobbs-Merrill, 1976. (documents)

ROGGE, JOHN, ed. *Refugees: A Third World Dilemma.* Savage, MD: Rowman & Littlefield, 1987.

United Nations High Commissioner for Refugees. *The State of the World's Refugees 1993.* New York: Penguin Books, 1993.

U.S. Committee for Refugees. American Council for Nationalities Service. *World Refugee Survey.* Washington, D.C., annually.

6 Agriculture, Fishing, and Forestry

These British trawlermen mend their nets before they can head out to sea again fishing for cod. The harvest of cod from the North Atlantic rose steadily until recently, but overfishing depleted them, and recent catches have dropped precipitously.

Courtesy of Bryn Campbell/Tony Stone Images.

Securing adequate food has been a constant human concern since the dawn of history, and a rising world population demands increasing supplies. The following pages will examine first the environmental and technological capability of the entire globe to produce enough food to supply the entire human population. This capability is quite secure today and for the future. Then, however, this chapter will map the disparities in the distribution of food production and supplies. Some countries are glutted with food; they either store surpluses or else **dump** the food on world markets (sell it for less than it costs to produce). Even in many of those countries, however, some people go hungry. In other countries overall food supplies are insufficient to provide the entire population with a nutritious diet. This problem of oversupplies of food in some places and insufficient supplies in others can be understood only as the result of a complex of environmental, technological, political, and economic factors.

The next topic examined in this chapter is world supplies of fish. Animal or vegetable resources that naturally renew themselves, such as fish or forests, are called **renewable resources**. They can be harvested at the rate at which they reproduce themselves through time, although supplies will diminish if they are harvested faster than they reproduce. In some cases, however, humankind can increase that replacement rate. In the case of the world's harvest of fish, for example, the harvest from the seas has quintupled in the past 40 years, significantly increasing world per capita food supplies. Increasing future supplies from the seas is dubious, but humankind may turn to raising more fish under controlled circumstances. The end of the chapter looks at another rich renewable resource—the raw material of the world's forests.

THE WORLD POPULATION: FOOD RATIO

Thomas Malthus and the Malthusian Equation

Discussions of world population growth inevitably invoke the ideas of Thomas Robert Malthus (1766–1834), who asked whether humankind would always be able to feed itself. Malthus's statement of the relationship between population and food supply still demands our attention. An understanding of Malthus's intellectual milieu will also help us understand his argument.

Thomas's father, Daniel, was a disciple and personal friend of both the French philosopher Jean Jacques Rousseau (1712–1778) and the English theorist William Godwin (1756–1836). Daniel Malthus and Godwin spread in England the new, optimistic "enlightened" idea that progress is achievable through human action. Godwin's wife, Mary Wollstonecraft, (1759–1797) was one of the first feminists, and her essay *A Vindication of the Rights of Women* (1792) raised a storm of opposition. Later, William and Mary's daughter, Mary Godwin (1797–1851; wife of the poet Percy Bysshe Shelley) was less optimistic about the future of humankind. She would write the novel *Frankenstein* in 1818. In this classic, Dr. Frankenstein's confidence in science leads him to create a living creature which he hopes he can control, but that turns on him and kills him. This idea survives in our imagination that science can be a "Frankenstein's monster" that creates more serious problems than it solves.

In these debates between people who were optimistic about humankind's future and their pessimistic children, Thomas Malthus was a pessimist. When his father challenged Thomas to put his ideas of the future into print, Thomas produced, anonymously, his *Essay on Population* (1798). Thomas was by that time a professor of political economy and also a clergyman. Stated most simply, Thomas Malthus argued that food production increases arithmetically: 1-2-3-4-5…units of wheat. Population, however, increases geometrically: 2-4-8-16-32…people. Thus, the amount of food available per person must decrease as the population increases.

The human population can be kept in balance with food supplies only through checks on population increase. Malthus defined two types: positive and preventive checks. **Positive checks** refer to premature deaths of all types, such as those caused by war, famine, and disease. The alternatives to these positive checks are human actions designed to limit population growth. Malthus called these behaviors **preventive checks**. An example of a preventive check is a decision by a young couple to delay marriage and childbearing. According to Malthus, couples should have a sense of responsibility for the economic welfare of the children they might produce; at the least, couples ought to fear their own inevitable social and economic decline if they produce too many children. "There are perhaps few actions that tend so directly to diminish the general happiness as to marry without the means of supporting children," he thundered. "He who commits this act, therefore, clearly offends against the will of God."

Malthus did not have much faith in people's ability to restrain themselves, so he did not believe

FIGURE 6-1

This map divides the Earth into ten categories of agricultural land use. The success of agriculture depends largely upon the possibilities of the physical environment, so we should not be surprised that the categories mapped here largely coincide with the climates mapped in Figure 1-20 and the biomes mapped in Figure 1-23. Categories 9 and 10 (low-technology subsistence farming and nomadic herding) cover impressive areas on this map, but their productivity is low, and comparison of this figure with Figure 4-1 reveals that these areas are generally sparsely populated.

that preventive checks could effectively control population growth. Therefore, he predicted that the future of humankind would consist of endless cycles of war, pestilence, and famine. This *Malthusian theory* that overpopulation is a constant threat contrasts sharply with the *Marxist theory* discussed in Chapter 2, that technological advances can outpace population growth and provide humankind with increasing material welfare.

DEVELOPMENTS SINCE MATLTHUS
▼▼▼▼▼▼▼▼▼▼▼▼▼▼▼▼▼▼▼▼▼▼▼▼▼▼▼▼

Since Malthus's theory was first published, the human population has increased from 1 billion to over 5 billion. The mass starvation predicted by Malthus has not occurred, however, for several reasons.

New Crops and Cropland

Vast areas of the planet that were scarcely utilized during Malthus's lifetime have been opened to productive agriculture (see Figure 6-1). The greatest agricultural surplus areas of both North and South America, as well as surplus regions in Australia and South Africa, have been developed since Malthus published his theories.

Since the middle of the last century alone, over 3.5 million square miles (9 million square km) of the Earth's surface have been converted to permanent croplands. The United Nations reports that in the 1980s the world cropland area increased by another 2 percent. Most of these lands were opened by irrigation.

Not only have new croplands been planted, but many food crops have been transplanted to new areas where they have thrived, in some cases better than in their areas of origin (Figure 6-2). Even before Malthus wrote, the Western Hemisphere had contributed important food crops to the Eastern Hemisphere. For example, the potato, which yields the second highest number of calories per acre of any crop, is native to the Andes region of South America.

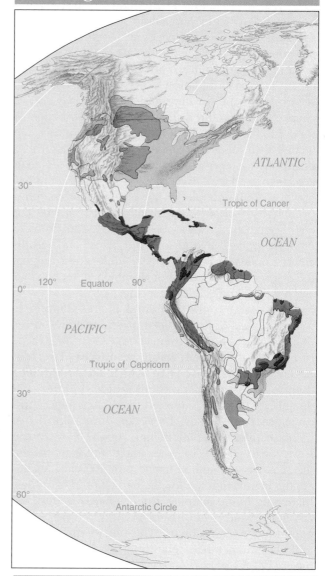

World Agriculture

By Malthus's day it had already become a major food in Northern Europe (Figure 6-3, p. 197). Today it is the world's fourth most important food crop, measured by total tonnage harvested, after wheat, rice, and corn. China has recently recognized the potato's versatility and is today, after Russia, the world's second largest producer. India is not far behind, and the crop is becoming a mainstay throughout Africa and Asia. New genetically engineered plants are being introduced to the tropics, where they are ready to harvest only 40 to 90 days after planting. Potato planting continues to spread, and the value of the world's potato crop increases each year. The true treasure of the Andes was not the gold which the Spanish conquerors sought, but the potatoes they trampled.

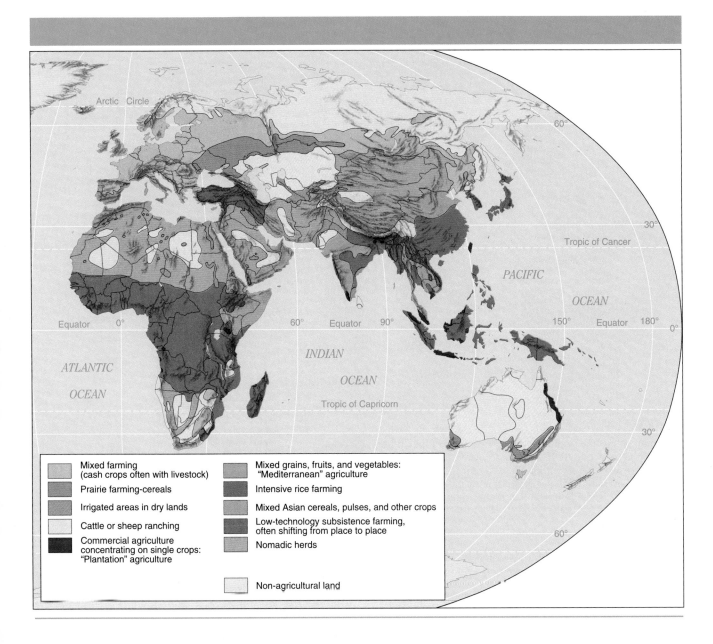

Mixed farming (cash crops often with livestock)

Prairie farming-cereals

Irrigated areas in dry lands

Cattle or sheep ranching

Commercial agriculture concentrating on single crops: "Plantation" agriculture

Mixed grains, fruits, and vegetables: "Mediterranean" agriculture

Intensive rice farming

Mixed Asian cereals, pulses, and other crops

Low-technology subsistence farming, often shifting from place to place

Nomadic herds

Non-agricultural land

Maize was another Western Hemisphere native, and still another was manioc, or cassava. In the sixteenth century, Portuguese traders introduced manioc into Africa's Congo River delta area, and it rapidly became a dietary staple there. The introduction of manioc to Africa improved Africans' diet so much that it is one reason why Africa's population did not decline over the following period, during which millions of Africans were taken off into slavery. Manioc is still a principal source of calories for Africans. Successful transplantation of additional crops continues around the world today.

Transportation and Storage Improvements in worldwide transportation have allowed regional specialization in food production, and specialization can multiply productivity. Today railroads, trucks, and cargo ships—many of them refrigerated—move quantities of food with a speed and efficiency that could not have been imagined in Malthus's day (Figure 6-4).

Today transportation also allows the shifting of food from surplus to deficit areas, so fewer people need die from local famines. In the past it was not uncommon for surplus food to rot in the sun 100 miles from a starving population because the food could not be transported.

The technology of food storage has also continually improved, decreasing both spoilage and loss to pests (Figure 6-5, p. 198). Improvements have been continuous from the introduction of the silo in the nineteenth century to freeze drying—actually, the

Agriculture, Fishing, and Forestry   **195**

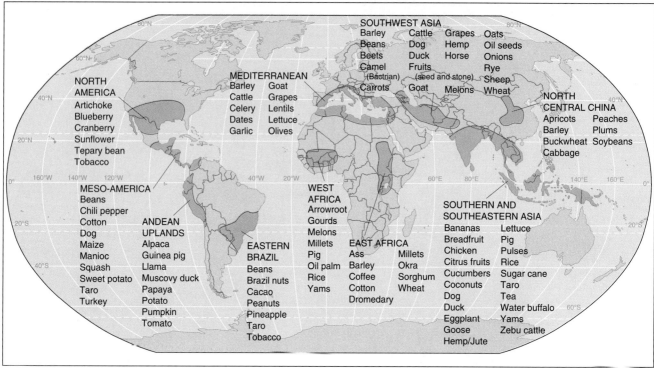

FIGURE 6-2

This map shows the origins of the world's food crops and domesticated animals. It is a tribute to human ingenuity that the people in so many different regions and biomes devised diets and lifestyles by learning to exploit the different plants and animals they found locally. Many of these plants and animals, however, have been widely redistributed, so the area that leads production today is not necessarily where that plant or animal was first domesticated.

Inka did this 500 years ago, but the rest of the world learned how only recently—and antiseptic packaging.

The Green Revolution The introduction of higher-yielding and hardier strains of crops through advances in botanical science is known as the **Green Revolution**. For example, since the 1970s specialists at Texas A&M University and at the International Rice Research Institute in the Philippines have produced higher-yielding strains of rice that are more resistant to disease and pests and that mature in only 110 days. This allows for two or even three crops per year. New potato strains are developed at the International Potato Center in Lima, Peru. Wheat yields have multiplied from new hybrids. India, where 1.5 million people died in a 1943 famine, became a grain exporter in 1977, even though its population doubled between those years.

Other Technological Advances The Green Revolution is just one aspect of a complex of factors known as the **Scientific Revolution in Agriculture**.

New pesticides save crops from insects that once wiped out entire harvests, and new fertilizers multiply yields per acre. Parallel scientific research has been directed to livestock. As with crops, the developments include world redistribution, increases in total numbers, improved breeding, and even "engineering" for greater hardiness and higher yield from each animal. In 1950, the average U.S. dairy cow gave 618 gallons (2,339 L) of milk per year; by 1992, that number had more than tripled to 1936 gallons (7,328 L). So far, however, only a fraction of the world's livestock herds is this productive.

Farm machinery invented since Malthus's day reduces the number of fieldworkers and at the same time increases yields by improving the regularity of plant spacing and the efficiency of harvesting. Heaters rescue many crops threatened by freezes. During the past 200 years humans have literally re-formed the Earth with drainage projects where there was too much water and with irrigation projects where there was not enough. Projects such as these are not new in theory, but they are achieved in scale beyond anything that Malthus could have foreseen.

Current Production of Potatoes and Rice

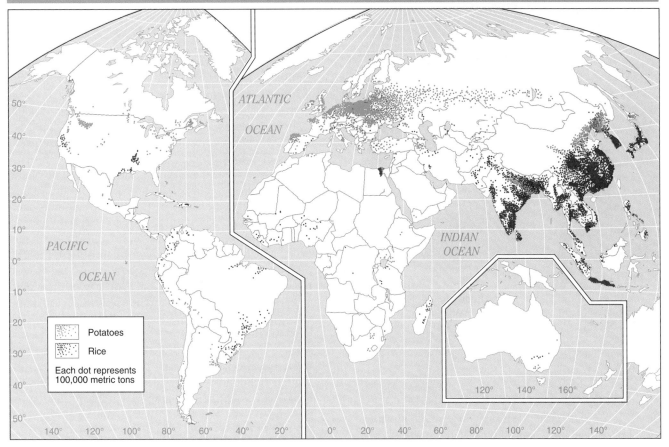

FIGURE 6-3

The potato is native to South America, but its greatest production today is in Europe. Rice remains the great staple food of Asia.

FIGURE 6-4

Modern techniques of food transportation and storage allow Alaskans to enjoy fresh produce even when outside temperatures read −40°F (−40°C).
(Courtesy of Carr Gottstein Foods Company, Anchorage, Alaska.)

THE WORLD'S PRINCIPAL FOOD CROPS

Today about 3.75 billion acres (1.5 billion hectares) of land are devoted to agriculture. Cereals (grains) and potatoes are the world's basic foods. Cereals have been grown since the dawn of history, and every major civilization has been founded on them as the principal source of food. Wheat, rice, maize, barley, sorghum, oats, rye, and millet are all members of the grass family of plants. They yield more food, both in bulk and in nutritive value, than most other crops. The preeminence of cereals as a food is also partly explained by the ease with which they can be produced, stored, and transported. A number of storage and transportation devices developed in the nineteenth century, such as grain elevators and

FIGURE 6-5

This officer of the U.S. Agency for International Development (AID) is overseeing the construction of a silo in Peru. Many countries produce enough food to nourish their people, yet much of the crop is wasted or lost to pests or spoilage because of inadequate storage and distribution facilities.
(Courtesy of the Agency for International Development.)

devices for mechanical transporting and shipping, handle grain as if it were a liquid (Figure 6-6).

Tables 6-1 and 6-2 list the largest importers and exporters of the major foods in world trade. In each case, a few countries that are not among the major producers of a specific crop do number among the major exporters. Argentina, for example, is the fifth most important exporter of wheat; Thailand, Vietnam, and Pakistan are the first, third, and sixth leading exporters of rice. This indicates that in those countries food production for export is a highly commercialized business.

Wheat Wheat was first domesticated in the Near East or in Egypt's Nile Valley by 5000 B.C., and today it is the most important cereal in world food production. It supplies about 20 percent of the total calorie consumption of the human species (Figure 6-7, p. 201) and provides a staple for over one-third of the world's population. The protein content varies between 8 and 15 percent.

Wheat grows best where the weather is cool early in the growing season. Modern wheat varieties are usually classified as winter wheats (fall-planted and unusually winter-hardy) and spring wheats (spring-planted). In North America, for example, the production of winter wheat is geographically focused on

Kansas, whereas spring wheat production is centered farther north, along the U.S.–Canadian border. Hard-kerneled varieties yield flour with a high content of *gluten* (a protein substance), which are used to make breads; soft-kerneled ones are starchier and are used to makes cakes and biscuits. *Durum* wheat is an especially hard-kerneled wheat used in pasta products. Wheat grain, *bran* (the fibrous outer coat of the wheat kernel that remains as a residue after milling), and the rest of the plant are all valuable livestock feed.

As much as one-sixth of world wheat production enters into world trade. This development of the past 125 years followed the agricultural settlement of the plains of the United States, Canada, Argentina, and Australia, the extension of the world's railroad network, and the building of ocean-going steamships.

In 1991, China was the world's leading producer of wheat (17 percent of world production). The next leading producers of wheat were the Soviet Union (15 percent, most of which was grown in today's Russia, Ukraine, and Kazakhstan); India (10 percent); the United States (10 percent); France (6 percent); and Canada (6 percent). The United States, France, and Canada were major exporters of wheat in 1991, but the Soviet Union and China imported substantial quantities. The successor states of the Soviet Union have continued to import significant quantities.

FIGURE 6-6

In the mid-nineteenth century, grain was still stored and transported in individual bags. This required enormous amounts of labor, and a great amount of grain was lost to spillage. Late in the nineteenth century, however, methods were devised to store and ship grain as if it were a liquid. These techniques greatly reduced the cost and improved the efficiency of transportation.
(Courtesy of the Kansas Wheat Commission.)

TABLE 6-1
Wheat, Rice, and Corn: Top Ten Exporters and Importers, 1990

Exporters (Millions of dollars)		Importers (Millions of dollars)	
Wheat			
United States	3,887	Soviet Union	2,490
France	3,296	China, Mainland	2,319
Canada	2,863	Italy	1,217
Australia	1,971	Japan	1,019
Argentina	871	Egypt	965
United Kingdom	760	Iran	707
West Germany	503	Netherlands	595
Netherlands	406	West Germany	429
Denmark	229	Algeria	422
Saudi Arabia	211	South Korea	419
Rice			
Thailand	1,086	Saudi Arabia	226
United States	804	Iran	216
Vietnam	375	France	216
Italy	357	United Kingdom	209
India	246	West Germany	165
Pakistan	242	Hong Kong	150
Belgium-Luxembourg	168	Brazil	147
Spain	108	Philippines	128
Uruguay	104	Iraq	124
China, Mainland	98	United Arab Emirates	120
Corn			
United States	6,206	Japan	2,295
France	1,854	Soviet Union	1,690
China, Mainland	404	South Korea	837
Argentina	329	China, Mainland	750
South Africa	176	Netherlands	538
Thailand	162	Mexico	473
Zimbabwe	87	United Kingdom	396
West Germany	71	West Germany	375
Hungary	51	Italy	363
Spain	41	Spain	307

Source: *Statistical Abstract of the United States*, 1993–1994, #1476–1419, p. 867.

Rice About half the world's population subsists wholly or partially on rice. It remains the leading crop in tropical Asia, where it was first domesticated. Rice cultivation requires warmth and abundant moisture, but it allows for continuous harvesting, so it supports high population densities. Rice has a protein content lower than that of wheat—8 to 9 percent.

Thousands of rice strains are known, both cultivated and wild. Brown rice has greater food value than white, since the outer coatings that are polished away to yield white rice contain protein and minerals, while the white endosperm is chiefly carbohydrate.

In 1991, the top rice producers were China (36 percent); India (21 percent); Indonesia (9 percent); Thailand (4 percent); Myanmar (3 percent); and Japan (2 percent). Of these leading producers, Thailand, India, and China exported rice. Only 3 or 4 percent of world rice production enters international trade. The United States produces only about 1 percent of the total world crop, but it accounts for 18 percent of all rice exports, a proportion second only to Thailand's.

TABLE 6-2
**Top Ten Importers and Exporters of Pulses (Lentils, Peas, and Other Dried
Leguminous Vegetables), 1989–91**

Exporters		Importers	
France	1,096,591	Netherlands	802,256
China	588,047	India	583,537
Turkey	522,694	Italy	434,065
Australia	372,197	Belgium	405,156
Poland	249,475	Spain	301,238
Denmark	235,166	Germany	254,317
Hungary	210,221	Japan	180,899
Thailand	159,726	Algeria	135,803
Argentina	149,255	Pakistan	115,033
Myanmar	127,267	Bangladesh	90,642

Source: World Resources Institure, *World Resources 1994–95.* New York: Oxford University Press, 1994 pp. 298–99.

Maize *Corn* is the name English-speaking people have traditionally given to the major cereal crop of any given region. Therefore, in England "corn" means wheat; in Scotland and Ireland, oats, but in North America, it means maize, or Indian corn. Maize was cultivated in the Western Hemisphere long before the arrival of Europeans, and today almost half of the world's maize is grown in North America (Figure 6-8). Maize is a staple for people in South America and Africa, but about 70 percent of the U.S. crop is fed to livestock. Maize is eaten fresh or ground for meal, and the average protein content is 10 percent.

Maize is susceptible to frost, so the length of the frost-free period limits its distribution. Yields are highest when the growing period is at least 140 days, with mean summer temperatures of about 75°F (24°C). Most crops are grown in areas that receive between 25 and 40 inches (63-102 cm) of rainfall, but there must be adequate water during the midseason growth period. Therefore, the geographic focus of the U. S. "Corn Belt" stretches across Illinois and Iowa.

In 1991, the world's top maize producers were the United States (40 percent); China (19 percent); Brazil (5 percent); Mexico (3 percent); France (3 percent); and the Soviet Union (2 percent—mostly in Ukraine). The United States, China, and France were

major exporters, but the Soviet Union imported substantial amounts. Russia, Ukraine, and the other successor states of the Soviet Union continue to import maize.

Potato The potato, or white potato, is a member of the nightshade plant family, as are tomatoes and eggplants. Its swollen underground stem, or tuber, is a vegetable. Potatoes grow successfully in cool, moist regions, and they are a staple carbohydrate in many developed countries. In some parts of Europe about 40 percent of the crop is used to feed livestock. Potatoes' protein content is about 6 percent.

Because of potatoes' great bulk and relatively low value, international trade in potatoes is small. Major exporters are the Netherlands, France, and Italy; other Western European countries are the main importers.

Barley Barley is the world's fourth most important cereal crop (Figure 6-9, p. 203). It has the shortest growing season of all cereals and can be grown farther north, at higher altitudes, and in more arid regions than any other cereal. Barley was probably first domesticated in northern China, and today it is a crop of both woodland and grassland climax vegetation areas. In rich countries it is used

World Wheat Production

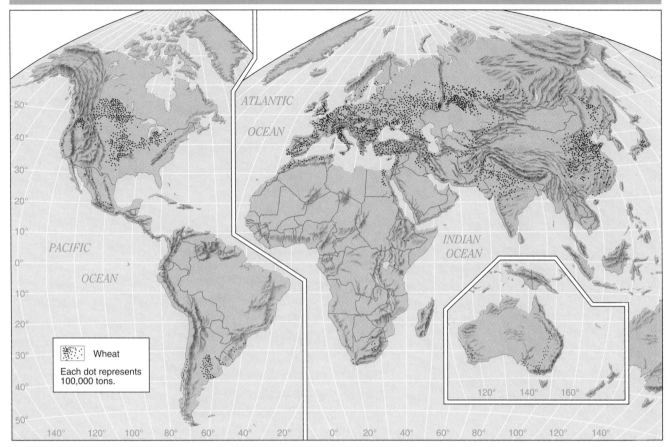

FIGURE 6-7

Today wheat is grown almost everywhere except in the tropics, and it is a major commodity in international trade.

mainly for animal feed and for malting beer and whiskey, but it is an important food crop in parts of Asia and in Ethiopia.

Sweet Potato The sweet potato is a member of the morning glory family and is native to Western Hemisphere tropical regions, but it is no relation to the white potato. It is rich in vitamins, and today it is an important tropical root crop around the world.

Cassava Cassava, or manioc, which is native to Brazil, remains a major staple across tropical lowland Africa. It is extremely resistant to drought. It has a very low protein content (0.9 percent) and should be supplemented with other high protein foods.

Sorghum and Millet Sorghum and millet are seldom mapped because they are usually grown as subsistence crops, and there is little reliable information on their production. Both are tropical cereals, and

although they provide staples in drier parts of Asia and Africa, the grains lack gluten and cannot be used for making bread.

Oats and Rye Both oats and rye prefer cool, damp climates. Oats, native to Eurasia, are grown mainly for feeding livestock, while rye remains popular in Central and North Europe for bread flour. Rye grows well where the soil is too poor and the climate too cool for wheat.

Pulses Pulses is the botanical name for a large family of herbs, shrubs, and trees also called the pea or legume family. The seeds, which include beans, peas, lentils, peanuts, soybeans, and carob, are rich in protein and may form the principal source of dietary protein in poor regions or in regions where religion prohibits the consumption of meat. Table 6-2 also reveals the importance of pulses (principally alfalfa, soybeans, and vetch) as animal feed in European importing countries. The root nodules of these plants

Agriculture, Fishing, and Forestry **201**

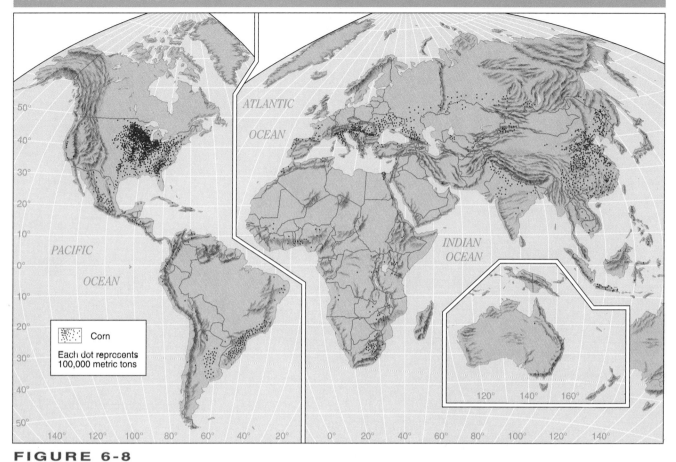

Corn

Each dot represents
100,000 metric tons

FIGURE 6-8

Maize was first domesticated in Central America, but world production today is concentrated in North America.

have bacteria that absorb nitrogen out of the air and release it into the soil (a process called *fixing* nitrogen). This ability makes these plants valuable as crops grown temporarily to reduce erosion and build up a soil's organic content.

LIVESTOCK

▼▼▼▼▼▼▼▼▼▼▼▼▼▼▼▼▼▼▼▼▼▼▼▼▼▼▼

Humans began to domesticate animals about the same time as they did plants, some 8,000 or 10,000 years ago. Dogs were probably domesticated first, independently at a number of places around the world. Dogs helped in hunting, and they were themselves eaten for food. As humans learned to domesticate plants, they found it convenient to herd plant-eating animals and to pen them close to settlements. From then on, growing crops and raising livestock advanced side by side.

A wide range of animals have been domesticated as livestock, and their numbers total over 15 bil-

lion—about three times the human population. These numbers include about 10 billion chickens, 3 billion ruminants (mainly cattle, sheep, goats, buffalo, and camels), and about 850 million pigs (see Figure 6-10 and look back at Figure 3-1). Domesticated livestock provide high-quality protein, either as meat or milk. They also supply hides, wool, and other raw materials, and they serve as draft power; large numbers of people in poor regions still depend on oxen and buffalo. In India alone, there are over 80 million draught animals. Among pastoral peoples, livestock represents wealth on the hoof (see Figure 6-11, p. 205).

The Direct and Indirect Consumption of Grain

Many domesticated animals eat grass, foliage, and other cellular plant material that humans cannot digest. Some grains, however, called *feed grains*, are fed to livestock. Eventually the livestock or their

Production of Barley, Oats, and Rye

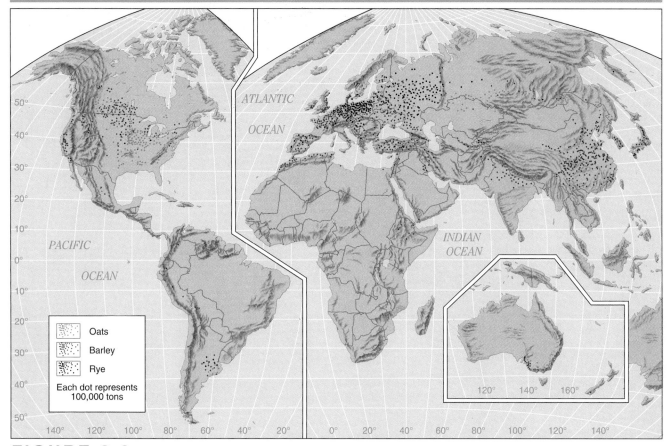

FIGURE 6-9

Barley, oats, and rye can serve either as human food or as feed for livestock.

products (dairy products, eggs) are consumed, so we say that feed grains are consumed by humans *indirectly*. Roughly 40 percent of the world's total grain harvest is fed to livestock. This percentage is higher in the richer countries. In the United States, for example, the figure is about 70 percent.

The richer a nation becomes, the more total grains it consumes, but the higher the percentage of its consumption which is in the indirect form of meat and dairy products. The people of the United States and Canada, for example, consume as much as one ton of grains per person per year. Only about 150 pounds of this, however, is consumed directly as bread or cereal. The rest is consumed indirectly. Some nutritionists argue that today the populations of the richest countries consume too much meat and dairy products, leading to increases in debilitating diseases such as heart disease. Nevertheless, most people enjoy meat and dairy products, and these dietary items are generally viewed as desirable. In poor countries, on the other hand, only about 400 pounds

of grain is available for each person per year, and the people consume most of it directly. In other places the people avoid meat and dairy products because of religious prohibitions. Chapter 3 noted the absence of pigs, for example, in Jewish and Muslim regions. These people must meet their protein requirements from fish, grains, or pulses (see Figure 6-12).

In 1991, humankind consumed about 175 million tons of meat, or about 70 pounds (32 kg) per person. Only in the richest countries, however, is meat the center of the diet. Citizens in the United States consume 246 pounds (112 kg) per year, compared with 196 pounds (89 kg) in Germany and 156 pounds (71 kg) in the United Kingdom. In many of the world's poor countries, by contrast, per capita consumption is 10 to 20 pounds (22 to 44 kg) per year.

The per capita consumption of beef in particular is closely related to the location of the world's grasslands. Argentina ranks first in beef consumption (153.9 pounds—70 kg—per person), followed by Uruguay (122.8 pounds—56 kg), the United States (97

Agriculture, Fishing, and Forestry   **203**

World Distribution of Cattle, Sheep, and Goats

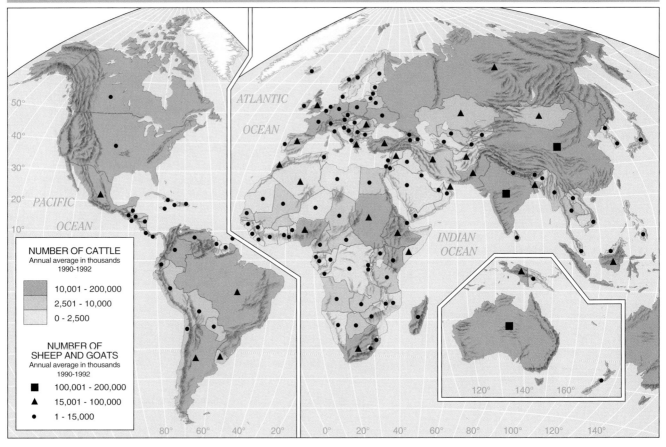

FIGURE 6-10

Cattle are widespread around the world, although the greatest number—almost 16 percent—are in India. Note the virtual absence of sheep and goats from the United States and Canada. Can you hypothesize why? The answer does not lie in physical-environmental conditions. (Data from World Resources Institute, *World Resources 1994–95.* New York: Oxford University Press, 1994 pp. 296–97.)

pounds—44 kg), and then Australia, Canada, and New Zealand. In the United States, annual per capita consumption of beef almost tripled between 1940 and 1976, from 55 pounds (25 kg) to about 155 pounds (71 kg). Since 1976, however, U.S. consumption of beef has fallen a full third. This is largely the result of popular concern about fat. From 1940 to 1990, U.S. per capita consumption of poultry nearly tripled. The figures for the consumption of both beef and chicken suggest the higher standards of living achieved in the United States from 1940 to today.

Some kinds of animals transform grain into meat more efficiently than do others, and humankind overall could greatly improve its food supply by concentrating on raising the most efficient animals. Chickens are the most efficient. They yield 1 pound (.45 kg) of edible meat for every 4 pounds (1.8 kg) of grain they consume. Pigs produce 1 pound for every 7 pounds (3.2 kg) of grain, and beef cattle 1 pound for every 15

pounds (6.8 kg) of grain. In addition, chickens reach maturity—and therefore can be consumed—much faster than do pigs or cattle. There are already twice as many chickens on Earth as people, yet a still greater emphasis placed on raising chickens could immensely improve the human diet (Figure 6-13). China is widely replacing swine with chicken farming, and several other countries have established programs to multiply their chicken populations. In many societies, however, cattle are viewed as a status symbol, and this preference delays the switch into more productive livestock.

Problems Associated With Cattle   There are about 1.3 billion cattle on Earth today, and recently some scientists have begun to criticize these herds as inefficient and even harmful. The indictment against them is long. In the United States, runoff from mammoth feedlots is despoiling streams and underground aquifers. In sub-

FIGURE 6-11

Cattle represent wealth among the Masai people of Kenya. As the government requires these traditionally nomadic people to settle, multiplying cattle overburden the environment. (Photo by Eliot Elisofon, National Museum of African Art, Eliot Elisofon Archives, Smithsonian Institution.)

The World Distribution of Diets

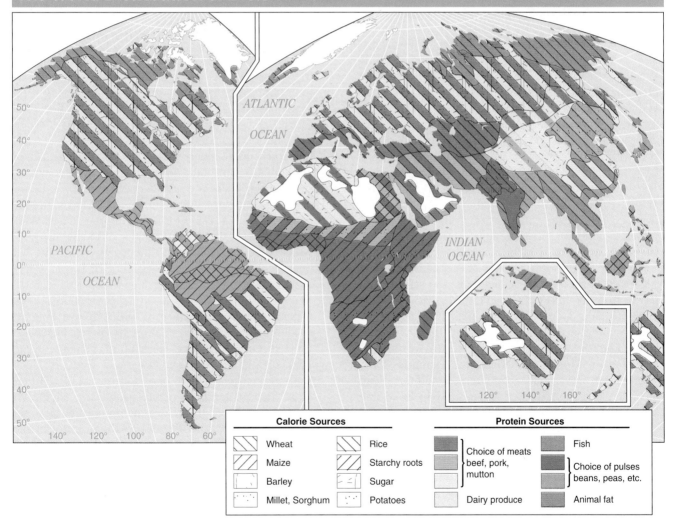

FIGURE 6-12

The meat and dairy diet of Europe and Europeans' settlements around the world stands out sharply on this map of sources of calories and protein. East and Southeast Asians rely more on fish, while pulses are important in South America, Africa, and in Southwest and South Asia.
(Recreated from *The Times Atlas of the World*, Times Book Division of The New York Times Book Co., 1983 by John Bartholomew & Sons Limited and Times Books Limited, London.)

FIGURE 6-13

Chicken production in advanced countries has moved from the barnyard into virtual factories in which chickens are mass-produced. The Earth's total chicken population is about twice its human population.
(Courtesy of U.S. Department of Agriculture.)

Saharan Africa, cattle contribute to desertification by denuding arid lands of fragile vegetation. In Central and South America, ranchers are felling rainforests for pasturage for cattle. Some agronomists argue that the millions of acres of land dedicated to feedstocks should be rededicated to crops for direct human consumption.

In addition, cattle contribute to global warming. The bacteria that live in every cow's gut enable the animal to break down the cellulose in grass. As a byproduct, however, these bacteria produce methane, which is one of the most important greenhouse gases. The amount of methane produced by livestock is annually estimated to be almost double the amount produced by all termites (which have the same bacteria) or from landfills (which give off gas from decomposition), and 50 percent more than from all worldwide burning of vegetation.

Other livestock contribute to environmental devastation. Sheep and goats, for example, have overgrazed substantial areas in the Mediterranean basin, as well as parts of Africa and India. Some scholars single out cattle, however, as a domesticated animal with an overall negative impact on the Earth.

Dairy Farming Nearly 90 percent of the world's supply of cow's milk is produced in developed countries. In the past, fresh milk was available only on farms and in nearby towns, but in rich countries urbanization created a demand for large-scale commercial milk production. Milk is heavy and spoils quickly, so the cost of transporting it is high. Therefore, most cities receive their supplies from dairy farms in surrounding regions referred to as the cities' *milksheds.*

Some areas that are remote from urban centers but have cool damp climates unsuited to grain farming may specialize in dairy farming. Wisconsin in the United States and Switzerland in Europe are examples. Much of the milk from these regions, however, is first transformed into butter, cheese, or dried, evaporated, or condensed milk. These products are not only lighter and less perishable than milk, but they are more valuable. The difference between the value of a raw material and the value of a product manufactured from that raw material is called the **value added by manufacturing**. Adding value to raw materials produces wealth. Wisconsin cheese and Swiss cheese and milk chocolate are exported worldwide, providing incomes to those two regions. In Chapter 11 we will see how the concept of value added applies to manufacturing from mineral raw materials.

New Crops Offer New Potential

The Green Revolution has so far focused attention on the improvement of just a few of humankind's most important crops, and the success in improving these crops has raised concern that humankind is becoming too reliant on too few crops (Figure 6-14). Wheat responded to scientific yield enhancement so well that today humankind relies more on wheat than on any single crop in the past. If a new wheat disease suddenly appeared today, it could destroy a significant percentage of humankind's total food supply. A model of such a disaster occurred in the United States corn crop in 1970. A new fungus suddenly appeared which was well-matched to the T-cytoplasm that had been incorporated into 80 percent of the country's seed corn, and U.S. corn production fell 15 percent.

A diversity of crops offers protection against catastrophe in case any one crop should fail, but modern commercial agriculture is increasingly specialized. Only "primitive" agriculture, which is in fact very sophisticated, has preserved diversity (look back at Figure 3-14). Ethnobotanists have reported more than 50 kinds of potato grown in a single village in the Andes. Therefore, paradoxically, new scientific appreciation of "primitive" agriculture may multiply world harvests. Scientists are just now rediscovering valuable food crops which were cultivated by the Inkas 500 years ago. When archaeologists returned some Bolivian fields from the most modern farming techniques to the methods used by the Tiahuanaco people over 1,000 years ago, yields multiplied by seven times. Modern Mexico City's Xochimilco Park preserves an example of the cultivation system of canals and artificial islands that

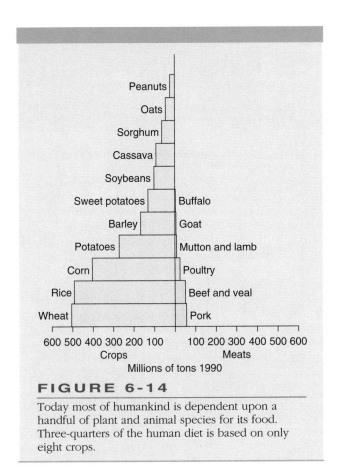

FIGURE 6-14

Today most of humankind is dependent upon a handful of plant and animal species for its food. Three-quarters of the human diet is based on only eight crops.

allowed the Aztecs to feed the 200,000 people living in the city in 1520 (Figure 6-15).

Today humans rely on only 20 plant species for almost all of our food, but in the course of human history about 7,000 have been utilized, and at least 75,000 species have edible parts. Many of them are superior to the plants cultivated today. The New Guinea winged bean, for instance, is entirely edible—roots, seeds, leaves, stems, and flowers. It grows rapidly, up to 15 feet (4.5 m) in a few weeks, and it offers a nutritional value as high as that of any known crop. The American Academy of Sciences has recommended the cultivation of 36 other crops, including amaranthus, buffalo gourd, tamarugo, guar, mangosteen, and soursop. Most of us have undoubtedly never heard of these, yet each offers enormous potential for food. The potential of raising food from more and different plants is a reason why the extinction of plant species threatens humankind's welfare. The steady loss of genetic diversity before botanists can study each species sacrifices potential new crops or genes for interbreeding.

Agricultural technology offers many alternative foods, but what people eat, or refuse to eat, is to a great degree *cultural* (Figure 6-16). Culinary imagination could probably make a wide variety of alternative crops or animals appealing and palatable. For exam-

FIGURE 6-15

Mexico City's Xochimilco Park maintains the system of cultivation practiced by the Aztecs in the lakes surrounding their capital 500 years ago. Grass and other organic material were woven together into thick mats covered with lake bottom mud. At first the mats float, and whole fields can be towed from one place to another. In time, the plants send down roots that take hold in the shallow waters, and the farmers add more layers of organic material. This system supported a city that Spanish conquistadors described as surpassing the glory of any in Europe. The park was declared a World Cultural Heritage Site by UNESCO in 1987. (Courtesy of Professor Milton Rafferty, Southwest Missouri State University.)

FIGURE 6-16

Peasants bring live iguanas into downtown Hanoi, Vietnam, for sale for Saturday night dinner. Many people around the world find nutrition in and enjoy eating things most Americans do not find tempting. (Courtesy of Julia Nicole Paley.)

Agriculture, Fishing, and Forestry **207**

How Farmers Decide How to Use Their Land: Von Thünen's "Isolated City" Model

Johann Heinrich von Thünen (1783–1850) was a Prussian aristocrat who wondered what product he could grow on his suburban estates and market most profitably. Thinking about this problem led him to question how various crops become distributed across any countryside. To answer his question, von Thünen devised one of geography's earliest examples of deductive reasoning. He began with the idea of an imaginary city market in an isotropic plain, and, applying the assumptions of the theory of economic man discussed in Chapter 2, he deduced a model pattern for the distribution of different land uses.

Von Thünen noted that the different uses to which parcels of land are put results from the different values placed upon the land, called the *rent value* of land. On an isotropic plain, transport costs are the only variable, so differences in rent reflect the transport costs from each farm to the market. The greater the cost, the lower the rent that can be paid if the crop produced is to be competitive in the market. In addition, perishable products such as milk and fresh vegetables need to be produced near the market, whereas less perishable crops such as grain can be produced farther away.

Von Thünen deduced that a pattern of concentric zones of land use will form around a city market (Figure A). The *intensity* of cultivation, that is, the amount of labor and capital applied, will decline with distance from the market. Perishable crops that have the highest market price and the highest transport costs per unit of distance, such as vegetables, will be grown closest to the market. Today a hint of this zone around a city survives in New Jersey, "The Garden State." New Jersey farms historically supplied the cities of Philadelphia and New York. Today New Jersey is heavily industrialized, yet it also still specializes in greenhouse products, dairy products, eggs, and tomatoes, which are typical market garden commodities. In Von Thünen's model, fields farther away from the market will be dedicated to less perishable crops with lower transport costs per unit of distance. In the rings farthest outward from the city, livestock grazing and similar extensive land uses predominate.

Von Thünen elaborated his model by adding a navigable river flowing through the town (Figure B). He noted that the areas along the river's banks would enjoy greater accessibility to the city market. In other words, the cost-distance to the city market was lower along the riverbank, so each of the zones would extend out along the river.

Some scholars have tried to apply Von Thünen's model to entire countries. They have plotted the use of cropland in the United States, for instance, as if the eastern seaboard cities or Chicago were a central market. Trying to apply Von Thünen's model to vast territories under contemporary conditions, however, strains the model's usefulness. The model focuses on

---

ple, the menu for the New York Entomological Society's 100th anniversary banquet in 1992 featured live honey-pot ants as an appetizer, cricket and vegetable tempura, mealworm balls in zesty tomato sauce, roasted Australian kurrajong grubs, waxworm fritters in plum sauce, giant Thai waterbugs sautéed in olive oil, and, for dessert, chocolate cricket torte. All of these insects are very nutritious, and they are common fare in many countries.

What Determines Agricultural Productivity?

 Figure 6-17 maps the trememdous variation in agricultural yields around the world. This variation is partly attributable to variations in the physical environment. For example, Western European countries enjoy mild climates, and their agriculture is highly productive, whereas the Sahara Desert region of northern Africa is generally unproductive. Surprising exceptions, however, can also be found. For example, Saudi Arabia records the highest yields anywhere on Earth.

When we compare Figure 6-17 with Figure 6-18 (p. 211), however, we see considerable coincidence between the variations in productivity and the use of tractors and the application of fertilizers. The countries that enjoy the highest yields rank high in at least one of these two investments. Investment in agriculture seems to be a principal factor determining productivi-

distance from city markets, but in the real world proximity to the market is only one consider- ation. Physical environmental conditions, governmental regula- tions, the economic system, the pattern and the regulation of the transport system, and still other factors must be considered.

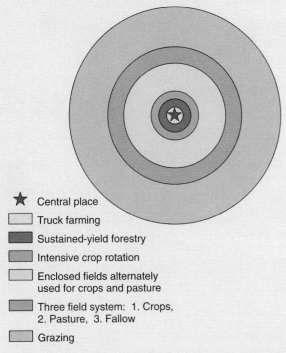

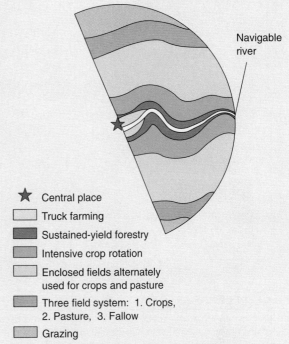

Central place
Truck farming
Sustained-yield forestry
Intensive crop rotation
Enclosed fields alternately used for crops and pasture
Three field system: 1. Crops, 2. Pasture, 3. Fallow
Grazing

(A) Von Thünen's model shows rings of decreasingly intensive land use concentrically outward from the city. The ring of sustained yield forestry may be sur- prising, but forestry was then an intensive land use. The forest provided grazing for livestock, fuel, and raw materials for building and many other purposes.

Navigable river

Central place
Truck farming
Sustained-yield forestry
Intensive crop rotation
Enclosed fields alternately used for crops and pasture
Three field system: 1. Crops, 2. Pasture, 3. Fallow
Grazing

(B) If a navigable river flows through the town, the activity characteristic of each zone extends outward along the river.

ty. Many of the rich places that are able to invest in tractors and fertilizer can also be assumed to be places where farmers have the best *hybrid* seeds. These are the product of careful interbreeding of the best strains of a crop through generations. These countries also have the best irrigation, pesticides, and all of the other modern technological apparatus of agriculture. Therefore, we can conclude that capital input is a principal determinant of agricultural produc- tivity today. Capital investment may be as important as the natural environment, or even more important. If investment in agriculture could be increased in the regions where there is little investment today, total world agricultural output would undoubtedly soar (see Figure 6-19).

FOOD SUPPLIES IN THE FUTURE

Until now all the factors just discussed have held off the specter of worldwide starvation. Is it possible that humankind is now, at last, at the end of its ability to increase food supplies?

The answer to this question is no. There is no reason to fear that humankind is, technologically, in danger. If demographers are correct in their projec- tions of the Earth's future population that we read in Chapter 4, the population can be fed. Farmers now utilize almost all potential cropland, and the amount of cropland per capita is falling, but humankind has scarcely begun to maximize productivity even with

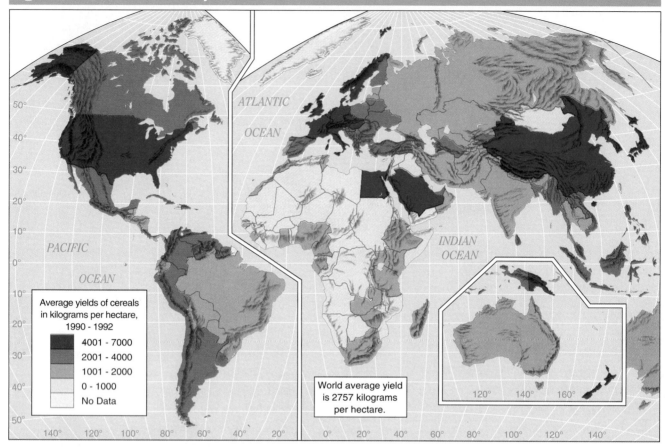

Average yields of cereals in kilograms per hectare, 1990 - 1992
- 4001 - 7000
- 2001 - 4000
- 1001 - 2000
- 0 - 1000
- No Data

World average yield is 2757 kilograms per hectare.

FIGURE 6-17

Yields per hectare vary greatly around the world, and these variations are not always clearly related to natural environmental conditions. This map does not reproduce the pattern of precipitation (Figure 1-18), climates (Figure 1-20), or biomes (Figure 1-23). Factors other than natural environmental conditions must be found to explain agricultural productivity. (Data from World Resources Institute, *World Resources 1994–95*. New York: Oxford University Press, 1994, pp. 292–93.)

present-day technology. The leading technology has been applied to only a small portion of the Earth, and emerging technology offers still greater possibilities. A study published in 1994 by the Council for Agricultural Science and Technology even suggests the possibility of feeding double the Earth's present population while still being able to return significant areas of cropland to natural states or wilderness.

The Green Revolution and the Scientific Revolution in Agriculture Continue

The Green Revolution and the Scientific Revolution in Agriculture are not over. Some scientists argue that they have just begun. New scientific advances promise to multiply future food yields. Biotechnology, for example, offers genetically altered crops that can be custom-designed to fit the environment and produce bountiful

harvests. In 1988, scientists mapped the *genome* of rice—the set of 12 chromosomes that carries all the genetic characteristics of the plant. This development could enable geneticists to produce improved strains of rice. Biotechnology can also replace chemical pesticides and fertilizers, whose biological or even genetic impact on our own bodies is not fully understood (see Figure 6-20, p. 213). So far, most advances in the Green Revolution have addressed improving yields of wheat, potatoes, and rice. Many other traditional crops, including the tropical crops such as sweet potatoes and cassava, are winning new attention (Figure 6-21, p. 213).

Scientists have also conducted research on *halophytes*, plants that thrive in salt water. Interbreeding halophytes with conventional crops has made these crops more salt-resistant, which means that they can grow in more diverse environments. Farmers are today harvesting lands once thought too salt-soaked to support crops in Egypt, Israel, India,

210 *Chapter 6*

The Use of Tractors and Fertilizers

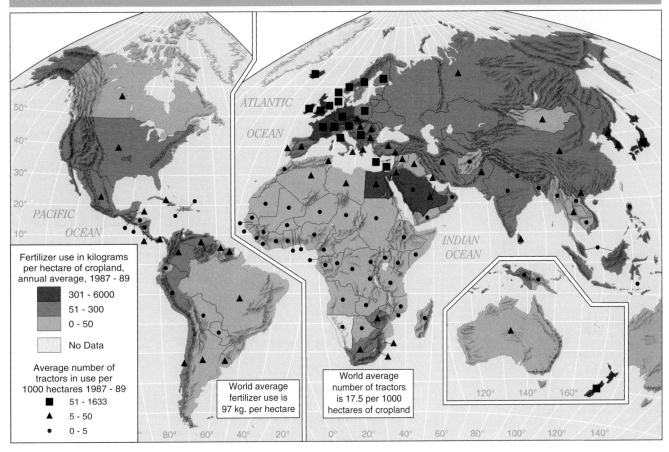

Fertilizer use in kilograms
per hectare of cropland,
annual average, 1987 - 89

- 301 - 6000
- 51 - 300
- 0 - 50
- No Data

Average number of
tractors in use per
1000 hectares 1987 - 89

- ■ 51 - 1633
- ▲ 5 - 50
- • 0 - 5

World average
fertilizer use is
97 kg. per hectare

World average
number of tractors
is 17.5 per 1000
hectares of cropland

FIGURE 6-18

Capital inputs in agriculture such as the amount of fertilizer used and the number of tractors available vary tremendously. Does this map reproduce the pattern of Figure 6-17? (Data for World Resources Institute, *World Resources 1994-95*. New York: Oxford University Press, 1994, pp. 29–95.

and Pakistan. Conventional crops may someday be grown in salt water.

All of these possibilities justify optimism. The economist Henry George (1839–1897) succinctly contrasted the rules of nature with the multiplication of resources through the application of human ingenuity. He said, "Both the jayhawk and the man eat chickens, but the more jayhawks, the fewer chickens, while the more men, the more chickens." This principal is key to understanding and counting all resources.

FIGURE 6-19

These two photographs illustrate the extremities of capital investment in agriculture. If the Senegalese peasant (a) had as much capital to invest in farm machinery, fertilizer, and improved seeds as the Nebraska farmers (b) do, who knows how much food he could raise?
(Courtesy of (a) the Agency for International Development/Carl Purcell and (b) U.S. Department of Agriculture.)

Agriculture, Fishing, and Forestry   **211**

Continuing Crop Redistribution

Questions

1. Can you think of the first time you tried some new food? How about a kiwi or a Japanese buckwheat noodle? Find "exotic" foods in local markets. Why were these first imported to your local store? For a local immigrant population?

2. New foods from around the world are introduced almost daily in heterogeneous urban centers. How has the change in the source areas of immigrants to the United States discussed in Chapter 5 caused the introduction of new foods to U.S. markets?

Different peoples still harvest different indigenous crops and enjoy different diets, and food crops continue to migrate around the world. *Chenopodium quinoa*, for example, is a native of South America. It is an annual, broad-leafed herb usually standing about 3 to 6 feet high. Leafy flower clusters rise from the top of the plant. The dry, seedlike fruit is about 2 mm in diameter and enclosed in a hard, shiny, four-layered fruit wall. Quinoa seeds are flat and pale yellow, and they can be steamed, ground into flour, or fermented to make a mildly intoxicating beverage. The leaves are highly edible and often used for livestock feed.

Quinoa has recently been appearing in health food stores in the United States. Of all the world's grains, it is the highest in protein content, and it has an amino acid profile that parallels the ideal standard set by the

This experimental quinoa field in Colorado's San Luis Valley is where Colorado State University scientists develop new strains of this South American native crop.
(Courtesy of Professor Dwayne Johnson, Colorado State University.)

United Nations Food and Agriculture Organization (FAO). New varieties of this member of the genus goosefoot (*Chenopodium*) are being bred that are better-suited to North American conditions, but significant U.S. production has begun in Colorado.

Technological solutions to problems, however, can still trigger unexpected new problems. Some people fear that bioengineered plants may have unforeseen effects in nature (outcompeting and destroying natural species, for example) or on our own bodies. Chapter 2 noted that many people fear the results of opening even human biology to manipulation. These fears remind us of "Frankenstein's monster" referred to at the opening of this chapter.

In 1968, Professor Paul Ehrlich of Stanford University wrote in a book that became a bestseller: "The battle to feed humanity is already lost....we will not be able to prevent large-scale famines in the next decade." Local famines did occur, but they were not caused by a lack of food on Earth. From 1968 to 1992, world grain production rose 60 percent. For 30 years world excess food stocks relative to consumption have grown faster than population. Nevertheless, the debate between the optimists and the pessimists continues.

WORLD DISTRIBUTION OF FOOD SUPPLIES AND PRODUCTION

The amount of nutrients needed per capita in each country varies greatly, depending, for the most part, on the climate and on the median age of the population. A country with a low median age needs less food per capita than one with a higher median age. The amount of food intake is measured in calories, which are units of fuel the body needs for energy production. Figure 6-22 maps the calories available as a percentage of the need in each country.

Adequate nutrition is not only a matter of calories. A balanced diet includes carbohydrates (derived from staples such as rice, corn, wheat, and potatoes), proteins (from meat, fish, poultry, eggs, dairy products, or pulses), vitamins (from fruits and vegetables as well as other sources), fats, and minerals. Overall,

FIGURE 6-20

The rows of potato plants on the left and right were devoured by the Colorado potato beetle. The plants in the center row, however, had a gene introduced into them from a common soil bacteria. Those plants produce a protein that acts as a natural insecticide. Biotechnology offers the promise of protecting many crops in this way. (Courtesy of Monsanto Company.)

FIGURE 6-21

Agricultural colleges, such as this one in Tanzania, instruct local farmers in the latest farming techniques. Most agricultural research, however, has been done in the midlatitude areas which are already rich and advanced. More is known about the physical conditions for agriculture in Iowa than in Tanzania. There is still much to be learned about the physical conditions of farming, such as soils, in many countries, as well as about the possibilities of enhancing yields of tropical crops. (Courtesy of the Agency for International Development.)

Per Capita Food Supplies

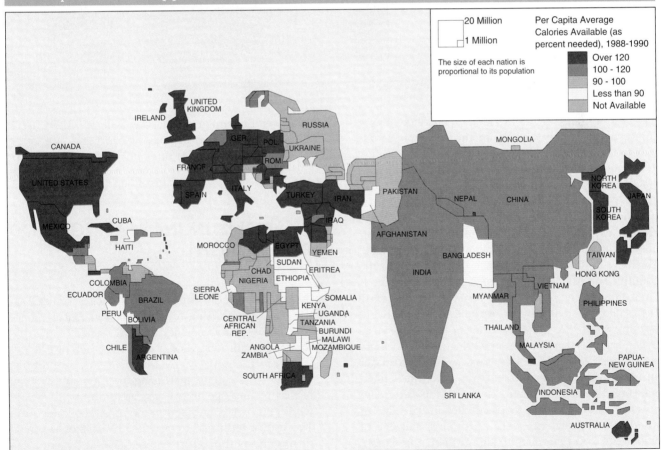

FIGURE 6-22

The number of calories available as a percentage of the need is here mapped on a population cartogram. This suggests the percentage of the Earth's population in each category of supply. (Data from World Resources Institute, *World Resources 1994–95.* New York: Oxford University Press, 1994, pp. 272–73.)

FOCUS
The Aswan Dam

Egypt's Aswan Dam, built between 1960 and 1970, illustrates how technological innovation can backfire. The dam allowed Egyptians for the first time to control the flooding of the Nile River for hundreds of miles to the north, and Lake Nassar behind the dam, covering about 2,000 square miles (5,180 km²) of Egypt and the Sudan, provides an assured supply of water to irrigate new fields.

Since the dam was built and the river was controlled, however, the Nile's floodwaters no longer fertilize the valley, and this has increased demand for fertilizers. The sardine fisheries of the Mediterranean Sea have lost the nutrients that the river formerly provided, so they are suffering. The Nile delta, deprived of new silt, is eroding badly, and the incidence of parasitic diseases is increasing in Egypt.

The dam may have contributed to ending Egypt's wars with Israel. About 95 percent of Egyptians live within 12 miles of the Nile or in its delta. It would be fairly easy to destroy the dam with just one or two bombs and quickly drown millions of people. Israel announced that the Israeli Air Force had the capability to do that.

Egypt's Aswan dam produces enough electricity to provide half of Egypt's needs.

the food supplies in some countries fall short of per capita needs. Many people in those countries are well-fed, but many others must be hungry. Even in the countries in which overall per capita food supplies are adequate, however, some people are hungry.

No country is autarkic in food. Most countries both import and export food, and a few countries are net exporters of food despite the fact that portions of their own populations are undernourished. This may be because of injustice or civil strife in the country. The African country of Zaire, for example, is environmentally richly endowed, yet in the early 1990s, a great many of its 40 million people were starving. In 1992, the U.S. Catholic Relief Services Organization sorrowfully noted, "In our view, Zaire is not faced with a food shortage per se, but rather an acute food access problem." In 1992, the United States sent troops to Somalia to ensure the safe delivery of emergency relief food supplies in that nation, where warring factions were preventing delivery of emergency food supplies to the civilian population. The question of whether one nation, or the United Nations, has the right to intervene in another nation

even for humanitarian reasons will be taken up in Chapter 13, but food supply problems are often problems of economics or politics, not geography or technology.

PROBLEMS IN INCREASING FOOD PRODUCTION

▼▼▼▼▼▼▼▼▼▼▼▼▼▼▼▼▼▼▼▼▼▼▼▼

 Technology creates the potential for increasing food production, but it cannot guarantee that this will occur or that the food will be produced or distributed where it is needed. Unfortunately, many practices and conditions actually work against increasing the food supply. One is that technology is not spread around the whole world, nor is it distributed in the most economic way. The use of fertilizer exemplifies this. Other potential problems in the effort to increase food supplies include lack of financial incentives, unsuccessful land redistribution, and inequitable land ownership.

Fertilizers: Overuse and Diminishing Returns

Fertilizers can improve crop yields significantly when used in appropriate amounts. For many years, however, the quantities of fertilizer applied in the advanced countries have far exceeded the point of **diminishing returns**. Diminishing returns exist when, in adding equal amounts of one factor of production, such as fertilizer or labor, each successive application yields a smaller increase in production than the application just preceding. Table 6-3 applies this concept to a hypothetical example of fertilizer. In this table the point of diminishing returns is 4 pounds of fertilizer. This is true because adding the second and third pounds of fertilizer increased the yield by 11 bushels, but adding the fourth increases the yield by only 10 bushels. At some point, then, the diminishing increases in crop yields no longer justify the application of greater amounts of fertilizer at one place. That fertilizer could more efficiently be applied in another place that has not reached the point of diminishing returns.

Today, U.S. agriculture uses about 160 pounds (73 kg) of fertilizer for each man, woman, and child in the country. If each U.S. farmer's last 10-pound bag of fertilizer were given to a farmer in a poor country, the world food supply would increase dramatically. In other words, the same fertilizer that yields only diminishing returns in the United States could be used much more productively in a poor country.

Many Farmers Lack Financial Incentives

If a market exists for any surplus that farmers can produce, and if the political and economic system sufficiently rewards farmers for their effort, produc-

tion of a surplus is all but assured (assuming adequate technology is available). Chapter 3 referred to the agronomists' dictum: "The market produces the surplus."

In many countries, however, farmers have little incentive to increase food production. Many governments buy farmers' crops at fixed low prices, charge export levies, support high exchange rates, and impose high import duties on the tools and agricultural chemicals. These are all ways of taxing farmers. The wealth that is squeezed from the farmers in these ways is often used to subsidize urban food supplies or to support urban civil service bureaucracies, military spending, or unprofitable state-run industries. In some countries, deteriorating conditions in the countryside trigger migration to the cities (a topic we shall examine in Chapter 9), and swelling urban populations increase the demand for subsidized food. Many farmers can produce more food if given an incentive to do so. Since 1991, when India lifted some agricultural price controls, grain production has risen about 4 percent each year, whereas population has been rising at under 2 percent per year.

Systems of Land Ownership Can Discourage Production

In many countries, systems of land ownership retard food production while increasing political and social inequalities. In Brazil in 1990, for example, 70 percent of the rural population lacked title to land, while 0.7 percent of farms occupied 43 percent of the farmland. Many of the large landholdings were not worked to maximum productivity, while the concentration of land ownership spurred the landless to attempt to farm new claims in the rainforest regions. As noted in Chapter 2, this expansion degrades the environment while producing only disappointing yields.

In many countries the landholding systems are holdovers from colonial periods. One widespread consequence of colonialism was to replace precolonial communal systems with private property. The distribution of the property under colonialism was highly inequitable, creating one class of large estate holders and another, much larger, class of landless poor. The large landowners often moved to the cities, becoming absentee landlords unwilling to make the investments necessary to maximize the return from their own agricultural holdings.

In contrast, individual farmers of small personal holdings are more likely to maximize their productivity. In the Philippines, for example, large blocks of land were granted to the Spanish colonial elite, and

TABLE 6-3
The Diminishing Returns of Fertilizer Use

Fertilizer (pounds)	Total yield (bushels)	Production increase (bushels)
1	10	10
2	21	11
3	32	11
4	42	10
5	51	9
6	59	8

the natives on these land grants became tenant farmers. After a few generations, the landlords were mostly mestizos (persons of mixed native and colonial ancestry), and their descendants owned the best agricultural land. This land-grant system prevailed in all areas colonized by Spain, and it is at least partly responsible for political unrest in Central and South America today.

Indonesia presents a contrast to the former Spanish colonies. It is much larger than the Philippines, but much like it in geography and population type. For more than 300 years under Dutch rule, however, only native Indonesians could own land, except for city lots. Dutch, Arabs, Chinese, and English could rent agricultural land, but they could not acquire title to it. Thus, land ownership among Indonesian peasants remained much more widespread than it was in the Philippines. Indonesian peasants who have inherited their land and expect to leave it to their children are much less likely to join peasant guerrilla movements than landless Filipinos are.

When land is leased to those who actually work it, the conditions of tenancy and of payment determine whether the farmers are encouraged to produce significant surpluses. If the landlord takes too large a share, surpluses will be small.

The collectivization of agriculture under Communist regimes crippled farm production in several countries, but today private farming is being encouraged. Poland ended food subsidies and restrictions on direct marketing in 1989; farmers' markets sprang up overnight, food shortages eased, and prices fell. The other formerly Communist Eastern European countries are today breaking up large communal landholdings and encouraging private farmers to increase food production. Giving or selling government assets to private individuals or investors is called **privatizing** an activity. Small private market gardens were always allowed in the former Soviet Union, and although these accounted for less than 2 percent of total Soviet farmland, they yielded 60 percent of the country's potato crop and around 30 percent of its vegetables, meat, eggs, and milk. By the summer of 1993, Russia reported having 260,000 private farms and hoped to have 650,000 by 1995. China privatized its previously collectivized agriculture in 1978, and within 10 years farm output rose 138 percent. China turned from a net importer of food products into a net exporter. (Evidence suggests, however, that a side effect of privatization in China may be to increase the birth rate by intensifying the traditional urge to have sons to carry on the family farm.)

In many other countries, title to land is unclear. Generations of peasants may have tilled small plots,

but families have never registered title, and bureaucratic complications make it difficult to establish title today. In Peru, for example, only about 20 percent of all land has clear title. Lack of clear title discourages the occupants from making improvements on the land. Furthermore, as we have seen, successful farming today requires capital investment in irrigation pumps, chemicals, improved seeds, and machinery. Farmers can borrow money to make these investments only if the land can serve as collateral (a thing of value that can be seized by a creditor in case of nonpayment of a debt). Farmers without clear title to the land lack collateral. New programs to register landholdings have been launched in El Salvador, Brazil, Bolivia, India, Indonesia, and Peru, as well as in formerly communist countries. Establishing title to farmers' land throughout the world could trigger investment and improved productivity.

Unsuccessful Land Redistribution: Mexico

Communal holdings cannot serve as collateral, so communal ownership does not necessarily increase productivity. Mexico demonstrates this phenomenon. More than one-half of the country's arable land is held in **ejidos**, a form of land tenure in which a peasant community collectively owns a piece of land and the natural resources and houses on it (Figure 6-23). Mexican law states that ejidos are "inalienable, nontransferable and nonattachable." They cannot be used as collateral, and

FIGURE 6-23

This ejido represents the traditional Mexican form of communal landholding. As the rural population has risen, the amount of land per farmer has dropped, and capital inputs are meager.
(Courtesy of Carl Frank/Photo Researchers.)

this illiquidity discourages possible lenders. This system, combined with a government tradition of paying farmers low prices for their crops while at the same time subsidizing food for urban consumers, has kept farmers poor and productivity low and, in turn, intensified migrations of peasants to the cities. Mexico imports 35 percent of its total food supplies, including nearly half of its staples of corn, wheat, and beans, from the United States. Mexico is also one of the world's largest importers of dairy products.

In 1991, the government launched an effort to overhaul the ejido system. Official recognition of ejido ownership was, however, a principal issue of the Mexican Revolution (1910–1920), in which over 1 million Mexicans lost their lives, and many Mexicans remain attached to the idea. Changing the system will require an emotional political battle, but the government hopes that it will substantially increase food production.

Barriers to Increasing Production in Africa

The return of land to communal holdings in some African countries has repeated the Mexican experience. Successful farmers cannot borrow money for investment to expand their holdings. These conditions promote increased migration to the cities. As long as tenure depends on occupancy, migrants to the city leave their families in the villages. In some parts of rural Zimbabwe, 40 percent of the households have lost the father to the town. The overworked women, left to tend children as well as crops, can scarcely rise above subsistence. The United Nations Children's Fund (UNICEF) estimates that women grow 80 percent of Africa's food.

In Zimbabwe the government has seized numerous large farms—from both white and black owners—that were profitably exporting foods. Rather than redistribute this land to peasants, however, the government has resettled the farms but retained ownership. Therefore, the farmers cannot use the farms as collateral to borrow for investment, nor can they sell their produce except through state marketing boards that pay them little for their effort. In Nigeria, too, the government owns almost all land, leases it to farmers, and buys all output at fixed low prices.

Civil wars have reduced food production in some areas of the world, again, particularly in Africa. The continent was feeding itself in 1960, but by 1990 Africa imported 40 percent of its food supply. In 1976, the U.S. Central Intelligence Agency (CIA) estimated that Angola could produce food enough for 250 million people, but in 1990, after years of civil

war, that country had to import half of the food for its 10 million people.

All of these complicated factors inhibit food production in many countries. Therefore, it is virtually impossible to map *potential* food production and supplies. Food is available in world markets for countries that can afford to import it and choose to do so. Nevertheless, people are going hungry. The United Nations estimated that about 500 million people were undernourished in 1991, about 9 percent of the Earth's population. Many of these people could be fed if their homelands could stop discriminating against farmers (particularly women), end their civil wars, stop subsidizing urban populations at the expense of farmers, let their farmers import the tools and chemicals they need, and reform landholding systems. The number of instances of starvation around the world is still high. Most of these, however, are triggered by civil disruption. They do not indicate deficiencies in the environment or in human technical capabilities.

Commercial Crops in the Poor Countries

 Today many farmers in poor countries do not concentrate on growing staple crops for local consumption. Instead they raise *cash crops* for sale. Senegal, for example, exports peanuts, the Ivory Coast cocoa, Angola coffee, Zimbabwe tobacco, and Kenya tea and flowers, while food availability in those countries is low (Figure 6-24). We must investigate whether these countries would not be better off growing food for their own consumption.

FIGURE 6-24

Colombia's exports of cut flowers tripled from 1982 to 1992, to a total of $315 million. Today Colombia provides the United States with four-fifths of its carnations and a third of its roses. Would Colombians be better off raising food for themselves?
(Courtesy of the Colombian National Tourist Agency.)

If any country chooses to dedicate its land to specialty crops, it can buy basic foodstuffs on world markets. In that way the country can achieve what is called *economic self-sufficiency* in food. Farmers who choose to depend on cash crops, however, face two risks. One is that the price for their product could collapse because of overproduction elsewhere. The second is that prices for the foods they will have to buy could rise faster than their own incomes do.

Some observers have argued that the production of cash crops competes with food crops for scarce resources of fertile land, water, and capital. In most countries that are expanding their production of cash crops, however, the production of food crops is growing too. Farmers have often been able to borrow techniques from their production of cash crops and use these to increase their production of food crops, or they have brought fallow land into cultivation in crop rotation systems. Unfortunately, the reverse is also often true. Parts of Africa have seen a decline in the production of both kinds of crops.

Raising cash crops may increase farmers' disposable incomes rapidly, while maintaining food production. Guatemalan farmers raising snow peas for New Yorkers, for example, can make 14 times as much per acre as they would raising corn. Even after taking into account the twelve-fold increase in their costs, their return for each day of work more than doubles. A great share of the increased costs are for hired labor, demand for which increases.

The rise in income may not, however, improve the lives and health of farm families. In a comparative study of Kenya and the Philippines, the average household's income doubled when farmers turned to cash crops, but the average caloric intake of young children rose by only 4 to 9 percent. The reason seems to have been sex discrimination. In many poor countries women control the production of food crops, but men control the production of cash crops. The men often spent "their" money on things other than family needs.

In order for farmers to turn to cash crops, governments must provide the necessary infrastructure, including clean water and medical facilities, roads to markets, simple financial services to even out the farmers' cash flow, and information services to help ensure the parallel improvement of methods used to raise staple crops. The money for this needed investment might come from the higher earnings gained from agricultural exports.

Unfortunately, the world's most successful cash crop remains the one most governments try hardest to discourage—illegal drugs (see Figure 6-25). Several Caribbean countries have found it profitable to surrender self-sufficiency in food and to dedicate their crop-

land to another kind of grass—fairways for golf courses that attract tourists. The role of tourism in national economies will be examined in Chapter 12.

THE RICH COUNTRIES SUBSIDIZE PRODUCTION AND EXPORT OF FOOD

World production and distribution of food is made even more complicated and perplexing by the system of world trade. The percentage of the world's food that is traded internationally is rising, so the prices of commodities and the conditions of trade in food are increasingly important. Furthermore, as noted in Chapter 3, improvements in connectivity continue to unify world markets, so that prices for many basic foods are seeking common world levels. International trade in manufactured goods has been regulated since 1945, but trade in agricultural products, which is approaching $500 billion per year, will only begin to be regulated in 1995. Up until today, many rich countries have erected walls of high tariffs (taxes on imported goods) to protect their markets against food

FIGURE 6-25

This opium poppy field was discovered in a forest clearing by tourists hiking in the vicinity of the border between Thailand and Laos. Opium poppies are illegal in most places, but they are one of the most valuable of all field crops, so farmers persist in growing them. The region where Thailand, Laos, and Myanmar come together (the "Golden Triangle") produces an estimated 60 percent of the heroin that reaches the United States. (Courtesy of Julia Nicole Paley.)

imports from the poor countries. They also subsidize their own farmers by guaranteeing high prices. As a result, farmers in these countries produce enormous food surpluses, which the countries then export or dump.

In the 1960s, Western Europe was a net food importer, but by the late 1980s it was an exporter and had begun to stockpile surpluses of cereals, fruits, vegetables, dairy products, wine, olive oil, and sugar. In 1992, the governments of the twelve Western European members of the European Union took about $60 billion from general tax revenues, added $100 billion paid by consumers as higher prices, and gave it all to their farmers. Western Europe has about 10 million farmers, compared to only 2 million in the United States. Despite such subsidies, agriculture is still only 3 percent of Western Europe's total economic output. In 1992, the United States spent about the same amount as Union members on farm subsidies, but consumers paid only about $31 billion more as higher prices. Japan's 1992 subsidies to its farmers totaled about $74 billion.

These subsidies distort the world geography of agriculture. They contribute to production of agricultural surpluses in many rich countries. Furthermore, as long as these countries dump food, the governments of many poor countries neglect to invest in their own rural areas. Farmers in the poor countries cannot compete, and they may be reduced to subsistence or driven off their farms entirely.

Why Do Some Rich Countries Subsidize Agriculture?

 Many rich countries subsidize their farmers and erect import barriers for several reasons. First, some countries pursue self-sufficiency in food production as a national security measure. The governments want some defense against grain embargoes and crop failures in those countries that normally have surpluses to market. Subsidizing a country's own farmers offers some protection against these threats. Following U.S. threats to halt grain shipments to the Mideast in the 1970s, for example, Saudi Arabia spent heavily to improve its agriculture (Figure 6-26). Iraq failed to do so, and by the time it occupied Kuwait in 1990, Iraq depended on imports for almost 80 percent of its nutritional needs.

Second, many rich countries, particularly in Europe, subsidize agriculture in order to keep their farmers from migrating to the cities in search of work. Rural-urban migration can overwhelm the ability of cities to absorb the new labor force, and it threatens the quality of life in the cities.

FIGURE 6-26

Saudi Arabia has invested some of its oil income in irrigating the desert to achieve agricultural self-sufficiency. In 1991, wheat cost Saudi Arabia about $600 per ton to grow, yet in that year Saudi Arabia raised so much that it sold wheat to New Zealand for $80 per ton, $45 per ton below the world market price. This pattern of world production and trade was obviously not the result of physical conditions for agriculture, but it bankrupted wheat farmers in regions more naturally suitable to wheat growing than Saudi Arabia.
(Courtesy of Aramco World Magazine.)

Third, rich countries may subsidize farming as part of national land-use plans. To them it is valuable in itself to have a farming countryside and to preserve green areas around cities, called *greenbelts*. Their urban citizens enjoy driving through or visiting convenient rural landscapes, and they are willing to subsidize farms in order to preserve them.

Fourth, many rich countries subsidize their farmers because their systems of political representation favor farmers. The farmers in the United States, Japan, and Europe constitute a decreasing percentage of the national populations, yet, particularly in Europe and Japan, the electoral districts have never been readjusted to the relative depopulation of the countryside and the urbanization of the populations. Thus, a rural vote may heavily outweigh an urban vote. Farmers enjoy disproportionate political power over urbanites. Most urban consumers remain unaware that food prices subsidize their nation's farmers, because the cost of food as a percentage of their incomes is falling anyway. Electoral reform in the rich countries might be necessary to encourage food production in the poor countries.

Thus, the fact that farmers in many rich countries enjoy disproportionate political power and eco-

FOCUS
Subsidizing Agriculture in Japan

Japan exemplifies a rich country that subsidizes its agricultural sector. Japan is a mountainous archipelago with limited amounts of flat fertile land, but about half of Japan's nonmountainous land is farmed, so the Japanese are forced to live in tiny and expensive quarters. The average size of a Japanese farm is about 3 acres (1.2 hectares), 0.5 percent of the U.S. average, but the 9 percent of all farms that are larger than this produce 44 percent of the crop. Japan had 30 million farmers as recently as 1965, but today agriculture is the sole source of income for only about 450,000 Japanese.

Japan does not have a one-person, one-vote system of representation. The Japanese courts have upheld a 3-1 disparity in population of districts represented in the Japanese parliament (the Diet). In 1994, the actual disparity was as high as 6.5-1 in favor of agricultural districts, but the government is considering shifting more political power to urban regions. The Diet imposes import quotas and high tariffs on food products, ostensibly as a national security measure, and it has allowed the import of rice only in the very occasional years when Japan's own harvest is poor. Thus, Japanese must pay six to ten times more for domestic rice than they would pay for rice imported from Thailand, Australia, or California. In some years Japan has even dumped rice on world markets.

The Japanese government is today investing in farming technology, including livestock raising. On the northern island of Hokkaido, the Ministry of Agriculture's experimental stock-breeding farm controls herds by remote control radio, monitoring their health, moving them to new pastures, and even fencing them in electronically. Japan has the genetics techno-

Agriculture in Japan is carried out on tiny plots, but it is productive because it is capital intensive. The Japanese market is protected by tariffs. (Courtesy of Japan National Tourist Organization.)

logy and the resources to revolutionize agriculture from an inefficient industry to a capital- and technology-intensive industry, like electronics. Today Japan is a major food importer, but Japan hopes to export food by the year 2000.

nomic subsidies contrasts sharply with the situation in most poor countries. In many rich countries, the national governments drain wealth from the cities to support farming. In the poor countries, by contrast, governments often heavily tax farmers and subsidize the urban populations. Chapter 3 explained why this political and cultural imbalance is partly a legacy of political and cultural imperialism. It hinders economic development and the achievement of food self-sufficiency, and it triggers migration to the cities.

If all rich countries opened their markets to crops from developing countries, and if the farmers of the poor countries were allowed to compete fairly, then they might be able to sell more food to the rich countries and earn foreign exchange to buy seeds, fertilizer, pesticides, and farm machinery.

A Few Product Case Studies

The world sugar market exemplifies how subsidies and trade redistribute world production. Neither the United States nor Europe is the most efficient place in the world to produce sugar, yet both restrict imports and maintain artificially high domestic sugar prices to subsidize local producers. United States producers do not fill total U.S. market demands, so the United States does import sugar. The government assigns quotas to individual countries, and when it allows a country to sell a large quota in the U.S. market at the inflated U.S. market price, the U.S. government is, in effect, granting that country an indirect form of foreign aid. This system, however, does not consider the country's producing capacity or economic needs. It forces U.S. consumers to pay between $3 and $5 billion more per year for sugar than they would if sugar were bought on world markets and imported freely. Sugar seems inexpensive to U.S. consumers, but the U.S. domestic price for sugar hovers at about twice the world market price. World sugar prices can be found any day in the business section of a newspaper.

Sugar from sugar beets competes with sugar made from sugarcane, and Europe subsidizes its

FOCUS
U.S. Production of Oats

The production of oats illustrates the many forces at work in U.S. agriculture today. The number of acres of oats planted in the United States has fallen steadily, and between 1955 and 1993 production fell from 1.5 billion bushels to 300 million bushels. This decline has come at a time when consumer demand for oats in cookies, cereals, and other foods is rising. Following reports in the late 1980s of health benefits of oat bran, demand for oats in food leaped from less than 50 million bushels in 1987 to about 107 million bushels in 1993.

The reasons why U.S. farmers are no longer raising as many oats involve agribusiness, technology, the role of the federal government, and international trade. Except for human consumption, the market for oats has been crumbling for decades. The biggest single reason for this decline is the decreasing role of the horse, the principal consumer of oats, in the domestic economy. More recently, changes in farm equipment have played a role as well. Oats typically mature in July, and many Midwest farmers used to devote some land to oats because they had land left over after allocating as much to corn or soybeans as their machinery could handle during the fall harvest. Today's combines, however, cover far

more ground, so farmers do not need to spread the workload. In addition, farmers used to control weeds and insects by alternating oats with corn or soybeans, but farmers now rely on chemical pesticides. Fewer seed dealers bother to carry oats, so it is more difficult for farmers to get seeds. Many grain elevators have stopped setting aside space for oats, so it is more difficult to market the crop.

U.S. government farm-support policies are skewed against oats. The support levels were set by comparing the energy content (calories) of different grains in relation to corn as animal feed. Oats are relatively high in protein and fiber, which makes them good for people and for horses, but they are low in calories, so they cannot compete with corn, soybean meal, and other high-calorie feeds that are used to fatten cattle and hogs. Thus, oats get government subsidies based on their value in a market for which they are poorly suited. Farmers are reluctant to experiment with switching from corn and other more subsidized crops to oats, because if the farmers do not like the results of the switch, the government will not let them bring their land immediately back into the richer subsidy programs.

It is unlikely that many farmers will return to planting

oats. Oats are not even interesting anymore to the companies that develop agricultural machinery, pesticides, herbicides, and fungicides. Some chemicals used on other grains would be effective on oats, but they cannot be advertised as such because no manufacturer will make the substantial investment required to get the necessary permission from the Environmental Protection Administration.

As domestic production has fallen, imports have risen and kept prices down. Oats from Scandinavia and Canada are competitive in the U.S. market, but this is only because those countries subsidize oat production. Many foreign subsidies are, however, being lowered, so American processors are afraid that foreign farmers will shift to other crops. Companies such as Quaker Oats and General Mills worry about the risk of price and supply fluctuations stemming from growing reliance on imports. The companies could guarantee domestic supplies by contracting with individual suppliers, as is widespread in hog and poultry operations. The companies would, however, have to offer costly incentives, and if companies make commitments to buy crops before they are planted, the companies would be exposed to fluctuations in quality and output.

sugar beet industry so extravagantly that Europe exports sugar, to the dismay of Australia, the Philippines, and Caribbean cane sugar producers.

Because the price of sugar is inflated in the rich countries, the manufacturers of alternative sweeteners can inflate their prices and still compete in the manufacture of soft drinks and candies. These alternatives

include corn-based or even purely chemical sweeteners, about which there are health concerns. Sugar's share of the U.S. market for sweeteners fell by 50 percent between 1970 and 1992.

World sugar markets are not the only agricultural markets that are distorted by national subsidies and protectionism. In 1991, the European Union sold 100,000 tons of beef to Brazil at a subsidized price so low that it bankrupted Brazilian cattle ranchers, who can actually produce beef at prices lower than Europeans can. The U.S. market consumes 3.3 billion pounds of peanuts per year, but only 1.7 million pounds are allowed to be imported. This inhibits the production of inexpensive peanuts in Argentina, Ghana, Sierra Leone, and Senegal. Through the 1980s, the U.S. government-supported domestic price of peanuts averaged $714 per ton, while world market prices for peanuts averaged about $450 per ton. Therefore, the U.S. price support system for peanuts alone cost U.S. consumers $513 million per year ($2 per year per U.S. citizen). Numerous additional examples of distorted world markets in foods could be cited.

The Politics of Subsidies

How do the agricultural interests, who represent only a small minority of the U.S. population, exercise such influence over policy? Part of the answer lies in the politics of the U.S. Congress. The agriculture committees of both the Senate and the House of Representatives are dominated by farm interests. Urban members of Congress generally vote in favor of farm bills as long as they also contain provisions for food stamps and school lunches. Furthermore, the sparsely-populated agricultural states retain disproportionate representation in the U.S. Senate. We should note that farm subsidies are not all going to struggling family farms. In the United States in 1990, the 15 percent of "farming concerns" that had sales of over $100,000 received over 60 percent of all farm support payments.

In 1988, the United States called for a worldwide end to subsidies within 10 years, but the rich countries cannot agree on what constitutes a subsidy. The United States subsidizes water for California farmers but denies that that is a "farm subsidy." Throughout California's Central Valley, sprinklers irrigate fields of alfalfa, cotton, and rice—crops more suited to monsoon lands than to a desert, and more than 50 percent of federally irrigated land in California is devoted to crops that are already in surplus. Competition for this water, however, has been intensifying among farmers, urbanites, and environmental needs, and the 1992 Federal Omnibus Water Act, which will be discussed in Chapter 11, might change the balance of water uses in California.

The issue of farm subsidies is complicated, and in each country it is a volatile domestic political issue. The overall result, however, is to increase food surpluses in the rich countries and to discourage food production in the poor countries.

Food Aid to the Poor Countries

During the 1980s, Western donors gave $1 billion of food aid to Africa each year. This aid saved many lives, yet at the end of the decade more people were worse off, and per capita production of food in Africa was down. There are reasons why the aid itself might have been part of the problem and why it might have created long-term dependency.

When there is a famine, aid is needed urgently. When the worst of the famine has been alleviated, however, the availability of free food destroys the livelihood of local farmers and hinders the development of local agriculture. The timing of the gift of food must be just right. Unfortunately, in many famine situations by the time Western TV viewers have been galvanized by television footage of starving people and food aid has been sent, the local crops are ready for harvest. The free food is delivered just in time to wreck the local agricultural economy by bankrupting local farmers. Some agronomists have suggested that the rich countries stop sending food and start buying the local farmers' crops instead. That, however, would harm the donor countries' own agricultural economies.

Western food gifts and Western dumping of food also disrupt traditional trading patterns. For example, in 1975 cattle from the Sahel accounted for two-thirds of the beef eaten in the Ivory Coast. In the 1980s, however, Western European countries began to dump frozen surplus beef in West Africa. The Ivory Coast came to depend upon this cheap new source of beef, and by 1993 the Ivory Coast imported only 25 percent of its meat from the Sahel. This impoverished the 4 million Sahelians. At the same time, however, the Europeans were providing substantial financial aid to improve cattle breeding techniques in the Sahel! In addition, the dumping of frozen meats in West Africa posed a health risk to the purchasers, many of whom did not fully understand the risks involved in defrosting frozen food.

The survey of world food production and availability in the past few pages illustrates so many ways that the world's agricultural economy is manipulated and distorted that we cannot know what quantities of food the world's farmers would be capable of producing or what the geography of food production would be if these conditions were eliminated. In 1992 George Demko, a former Geographer of the United States, wrote, "All of us could be fed adequately if the world's supply of food were equitably distributed irrespective of cost, politics, hatreds, and corruption." He concluded, however, that "straightening out the system and ending political infighting, disparities in income, and other problems will not occur in our lifetime." Let us hope that this pessimistic conclusion is not accurate.

FISH

▼▼▼▼▼▼▼▼▼▼▼▼▼▼▼▼▼▼▼▼▼▼▼▼

Seafood represents about 2 percent of the total human diet but a higher percentage of our protein intake. About 7 percent of humankind's total protein comes directly from fish and shellfish, and another 5 percent comes indirectly from fish meal fed to livestock. In some coastal and island countries, such as Japan and the Philippines, fish and shellfish supply as much as 20 to 30 percent of the population's total protein.

Humans harvest four main groups of marine species. *Demersal fish* (primarily bottom dwellers) include cod, haddock, sole, and plaice. These fish tend to concentrate on broad continental shelves, especially of the North Atlantic. A second group is the *pelagic fish* (surface dwellers) include herring, mackerel, anchovy, tuna, and salmon. These two groups together make up about 72 million tons of catch per year. *Crustaceans*, such as lobsters, shrimp, and other shellfish, provide more than four million tons; and *cephalopods*, including octopus and squid, yield about 2 to 3 million tons. The Japanese harvest more cephalopods than any other nation, but cephalopods are important in the diets of Mediterranean and many poor countries.

Fishing may be divided into two categories: traditional, or artisanal; and modern.

Traditional Fishing

Traditional fishing is concentrated mainly in the developing countries. Little capital is invested, and incomes are low, but these activities employ about 80 percent of the world's fishermen. Artisanal fishing is important in the lives of the majority of the populations of islands and coastal areas, and it directly supports an estimated 40 million people. Particularly in Asia and the Pacific regions, artisanal fishing predominates in human labor and in the number of vessels involved.

Fishing is frequently a dangerous occupation, and it demands skills and local knowledge. Physical risks are great, and so are the risks in terms of income, which can fluctuate greatly. Many traditional fishing areas are characterized by close-knit communities with distinct customs and rituals. Traditional fishermen from Canada to Madagascar to Indonesia share certain hardships and experiences. Such a traditional fishing life can be designated as a culture. Yields of fish from these traditional ways feed substantial numbers of people, but they are only a small fraction of the global catch. This way of life is rapidly being undermined and replaced by modern, highly capitalized fishing.

Modern Fishing

About 87 percent of the annual commercial catch of fish and shellfish comes from the ocean, and the rest from fresh water. Concentrations of aquatic species suitable for commercial harvesting are called **fisheries**. The world's major commercial marine fisheries are shown in Figure 6-27. Between 1950 and 1989 the annual harvest from ocean fishing rose from 22 million tons per year to 100 million, although it slipped to 97 million tons in each of the following 3 years. Each country or individual fishing vessel harvests as much as it can from international waters, but the world's fisheries are dominated by a few nations whose enormous and efficient fleets take the bulk of the harvest. In 1991, the Soviet Union, China, and Japan accounted for a full third of the total global catch, and the top six nations accounted for over half. Russia was responsible for almost all of the Soviet catch, and it remains a major producer.

Demand continues to rise for fish and fish products. As people eat more meat, they create a demand for animal-feed supplements, including fishmeal. Almost one-third of the global fish catch now goes into meal and oil, mostly to feed the rich countries of Western Europe and the North America. The Japanese, by contrast, want more fish to eat directly.

Overfishing and Depletion of the Seas

One hundred million tons per year is currently thought to be the limit of sustainable harvest from the sea. By the late 1980s some traditionally rich fishing grounds were relatively empty, and scientists warned that some species were near extinction. If this is true, depletion of the world's fisheries has implications for future food production and for the economic stability of countries dependent on fishing.

> Several major regions are already thought to be fished to their sustainable limits or beyond, and virtually every other fishing region is in peril.

The World's Major Fisheries

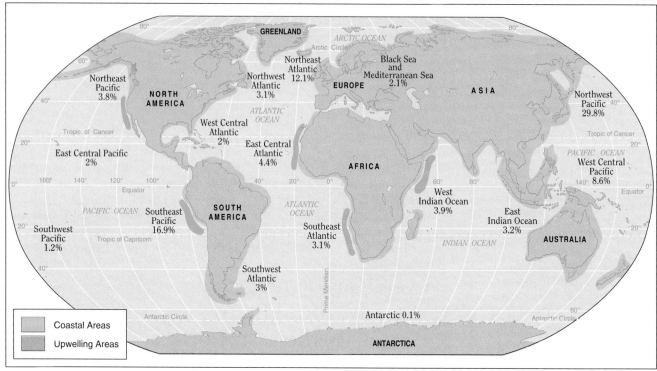

FIGURE 6-27

The figure shows the percentage of total world yield from each fishery. Many of the world's major fisheries are currently fished to their maximum sustainable yield. (Courtesy of the United Nations Food and Agriculture Organization.)

Several major regions are already thought to be fished to their sustainable limits or beyond, and virtually every other fishing region is in peril. The most dramatic depletions have been in the Atlantic, where commercial quantities of cod have all but vanished from the fabled Grand Banks, triggering the layoffs of 30,000 people in the Atlantic fishing communities of Canada. The collapse of stocks of pollock fish off Russia's Pacific coast has brought Russia into dispute with the governments of several other countries that fish there. Russia has demanded a moratorium on fishing. In July 1993, 150 nations attended a United Nations conference devoted to threats to the world's fisheries. The FAO announced that virtually every commercial species has been "depleted," "fully exploited," or "overexploited."

We do not know as much as we wish about fish populations and their migrations. Systematic fisheries research began only in 1902, with the formation of the International Council for Exploration of the Sea, which studies North Atlantic species. Today a great many international fishery commissions and advisory bodies exist. Some only consult, but others set quotas as well as conduct research. Some factors in changing fish populations are anthropogenic, but some are natural. Changes in water temperature and salinity can wipe out certain stocks, especially the small pelagics such as sardines, anchovies, pilchard, and capelin.

The major fishing nations have greatly enlarged their fleets and developed new technologies that multiply the yield from the sea. New electronic devices, for example, have discovered and tracked wholly new groups of fish. Some so-called *factory ships* are equipped with vast drag nets that virtually vacuum the sea.

Overfishing is not the only problem. Marine animals rely on coastal wetlands, mangrove swamps, or rivers for spawning grounds, and the world's wetlands and coasts are being destroyed by pollution and overdevelopment. About a third of the world's urban population lives within 100 miles (60 km) of the sea, contributing to the pollution that reaches the seas. Heavy metals have polluted fish and damaged the health of people who eat them. Sewage, fertilizers, and runoff from agriculture have overfed algae (tiny marine plants), causing them to grow so rapidly that they use up the oxygen that fish need to breathe.

The depletion of atmospheric ozone lets in ultraviolet radiation that harms life in the sea. The photosynthesis and growth of plankton, which forms the lowest link of the food chain, have decreased by

224 *Chapter 6*

as much as 20 percent near the surface of Antarctic waters, and the surface is where most marine growth and reproduction take place. Global warming might also alter ocean currents.

One wholly-new and scarcely-tapped potential source of seafood might be the vast swarms of tiny shrimplike krill found in the oceans around Antarctica. During the summer they feed on phylo- and zooplankton (one-celled plants and animals), and summer swarms of krill have been estimated at 650 million tons—a huge amount of living matter for only one species. At present, less than one-half million tons are harvested per year. An understanding of their vital role in the global oceanic ecosystem is essential before we harvest more, but Antarctic krill might become an important new source of protein.

International Law and Regulation of Fishing

About 90 percent of the world's marine fish harvest is caught within 200 nautical miles of the coasts, and therefore many coastal countries have recently extended their claims out into the sea. The 1982 United Nations Convention on the Law of the Sea authorizes each coastal state to claim a 200 mile **exclusive economic zone** (**EEZ**), in which it controls both mining and fishing rights (see Chapter 13). Countries as distant as Canada, Namibia, and the South Pacific Forum Fisheries Agency (a consortium of eight island nations) have already restricted foreign fishing vessels in their waters.

International talks focus on two categories of fish: species like pollock in the Bering Sea and cod off Canada's eastern coast, whose wanderings cause them to straddle territorial and international waters; and fish like tuna, swordfish and billfish, whose seasonal migrations cover thousands of miles. Coastal states dispute rights to these species with countries that have long range distant-waters fleets. The coastal states claim that factory ships in international waters are destroying the stocks in their territorial waters. Others argue that the crisis stems from coastal countries' mismanagement. Coastal countries want binding rules to regulate catches in international waters; distant-water countries want nonbinding guidelines drawn up regionally.

In 1993, U.S. interests set a precedent for international negotiations. The United States agreed to buy Greenland's commercial salmon fishing rights for 2 years in order to restore runs of the Atlantic salmon in North America. Scientists had estimated that 70 to 90 percent of salmon were being caught as they journeyed to the rich feeding grounds off the west coast of Greenland. Most of the money to purchase Greenland's rights came from private conservation groups.

Whales

Whales are not fish, but mammal inhabitants of the sea. The first whale species to attract human attention was the right whale, so called because it was easy to catch and a rich source of oil and whalebone—the "right whale" to hunt. The right whale was driven to the edge of extinction by the end of the nineteenth century. The western gray whale followed, and since 1900, the humpback, the blue, the fin, the sei, and then the minke in turn were overhunted.

In 1946, the International Whaling Commission (IWC) was established to regulate the whaling industry, but its quotas lacked a sound scientific base, and they were ignored. As worldwide concern over whales mounted and more nonwhaling nations joined the IWC, however, it finally imposed a moratorium on commercial whaling in 1986, even though it has no power to enforce its rules.

Even the IWC estimates that there are 87,000 minke whales in the Northeast Atlantic, and that 800 could be taken per year. Therefore, Japan, Norway, and Iceland protest IWC's moratorium, and they continue to hunt. The U.S. 1967 Fishermen's Protective Act requires the United States to impose sanctions on countries that hunt whales.

Aquaculture

Mechanized fishing on technologically-advanced ships has already multiplied yields from the sea, but humankind has scarcely begun the shift from hunting and gathering seafood to **aquaculture**, which involves herding or domesticating aquatic animals and farming aquatic plants (Figure 6-28). Humankind took this step with agriculture and livestock herding on land thousands of years ago.

The attractiveness of aquaculture stems from both demand and supply considerations. Global seafood consumption is on the rise, while the catch from the oceans may be peaking. In rich countries, fish has become the food of choice for many health-conscious people because it is filled with high-quality protein, is generally low in calories, and may reduce cholesterol levels.

In 1989, about 15 percent of American seafood consumption came from aquaculture (60 percent of that catfish), and that figure is expected to rise to 25 percent by 2000. World aquaculture production is expected to double to 25 million tons by 2000. Technology is advancing (computer-controlled feeding, for example) and new species are being introduced. Aquaculture protects both the purity of the aquatic environment and allows people to monitor what the fish are fed. Aquaculture presents another promising frontier for the multiplication of human food supplies.

Agriculture, Fishing, and Forestry **225**

FIGURE 6-28

The fish in these tanks in Idaho are not "caught," but raised under controlled conditions. The latest technological innovations in aquaculture improve the recycling of the water and allow the harvest of fertilizer nutrients from the fish waste.
(Courtesy of the U.S. Department of Agriculture.)

FORESTRY

▼▼▼▼▼▼▼▼▼▼▼▼▼▼▼▼▼▼▼▼▼▼▼▼▼

Forests once may have covered as much as 45 percent of the Earth's land surface outside Antarctica. Humankind has cleared forests for agriculture and fuelwood, and, more recently, for commercial lumbering and cattle ranching, but forests still range across some 30 percent of the Earth's land surface. About 55 percent of all wood harvested comes from broad-leafed trees, and about 45 percent from conifers. Chapter 2 discussed tropical deforestation and noted that the world's temperate forest area, by contrast, has been increasing, largely as the result of reforestation.

Forests sheltered and supported early primitive societies that subsisted on gathered fruits, nuts, berries, leaves, roots, and fibers, which they collected from trees, shrubs, and other plants. Few such primitive cultures remain, although many subsistence agricultural societies still gather forest products.

Today the most important forest resource is wood. Wood plays a part in more activities of a modern economy than any other commodity, and almost every industry depends on forest products in at least one of its processes. Wood serves for the construction of houses, but also for plywood, veneer, hardboard, particleboard, and chipboard. Furthermore, we use much wood in the form of paper. An average American citizen uses twice as much wood as all metals put together. The world's total consumption of wood is about 3.5 billion cubic meters per year, about half of which is fuelwood, half industrial timber. The use of industrial timber rose 50 percent between 1965 and 1990.

In Africa, South America, and Asia, most wood is used for fuel. In North America, Europe, Oceania, and the lands of the former USSR, by contrast, wood is produced for industrial use. This contrast is clear in the contrasts among the world's six leading producers of *roundwood*, that is, wood in the rough, whether destined for industrial or fuel uses. Canada, the United States, and the countries of the former Soviet Union harvest most of their wood for industrial purposes, and they are traditionally net exporters of roundwood. China, Brazil, and India, by contrast, each use most of their wood for fuel. These six countries together traditionally remove roughly half of the world's timber from their own territories.

TABLE 6-4
The Top Ten Importers and Exporters of Roundwood

In thousands of cubic meters, annual average 1989–1991			
Top 10 importers		Top 10 exporters	
Japan	48,744	United States	27,129
China	9,913	Malaysia	20,125
South Korea	8,417	USSR (former)	15,898
Italy	6,982	Australia	6,004
Finland	5,614	Chile	5,162
Sweden	4,862	Germany	3,949
Austria	3,855	France	3,908
Spain	2,227	New Zealand	2,916
Belgium	2,186	Papua New Guinea	1,424
Thailand	1,444	Indonesia	1,245

Source: World Resources Institute, *World Resources 1994-95.* New York: Oxford University Press, 1994, pp. 310-311.

Table 6-4 indicates world trade in roundwood. Several of the successor states of the USSR, most notably the Central Asian Republics, are traditional importers of wood and wood products, while others, most notably Russia, were substantial exporters. That pattern has probably continued into the 1990s.

SUMMARY

The environmental and technological capability of the entire globe to produce enough food to supply the entire human population is secure today and for the future. There are, however, great disparities in the world distribution of food production and supplies. The fact of oversupplies of food in some places and insufficient supplies in others can be understood only as the result of a complex of environmental, technological, political, and economic factors.

Thomas Malthus argued that food production increases arithmetically, but that population increases geometrically. Therefore, either preventive or positive checks on human population were needed. The human population has increased, but mass starvation has not occurred. This has been because of the farming of more areas of the planet, the transplantation of many food crops, improvements in worldwide transportation and the technology of food storage, the introduction of higher-yielding and hardier strains of crops, new pesticides, improvements in livestock, and farm machinery. These developments are called the green revolution and the scientific revolution in agriculture.

Today, cereals (grains) and potatoes are the world's basic foods, and a wide range of animals have been domesticated as livestock. The richer a nation becomes, the more total grains it consumes, but the higher the percentage of its consumption which is usually in the indirect form of meat and dairy products. Humankind overall could greatly improve its diet by concentrating on raising the most efficient animals. Some areas that are remote from urban centers may specialize in dairy farming. These areas often add value to milk by producing dairy products.

Johann Heinrich von Thünen questioned how various crops become distributed across any countryside, and he deduced a model pattern for the distribution of different land uses.

Variation in agricultural yields worldwide are partly attributable to variations in the physical environment, but today capital input is also important. Some scientists argue that the Green Revolution and the Scientific Revolution in Agriculture have just begun.

Technology creates the potential for increasing food production, but technology is not spread around the whole world, nor is it distributed in the most economic way. Problems in the effort to increase food supplies include lack of financial incentives for farmers, unsuccessful land redistribution, and inequitable land ownership.

World production and distribution of food is made even more complicated and perplexing by the system of world trade. Many rich countries have erected tariff walls against food imports, and they subsidize their own farmers. These subsidies contribute to production of agricultural surpluses in many rich countries, and discourage production in the poor countries. The world's agricultural economy is manipulated and distorted in so many ways that we cannot know what quantities of food the world's farmers would be capable of producing or what the geography of food production would be if these conditions were eliminated.

The world's harvest of fish has quintupled in the past 40 years and significantly increased the Earth's per capita food supplies. Although future increases from fishing are dubious, aquaculture offers new possibilities. The world's forests are another rich resource, but in many areas they are being harvested at unsustainable rates.

KEY TERMS

dump (p. 193)
renewable resources (p. 193)
positive checks (p. 193)
preventive checks (p. 193)
Green Revolution (p. 196)
Scientific Revolution
   in Agriculture (p. 196)
value added
   by manufacturing (p. 206)

QUESTIONS FOR INVESTIGATION AND DISCUSSION

1. What food crops are raised in your local region? What quantities are exported? Where?
2. What livestock is raised in your local region? What quantities are exported? Where?
3. Are there any significant food processing plants in your region?
4. Where did the materials in your breakfast come from?
5. Analyze the positives and the negatives of the quotation from Henry George.
6. What are possible ways of attacking hunger where it exists?

Agriculture, Fishing, and Forestry   **227**

diminishing returns (p. 215)
privatize (p. 216)
ejidos (p. 216)
fishery (p. 223)
exclusive economic
    zone (EEZ) (p. 225)
aquaculture (p. 225)

7. The text mentions that the global distribution of specific crops is always changing. How will it change if the globe heats up?

8. Go to your grocery store and investigate the latest storage and preservation packaging techniques. How have these allowed food to be preserved longer and to be more widely distributed?

9. How many of the food crops first domesticated on other continents (shown on Figure 6-2) have not yet found their way into your diet?

ADDITIONAL READING

Agricultural Biotechnology: The Next "Green Revolution"? World Bank Technical Paper 133, Washington, D.C. 1992.

HOBHOUSE, HENRY. *Seeds of Change: Five Plants That Transformed Mankind.* New York: Harper and Row, 1986.

KRUEGER, ANNE, with MAURICE SCHIFF and ALBERTO VALDÉS, eds. *The Political Economy of Agricultural Pricing Policies.* Vol. I *Latin America* (1991); Vol. II *Asia* (1991); Vol. III *Africa and the Mediterranean* (1992); Vol IV *Synthesis of the Economics in Developing Countries* (1992); Vol. V *A Synthesis of the Political Economy in Developing Countries* (1992). Baltimore, MD: Johns Hopkins University Press for the World Bank.

MACRAE, R., R. K. ROBINSON and M.J. SADLER, eds. *Encyclopedia of Food Science, Food Technology and Nutrition.* San Diego, CA.: Academic Press, 1993, 8 vols.

PRASAD, B. N., G.P.S. GHIMIRE and V.P. AGRAWAL, eds. *The Role of Biotechnology in Agriculture.* Published jointly by Oxford University Press, New Delhi, India, and International Science, Lebanon, NH, 1992.

Sharma, Narendra P., ed. *Managing the World's Forests.* Dubuque, Iowa: Kendall/Hunt for the World Bank, 1992.

U.S. Department of Agriculture. Economic Research Service. *World Agriculture: Situation and Outlook Report.* Washington, D.C., quarterly.

World Resources Institute, in collaboration with the UN Environment Program and the UN Development Program. *World Resources: A Guide to the Global Environment.* New York: Oxford University Press, 1986, 1987, 1988–1990, 1990–1991, 1992–1993, 1994–1995.

ZOHARY, DANILE and MARIA HOPF. *Domestication of Plants in the Old World: The Origin and Spread of Cultivated Plants in West Asia, Europe, and the Nile Valley.* New York: Oxford University Press, 1988.

7 | The Geography of Language

According to Genesis, everybody on Earth spoke the same language until King Nimrod ordered the construction of a tower to reach heaven. The Lord caused the workers suddenly to speak different languages so they couldn't work together. The tower came to be known as "Babel," which means "he confounded." (The Tower of Babel, by Pieter Bruegel, 1563.)

Courtesy of Erich Lessing/Art Resource.

*L*anguage and religion are two of the most important forces that define and bond human cultures. Their influences are so pervasive that many people take them for granted and cannot objectively observe their influences in their own and others' lives. Nevertheless, peoples who share either a language or a religion often demonstrate consistencies in other aspects of behavior. Furthermore, peoples who share either of these two cultural attributes can often more easily cooperate with one another in other areas, such as in international affairs. Sharing these attributes may help them understand one another in unique and profound ways.

Each language and each religion originated in a distinct hearth, and although the carriers of each of these cultural attributes have diffused around the world, each language and each religion still predominates within a definable realm. These realms of language and of religion are two of the most important of all types of culture regions.

This chapter will examine the world's great variety of languages, their development, the many cultural, political, and economic forces at work in their diffusion, and the ways in which competition among them still today reflects how the diffusion of some cultures and the dissolution of others is redrawing the world map. Religion will be the subject of the next chapter.

LANGUAGES AND DIALECTS

Any group of people who communicate exclusively with one another will soon develop "their own language," whether they are nuclear physicists or a social clique at a high school. If human beings were to appear suddenly in today's highly interconnected world, one worldwide language might develop. The great variety of languages spoken today testifies to the relative isolation of groups in the past. The distribution of any language illustrates the pattern of dispersal of its original speakers or their cultural impact on others.

Many social scientists believe that language is the single most important cultural index. A **language** is a set of words, plus their pronunciation and methods of combining them, that is used and understood to communicate within a group of people. The study of languages' vocabularies and rules of construction is called *grammar*, and the study of each language's sounds is called *phonology*. A **standard language** is the way any language is spoken and written according to formal rules of diction and grammar, although many regular speakers and writers of any language

may not always follow all of those rules. A country's **official language** is the one in which government business is normally conducted, official records are kept, signs are posted, and so forth.

The term language is usually reserved for major patterns of difference in communication. Minor variations within languages are called **dialects**, but scholars do not agree on the amount of distinctiveness necessary for a pattern to be considered a language. One suggested criterion is that in order for two dialects to be recognized as separate languages they must not be mutually intelligible. That is, a speaker of one of them cannot be understood by a listener who normally speaks the other. This single criterion does not always work, because some spoken dialects of one language are not always mutually intelligible. For example, a San Franciscan might not always be able to understand a Dubliner, a Glaswegian, or a resident of Lagos or New Delhi, yet what they all speak would be considered the English language, and the written form would be mutually intelligible. Sometimes the decision about what is a language is made on historical or political grounds, rather than linguistic grounds. When one crosses a certain political border, for example, "Swedish" becomes "Norwegian." Other examples of distinctions made at least partly for political reasons include the distinction between Twi and Fante (the former spoken in the Ivory Coast; the latter in neighboring Ghana), or, in southern Africa, Xhosa and Zulu.

DEFINING REGIONS OF SPOKEN LANGUAGE OR DIALECT

The systematic study of regional dialects is called **dialectology**. A dialectologist might, for example, study the local dialect in a city, or even one part of a city. The study of different dialects across space, by contrast, is called **dialect geography**, or **linguistic geography**.

Dialects usually diverge more in the way they are spoken than in the way they are written. This is because writing is often widely dispersed, but sounds are localized only among a group of people who speak together, called a **speech community**. Probably no dialectologist has ever developed the ability of the fictional Henry Higgins in G. B. Shaw's play *Pygmalion*: "I can place any man within six miles. I can place him within two miles in London. Sometimes within two streets." To this day in New York, however, some speech differences can be heard in various parts of the city, and there are distinct

Dialect Regions in France

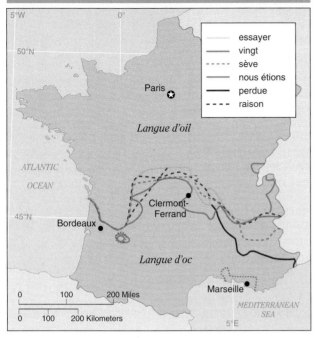

FIGURE 7-1

These six isoglosses represent the dividing lines between regions to the north and to the south in France where the people pronounce individual words differently. Clearly this bundle of isoglosses represents a significant border between two spoken dialects. (Recreated from the *Atlas linguistique de la France*)

dialect regions across the United States. Nevertheless, on Brooklyn's famous corner of "Toity-toid and toid," visitors from "Bawlmer" (a city in Maryland) and from "Nyew Awluns (a city in Louisiana) could probably make themselves understood.

Sometimes researchers survey speech and draw lines around places where speakers use a linguistic feature in the same way. These boundary lines are called **isoglosses**. Figure 7-1 reproduces a map from the early twentieth century *Atlas linguistic de la France*. The atlas's makers identified a bundle of isoglosses running across France from east to west, dividing the country into two major dialect areas. These two areas are known as langue d'oil (the North) and langue d'oc (the South) based on the words used for "yes" in these areas during the thirteenth century, when the division was first recognized.

Isoglosses frequently parallel physical landscape features, because physical features often act as barriers to human migration and diffusion. For example, the Pyrenees mountains divide Spain from France, and the Pripet marshes separate Belarus from Ukraine. By contrast, languages often diffuse quickly across broad lowlands, and languages have historically diffused along river valleys or other routes of trade and transportation.

Speakers along the border between two languages that are not historically related will often share features of pronunciation that are traditional in only one of the languages. Neither the vocabularies nor the grammars of the two languages blend along the border, but the phonics do. For example, the Khoisan languages of southern Africa (see Figure 7-5) share a clicking sound with the adjacent but unrelated Bantu languages Zulu and Xhosa. Another example is provided by the use of front-rounded vowels, such as in German *müde* (tired) or French *soeur* (sister) (see Figure 7-2). Front-rounded vowels occur along an axis which runs diagonally across northern Europe, and they are heard in French, Dutch, German, Danish, Norwegian, Swedish, and Finnish. They do not, however, occur in English, which is closely related to German, or in Spanish, which is closely related to French. The main factor explaining the occurrence of front-rounded vowels, therefore, must be proximity, not history.

Researchers often map what is called a **geographical dialect continuum**. This is a chain of dialects or languages spoken across an area. Speakers at any point in the chain can understand people who live in adjacent areas, but they find it difficult to

The Region of Front Rounded Vowels

FIGURE 7-2

Front-rounded vowels are heard along the axis on this map, but they are not used in neighboring languages that may be historically related to these. Front vowels do not occur in south German dialects, but they can be heard in the Italian of Northwest Italy. The only hypothesis of explanation for this distribution is proximity.

The Geography of Language **231**

FOCUS
Differences in Pronunciation

Different pronunciations of one language have long been of interest. The biblical Book of Judges records one of the earliest known uses of pronunciation to differentiate friend from enemy. The Gilieadites cut the Ephraimites off from the fords of the Jordan River, and whenever an Ephraimite fugitive tried to cross, he was required to pronounce the word "Shibboleth." The Ephraimites would say "Sibboleth," because they could not pronounce the word "correctly," so they were put to death. As a result of this story, "shibboleth," which originally meant a flowing stream, has in modern times come to mean any distinguishing mark. During World War II, the Dutch routinely tested suspected spies by asking the strangers to pronounce the name of the city Scheveningen. Only a native Dutch person could do it properly.

Varying pronunciations have also been a source of humor. Afferbeck Lauder used standard spellings to represent the impression of an Australian accent in his popular book *Let Stalk Strine* (1965). Definitions included: "Egg Nishner: A mechanical device for cooling and purifying the air of a room," and "Scone: A meteorological term. As in 'Scona rine.'" Probably no one raised in the United States could understand a speaker of "Strine."

understand people who live farther along the chain. An example of such a chain is the North Slavic continuum, which links Czech, Slovak, Ukrainian, Polish, and Russian.

Pidgins and Creoles

A **pidgin** language is a system of communication which has grown up among people who do not share a common language, but who want to talk with each other. Pidgins are marginal or mixed languages, and they usually disappear after a few years or else they evolve into a **creole**. A creole is a pidgin language that has survived long enough to become a mother tongue. That usually takes a generation or two. Examples of creoles include Gullah, the English-based language used among some African Americans living along the U.S. Southeast coast; Haitian French Creole; Guyanese Creole; the Krio language of Gambia; and Seychellois, a French-based creole spoken on the Seychelles Islands.

Changes in Languages

Individual languages change through time, but religious classics or classics of literature can exert a powerful force for stabilization. In English, for example, the works of William Shakespeare and the 1612 King James translation of the Bible have molded the language. Nevertheless, vocabulary and even grammar have changed so much that parts of even these works are difficult for many English speakers to read today.

The sounds of any language also change through time. Shakespeare's plays are full of *puns*, jokes based on the fact that two different words are pronounced the same. These puns cannot be conveyed in contemporary productions of the plays, however, because audiences cannot understand them. The words "hour" and "whore," for example, were pronounced the same in Shakespeare's day.

THE WORLD'S MAJOR LANGUAGES

Most scholars agree that there are more than 3,000 distinct languages, at least 30 of which are spoken by 20 million people each. About 60 percent of all the world's people, however, speak 1 of the 12 languages listed in Table 7-1. The language with the largest number of speakers is Mandarin Chinese (Guoyo), spoken by over 800 million people. Mandarin is one of eight main dialects of Chinese that are together considered to be one language because they have one written form. The spoken dialects are not entirely mutually intelligible.

Although more people speak Chinese than any other language, it is spoken by very few non-Chinese. English, by contrast, is the primary language of about 450 million people worldwide, and it is either the only official language or one of several official languages in about 50 countries. English serves as

TABLE 7-1
The World's Leading Languages and the Number of Speakers of Each in Millions

Language	Native speakers	Total speakers
Mandarin	817	907
Hindi	321	383
Spanish	320	362
English	316	456
Bengali	180	189
Arabic	178	208
Russian	173	293
Portuguese	165	177
Japanese	125	126
German	98	119
French	71	123
Malay-Indonesian	48	148

Source: U.S. Department of State

a second language for hundreds of millions of people, so it is the world's **lingua franca**, that is, a second language held in common for international discourse. Other languages have served as lingua francas in the past. Latin long served this role for both scholarship and business throughout Western civilization, partly as a legacy of the Roman Empire. Swahili, which developed among black peoples in eastern Africa who were in communication with Arab traders, has many Arab words. Today it is an official language in Tanzania and Kenya, and it is widely spoken or understood in Uganda, Zaire, Rwanda, Burundi, Mozambique, and in parts of Somalia, Sudan, and Malawi. Arabic derives special transnational importance as the language of the Koran, the sacred scriptures of Islam. In the past, people of many different ethnic and racial groups adopted the Arabic language when they converted to Islam, and they came to be known as Arabs. The Koran has been translated, but Muslims are still encouraged to study the original version. Arabic is the official language in 22 countries today.

Efforts have been made to develop wholly artificial languages which could serve as international lingua francas. Most have been based on elements drawn from natural languages—invariably one or more Western European languages. The best known artificial language, **Esperanto**, which means "hopeful," was invented by the Polish scholar Ludwig Lazarus Zamenhof (1859–1917). Today several journals and newspapers are published in Esperanto, and many books have been translated into it. There is even an extensive bibliography of original work in Esperanto, but it has never achieved official status as an international language.

Local and "Primitive" Languages

In contrast to these widespread languages, some languages are extremely local. Linguists have discovered fully developed languages in New Guinea spoken by only a few hundred people in certain valleys. These languages have developed in total isolation over long periods of time. They are utterly incomprehensible to people just 20 miles away in the next valley.

In the nineteenth century it was widely assumed that primitive cultures had primitive languages—languages with only a simple grammar, a few sounds, and a small vocabulary. Every culture that has been discovered and investigated, however, no matter how primitive it may be, has turned out to have a fully developed language with a complexity comparable to those of the so-called civilized peoples. There are no "bronze age" or "stone age" languages. Nor have any languages been discovered to correlate with other sorts of cultural groupings—nomadic, or pastoral peoples, for example. All languages have developed to express the needs of their users, and in that sense all languages are equal.

The vocabularies of different languages, however, do reflect the needs and the historical experiences of the people who use them, including descriptive landscape terms. Perception of landscape, as represented in vocabulary, is part of any culture's adaptive strategy. The more specialized your need is, the more specialized your vocabulary will be. It should not surprise us to learn that German has more words for different kinds of trees than Arabic does, but that Arabic has more words for different kinds of sand dunes. Vocabulary specialization occurs even among users of any one language. For example, most of us know what a horse is. The word horse, however, is not very specific. If we are a nomadic people who rely on horses, or if we are modern equestrians, we might want to know each horse's sex, age, and color. Therefore, we specialize our vocabulary with words such as mare, stallion, gelding, foal, roan, dapple, and bay. Similarly, the Hanunóo people of the Philippines distinguish and name more than 1,600 different types of plants—far more distinctions than even systematic botanists make.

All languages, however, share certain developmental sequences. One concerns recognition of life forms, and this sequence may contain some profound clues as to how humans relate to the rest of the natural world. In any language, the first distinction made among life forms is between plants and animals. After a language develops a word for "plant," it next develops a special word for "tree," then one for "grerb" (a small, green, leafy, nonwoody plant), then "bush" (plants between trees and grerbs in size), then "grass," then "vine." We cannot listen as vocabularies develop this sequence, but we infer the sequence from the fact that if a language has a word for bush it will have a word for tree, whereas having a word for tree does not mean that a language will have a word for bush. The differentiation of animal types also follows a general developmental sequence in all languages. First each language develops a word for "animal," then a word for "fish," then "bird," then "snake," then "wug" (small creatures other than fish, birds, or snakes—for example, worms and bugs), then "mammal."

Vocabularies also develop the recognition of colors in a definable sequence. A language that has only two basic color words will differentiate "dark" from "light." Languages that have three or more basic color words will next differentiate "red," then "yellow," then "grue" (green/blue), then different words for "green" and "blue," and so on. Within any given language, however, the number of colors any individual recognizes depends upon his or her needs. A painter's vocabulary has more words for colors than mine does, and the nuns who repair tapestries in the Vatican City have threads in 50 shades each of 350 colors.

LANGUAGE AND PERCEPTION
▼▼▼▼▼▼▼▼▼▼▼▼▼▼▼▼▼▼▼▼▼▼▼▼▼▼▼

Most linguists agree that all aspects of culture influence language, but there is less agreement about the extent to which language influences other aspects of culture. American linguist Edward Sapir (1884–1939) and his pupil Benjamin Lee Whorf (1897–1941) articulated a theory that language is a force in its own right. This is known as the Sapir-Whorf hypothesis.

This hypothesis combines two principles. One is *linguistic determinism*, which states that language determines the way we think. The second principle

of *linguistic relativity* states that the distinctions encoded in one language are not found in any other language. Whorf wrote:

> We dissect nature along lines laid down by our native languages. The categories and types that we isolate from the world of phenomena we do not find there because they stare every observer in the face; on the contrary, the world is presented in a kaleidoscopic flux of impressions which has to be organized by our minds —and this means largely by the linguistic systems in our minds. We cut nature up, organize it into concepts, and ascribe significances as we do, largely because we are parties to an agreement to organize it in this way—an agreement that holds throughout our speech community and is codified in the patterns of our language.

Whorf illustrated this hypothesis with examples from several languages. The language of the Hopi Indians, for example, lacks a concept of time as a dimension—there are no forms corresponding to English verb tenses. Instead, the language allows the speaker to talk about various durations from the speaker's point of view. It is difficult to follow some of the ideas that Sapir and Whorf presented in their examples precisely because the examples force us to think in totally new ways. Sometimes we can get only a glimpse—a fascinating glimpse—of a totally new way to see the world.

Critics of the Sapir-Whorf hypothesis point out that languages can make up new words for new needs or adopt another language's terminology. Successful translations among languages can be made. Just because a language lacks a word, it does not mean that its speakers cannot grasp that concept. The conceptual uniqueness of a language such as Hopi can be explained (more or less) in English. Some conceptual differences between cultures may be attributed to language, but most scholars today believe that mutual comprehension is indeed possible. One language may take many words to say what another can express in a single word, but the circumlocution can make the point.

Today most scholars accept a weakened version of the Sapir-Whorf hypothesis. Language cannot determine the way we think, but it can influence the way we perceive and remember. Experiments have shown that people find it easier to make a conceptual distinction if it neatly corresponds to words available to them. Language may also affect the ease with which we perform mental tasks. For example, Chinese students have been demonstrated to be better at arithmetic than American students, and one explanatory hypothesis is that the Chinese words for numbers are easier to remember. Compared to students in Columbia, Missouri, kindergarten-aged children in Hangzhou, China, could hold more digits in short-term memory (6.7 vs. 4.1). The Chinese were much more likely to use verbal counting, a faster and more sophisticated strategy than finger counting. These differences show up too early to be explained by differences in formal instruction. The researchers hypothesize that Chinese children do better in math because their words for numbers take less time to say. That enables the Chinese children to hold more numbers in their heads and work with them quickly. Studies such as these are a part of the field of *psycholinguistics.*

Today all social sciences are scrutinizing language carefully to find subconscious assumptions and prejudices in it. This search carries over into our everyday speech. It has long been traditional, for example, to use the masculine gender when referring to people in general. Some scholars have argued that this suggests that men do everything and, thus, unconsciously restricts the opportunity available to females. Therefore, you will notice that the third paragraph of the first chapter of this textbook referred to the first geographer as "he or she."

Changing the words we use can redirect our sensitivity and make us think anew. Chapter 1 noted that map conventions such as putting North at the top conceal information because we are not really looking at the map, but only being reminded of something we have "seen" often before. When something is always the same, we cease really seeing it. The same is true with hearing words. Only by being startled can we be made aware and budged out of our conventions, prejudices, and assumptions. Some philosophers argue that we are so enmeshed in the assumptions of our words that there is in fact no such thing as "objective" science. Words betray us. At a time when the words "negro" and "colored" yielded to African American, what does it mean that a popular rap group co-opts a hated term to call itself "Niggers with Attitude," or that parading gay activists call themselves the "queer nation"? Everyday speech reveals many examples of changing our sensitivity and our attitudes by changing the words we use. Poets explore these ideas and thus claim the name of scientists.

NATIONAL LANGUAGES

There is no exact correspondence between languages and the countries of the world. A few languages, such as Icelandic and Japanese, are associated almost exclusively with one country, but several languages are shared by many countries. The relationship between languages and nationalism is very complicated. In European history, language has been interpreted as the basis of nationalism. As early as 1601, King Henri IV of France seized French-speaking territories from the Duke of Savoy and declared to his new subjects, "It stands to reason that since your native tongue is French, you should be subjects of the King of France." This logic, however, never stopped the French from seizing the territory of non-French speakers whenever they could. The notion of language as a basis for nationality eventually spread among the Germans. Martin Luther's translation of the Bible into German fixed the common language, as the Koran had fixed Arabic (Figure 7-3). Thus, among the Germans as among the Arabs, language, religion, and national self-consciousness became intertwined.

FIGURE 7-3

This is the title page of the last edition of Martin Luther's German translation of the full Bible printed before his death in 1546. This book launched the Protestant Reformation and fixed the German language. It can still be read by German speakers today. The page's rich decoration includes references to Christian symbols, including the rewards of heaven and the terror of hell.
(Courtesy of the Pierpont Morgan Library, New York. PML19474.)

A series of International Statistical Congresses held in the nineteenth century legitimized the idea that language defines nationality. The intellectual field of statistics was born in efforts to count and compare numerical values such as population, territory and economic output among *states*. Participants at the first International Statistical Congress in 1853 argued whether countries' censuses should include a question of language and whether language had any bearing on nationality, but they were unable to come to any decisions. The Congress of 1860 decided that a language question should be optional and that each political unit could decide whether language was a significant determinant of nationalism. In 1873, however, the Third Congress recommended that all censuses henceforth include a question of language. The language that a person speaks at home was thought to be the only aspect of nationality which could objectively be counted and tabulated, and, therefore, a person's language was "definitive" of that person's nationality. In other words, statisticians proposed that what could most easily be counted is what really counts. This demonstrates how social scientists can *affect* a phenomenon they are trying only to *reflect*. These decisions made by statisticians for the single purpose of simplifying counting techniques played a key role in augmenting feelings of competitive, self-conscious nationalisms that eventually contributed to causing World War I.

Language may serve as a basis for national identification, but, as we shall see in Chapter 10, it does not always necessarily do so. Some Frenchmen speak German, and the Irish speak English.

Language and Nation Building

Philological nationalism is the idea that "mother tongues" have given birth to nations. This idea persists despite the fact that standard languages usually result from self-conscious efforts at nation building by centralized governments. Standardized languages cannot emerge before a government imposes mass schooling and mass literacy or, alternatively, universal service in a national army. Usually these centralizing pressures transform the language of a small percentage of the population, the political or cultural elite, into the national language. Thus, today's world map of languages is partly a result of the world political map. In 1789, 50 percent of the French population did not speak French at all, and only 12 to 13 percent spoke it "correctly." In 1860, only 2.5 percent of the population of the new Italian state spoke Italian for everyday purposes.

Several national languages have appeared still more clearly as the result of deliberate political pressure. Some languages were virtually invented, such as Romanian in the nineteenth century. In 1919, Hebrew was spoken by no more than 20,000 people, but the Israeli state has nurtured it for nationalistic reasons. Other new states have cultivated new national languages to unite diverse populations. The Indonesian government, for example, has succeeded in codifying and establishing one common language, Bahasa Indonesian ("bahasa" means language). Slovak is the official language of the new country of Slovakia, but few people use it. If the Slovaks want to differentiate themselves linguistically from their neighbors, they will have to nurture this language.

Not all countries succeed in reviving or creating national languages. When Ireland achieved independence in 1922, the government tried to enforce the use of Gaelic rather than English. The people themselves did not accept it, however, but continued to use English.

A minority people in a country often refuses assimilation and clings to its language as a gesture of cultural independence. The Kurdish minority in Turkey, an estimated 10 million people, clings to Kurdish, although its public use was outlawed from 1984 to 1991. Kurdish minorities in Iraq and Iran have also suffered political and linguistic discrimination.

Official Languages Contrasted with Indigenous Languages

Figure 7-4 shows the official languages of the countries of the world. One of the most notable features of the map is the widespread diffusion of the major European languages. European colonial powers forced their languages on their colonies, regardless of how many native languages were already spoken and whether many Europeans actually settled in any given colony. The conqueror's language became the language of government, law, and usually education. When the colonies gained their political independence and had to choose official languages, many kept their former ruler's language.

Several forces compete to stabilize or to change that map today. International trade and travel reinforce the usefulness of widespread international languages. A competing force, however, is growing national pride in new countries. Chapter 3 noted how some countries are emphasizing or even resurrecting traditional folk cultures as a way of enhancing national identity and community. The governments may grant official recognition to indigenous languages. Therefore, more and more countries today recognize two or more official languages—one international language, plus one or more local indigenous languages. Bolivia, for example, now recognizes the native American languages Aymará and Quechua, as well as Spanish, as official. In Kenya the backlash against colonialism was so strong that the country terminated the recognition of English as an official language in 1974. Many individuals find it useful to be bilingual, or even trilingual, in their own home countries.

Figure 7-5 maps the language regions of indigenous peoples, and this map differs markedly from Figure 7-4. In most of the Western Hemisphere, indigenous peoples are outnumbered by migrants and their descendants. In many other areas, however, most notably Africa, most of the local population does not actually speak the official language. The map of official languages shows six international languages (English, French, Spanish, Portuguese, Arabic, and Russian) covering most of the globe, yet a majority of the Earth's people have as their native tongue a language spoken by fewer than 1 percent of the world's population.

Furthermore, some of the languages listed in Table 7-1 as having among the greatest numbers of speakers are spoken only in fairly restricted, although densely populated, areas. Bengali, for example, the world's seventh leading language in terms of number of speakers, is spoken only in Bangladesh and adjacent parts of India. The languages of those European countries that explored and conquered much of the world, by contrast, are impressively widespread on the map.

Why Did New Countries Retain Their Former Rulers' Languages?

When many former colonies won their independence, their populations consisted of different peoples who spoke many different languages. These countries often retained their former rulers' languages to avoid having to choose one of many native languages as the new national language. The former ruler's language was the lingua franca. Many African countries have retained English or French for this reason, and attempts to change the official language can trigger internal strife. In Algeria, for example, the majority of the population speaks Arabic, but the Berber minority, which speaks a Berber language, protested violently when the government replaced French with Arabic. Over 250 languages are spoken in Nigeria, and advocates of a national language could not decide among Hausa, Yoruba, or Igbo, the three major native tongues. Therefore, English was chosen, although only about 20 percent of the population speaks it well, and still fewer use it as their first

language. Another reason for the continuing preferences for European languages is that the governments cannot afford to offer schooling in the many indigenous languages.

Tanzania repudiated English and chose Swahili as its official language, and that decision resulted in both advantages and disadvantages. The choice allowed Tanzania to bridge the linguistic divisions among its many peoples. Some observers give the language credit for Tanzania's having achieved 30 years of relative political stability and a sense of national unity. Books in Swahili are expensive, but the Canadian government has subsidized the publication of many children's books in Swahili, and Tanzania claims to have achieved one of the highest literacy rates in Africa, 90 percent. Unfortunately, the choice of Swahili has caused the decline of English. Tanzania's first president and founding father Julius Nyerere himself translated several of Shakespeare's plays into Swahili, but a lack of contemporary books in the sciences means that higher education must be conducted in English. This forces a language transition on students who go on to college. The slow decline of English-language skills also isolates Tanzania from the international community, and this hinders international trade and other exchanges—even with Canada.

India is another country in which linguistic rivalries maintain English as an official language, although it is spoken by a small fraction of the total population. The leaders of India's independence movement intended Hindi one day to be India's official language, and Hindi is spoken by at least 400 million Indians. Hindi speakers, however, are concentrated in northern India, and other Indians are unwilling to accept the domination of Hindi. Whenever Hindi has been pressed on India's southern states, they have threatened to secede. India recognizes English as one of 16 official languages, and English remains the preeminent language of the upper classes and the upwardly mobile. Chapter 5 noted how knowledge of English is an asset for Indian migrants around the world.

Language Innovation in Former Colonies

A former colony can rise to greater wealth and influence than its former ruler, and when this happens, the colony's dialect can become the international standard for its language. This has happened, for example, with the U.S. version of English. Noah Webster published his *American Spelling Book*, which was self-consciously different from English, in 1783,

The Geography of Language **237**

The Official Languages of the Countries of the World

FIGURE 7-4

the same year that the United States won its political independence. Today American English competes with British English for precedence worldwide, although some scholars argue that American is no longer a dialect, but a language in its own right. In 1992, the prestigious worldwide radio broadcasting service of the British Broadcasting Corporation acknowledged the worldwide appeal of the American

language, and it hired its first native American speaker for its broadcast English language lessons. Some observers compared this linguistic surrender to Britain's military surrender at Yorktown in 1781.

Portuguese boasts 180 million speakers, and at a 1990 international conference the governments of Brazil, Angola, Mozambique, São Tomé and Principe, Guinea-Bissau, and Cape Verde—all former

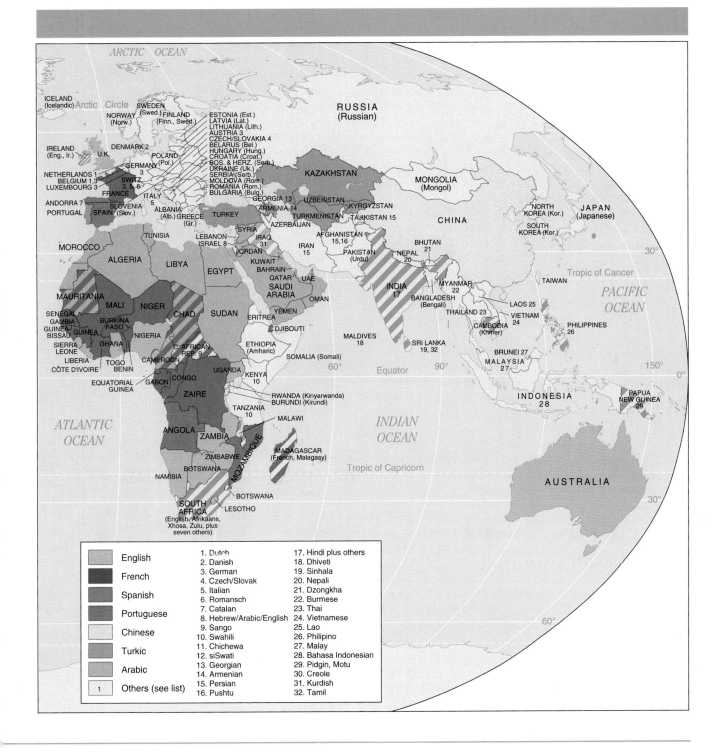

Legend:

English
French
Spanish
Portuguese
Chinese
Turkic
Arabic
1 — Others (see list)

1. Dutch
2. Danish
3. German
4. Czech/Slovak
5. Italian
6. Romansch
7. Catalan
8. Hebrew/Arabic/English
9. Sango
10. Swahili
11. Chichewa
12. siSwati
13. Georgian
14. Armenian
15. Persian
16. Pushtu
17. Hindi plus others
18. Dhiveti
19. Sinhala
20. Nepali
21. Dzongkha
22. Burmese
23. Thai
24. Vietnamese
25. Lao
26. Philipino
27. Malay
28. Bahasa Indonesian
29. Pidgin, Motu
30. Creole
31. Kurdish
32. Tamil

Portuguese colonies—accepted the Brazilian version as the international standard. Brazilian Portugese is much simpler in spelling than is the Portuguese of Portugal, but it incorporates many native American, African, and even American words. The Portuguese were humiliated. Mushrooming Brazilian influence in Portugal, through investment, trade, and media products, threatens to inundate Portugal's own culture.

Brazil has 15 times as many inhabitants as Portugal and an economy more than 8 times larger.

Polyglot States

Figure 7-4 shows that many countries grant legal equality to two or even more languages. These are

The Geography of Language   **239**

The World's Indigenous Language Families

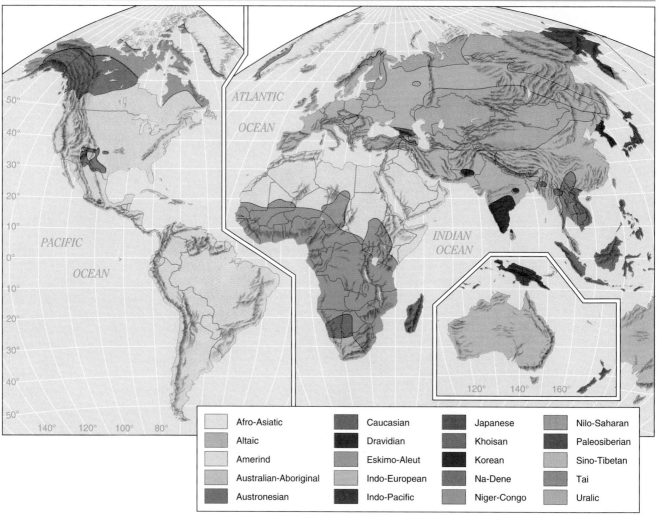

Legend:
- Afro-Asiatic
- Altaic
- Amerind
- Australian-Aboriginal
- Austronesian
- Caucasian
- Dravidian
- Eskimo-Aleut
- Indo-European
- Indo-Pacific
- Japanese
- Khoisan
- Korean
- Na-Dene
- Niger-Congo
- Nilo-Saharan
- Paleosiberian
- Sino-Tibetan
- Tai
- Uralic

FIGURE 7-5

This map may be said to "underlie" Figure 7-4, the map of official languages. (Western Hemisphere adapted from *Language in the Americas* by Joseph H. Greenberg with the permission of the publishers, Stanford University Press. 1987 by the Board of Trustees of the Leland Stanford Junior University. Eastern Hemisphere adapted with permission form David Crystal, *Encyclopedia of Language.* New York: Cambridge University Press, 1987.)

called **polyglot states**. Switzerland, for example, recognizes German, French, Italian, and Romansch. If various languages predominate in distinct regions of a country, the country might accept each language as official in its region. Belgium, and even its capital city of Brussels, is legally divided into French and Flemish zones. In the center of Brussels, all signs are in both languages (see Figure 7-6). The precedence of languages is hotly contested, and several Belgian governments have fallen because they have been seen as giving preference to one language or the other.

Canada is also officially polyglot, and language is a source of friction between the Anglophonic (English-speaking) majority and the Francophonic (French-speaking) minority (Figure 7-7). Disputes over the rights and privileges of the two languages in Canada have embroiled Canadian politics for many years. Quebec has even threatened to secede from Canada.

Polyglot states usually select one language as official for the federal center and for communications among the states, and whatever the country's constitution or laws might say, one language will generally be the "preferred" language of the country. Those who do not use it may find their opportunity restricted or their upward mobility blocked. Russian, for instance, was long in general use throughout the Soviet Union. Native speakers of the minority languages could rise in the national power structure only if they also spoke Russian.

FIGURE 7-6

All street signs in Brussels, the capital of polyglot Belgium, must be in both official languages: French and Flemish. The name Häagen-Dazs, which is advertised in the window, is a linguistic curiosity. The name means nothing; it was concocted by a New York businessman to sound vaguely Scandinavian and to profit from associations of Scandinavia with cleanliness and health.
(Courtesy of Robert Mordant.)

LANGUAGE COMMUNITIES AND SOCIAL CLASS

▼▼▼▼▼▼▼▼▼▼▼▼▼▼▼▼▼▼▼▼▼▼▼▼

Different language communities within one country may be defined not only geographically, by region, but also socially, by class. In newly independent countries that retain official European languages, those who speak only the local language are at a great disadvantage, as they find at an early age when they go to school. Appropriate teaching materials are often inferior or unobtainable. For example, Tagalog is the language of the people native to the Philippine Islands. The islands were colonized, however, first by Spain and then by the United States, so Spanish and English are the languages of the upper classes. Today upper-class English- or Spanish-speaking Filipinos have been heard to sneer, "We speak Tagalog only to servants." Dutch remains the lingua franca of the elite throughout Indonesia, which was a Dutch colony for hundreds of years.

Throughout Central and South America the native American populations and their cultures were smothered under Spanish and Portuguese rule, and vestiges of this cultural imperialism remained even after independence. For example, the original Constitution of Bolivia, written by Simon Bolívar, accepted as citizens only those who could speak Spanish. Today Bolivia, Peru, and Paraguay all recog

nize native American languages in addition to Spanish. Official recognition of these native languages reveals the political awakening of the native American population. You may have noticed that this book has consistently referred to Peru's historic "Inka" empire, rather than the traditional "Inca." "Inca" is Spanish, whereas "Inka" is Quechua, and today the government of Peru and scholars prefer the native American spelling.

Linguistic geographers have noted that in Britain the upper classes from throughout the country speak the same English. This is variously called *the King's English, the Received Pronunciation* (received, presumably, down from royalty), or *Oxbridge English,* a term that combines the names of England's two leading universities, Oxford and Cambridge. Regional differences become greater and greater, however, as you descend the social scale, so that sometimes lower-class people from different regions of Britain cannot understand one another. In the United States, by contrast, there is no "received American pronunciation" shared among the upper classes of New Orleans, Boston, and Seattle. Regional variations are usually shared by speakers of all social classes. There are several hypotheses to explain this observation. One is that it results from the country's great size, a second is that it results from the nationwide distribution of prestigious institutions of higher education, and a third suggests that the U.S. class structure is less geographically based or even less

FIGURE 7-7

Canada is officially a bilingual country, so government signs must be posted in both French and English.
(Courtesy of Michael Evans/Leo de Wys.)

rigid than that of Great Britain. It has also been observed, however, that when speakers move from one region to another they often quickly adopt their own speech to the new speech community. Even a casual traveler might quite readily and unconsciously pick up local speech patterns.

In almost all countries, mastery of the standard written language is a sign of upper-class status. This mastery can be achieved by anyone in the society, no matter what that person's birth or origins. Therefore, many people believe that standard languages are in fact the most truly democratic aspect of any society.

THE GEOGRAPHY OF ORTHOGRAPHY

▼▼▼▼▼▼▼▼▼▼▼▼▼▼▼▼▼▼▼▼▼▼▼▼▼▼

The geography of languages is complicated by the geography of **orthography**, which is a system of writing. Most languages are written in **alphabets**, which are systems in which letters represent sounds.

Alphabetic Languages

There are several alphabets in use today, and the alphabet in which a language is written can reflect a historic diffusion (Figure 7-8). The number of letters in existing alphabets range from 11 in Rotokas, the language of the Solomon Islands, to 74 in Khmer, the language used in Cambodia. Some alphabets, such as Hebrew and Arabic, are consonantal; vowel markings are optional. Many languages use *diacritics*, which are small marks added to a letter to indicate a special sound. The acute accent marks on the word résumé and the cedilla on façade are familiar examples.

In a perfectly regular alphabetical system, there is only one symbol for each sound. Some alphabets that have been devised by scholars to record languages that previously had no written form meet this criterion. Most languages actually used today, however, called **living languages**, fail to meet it. Languages vary greatly in their regularity. Spanish and Finnish, for example, have very regular systems, whereas English and Gaelic are very irregular.

The earliest known alphabet was the North Semitic alphabet, which developed around 1700 B.C. in Palestine and Syria. It consisted of 22 consonantal letters. The Hebrew, Arabic, and Phoenician alphabets were based on this model. Then around 1000 B.C., the Greeks added vowels to the Phoenician alphabet. The Greek alphabet in turn became the model for the Etruscan alphabet used in parts of what

is now Italy (ca. 800 B.C.), and the Etruscan alphabet eventually provided the letters for the ancient Roman alphabet.

Modern Western European languages are written in the Roman alphabet because Western Europeans were converted to Christianity from Rome. In Eastern Europe, by contrast, Russian, Belarussian, Ukrainian, Serbian, Bulgarian, and a few other languages used in Russia are written in the Cyrillic alphabet. This is the Greek alphabet as it was augmented and taught by the Greek missionary Saint Cyril (d. 869), who converted many of these peoples to Christianity. Serbian and Croatian are one spoken language, but the Serbs write it in Cyrillic, and the Croats write it in Roman. Thus, the line between Serbs and Croats demarcates a cultural borderline across Europe (Figure 7-9).

A similar cultural divide separates Pakistan from India. Spoken Urdu, the language of Pakistan, is closely related to the spoken Hindi of northern India. The Muslim Pakistanis, however, write it in Arabic script, while the Hindus of India write it in Devanagari script.

Sometimes a people can change the orthography in which they write their language as a self-conscious political act. Romania, for example, switched from the Cyrillic to the Latin alphabet early in the twentieth century. This change reflected a deliberate choice to be a "Western European" nation. In the same way, several languages of Central Asia, called Turkish, historically came to be written in Arabic when these people converted to Islam (see Chapter 8). The Turks of today's country

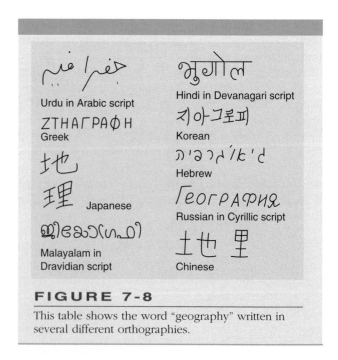

FIGURE 7-8

This table shows the word "geography" written in several different orthographies.

242   Chapter 7

The Distribution of Alphabetic Scripts in Eurasia

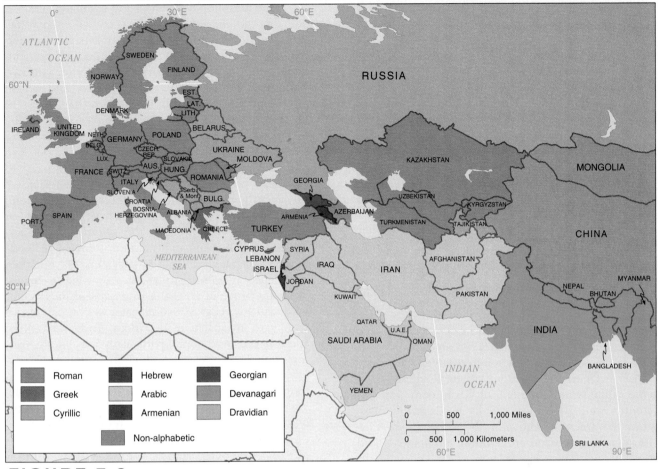

Legend:
- Roman
- Greek
- Cyrillic
- Hebrew
- Arabic
- Armenian
- Georgian
- Devanagari
- Dravidian
- Non-alphabetic

FIGURE 7-9

This map reveals the diffusion of religions and other cultural markers.

of Turkey, however, who are "distant cousins" of the Turks of Central Asia, replaced Arabic with the Roman alphabet early in the twentieth century to signal a deliberate turn to Western-style modernization (see Figure 3-31). The Turkish peoples in Central Asia, by contrast, came under the domination of Russia in the political framework of the USSR, and in 1939, Soviet dictator Joseph Stalin forced them to replace their Arabic script with Cyrillic script. His objective was to cut off these peoples from their cultural heritage and to intensify relatively minor linguistic differences among them. In 1992, however, with the collapse of the USSR, the newly independent Turkic republics of Kazakhstan, Kyrgyzstan, Azerbaijan, Uzbekistan, and Turkmenistan all chose to abandon Cyrillic script. They did not choose to revert to Arabic script, but to adopt Roman characters instead. Western observers hope that this reflects a commitment to join the cultural community of the West.

The language of Malta, the crossroads island of the Mediterranean, is an Arabic language, but it is written in the Roman alphabet. This reflects Malta's former status as a colony of Great Britain.

Many Islamic societies place a special importance on *calligraphy*, that is, highly stylized artistic writing. This is because Islamic orthodoxy generally spurned the creation of the figural artistic images that are so important in Western Christian art. Instead, Arabic writers developed calligraphic styles that are as distinct as the artistic styles of the Christian West (Figure 7-10).

Nonalphabetical Languages

Not all languages are written in alphabets. The major nonalphabetic form of writing is Chinese, in which each character represents a word or concept. These characters are called **ideograms**. The Chinese dictionary of K'ang Hsi (1662–1722) contains nearly 50,000

The Geography of Language   **243**

FIGURE 7-10

"There is no God but God, and Muhammad is his prophet," written in a variety of traditional artistic styles, can often be seen over doorways in Islamic lands, as here, in Morocco.
(Courtesy of the Moroccan National Tourist Office.)

characters, and basic literacy in the modern language requires knowledge of some 2,000 characters.

Japanese *kanji* script is a derivative of Chinese. Notice in Figure 7-8 that the written Japanese is the same as the written Chinese. In Japan today four different writing styles are actually used: *kanji, hiragana* (to express grammatical distinctions), *katakana* (for words borrowed from other languages), and some Roman letters, in addition to Arabic numerals. The Japanese Ministry of Education defines 1,850 characters as essential for everyday use. Of these, 881 are taught during the six years of elementary school.

The Koreans invented their own alphabet in the mid-fifteenth century to overcome the problem of adapting Chinese ideograms to record the spoken Korean language. Today Chinese continues to be the written language for scholars. South Koreans mix Chinese ideograms and the Korean alphabet, but North Koreans use only the Korean alphabet as a political gesture to emphasize independence and self-reliance.

The Vietnamese changed from Chinese writing to Roman orthography in the eighteenth century, when they were Christianized by Jesuit missionaries. Today, the demands of international communication are forcing all of these countries to adopt romanized models.

Many traditional native American, African, and Asian languages never had written forms until Christian or Islamic missionaries set out to translate the Bible or the Koran into these languages, using the missionaries' own othographies. Writing opens new possibilities of exact storage and communication of information, so these peoples may have benefited from having their languages written. Imposing *any*

orthography on a people whose culture had been oral, however, is an act of cultural imperialism.

The transition from an oral to a literate culture is arguably the greatest transition in human cultural history. To purely oral people, writing is strange and magical—a conversation with no one and yet with everyone. The contemporary U.S. writer Richard Rodriguez, who is of Mexican ancestry and whose mother tongue is Spanish, has written that "works of literature, while never intimate, never individually addressed to the reader—are so often among the most personal statements we hear in our lives....One can use *spoken* words to reveal one's personal self to strangers. But *written* words heighten the feeling of privacy. They permit the most thorough and careful exploration. ... The writer is freed from the obligation of finding an auditor in public." You might understand how confusing the transition from an oral culture to a literate culture is if you think of your own experience in learning to talk to answering machines. The experience of speaking to an absent interlocutor who will not hear you until later is comparable to writing to an absent reader. Both are learned skills. Still today many people, particularly the elderly, cannot bring themselves to speak to answering machines.

Many languages today still do not have written forms. More than 100 Chadic languages, for example, are spoken by about 50 million people across northwest central Africa from northern Ghana to the Central African Republic. Only one of these languages, however, Hausa, with about 36 million speakers, has a written form. Speakers of Hausa adopted Arabic script in the sixteenth century, but today these people are switching to Roman script.

The study of orthography reminds us that printing and distributing books is also a key factor in standardizing a language. The oldest printed book in existence is a copy of a Buddhist religious text carved on several wooden blocks and dated 868. Johann Gutenberg (c. 1397–1468) is credited with the invention of movable type in the West. Printing in an alphabetic language is easier than printing in an ideographic language with thousands of ideograms, and printing presses and printed materials quickly spread throughout Europe. William Caxton (c. 1421–1491) first printed books in English.

THE DEVELOPMENT AND DIFFUSION OF LANGUAGES

The way languages develop and diffuse demonstrates the basic principles of diffusion for any cultural innovation or attribute. Any isolated group of people develops a language of its own that enables them to describe everything they see or experience together. If groups of these people break away and disperse, then each group discovers new objects and ideas, and the people have to make up new words for them. After many years, the descendants of each of these breakaway groups have their own language. Each descendant language has a vocabulary of its own, but each also retains a common core of words from that earliest shared language. The ancestor that is common to any group of several of today's languages is called a **protolanguage**. The languages which are related by descent from a common protolanguage make up a **language family**.

The Indo-European Language Family

In 1786, the English philosopher Sir William Jones (1746–1794) first pronounced his theory that a great variety of languages spoken across a tremendous expanse of the Earth demonstrate similarities among themselves so numerous and precise that they cannot be attributed to chance and cannot be explained by borrowing. These languages, then, must descend from a common original language. The group of languages first identified by Sir William is called the

Indo-European family of languages, and languages in this family are spoken by about half of the world's population today (Figure 7-11).

Jacob Grimm (1785–1863), one of the brothers who collected childrens' fairy tales, formulated rules to describe the regular shifts in sounds that occurred when the various Indo-European languages diverged from one another. There is, for example, a regular relationship between words beginning with p in Latin and f in Germanic languages (as in pater and father). These rules are known as **Grimm's Law**.

Sifting through the vocabularies of all Indo-European languages yields a common core vocabulary, which is the common ancestor of these languages, **proto-Indo-European**. The vocabulary of proto-Indo-European tells us a surprising amount about how proto-Indo-European society was organized and how the people lived. It also hints at the language's hearth. Reconstructed proto-Indo-European has words for distinct seasons (one with snow), woody trees (including the beech and the birch), bears, wolves, beavers, mice, salmon, eels, sparrows, and wasps. These things can be found together around the Black Sea, but proto-Indo-European also includes words clearly borrowed from the languages of the Near East. The word wine, for instance, seems to descend from the non-Indo-European Semitic word "wanju." Thus, the hearth area

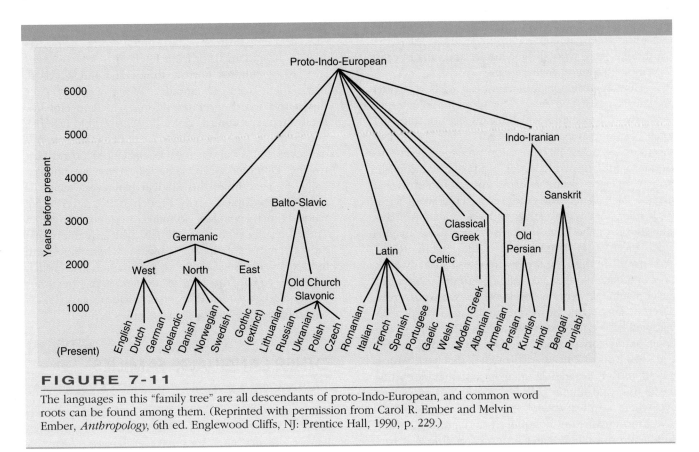

FIGURE 7-11

The languages in this "family tree" are all descendants of proto-Indo-European, and common word roots can be found among them. (Reprinted with permission from Carol R. Ember and Melvin Ember, *Anthropology*, 6th ed. Englewood Cliffs, NJ: Prentice Hall, 1990, p. 229.)

The Geography of Language   **245**

FOCUS
The Basque Language

The Basque language (Euskera) presents an anomaly on the map of languages. It is not related to any contiguous languages, and scholars now believe that it is much older than any of today's Indo-European languages. The Basques tell a joke about their own antiquity: When the Lord wanted to create the first human beings, the Lord got the bones to copy from an old Basque cemetery.

Basque also demonstrates how tenaciously a people can hold on to their language as a symbol of their distinctive nationality. From 1937 until the mid-1950s, the Spanish government tried to stamp out Basque. The government forbade its use in schools, church, the media, and all public places. Books in Basque were publicly burned. Basque names were not allowed in baptism, and all Basque names on official documents were translated into Spanish. Basque inscriptions were removed from public buildings and even from tombstones.

Official policy changed, however, in the early 1960s. Basque was permitted first in church services, then in schools and in the media. A 1969 government decree permitted its teaching in primary schools, and ten years later Basque could be used at all levels of education. In 1980 the first Basque Parliament was elected, with Euskera as a second official language in Spain's Basque provinces. One group of Basques, however, still fights for full independence for the Basque people, and acts of violence continue.

The French long discouraged the use of Basque in the Basque region of France, but the French government never acted against it as forcefully as the Spanish government did.

for Indo-European languages was probably in Anatolia, today's Turkey, some 10,000 years ago. Scholars hypothesize that Indo-European agricultural peoples spread across Europe and replaced the nomadic hunter-gatherers, whose sole direct survivors are the Basques, who are genetically and linguistically far removed from other Europeans.

Hunting common Indo-European roots of words provides a fascinating study. The proto-Indo-European root "aiw," for example, which means "life" or "the vital force," descended into Hindi as "ayua," "life," but it also shows up as "aetas" in Latin, "aion" in Greek, "ewig" in German, and the words "ever" and "age" in English. Words that are distant cousins appear in surprising places. "Maharajah," for example, a Hindi word for a great ruler, may seem exotic, and yet "maha" is a distant cousin of the English words "major" and "magnitude." "Rajah" is a distant cousin of the English "reign" and "royal." In this case the distant cousins clearly look or sound alike, so they are called **cognates**. One must never assume any connection between two words from different languages, however, until it can be proven that they have a historic common root. The study of the origin and history of words is called **etymology**.

Proto-Indo-European provided the basic stock for all Indo-European languages, but that does not mean that one ethnic or racial group spread out to live where all these peoples are today. Sometimes a few Indo-Europeans conquered and imposed their language on a much larger group of people that had previously developed a language of their own—as occurred throughout Latin America and Africa. Sometimes peoples adopted Indo-European just to be able to communicate with Indo-Europeans. Not everyone who speaks English in the world today necessarily has an ancestor from England.

Also, cultures borrow things and the words to name them from one another. Western culture gave Japan the word "erebata" along with the elevator itself. Japanese culture gave the West edible raw fish and a name for it: "sushi." Vocabulary borrowings can even tell us about the circumstances of the borrowing. Different social classes often borrow differentially. For example, the English language, which is Germanic, adopted many French words after 1066, when French-speaking Normans conquered England. Therefore, the names of animals are Germanic when they are hunted or in the farmyard, where they were tended by the peasants, but the names are French when served to the lord as dinner: a pig becomes pork, a cow or bull becomes beef, a chicken becomes poultry, and a deer becomes venison.

OTHER LANGUAGE FAMILIES

The classification of all of the Earth's languages into families is an enormous and difficult task. Comparative linguists do not agree whether it has yet been satisfactorily completed. Figure 7-5 reflects the most generally agreed-upon state of understanding of the world's language families. The Indo-European family has been studied more than any other, and researchers have benefited from written sources dating back 3,700 years—tablets in the extinct Indo-European language of Hittite. These sources give researchers considerable confidence in including individual languages in the family and in reconstructing proto-Indo-European.

The following discussion includes only the major families and the major languages within each family. It draws special attention to instances where language establishes or reinforces a threatened group identity—particularly in opposition to hostile neighbors. Linguistic distinctions fuel self-identity and often trigger political strife.

The Uralic Family

No written forms of any Uralic language survive from before the thirteenth century, but scholars are reasonably confident that proto-Uralic was spoken in the region of the north Ural Mountains about 7,000 years ago (see Figure 7-12). The language in this family with the most speakers today is Hungarian, or Magyar, with around 15 million speakers. The settlement of these people who were originally Asian nomads on the plains of Hungary exemplifies relocation diffusion, and their isolation in the midst of Indo-Europeans has occasioned discrimination—sometimes humorous, sometimes not. The Austrians, for example, who were united with the Hungarians in the Hapsburg Empire for hundreds of years, insist that "Asia begins on the Mariahilferstrasse"—the principal road leading to Budapest.

The Finns and the Estonians are also Uralic-speaking peoples who retain these languages as their official national languages. Hungarian, Finnish, and Estonian all fall within the Finno-Ugric subgroup of the Uralic family. The other subgroup, called Samoyedic, is today represented by only a few thousand people scattered across northern Russia.

The Caucasian Family

The small region between the Black Sea and the Caspian Sea contains a large number of languages. The distribution of languages and dialects from village to village throughout this rugged area is so complicated that dialectologists have made several different estimates of the numbers of languages and speakers. Some 40 languages have been assigned to a single Caucasian family with a total of about 5 million speakers. Georgian, the national language of Georgia, accounts for about 3 million of these, and it is the only one with a significant written form, dating from the fifth century. During the years that this region was part of the USSR, the Russians wrote down a few of the other minor languages in Cyrillic script, but Russian itself became a regional lingua franca. Today linguistic and political splintering continue.

Notice on Figure 7-11 that Armenian, the official language of the Republic of Armenia, spoken by about 6 million people there and in neighboring areas, is *not* a Caucasian language. It is a very old Indo-European one, dating back to about 1000 B.C. in its spoken form and to the fifth century A.D. in its

The Distribution of Uralic Languages

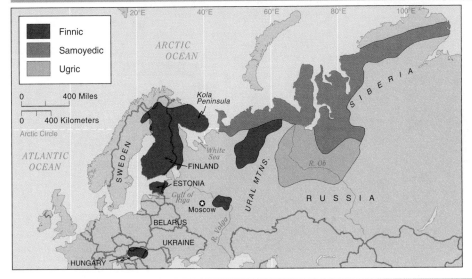

FIGURE 7-12

written form. The antiquity of this linguistic distinction provides a firm root for Armenian nationalism. The language of the neighboring Azeri is Altaic, as described in the next section.

The Altaic Family

The Altaic languages take their name from the Altai Mountain range in Central Asia, but today speakers of Altaic languages are spread from the Balkans to Northeast Asia. The oldest writing in any of them dates from the eighth century.

The largest subgroup of Altaic languages is the Turkic group (see Figure 7-13). Several Central Asian Turkic peoples have only recently won political independence since the collapse of the USSR. These include the Azeri or Azerbaijani (about 15 million), Turkmen (3 million), Uzbeks (13 million), Kyrgyz (5 million), and Kazakhs (7 million). Still other Turkic groups that number just one or two million each lack full political independence but are demanding it. These groups include the Tatars of Russia and Kazakhstan; the Bashkirs, mostly in Russia; the

Chuvash of the middle Volga region; and the Yakuts of the Yakut region of Siberia. Some 10 million Uighurs are spread across Central Asia into western China, and China fears that a resurgence of cultural pride and nationalism among Turkic groups could trigger demands for Uighur political independence. Whether these people all share a political affinity, and whether they share a political affinity with today's Turkey, is a matter of considerable international interest.

A second branch of the Altaic language family is Mongolian, the language of the country of Mongolia and of the Chinese province of Inner Mongolia (look ahead at Figure 10-16). The Mongolian people straddle this international border, and the Mongolians in Inner Mongolia probably envy their "cousins" who enjoy national autonomy. Government leaders in Beijing fear that Inner Mongolia might want to secede and join independent Mongolia.

Manchu is a third distinct Altaic language, spoken by about 20 million people today. The Manchu homeland is northeast of China, but the Manchu people have been assimilating and acculturating to the Chinese for about 200 years. The last ruling royal

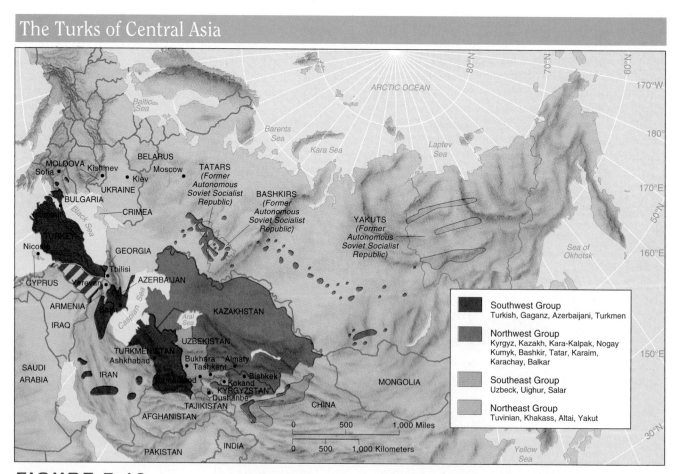

The Turks of Central Asia

FIGURE 7-13

Students of international political affairs are closely watching the emergence of political affinity among all of these Turkic peoples.

dynasty of China was not in fact Chinese, but Manchu. During World War II the invading Japanese separated Manchuria from China and established an independent Manchuria, but that was actually a puppet state of Japan. Today Manchuria is again a part of China, and so many Chinese have settled there that Manchu people are a minority in their homeland.

Korean and Japanese

The Korean language, spoken by about 75 million people in the two Koreas and the surrounding region, bears some similarity to the Altaic languages. Some scholars group it as part of the Altaic family, but others place it in a separate family. More than half of its vocabulary is from Chinese, and its earliest written form, dating from before the twelfth century, is written in Chinese characters.

Japanese too may either be classed alone or among the Altaic languages. Today it is spoken by around 130 million people, mostly in Japan, Brazil, and the United States. The earliest written form of Japanese, dating from the eighth century, uses Chinese *kanji* forms. The Tokyo dialect is considered standard today, but distinct dialects are spoken in outlying regions. The Ryukyu Islands, for example, which enjoyed 1,000 years of independence until they were conquered by the Japanese in 1879, have a distinct dialect.

The Dravidian Family

Languages of the Dravidian family are spoken in the southern part of the Indian subcontinent (see Figure 7-14). It is generally believed that Dravidian languages and peoples lived in the north of India about 4000 B.C., but that they were displaced to the south by Indo-European invaders from the region of today's Afghanistan. An enclave of the Dravidian language Brahui in Pakistan may be a remnant of Dravidian culture surviving there. The oldest written records in Dravidian are in the Tamil language, and they date from the third century B.C.

In the 1950s, the federal government of India established individual states in the southern region to serve as homelands for each of the major Dravidian languages. At the time, some critics of the policy feared that it would encourage political separatism, but, in fact, the recognition of individuality may have been necessary to keep India together. The state of Andhra Pradesh was created for the 60 million speakers of the Telegu language, Tamil Nadu for the 60 million speakers of Tamil, Kerala for the 30 million speakers of Malayalam, and Mysore for the 30 million speakers of Kannada.

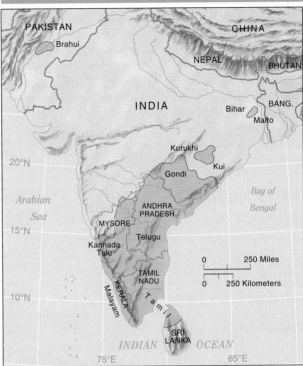

Distribution of Dravidian Languages

FIGURE 7-14

The fact that Tamil people live in both India and Sri Lanka has embroiled India in Sri Lanka's civil war. Sri Lanka's minority Tamil speakers battle the country's Indo-European Sinhalese-speaking majority.

The Tamil language has the greatest geographical spread. There are several million speakers of Tamil in Sri Lanka, as well as in Malaysia, Indonesia, Vietnam, and parts of East and South Africa.

The Sino-Tibetan Family

Classifications of languages within the Sino-Tibetan family are matters of great controversy, but certainly Chinese itself is, measured by number of users, the world's leading language. Most Chinese speakers are in China or on the island of Taiwan, but the Overseas Chinese, as noted in Chapter 5, have migrated widely.

Language has become a sharp political issue on the island of Taiwan. There, Mandarin Chinese, the language of the Chinese mainland, has been the official language since 1949, when the Chinese Nationalist government fled to Taiwan after the victory of the communist revolution on the mainland. The persistent use of Mandarin signaled that the island's government still claimed to be the legal government of all China. Today, however, the use of the Taiwanese dialect is rising. This suggests growing sentiment for a declaration of political independence for Taiwan. This point will be discussed in Chapter 12.

Other groups of languages in the Sino-Tibetan family include the Tibetan group (spoken almost exclusively in Tibet), a Burmese group (largely restricted to Myanmar), and the Miao-Yao languages spoken in parts of southern China, northern Laos, Thailand, and Vietnam. Each of these languages has only a few speakers, and scholars dispute the exact relationships among them.

THE LANGUAGE FAMILIES OF AFRICA

In Africa an extraordinary richness of distinct native languages—by some counts more than 1,000—is spoken. Figure 7-15 uses a finer classification system than that used on Figure 7-5. The great number reflects the minimal interaction of Africa's many native peoples before European conquest. Most African languages do not have a written tradition, and only 40 or so have as many as 1 million speakers. Some African governments are reviving the use and study of indigenous languages in order to emphasize their peoples' preconquest heritage.

The most widely accepted classification of African languages recognizes four main families, but the boundaries between them and even the members of each are matters of scholarly controversy. There are few historical records to help. Written records of African languages have existed only since the first Christian missionaries began their work on the continent about 150 years ago.

The Niger-Congo Family

The Niger-Congo family is the largest family of African languages. This one family is subdivided into six groups with several hundred members each, and its speakers are spread across sub-Saharan Africa.

The Bantu languages form the largest subgroup. These include Swahili, Kongo, Rwanda, Zulu, and Xhosa, and they are spoken by almost 100 million people. Modern scholars place the culture hearth of the Bantu-speaking peoples in the eastern part of today's Nigeria (see Figure 7-16). The proto-Bantu speakers probably kept goats and practiced some form of agriculture, and those technological achievements allowed them to spread at the expense of neighboring hunter-gatherers. As the Bantu speakers expanded, they began to cultivate cereal crops and to

herd sheep and cattle. At the same time, about 1000 B.C., they began to make and use iron tools, and this still further enhanced their technological superiority. By 1,500 to 2,000 years ago, Bantu speakers had spread throughout Central Africa and into the northern areas of southern Africa.

The Nilo-Saharan Family

Nilo-Saharan languages are spoken in two clusters: one along the upper Nile River in today's Sudan (the Nilotic group on Figure 7-15), and the other along the upper Chari River in today's Chad (the Kanuri). Most Nilo-Saharan languages are called *remnant languages*, because their few speakers occupy refuge areas after their hearth areas were occupied by rival groups.

The Khoisan Family

Khoisan is another remnant group. Its native speakers, the Bushmen and Hottentots of today's Namibia, are rapidly being acculturated to other languages. Their languages, and even the peoples themselves, are disappearing.

The Afro-Asiatic, or Semitic-Hamitic Family

Proto-Afro-Asiatic originated in Southwest Asia in the region generally known as the Fertile Crescent—one of the most important and earliest centers of human civilization— about 7000 B.C. Variants were spoken by the ancient Babylonians, Assyrians, and Phoenicians. A distinct language evolved in ancient Egypt before the third millennium B.C., but that language is extinct today. An echo of it survives only in the Coptic language that evolved out of it in the second century A.D. Coptic is still used by Egyptian Christians, who are called Copts.

Today Afro-Asiatic languages predominate from Arabia and the Fertile Crescent westward to the Atlantic. Arabic is the most widespread Semitic language, with about 200 million speakers. Arabic has many different dialects, but its written form is standard. It has a special status among all Muslims as the language of the Koran.

The second major Semitic language, with about 20 million speakers, is Amharic, which is spoken in Ethiopia. This language was carried there from the Near East about 3,000 years ago, and it has since developed largely in isolation. In the distant past, lively trade linked the peoples of today's Israel and Lebanon with today's Yemen, on the southwest tip of

The Languages of Africa

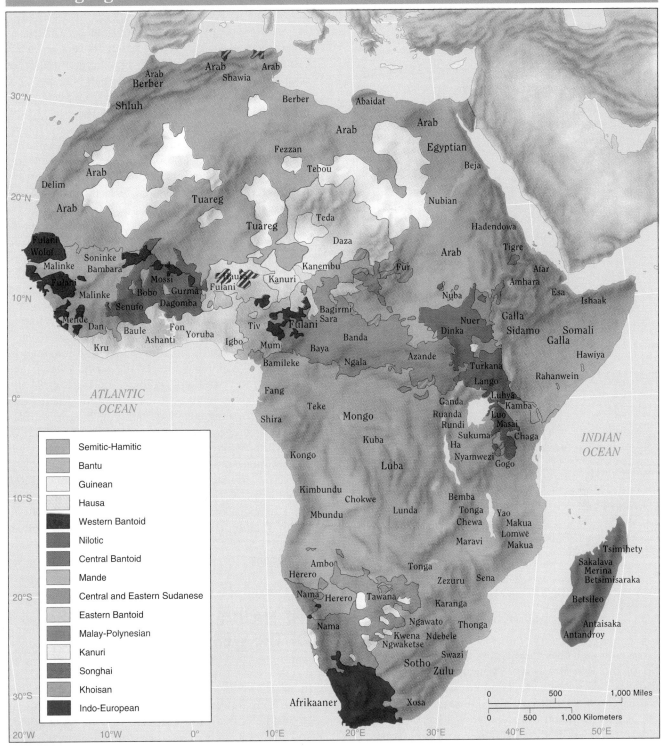

Legend:
- Semitic-Hamitic
- Bantu
- Guinean
- Hausa
- Western Bantoid
- Nilotic
- Central Bantoid
- Mande
- Central and Eastern Sudanese
- Eastern Bantoid
- Malay-Polynesian
- Kanuri
- Songhai
- Khoisan
- Indo-European

FIGURE 7-15

Most of the official languages in Africa are the languages of the former European rulers, but this map shows the extraordinary number of native languages, many of which are in fact spoken by a majority of the population.

the Arabian peninsula. The Bible records that the Queen of Sheba (today's Yemen) came to visit King Solomon in Jerusalem. Sheba also traded extensively across the narrow Red Sea with the peoples of

The Bantu Languages

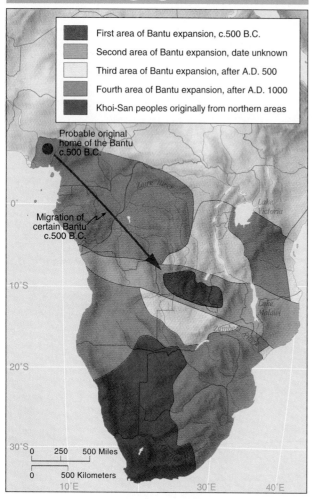

First area of Bantu expansion, c.500 B.C.

Second area of Bantu expansion, date unknown

Third area of Bantu expansion, after A.D. 500

Fourth area of Bantu expansion, after A.D. 1000

Khoi-San peoples originally from northern areas

FIGURE 7-16

The name Bantu was coined in 1862 by the linguist W. H. Bleek. He combined *ntu*, a common sound meaning "humankind," with *ba*, a plural indicator, to represent the close relationship of all African languages from the Cameroon Mountains east to Ethiopia and as far south as the Cape of Good Hope. Today scholars recognize about 350 Bantu languages in which a complex series of prefixes, infixes, and suffixes are added to unchanging roots.

Ethiopia. This connection across the narrow southern tip of the Red Sea is a very ancient and important route of cultural diffusion between Africa and Arabia.

Hebrew, a third Semitic language, dates from about 2000 B.C., and in recent times it has been virtually resuscitated by the government of Israel. Today it has about 4 million speakers.

Languages from the Hamitic group of the Afro-Asiatic family once dominated much of North Africa, but this region was blanketed by Arabic. Remnant Hamitic languages survive among the Berbers of Morocco and Algeria, the nomadic Tauregs of the Sahara, and the Cushites of Ethiopia, Somalia, and Kenya.

THE AUSTRONESIAN FAMILY

The Austronesian family of languages, sometimes called the Malayo-Polynesian family, covers an extraordinarily wide area as well as a great number of peoples (see Figure 7-17). Austronesian languages are spoken by about 300 million people from Madagascar to Easter Island, from Taiwan and Hawaii to New Zealand.

This immense family of languages probably developed in Southeast Asia, where the related Tai family also originated. Austronesian languages were undoubtedly diffused by daring seafaring peoples, but so far linguists have been unable to agree on systems of names or categories for them. For example, scholars have recorded over 70 names for the many dialects of the Dayak language of northwest Borneo, but several of these may be different names for the same dialect. In addition, a great many pidgins and creoles are widespread among these trading peoples, and even these have been overlain by new lingua francas, such as Dutch, English, Malay, Chinese, and French. About 150 million people speak Malay-Indonesian, making it one of the world's languages with the most speakers. Pilipino, or Tagalog, is the language of the natives of the Philippine Islands. No Austronesian language, but rather, Papuan or Indo-Pacific languages are today spoken on the island of New Guinea, directly in the stream of the expansion of Austronesian.

The language of Madagascar may seem curiously out of place on the world map. It is not related to the languages of nearby Africa, but to those of the South Pacific. This suggests that Indonesian traders migrated to the uninhabited island during the first millennium A.D.

THE NATIVE LANGUAGES OF THE WESTERN HEMISPHERE

In 1492, there may have been several thousand native American languages spoken in the Western Hemisphere, but today there are only a few hundred, and the total number of surviving speakers of those native languages is less than 500,000.

Linguists dispute the classification of these many languages, but one widely accepted scheme divides them all into three great families: Eskimo-Aleut, Na-Dene, and Amerind. Each of these three families shows a closer affiliation with Asian language groups than with either of the two other

The Diffusion of the Austronesian Family of Languages

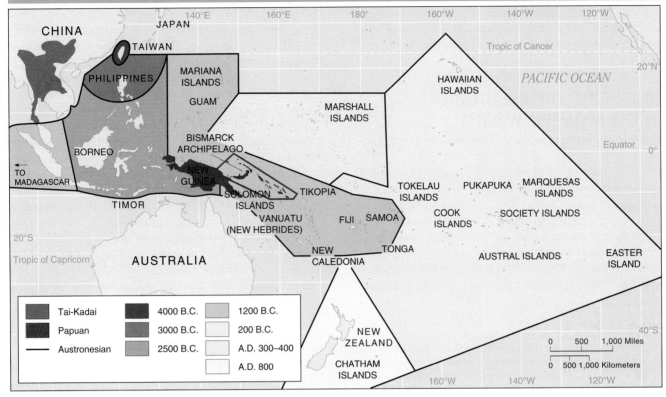

FIGURE 7-17

Austronesian languages are believed to have diffused from the Southeast Asian hearth area in seven stages. (Adapted from P. Bellwood, "The Austronesian Dispersal and the Origin of Languages," *Scientific American* (July 1991): p. 88. Copyright 1991 by Scientific American, Inc. All rights reserved.)

native American families. This suggests that each corresponds to a separate wave of migration from Asia. The first wave gave rise to the Amerind family. The ancestors of today's Na-Dene speakers came somewhat later, and the ancestors of the speakers of the Eskimo-Aleut languages arrived most recently—around 2000 B.C.

The Na-Dene family includes Apache and also Navajo, which is one of the few native American languages whose number of speakers has increased in recent years. During World War II an elite corps of Navajos provided an invaluable undecipherable code among U.S. military operations in the Pacific. They simply spoke Navajo supplemented by about 400 newly coined terms (grenades, for example, were "potatoes"). This was one of the few unbroken codes in military history.

The region of Amerind speakers stretches all the way from the Algonquin speakers of Canada and the eastern United States through the Quechua region from Colombia to Chile down to the tip of the continent.

SUPERFAMILIES

Some comparative linguists believe that after all languages have been classified into families, protolanguages will be constructed for each language family. In turn, study of those protolanguages will yield superfamilies. Such scholars argue that the Indo-European family, for instance, is only one of six branches of a larger group, which they call Nostratic. Reaching even further back into prehistory, Nostratic itself is only one descendant of a protolanguage, or "Mother Tongue," that was spoken in Africa 100,000 years ago and that then diffused around the globe. The further back into prehistory that scholars attempt to reconstruct these languages, however, the more controversial the work becomes. Some linguists believe that the time depth of such studies is too great to reconstruct more protolanguages, let alone one prehistoric Mother Tongue.

The language trees worked out by the most adventurous historical linguists, however, concur to some degree with the groupings of humankind based on the genetic evidence discussed in Chapter 5 (Figure 7-18). Geography, linguistics, anthropology, and biology each contribute to this fascinating but highly controversial research into prehistory.

LINGUISTIC DIFFERENTIATION IN THE MODERN WORLD

▼▼▼▼▼▼▼▼▼▼▼▼▼▼▼▼▼▼▼▼▼▼▼▼

The diffusion and differentiation of languages continues today. Brazilian Portuguese differentiated itself from the language of Lisbon just as American English grew apart from the language of London. Today an American, a Nigerian, and an Indian may all speak English, but they will not be able to understand one another completely without effort. As English has diffused, it has differentiated into these and other dialects. Sometimes even the

same words convey different meanings when they are used by people from such diverse cultures as the United States, Nigeria, and India.

In 1877, the linguist Henry Sweet (the model for the fictional character Henry Higgins in *Pygmalion* and *My Fair Lady*) predicted that by 1977 even the English, Americans, and Australians would speak mutually incomprehensible languages, because of their isolation from one another. These dialects of English are different today, but, despite jokes about "Strine," they have not differentiated to the degree Professor Sweet predicted. This is partly because English, American, and Australian cultures have not greatly differentiated from each other in their other aspects. Also, Sweet never could have foreseen the improvements in world communication, nor the role that nearly universal education would play in fixing language in England, America, and Australia.

Other instances of language differentiation continue. Both French and Spanish are widely used, and

FIGURE 7-18

This diagram illustrates the currently hypothesized genetic relationships among peoples as well as the relationships among language families. Research in one field seems to reinforce research in the other. Compare this chart with the language map (Figure 7-5) and the map of genetic similarities among the world's peoples (Figure 5-3).
(From Philip E. Ross, "Hard Words," *Scientific American* (April 1991): 145. Copyright 1991 by Scientific American, Inc. All rights reserved.)

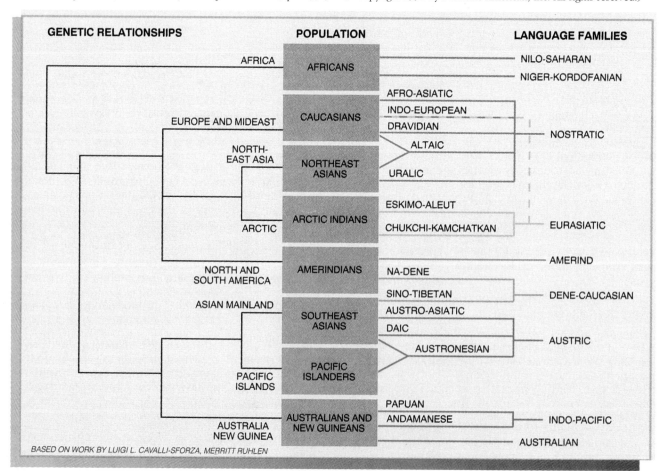

both the French and the Spaniards try to "purify" their languages of linguistic borrowing while also adding new words so that their languages stay useful in the modern world. The French Academy was founded in 1635 and ordered by the king "to labor with all possible care and diligence to give definite rules to our language, and to render it pure, eloquent, and capable of treating the arts and sciences." The Academy published its first dictionary in 1694, and still today it devises new French words for new concepts and items in international trade and discourse. Under the onslaught of foreign languages, particularly English, the French seem to be growing increasingly protective of their language. In 1977, Parliament passed a law banning the use of any foreign words in official contexts if an equivalent word exists in French. In 1991, a special joint session of Parliament convened in order to add to the French Constitution: "The language of the Republic is French." Despite these efforts, however, the French spoken in the 35 Francophone countries and in Quebec consists of many different dialects. In 1992, King Hassan of Morocco expressed concern about the future of French. Asked if he was referring to the spread of English, he replied scornfully, "Not at all. What appalls me is the way people in France speak and write our language."

Spanish has also experienced differentiation. The world's first modern grammar book was written for the Castilian Spanish of the royal court, and it was presented to Queen Isabella in 1492. Language has been called the greatest tool of imperialism, because it colonizes peoples' minds. Spain created a Royal Academy shortly after France did. Nevertheless, the various dialects spoken and written throughout Latin America and the Philippines today differ significantly from one another and from the Spanish of Spain.

When the province of Bengal was one state of the British Indian Empire, its people spoke one language, Bengali. That province was split, however, into West Bengal in India and East Bengal in Bangladesh. Since then Bangla, the dialect of Bangladesh, is becoming a distinct language. This is despite shared Bengali pride in the writings of Rabindranath Tagore, the first non-European to win the Nobel Prize for Literature (1913).

I n 1992, King Hassan of Morocco expressed concern about the future of French. Asked if he was referring to the spread of English, he replied scornfully, "Not at all. What appalls me is the way people in France speak and write our language."

COMPETITIVE EXPANSION AND SHRINKAGE

The map of the geography of languages is not static. The use of some languages is expanding because the speakers of those languages are diffusing around the world, are gaining greater power and influence in world affairs, or are winning new adherents to their ideas. The collapse of Russian-dominated Communist governments in the countries of Eastern Europe, for example, has practically terminated the study of Russian language and the use of Russian as a lingua franca throughout this region. Russian is being replaced by German, which indicates the repair of historic ties with Germany.

English is currently the world lingua franca, partly because much of the world's science and business uses it. Many multinational corporations have designated English their corporate language, whatever the languages of their home countries might be. In computer terminology there is virtually no language but English. Another reason for the spread of English is the phenomenal global appeal of U.S. popular culture, especially movies and music. A 1992 survey by *Readers' Digest* magazine confirmed the increasing use of English across Europe. The survey found that whereas only 40 percent of those over 24 years of age in the European Union spoke English, 70 percent of the population between the ages of 18 and 24 did.

The religion of Islam is spreading in Africa (a development we will examine in the next chapter), and some scholars believe that this will inevitably expand the use of Arabic and the definition of who is an Arab. The southern Sudan may become Arabized, and Mauritania may be completely Arabized. Will the Eritreans be Arabs a century from now? Ancient links across the Red Sea are being reforged. This might lead to a redefinition of the Arab world. Today we think of Africa as ending at the Red Sea, but perhaps in the future, culturally, Africa will end at the Persian Gulf.

Japan has recently won such tremendous wealth and economic power that Japanese might be expected to become an international language, but this has not happened. The reason is that the Japanese language developed over centuries in isolation, and non-

Japanese people find it so complicated and difficult that the Japanese actually suffer handicaps in international communication. Japan's Ministry of International Trade and Industry subsidizes efforts to develop computer software that can translate documents into or out of Japanese. In January 1993, the world's first computer-translated international phone call was placed from Kyoto to Pittsburgh. The two callers had to speak slowly, but the system worked. A speaker in Kyoto said, "Moshimoshi," and 12 seconds later, a computer in Pittsburgh said "hello." Automatic interpreting telephony and other speech synthesis and speech recognition devices offer fascinating possibilities. Nevertheless, for the present, the Japanese have been quick to adopt the languages of others, most notably English. Today Japanese power and influence actually contribute to the diffusion of English.

While some languages are growing and spreading, devising new ways of expressing new concepts and winning new speakers, others are dying out. For example, among the Celtic languages of Western Europe, both Manx, the language of the Isle of Man, and Cornish, the language of Cornwall in southwestern England, have been overwhelmed by English and have disappeared. The tombstone of Dorothy Pentreath (d. 1777) in Saint Paul's parish church, Cornwall, declares her to have been the last person to speak Cornish. Welsh survives and is even enjoying a revival in Wales, but it is not spreading.

Today most of the countries of Western Europe feel secure in their senses of national identity, and they no longer fear linguistic pride as separatism. Therefore, they have recently taken steps to revive and even restore small regional languages that have been neglected or persecuted for years. The French government, for example, suppressed languages other than French in 1789 and banned them completely in 1886. Today, even as the government defends the uniformity of official French, it has simultaneously taken steps to preserve pockets of non-French language within France. In January 1993, the government declared that some regions ought to prepare for regional bilingualism. France has eight distinct languages and more than a dozen dialects. Linguists fear that it may be too late to save Corsican, Provencal, Basque (in France), and Breton, but the French government does hope to save Alsatian German, Flemish (in the region around Dunkirk), and Catalan in the south.

Meanwhile, countless languages or dialects throughout Africa, South America, and Asia are disappearing before they are even recorded. In the nineteenth century there were thought to be over 1,000 native languages in Brazil; today there are fewer than 200. The loss of a language forecloses an opportunity to explore the workings of the human mind.

The Languages of Signs and Gestures

The study of nonverbal visual communication is called **kinesics**. Signing, for example, is using the hands in a conscious verbal manner to express the same range of meaning as would be achieved by speech. Several different sign languages exist, including American, French, British, Danish, and Chinese, and they are not mutually intelligible. They use different signs, different rules of sign formation, and sentence structures. Even within regions that use the same spoken language, different signers may not be mutually comprehensible. British Sign Language, for example, differs from American. One system adopted by the Unification of Signs Commission of the World Federation of the Deaf in 1975 is called Gestuno, but its use is not universal. Some of the hand movements of sign language can be plausibly interpreted by nonsigners because they reflect properties of the external world—they are *iconic*. After a sign's meaning has been revealed to a nonsigner, it may seem "obvious," but the meanings of the vast majority of signs are not apparent to nonsigners.

Gestures contrast to signs. There are fewer gestures, and these are used to express a small number of basic notions (see Figure 7-19). Other physical traits of behavior are also specific to certain cultures.

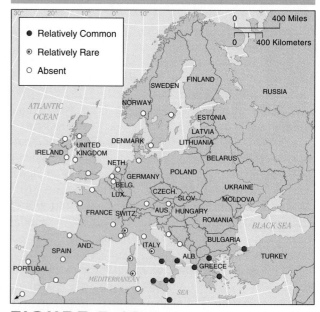

Where a Toss of the Head Means "No"

FIGURE 7-19

This map shows the region within Europe within which throwing one's head back answers a question or suggestion in the negative. (Adapted with permission from Desmond Morris, Peter Marsh, Marie O'Shaughnessy, *Gestures: Their Origins and Distribution*, New York: Stein & Day, 1979.)

Europeans, for example, generally convey food to their mouths with their forks in their left hands, whereas Americans usually hold their forks in their right hands. Guidebooks of gestures have been written to prevent travelers from unconsciously "saying" something rude in foreign lands.

Some aspects of facial expressions, eye contact, and body posture seem to be almost universal, while others vary from culture to culture. For example, when people greet each other from a distance, they raise their eyebrows for about one-sixth of a second. This is called an eyebrow flash. The gesture seems universal, although some cultures suppress it. The Japanese, for example consider it obscene. We are usually not aware of doing this ourselves, but it evokes a strong response. To receive an eyebrow flash from someone we do not know can be uncomfortable, embarrassing, or even threatening.

TOPONYMY

▼▼▼▼▼▼▼▼▼▼▼▼▼▼▼▼▼▼▼▼▼▼▼▼▼

Local place names are an important feature of any cultural landscape. If the cities we visit bear names like Saint Paul, we are in a Christian region; conversely, Islamabad ("the place of Islam," the capital of Pakistan) indicates the Islamic realm.

The study of place names, called **toponymy**, often reveals a great deal about what people in a particular location do or believe. The historic city of Saint Petersburg in Russia provides an example. When Russia fought Germany during World War I, the Russians thought that the name Saint Petersburg was too German, so they translated the name of the city to the more Russian "Petrograd." Following the Russian Revolution of 1917, the city's name was changed to Leningrad. In 1991, with the dissolution of the Soviet Union, the city's name reverted to Saint Petersburg. Since the fall of Communist governments all across Eastern Europe in 1991, places have reassumed names they had before the Communists came to power (Figure 7-20). Newly independent countries often change away from their old colonial names. In recent times in Africa, for example, Upper Volta (a simple geographical designation) became Burkina Faso ("The Republic of Upright Men"). Rhodesia (named for an English imperialist) became Zimbabwe (the name of an ancient native city there), and its capital, Salisbury (in honor of an English lord), became Harare ("one who never sleeps").

Toponymy can also reveal history when other evidence has been erased. It may tell us what people first settled a place (Figure 7-21). Much of the territory of Eastern Europe, for example, has long been dis-

Communist era name	Previous and now restored name
Andropov	Rybinsk
Brezhnev	Naberezhnye Chelny
Chernenko	Sharypovo
Frunze	Bishkek
Georgiu-Dezh	Lisky
Gorky	Nizhny Novgorod
Gotvald	Zmiev
Kalinin	Tver
Kuibyshev	Samara
Kirovbad	Gyanja
Leninabad	Khodjent
Leningrad	St. Petersburg
Mayakovsky	Bagdati
Ordzhonikidze	Vladikavkaz
Sverdlovsk	Yekaterinburg
Voroshilovgrad	Lugansk
Zhdanov	Mariupol

FIGURE 7-20

During the period of Communist government in the Soviet Union, all of these historic cities' names were changed to honor Communist leaders. Today the cities are reassuming their earlier names.

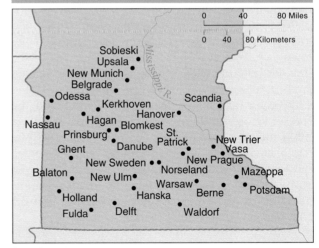

Minnesota Place Names

FIGURE 7-21

Some North American cities retain the names of explorers who just passed by, but these Minnesota place names supply a good clue to the origins of the immigrant settlers in the area. Each is a city in Europe or a hero to some national group. How many can you identify?

puted by the Germans from Central Europe and the Slavic peoples to the East. Today Eastern Germany has towns with names ending in "-ow," with "-in," and with "-zig." These are all Slavic suffixes, and their distribution describes quite accurately the geography of Slavs as late as 800 A.D. Similarly, places south of the Danube and west of the Rhine with names ending in "-weiler" (hamlet) were once part of the Roman Empire. This ending derives from the Roman "villare" (country estate). In Spain, "guada" (as in Guadalquivir or Guadaloupe) is a corruption of the Arabic "wadi," which means river. Guadalquivir is Wadi al Kabir, the great river. The English countryside retains place names left from Roman, Norman (French), and Danish rule. In Danish, for example, the suffix "-by" means village, as in Derby, and the suffix "-thorp" means hamlet, as in Althorp.

New Zealand place names reveal how European settlement came to overlay older indigenous settlements. The four largest cities in New Zealand have European names. Of the 20 regional centers of 10,000-100,000, people, 40 percent have Maori names. Almost 60 percent of the small towns with populations under 10,000 have Maori names.

Sometimes names are intended to attract settlers to a place. Eric the Red, who was from Iceland (not a particularly inviting name), tried to lure settlers to his new western real estate development by calling it Greenland. The climate may in fact have been a bit milder at the time, but Eric was the spiritual ancestor of every developer who builds a Paradise Valley in suburban American today. Place names recalling woods and streams are popular even where there are none.

Many names indicate what type of environment early settlers did in fact find. Chapter 1 noted that many towns in the North China plain are named "Wang," which means oak tree, even though the towns are in treeless plains today. Names may even indicate how the early settlers converted the natural landscape to a cultural landscape. From about 800 to 1300, for example, Germanic peoples cleared the forests across northern Europe. Local place names often indicate how this was done. German suffixes "-roth" and "-reuth" (as in Bayreuth) refer to digging out roots after cutting down the trees. In England the suffix "-ley" or "-leigh" (as in Woodleigh) means clearing, and place names ending in "-brind," "-brunn," or "-brand" refer to the use of fire.

Place Names across the United States

The Europeans who came to North America did not come to lands without languages and without names. Literally thousands of native American lan-

guages and dialects have been differentiated. These languages, however, were so different from European languages and from each other that a great deal of confusion remains about what the original native place names were, and how many of them came to be Europeanized in various regions. The predominant European languages were at first Spanish, French, and English.

Europeans sometimes ignored the local languages and place names, but in thousands of cases the native names remain. Sometimes two languages combined into different types of names. For example, today's Bogalusa, Louisiana, comes from the Choctaw for "black stream," and it was rendered in Spanish (Arroyo negro o Bos-holizà), French (Bogue Luca), and English (Black Creek), before Bogalusa resulted.

Native American names usually identified places. "The fishing place," "the falls," or "the point" are typical examples. The Europeans, by contrast, named places with respect to history, ancestral pride, or nostalgia, but without meaning. To name a place Chester, for example, after a place in England, may reflect homesickness, but it does not identify the new site. Chester, England, was the hometown of many settlers brought to America by William Penn, and the place name spread across North America in a typical example of relocation diffusion. Chester, Pennsylvania, where Penn landed in 1682, is echoed all the way out to various Chesters on the West Coast, many of which were settled by pioneers from Chester, Pennsylvania. Similarly, settlers from Wallingford, England, populated Wallingford, Connecticut, and Wallingford, Vermont, and settlers from these two Wallingfords relocated west to found new Wallingfords all the way out to the Wallingford neighborhood of today's Seattle, Washington.

There are several categories of place names. Descriptive names are common in all languages, and a descriptive name may be carried through several languages as different settlers occupy a place. Today's Red River was previously the French Rivière Rouge, and that descended from an earlier corresponding native name.

Names of battles, saints, and psychological conditions usually reflect incidents. Places such as St. Augustine, San Diego, or St. Lawrence, for example, were usually named by French and Spanish explorers for the saint on whose day the place was sited. Commemorative names in honor of historic personages include the many Washingtons, Lincolns, and Jeffersons. Alexander Von Humboldt, the great geographer mentioned in Chapter 1, is commemorated by counties in California, Iowa, and Nevada; towns in six states and one Canadian province; a bay in California, a peak in Colorado, and a National Forest, a mountain range, and a river in Nevada. Von Humboldt visited the United States and became a friend of President Jefferson in 1804. Karl Ritter, the other founder of modern geography, never visited the United States, but a mountain has been named for him in California.

In Canada and the United States today the majority of place names are actually possessives and personal names, such as Jones Creek. Many place names have two parts. The generic part tells what sort of a place it is—"gap," for example, a natural highway through mountains, or "ford," a shallow place where a river might safely be crossed. The other part of the name, the specific, identifies it as a particular place of that sort: the Cumberland Gap, for example, or Harrisford.

The areas explored or settled by the Spanish, French, or English can be traced by trails of place names they left behind. The arc of French settlement from Quebec to New Orleans, for example, can be traced through Montreal (Mount Royal); Detroit (the "straits" between Lakes Erie and St. Clair); Fond du Lac, La Crosse, and Prairie du Chien (on routes across Wisconsin from the Great Lakes to the Mississippi); down the Mississippi past Dubuque (named for settler Julien Dubuque); Saint Louis; and Baton Rouge ("Red stick" translated from the Choctaw "istrouma," a boundary mark between two tribes). Spanish-named settlements stretch across the South from St. Augustine (originally San Augustine), Florida, to San Antonio, Texas, to Sante Fe, New Mexico, to San Diego, California.

Three major dialects had developed in the thirteen English colonies along the eastern seaboard by the time of the American Revolutionary War: northern, midland, and southern American English. Each of these diffused, and one way that paths of settlers can be traced toward the west is by reading place names (Figure 7-22). Users of the northern dialect frequently named places Brook, Notch, and Corners. People moving westward from the midland areas commonly named places Gap, Cove, Hollow, Knob, and Burgh. Southern settlers used names including Bayou, Gully, and Store.

The U.S. Board on Geographic Names was created in 1890 to certify place names. Today there is a committee to certify the names of foreign places for American usage, as well as a Committee on Domestic Names. The Advisory Committee on Extraterrestrial Features surrendered its duties to the International Astronomical Union in 1979.

LANGUAGES IN THE UNITED STATES

The U.S. Constitution did not specify English as an official language, and many local governments through history have found it useful to provide ser-

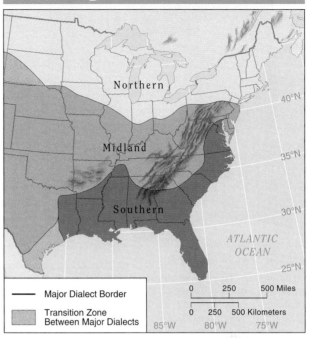

Dialect Regions

FIGURE 7-22

The three regional dialects of northern, midland, and southern American English developed already in the eighteenth century, and as pioneers headed westward from each dialect region, they took place names with them. (Adapted with permission from Kurath, Allen, Wood, and E. B. Atwood, *The Regional Vocabulary of Texas*, Austin: University of Texas Press, 1962.)

vices and even keep official records in other languages (Figure 7-23).

The population of the United States has always been composed of a great variety of peoples speaking a great variety of languages. English was the language of the principal colonial ruler and of the greatest number of European settlers, and it has always served as a lingua franca. A distinct American English nevertheless evolved. From the days of the earliest settlers, the American language adopted terms from native American languages. These included the names of native animals and plants unknown in Europe or Africa, products derived from them, and also place names. It has been said that the explorer Meriwether Lewis (1774-1809), who recorded many native words as he traveled across the continent from 1803 to 1806, added more words to English than anyone else has ever done. Each of the many immigrant groups has in turn learned American English, but each has also contributed to American English vocabulary, grammar, and diction.

Black English, an American dialect, demonstrates that the speech of parts of the African-American population retained and preserved elements of West African

FIGURE 7-23

The Florida State Constitution insists that "English is the official language of the State of Florida." In 1980, Dade County (metropolitan Miami) passed an ordinance prohibiting the use of any language other than English for government business. By 1990, however, more than half (50.1 percent) of county residents spoke Spanish at home, so in 1993, the county ordinance was repealed.
(Courtesy of Donald Dietz /Stock, Boston.)

languages: Yoruba, Fante, Hausa, Ewe, and Wolof. Most of us speak at least one word of Wolof, the language of Senegal: "degan," "to understand" (as in, "Can you dig it?"). Some people object to the name "Black English," because there is no simple correlation between color and language, so in dialect studies the term "Black English Vernacular" (BEV) characterizes the nonstandard English spoken by lower-class blacks in urban areas. Scholars have recently begun to study how the grammar of BEV diffused from West Africa to the southern United States and from there into the cities of the North. In a landmark 1977 court case involving the school system of Ann Arbor, Michigan, the court ruled that BEV was sufficiently distinct to require the school board to identify children speaking it and to use that knowledge to help those children learn standard English. Since that decision, several districts across the country have taken similar steps to recognize BEV and to help its speakers learn standard English.

As noted above, regional variations in American English existed by the time of the Revolution, and these regional variations have been mapped and traced along the various routes of westward expansion. Regional variations persist, both in grammar and in pronunciation, despite the nationwide reach of popular media. Television homogenizes national speech patterns much less than scholars originally thought it would.

In 1931, the scholar Hans Kurath began surveys for a linguistic atlas of the United States and Canada. He divided small regions into survey areas, and the first atlas to appear dealt with New England. The amount of work involved means that publication has been slow and irregular. The year 1991 saw publication of Volume Two of the *Dictionary of American Regional English*, a study of regional peculiarities begun in 1965. It got as far as the letter "H," and it informs us, among other things, that when you put on your best clothes you are "diking up" in the South but "dickeying up" in New England.

Numbers, Distribution, and Changes of Non-English Speakers in the United States

The number of U.S. residents for whom English is a foreign language jumped 38 percent from 1980 to 1990, from 23 million to 32 million (Table 7-2). This increase resulted largely from the new wave of immigration discussed in Chapter 5. About 14 percent of the nation's population over 5 years of age grew up or are growing up today speaking a language other than English.

The 17.3 million speakers of Spanish account for more than half of all people whose first language is not English. Languages showing the largest recent increases suggest the source areas of the newest immigrants. Aside from the French Creole of the Haitians, the greatest increases were all recorded among Asian languages: Mon-Khmer, spoken by Cambodians; the languages of Pakistan and northern India; Vietnamese; Thai; Korean; Chinese; and Tagalog.

The state with the highest percentage of residents speaking a language other than English was New Mexico (33.5 percent), followed by California (31.5 percent), Texas (25.4 percent), Hawaii (24.8 percent), and New York (23.3 percent). The greatest percentage changes, however, occurred in the South, just as we saw in Chapter 5 that the greatest increases in the percentage of the foreign-born population occurred in the South. Foreign-language speakers made up only 4.8 percent of Georgia's population in 1990, for example, but that represented the largest increase in any state, from 134,000 people to 285,000 people, or an increase of 112 percent. The greatest numeric change was in California, which added 3.6 million people whose mother tongue was a language other than English.

The Movement to Declare English the Nation's Official Language

During the 1970s and 1980s, a political movement to declare English the nation's official language gathered

TABLE 7-2
The Leading Non-English Languages in Use in the United States, 1990

Language used at home	Total speakers over 5 years old 1990	1980	Percentage change*
Spanish	17,339,000	11,549,000	50.1
French	1,703,000	1,572,000	8.3
German	1,547,000	1,607,000	− 3.7
Italian	1,309,000	1,633,000	−19.9
Chinese	1,249,000	632,000	97.7
Tagalog	843,000	452,000	86.6
Polish	723,000	826,000	−12.4
Korean	626,000	276,000	127.2
Vietnamese	507,000	203,000	149.5
Portuguese	430,000	361,000	19.0
Japanese	428,000	342,000	25.0
Greek	388,000	410,000	− 5.4
Arabic	355,000	227,000	57.4
Hindi, Urdu and related	331,000	130,000	155.1
Russian	242,000	175,000	38.5
Yiddish	213,000	320,000	−33.5
Thai	206,000	89,000	131.6
Persian	202,000	109,000	84.7
French Creole	188,000	25,000	654.1
Armenian	150,000	102,000	46.3
Navajo	149,000	123,000	20.6
Hungarian	148,000	180,000	−17.9
Hebrew	144,000	99,000	45.5
Dutch	143,000	146,000	− 2.6
Mon-Khmer	127,000	16,000	676.3

Source: Census Bureau *Calculations are from numbers before rounding.

strength. By 1994, 17 states had statutes or constitutional clauses declaring English the sole official language, and 7 more were considering such clauses. This movement may reflect a dedication to a national language as a bonding force of a diverse population, but some people fear that it may also reflect a resentment of the changing immigration trends. A storm of opposition arose in 1994 when the U.S. Internal Revenue Service first offered income tax forms in a language other than English—in Spanish.

In two contexts language has become a civil rights issue: education and voting. If, as is generally agreed, the inability to communicate in English is a handicap in the United States, then the teaching of English becomes a key route to equal opportunity for all children. In a case involving Chinese-American children in San Francisco, the U.S. Supreme Court ruled that "students who do not understand English are effectively foreclosed from any meaningful education" (*Lau* v. *Nichols*, 1974). This Court ruling triggered a national concern to introduce bilingual education where English was the students' second language. Arguments persist, however, over whether the only purpose of these programs is to ease the students' transition to English, or whether they also should preserve the languages and cultures that immigrant children bring to school. Some people argue that using the education system to acculturate all children to the English language denigrates the richness of their native inheritances, while others argue that preservation of other languages threatens to lock the youngsters into second-class citizenship.

Another volatile issue is equal access to the political process. The Voting Rights Act of 1965 sought to draw many disenfranchised Americans into the political process by banning the literacy tests that had been common in the South. Amendments in 1975 and 1982 mandated that voter registration districts (usually counties) whose populations contain a certain percentage of minority groups unskilled in English must provide registration and election materials in other languages. Today those electoral districts include vast areas of the country, including both inner-city neighborhoods and rural areas. Elections in these areas are observed by the Department of Justice.

U.S. Government Policy Toward Native American Languages

From the 1880s until the 1950s, federal policy discouraged or eliminated native American languages. In 1990, however, federal legislation was passed to "encourage and support the use of Native American languages as languages of instruction." In 1993, the state of Oklahoma, where over 700,000 native American children attend public schools, began providing Cherokee, Creek, Choctaw, Chickasaw, and Seminole as part of a program offering second languages to all students. Native Americans support 14 radio stations and several magazines and newspapers printed in native American languages. Many equate linguistic survival with cultural survival, and perhaps they are correct.

SUMMARY

The distribution of any language illustrates the pattern of dispersal of its original speakers or their cultural impact on others. The great variety of languages spoken today testifies to the relative isolation of groups in the past. There are more than 3,000 distinct languages, but about 60 percent of all people speak 1 of just 12 languages. The language with the most speakers is Mandarin Chinese. English is the world's lingua franca. In contrast with those widespread languages, some languages are extremely local. Written language is usually more widely dispersed than spoken language, because writing tends to standardize languages. Sounds are localized only among speech communities.

The vocabulary, grammar, and sounds of languages change through time. The vocabularies of different languages reflect the needs and the historical experiences of the people who use them. All aspects of culture influence language, and some psycholinguists believe that language is a force in its own right.

Most languages are written in alphabets, but major nonalphabetic languages include Chinese, Japanese, and Korean. Christian or Islamic missionaries provided systems of writing for many traditional native American, African, and Asian languages, but many languages still today do not have written forms.

A few languages are associated almost exclusively with one country, but several languages are shared by many countries, and polyglot countries officially recognize several languages. Standard languages usually are the product of efforts at nation building by central governments.

Languages can be categorized into families on the basis of common ancestry, and they continue to diffuse and differentiate. The use of some languages is expanding, while other languages are dying out.

Kinesics is the study of nonverbal visual communication, including signing and gestures. Toponymy often reveals a great deal about what people in a particular location do or believe; it can reveal history when other evidence has been erased.

The population of the United States has always been composed of a great variety of peoples speaking a great variety of languages. Regional differences persist, and some people think that English should be the official national language.

KEY TERMS

language (p. 230)
standard language (p. 230)
official language (p. 230)
dialect (p. 230)
dialectology (p. 230)
dialect geography (linguistic geography) (p. 230)
speech community (p. 230)
isogloss (p. 231)
geographical dialect continuum (p. 231)
pidgin (p. 232)
creole (p. 232)
lingua franca (p. 233)
Esperanto (p. 233)
polyglot states (p. 240)
orthography (p. 242)
alphabet (p. 242)
living language (p. 242)
ideogram (p. 243)
protolanguage (p. 245)
language family (p. 245)
Grimm's Law (p. 245)
proto-Indo-European (p. 245)
cognate (p. 246)
etymology (p. 246)
kinesics (p. 256)
toponymy (p. 257)

QUESTIONS FOR INVESTIGATION AND DISCUSSION

1. How many languages are spoken in your community? Does the local school system have bilingual or English as a Second Language (ESL) programs? Where do your community's non-English speakers come from?
2. The breadth of distribution of a language is not directly related to the number of people who speak it. Several Asian languages are geographically restricted, yet they number among the top languages in terms of numbers of speakers. Hindi and Urdu (which are almost the same spoken language) and Bengali are examples. Why are these languages not more widespread across the map (Figure 7-4)?
3. Is it a good thing that there are so many languages in the world? List advantages and disadvantages.
4. The map of the official languages of the world gets more complicated each year. The number of official languages is increasing as countries react against the legacy of colonialism and accept native languages. Should we regret this as a lack of international understanding, or celebrate it as a triumph of democracy in the world?
5. The Dutch have considered making English the official language for college-level education in the Netherlands. Can you consider why?
6. If you speak two or more languages, can you think of specific ideas or images that may be difficult to translate from one to another? Try to explain this difficulty to your fellow students.

ADDITIONAL READING

AXTELL, ROGER. *Gestures: The Do's and Taboos of Body Language Around the World.* New York: John Wiley & Sons, 1991.

CAMPBELL, GEORGE L. *Compendium of the World's Languages.* New York: Routledge, 1991. 2 vols.

CASSIDY, FREDERIC G. and JOAN HOUSTON HALL, eds. *Dictionary of American Regional English.* Cambridge, MA: Belknap Press of Harvard University Press. Volume 1 1985; Volume 2, 1991.

COOPER, ROBERT L., ed. *Language Spread: Studies in Diffusion and Social Change.* Bloomington: Indiana University Press, 1982.

CRYSTAL, DAVID. *The Cambridge Encyclopedia of Language.* Cambridge, England: Cambridge University Press, 1987.

KRETZSCHMAR, WILLIAM A., Jr., Virginia G. McDavid, Theodore K. Lerud and Ellen Johnson. *Handbook of the Linguistic Atlas of the Middle and South Atlantic States.* Chicago: University of Illinois Press, 1991.

MOSELEY, CHRISTOPHER, ed. *Atlas of the World's Languages.* New York: Routledge, 1993.

RUHLEN, MERRITT. *Guide to the World's Languages.* Vol. 1, *Classification.* Palo Alto, CA: Stanford University Press, 1987.

8 The Geography of Religion

Moscow's Saint Basil's Cathedral (1554) was built to commemorate Russia's victory over the Tatars, a victory that launched Russia's own empire. The communists turned the church into a museum, but it may be rededicated to Christian worship. The onion domes, designed by a Venetian architect, exemplify cultural diffusion of the Byzantine style.

Courtesy of David Sutherland/Tony Stone Images.

A religion is a system of beliefs regarding conduct in accordance with divine commands found in sacred writings or declared by authoritative teachers. Most religions involve personal commitment to worship a god or gods, but the teachings of several Asian systems of belief, including Confucianism, Taoism, and Shintoism, are basically ethical and psychological. That is, they focus on appropriate behavior (*orthopraxy*) rather than belief in a set of philosophical or theological arguments (*orthodoxy*). Several do not even address theological questions such as the nature of God or the gods, or life after death.

The world distribution of religions forms a mosaic even more complex than the distribution of languages (Figure 8-1). Each of the major religions originated in one place and diffused from there, but today communicants of various religions mingle around the globe, and religious affiliations cut across lines of politics, race, language, and economic status. Table 8-1 estimates the number of adherents of each of the world's principal religions. No worldwide religious census is taken, however, and these figures cannot indicate the depth of anyone's belief nor the degree to which a person's religion actually affects his or her behavior. The strictest adherence to traditional beliefs is called **fundamentalism**. **Secularism**, by contrast, is a lifestyle or policy that deliberately ignores or excludes religious considerations.

THE TEACHINGS, ORIGIN, AND DIFFUSION OF THE WORLD'S MAJOR RELIGIONS

This chapter will note the teachings of each of the world's major religions, but we cannot examine the fine points of each religion's message. A textbook in comparative theology would be needed to do that. Furthermore, there are great variations in belief and

The World's Major Religions

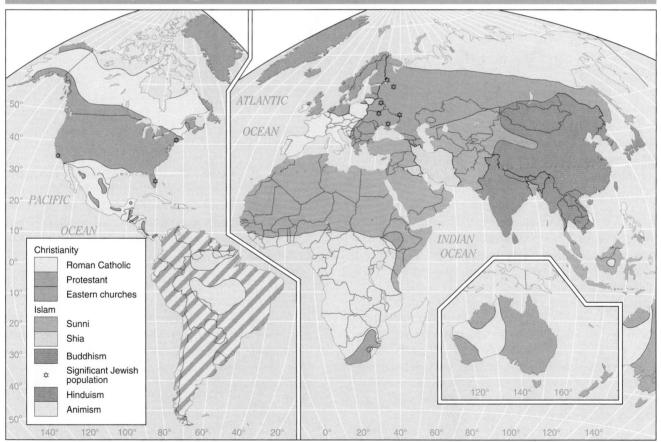

FIGURE 8-1

This map shows the predominant faith or faiths in each region, but it cannot show small minorities.

TABLE 8-1
The Number and Distribution of Adherents of the World's Major Religions

	Africa	Asia	Europe	Latin America	Northern America	Oceania	Former U.S.S.R.	World
Christians	327,204,000	285,365,000	413,756,000	435,811,000	239,004,000	22,628,000	109,254,000	1,833,022,000
Roman Catholics	122,907,000	123,597,000	262,638,000	405,623,000	97,022,000	8,208,000	5,590,000	1,025,585,000
Protestants	87,332,000	81,476,000	73,939,000	17,263,000	96,312,000	7,518,000	9,858,000	373,698,000
Orthodox	28,549,000	3,655,000	36,165,000	1,764,000	6,008,000	576,000	93,705,000	170,422,000
Anglicans	26,863,000	707,000	32,956,000	1,300,000	7,338,000	5,719,000	400	74,883,400
Other Christians	61,553,000	75,930,000	8,058,000	9,861,000	32,324,000	607,000	100,600	188,433,600
Muslims	278,250,800	636,976,000	12,574,500	1,350,500	2,847,000	100,500	39,229,400	971,328,700
Nonreligious	1,896,000	691,144,000	52,411,000	17,159,000	25,265,000	3,291,000	85,066,000	876,232,000
Hindus	1,475,000	728,118,000	704,000	884,000	1,269,000	360,000	2,000	732,812,000
Buddhists	21,000	313,114,000	272,000	541,000	558,000	26,000	407,000	314,939,000
Atheists	316,000	161,414,000	17,604,000	3,224,000	1,319,000	535,000	55,898,000	240,310,000

Source: Adapted with permission from 1993 Encyclopaedia Britannica Book of the Year, © 1993 Encyclopaedia Britannica, Inc.

practice within religions that have recorded histories of thousands of years and that claim today hundreds of millions of followers. These variations occur both through time and across space.

We will identify the hearth region of each religion and review the story of its diffusion, although it is difficult for scholars to identify precise reasons why any religion diffuses. Believers in a religion will say that their religion diffused because that was God's or the gods' will. This kind of argument cannot be criticized or analyzed. Some believe that it is true, and others do not, but it is grounded in a faith that cannot be tested. A few religious groups **proselytize**, that is, actively try to convert others to their religious beliefs. Other religions do not proselytize. Some scholars differentiate *universalizing religions*, which proselytize, from *ethnic religions*, which are identified with a particular group of people and do not proselytize.

Outside observers have suggested that cultural imperialism has often played a role in the diffusion of a religion. Many conquered peoples have converted to the religion of their conquerors. Some people may

convert because they hope to enjoy higher social or economic status either in this world or in a next world after death. Whole peoples have converted from one religion to another after their own ruler converted, but we cannot know if they did this out of love, fear, or in hope of reward. As we have noted before, the motivation for human behavior is seldom completely understood. The world geography of religion is not a static pattern today. Some religions are diffusing widely, and their continuing diffusion triggers changes in societies and in landscapes.

After we have inventoried the world's major religions and their distributions, we will investigate the impact of religion on other factors of geography. We saw already in Chapter 6 that different religions' teachings about food affect the geography of farming, livestock, and international trade in food. Religion is a major determinant of human behavior, so the geography of people holding and following a set of teachings affects political geography, economic geography, and other aspects of cultural geography. Most religions involve joining an organized body of fellow believers, and local religious administrative patterns, such as Roman Catholic dioceses, often organize and focus socializing, community spirit, and, in many cases, education. More Roman Catholic New Yorkers know what parish they live in than what city council district they live in. It is through the parish that they make friends and meet potential spouses. "Religion" means "rebinding" of people with their gods and of people with one another. The geography of such binding will always be a key consideration of human geography.

The third section of this chapter will investigate relationships between religions and environments and landscapes. Most religions include some aspect of mediation between humans and supernatural or divine powers that control environmental forces. Farmers of any religion may pray for rain. Religious teachings often influence how the adherents of any religion exploit, preserve, or abuse the natural environment, and followers of different religions often create strikingly different cultural landscapes.

Judaism

Judaism has only about 18 million adherents, but it was the first of the great **monotheisms**—religions that preach the existence of one God—to emerge in history. Many scholars suggest that **polytheism**, the worship of many gods, is the most primitive form of worship and that among some peoples polytheism slowly developed into *henotheism*—the worship of one god without denying the existence of others. Henotheism can be found in the older parts of the Bible (for example, Micah 4.5). Henotheism then evolved into monotheism.

Judaism rests on a belief in a pact between God and the Jewish people that they would follow God's law as revealed in the *Pentateuch*—the first five books of the Old Testament. This covenant was granted to Abraham, who was probably a historical person who lived in the city of Ur, in Mesopotamia (today's Iraq), about 2000 B.C. Both Jews and Arabs claim descent from Abraham—Jews through his son Isaac, and Arabs through another son, Ishmael. Today Judaism is divided among a great variety of Orthodox (fundamentalist) and Reform *sects*, or subdivisions of a religion.

Judaism developed in the Near East over many centuries. Jewish political power reached its height under Kings David and Solomon in a state roughly covering today's Israel about 1000 B.C., but the Babylonians destroyed that state in 586 B.C. Later, during the period of the Roman Empire, Jews diffused around the Mediterranean as farmers and traders. In A.D. 70, the Romans destroyed the Jews' Temple in Jerusalem, and the Jews scattered in the *Diaspora*. Jewish communities were established as far away as Yemen, Ethiopia, India, and even China. Jews migrated throughout Western Europe, but during the Middle Ages many European Christian rulers persecuted and expelled them. Ferdinand and Isabella of Spain, for example, expelled the Jews in 1492. Jews returned to Western Europe during the Enlightenment of the seventeenth and eighteenth centuries, but they were required to live in segregated communities called **ghettoes**. Legal emancipation came only in the nineteenth century. Millions of Jews came to the United States from Eastern Europe at that time, and migration from Russia and surrounding lands continues.

During the Diaspora many Jews visited Jerusalem if they could, but during the nineteenth century many Jews came to accept **Zionism**, the belief that the Jews should have a homeland of their own. The Jewish leader Theodor Herzl (1860-1904) originally sought to establish a Jewish state either in Argentina or in Uganda, but because of Judaism's historical roots, the Near East was a preferred site. When Israel came into existence in 1948, many Jews who had survived the Holocaust in Europe and many Mediterranean Jews migrated there (Figure 8-2). Israel has continued to welcome Jewish immigrants from around the world, although several Jewish sects repudiate Zionism.

Israel's Law of Return gives every Jew the right to live there as a citizen, but the definition of a Jew can be debated. Judaism does not generally proselytize today, but this may have been different in the past. There are many small groups with vaguely Jewish customs around the world. Israel has taken in groups from India, and also thousands of *Falashas*, black Jews from East Africa. Tens of thousands more from Ethiopia and Somalia are now claiming right of entry. Many Israelis question the true Jewishness of the Falashas, however, and doubt whether Israel can absorb them. The current studies of genetic markers discussed in Chapter 5 may prove whether Judaism is today more a matter of belief or of genes.

Israel was intended to be a homeland for the Jewish people, yet the balance of Jews and non-Jews in Israel has continuously shifted as a result of Israeli territorial aggrandizement, immigration to Israel, and the birth rates of the various groups in Israel. In 1993, the total territory occupied by Israel—including territory occupied during the 1967 war—contained about 3.7 million Jews, but also 2.5 million Arabs and Palestinians. That high number of non-Jews may have motivated Israel to surrender some territory inhabited by Palestinians in exchange for peace and diplomatic

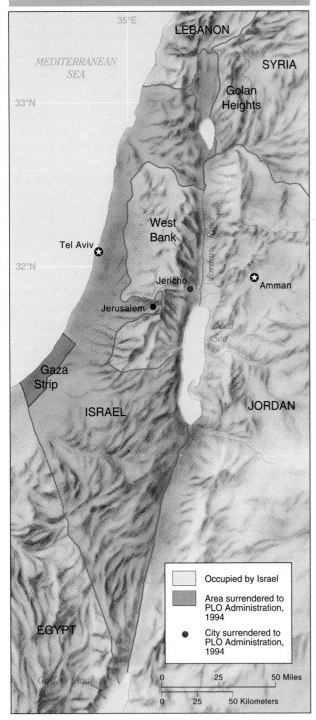

Israel

FIGURE 8-2

Israel was carved out of the Mideast in 1948, and it occupied larger areas during a war in 1967. Jewish settlers have occupied these territories, but their international legal status remains unclear. Israel has surrendered territories to Palestinian control and may return the Golan Heights to Syria. The United States does not recognize Israel's claim to Jerusalem as its capital.

recognition in 1993. Within just the territory occupied by Israel before 1967 plus Jerusalem, which Israel insists it will never surrender, Jews made up a full 81.9 percent of the population of about 5 million. In the 1993 pact, Israel agreed to yield authority in the Gaza Strip (occupied by Palestinians) and the city of Jericho (occupied largely by Palestinians) to the Palestine Liberation Organization. The surrender of territory begun in 1994 will increase the "Jewishness" of Israel's remaining territory. The total fertility rate among the Jews in Israel, however, has been falling steadily to 2.6 in 1992, while that of the Muslim population, by contrast, stood at 4.6. Many of the Jewish immigrants from Russia in the early 1990s were elderly, so their immigration increased the number of Jews relative to non-Jews in Israel, but it depressed the overall Jewish fertility rates. Israel cannot remain distinctly Jewish in population unless it surrenders territory occupied by non-Jews or expels non-Jews.

An estimated 4.3 million Jews lived in the United States in 1990. The majority live in New York State and in California, but Jewish communities are scattered throughout the country.

Christianity

Christianity, the belief that God lived on Earth as Jesus Christ, emerged from Judaism, but part of the widespread appeal of Jesus's teachings rests on his emphasis on God's love for all and on a prophetic urgency that stressed the nearness of the Kingdom of God. Jesus lived his entire short life in Judea, which was then a part of the Roman Empire, and his own journeys spanned little more than 100 miles. After Jesus's death, Saint Paul preached more widely (Figure 8-3), and his letters to early Christian groups make up much of the *New Testament*. Christianity is a proselytizing religion. Many Christians believe that conversion of others is a duty (Matthew 28.19).

The Bible records that the first non-Jew to covert to Christianity was an Ethiopian (Acts 8). Today many scholars believe that man was more probably from Makurra, one of three kingdoms in what is today the Sudan. The Christian church of Egypt, called the Coptic Church, was founded in Alexandria in A.D. 41, and Christianity did diffuse south from Egypt to today's Ethiopia by the fourth century. It has survived there despite having been cut off from the Mediterranean world by the later conversion of Egypt and the Sudan to Islam (see below). Jesus's disciple Thomas may have traveled to India, and while we cannot know for certain whether that happened, we do know that Thomas's followers had established a few Christian churches in India as early as the second century.

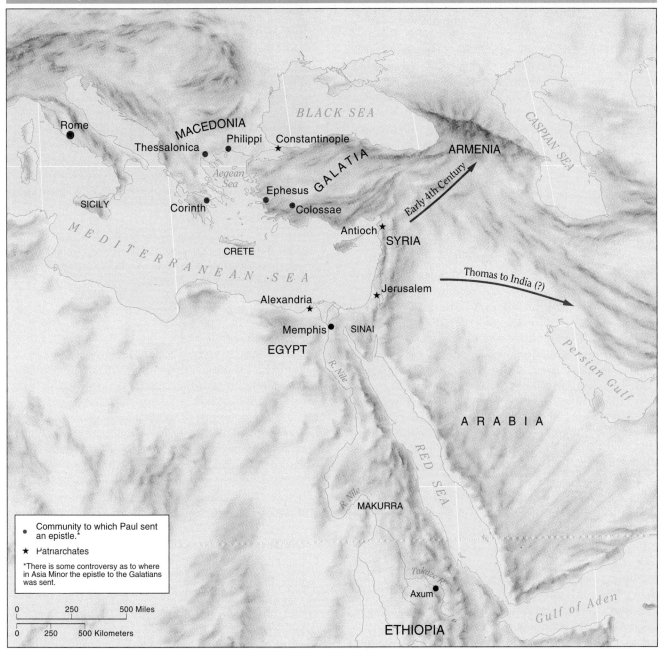

FIGURE 8-3

Saint Paul visited the Christian communities of major cities from Rome to Jerusalem. By A.D. 350 five patriarchs—co-equal leaders of the Church—had been recognized, with their seats at Alexandria, Jerusalem, Constantinople (today's Istanbul), Antioch (today's Antakya), and Rome. To the far South, a Cathedral of Saint Mary was built at Axum in 330, where the Christian emperors of Ethiopia were crowned until 1931.

Christianity also spread to the northeast, and very early in the fourth century Armenia became the first Christian nation. The Roman emperor Constantine (288?–337) converted to Christianity and favored the religion, and by the end of the fourth century Christianity was the Roman Empire's official religion. Christianity survived the Empire's disintegration and spread throughout Europe.

Later, European Christians often cited religious conversion as their purpose for exploring and conquering other lands. Throughout the colonial period, Christian missionaries were the chief agents of the partial Europeanization—of religion, language, social mores, and the acceptance of secular authority—that made many natives more tractable to European rule. The rapidity and the completeness of Christianity's diffusion, however, especially throughout the Western hemisphere, requires analysis. Certainly its message of all-embracing love carried strong appeal in lands in which many of the gods had been portrayed as terrifying and cruel. Furthermore, the Christian heaven after death is democratic and accessible to all classes of people. Some native Americans might have adopted Christianity because they were astonished and demoralized by the Spaniards' invulnerability to the diseases that were wiping out the native population. Neither the Spaniards nor the native Americans understood that those diseases had been brought by the Spaniards themselves. The Spaniards convinced the natives to acquiesce to Spanish secular authority and to seek both physical and spiritual salvation in the Spaniards' church. Similar beliefs prevailed among both the native Americans and the whites in North America.

Christian missionaries often adapted native practices. For example, we do not know the actual date of Jesus's birth, but in the fourth century the Christian churches of Western Europe timed the celebration to replace the celebration of the winter solstice—the shortest day of the year, and the day many Northern hemisphere peoples traditionally celebrated the "rebirth of the sun." Christians still demonstrate willingness to accept indigenous practices in order to bring peoples to Christianity. The final document of the 1992 Latin American Conference of Roman Catholic bishops acknowledged the values in the distinct cultural and religious traditions of Latin America's native Americans and blacks. The bishops pledged to incorporate their symbols, rituals, and cosmologies into Catholic practice whenever "compatible with the clear sense of the faith" and "the general discipline of the church."

During the period of European exploration, Christian missions to Asia were generally less successful than those to the Western hemisphere. In 1542, exactly 50 years after Christians conquered Granada—the last Arab Muslim outpost in Europe—and Columbus first voyaged to the West, Saint Francis Xavier arrived in Goa, in India, to continue the conversion of Asia (Figure 8-4). Saint Francis went on a pilgrimage to the shrine where Saint Thomas is said to have preached and met martyrdom, São Tomé de Meliapur, today in Madras. By 1549, Saint Francis was

FIGURE 8-4

The Basilica of Bom Jesus in Goa, India, is the final resting place of the missionary Saint Francis Xavier. Saint Francis is today the patron saint of Roman Catholic missionaries. (Courtesy of Father Vasco do Rego.)

in Japan, which nearly converted to Christianity in the early seventeenth century but then expelled or persecuted Christians when it closed its doors to the world in the mid-seventeenth century. Christian missionaries had some initial success in China, but in 1723, Christianity was banned there as well.

When people convert to Christianity, and, as will be seen later, to Islam, they break the continuity of their cultural inheritance. This is because both Christians and Muslims believe that conversion is a spiritual rebirth, a repudiation of much of what came before, so it can destroy a people's earlier culture more completely than can political conquest.

Christian Sects and Their Distributions Roman Catholics make up the largest single sect or denomination among Christians today. They believe that the only true Christian church is that headed by the Pope in Vatican City (Figure 8-5). The number of Roman Catholics counted in Table 8-1 is the number of people baptized Roman Catholic; the Church insists that once a person has been baptized as a Roman Catholic, that person is always Roman Catholic, even if the person converts to another denomination or is *excommunicated* (expelled from participation in church ceremonies). The great many Protestant denominations (Lutherans, Anglicans, Baptists, Methodists, and others), were named for their protest against the Church of Rome. Still other Christian sects include the Georgian, Armenian, Ethiopian, and Coptic churches. Orthodox Christians make up another group of churches. They recognize the spiritual leadership of the Patriarch of Constantinople (today's Istanbul), but they are subdivided into national churches, such as the

FOCUS
Christianity Overlying Indigenous Religions

When Christianity spread, its teachings often overlay native practices, and, to emphasize this, churches were often built right on top of those of indigenous or older religions. Throughout Italy, for example, churches dedicated to the Virgin Mary were regularly built on top of temples dedicated to Minerva, the Roman goddess of wisdom. These churches are called Santa Maria Sopra Minerva—Saint Mary on top of Minerva.

This practice was copied throughout Latin America, so many Christian churches stand on top of ruins of pre-Christian native American temples. In Mexico City, for example, the ruins of the greatest Aztec temple are beside and partly under the cathedral. Today, archeologists often join local peoples eager to reclaim and study their pre-Christian heritage in wishing to pull down some of the Christian churches that may be redundant and of little or no historic or aesthetic interest. In Cuzco, Peru, for example, the historic capital of the pre-Columbian Inka empire, an undistinguished seventeenth-century Christian church and

monastery sit on top of the holiest shrine of the Inka culture—the Temple of the Sun. The church's stone foundations and supporting walls were so exquisitely carved by the Inkas that no mortar was needed, and still today a knife blade will not fit between the joints. The Inka masonry supports inferior Spanish construction of roughly cut stones and mortar. Much of the original temple lies hidden beneath and behind walls of the church, and the most important part of the temple, the great

central chamber, lies concealed directly beneath the church sanctuary. The mayor of Cuzco, archeologists from Peru and from the United States, and many local citizens want to disassemble the church and monastery complex. This is partly for historic reasons, but also partly to enhance tourist income in this poor region. Four lone monks, however, inhabit the church and monastery, and their director insists "We will defend the church to the very end."

This perfectly joined wall, part of the Inka Temple of the Sun in Cuzco, Peru, today supports the Church of Santo Domingo. (Courtesy of American Museum of Natural History.)

Serbian, Greek, Bulgarian, Ukrainian, Belarussian, and Russian Orthodox churches. Orthodox Christians in the United States and Canada have never formed national Orthodox Churches but have clung to their traditional affiliations.

Roman Catholicism dominates in Southern Europe, throughout Central and South America (with the exceptions discussed below), Quebec in Canada, and wherever else Mediterranean peoples colonized or converted, as in former French or Portuguese African colonies, the Philippines, and Vietnam (Figure 8-6). Protestant sects dominate in Northern Europe and wherever Northern Europeans have settled or converted: North America, Australia, New Zealand, and the

parts of Africa that were either English colonies (today usually Anglican) or German colonies (Lutheran).

Orthodox Christianity predominates in Eastern Europe because these peoples were converted by missionaries from Constantinople. The people of today's Ukraine converted first, and Kiev became the capital of a Christian civilization that lasted until it was destroyed by the Mongols in the thirteenth century. By then, however, Christianity had diffused from Ukraine northward to the Russians. Meanwhile, in 1054, the Pope and the Patriarch of Constantinople had excommunicated each other. Thus, Roman Catholic Western Europe came to be culturally divided from Orthodox Eastern Europe. Later the Uniate

FIGURE 8-5

This is one of the world's smallest principalities, the Vatican City. It is entirely within the city of Rome, but it is ruled by the Pope according to a treaty signed between the papacy and Italy in 1929. The official language here is Latin.
(Courtesy of Louis Renault/Photo Researchers.)

Church developed in Ukraine; it bridges the East-West gap by using the Cyrillic alphabet and following the Eastern rite, but recognizing some spiritual authority of the Pope.

The Changing Geography of Christianity Today the world map of Christianity is being redrawn. This is primarily the result of two struggles within Christianity. These struggles are on the surface theological, yet they affect many other aspects of life in the societies in which they are being fought. The outcomes of these struggles will change world cultural geography, politics, population geography and demography, and economics. One is the struggle between Evangelical Protestantism and Roman Catholicism, and the other is the challenge of liberation theology within the Church of Rome itself.

Evangelical Protestantism Several theological points differentiate Protestantism from Roman Catholicism, but among the most important is that Roman Catholics believe that Jesus gave Saint Peter unique responsibility for founding the Christian Church, that Saint Peter became the first bishop of Rome, and that the continuing line of bishops of Rome (the Popes) retains this special responsibility. The Vatican flag displays crossed keys, representing the belief that Saint Peter was given the keys to heaven and that there is no salvation outside the Church of Rome. This belief that a church or priest intercedes between God and humankind is called **sacerdotalism**.

Protestants deny sacerdotalism. They emphasize salvation by faith through personal conversion, the authority of Scripture and each individual's responsibility to read the Scriptures, and the importance of preaching as opposed to ritual. **Evangelical Protestants** (the word "evangelical" comes from the Latin for "bringing good news") are Protestant groups that most actively spread faith in individuals' ability to change their lives (with God's help). This message offers to many people a new sense of personal empowerment. Thus, Evangelical Protestantism frequently brings a revolutionary force into traditionally rigid or stratified societies.

Today Evangelical Protestantism is growing and spreading throughout the world. In some places it is replacing other non-Christian religions, but in many places it is replacing Roman Catholicism. It is spreading throughout Asia (especially into South Korea and the Philippines), into Africa, and throughout Latin America. In Latin America, at least one-fifth of the total population was Protestant by 1990. This percentage was even higher in Brazil, which counts from baptism the world's largest Roman Catholic population, but where today full-time Protestant pastors outnumber Roman Catholic priests. Evangelical Protestantism has replaced Roman Catholicism in several countries as the most widely practiced faith.

This conversion brings profound changes to these societies. In most Latin American states, the Church of Rome has traditionally enjoyed special privileges and a role in education, so its teachings have been enacted into law. Laws regulating marriage and divorce, for example, followed Catholic teachings. Today, however, a rising share of elected officials throughout Latin America is Protestant, and the press refers to elections outright as "Holy Wars."

The diffusion of Evangelical Protestantism is at least partly responsible for changing the dynamics of population growth. Evangelical Protestants are often as opposed to abortion as Roman Catholics are, but they are not always so adamantly opposed to other forms of birth control. As Roman Catholicism has lost its political power in Latin America, Quebec, Spain, Italy, and elsewhere, governments have begun sponsoring family planning, distributing birth control devices, and introducing sex education into school curricula. The Philippines, for example, elected its first Protestant president, Fidel Ramos, in 1992, and soon his new health minister, Juan Flavier, began touring the country discussing birth control and safe sex. The government hopes to avoid the spread of AIDS and to curb one of the

The Distribution of the Major Christian Sects in Europe

FIGURE 8-6

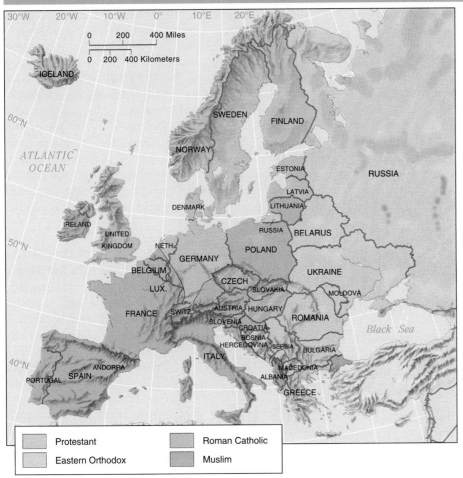

Legend:
- Protestant
- Eastern Orthodox
- Roman Catholic
- Muslim

world's highest rates of population growth. Surveys have shown that 80 percent of Filipinos approve of artificial contraception.

Peru's president Alberto Fujimori was baptized a Roman Catholic, but he was elected on a program dedicated to population planning. He said, "The populace knows how to distinguish between religious and social affairs....A sector of the Church is restricting the freedom of citizens with its medieval opinions and recalcitrant positions." This is a powerful criticism of Church leaders in a country in which the Catholic church claims 90 percent of the population and that has one of the highest annual population growth rates in Latin America.

Scholars have hypothesized other possible results from the spread of Evangelical Protestantism that bear watching through coming years. One is the spread of literacy. This follows from Protestants' emphasis on individual responsibility to read and study Scripture. Literacy has risen in areas where Evangelical Protestantism has spread, but the exact cause-and-effect relationship is difficult to measure. A small analogous example of the effect of religion on literacy, however, might be seen in the United States. Utah ranks low among American states in spending on education, yet it boasts one of the nation's highest literacy levels. Several factors probably combine to explain this paradox, including the fact that few immigrants settle in Utah. Another factor, however, is probably the dominance of the Mormon religion there. Mormonism, officially called the Church of Jesus Christ of Latter-Day Saints, is based on the Bible plus several unique scriptures that Mormons are committed to read and study. Therefore, literacy is virtually universal among Mormons. Protestants, similarly, are committed to reading and study of scripture. If the hypothesis is correct that Evangelical Protestantism raises literacy rates as it diffuses, this effect may carry further ramifications—greater political participation, for example.

Still other possible consequences of the spread of Protestantism that have been hypothesized in the areas of women's rights and even economics are discussed later in this chapter.

Liberation Theology The Roman Catholic church in Latin America is itself divided over the issue of **liberation theology**. This term comes from a 1968 book, *The Theology of Liberation*, by the Peruvian Father Gustavo Gutiérrez. Liberation theology puts the problems of overcoming poverty at the heart of Christian theology. It recommends political activism, using the Church's institutional framework to organize a popular struggle for social justice and equality. As a result, many Latin American bishops proposed national redistributions of land and income. The Church began to provide legal services for the poor, and some priests attacked elite power structures from their pulpits.

The diffusion of liberation theology has provoked political repercussions: spiritual disarray and political polarization in Argentina and Chile, a new sense of self-confidence in Brazil, and conflict between the "popular Church" and the Church hierarchy in Nicaragua. It also triggered militant opposition among conservatives threatened by changes on either religious or political grounds. For example, in El Salvador during the 1980s nuns, priests, and even Archbishop Oscar Romero were assassinated by right-wing elements of the military regime.

Pope John Paul II has rebuked liberation theologians. He has instructed Father Gutiérrez to refrain from writing or teaching, and he has redrawn maps of dioceses and appointed conservative bishops who have closed Church legal services offices, land rights offices, and liberal seminaries. Nevertheless, during a visit to the Dominican Republic in 1992, the Pope declared that "the Church reaffirms the preferential option for the poor." At the same time, however, he criticized Protestants' "fanatic and growing proselytism," and he urged Latin American bishops to defend their flocks against these "rapacious wolves."

The most conservative Latin American governments have favored evangelicals as a counterforce to liberation theology. In Guatemala, for example, government death squads murdered politically active priests during the 1980s, and one Roman Catholic bishop was forced into exile. Meanwhile Protestant sects, with government encouragement, converted more than 30 percent of the population. Jorge S. Elias, elected president in January 1991, was the first Protestant ever elected president of a Latin American country. The Roman Catholic bishops of neighboring Honduras accused the U.S. CIA of subsidizing Protestantism as a way of undermining the Church of Rome and pacifying the people. In Mexico, Protestants denounce Catholics as agents of the Vatican, and Catholics denounce Protestants as "Yankee" agents out to sabotage traditional Mexican culture. The interplay between theology and politics demonstrated in these examples may help us understand important social and political movements through coming decades.

T he interplay between theology and politics demonstrated in these examples may help us understand important social and political movements through coming decades.

The Future of Christianity
Christianity, whether Roman Catholic or Protestant, is still spreading rapidly. Evangelical Protestantism is winning new converts worldwide, and a 1991 papal encyclical called for renewed missionary efforts by Roman Catholics. The observance of Christianity will probably increase throughout Eastern Europe as political liberalization proceeds there, but overall the European core area of Christianity is of diminishing importance in relative numbers (Figure 8-7).

The movement of Christianity's center of gravity to the expanding populations outside Europe may inspire changes in Christian theology and practice. In 1994, Bishop Bonifatius Hauxiku of Namibia demurred, "Our African people have accepted Christ, but this Christ walks too much among them in a European garment." Several examples of challenges to tradition are listed below.

- European Christians spread grape and wheat crops worldwide to provide wine and bread for the mass. Today, however, local churches around the world are adopting local foods for this sacred rite.
- At a world convocation of Roman Catholic bishops in 1990, a Brazilian bishop advocated ordaining married men in his country. The Vatican remains opposed to this, but the bishop of Nassau, the Bahamas, emphasized that on such questions "the Church should not be tied to cultural vestiges typical of the European

The Distribution of the World's Christians

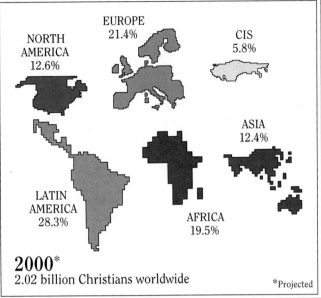

FIGURE 8-7

This cartogram shows that in terms of numbers of communicants, the European core area of Christianity is losing its predominant status. Converts elsewhere have challenged traditional Christian practices that are rooted in European culture. (Adapted with permission from the *Christian Encyclopedia* as printed in *The Economist*, Dec. 24, 1988, p. 62.)

experience." It is revolutionary that such initiatives should come from non-European Christians.

- Efforts are being made in India to incorporate pre-Christian cultural and Hindu concepts into Roman Catholic practice. These efforts, called **interculturation**, have provoked tremendous controversy. Elsewhere **syncretic religions** that combine Christianity with traditional practices are winning converts. Examples include Candomblé and Umbanda in Brazil, Cao-Dai in Vietnam, and Chondo-Kyo in Korea.

- At the 1991 meeting of the World Council of Christian Churches, a Korean priest, accompanied by gongs, drums, and clap sticks, summoned "the spirit of the Amazon rainforest" and "the spirits Earth, air, and water raped, tortured and exploited by human greed." Representatives of hundreds of millions of Christians were offended, and they threatened to abandon the World Council. A performance such as this underscores the problem of how to express Christian belief in terms meaningful to a great diversity of cultures without abandoning essential Christian distinctiveness.

Islam

Through the centuries after Abraham and Ishmael, the Arabs fell away from monotheism to polytheism, but they were brought back to monotheism by Muhammad (570?–632) He founded the religion called *Islam*, which means "submission [to God's will]." "One who submits" is a *Muslim*. The Arabic word for the one God is "al-elah," or Allah. This word is cognate to the Hebrew *eloh*, "god," and *elohim*, "the God." Muslims believe that Muhammad was the last of God's prophets, who also include Adam, Noah, Abraham, Moses, and Jesus. Muslims accept five essential duties, called the Five Pillars, which are belief in the one God, five daily prayers, generous giving of alms, recognition of one month (called *Ramadan*) as a fast, and, if possible, a pilgrimage (**hajj**) to Mecca at least once in one's lifetime.

Christians believe that their religion builds on Judaism, adding the New Testament to the Old. Muhammad envisioned his teachings as a continued evolution of monotheism. His holy book, the Koran, instructs Muslims to honor Christians and Jews as "people of the Book," that is, the Bible. Muhammad expected Jews and Christians to embrace Islam. He directed Muslims to face Jerusalem in prayer, as worshippers in Christian churches traditionally do, but he later changed the direction to face toward Mecca (Figures 8-8 and 8-9).

The Diffusion of Islam Muhammad was forced to flee from his native city of Mecca in 622, the year from which Muslims date their calendar, but by his death he had converted and united most Arabs. As Islam diffused, diverse peoples throughout the Near East and across North Africa adopted the language of the Koran, and all came to be known as Arabs. To the East, Persia (today's Iran) had its own ancient culture, and although the Persians converted to Islam, they retained their own Indo-European language, Farsi. Therefore, today's Iranians are not Arabs, nor are any of the peoples to the East of Iran, even if they are Muslims.

Within slightly more than a century after Muhammad's death, Islam had greatly expanded both East and West (Figure 8-10). To the East, Arab Muslim armies conquered and converted the peoples across today's Iran and Afghanistan. Many Central Asian Turks, who then paid tribute to the Chinese, welcomed the Muslims, and in 751 an Arab army battled a Chinese army in the Talas River Valley, in today's Kyrgyzstan. This was the westernmost extension of Chinese power. Islam later diffused into China itself through trade links, so that by the thirteenth century Marco Polo would describe peaceful groups of Arab and Chinese Muslims living in China. From about 1150 to 1550, new groups of Muslim warriors from Central Asia penetrated and conquered India, and Islam later expanded down into South and Southeast Asia.

To the West, Muslims swept across North Africa, crossed the Straits of Gibraltar ("the rock of Tariq," named for a Muslim general) marched northward through Iberia and reached central France before they were stopped at Tours in 732. Within this vast realm from China to France, the conquering Arabs eventually lost their dominance to the non-Arab majority, including the Persians and the several Turkish peoples. For hundreds of years the Muslim culture realm greatly exceeded the Christian culture realm in extent, power, and riches. Learned travelers crossed its length and breadth, and their descriptive writings constitute some of the greatest works of historical geography.

FIGURE 8-8

This shrouded building is the Kaaba in Mecca, where, Muslims believe, the biblical patriarch Abraham and his son Ishmael built the first sanctuary dedicated to the one God. Through millenia the temple was given over to the worship of idols, but Muhammad threw out the idols, rededicated the Kaaba, and had its walls covered with paintings of Abraham, Mary, and Jesus. That building has since been replaced, but this is probably the longest continuously revered spot on Earth. To the left of the Kaaba a small stone canopy covers what believers accept as the well of Zamzam (Genesis 21.19), where Abraham's concubine Hagar and their son Ishmael quenched their thirst. Ishmael is buried just in front of the Kaaba.
(Courtesy of Aramco World Magazine.)

FIGURE 8-9

Many sites in Jerusalem are sacred to Jews, Christians, and Muslims. All believe that God prevented Abraham from sacrificing his son Isaac on this rock (Mount Moriah), thus ending human sacrifice. King David later built the Jews' Temple here. Muslims believe that Muhammad ascended to heaven from this spot. The beautiful building standing here now, the Dome of the Rock (begun in 643), is the oldest existing monument of Muslim architecture. Christians believe that Jesus ascended to heaven from the Mount of Olives in the background. Just in front of and below this picture is the only wall remaining of the Jewish Temple, which was destroyed by the Romans in A.D. 70, triggering the Jewish diaspora.
(Courtesy of the Israel Ministery of Tourism.)

The Diffusion of Islam

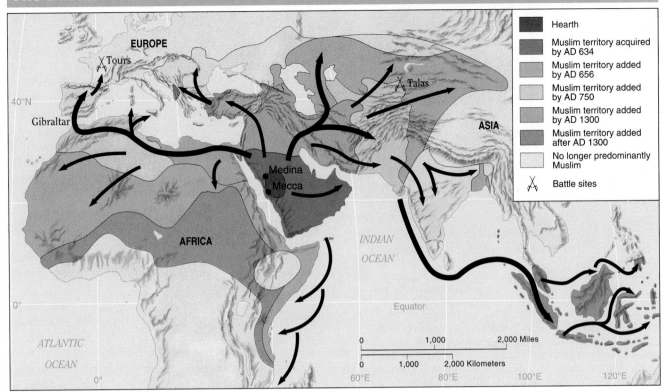

Legend:
- Hearth
- Muslim territory acquired by AD 634
- Muslim territory added by AD 656
- Muslim territory added by AD 750
- Muslim territory added by AD 1300
- Muslim territory added after AD 1300
- No longer predominantly Muslim
- Battle sites

FIGURE 8-10

The first nine centuries of Islam were a period of almost continual expansion, and by 1800 Arab and Indian traders had spread the faith through much of island Southeast Asia.

Muslims renounce their earlier cultures, just as converts to Christianity do. Everything before Islam is, in Arabic, *jahiliya*, "from the age of ignorance." This leaves little room in these peoples' historical consciousness for their pre-Islamic past. They may view their own ancient cultural landscapes without interest and even discourage tourists from viewing pre-Islamic ruins. For example, despite Persia's brilliant antique history, contemporary Iranians believe that their glory began with the coming of Islam. The twentieth-century state of Pakistan also contains ruins of civilizations thousands of years old, yet contemporary Pakistanis disdain them. Similarly, the study of the history and art of pharaonic Egypt is the result of Christian European historical interest. Today foreign tourists visiting pharaonic ruins provide Egypt's primary source of foreign exchange, yet Muslim fundamentalists regard this tourism as sacreligious. Some Egyptian Muslims, however, take nationalist pride in Egypt's antique past. Thus, nationalism can compete with Islam for individuals' primary loyalty.

Islam filtered south across the Sahara Desert from North Africa to reach the black peoples of the Sahel and savanna, and there it competes still today with Christianity, which entered sub-Saharan Africa from the Europeans' coastal incursions. From Senegal to the Congo the coastal areas are Christian, and the inland areas are Muslim (Figure 8-11). Many contemporary black African writers and artists, such as the Senegalese filmmaker Ousmane Sembène, have in their works noted how Africa suffers from being a cultural battleground between two religions, neither of which is native to the continent.

Within Europe, the population of Albania, long ruled by Turks, is predominantly Muslim. Bosnia-Herzegovina had a substantial Muslim population when Yugoslavia disintegrated in 1991 and war broke out among the successor states. The war was fueled partly by Serbian territorial claims and aggression against the Muslim inhabitants of Bosnia-Herzegovina. Serbians spoke of the expulsion of Muslims as "ethnic cleansing." The leaders of two powerful Muslim countries—Pakistan's Prime Minister Benazir Bhutto and Turkey's Prime Minister Tansu Ciller—visited the beseiged Bosnians together and offered Muslim support, but diplomacy seemed to paralyze action. Serbia itself has a Muslim Albanian minority.

FIGURE 8-11

The Basilica of Our Lady of Peace in Yamoussoukro, the Ivory Coast, is the world's largest Christian church. The country's first president, Félix Houphouët-Boigny (1905?-1993), converted to Roman Catholicism as a teenager, and he built the basilica as a pilgrimage center for Africa's Catholics and as a bulwark against Islam and native religions in his country, which in 1993 was about 15 percent Christian, 20 percent Muslim, and 65 percent animist. The air-conditioned Basilica can hold 18,000 people. (Courtesy of Gideon Mendel/Magnum.)

Today Muslims are distributed from Morocco to Indonesia, north to the frontiers of Siberia and south to Zanzibar, with outposts throughout the world. Even Rome boasts a splendid new mosque. The leading Muslim states, in numbers of adherents, are Indonesia, Bangladesh, India, and Pakistan. Indonesia's actual status is dubious. More than 90 percent of Indonesians put "Muslim" on their identity card, thus making up a full 20 percent of the world's Muslims, but this may reflect the fact that it is illegal to have no religion. This law is a relic of the government's intense anti-communism.

Many Americans tend to equate Islam with Arabs, but Arabs constitute only a fraction of Muslims. There are more Muslims in Indonesia and Bangladesh than in all the Arab countries combined. Americans' association of Islam with Arabs is a political preconception, caused by the distinctive role of Islam in Arab countries and by U.S. news media focus on Arab lands as sources of oil and as antagonists of Israel. The next time you think of a Muslim, think of a 17-year-old girl living in the equatorial forest of Indonesia, an 11-year-old boy living on the tropical mudflat of Bangladesh, or perhaps the family down the block, rather than an Arab on a camel in a desert.

Islamic Sects and Their Distributions Theoretically, there is no distinction between temporal (political) and spiritual rule in Islam, and the religion's two principal sects date back to a struggle over rule of the Islamic world that occurred shortly after Muhammad's death (Figure 8-12). **Sunni Muslims** accept the tradition (*Sunna*) of Muhammad as authoritative and approve the historic order of Muhammad's first four successors, or *caliphs*. About 85 percent of Muslims worldwide today are Sunni. The ruling family of Saudi Arabia is of the Sunni Wahabi sect, a Puritanical movement that arose in the early eighteenth century and is the only modern separatist Sunni sect. Most of the other 15 percent of Muslims are **Shia Muslims**, or **Shiites**. They believe that Muhammad's son-in-law Ali should have been the first caliph, and they commemorate the martyrdom of Muhammad's grandsons in the battle of Karbala. Through the centuries, differences in ceremony and in law have further differentiated the Sunnis from the Shiites.

Shiites form the majority in Iran, Azerbaijan, and in southern Iraq, and important minorities in Kuwait, Lebanon, Syria, Yemen, Saudi Arabia, and Pakistan. Animosity between Sunnis and Shiites can be fierce. Countries that have Shiite majorities often refuse to cooperate in international affairs with countries in which the majority is Sunni. In countries that contain significant shares of both sects, the sects quarrel over the interpretations of Muslim teachings that should be enacted into national law. Several countries, including Lebanon and Pakistan, have suffered civil disturbances between the two groups. Even the hajj to Mecca has regularly been disrupted by violent clashes between Sunnis and Shiites.

During the war between United Nations troops and Iraq in 1991, the Shiites of southern Iraq rose up against the Sunni government in Baghdad. United Nations troops drove the Iraqi army out of Kuwait, but they did not advance to aid the rebelling Shiites, as the Shiites had expected them to do. When the government in Baghdad felt certain that the United Nations troops would not advance, it counterattacked against the Shiites. The United Nations then interceded in 1992 and established a zone of Shiite safety from Iraqi air force interference south of the 32nd parallel. Nevertheless, Iraqi tanks rolled across rebelling Shiite areas emblazoned, in Arabic, "No more Shiites after today." The city of Najaf, a focus of 1,000 years of Shiite cultural history, mosques, libraries, and seminaries, was virtually destroyed. Since then, the Iraqi government has diverted waters of the Tigris and Euphrates Rivers to drain Shiite-populated marsh regions, driving the Shiites into Iran. This is an ecological disaster as well as a human disaster.

The Distribution of Sunni and Shiite Muslims

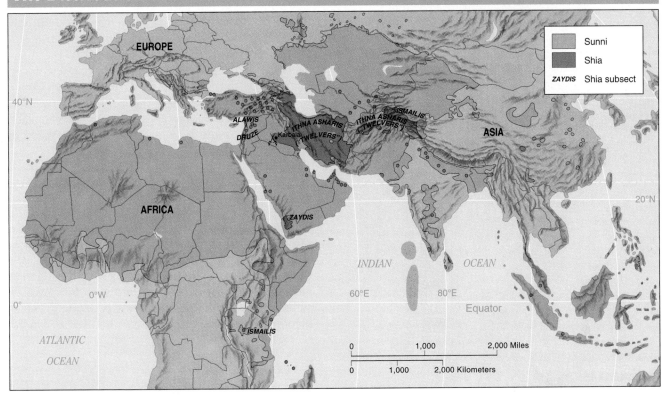

FIGURE 8-12

Karbala in today's southern Iraq witnessed a great battle among family members and followers of Muhammad in 680. That battle split Muslims into the Sunni and Shia sects.

Hinduism and Sikhism

Hinduism is the most ancient religious tradition in Asia. The oldest Hindu sacred texts (the *Vedas*) date to about 1800 B.C., but the religion originated somewhere in Central Asia long before that. It entered the Indian subcontinent with the coming of Central Asian peoples about the time of the writing of the Vedas (Figure 8-13). Hindus believe in one Supreme Consciousness, Brahman, whose aspects are realized in three deities: Brahma, the creator; Vishnu, the preserver; and Siva, the destroyer. These are coequal, and their functions are interchangeable. All other Hindu "gods," saints, or spirits are emanations of Brahman.

Hinduism classifies people in a hierarchy of **castes**. The four main castes are (1) the Brahman, or priestly caste; (2) the Kshatriya, or warrior caste; (3) the Vaisya, or tradesman and farmer caste; and (4) the Sudra, or servant and laborer caste. Each of these is split into hundreds of subcastes. Many are defined by occupation, although it is difficult to fit many new occupations into these old categories. Many white-collar professionals, for example, would seem to fall

into the Brahman class, because they require special education and training, but others contend that they are just Vaisyas, because they work in order to make money. Nevertheless, people are expected to socialize, marry, and stay in the caste into which they were born. A group called *untouchables* is considered so low that their status is below the formal caste structure. Caste discrimination was abolished in the Indian Constitution of 1950, but it still structures Indian life.

In all of India, only about 20 percent of the total population belongs to the three top castes. Almost 50 percent are Sudras, and another 20 percent untouchables. The rest of the population is made up of members of primitive peoples called "tribals," and non-Hindus outside the caste system. The geographic distribution of the castes reinforces the theory that Hinduism was brought to India by a people who invaded from the north. In southern India only 4 percent of the people are Brahmans, and there are virtually no Kshatriyas or Vaisyas.

The members of many subcastes lobby the central government to be declared "backward," because they would thereby become eligible for special job quotas, development grants, and other privileges.

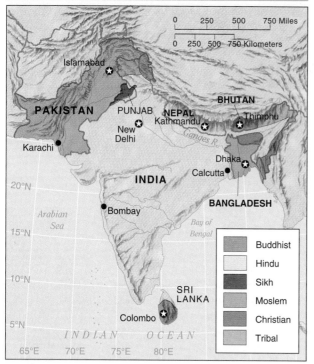

Religions in South Asia

FIGURE 8-13

Hinduism is today restricted almost exclusively to India, Nepal, and to places where Indians have migrated, such as a few countries around the Caribbean Sea and Fiji in the Pacific Ocean. Sikhs form a majority in the Indian state of Punjab, but smaller Sikh communities can be found throughout India.

India has steadily increased the number of groups labeled backward. Critics argue that this form of affirmative action perpetuates the caste system, and when job guarantees are not accompanied by intensive educational efforts, they close the most desirable positions in the country to the people best able to fill them.

Hindus believe in *reincarnation*, that is, individual rebirth after death. The caste into which you are born is not haphazard, but results from your behavior in an earlier life. Thus, each individual has earned different rights and duties and will enjoy due happiness or suffer unhappiness in this life. Some people believe that this Hindu teaching, called *karma*, discourages ambition. Only by docilely keeping to your place in this life can you hope to enjoy a better position in your next life. The goal of Hindus is liberation from the cycle of death and rebirth.

Sikhism is an offshoot of Hinduism based on the teachings of Guru (teacher) Nanak (c. 1469-1539). Guru Nanak tried to reconcile Hinduism and Islam. He went on the hajj, and he taught monotheism and the realization of God through religious exercises and

meditation. Nanak opposed the maintenance of a priesthood and the caste system. Guru Nanak was followed by Guru Arjun, who compiled the Sikh holy book, *Abu Granth Sahib*, and built in Amritsar a Golden Temple to house it (Figure 8-14). For many years the Sikhs had their own state in northern India, but in 1708 the last Guru died, and the Sikhs were eventually conquered by the British. Many Sikhs dream of the restoration of an independent Sikh state, to be called *Khalistan*, which is Punjabi for "Land of the Pure."

Buddhism

Siddhartha Gautama (c. 563-483 B.C.) was a Hindu prince in today's Nepal who, through meditation, achieved the status and title of Buddha, or "Enlightened One." Buddha was not a god, but a man who taught the Four Noble Truths: (1) life involves suffering; (2) the cause of suffering is desire; (3) elimination of desire ends suffering; (4) desire can be

FIGURE 8-14

The Sikhs' 400 year-old Golden Temple in Amritsar, India, stands in the middle of a sacred lake (in Sanskrit *amrita saras*, "pool of immortality"). The building was occupied by armed Sikh extremists in 1983, and the following year Indian troops stormed it and drove out the occupants in a bloody encounter. Sikh fanatics retaliated by assassinating India's Prime Minister Indira Gandhi.
(Courtesy of Jehangir Gazdar/Woodfin Camp.)

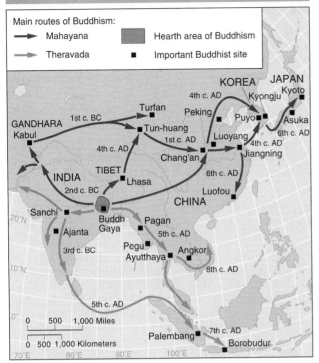

The Early Diffusion of Buddhism

Main routes of Buddhism:
- ➡ Mahayana — ▨ Hearth area of Buddhism
- ➡ Theravada — ■ Important Buddhist site

FIGURE 8-15

Buddhism was localized near its hearth for hundreds of years, but it spread widely through the support of the third century Indian emperor Asoka. It has largely been abandoned in India.

eliminated by right thinking and behavior. This cessation of suffering is called *nirvana*, which means the "blowing out" of greed, hatred, and delusion.

As Buddha's teachings diffused, sects and schools arose. Chapter 7 noted that the oldest surviving printed books are copies of the Buddhist texts, the *Pali Canon*. The Theravada school of Buddhism ("doctrine of the elders") diffused to the South (Figure 8-15). Today it is the state religion in Thailand and Sri Lanka. This school centers around the ideal of a monk striving for his own deliverance. Mahayana Buddhism (the "great vehicle," because it carries more people to nirvana), in contrast, which diffused to the north, idealizes the concept of the *bodhisattva*, someone who helps others achieve enlightenment. Mahayana thus recognizes numerous minor divinities. Mahayana monasteries arose in what is now Afghanistan, and the religion diffused out from that focus of Central Asian trade routes (look back at Figure 3-9). Buddhism had followers throughout India for a thousand years, but it was eventually overwhelmed by Hinduism and Islam. In Bhutan, Tibet, and Mongolia, Buddhism evolved a special form called *Lamaism*, which is known for its elaborate ritu-

als and complex priestly hierarchy (Figure 8-16). Chinese Buddhists produced a new theory of spontaneous enlightenment, called *Ch'an*, which diffused into Japan as *Zen*.

Today Buddhism has several hundred million followers, but its adherents are hard to count because it is not an exclusive system of belief (Figure 8-17). Buddhist philosophy has won considerable influence in the modern Western world. Much popular New Age philosophy derives from Buddhism.

Other Eastern Religions

Confucianism is a philosophical system based on the teachings of K'ung Fu-tzu (c. 551-479 B.C.). He taught a system of "right living" preserved in a collection of sayings, *The Analects*. Confucianism was the state religion of China from 136 B.C. until A.D. 1911. Confucianists hold that people may attain heavenly harmony by cultivating knowledge, patience, sincerity, obedience, and

FIGURE 8-16

The Dalai Lama, the Buddhist spiritual leader of the people of Tibet, was forced into exile by the government of China. He has won worldwide acclaim—including the Nobel Peace Prize in 1989—for his efforts to promote religious freedom. Hollywood stars, including Richard Gere, have become devoted to him.
(Courtesy of Rick Maiman/Sygma.)

FIGURE 8-17

This 112-foot high statue was made in China and erected and dedicated near the Lo Pin monastery in the mountains of Lantau Island, Hong Kong, in December 1993. Images of Buddha try to capture the serenity associated with this religion. The lotus on which he sits symbolizes the Dharma (truth) in his teachings.
(Courtesy of C.K. Tsui/Pro-file Photo Library.)

the fulfillment of obligations between parents and children, subject and ruler. These moral precepts permeate life in many Eastern societies, and today the word *Confucian* is often popularly used as a synonym for these qualities. Confucianism influenced Western philosophy and political theory at the time of the Enlightenment. Many Westerners saw it as the realization of Plato's ideal of a state ruled by philosophers.

Taoism is derived from the book *Tao-te Ching* attributed to a legendary Chinese philosopher Lao-tze. It advocates a contemplative life in accord with nature. It is uncertain, however, whether Lao-tze was an historical figure. The name simply means "old philosopher," and modern scholars date the *Tao-te Ching* from the fourth century, B.C. Nevertheless, the image of a kind elderly man wandering in communion with nature has long exerted powerful appeal (Figure 8-18).

Shinto is the ancient native religion of Japan. It includes a set of rituals and customs involving reverence for ancestors, celebration of festivals, and pilgrimage to shrines, but there is no dogmatic system or formulated code of morals (Figure 8-19). It is specifically nationalistic, and it traditionally recognized the emperor as divine. The Emperor Hirohito renounced this divinity in 1946, following Japan's defeat in World War II.

Animism and Shamanism

Animism is a belief in the ubiquity of spirits or spiritual forces (Figure 8-20). Animistic religions may be

basically monotheistic, or they may recognize hierarchies of divinities who assist God and personify natural forces. No single set of animist beliefs has diffused widely, but millions of Africans believe in animism, and animists are also found among native Americans in the Amazon basin and among native peoples in the interior mountains of Borneo and New Guinea.

Animism is frequently accompanied by **shamanism**. A *shaman* is a medium who characteristically goes into autohypnotic trances, during which he or she is thought to be in communion with the spirit world. Shamanism is noted among the peoples

FIGURE 8-18

Lao-tze (on the left) and Confucius protect the infant Sakayumi, the future Buddha, in this fourteenth-century Chinese painting. Lao-tze's appearance reflects his legendary life spent wandering and communing with nature, in contrast to the courtly scholar Confucius. Images such as this, emphasizing the compatibility of Confucianism, Taoism, and Buddhism, became popular as Buddhism spread across East Asia.
(Courtesy of the British Museum.)

FIGURE 8-19

These beautiful buildings make up part of the Toshogu Shinto shrine in Nikko, Japan, about 75 miles (120 km) north of Tokyo. Millions of Japanese visit this city of shrines and memorials every year to learn about their history as well as to perform rituals.
(Courtesy of Japanese National Tourist Organization.)

FIGURE 8-20

This animist altar stands in a Dogon village in the West African country of Mali. It is covered with a patina of millet gruel laid on as an offering to the spirits of the harvest. (Photo by Eliot Elisofon, National Museum of African Art, Eliot Elisofon Archives, Smithsonian Institution.)

of Siberia, some native American peoples, and in Southeast Asia and on East Indian islands. Almost everywhere both animism and shamanism are yielding to Muslim and Christian proselytizing. As noted in Chapter 3, however (see Figure 3-14), the world's scientists and pharmaceutical companies are racing to tap shamans' ethnobotanical knowledge before that knowledge is lost.

RELIGIOUS INFLUENCES ON GOVERNMENT

▼▼▼▼▼▼▼▼▼▼▼▼▼▼▼▼▼▼▼▼▼▼▼▼

Religion so deeply affects what people assume to be natural or desirable human behavior that religious teachings can be translated into laws unconsciously. To determine the degree of influence that a religion exercises on a legal system, we might explore whether regions where different religions prevail demonstrate different legal systems. For example, some studies of traditional African justice have suggested that traditional law in some African cultures focuses on restoring harmony to the community. This can be contrasted with the Old Testament tradition, which includes a notion of exacting retribution.

Almost everywhere, organized religion is an institutional, economic, and political force. Any organized religion has its own bureaucracy and sources of income, and its religious services and preaching offer a channel of communication that often reaches more of the population than government information does. Relations between the government and the leading religious organizations constitute a major political issue in most countries (Figure 8-21). Religious leaders may lend legitimacy to rulers, or, conversely, they may provide a rallying point for political opposition. For example, in the Philippines in 1986 the Archbishop of Manila refused to bless the conspicuously fraudulent re-election of President Ferdinand Marcos. That act triggered nationwide protests against the Marcos regime, the army refused to obey the president's command to put down the insurrection, and the Marcos family fled the country. The Roman Catholic Archbishop of Benin played a similar key role in bringing down a military government there in 1991, and he chaired the following constitutional convention.

Even in the United States, politicians court religious leaders and organizations, and the religious leaders themselves make strong political gestures. For example, in 1993 the annual convention of the Southern Baptist church, the largest denomination in the United States, invited both former President

Reagan (a Presbyterian) and former President Bush (an Episcopalian) to speak, but it did not invite either President Clinton or Vice President Gore, both of whom are Southern Baptists. Church leaders were angered by the administration's defense of abortion rights and equal rights for homosexuals.

In some countries a community of citizens might be alienated from their national government if coreligionists in an adjacent country enjoy more complete religious freedom. For example, the Islamic vigor of Iran provided religious inspiration to the Muslim populations of the Central Asian Republics when they were under the rule of the Soviet Union.

Official National Religions and Churches

Most countries ostensibly guarantee freedom of religion and observe a form of secularism. Many governments nevertheless favor one religion over others implicitly or explicitly. Therefore, the world political map partly fixes or stabilizes the map of world religions, just as it fixes the map of world languages.

Today's world contains few **theocracies**, a form of government where a church rules directly. The Vatican is one. Morocco may be a modified theocracy, because the king's legitimacy derives partly from his descent from Muhammad (Figure 8-22). Many theocracies have existed in history: Tibet, for example, and Massachusetts Bay Colony in British North America. Utah was once a Mormon theocracy, and although it is today increasingly cosmopolitan, no Utah politician can ignore Mormon church leaders (Figure 8-23).

Israel has grown more observant of Jewish law since its founding, and more religious strictures have

been enacted into national law. In 1991, for example, laws were passed banning the production or sale of pork. Laws prohibit activities on the Sabbath.

Christianity from its beginning distinguished between religion and politics. Jesus said that there are "things that are Caesar's"—legitimately due to Caesar, as ruler of an autonomous political and social

FIGURE 8-23

This is the citadel of Mormonism in Salt Lake City. In 1847, this site was surrounded by hundreds of miles of wilderness, but Mormon leader Brigham Young recognized it as an oasis which Mormon industriousness could make bloom and where Mormons might escape the persecution they had suffered in the East. The building in the form of an overturned hull of a ship (it was in fact built that way) is the Tabernacle, and the neo-Gothic building behind it is the Temple. Today Salt Lake City is a metropolitan area of over 1 million people, and the Utah state capitol can be seen in the right background.

Mormonism is one of the fastest growing religions in the world. In 1950, there were 1 million Mormons, nearly half of whom lived in Utah. Today the religion counts almost 9 million members around the world, fewer than 15 percent of whom live in Utah.
(Courtesy of the Church of Jesus Christ of Latter-day Saints. Used with permission.)

order (Luke 20.25). This distinction allowed Europe to develop secular government, secular knowledge, and a secular culture. Nevertheless, several countries are today explicitly Christian or at least support various Christian sects with public funds. The countries with what are called *established* Christian churches include the United Kingdom, Germany, Ireland, Sweden, Norway, Denmark, Iceland, Finland, Argentina, and Peru. The Norwegian Constitution, for example, specifies Lutheranism as the state religion; pastors are government officials.

Many countries are officially Islamic, including Mauritania, Afghanistan, Libya, Saudi Arabia, Yemen, Oman, Qatar, Bahrain, Iraq, Iran, Comoros, Maldives, Pakistan, Bangladesh, and Malaysia. Islamic states enact Islamic teachings into law, called *Sharia*, and they establish Sharia courts to rule whether secular law conforms to Islamic teaching. If the Sharia court rules against a secular law, that law is usually repealed. In Islamic lands, laws against blasphemy (irreverent or impious action) may be strictly enforced. Iran's religious leaders have even tried to extend the reach of punishment for blasphemy into foreign countries. They have offered a reward for anyone who would murder the Pakistani writer Salman Rushdie for his 1989 book *Satanic Verses*, which, they insist, blasphemes Muhammad. Rushdie has been burned in effigy throughout the Muslim culture realm.

When Muslims migrate to Western countries, they may join with fundamentalist Christians as the most conservative element of the Western nation's political complexion. Muslims strongly favor "family values" issues, including opposition to abortion, pornography, and sexual promiscuity. These "family values," however, do not include women's rights, as we shall see later in this chapter.

Religion and National Identity

Religion and nationalism are two powerful cultural forces that reinforce one another in some countries, but compete in others. We have already noted how some Egyptians' sentiments may be divided between pride in Egypt's antique past and disdain for that pre-Islamic period. Many people today, in a great variety of contexts all around the world, are testing their loyalties to their religions against their loyalties to secular governments. The results of these tests will carry consequences throughout the world's political, social, and cultural organization.

Disestablishment in Some Christian Countries

Several Christian countries are disestablishing their national churches. In the 1960s, the Canadian province of Quebec secularized welfare, health, education, and trade union organization, all of which had been controlled by the Roman Catholic church. Italy disestablished the Roman Catholic church in 1984, and Spain followed in 1988. Colombia's 1991 Constitution terminated its 1887 Concordat (understanding) with the Church of Rome by which church institutions had been tax exempt, religious instruction had been compulsory in public schools, and priests had been subject only to ecclesiastical courts, even for criminal acts. Furthermore, only marriages sanctioned by the Church had been legally recognized.

The Geography of Religion **285**

England broke off from the Church of Rome in 1534 and established its own church, the Church of England (the Anglican Church), but many people in England now want to disestablish that. The Church of England still gets free time on the national media, study of the Christian religion is compulsory in school, and 24 Anglican bishops sit in the House of Lords, one of two houses of Parliament.

The states of western Germany collect "church taxes" for both Roman Catholic and Protestant churches. Christian taxpayers must declare their affiliation and pay the tax, or else they must formally leave the church into which they were baptized.

One popular hypothesis holds that as many of today's Christian societies have gotten richer, the role of religion has dwindled, and that other societies currently practicing other religions will follow this example. In other words, religiosity is in itself a sign of backwardness, and the secularized West presents an inevitable model. Western intellectuals have defended this hypothesis since the eighteenth-century Enlightenment. Thomas Jefferson and later the French observer Alexis DeToqueville both assumed that all Americans would eventually become Unitarians or *pantheists*, worshippers of the whole universe *as* God. Many Americans assume (often unconsciously) that all societies evolve toward wealth, secularization, and democracy. But perhaps these are, after all, choices, or happenstance, not inevitable occurrences.

Even in the rich, historically Christian countries, the belief and practice of Christianity remains vital. In 1991 and 1992, over 1,000 people were polled in each of 14 countries. Table 8-2 reveals that a majority in all but the former East Germany said they believed in God, and that the majorities in 7 of the 14 believe in heaven. Substantial percentages attend church regularly. Levels of belief and practice vary, but overall the level of religious devotion in these traditionally Christian countries remains high.

National Churches are Reviving in Other Christian Countries

Many countries in East Europe have revived their national churches as a means of defining and emphasizing their national identities since the fall of Communist governments. The Uniate Church of Ukraine, for example, survived persecution by the Communists and revived in the late 1980s.

Eastern Orthodox national churches traditionally cooperate with whatever government is in power. Therefore, the Communist governments in Serbia, Bulgaria, Belarus, Ukraine, and Russia hampered the Orthodox churches but did not ban them. All priests, for example, had to be screened by the state. After the fall of communism, some people felt that these churches were tainted by having cooperated with the

TABLE 8-2
Religious Beliefs and Practices in Fourteen Countries

	Percentage Theist	Percentage believe in heaven	Percentage regularly attend services[1]
W. Germany	67	43	15
E. Germany	26	19	4
Britain	70	54	17
USA	94	86	44
Netherlands	50	40	21
Hungary	64	27	19
Italy	85	58	49
Ireland	92	87	76
N. Ireland	94	90	57
Poland	88	64	67
Norway	59	47	11
Israel	71	43	—
Slovenia	61	32	—
N. Zealand	70	59	20

[1]Two or three times a month or more.

Source: The International Social Survey Program, p. 674. Reprinted by permission.

Communist governments. Most of the new governments have nevertheless honored the Orthodox churches, although this may be partly an effort to borrow respectability and legitimacy from them. Even before the collapse of the Soviet Union, for example, Boris Yeltsin invited the Patriarch of Moscow to his swearing-in as President of Russia. Eventually Yeltsin was baptized and started attending church services (Figure 8-24). In the early 1990s, thousands of Orthodox churches and monasteries opened or reopened in Russia and other Orthodox countries.

Missionaries of every denomination have tried to win Eastern European converts, but the leaders of the indigenous Orthodox churches are outraged at the presumption that their countries are religious vac-

FIGURE 8-24

In October 1993, Russian president Boris Yeltsin attended a religious ceremony honoring the memory of the eleventh-century Prince Yaroslavl the Wise. Yaroslavl had given the first written legal code to the peoples of today's Russia, Ukraine, and Belarus. Politicians often try to associate themselves with historic national and religious values, but, in this case, the fact that Yaroslavl ruled over two peoples other than the Russians suggests, perhaps, a continuing sense of Russian hegemony.
(Courtesy of Reuters.)

uums. In 1993, the Russian Parliament required foreign missionaries to obtain licenses to preach, broadcast, or publish in Russia, but these rules have been assailed as a retreat from the religious freedom guaranteed in 1990.

Several Communist countries had been traditionally Roman Catholic, including Poland, Lithuania, the Czech and Slovak lands, Hungary, Slovenia, and Croatia. In these, church leaders often spoke out againt Communist governments when they defied Roman Catholic teachings. For example, Communist governments guaranteed women's rights to abortion. Pope John Paul II, who is Polish, played a role in negotiating the transition from communism to democracy in several Eastern European countries, and today in these countries freedom for the Catholic Church has become a symbol of national sovereignty. This freedom, however, revives the Church's political

agenda, which includes, at least, compulsory religious education, a ban on abortions, and financial support for the Church and its activities. The new democratic political leaders may not support these causes any more than the former Communist rulers did. Therefore, in each of these countries the Church may conflict with new democracies. In Poland, for example, the Roman Catholic Church has exercised new influence and won a ban on abortions, a requirement for religious teaching and compulsory prayers in all schools, and a requirement that television and radio broadcasts "reflect Christian values." This law banned sex education and ads for condoms from television, and epidemiologists have suggested that it may partly explain the rapid rise in AIDS in Poland. Meanwhile, church attendance has fallen sharply.

When Lithuania was annexed by the Soviet Union in 1940, many of its churches were closed and its priests sent to concentration camps. In September 1993, Pope John Paul II arrived to visit on the same day that the last occupying Russian soldier left Lithuanian soil. As in Latin America, the Pope expressed his fear of Evangelical Protestants, whom he calls "sects": "You will have to respond in particular," he warned, "to the disturbing phenomenon of sects, whose proliferation is abetted by widespread misconceptions about religion." Evangelicals have established substantial congregations throughout East Europe, and Lutherans and Lutheran church organizations, who were a major element of Eastern European religious life before communism, are re-establishing their religious missions.

Neither of the two most strongly Roman Catholic countries in Europe today, Ireland and Poland, was ever part of the territory of the Roman Empire. It has been hypothesized that these two peoples have clung to their Roman Catholicism partly in order to differentiate themselves from their powerful non-Roman Catholic neighbors—the Irish to differentiate themselves from the English, and the Poles to differentiate themselves from the Protestant Germans and the Orthodox Russians. Thus, despite the international nature of the Catholic Church, it has in many cases defined and sustained nationalism.

The Struggle between Fundamentalism and Secular Government in the Islamic Realm

Islam recognizes no distinction between religion and politics, so Islamic fundamentalism often challenges secular national governments. National parliaments, bureaucracies, and military organizations become the "enemies of the people" if they fail to rule according to divine law. The Ayatollah (a religious title) Khomeini, who ruled Iran from 1979 to 1989, wrote, "We don't say that the government must be composed of the clergy, but that the government must be directed and organized according to the divine law, and this is only possible with the supervision of the clergy." Today the Iranian Constitution guarantees civil rights, but they are "subject to the fundamental principles of Islam." All legislation in Iran must be approved by the Guardian Council, a body of 12 religious judges.

Fundamentalism is rising within several countries across the Islamic realm, and there are many hypotheses of explanation for this. One is the impoverishment of traditional rural areas discussed in Chapter 6. This triggers rural-urban migration. Many displaced rural people are alienated and disoriented by urban life, and they find appeal and solace in religious fundamentalism. Secular governments must choose either to battle Islamic fundamentalism or else to acquiesce to its demands. Algeria presents an example of a national government that has fought fundamentalism. Early in 1992, the Algerian army canceled scheduled elections and seized power in order to prevent an expected political victory by Islamic militants. Turmoil continued into the mid-1990s.

In Egypt, the most populous Arab land, fundamentalists have accused the government of being insufficiently Islamic, and in 1981, extremists assassinated President Anwar al-Sadat for that reason. Sadat had shared the 1981 Nobel Peace Prize for negotiating a peace with Israel. Assassinations of Egyptian officials continued into the mid-1990s. Spokesmen for the militants argue that their targets are not just un-Islamic, but thoroughly corrupt by any standard and that only the Western press interprets the assassins' actions as unjustified. Muslim extremists have assassinated intellectuals across the Arab world. The prominent Egyptian writer Farag Fodah, for example, was killed in 1992. At the trial of his murderers, Sheik Mohammed al-Ghozali, who preaches on Egyptian government television and whom the Western press had labeled "a moderate," testified for the defense. "A secularist," he said, "represents a danger to society and to the nation that must be eliminated. It is the duty of the government to kill him." In 1993, 100 Muslim intellectuals from several countries contributed to a book in honor of Salman Rushdie's right to intellectual freedom, and today those intellectuals feel that their lives are in danger. The Egyptian government has caught and executed several fundamentalist assassins, but at the same time it has tried to mollify fundamentalists by enforcing laws prohibiting the construction or repair of Christian churches. The Egyptian

constitution guarantees freedom of religion, but it prohibits proselytizing any religion other than Islam.

In 1986 Nigeria joined the Islamic Conference Organization (OIC), a worldwide association of Muslim states. This action outraged Christian leaders. Anthony Olu Okogie, the Catholic Archbishop of Lagos, said, "The OIC issue showed us that they want to make Nigeria an Islamic state. Religion should be a private affair. We want to preach peace, but if the worst comes to the worst, and they want us to burn down this country, we will." Street battles between Christians and Muslims have become commonplace throughout Nigeria.

In Indonesia and Malaysia, Islamic fundamentalism could threaten the security of the Chinese minorities. Indonesia has struggled to establish a secular government, but in 1992 the government granted official status to Sharia rulings on family law. Even in the Philippines, Muslim groups on several islands have fought against the central government.

Liberal democracy, as it is known in the West, was shaped by a thousand years of European history, and, beyond that, by Europe's double heritage of Judeo-Christian religion and ethics and Greco-Roman statecraft and law. No such system ever originated in any other cultural tradition, and it remains to be seen whether such a system can survive when transplanted and adapted in another culture.

Nationalism and Politics in the Buddhist Realm

Buddhism has revived in Mongolia since the fall of the Communist government, and it may become the state religion. This worries the Chinese, because that action could revive Mongolian nationalism in Inner Mongolia, which is a province of China.

In Japan the Soka Gakkai, an evangelical fundamentalist Buddhist organization with more than 11 million members, may be winning political power. During World War II Soka Gakkai leaders were imprisoned for refusing to follow state-sponsored Shinto, but after the war they founded Komeito (Clean Government Party). In 1993, this party won 51 seats in Parliament and even a seat in the cabinet. Komeito and Soka Gaikkei are not today formally linked, but some Japanese worry that the constitutional clause separating church and state is being breached.

States Split by Two or More Religions

When different religious communities within a national population compete to write national laws, civil strife may result. In the Sudan, for example, attempts to enforce Sharia have intensified conflict between the Arab Muslims in the north and the black Christians and animists in the south.

Muslims believe in proselytizing, as Christians do, and this necessarily creates an element of competition between the two religious communities. Mosques exist throughout the Christian realm, while a few Arab Muslim countries, by contrast, ban Christian worship altogether. This fact may testify to which religion teaches greater tolerance, or else it may suggest which culture realm is more secular. The competition for converts between Islam and Christianity can spill over into the writing of laws, and it has triggered murderous street rioting in many countries, as in Nigeria. The constitution of Nepal tries to avert religious violence by guaranteeing freedom of religion but forbidding proselytizing. This satisfies neither Christians nor Muslims; instead, it frustrates and angers both groups.

Sometimes states will enforce two separate systems of law for citizens of different religions. Senegal and several other African states allow Muslim men, but only Muslim men, more than one wife. In India Muslims are allowed to follow Muslim law in many matters of education, marriage, divorce, and property. For example, in Islamic law, no divorced woman can claim support from her former husband. Therefore, in India, a divorced Muslim woman can claim financial support only if she herself renounces Islam.

In some countries struggles that are ostensibly religious actually conceal other social divisions. Fighting in Northern Ireland is, on the surface, a conflict between a Protestant majority and a Roman Catholic minority. That split, however, also involves competition for jobs and opportunity, and it is complicated by the fact that both sides draw on outside support.

Religious Tensions on the Indian Subcontinent

The Indian subcontinent provides an example of religious conflict both within individual countries and among countries (Figure 8-25). The British ruled the subcontinent (today's India, Pakistan, and Bangladesh) as one colony, "India," and they intended to grant it independence as one secular state. Many of the Muslims within this immense territory, however, demanded that the colony be divided so that they could create a Muslim state. When independence came in 1947, the Muslim population established the country of Pakistan. It originally included a western territory (today's Pakistan) and a separate eastern territory. In the months before the partition, millions of Muslims fled toward areas that were designated to become Pakistan, and millions of Hindus to regions scheduled to become parts of India. Tens of thousands died in terrible acts of violence.

Border Disputes in South Asia

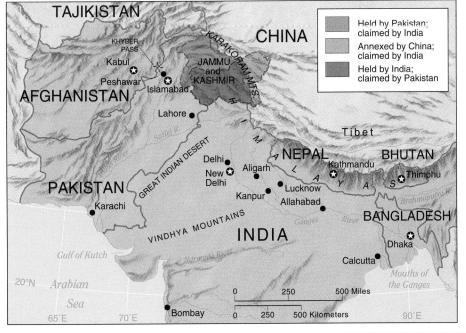

FIGURE 8-25

Antagonism and border disputes persist between India and Pakistan and between India and China. These three nuclear powers contain about 40 percent of the human population.

The Geography of Religion **289**

In 1971, eastern Pakistan broke off from Pakistan to form a separate Islamic country, today's Bangladesh. Meanwhile, religio-political turmoil continues in what is left of Pakistan. Islam has not proven a sufficient bond to unite Sunnis (about 70 percent of the population) and Shiites (about 20 percent), and tribal loyalties and identifications subdivide the population still further. About one-tenth of the population is not Muslim, and a system of separate electorates grants non-Muslims control just 10 of the 307 seats in Parliament. These seats do not represent districts, so candidates for them must campaign across the entire country.

India itself emerged as a secular state with a majority Hindu population. Between 10 percent and 13 percent of India's population, however, is Muslim. The Congress Party, which has ruled India for most of its independent existence, is dedicated to a secular India, but in the early 1990s that party, and the secular ideal, came under increasing pressure from the Bharatiya Janata Party, which wants Hinduism to be India's state religion.

Several considerations exacerbate Muslim-Hindu tensions. One is Hindus' historical consciousness. For hundreds of years Muslim minorities ruled Hindu majorities in many areas (Figure 8-26). Today many Hindus object to the continued use of Muslim names for North Indian cities. Aligarh, for example, would

FIGURE 8-26

The Taj Mahal in Agra, India, completed in 1648, is the mausoleum a Muslim ruler built for his beloved wife. India's Muslim rulers were so rich and powerful that their name, Mughals, has come down to us in English as a word for a business executive, a *mogul*.
(Courtesy of the Agency for International Development.)

be, in Hindi, Harigarh. Second, many Hindus resent the laws mentioned above that recognize Muslim practices. Third, many Hindus believe that since both Pakistan and Bangladesh are Muslim states, the Hindus should have a Hindu state. Fourth, India and Pakistan remain bitter political rivals, and many Hindus doubt the patriotism of India's Muslims. India's Muslims suffered persecution during India's wars with Pakistan in 1965 and again in 1971, and conflict continues on the Indian-Pakistani border. In 1947, India and Pakistan fought over the border state of Jammu and Kashmir. The population of that state was mostly Muslim, but the territory was eventually split. The population of Indian Kashmir today is about 60 percent Muslim. India exempts this state from many acts of federal legislation and grants Kashmir other special rights, but India refuses to allow a plebiscite to determine the population's own political preferences. Many Indians feel that it is important to India's identity as a secular country to include a state with a Muslim majority.

Muslims and Hindus battled regularly on Indian streets through the early 1990s, and Hindu extremists destroyed Muslim mosques. Thousands of people died. These religious tensions are further complicated by the fact that many Hindu untouchables are converting to Islam or Buddhism, and Christian missionaries are proselytizing all groups. Even the Basilica of Bom Jesus in Goa has been threatened with destruction, for it may stand on top of a Hindu temple.

RELIGION AND WOMEN'S RIGHTS

Religions differ in their teachings about women, and therefore we may investigate whether the geography of religions affects the geography of women's rights. The actual practice of any religion, however, varies from place to place, and these variations reflect variations in local cultures in addition to religion. Therefore, it is difficult to determine exactly how much variation can be attributed to the predominant religion of a place. The essay on the scientific method discussed such problems in apportioning explanatory power among variables.

For example, the Catholic church maintains a worldwide prohibition against the ordination of women. By contrast, the official head of the Anglican church, the Archbishop of Canterbury, has approved the ordination of women. Local Anglican communities around the world, however, hold different attitudes, and these reflect local variations in general social attitudes toward women's rights. In U.S. culture, for example, women's rights are protected in many areas, and the U.S. Episcopal church, which is the American province of the Anglican communion, has consecrated female bishops. Anglican communicants from some other world regions, however—most notably Africa, where women's rights are not always assured—have refused to recognize U.S. female bishops. In the case of the Anglican church, therefore, local cultural considerations seem more important than church teachings.

Variations in practice can also be found across the Islamic realm. Islam originated in an Arab culture that granted women few rights, but the list of women's rights in the Koran was liberal for the time and place. Islam outlawed female infanticide, encouraged the education of girls, and established a woman's right to own and inherit property. At the same time, Islam also fixed certain discriminatory practices. It teaches that a woman's testimony in court is half as valuable as that of a man and that men are entitled to four spouses, whereas women may have only one. In practice, women's rights vary throughout the countries where Islam predominates. Women are frequently secluded, or veiled in public, in Islamic countries, but this is compulsory only in Saudi Arabia and Iran. Also, only in the most fundamentalist Arab societies are women generally forbidden to work outside the home.

Future struggle is nevertheless almost inevitable between the expansion of women's rights in Islamic societies and an Islamic fundamentalist backlash that demands the repeal of laws that had banned polygyny (having more than one wife), permitted birth control, and given women the right to divorce in some countries. The Iranian Revolution of 1979 rescinded women's rights granted under the Shahs. In Egypt the Supreme Court in 1985 struck down a 1975 law that gave a woman the right to divorce her husband should he take a second wife. Sudan's military regime, which seized power in 1989, refused to allow women the right to leave the country unaccompanied or without the permission of a father, husband, or brother. Algeria's 1984 Family Code gives a husband the right to divorce his wife and eject her for almost any reason.

The Old Testament contains restrictions on women's rights that are comparable to those found in the Koran, but few Jews or Christians today observe them. In Israel, however, Jewish law denies women equality with men. For example, Israel has no civil marriage or divorce, and in Jewish law only a husband can initiate a divorce. A man stuck in an unhappy marriage may go off and live with another (unmarried) woman, and even have children with her who are not considered illegitimate. Women, on the other hand, are trapped. Thousands of Israeli women live apart from their husbands, wanting divorces, but unable to get them.

It has been hypothesized that religious teachings about women's rights may affect women's role in politics, but this hypothesis is difficult to test. What sort of evidence could be sought to investigate this hypothesis? Reading scriptures alone does not prove whether those teachings are obeyed. Could we draw any definite conclusions from an examination of the percentages of national legislators who are female? the percentages of voters who are female? The two Muslim Prime Ministers who in 1993 visited Bosnia together—Bhutto of Pakistan and Ciller of Turkey—are both female, and women have also served as head of government in countries which are officially Christian (only Catholic and Protestant), Jewish, and Buddhist. National contrasts in granting women civil rights will be discussed in Chapter 10.

THE GEOGRAPHY OF RELIGION IN THE UNITED STATES

The U.S. Census Bureau has never surveyed the distribution of religions within the United States, but Figure 8-27 maps the variations of church membership across the country. The map partly reflects the national origins of many of the settlers in distinct regions. The City University of New York polled Americans nationwide on their religious affiliation during 1989 and 1990. The poll did not ask about baptism, church contributions, or regular worship. It only asked about religious self-identification. A full 86.4 percent of Americans identified themselves as Christian. This is a degree of religious commitment, and of religious homogeneity, almost unique in the world. Only 3.7 percent of Americans claimed any religion other than Christianity. Only 7.5 percent of the U.S. population claimed to have no religion, and 2.3 percent refused to answer. Some regional variations in religious affiliation revealed by this poll are shown in Table 8-3. The largest percentages of the population claiming no affiliation were concentrated in the West. Can you suggest any hypotheses of explanation for this?

Religion in U.S. Politics and Government

Several of the American colonies were founded as theocracies, and most retained established churches until the War for Independence. In forging the new United States, however, James Madison included in the Bill of Rights a prohibition against the federal govern-

The Geography of Religion　　**291**

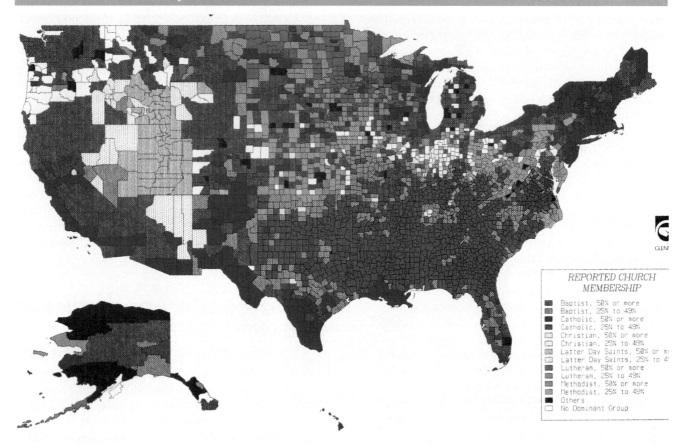

REPORTED CHURCH
MEMBERSHIP

Baptist, 50% or more
Baptist, 25% to 49%
Catholic, 50% or more
Catholic, 25% to 49%
Christian, 50% or more
Christian, 25% to 49%
Latter Day Saints, 50% or m
Latter Day Saints, 25% to 4
Lutheran, 50% or more
Lutheran, 25% to 49%
Methodist, 50% or more
Methodist, 25% to 49%
Others
No Dominant Group

FIGURE 8-27

This map of Americans' church affiliations reveals the predominance of Baptist membership in the Southeast and of Mormons in Utah and adjacent regions. The predominance of Lutherans in the northern Midwest reflects the Scandinavian and North German backgrounds of that region's pioneer settlers, and the distribution of Roman Catholics reflects, among other factors, the migration to the United States of Hispanics and Southern Europeans. (Reproduced with permission from Martin B. Bradley, Norman M. Green, Jr., Dale E. Jones, Mac Lynn, and Lou McNeil, *Churches and Church Membership in the United States: 1990.* Atlanta, GA: Glenmary Research Center, 1992.)

TABLE 8-3
Percentage of U.S. Population Claiming Certain Religious Affiliations and the Five States Within Which the Largest Percentage Claimed that Affiliation

Catholic	26%	Baptist	19%	Methodist	8%	Lutheran	5%	Jewish	2%	No religion	8%
R.I.	62	Miss.	55	Del.	27	N.D.	37	N.Y.	7	Ore.	17
Mass.	54	Ala.	51	Iowa	16	Minn.	34	N.J.	4	Wash.	14
Conn.	50	Ga.	51	S.C.	16	S.D.	30	Fla.	4	Wyo.	14
La.	47	N.C.	47	W. Va.	15	Wis.	26	Mass.	4	Calif.	13
N.J.	46	D.C.	47	Kan.	15	Neb.	16	Md.	3	Ariz.	12

Source: City University of New York

ment's establishing any religion—the world's first such clause. This was intended to protect the churches from government interference. The Supreme Court regularly rules on whether certain acts carried out by individuals in the practice of their faiths can be banned by local governments. In 1993, for example, the Court ruled that the First Amendment forbade one city's ban on the ritual sacrifice of animals by followers of the Afro-Cuban syncretic religion Santería (*Church of the Lukumi* v. *Hialeah, Florida*). The 1993 Religious Freedom Restoration Act requires all levels of government to meet stringent standards before instituting measures that might interfere with religious practices.

George Washington insisted, "The government of the United States is not, in any sense, based on the Christian religion." Nevertheless, U.S. culture and the U.S. legal system have traditionally favored Christianity. Until 1923, Webster's American Dictionary refused to recognize a plural form for the word "religion"; only Christianity was accepted as "religion;" all other forms of worship were considered to be heathenism. Despite the Constitutional prohibition against the "establishment of religion," Congress added the words "under God" to the pledge of allegiance in 1954, and the Supreme Court has upheld the writing into law of "Judeo-Christian moral and ethical standards" (*Bowers* v. *Hardwick*, 1986).

The U.S. Protestant majority has often feared that Roman Catholic citizens would follow political dictation from Rome, and U.S. Roman Catholics have suffered discrimination. John F. Kennedy was the first—and only—Roman Catholic to win the presidency, and during his campaign he felt it necessary explicitly to state that he would obey national laws. A Roman Catholic priest sat in Congress in the 1970s, but the Pope ordered him to resign. Individual U.S. Catholic bishops have excommunicated state and local legislators who have defended the prochoice position on abortion, and prochoice legislators nationwide have been barred from speaking on Catholic church property. Meanwhile, many Protestant ministers have organized political campaigns, often using tax-exempt publications and other church properties too. Many ordained Protestant ministers have held political office, including congressional seats, and they have worked to win passage of explicitly religious platforms.

The first mosque in the United States was built by Syrian and Lebanese settlers in Ross, North Dakota, in 1929, and today Islam is growing so rapidly that U.S. Muslims will soon outnumber U.S. Jews. On June 25, 1991, a Muslim Imam (preacher) gave the opening invocation for the U.S. House of Representatives for the first time; this occurred on February 6, 1992, in the Senate.

RELIGIOUS INFLUENCES ON THE GEOGRAPHY OF DIETS

Different religions teach different lessons concerning the sacredness of plant and animal life. Therefore, the geography of religion affects the geography of agricultural practices and of diets. Buddhists, for example, are generally vegetarians. Hindus believe that cattle are sacred; therefore, they will not eat beef, and they use cattle only as draft animals and as a source of milk. Outside observers have tried to measure the advantages and disadvantages of preserving cattle in economic terms. Aging and unproductive cattle may overgraze the land or compete with humans for limited food supplies. Sacred cows roam the streets of Indian cities unmolested, befouling and congesting them (Figure 8-28). On the other hand, cattle dung provides a fuel and fertilizer, and the cattle that die of natural causes are eaten by non-Hindus anyway. Whatever the balance of these economic arguments might be, to a religious Hindu this matter is not open to economic debate.

Altogether, a significant share of humankind limits its diet because of religious beliefs. These religious dietary strictures affect world trade in food. Table 6-2 shows that several of the countries that limit their consumption of meat for religious reasons are the greatest importers of pulses, which provide an alternative source of protein. Annual per capita consumption of

FIGURE 8-28

In India cows are sacred and cannot be molested, no matter how much of a nuisance they become. McDonald's restaurants in India do not offer beef hamburgers. (Courtesy of Carl Wolinsky/Stock Boston, Inc.)

meat in India, for example, is only 4.4 pounds (2 kg), and India is a major importer of pulses.

RELIGION AND ECONOMIC DEVELOPMENT

▼▼▼▼▼▼▼▼▼▼▼▼▼▼▼▼▼▼▼▼▼▼▼▼▼▼▼

The religion that predominates in any region may have a significant impact on that society's economic growth. Elaborate burial practices, for example, can drain a society of capital. The tombs of Egypt's pharaohs absorbed immeasurable amounts of labor to build, and the rulers were buried with heaps of rich possessions (Figure 8-29). Many peoples have buried their dead along with their possessions. The North American Pawnee and Comanche peoples, for example, buried warriors with their horses; the natives of Patagonia in today's Argentina customarily buried people with all their ornaments, weapons, and livestock.

Some societies lavish expenditure on houses of worship and other religious institutions. Europe's great cathedrals absorbed a good share of the sweat and treasure of whole societies—sometimes for decades or even longer—and their treasuries dazzle with the accumulated riches of centuries. Roman Catholics, when possible, build their houses of worship with a greater degree of ornateness than Protestants do, because for Catholics a church is the house of God, whereas for Protestants it is only a place of assembly. In Muslim lands, great mosques are the pride of many communities that are otherwise poor. In Theravada and Lamaist Buddhist societies high percentages of the male labor force spend several years, or even their whole lives, as monks.

Furthermore, religious teachings can affect a society's economic growth because religious teachings can influence economic behavior. The exact cause-and-effect linkages are conjectural, but they present coincidences that must be explored as strong hypotheses. One hypothesis is that religions can affect the way people view moneymaking. For example, the scriptures of most religions bar the charging of interest as unfairly taking advantage of another person. Christian teachings, however, have evolved to differentiate between two reasons for borrowing money: for needs, and for investment. It may be sinful to charge interest on money borrowed to buy food, but it may not be sinful to charge interest if the borrower invests the money for his or her own profit, as, for example, when the borrower buys a truck to increase the profit in his or her own business.

The religious leaders in some societies do not accept such a distinction. Islamic fundamentalists, for

FIGURE 8-29

This gold sarcophagous of Egypt's Pharaoh Tutankhamen (fl. c. 1350 B.C.) was among the rich objects found when his tomb was discovered and opened in 1922. Followers of many religions are often buried with their possessions. Some kings' servants have even been killed to accompany them in the afterlife.
(Courtesy of Superstock.)

example, condemn charging interest. Pakistan's Federal Sharia Court ruled in 1992 that interest is religiously unacceptable and therefore illegal. This ruling threatened Pakistan's entire modern financial system, as well as the country's international trade and financial links. Islamic teachers have held that loaning money without taking an active interest in what the borrower does with the money is irresponsible on the part of the lender and invites irresponsibility on the part of the borrower. Therefore, they have insisted, a capitalist must buy and own part of a business; this should encourage the lender to take a more active interest in its success. Capitalists call this holding an equity position. These ideas are not impractical; Muhammad was himself a successful businessman. Limiting capitalist participation to equity positions does, however, restrict the flexibility of an economic system.

The government of Malaysia wants to encourage capitalist development, especially among the

Muslim Malays, so it is devising new religiously acceptable frameworks for commercial life. The Malaysian Institute of Islamic Understanding brings together scholars to debate the fine points of stock-broking and money management. In practice, new financial structures and compromises have to be defined in each Islamic country.

In many countries a religion can become a major landowner or financier, and religious organizations do not always maximize the productive capacity of their property. Real property (land) that is restricted in ownership or in the purposes for which it can be used is called *mortmain*, literally "dead hand." Church property may also enjoy preferential tax rates or escape taxation entirely. Throughout Latin America, the Catholic church is a principal landowner. In the United States, church-owned properties are usually concentrated in the cities. They cannot be taxed, so their concentration reduces any city's ability to raise property-tax income. In turn, however, the churches often provide a wide range of services (such as soup kitchens and hospitals) to a public beyond their own communicants.

Studies of comparative economic development are often challenged to identify the reasons for the poverty of Haiti, which contrasts sharply to the countries around it. Some observers seeking to explain this poverty have argued that the most important factor differentiating Haiti from its neighbors is Haiti's traditional religion, called voodoo or Vodun. Some scholars argue that this religion encourages its believers to externalize guilt onto spiritual forces. This limits the person's responsibility, but it also limits the person's potential to improve his or her condition.

Weber and the Protestant Ethic

German sociologist Max Weber linked Protestantism and capitalism in *The Protestant Ethic and the Spirit of Capitalism* (1904). According to Weber, Protestantism encourages individualism. With the rise of Protestantism in Western Europe acquisitiveness, as the result of actively exercising individual ability in the marketplace, became a virtue. Today this ethic seems to characterize adherents of many different religions, so it is usually referred to as the *work ethic*.

If there is any advantage in the Protestant attitude toward money making, then it might be hypothesized that the worldwide expansion of Protestantism will encourage individualistic capitalism. This explains why the spread of Protestantism is reported and happily received in America's business press, which clearly accepts the link. *The Wall Street Journal, Business Week, Forbes,* and other business publications devote extensive coverage to the spread of Protestantism as economic news, not religious news.

If the corporate managers who read these publications believe that Protestantism is hospitable to capitalism, then they will direct their corporations' investment to areas undergoing conversion. Thus, a belief can launch its own fulfillment, whether or not it can objectively be shown to be true.

The Catholic Church and Capitalism

In 1991, Pope John Paul II issued an encyclical on economic matters, *One Hundred Years*. This document asserted a superiority of capitalism over socialism. "On the level of individual nations and of international relations," wrote the Pope, "the free market is the most efficient instrument for utilizing resources and effectively responding to needs." This document was widely interpreted as encouraging the East European governments that were then turning from communism to democratic and capitalist systems. The Pope insisted, however, that capitalism be tempered with concern for social justice and human dignity.

In 1992, the Church released the first universal catechism (compilation of church teachings) since the Council of Trent (1545–63). In its elaboration of the Seventh Commandment ("Thou shalt not steal"), this document questions the profit motive, sees speculation as preying on the ignorance and distress of others, points out that markets do not always meet human needs, and insists that governments must regulate markets "according to a just hierarchy of values." It demands that every worker receive a "just salary," and it rejects as a "sin against the dignity of persons" the concept that workers shall be viewed as essentially "a source of profit." Thus, while the Catholic Church praises capitalism's demonstrated ability to produce wealth for a society, it worries about what capitalism does to individuals. This concern complements the insight of Adam Smith (1723-1790), who is often credited as the founder of modern capitalism. Smith noted that a whole community can be enriched when individuals pursue their own self-interest. "Economic man," whom we met in Chapter 2, is driven to maximize his financial return, and this urge may not necessarily lead to virtuous behavior or individual happiness. The attitudes the Church's new catechism expresses toward markets, speculation, and worker-management relations may influence the behavior of individual Roman Catholics and of governments in Roman Catholic lands around the world.

Religion and Economics in Asia

The Confucian tradition in East Asian societies may exemplify another religious influence on economic development. Diligence, obedience, and high savings rates characterize the peoples of neo-Confucian Japan, South Korea, Taiwan, Hong Kong, and Singapore, and of every Chinese settlement in the world. It is risky to hypothesize a cause-and-effect relationship, but the coincidence must be noted.

In addition, the Confucian view of leadership by an intelligent elite with the moral obligation to guide the people enhances the status of jobs in government bureaucracies. In 1991, the Korean scholar Jung-en Woo wrote, "The peoples of China, North Korea, South Korea, Japan, and other Confucian cultures deeply believe that the state ought to provide not only material wherewithal for its people, but moral guidance. By and large, Westerners have no way to understand this point except to assert that the Asian countries suffer from a series of absences: no individual rights, no civil society, no Enlightenment, and thus a weak or absent liberalism." Ambitious young graduates compete for positions with the Japanese Ministry of Finance or the Korean Economic Planning Board the way Americans compete for jobs at investment banks. The ability to attract talent gives these Asian governments more legitimacy and more competence in dealing with businesses than the U.S. government can bring to bear. Perhaps this has helped to build successful economies.

Several East Asian leaders have praised their own societies' traditional communitarianism and decried Western Christian individualism. Zealous Evangelical Protestant missionaries have even been arrested in Singapore for interfering with that nation's Confucian traditions. The missionaries' intolerance caused them to be branded, ironically, as Marxists. Singapore's long-term Prime Minister Lee Kuan Yew expressed many Asians' thoughts about the relationship between religion and economics as follows: "If we follow the West in our social relations and family structures, we will be in deep trouble. In the West the Christian religion used to instill fear of punishment in hell or reward in heaven. Science and technology have eliminated that fear, so the controlling mechanism has gone awry. I am hoping that because Asian moral control is based on what is good in this secular world—not a spiritual afterworld—we will not lose our moral bearings." The hypothesis that Asian philosophies carry advantages for economic development, however, cannot explain why the Christian realm developed so much faster than did Asian societies from 1800 to the present.

Hypotheses about religion and economics bear significant implications for governmental policy. If it is true that different religions' teachings can influence peoples' attitudes toward economic growth and, therefore, a society's ability to achieve economic growth, then economists' arguments that the solutions to some countries' poverty depend largely on economically correct policies are false, or at least inadequate. Sending the right market signals to trigger economic growth may work in China, where people are imbued with the Confucian values of education, work, merit, and saving for the future. Those signals, however, are not going to do much good in Haiti, where the ability of people to receive and act on them may be minimal.

RELIGIONS AND THE ENVIRONMENT

▼▼▼▼▼▼▼▼▼▼▼▼▼▼▼▼▼▼▼▼▼▼▼▼▼

Most religions address the questions of how the world came into existence, what power created it, and what, then, humankind's relationship should be to other living things and to the physical environment. Many religions mediate between heavenly powers and humanity. This mediation may include services to praise and thank the divine powers for their blessings and gen-

FIGURE 8-30

Volcanic Mount Fuji, Japan's highest peak at 12,389 feet (3776 m), is sacred to many Japanese. It has been celebrated for centuries in Japanese paintings and verse. Today the mountain and Lake Ashi ("foot"), shown in the foreground, are within the Fuji-Hakone-Izu National Park. (Courtesy of the Japanese National Tourist Organization.)

erosity, or to propitiate gods who may otherwise bring ruin to human undertakings.

Many major faiths not normally classified as animist nevertheless designate specific features of the environment as holy. Some religions venerate rivers (Hindus venerate the Ganges; Christians the Jordan); some venerate high places (the Shinto venerate Mount Fuji; Jews and Christians venerate Mount Ararat) (Figure 8-30). The Christmas tree is a relic of the pre-Christian tree worship that covered much of heavily forested Northern Europe. The oak tree retains associations with Germanic folklore, nationalism, and military virtues. In Northern Europe the oak is the equivalent of the southern European laurel wreath, associated, in classical religions, with the god Apollo.

Early religious beliefs often formed as attempts to explain natural phenomena, such as the changes in the seasons or floods. Many religions celebrate natural events, time their annual ceremonies in accordance with astronomical events, or celebrate human affairs connected to the environment—most religions celebrate harvest festivals of thanksgiving, for example. Many religions retain forms of sun or moon worship. The Woman of the Apocalypse (Revelations 12.1) comes "robed with the sun, beneath her feet the moon, and on her head a crown of twelve stars." The synoptic gospels relate that an eclipse of the sun occurred during the crucifixion of Jesus, but Saint Augustine explained that the sun and moon represent the two testaments: The Old (the moon) is only to be understood by the light shed upon it by the New (the sun). Ancient Romans worshipped the moon goddess, and the moon continues to symbolize chastity when shown under the feet of the Virgin Mary in Christian art (Figure 8-31).

Different religions recognize different holidays and pace human activities through the year differently, and this is often reflected in national laws across different religious realms. Most Christian countries use the solar calendar. Jews use a calendar of 12 months plus an additional month 7 times in 19 years. Muslims use the lunar calendar. Muslims celebrate the Sabbath on Friday, Jews on Saturday, and most Christians on Sunday. Many governments prohibit work or other activities on that day. Muslims recognize an entire Holy Month and accordingly curtail their activities.

Some scholars have sought to explain the origins of monotheism through environmental factors. The Judeo-Christian-Islamic tradition evolved in desert lands. The geographer Ellen Churchill Semple argued that such desert-dwelling peoples "receive from the immense monotony of their environment the impression of unity." These peoples' view of the regularity of the stars and planets, she suggested, gave birth to the idea that there must be one thing causing it. Therefore,

FIGURE 8-31

The doctrine that the Virgin Mary was born without sin (called the Immaculate Conception) is portrayed in art by showing her astride the moon, traditional symbol of chastity. This *Assumption of the Virgin* was painted by the Flemish painter Miguel Sithium about 1500.
(Courtesy of the National Gallery of Art, Washington, DC.)

she wrote, desert dwellers "gravitate into monotheism." This environmental-religious determinism is poetic, but it cannot be defended. We do not know when, where, or why monotheism arose. There were and still are many polytheistic desert-dwelling peoples and, conversely, monotheists who live in jungles.

Religions and Environmental Modification

Most religious traditions award humankind a special role in the world, but, it has been argued, some religions suggest more adaptive approaches to the physical world, while others suggest more exploitative approaches. In the Judeo-Christian-Islamic tradition, for example, God promised Noah rule over all living things after the flood (Genesis 9. 2-3), and, later, "dominion over the works of [God's] hands" (Psalms 8.6). Psalm 115, Verse 16 emphasizes that "the heavens are the Lord's heavens, but the Earth he has

The Geography of Religion **297**

F O C U S
The Bishnois

The Bishnois of India illustrate how, in a precarious environment, a religion may codify essential feelings in favor of environmental preservation. The Bishnois are a reformist Hindu sect founded in the fifteenth century. About 600,000 Bishnois live in India's most arid zone in a belt from southwest Rajasthan to the state of Haryana, north of New Delhi. This fragile zone acts as a shelter belt to prevent desertification and expansion of the Thar Desert. The Bishnoi are vegetarians. Their creed teaches compassion and protection of all living things, and they do not allow the killing of wildlife or the chopping of trees. In the eighteenth century, a king who was building a new palace sent soldiers to cut down the trees in the Bishnoi territory. The Bishnois hugged the trees, and, the story goes, over 300 were slaughtered before the king ordered the killing to stop. The king ordered all people to respect the Bishnois in their territory forever. They still enjoy special rights as quasi-environmental rangers in this region where perhaps no one else could live and where any ill-considered attempts to exploit the environment would create severe disruption.

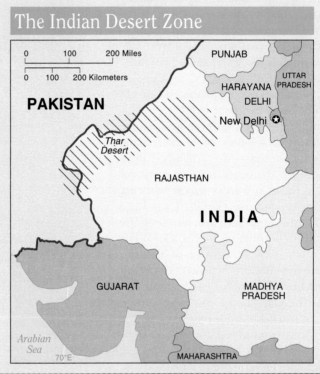

The Indian Desert Zone

Desertification threatens to expand in western India if environmental limitations are not carefully respected.

given to the sons of men." Some scholars have argued that this constitutes a divorce between humankind and nature and that this divorce has resulted in the present environmental problems.

Other scholars disagree. The Bible admonishes humankind to treasure the Earth and its bounty, and all religions exhibit a difference between what they preach and what their followers actually do. Some animists make offerings to propitiate the forests, but then they chop down the trees.

Scholars have surveyed different populations to test their attitudes toward nature. Anthropologist Florence Kluckhohn, for example, surveyed different groups in the U.S. Southwest to uncover differences in their attitudes toward the relative powers and responsibilities of God, of nature, and of humankind. Professor Kluckhohn found that most Hispanic Roman Catholics felt that humans are subject to nature, whereas most Mormons view humans as living in harmony with nature. Protestant Anglo-Texans generally believe that humans control nature, and the local native American groups, most of whom are traditionally animists, favored the idea of harmony with nature. These beliefs may or may not concur with the "official" teaching of any of these groups' religions. All are affected by the surrounding culture.

Other scholars have suggested that Hinduism and Buddhism emphasize resignation before the forces of nature to a greater extent than do other traditions. This is hard to reconcile with the vast works of environmental modification that Hindu and Buddhist societies have accomplished.

Environmental degradation today afflicts areas of all religious persuasions. The Worldwide Fund for Nature is translating all the world's major religious texts into English specifically to allow scholars to compare what the various scriptures say about conservation. Muslim, Christian, Buddhist, Hindu, Muslim, Jewish, Sikh, Shinto, and Taoist organizations are all cooperating.

RELIGIOUS LANDSCAPES

▼▼▼▼▼▼▼▼▼▼▼▼▼▼▼▼▼▼▼▼▼▼▼▼▼▼

Most landscapes reveal the predominant local religion. Chapter 3 noted the importance of the architecture of houses of worship in many landscapes, but religious buildings include religious schools or retreats for clergy, monasteries, and similar institutions. Also, as noted in Chapter 7, local place names—toponyms—often reflect the local religion.

Places of worship may become goals of pilgrimages, and these attract visitors and play a major role in cultural diffusion (Figure 8-32). These places may be associated with the life, death, or burial of the religion's founder or respected teachers, or they may be places where possessions of these respected teachers are venerated. Fairs and markets grow up at these sites. Muslims' hajj pilgrimages bring together people from around the world, and the Saudi royal family derives political prestige as protector of the holy places. In the Buddhist realm, the Anuradhapura temple in Sri Lanka houses what is regarded as the oldest historical tree in

FIGURE 8-33

The eyes on this great stupa in Nepal emphasize that God sees everything. This symbolism is virtually universal. It can even be found on the back of the U.S. dollar bill. (Courtesy of Julia Nicole Paley.)

FIGURE 8-32

The Cathedral of Santiago, in the northwest corner of Spain, is believed to hold the relics of Saint James. Pilgrims from throughout Europe and the world have sought it out. In medieval times these traditional routes and staging points played important roles in trade and in the diffusion of ideas. The routes are marked by a chain of historic churches, hotels, universities, and trade fairs. Still today a Pilgrim's Certificate is given to anyone who walks at least 100 km (62.5 miles) along the medieval pilgrim's trail, and it must be stamped by parishes along the way. In 1993, 100,000 people earned these certificates.

existance. It was planted in the third century B.C. as a cutting from the original tree under which Buddha found enlightenment. Many sacred temples throughout India attract millions of Hindu worshippers.

Different religions' burial practices create different mortuary landscapes, and these can reveal the philosophies and commitments of the people who build them. Hindus and Buddhists cremate their dead. The small sect of Parsis in India leave their dead on high scaffolds to be eaten by vultures to avoid contaminating sacred Earth, air, or fire. Christians and Muslims usually bury their dead and erect monuments to them. Buddhist landscapes are marked by stupas (called pagodas east of India). These serve as temples today, but the form was originally devised as a funeral mound to contain the remains of a revered bodhisattva (Figure 8-33). The people of Okinawa buried their dead in family tombs on the island's hillsides, where the spirits of the newly deceased joined those of their ancestors. During World War II the Japanese fortified these tombs, so U.S. troops destroyed them all. The destruction of these family tombs devastated Okinawan culture. The type of burial ground most favored in the United States since about 1840 is called the rural-style cemetery. "Cemetery" means "place of sleep," and rural-style cemeteries are pastoral gardens of graves. Nature predominates, and the landscape suggests that humankind fits into nature, as death is a part of life. This style of cemetery was developed during the Romantic cultural movement.

Religions codify views about human relationships to each other, to this world, and to a possible next world. They affect almost all other aspects of human culture and behavior, and the study of their teachings, distributions, and impact will always be a key aspect of human and cultural geography.

SUMMARY

Each religion originated in one place and diffused from there, but it is sometimes difficult to explain why people convert to one religion from another. A few religious groups proselytize.

Judaism, the first of the great monotheisms, rests on a belief in a pact between God and the Jewish people. This covenant was granted to Abraham. Both Jews and Arabs claim descent from Abraham. Modern Israel was intended to be a homeland for the Jewish people, yet the balance of Jews and non-Jews in Israel is continuously shifting as a result of Israeli territorial aggrandizement, the number and nature of immigrants to Israel, and the birth rates of the various groups in Israel.

Christianity emerged from Judaism. The Roman Emperor Constantine favored the religion, but Christianity diffused into Armenia, spread south into Africa and even reached India before it was accepted throughout Europe. Christianity is a proselytizing religion, and European Christians often cited religious conversion as their purpose for exploring and conquering other lands. Conversion to Christianity or to Islam usually weakens the continuity of a people's cultural inheritance.

Roman Catholics make up the largest single Christian denomination today, but other sects include the great many Protestant denominations, the Georgian, Armenian, Ethiopian and Coptic Church of Egypt, as well as several Orthodox churches. Roman Catholicism dominates in Southern Europe, throughout Central and South America, Quebec in Canada, and wherever else Mediterranean peoples colonized or converted. Protestant sects dominate in Northern Europe and wherever Northern Europeans have settled or converted: North America, Australia, and New Zealand, and the parts of Africa that were either English colonies or German colonies. Orthodox Christianity predominates in Eastern Europe.

Today Evangelical Protestantism is diffusing widely—often in formerly Roman Catholic areas. Liberation theology splits the Roman Catholic church itself. Christianity continues to win new converts, but practices and doctrines that derive from European culture are under pressure for change.

Muhammad revitalized montheism among the Arabs and founded Islam. Today Islam predominates from Morocco to Indonesia. Its two principal sects are the Sunni and the Shiite Muslims.

Many Asians follow the tradition of Hinduism and its offshoots of Buddhism and Sikhism. Hindus believe in one Supreme Consciousness, Brahman, whose aspects are realized in three deities. Hinduism classifies people in a hierarchy of castes, and Hindus believe in reincarnation depending on one's behavior in life. The Sikh Guru Nanak tried to reconcile Hinduism and Islam. Many Sikhs dream of the restoration of an independent Sikh state.

Buddha was a Hindu prince who, many believe, achieved understanding. Theravada Buddhism is the state religion in Thailand and Sri Lanka, whereas Mahayana, in a form called lamaism, predominates in Bhutan, Tibet, and Mongolia. Chinese Buddhists produced a theory of spontaneous enlightenment, called Ch'an, which diffused into Japan as Zen. Today Buddhism has several hundred million followers, and much popular New Age philosophy derives from Buddhism.

Buddhism is compatible with the ancient Chinese philosophies of Confucianism and Taoism, and Shinto is a Japanese nationalist tradition. Their teachings are more exclusively ethical and psychological. Many people around the world follow animism, often accompanied by shamanism.

Religion is a key factor in any human culture and influences almost every other aspect of human motivation and behavior: politics, women's rights, diet, economics, and environmental modification, calendars, travel and pilgrimage, and landscape transformation.

KEY TERMS

fundamentalism (p. 265)
secularism (p. 265)
proselytize (p. 266)
monotheism (p. 267)
polytheism (p. 267)
ghettoes (p. 267)
Zionism (p. 267)
sacerdotalism (p. 272)
Evengelical
   Protestantism (p. 272)
liberation theology (p. 274)
interculturation (p. 275)
syncretic religions (p. 275)
hajj (p. 275)
Sunni Muslims (p. 278)
Shiite Muslims (p. 278)
castes (p. 279)
Confucianism (p. 281)
Taoism (p. 282)
Shinto (p. 282)
animism (p. 282)
shamanism (p. 282)
theocracy (p. 284)

QUESTIONS FOR INVESTIGATION AND DISCUSSION

1.   How many religions have temples, churches, or synagogues in your community? What, roughly, is the statistical breakdown of religious affiliations of the people of your community? Does this reflect the source areas of settlers to your region? Does each place of worship anchor a residential neighborhood of that religion's communicants? Examine a few different houses of worship to estimate the amount of money expended on their construction and maintenance. What does this say about the importance of religion in people's lives?

2.   How many religiously oriented parochial schools exist in your community? With what religions are they associated?

3.   The fourteenth century Islamic scholar Ibn Batuta could travel from Morocco to China unmolested—even welcome across most of the region. What political, religious, and other criteria of territorial organization have fragmented that region today, so that a contemporary traveler might have trouble retracing Ibn Batuta's travels?

ADDITIONAL READING

BECHERT, HEINZ, and RICHARD GOMBRICH. *The World of Buddhism.* London: Thames and Hudson, 1984. U.S. edition 1993.

CHADWICK, HENRY and G. R. EVANS, eds. *Atlas of the Christian Church.* New York: Facts on File, 1987.

AL-FARUQI, ISMA'IL R. and DAVID E. SOPHER, eds. *Historical Atlas of the Religions of the World.* New York: Macmillan, 1974.

JUERGENSMEYER, MARK. *Religious Nationalism Confronts the Secular State.* Berkeley: University of California Press, 1993.

LEWIS, BERNARD. *The World of Islam.* London: Thames and Hudson, 1976. U.S. edition 1992.

MCMANNERS, JOHN, ed. *The Oxford Illustrated History of Christianity.* New York: Oxford University Press, 1990.

MARTIN, DAVID. *Tongues of Fire: The Explosion of Protestantism in Latin America.* London: Basil Blackwell, 1990.

MARTY, MARTIN E. and R. SCOTT APPLEBY, eds., four volumes: *Fundamentalisms Observed. Fundamentalisms and Society: Reclaiming the Sciences, the Family, and Education. Fundamentalisms and the State: Remaking Polities, Economies, and Militance. Accounting for Fundamentalisms: The Dynamic Character of Movements.* All Chicago: University of Chicago Press, published, respectively, in 1991, 1992, 1993, and 1994.

REGENSTEIN, LEWIS. *Replenish the Earth: A History of Organized Religion's Treatment of Animals and Nature.* New York: Crossroad, 1991.

ROBINSON, FRANCIS. *Atlas of the Islamic World Since 1500.* New York: Facts on File, 1982.

SCOTT, JAMIE and PAUL SIMPSON-HOUSLEY, eds. *Sacred Places and Profane Spaces: Essays in the Geographics of Judaism, Christianity, and Islam.* New York: Greenwood Press, 1991.

STOLL, DAVID. *Is Latin America Turning Protestant?: The Politics of Evangelical Growth.* Berkeley: The University of California Press, 1991.

World Wide Fund For Nature. A series of volumes "World Religions and Ecology," including volumes on Islam, Christianity, Judaism, Buddhism, and Hinduism. New York: Cassell, 1992.

9 | Cities and Urbanization

Rio de Janeiro is no longer the capital of Brazil, but its breathtaking beauty and flourishing cultural and economic life keep it lively. The Third United Nations Conference on the Global Environment met here in 1992.

Courtesy of Sue Cunningham/Tony Stone Images.

A **city** is a concentrated nonagricultural human settlement. Every settled society builds cities, because some essential functions of society are most conveniently performed at a location that is central to the surrounding countryside. Cities provide a variety of services, including government, religious services, education, trade, manufacturing, wholesaling and retailing, transportation and communication, entertainment, business services, and defense. The region to which any city provides services and upon which it draws for its needs is called its **hinterland**.

In ancient Egyptian hieroglyphics, the earliest writing that we can read, the ideogram meaning "city" consists of a cross enclosed in a circle. The cross represents the convergence of roads which bring in and redistribute people, goods, and ideas. The circle around the hieroglyph denotes a moat or a wall. Few modern cities have walls, but cities do have legal boundaries, and within those boundaries they usually exercise a degree of self-government. The process of defining a city territory and establishing a government is called **incorporation**.

Sometimes several cities grow and merge together into vast urban areas called *conurbations*. In the northeastern United States, for instance, one great conurbation stretches all the way from Boston to Washington, D.C. This conurbation has been called *Megalopolis*, which is Greek for "city of millions." The world's largest urban areas are generally called *metropolises*, Greek for "mother cities."

In several countries one large city concentrates a high degree of the entire national population or of national political, intellectual, or economic life. These cities are called **primate cities**. Paris, for example, is the primate city of France, and Bangkok is the primate city of Thailand. Not all countries, however, have a primate city. The United States does not. Whether a country has a primate city depends on national history and social and economic organization.

Today in all countries urban populations are growing faster than rural populations. This process of concentrating people in cities is called **urbanization**. The United Nations predicts that about one-half of the world's population will live in urban areas by the year 2000, compared with 30 percent in 1950. The rate of urbanization, however, is not the same in all countries (Figure 9-1). One of the purposes of this chapter is to explain why this is so.

The geographic study of cities, **urban geography**, includes three topics:

1. The study of the functions of cities and their economic role in organizing territory;

2. The comparative study of urbanization as it occurred in the past and as it is continuing in different countries today;

3. The study of the internal geography of cities, that is, the internal distribution of housing, industry, commerce, and other aspects of urban life.

This chapter will review each of these topics and then examine the growth and internal geography of U.S. metropolitan areas. This examination will illustrate many of the principles of urbanization in situations which will probably be familiar to you. The chapter ends with a brief examination of key ideas of urban and regional planning and a look at non-Western cities.

URBAN FUNCTIONS

▼▼▼▼▼▼▼▼▼▼▼▼▼▼▼▼▼▼▼▼▼▼▼▼▼▼

Cities first developed some 4,000 to 5,000 years ago, and there are two theories of their beginnings. One argues that improvements in agricultural productivity created surpluses, and the surpluses allowed the appearance of cities and their specialized services. The other argues that two urban groups—priests and soldiers—intimidated or forced the peasantry to surrender food to support them. The theories are complementary, and each is probably partly true.

The first cities appeared in today's Turkey and Iraq, about 4000 B.C. Cities developed in Egypt's Nile Valley about 3000 B.C., in the Indus Valley of modern Pakistan about 2500 B.C., and in the Yellow River Valley of China by 2000 B.C. Mexico and Peru supported cities by about A.D. 500. Some early cities reached considerable populations. Ur, which flourished in Mesopotamia by 3500 B.C., may have had 200,000 people, and Thebes, the capital of ancient Egypt, may have had 225,000 in 1600 B.C. Later, Rome probably reached 1 million by the second century A.D. Kaifeng, China, housed 1 million by 1150, and Hangzhou held that many by 1275. In modern times, London reached 1 million by the early nineteenth century.

In the beginning, cities' ritualistic and political importance far outweighed their economic functions. Cities were the seats of government officials, or even of the gods themselves, so they were appropriately embellished (Figure 9-2). Craft workers flourished within their walls, and specialized occupations emerged, sustained by the peasantry of the hinterland.

As societies develop, production and trade join services as the paramount activities in most cities.

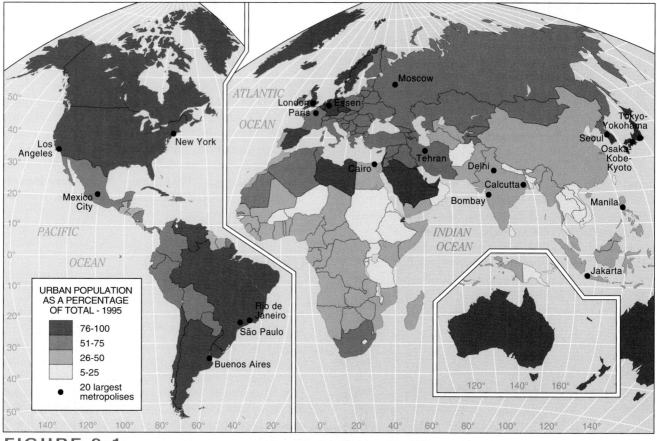

FIGURE 9-1

This map shows the varying degree of urbanization of countries and the world's largest metropolises. Different countries use different definitions of "urban area," ranging from settlements as small as 200 persons to as large as 30,000, but most countries use a minimum of 2,000 to 5,000 people. The U.S. Bureau of the Census accepts a minimum of 2,500 people. (Data from World Resources Institute, *World Resources 1994–95.* New York: Oxford University Press, 1994, pp. 286–87.)

FIGURE 9-2

The Todaiji Temple in Nara, Japan, is the world's largest wooden structure. It houses a colossal statue of Buddha 53 feet (16 m) high. Both temple and statue were originally made in the eighth century, when Nara was Japan's capital city. Today Nara is not a large city or important in Japanese economic life, but much of it has been preserved as a treasure of Japanese history and culture.
(Courtesy of Japanese National Tourist Office.)

Cities bring people and activities together in one place for greater convenience. This is called **agglomeration**. Agglomeration promotes the convenient division of labor, which is the separation of work into distinct processes and the apportionment of work among different individuals in order to increase productive efficiency. Cities also promote and administer the regional specialization of production throughout their hinterlands. This regional specialization may be thought of as a territorial division of labor. For example, the von Thünen model of land use introduced in Chapter 6 deduced how the city market encourages the specialization of land use around the city. With industrialization, cities become center of production.

The Three Sectors of an Economy

Economic activities are generally divided into three sectors: primary, secondary, and tertiary. The names used for the three sectors indicate the degree to which each sector is removed from direct involvement with the Earth's physical resources.

Workers in the **primary sector** extract resources directly from the Earth. Activities in this sector include agriculture, fishing, forestry, and mining. Agriculture usually provides the bulk of primary sector employment. Workers in the **secondary sector** transform raw materials produced by the primary sector into manufactured goods. Construction is included in this sector. All other jobs in an economy are within the **tertiary sector**, sometimes called the **service sector**. The priests, soldiers, and government officials mentioned above were early examples of workers in the tertiary sector, and today the tertiary sector includes a great range of occupations, from a store clerk to a surgeon, from a movie ticket seller to a nuclear physicist, from a dancer to a political leader. Because this sector includes so many occupations, some scholars have tried to split the tertiary sector into a tertiary services sector and a *quaternary sector*, called an *information sector*. There is no consensus, however, as to what this sector includes, and no agencies of the U.S. government or of the United Nations recognize this sector in statistics. Therefore, it is not used in this book.

As economies grow, the balance of employment and output shifts from the primary sector toward the secondary and tertiary sectors. Fewer people work on farms, and more people work in factories and offices. These activities normally require urban settings, and therefore, as an economy's secondary and tertiary sectors grow, its cities grow. This growth of the secondary and tertiary sectors is called *sectoral evolution*, and it will be examined in detail in Chapter 12.

Western culture often contrasts the countryside with the city—the former "natural" and the latter human-made (artificial) and "unnatural." In fact, fields and pastures are plagioclimax environments, and they are as human-made as factories or city streets are. Both rural and urban landscapes are cultural landscapes, products of human transformation of natural landscapes. Economically, cities and the countryside depend on each other.

The Economic Bases of Cities

Cities depend on their hinterlands for, at the very least, their food. The cities must, in turn, provide services or "export" something to the outside. Many cities produce and export manufactured goods, but the exports of a city are not necessarily things that leave it. They may be things or services that people come to the city to buy. If people go to Houston for heart surgery, for example, then heart surgery is counted as an export of Houston. Vacations are an export of Miami Beach; gambling is an export of Las Vegas. In this economic sense, capital cities export government.

Some of the workers in any city produce the city's exports, but others serve the needs of the city's own residents. The part of a city's economy that is producing exports is called the **basic sector**, and that part of its economy serving the needs of the city itself is called the **nonbasic sector**. A city's basic and nonbasic sectors may be in the primary, secondary, or tertiary sectors of the economy as a whole.

Jobs in the basic sector of a city create jobs in the city's nonbasic sector. For example, if a local factory makes a product that is sold around the world, the factory workers will spend their earnings by shopping locally, getting their hair cut locally, and purchasing other local goods and services. Each job in the basic sector actually supports several nonbasic sector jobs, because earnings from exports circulate and recirculate through the local economy. When a factory worker buys a shirt, the store clerk can get a haircut; the barber, in turn, might eat at a local restaurant, and so on. Thus, jobs in the basic sector have a **multiplier effect** on jobs in the nonbasic sector.

Cities can be classified economically by examining each city's basic and nonbasic sectors and by comparing these sectors among different cities. New York, Detroit, and Hollywood, for instance, each contain a number of dry cleaners and doctors. These people, for the most part, work within the nonbasic sectors. In terms of exports, however, workers in New York provide specialized financial services, workers in Detroit export cars, and workers in Hollywood make movies.

There are two ways of measuring economic specialization. It is possible to analyze either (1) a city's employment structure or else (2) the flow of money through the city's economy. The former approach assumes that any city that has an unusual concentration of workers in a specific job category must be exporting that product or service. A city that has a concentration of auto workers is presumably an auto-assembly site, and so on (Figure 9-3). Methods of analyzing the flow of money through economies are called input-output models. Advanced urban geography courses study both methods in detail.

THE LOCATIONS OF CITIES

The location of any city depends on a balance of the site and situation factors. A mining town is obviously the result of its site. Some mining towns virtually sit on top of valuable ore deposits but are otherwise isolated (Figure 9-4). Cities often developed at waterfalls to exploit hydropower for industry. America's first planned industrial city—Paterson, New Jersey—was built at the Great Falls of the Passaic River in 1791.

U.S. Cities with Specialized Economies

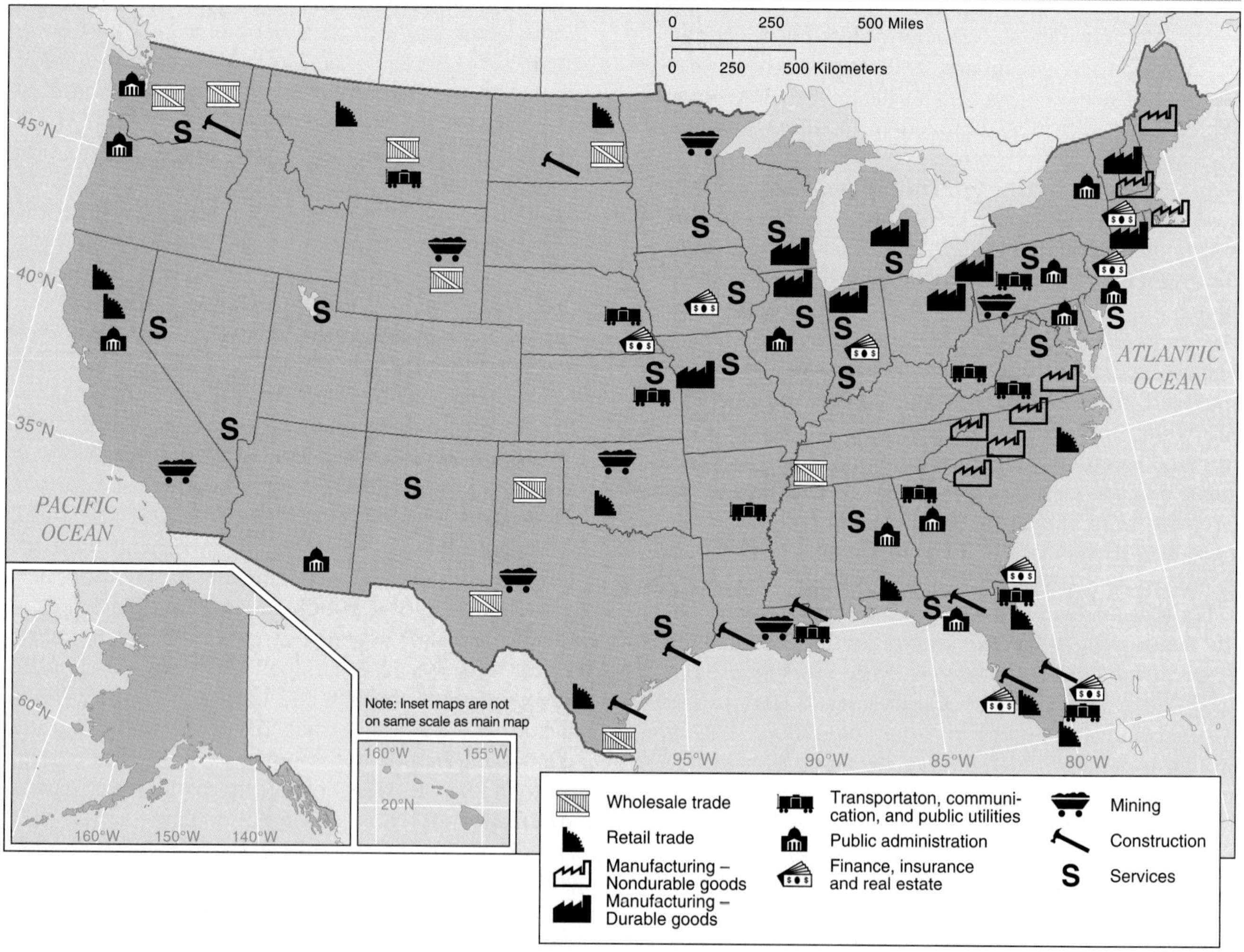

0 250 500 Miles
0 250 500 Kilometers
45°N
40°N
35°N
60°N
PACIFIC OCEAN
ATLANTIC OCEAN
95°W
90°W
85°W
80°W
160°W
150°W
140°W
160°W
155°W
20°N
Note: Inset maps are not on same scale as main map
Wholesale trade
Retail trade
Manufacturing – Nondurable goods
Manufacturing – Durable goods
Transportaton, communication, and public utilities
Public administration
Finance, insurance and real estate
Mining
Construction
S
Services

FIGURE 9-3

This map shows cities that have specialized economies; it does not indicate the cities' individual ranking in each of the activities. Los Angeles, for instance, is America's greatest manufacturing center, but it does not appear on this map because its economy is diverse. (Adapted from J. Clark Archer and Ellen R. White, "A Service Classification of American Metropolitan Areas," *Urban Geography* 6 (1985): 122–51.)

FIGURE 9-4

Butte, Montana, is an example of a city whose location was determined by site factors, in this case its proximity to a rich copper mine.
(Courtesy of Lawrence B. Dodge, Butte, MT.)

Other cities are more the result of their geographic situations. Cities frequently grow up at places where two different physical areas meet or along the border between two cultures. Tombouctoo was located on a key trade route, on the border between two physical environments and also two cultures (see Chapter 3). Many cities spring up as transportation hubs or at bottlenecks, such as at a bridge across a river, or where two political jurisdictions funnel trade through border checkpoints. Some cities grow at sites where the method of transportation necessarily changes. These are called *break-of-bulk points*. A seaport is an example. Another is the head of navigation of a river. If there is a waterfall on a navigable river, cargo has to be off-loaded and reloaded beyond the waterfall up or down river, or else the cargo might be shifted to rail or truck.

Louisville, Kentucky, for example, was laid out in 1773 at the falls of the Ohio River. The river is navigable both up and down river from Louisville. The opening of the Portland Canal in 1830 allowed ships to pass around the falls, but by then the city was well developed. It provided many services to its rich developing hinterland and to westward-moving pioneers. The bridge across the Ohio River focused north-south traffic, and later the railroad lines also focused on the city.

If a situation is favorable, a great city may arise on an unfavorable site. Many of Asia's coastal cities were built by European merchants or conquerors at sites that provided access to the sea and that may have been defensible, but they sit on deltas or the swampy foreshores of tidal rivers. Karachi, Pakistan; Madras and Calcutta, India; Colombo, Sri Lanka; Yangon, Myanmar; Bangkok, Thailand; Ho Chi Minh City, Vietnam; and Guangzhou, Shanghai, and Tianjin, China, all present to this day formidable problems of drainage, water supply, construction, and health.

One of human geography's great paradoxes is that one of the world's largest cities, Mexico City, is located on one of the world's most unfavorable places to build: a drained lake bed in an earthquake zone in a basin of interior drainage at a high elevation in a dry climate. These conditions combine to cause physical instability, alternating flooding and lack of water, and the world's worst air pollution. Both human lungs and internal combustion engines are inefficient when high altitude reduces oxygen levels by 23 percent, and air pollution is aggravated by local windstorms. Today's Mexico City, however, was the site of the Aztec capital Tenochtitlán at the time of the arrival of the Spaniards. The Spaniards maintained the site as their capital, and so have modern Mexicans. Thus, this great city testifies to the power of history and geographical inertia.

URBAN ENVIRONMENTS

▼▼▼▼▼▼▼▼▼▼▼▼▼▼▼▼▼▼▼▼▼▼▼▼

Cities make special demands on physical environments, and this can create problems for their inhabitants. Chapter 1 hypothesized an urban biome, and among its characteristics was a lack of natural vegetation. This is unfortunate, because plants absorb air pollutants, give off oxygen, help cool air as water evaporates from their leaves, muffle noise, provide habitats for other animals, and satisfy an important psychological need for nature.

Chapter 2 listed some urban air pollutants, including ground-level ozone, sulfur oxides, nitrogen oxides, carbon monoxide, and carbon dioxide. These result largely from burning fossil fuels, and they all cause respiratory problems and impair human motor functions. Lead poison is produced mainly by emissions from combustion engines, and it adversely affects metabolism, blood, and kidney functions. The United States has made great efforts to clean the air in its cities, but as of October 1991, more than 86 million people lived in U.S. metropolitan areas where the air was still officially labeled "unhealthy" (Figure 9-5). Air pollution can increase mortality rates from lung and heart diseases.

Large cities also create local climatic conditions, called **microclimates**, that are different from those

Cities and Urbanization   **307**

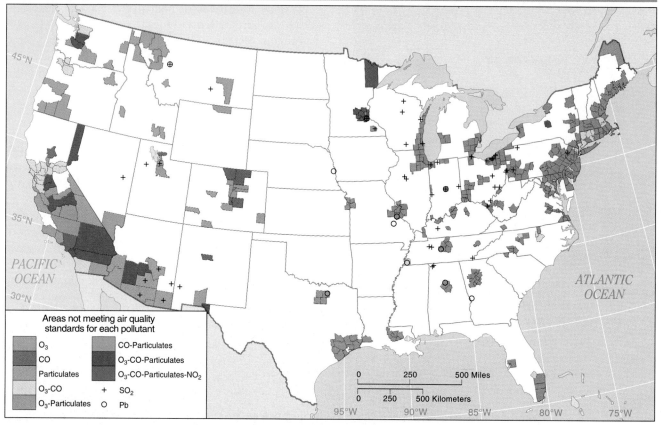

FIGURE 9-5

This map shows counties that (as of October 1991) did not attain the standards for healthy air as defined under the Clean Air Act Amendments of 1990. (Courtesy of the Environmental Protection Agency, *Air Quality Atlas*, May 1992: p.9.)

of surrounding areas. In urban microclimates average temperatures, precipitation, fog, and cloudiness are generally higher than in nearby rural areas. The machinery in cities generates heat, and air conditioning units pump heat out into the streets. Buildings, paved streets, and parking areas absorb incoming insolation. Rainfall runs off quickly so that little standing water is available to cool the air by evaporation. The cumulative effect of these factors is to create an **urban heat island**. This dome of heat over a city traps pollutants, which winds may carry out over areas even hundreds of miles away.

Cities are also noisy. Even the Emperor Julius Ceasar complained about the noise in ancient Rome. Noise interferes with communication and sleep, raises blood pressure, triggers muscle tension, migraine headaches, elevated cholesterol levels, gastric ulcers, and psychological stress. It diminishes people's ability to concentrate, and, if it is loud enough for long enough periods of time, it can damage hearing capability.

Water running off pavement can overload sewers and drains, contributing to water pollution and flooding. Many cities are built on floodplains, which are naturally susceptible to flooding. As cities grow and their water demands increase, it becomes difficult to assure adequate quantity and quality of supply. In the United States in 1990, for example, per capita water use in urban areas was 375 gallons per day. The problems associated with meeting these demands, as well as the problems of cities' production of garbage and solid waste, are discussed in Chapter 11.

HOW CITIES ORGANIZE SPACE

▼▼▼▼▼▼▼▼▼▼▼▼▼▼▼▼▼▼▼▼▼▼▼▼▼

The existence of a city and the task of supplying any city with its needs result in the organization of the hinterland. The relationships between cities and their hinterlands have inspired two models of understanding why cities are distributed across territory the way they are. Johann Heinrich von Thünen's isolated city model was analyzed in Chapter 6. The other model is Walter Christaller's central place theory. Both von Thünen and Christaller started with the simplest imaginary landscape—an isotropic plain on which transportation cost is determined according to straight-line distance.

Von Thünen asked the question "If there were a city on such an isotropic plain, how would the territory around the city be organized?" Christaller asked the complementary question, "How would cities be distributed on such a plain?"

Christaller's "Central Places"

Walter Christaller (1873-1969) asked: If cities are to be distributed to serve as convenient centers for exchange and other services across an isotropic plain, how will cities be distributed? What will be the pattern of towns and their hinterlands?

To answer these questions, Christaller developed his **central place theory**. This model has three requirements: (1) the hinterlands must divide the space completely, so that every point is inside the hinterland of some market; (2) all markets' hinterlands must be of uniform shape and size; and (3) within each market region, the distance from the central place to the farthest peripheral location must be minimal. The only pattern that fills these three requirements is a pattern of hexagons—six-sided figures (Figure 9-6). Therefore, market towns will be distributed across an isotropic landscape as foci of a grid of hexagons. The distributions of towns in several real landscapes have been shown to approximate this model.

Hexagons appear repeatedly in both nature and in industrial design. Honeycombs are hexagonal, and so probably is the pencil in your drawer. A hexagonal shape allows the greatest possible number of pencils to be cut out of a block of wood.

Urban Hierarchies

Chapter 3 discussed diffusion up or down a hierarchy of cities. There are many small towns, fewer and more widely spaced medium-sized cities, and still fewer big cities. A hierarchy of cities exists because the more specialized a service or product is, the larger the number of potential customers that is needed for that product or service to be offered. No product or service can be offered without a minimum number of customers. This minimum demand is called the **threshold** for that product or service. A dry cleaner,

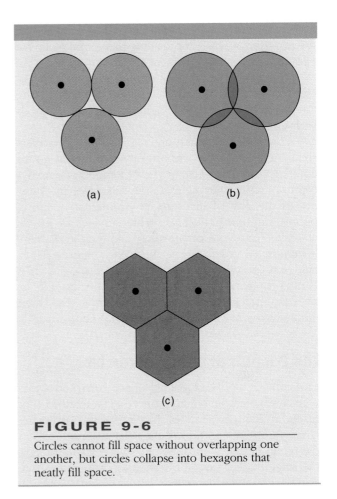

FIGURE 9-6

Circles cannot fill space without overlapping one another, but circles collapse into hexagons that neatly fill space.

for example, needs a minimum number of customers to earn a living. Most people periodically need a dry cleaner, so the threshold of demand for a dry cleaner is low. Each neighborhood or small town can support one. Similarly, many people frequently buy fresh bread and milk. Therefore, small groceries can be found in each neighborhood.

Few people, however, take tuba lessons or buy diamond bracelets. The threshold of demand sufficient to support tuba teachers or expensive jewelers is high, so only larger towns will be able to support these enterprises. On an isotropic plain on which people and buying power are equally spaced, dry cleaners and grocers will be closely spaced; tuba teachers and jewelers will be widely spaced.

In real landscapes, factors in addition to distance must be considered. If one area has a concentration of rich people, for example, then the density of jewelers will be greater there. Quantity of purchasing power substitutes for numbers of people. If another area has a high population density, then grocery stores would be denser there.

A provider of a service can do one of two things to reach his or her necessary threshold of customers. First, the provider can be *itinerant* (travel from place to place). Alternatively, the provider can set up shop at one convenient place and wait for people to come. Convenience and accessibility are the principal purposes of cities. Agglomeration of services in cities saves travel costs, and it allows the thresholds to be reached for more specialized goods and services.

The range of goods and services in a city offers small businesses the opportunity to rent pieces of equipment and to hire services or temporary employees only when they need them. This saves the businesses the cost of investing in equipment or hiring people full-time. These available services or goods are called **external economies**, and their availability lowers initial costs for new business ventures. Thus, cities are called incubators of new businesses. If a company grows large enough to justify buying its own equipment or hiring full-time employees, then the company has achieved an **internal economy**.

The Patterns of Urban Hierarchies

Christaller's model of evenly spaced market towns with hexagonal hinterlands can be elaborated to represent a hierarchy of cities. In the isotropic plain the larger cities that are "above" the small market towns in the urban hierarchy must be more widely spaced than the small market towns, but they too must be evenly spaced, and their hinterlands must also be hexagonal. Therefore, the distribution of the larger cities is represented by a grid of larger hexagons superimposed on the grid of small hexagons (Figure 9-7).

Many factors in the real world disrupt Christaller's model; the world is not an isotropic plain. The model has proven useful, however, in planning new cities in countries with unsettled lands. The land that the Dutch have claimed from the sea behind new dikes is similar to an isotropic plain. There Dutch geographers have planned new market towns on hexagonal grids. Brazil has adapted the model to settle territories in Amazonia. The government built what it calls an "Agrovila" every 6 miles (10 km), with a school, a health-care center, and a post office. Every 25 miles (40 km) the government placed an "Agropolis;" these offer the services of an Agrovila, plus sawmills, stores, warehouses, banks, and other commercial services. Each 85 miles (136 km) Brazil built a "Ruropolis," a city, and established light industry there. The ratio of 6, 25, and 85 miles cannot be reconciled with the proportions of a "pure" geometric grid, but they have been found useful by the Brazilian planners. The planned towns serve as growth poles, foci for economic development, as discussed in Chapter 3.

When improvements in transportation allow people to travel farther to obtain services or goods, the

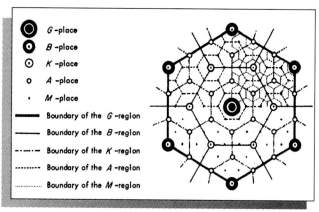

FIGURE 9-7

The distribution of cities on an isotropic plain could be represented by a grid of large hexagons superimposed upon the grid of the small hexagons that represent the hinterlands of the small market towns. Geometry defines at least three ways of doing this. The pattern illustrated is called the K3 model because in it the hinterland of each city includes a hinterland that the city serves as a market town plus one-third of the hinterlands of each of the six surrounding contiguous market towns. When a resident of a market town needs a higher-order service available only in a city, he or she finds each of the three cities equally convenient. Therefore, that person might visit each city occasionally, so news travels rapidly across a landscape organized in this way.
(Reprinted with permission from Walter Christallen, *Central Places in Southern Germany,* trans. Carlisle Baskin. Englewood Cliffs, NJ: Prentice Hall, 1966, p. 61.)

smallest central places may lose their reason for existence and disappear. This is happening in the United States. Across the farm states of Kansas, Nebraska, Iowa, and the Dakotas most small towns developed as commercial centers for farmers. They offered grocery stores, banks, hardware stores, farm implement dealers, automobile dealers, and feed stores. Technological advances, however, starting with the tractor, as well as advances in transportation and communication and other economic forces, have increased the size of farms and cut the number of farmers. As a consequence, there is less need for the small-scale central places. Farm states are today punctuated by wholly abandoned towns, and the number of rural towns is expected to continue to decline. Only small towns within a reasonable distance of a metropolitan area can survive, or even prosper. These become **exurbs**, the name given to settlements that make up the outermost ring of expanding metropolitan areas.

WORLD URBANIZATION
▼▼▼▼▼▼▼▼▼▼▼▼▼▼▼▼▼▼▼▼▼▼▼▼▼

Urban geographers compare urbanization as it occurred in the past and as it is continuing today. The rapidity of worldwide urbanization today presents many nations with both challenges and opportunities for the welfare of their populations.

Early European Urbanization

In the seventeenth century, Holland became the first modern urban country, with over one-half of its population living in towns and cities. The principal economic functions of the cities were tertiary sector functions. The cities arose as administrative and commercial foci of Holland's global shipping, banking, and trading activity, and the urban populations could be supported by Holland's highly productive agriculture. This urbanization occurred before the Industrial Revolution.

In Britain, a larger country with a more varied economy, urbanization occurred with industrialization. Over one-half of Britain's people lived in cities and towns by about 1900. Several developments over the previous 200 years had resulted in the concentration of Britain's population. These included the following:

1. Improvements in agricultural technology—part of the Agricultural Revolution—reduced the number of necessary agricultural workers. Land owners found it profitable to release employees and to evict tenants. The resulting rural depopulation is described in Oliver Goldsmith's poem "The Deserted Village" (1770), which laments that "rural mirth and manners are no more."

2. These displaced people migrated to the cities. There, many were absorbed by the concurrent **labor-intensive** stage of the Industrial Revolution. An activity is labor-intensive if it employs a high ratio of workers to the amount of capital invested in machinery. Other newcomers to the city found work in the tertiary sector. The largest class of urban workers was actually domestic servants—a tertiary sector job. In 1910, they formed more than one-third of the total British labor force. A similar pattern could be found in the United States and other rich Western countries.

    Rural-urban dislocation caused appalling hardship. The history of urban life in nineteenth-century England records overcrowding in dreadful slums, malnutrition, starvation, crime, attempted assassinations of political figures, and death. The descriptions of the miseries of England's slum populations in Charles Dickens's novels still haunt our imaginations. Cities have never easily absorbed all those who have flocked to them.

3. Population pressures were somewhat relieved by emigration, or else by the forcible exportation of criminals and debtors throughout the British Empire. The colonies of Georgia and Australia absorbed many of these deported people.

This British experience provided the world's first model for modern urbanization, and most of today's developed countries experienced similar histories. In 1800, the 21 European cities with populations of 100,000 or more held about 4.5 million people, or one thirty-fifth of the total European population. By 1900, there were 147 such places with a total population of 40 million, or one-tenth of the total population.

The word *model* as used here does not mean that the British experience of urbanization was perfect or that it should be imitated. It imposed terrible hardship on millions of people. Britain provides a model only in the sense that Britain experienced these forces of urbanization first, so that experience can be compared with the current situation. In British history the push of rural displacement and the pull of urban job opportunity were not coordinated. Governments have since learned better how to manipulate both forces in order to ease the process of urbanization, and yet some aspects of contemporary urbanization are causing as much hardship as the experience did in Britain.

URBANIZATION TODAY

Today urbanization is occurring in many places without concomitant economic development, especially in the world's poor countries. Burgeoning populations swamp the cities' ability to absorb the people and to put them to work. The populations overload the cities' infrastructures of housing, water supply, sewerage, internal transportation, and education. Reliable water, electricity, and telephone services are becoming increasingly rare. From the tops of new skyscrapers in many modern cities, the view presents vast shantytowns of the desperately poor (Figure 9-8). Living conditions in these pullulating cities are no better, for the majority, than those which existed in Europe in the nineteenth century.

Rapid urbanization is a problem, but problems can also be opportunities. New urban populations present concentrated labor forces that might be put to productive work in ways that could raise the living standards of all people. To do so, however, presents a tremendous challenge.

The rapid urbanization often reflects deteriorating conditions in the countryside. To that degree

Cities and Urbanization **311**

FIGURE 9-8

The immigration of Bolivian peasants into La Paz has inflated the city's population to over 1 million and surrounded the modern downtown skyscrapers with rings of slums. Sights like this are typical of the growing cities in the poor countries.

(Courtesy of Agency for International Development, John Metelsk.)

urbanization is a rural problem, but in many poor countries it is exacerbated by sharp cultural differences between urban and rural populations. Many cities in today's poor countries evolved as outposts of international commerce grafted onto the local societies, and their populations may even be cosmopolitan mixes of ethnic groups that are not native to the region. Ethnic animosities may intensify economic problems. Throughout Latin America, for example, the cities are predominantly white or mestizo, and measures intended to restrict urban migration are interpreted, often correctly, as racist discrimination against native Americans. In East Africa, as noted in Chapter 5, urban populations were often Asian, and in Southeast Asia urban populations were Chinese

and Indian. These cultural and ethnic contrasts make it harder to deal with rapid urbanization.

At least six other circumstances differentiate urban growth today from the historic British experience.

1. The commercialization and mechanization of agriculture has accelerated, accelerating the displacement of rural workers. Furthermore, modern agricultural technology requires capital investment, so the income gap between rich and poor widens. At the same time, rising rural populations multiply pressures on the land. National radio and television penetrate the countryside, spreading an image of the city as a place of opportunity.

2. In the past, disease and starvation among the poor kept urban death rates above urban birth rates. Without a steady influx of rural people, the town populations would have decreased. Today, however, the introduction of medical care and hygiene lowers urban death rates and triggers natural population increases. These compound the population growth that results from in-migration from the countryside.

3. Urban economies have changed in ways that make cities less able to employ the displaced rural population. Historically, urban economies offered entry-level job opportunities for the unskilled—jobs available for anyone with a strong back and a willingness to work. These jobs concentrated in domestic servitude, manufacturing, and construction. Today, however, domestic servitude has declined, there may be fewer jobs in manufacturing and construction, and many of the jobs that exist in these industries demand skills. Manufacturing and construction are less labor-intensive than they used to be; they are increasingly **capital-intensive**. Machinery has replaced workers. Tertiary sector jobs other than domestic servitude have multiplied, but most of these require job skills.

4. For most societies, the safety valve of emigration has been stopped. Migration opportunities for the unskilled and for rural workers are decreasing.

5. The world economy is increasingly interdependent, so local governments have less control over local economies. Factory openings or closings in Bangkok are determined in Tokyo. Worker opportunity in Caracas, Venezuela, is regulated from New York.

6. Many governments continue to favor urban projects over rural projects. One reason for this may be status. A shiny new hospital in the capi-

312   *Chapter 9*

tal city, for example, may seem more impressive than a thousand new water pumps in poor villages. Another reason may be that the government fears urban rioting. Urban food riots have occurred in several African countries, including Nigeria, Egypt, and Zambia. Even in relatively well-off Venezuela, urban riots over the price of food left over 250 dead in March 1989 and over 300 dead in November 1992. As noted in Chapter 6, many governments subsidize urban food supplies by underpaying the farmers or buying cheap food that is dumped on world markets. This discourages their own farmers. If countries showcase urban projects and subsidize urban food supplies while neglecting rural areas, their actions enhance the perceived opportunity in the cities without providing real opportunity. Young farm people who migrate to the cities only exchange poverty for destitution.

Government Policies to Reduce Urban Migration

People migrate from the countrysides to the cities because of the balance of push and pull factors. Governments might regulate this migration, therefore, either by reducing the attractiveness of the cities or else by improving rural life.

Reducing the Pull of Urban Life Forceful measures to limit urbanization are not new in human history. The Russians have been required to have internal passports since the days of Czar Peter the Great over 200 years ago. Recently the world has even witnessed brutal instances of compulsory "ruralization." When North Vietnam absorbed South Vietnam in 1976, the new rulers relocated millions of urbanites to rural areas to raise food using labor-intensive methods. The government argued that the cities had bloated on U.S. financial assistance. The government of Kampuchea (Cambodia) similarly relocated urban populations in the late 1970s. Some governments use media to discourage migration to the city. Ghana, for instance, broadcasts to the villages films suggesting that life is better there than in the cities.

Many cities try to discourage newcomers by restricting housing and economic opportunity. Building codes that ban substandard housing, for instance, are often ineffective in stopping the growth of slums, but they make squatters' settlements illegal. Therefore, their residents do not get city services and will not risk investing to improve the property. The city, in turn, cannot collect property taxes. These "illegal" communities hold 30 to 60 percent of the total populations—70 to 95 percent of newcomers—in the cities of the poor countries. Cities also try to restrict small businesses in residential areas, but backyard workshops may thrive anyway. Cities may discriminate against new urbanites in education, housing permits, business licenses, or other job opportunities. Hawkers are chased from city centers. In some countries frustrated city authorities have even bull-dozed squatters' settlements with only 1 or 2 days of warning. This has happened in major cities throughout Latin America, Africa, and Asia, leaving tens of thousands of people absolutely homeless.

In the extreme, the urban governments may round up squatters and drive them out. Tanzania's Human Resources Deployment Act declares that anyone who cannot show proof of employment in Dar es Salaam can be deported to a state farm; thousands have been moved.

Chinese are not supposed to move anywhere without getting their household registration changed, but people move anyway. In 1990, China counted 80 million "surplus" rural workers and an additional 50 million migrant laborers, called *floaters*. In major cities they make up more than 10 percent of the population, freely coming and going. Because their residence is technically illegal, they have no right to education, medical care, and other government services. Many urban residents find the floaters useful. They bring produce to the city or provide services not provided by central planning.

Improving Rural Life Instead of trying to drive people out of the cities, governments might invest in rural health care and education, and housing, roads, and other infrastructure to raise rural standards of living, keep people in the countryside, and also raise national food production.

If the government is poor, it may nevertheless organize labor-intensive projects. Agricultural terraces, roads, and dams can be built by hundreds of thousands of workers equipped with shovels and wicker baskets just as well as by relatively few workers equipped with expensive earth-moving machinery (Figure 9-9). During the Depression in the 1930s, even the U.S. government Civilian Conservation Corps, Civil Works Administration, Public Works Administration, and Works Progress Administration all used labor-intensive methods to alleviate unemployment and at the same time build or repair national infrastructure.

Cities' Economic Vitality

 Most of the previous paragraphs about urbanization in the developing countries

FIGURE 9-9

Construction of even massive infrastructure projects such as this dam in India can begin with women carrying baskets of dirt on their heads. In the rich countries, however, labor-intensive construction methods have yielded to capital-intensive methods.
(Courtesy of the Agency for International Development.)

may have seemed pessimistic. They presented urban growth as a problem for which solutions were needed. There is another side to the coin, however—it is possible to view the backyard shops, the street hawkers, and the Chinese floaters as examples of opportunity and growth. These activities hint that urban migration, balanced between management and liberty, can provide a reservoir of economic vitality that can be harnessed for national growth.

The book *The Other Path* (1986) is a worldwide bestseller about urbanization in Peru by the economist Hernando De Soto. De Soto noted that in many cities the productive activities of a substantial share of the population do not appear in official accounts. The people may not have licenses to do what they are doing, they may be avoiding taxes, or for some other reason their activity escapes official notice. These activities make up the **informal** or **underground sector** of an economy. Every city in the world has such a sector. In the rich countries, immigrants, often illegal immigrants, often activate this sector. Studies suggest, however, that in the cities of the poor countries the informal sector may be particularly important.

De Soto estimated that in Lima, Peru, the informal economy employed fully 60 percent of the population and produced 40 percent of all goods and services. The poor own and control 95 percent of the

public transportation network of private taxis and vans. This represents capital of at least $1 billion. De Soto estimated that the urban land and housing occupied by the poor was worth an additional $17 billion. None of this, however, was legal, so it could not be taxed by government or used as collateral by business owners. If the government simply legalized this activity, the business operators could borrow additional investment capital, their businesses could grow, and government could tax the activity. Therefore, the government's refusal to recognize what was happening handicapped the country's economic growth and vitality.

In more recent writings in the mid-1990s, De Soto insists that economic growth will occur most rapidly in the developing countries that ensure property rights. A modern market economy fosters specialization and greater productivity, but it cannot develop unless property rights are acknowledged and protected; therefore, he argues, governments must formalize the spontaneous emergence of informal property. Chapter 6 already noted how productivity in agriculture can be lifted by guaranteeing property rights. The same guarantees, De Soto insists, must develop in the cities.

If governments view urban migration as a problem, they cannot see how urban immigrants' industriousness could be an asset for economic growth. Millions of people continue to choose to migrate to the cities, where they do survive, or even thrive. There is something terrifically dynamic going on, and geographers, economists, and government officials at all levels are challenged to understand and measure it.

THE INTERNAL GEOGRAPHY OF CITIES

Urban geographers study not only the distribution of cities in the landscape, but also the distribution of activities and of housing within cities. This distribution may be caused by economic forces, by social factors, or by deliberate actions of the government.

The Role of Economic Factors

Certain locations in cities are more desirable than others, either for businesses or for residences, and these can demand a higher rent. Many factors influence desirability. Some areas are considered more prestigious than others, as mental maps and rents reveal. Residential neighborhoods can generally be classified by the income level of their residents. For businesses, accessibility is usually a principal determinant of a location's rent. The success of a department store, for instance, will depend partly on whether customers can reach the store easily. A city's most convenient and busiest intersections are most valuable for commerce.

Scholars have devised three common models of North American urban growth and land usage: the concentric zone model, the sector model, and the multiple-nuclei model.

Figure 9-10 illustrates the *concentric zone model* of urban growth and land use. The core of the city, called the **central business district (CBD)**, concentrates office buildings and retail shops. Land owners usually maximize the density of use on this valuable land by building up, so a traditional CBD is identifiable by tall buildings as well as crowded streets. Even within the CBD, clusters of functions appear. Lawyers' offices, for instance, cluster near courts or near the offices of their client firms. Retail stores of one type, such as jewelry stores, may cluster so that consumers can comparison shop. The CBD is surrounded by less intensive business uses such as wholesaling, warehousing, and even light industry—that is, nonpolluting industries that require relatively small quantities of raw materials. Residential land use surrounds this urban core.

Sociologists devised the concentric zone model to describe the patterns of housing density and quality in U.S. cities early in the twentieth century. The model proposed that the city's inner zones were con-tinuously expanding. They would "invade" the contiguous outer zones and eventually replace those land uses, or "succeed" them, by pushing those land uses still farther out from the expanding CBD. Cycles of invasion and succession described the replacement of populations of one income level by another, and also the observed replacement of one ethnic group by another among the various urban immigrant communities. One section of New York's Lower East Side, for example, was an Irish neighborhood in the 1860s; by 1900, immigrant Italians had replaced the Irish, and today immigrant Chinese are invading and succeeding the Italians.

The concentric zone model can be modified by considering the effect of transportation routes, which affect accessibility. New means of transport—first canals, then railroads and tramways—spread out radially from the heart of the city, although their paths were modified by topography. Industrial and residential growth took place in ribbons or fingers along these radial routes, and wedges of open land were usually left between these radial routes. This model compares with von Thünen's model of agricultural land use.

A second classic sociological model of urban housing, *the sector model* (Figure 9-11), assumes that high-rent residential areas expand outward from the city center along the new transportation routes. Middle-income housing clusters around high-rent housing, and low-income housing lies adjacent to the wholesale and light manufacturing areas.

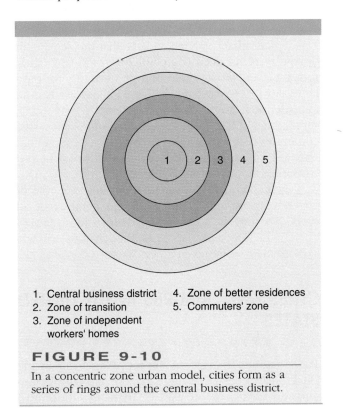

1. Central business district
2. Zone of transition
3. Zone of independent workers' homes
4. Zone of better residences
5. Commuters' zone

FIGURE 9-10

In a concentric zone urban model, cities form as a series of rings around the central business district.

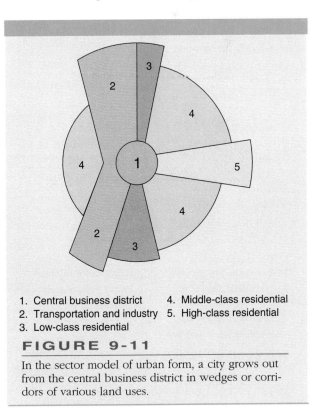

1. Central business district
2. Transportation and industry
3. Low-class residential
4. Middle-class residential
5. High-class residential

FIGURE 9-11

In the sector model of urban form, a city grows out from the central business district in wedges or corridors of various land uses.

The *multiple-nuclei model* shown in Figure 9-12 recogizes the development of several nodes of growth within an expanding city area. These nuclei may each concentrate on a different special function. This model best describes the expansion of U.S. metropolitan areas, as will be discussed in detail below.

Social Factors in Residential Clustering

Social considerations play a role in urban residential clustering. Some people want to live with people like themselves, and this causes a clustering called **congregation**. Ethnic groups or immigrants of a common background, for example, may want certain services, such as grocery stores offering their traditional foods.

In other cases, however, people live together because discrimination forces them to do so. These people suffer **segregation** from others. In any specific case it may be difficult to determine the degree to which people are congregating or are victims of segregation. Chapter 8 noted that in the past Jews were legally segregated in ghettoes, but today the word *ghetto* means any residential concentration of any one kind of people. The factors causing residential clus-

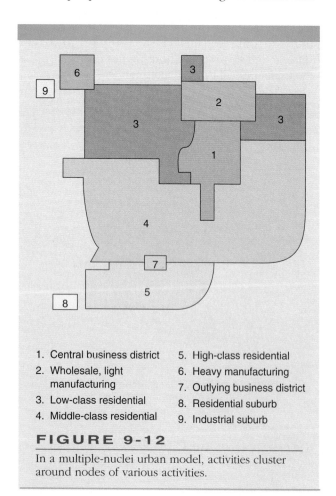

1. Central business district
2. Wholesale, light manufacturing
3. Low-class residential
4. Middle-class residential
5. High-class residential
6. Heavy manufacturing
7. Outlying business district
8. Residential suburb
9. Industrial suburb

FIGURE 9-12

In a multiple-nuclei urban model, activities cluster around nodes of various activities.

tering of any group—of Chinese Americans in "Chinatowns," Italian Americans in "Little Italys," Hispanic Americans in "barrios," or of African Americans discussed later in this chapter—must be evaluated carefully in each case. Sometimes the words placed in quotation marks above are considered insulting, but they usually are not meant to be.

Religion is another social consideration that frequently causes clustering. People who share a religious faith may cluster around their house of worship, and immigrants of that faith will seek that neighborhood. Language communities frequently form, as do communities of the elderly and communities of gays and lesbians. Almost any factor of social bonding can encourage the creation of an identifiable residential neighborhood.

The Role of Government

Government may determine land use. Restricting or prescribing the use to which parcels of land may be put is called **zoning**. In U.S. history, local government has often been averse to planning, so zoning has more often been restrictive than prescriptive. Industrial and commercial districts, for example, are usually kept away from residential neighborhoods. Land can even be set aside for different types of industries, and residential zoning can specify the type of housing allowed, such as single-family homes or apartment buildings. Each incorporated jurisdiction across vast conurbations exercises independent zoning power. This creates problems that will be discussed below.

Some urban planners believe that the separation of land uses has been overemphasized in many cities. It may have been desirable or even necessary for public health to separate industry from housing when all industry was noisy, polluting, or smelly, but today, separating homes from jobs may require excessive commuting. Integrating light industry, commercial activities, and housing might revitalize downtowns and encourage the use of public transit—or even walking.

Economic factors, social factors, and the role of government influence the internal geography of all cities. We turn now to a study of how these factors operate in U.S. cities. Later in this chapter we will explore their workings in a non-Western context.

CITIES AND SUBURBS IN THE UNITED STATES

The dominant feature of metropolitan form in the United States has been the explosive growth of cities out across the countryside. Growing cities have spilled over their legal boundaries into areas called *suburbs*. Some suburbs are entirely residential, but others offer services for the surrounding residential population. Some suburbs are older cities that have been engulfed by the growth of a larger neighbor, but others are newly-incorporated settlements. What defines an area as a suburb is its economic and social integration with a larger population nucleus nearby. *Town* and *village* are inexact terms that generally designate settlements smaller than cities, but that may be incorporated.

Many of the developments described in the following discussion are now occurring elsewhere around the world, but they occurred first in the United States—largely because of the nation's prosperity.

Early Suburbs

Most large U.S. cities included manufacturing districts by the late nineteenth century. These were noisy and dirty, and they often attracted a working class, largely made up of immigrants, whom many long-established residents found to be unpleasantly "different." These biases pushed those who could afford to leave to the suburbs.

At the same time, a cultural preference for rural or small-town life pulled many people out of the city. This predilection for a return to nature, to the land or to open spaces—even if only a suburban yard—is another manifestation of Romanticism. Many Americans fell in love with the idea of "the country." Therefore, when the railroads brought older rural communities within commuting distance of the city, many people who had the time and the money necessary to live outside the city and to commute to work began to do so. In other cases, the wealthy built new towns (Figure 9-13). "Streetcar suburbs" sprang up when that new means of transportation appeared. Some of these planned suburbs eventually became completely built up, merged into other settlements, and lost their identities.

The automobile ultimately opened the nation's landscape to suburban growth. For those who disliked urban life, the suburb was the solution. "We shall solve the city problem," wrote Henry Ford in 1922, "by leaving the city."

Government Policies and Suburban Growth

The dispersion of housing in suburbs was slowed by the Great Depression in the 1930s and by World War II in the early 1940s. Following the war, however, government policies established a new balance of push and pull forces that encouraged the movement of investment, residents, and jobs out of the central cities into the suburbs.

The Federal Housing Administration (FHA) guaranteed loans, so down payments shrank to less than 10 percent of the house price. Suddenly, thousands of families could afford new houses. FHA benefits did not apply equally everywhere or to to everyone. The FHA favored the construction of new single-family houses in the suburbs over the rehabilitation of older houses or apartment buildings in the central cities. Also, the FHA opposed what it termed "inharmonious racial or nationality groups." In some places the presence of one nonwhite family on a block was enough to label the entire block "Negro" and to cut it off from FHA loans. Government policy helped segregate the suburbs.

The government also granted tax and financial incentives to homeowners. These included the deductibility of both mortgage interest payments and of local property taxes from gross taxable income. These two benefits alone often made buying a new house cheaper than renting. By 1993, the annual tax loss to the U.S. Treasury of these two benefits totaled over $80 billion. This is, in effect, a subsidy to homeowners that is many times greater than government expenditure on public housing.

The government also protected homeowners from capital gains taxation, that is, taxes on an increase in the value of the house, and savings and loan institutions were allowed to pay higher interest on savers' money than commercial banks were, because money in savings and loans was directed into housing. (This last condition was true only until 1980.)

The Veterans Administration Housing Program, begun in 1944 and called "Homes for Heroes," pumped additional billions of federal dollars into housing programs. By 1947, the Levitt Company was completing 30 new single-family homes in Levittown, formerly a Long Island potato field, each day, and similar developments were springing up on the outskirts of every major U.S. city (Figure 9-14). Nationwide housing starts jumped from 114,000 in 1944 to 1,696,000 by 1950.

The suburbs brought fulfillment of the dream of home ownership to an increasing share of U.S. families. The percentage of U.S. housing that was owner-occupied rose from 44 percent in 1940 to 62 percent in 1960, signaling middle-class status for a rising share of the population. Expanding home ownership increased the number of citizens who profited from the many homeowner subsidies. Therefore, it has reduced the political possibility of rescinding them.

Cities and Urbanization   **317**

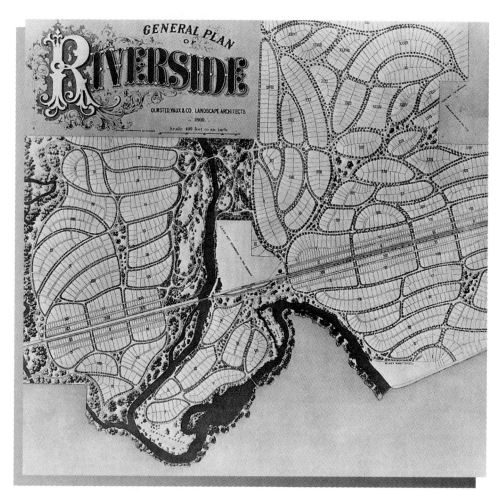

New suburbs incorporated as the population spread out, and the Census Bureau devised a new term for these sprawling conurbations: **metropolitan statistical area** (**MSA**; originally standard metropolitan statistical area, *SMSA*). The Bureau defined a metropolitan area as "an integrated economic and social unit with a recognized large population nucleus." Thus, MSAs are the principal central cities and their suburban counties (except in New England, where the definitions are in terms of cities and towns). Two or more contiguous MSAs form a **consolidated metropolitan statistical area** (**CMSA**). By 1990, the nation's 284 MSAs contained 77.5 percent of the total population, 31.3 percent in the central cities and 46.2 percent in their suburbs. These MSAs covered only 17 percent of the country's land surface (Figure 9-15).

MSAs and CMSAs are regions defined for statistical measurement. They are not governmental jurisdictions, yet their definition is a matter of tremendous political and economic importance. The federal government calculates regional wage rates within MSAs, and then it pays wages and awards contracts based on those values. Therefore, the inclusion or exclusion of particular suburbs may raise or lower the wages

for thousands of workers in any MSA by several percentage points.

The Suburban Infrastructure

Unplanned sprawl of single-family homes is expensive. It requires, first, roads. Individualized transportation also demands energy, and so does heating and cooling individual homes, even if they are well insulated. Dispersed housing also requires enormous investment in sewerage, water pipelines, telephone lines, and electrical wiring. The cost of providing electric wiring for 100 people in an apartment building is much less than the cost of wiring for the same 100 people spread out in 20 single-family homes over 4 or 5 acres (1.5 to 2 ha).

Infrastructure costs are further inflated by the fact that U.S. suburbs have not expanded contiguously outward from the city, like the waves from a stone tossed into a pond. Each developer wants to buy land as cheaply as possible, so the developer buys land beyond the edge of growth. This is called *leapfrogging*. The infrastructure network cannot be

FIGURE 9-14

In building their first Levittown, 25 miles (40 km) east of Manhattan, the Levitt family changed U.S. homebuilding techniques. The land was bulldozed and the trees removed, and then trucks dropped building materials at precise 60-foot (19-m) intervals. Construction was divided into 27 distinct steps. At the peak of production more than 30 houses were completed each day. Through the years owners have personalized them so much that few visitors today can see that the houses were originally identical. (Courtesy of UPI/Bettman.)

advanced in a regular pattern. This increases initial costs. Later, the leapfrogged areas are filled in, and then some infrastructure has to be rebuilt. Many suburbs, for example, originally relied on individual home wells for water, and their sewage was treated in individual home septic tanks. As the suburbs matured and density increased, water supplies became polluted. Homeowners had to pay for wholly new public water mains and sewers.

Suburbs are spacious, and between 1945 and 1990, U.S. urban areas spread at the rate of about 1 million acres (405,000 ha) per year. This required that a good share of the country's most productive farmland be paved over, such as the Long Island potato fields covered by Levittown. The reason for the original location of many cities had been the agricultural productivity of their immediate hinterlands, but that agricultural productivity disappeared under concrete and crabgrass. As a result, a rising percentage of the nation's food is now grown in suboptimal conditions, requiring expensive fertilizer or irrigation. In addition, the food has to be transported farther, consuming still more fuel, perhaps requiring refrigeration or special handling, and further boosting food prices.

Farmers were virtually forced to sell their land to developers because property taxes are calculated on land's potential value, not its current-use value. A farmer who was making a small profit by farming could not afford to pay property taxes calculated on the land's potential value as housing. Even if a farmer could somehow meet the annual property tax bill, then the farmer's heirs would eventually have to sell the farm to pay inheritance taxes, which are also based on land's potential value. Only recently have these tax laws been changed to preserve greenbelts around some cities.

In the 1950s and 1960s, U.S. citizens did not worry about the costs of creating new suburbs. Between 1950 and 1973, median family income doubled in real terms. Demands for housing mushroomed with the baby boom and the splintering of families into separate households, partly as the result of a rising divorce rate. The average number of occupants of a U.S. household shrank from 3.67 in 1940 to 2.63 in 1990.

Today a greater share of U.S. wealth is invested in housing and the necessary infrastructure than in any other nation. This tremendous investment succeeded in bringing a sense of well-being and a higher quality of housing to the U.S. population than is arguably available in any other country. A high proportion of the nation's housing stock is substantially new, and a high proportion of Americans enjoy private ownership of spacious, free-standing, well-equipped homes. It is becoming clear, however, that this development has brought with it some steep social costs.

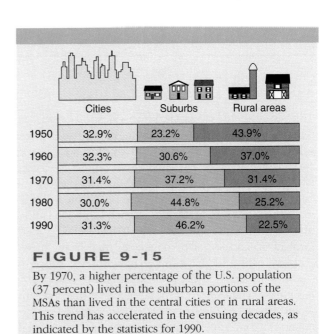

FIGURE 9-15

By 1970, a higher percentage of the U.S. population (37 percent) lived in the suburban portions of the MSAs than lived in the central cities or in rural areas. This trend has accelerated in the ensuing decades, as indicated by the statistics for 1990.

Cities and Urbanization　　**319**

GEOGRAPHIC REASONING
The Question of Private Property

Questions

1. If you shop at a mall, are political candidates allowed to campaign there?
2. Would you be allowed to set up a table and gather signatures to run for mayor yourself? Where else could you go in your community?
3. Are beggars or charity fundraisers found either in your downtown or at the mall? Might their presence distract you from shopping?
4. Does the public function doctrine cover malls in your state?

One significant difference between traditional downtowns and most new villages and suburban malls is that the latter are private property. People cannot be banned from a traditional downtown, and the right to petition on a public sidewalk is constitutionally protected. People may, however, be banned from private property, and constitutional rights, such as political pamphleteering, may be restricted. In many communities today, malls are the only public gathering places, so if the mall owners are allowed to decide who may speak in them, mall owners can determine the public's access to ideas. Candidates for political office have been banned from busy malls owned by their opponents, and so have labor union organizers.

U.S. law recognizes that private properties can perform functions traditionally associated with government. This is called the *public function doctrine*. The U.S. Supreme Court has ruled that the Constitution does not protect citizens' access to shopping centers against the wishes of the owners (*Pruneyard v. Robins*, 1980). Some states, however, have upheld the right of public access under their state constitutions. Each state has balanced public and private rights differently.

Today in the United States increasing numbers of well-to-do people live, work, and shop in private environments, enjoy private recreational facilities, send their children to private schools, and travel by private means. This has been called "the secession of the successful." These private environments are patrolled by private security officers, one of the country's fastest-growing occupations. (Security guards constitute almost 2 percent of the nation's total labor force, triple the number of public police officers.) The privatization of space is a new form of economic and social segregation that carries profound influence throughout U.S. political and social life.

Social Costs

The suburan lifestyle has imposed social costs on those who enjoy it. Americans generally sort themselves out residentially so distinctly that sociologists and mass marketers can confidently predict an astonishing amount of information about people on the basis of their address alone. The selection of "junk mail" that arrives in mailboxes in each Zip code is fine-tuned. In low-density suburbs, it has been argued, local racial and social homogeneity may have caused conservatism and conformity. In addition, property owners in many suburbs established *restrictive covenants*. These were legal agreements that the land would never be sold to people of a designated race or religious group. Such convenants are no longer legal, but they were common as late as the 1970s.

In 1993, more than 32 million Americans, nearly one in eight, were living in communities regulated by some form of homeowners' association, and that number is growing. Most of these associations require not only that homeowners maintain their yards, but that major changes in a home's structure or exterior appearance be approved by the association. The purpose of these rules is to maintain neighborhood property values, but some Americans find them stifling. In 1992, one young suburban couple went to jail in Washington State for painting their house mauve.

The suburbs overall are diverse—there are poor suburbs as well as rich ones, and there are suburbs of every ethnic and racial group. Individual suburban communities, however, are usually homogeneous. In 1952, *The Los Angeles Times* became America's first metropolitan newspaper to create separate zone editions. This pleased local neighborhood advertisers, but it fragmented the population's sense of being in one city.

The Movement of Jobs to the Suburbs

The suburbs first expanded as bedroom communities for the middle-class workers who left their suburban families each morning to go to work in the city. Soon, however, the completion of metropolitan expressways reduced the geographical advantage of the central city CBD. Many locations on expressways enjoy accessibility, and the best locations are at the intersections of two expressways. Retailers soon took advantage of these new crossroads to build giant shopping malls. Developers put up suburban office buildings, and corporations built spacious office parks.

Manufacturing establishments abandoned the central city too, for several reasons. New light industrial facilities proliferated in the suburbs because new technology favors horizontal buildings, rather than the vertical ones characteristic of the central cities. In addition, light industries relocated to escape central-city congestion, as well as higher costs for energy, taxes, wages, and rent. Warehousing also relocated out from the inner-city railroad yards to the suburban highway interchanges. Suburban airports grew to provide jobs in both freight and passenger services.

In the 1950s, suburban growth was fed by young married couples who wanted to raise children away from the cities. By the early 1970s, however, the proliferation of jobs in the suburbs became the driving force for new housing. The suburbs surpassed the central cities in employment, and job opportunity became a pull factor for continuing suburbanization (Figure 9-16). Exurbs grew farther and farther out from the central city, and their skilled labor pools and cheap land draw the next wave of development: light industrial and back-office jobs.

Redistributions of Population and Jobs in Metropolitan Atlanta, Georgia

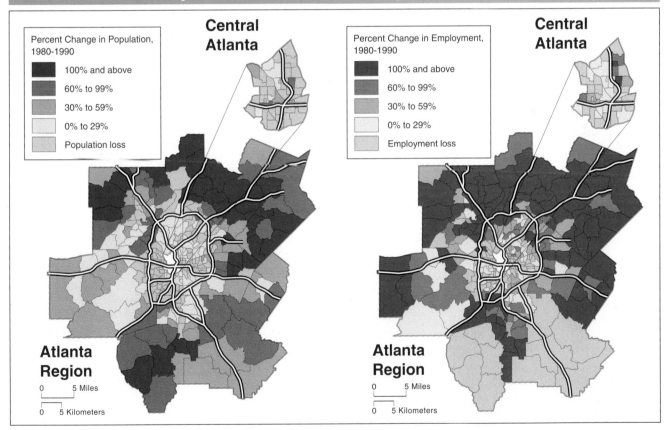

FIGURE 9-16

In Atlanta, as in most U.S. metropolitan areas, the greatest population growth (a) and job growth (b) are occurring in the regions located the farthest from the central city. (Adapted with permission from *Atlanta Region Outlook May 1991*, Atlanta Regional Commission.)

Some of the new exurbs, called satellite cities or edge cities, boast greater volumes of retail sales and contain more office space than the old central cities. They even offer amenities formerly found exclusively in the central cities: art galleries, theaters, sports teams, and fine restaurants. Some of these suburban centers are well designed, and the newest balance local housing and employment. Residents can walk to work or ride local public transit. They form self-contained villages (Figure 9-17). New satellite developments may even allow mass transit into the central city (Figure 9-18), thus reproducing the advantages of the original planned suburbs such as Riverside.

Changing Commuting Patterns

When suburban workers commuted in and out of the city, radial mass transit systems focused on the CBD could serve transport needs tolerably well. Today most commuting is no longer from the suburbs into the central city, but from one suburb to another (Figure 9-19). It relies on individualized transportation. Mass transit cannot serve populations that are spread out at low densities.

As more family members have gone to work, the number of cars on suburban highways has increased faster than the local populations. This is true even in the older, closer-in suburbs that are los-

ing total population to exurbs. Nassau County, a suburban county contiguous to New York City, reported a 2.6 percent population decline between 1980 and 1990, but a 15 percent rise in passenger car registrations. Suffolk County, which is farther from New York City than Nassau, experienced a 3 percent population increase but a 29 percent increase in passenger cars. Everywhere rush hour starts earlier and eases later each day, and even Saturdays now generate traffic jams of shoppers.

All those cars generate air pollution, especially when they are idling, and in 1991, 70 percent of urban interstate highways were officially congested during peak hours, up from 40 percent in 1975. Improving air quality in metropolitan areas will require improving transportation. Technological innovations include traffic monitoring, new methods of informing drivers of traffic jams, and the computerized dashboard navigation systems shown in Figure 1-5. Electric cars are being developed. Several cities

FIGURE 9-17

Rancho Santa Margarita, a new community 50 miles (80 km) southeast of Los Angeles, has won considerable national attention. It will not be fully developed until at least the year 2000, but by 1994 it already had some 7,300 homes. Approximately 85 percent of Rancho Santa Margarita homebuyers who are employed work in the town itself or somewhere within Orange County. Many of these new suburban communities around the country are getting some people out of their cars and back on their feet, and re-integrating home, work, and recreation. (Courtesy of Santa Margarita Company.)

FIGURE 9-18

Through the 1990s, the opening of the Virginia Railway Express System, a new rail system to the suburbs of Washington, D.C., will redistribute development and might stimulate new planned railroad communities there. (Map by Dave Cook of *The Washington Post.* © 1991 The Washington Post. Reprinted with permission.)

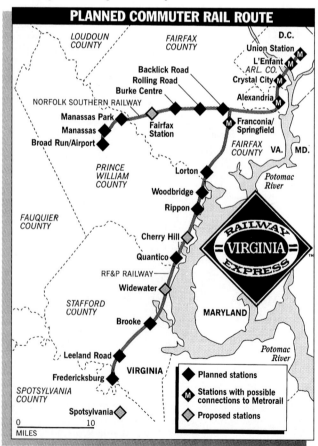

Changing Patterns of Suburban Commuting

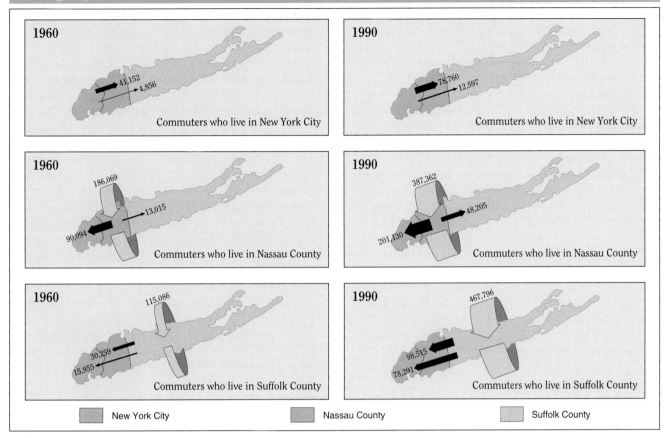

FIGURE 9-19

These maps of commuters' journeys to work on New York's Long Island show that greater numbers and proportions of suburban residents work within their own home counties. Individual automobile transportation necessarily replaces mass transit into the city.

are rediscovering mass transit. Los Angeles opened a rudimentary subway system in 1993.

The Geography of Retailing

Highway congestion, women going to work, fear of crime, 1-800 telephone numbers, credit cards, home shopping television, and other factors are changing the geography of U.S. retailing. More Americans are shopping from home by telephone, catalog, or computer. Elaborate catalogs, called catazines or magalogs, fill mailboxes and compete with fashion magazines, and the number of catalog shoppers doubled from 1980 to 1990. In 1992, some 55 percent of the adult population bought $52 billion of merchandise from home, an increase, in constant dollars, of over 50 percent from 1988. Consumers may soon buy by modem, without even talking to a person, or perhaps by punching numbers into their telephones. These developments introduce *virtual shopping* (see Chapter 3), which reduces the length of time before orders are filled, the quantity

of goods sitting in inventories, and the number of middlemen standing between a product and the consumer.

The shift to such efficient shopping will affect traditional retail outlets. In the 1980s, U.S. retailers built enormous quantities of new retail space, so that by 1993 there were 39,000 shopping centers offering 18 square feet of retail space per capita, up from 14.6 square feet in 1986. As shopping at home increases, stores will suffer. In 1982, shoppers spent 90 minutes on each visit to a mall; in 1992, an average visit lasted only 72 minutes. The number of stores visited per trip dropped from 3.6 to 2.6. Many of the 19 million jobs in retailing in the United States in 1993 are vulnerable to elimination.

Telecommuting

It is easier to move information than to move people, so more people are working at home at computer terminals. This is called **telecommuting**. Clerical work or other work that does not require face-to-face contact might abandon cities within the next few decades. In 1994, an estimated 6 million employees of private companies in the United States were already working from home on computers, and the state governments of California and Washington had telecommuting programs. California has distributed telecommuting centers throughout the suburbs.

Telecommuting eases the pressure on transport facilities, saves fuel, reduces air pollution, reduces the demand for office space, reduces absenteeism, lowers the fixed costs of a business enterprise, and has been shown to increase workers' productivity. It also allows employers to accommodate employees who want flexible work arrangements, thus opening employment opportunity to more people. Women who still have traditional household responsibilities in many families can take on an "inside-outside" job. Enterprises that provide paths of information movement (providers of E-mail, vendors of telecommuting software, developers of video conferencing and data compression systems, manufacturers of home office electronic equipment, and delivery firms) will probably thrive, but we do not know what increased telecommuting will mean for the future geography of large cities.

Developments in the Central City

The drain of jobs from the central city and the concomitant development of the suburbs have hollowed out many major U.S. metropolitan areas. They have been called "economic doughnuts." Many CBDs have lost their purpose. Commercial, professional, and financial offices have relocated to the suburbs, and upscale retailing has followed. Many U.S. downtowns consist of only a government center, a convention center, and a few hotels; the streets are deserted after 6 P.M.

Economic Decline Central cities can thrive as long as (1) their economies offer a complete range of job opportunities from entry-level jobs for the unskilled up to specialized jobs for skilled workers, and (2) family stability, education, and other social systems help people ascend the socio-economic ladder. In other words, cities do not have to retain their middle classes, but they have to offer the lower classes opportunity to become middle class. Unfortunately, the departure of the middle classes to the suburbs occurred at the same time as two other developments.

First, urban economies were transformed by the out-migration of entry-level jobs, particularly in manufacturing, construction, and warehousing. This out-migration broke the rungs of the ladder of upward mobility. In 1968, economist John Kain first suggest-

ed that the removal of manufacturing jobs to the suburbs and the concentration of the poor in the central cities created a spatial mismatch between the suburban job opportunity and the central city low-income housing. This spatial mismatch, he argued, could explain the high unemployment found in the central cities. This **spatial mismatch hypothesis** has until today dominated analyses of inner-city unemployment—and, therefore, the formulation of potential solutions.

Second, just as many unskilled jobs were leaving the central cities, new waves of unskilled workers were pouring into them. The stream of new migrants to the cities included African Americans from the rural South, Hispanics, and other immigrants. This influx continued long after the numbers of entry-level job opportunities began to shrink.

Much of the inner-city housing stock began to deteriorate. This was because the government incentives still made it profitable to give up a house in the city and to move to the suburbs. Policymakers thought that if new suburban houses were available to middle-class people from the cities, the urban poor could move into the older city dwellings. This succession, called *filtering*, would solve the housing problem for lower-income families. Much of the central-city population remaining behind, however, was financially incapable of maintaining the inherited housing stock. Therefore, many central-city neighborhoods deteriorated. The term "inner-city neighborhood" became a euphemism for slum.

The Service Economies Some downtowns have enjoyed spectacular growth in financial, information, and specialized technical services. In these areas skyscrapers have replaced rusty factories (Figure 9-20). New York's leading export, for instance, which for decades was garments, by the late 1980s was legal services. Between 1950 and 1987, the number of manufacturing jobs in New York fell from over 1 million, representing 30 percent of all jobs, to 383,000, or 11 percent of all jobs. Jobs in services jumped from 507,000 (14 percent) to 1.1 million (31 percent), and in finance, insurance, and real estate from 336,000 (10 percent) to 547,000 (15 percent). This shift in job opportunity is often called a switch from *blue-collar jobs*, performed in rough clothing and usually involving manual labor, to *white-collar jobs*, salaried or professional jobs that do not involve manual labor.

The people who held the best new jobs, notably the young, urban professionals ("yuppies"), occupied and restored select older residential neighborhoods in a process known as **gentrifica-**

FIGURE 9-20

No U.S. city illustrates a transformed economy better than downtown Pittsburgh, which was transformed from a dirty and smokey industrial area (a) into a new park and gleaming service center, a "Golden Triangle" (b). This is where the Allegheny River (upper left) meets the Monongahela (at right) to form the Ohio. Three Rivers Stadium is off to the left.
[(left) From the Pittsburgh Gazette-Times, *Story of Pittsburgh*, 1908. Courtesy of Professor Arthur G. Smith, University of Pittsburgh.(right) Courtesy of Professor Arthur G. Smith, from *Pittsburgh Then and Now*. Pittsburgh: University of Pittsburgh Press, 1990.]

tion. In some cities, such as Seattle and San Francisco, the demand for new housing for these white-collar workers spurred the governments to limit construction of office towers or to require builders of commercial space to construct residences as well.

Some immigrants have contributed to central-city rejuvenation. Some bring job skills (cobblers, for example), and some entrepreneurs open small shops. For them, the city's traditional advantages of agglomeration and external economies confirms the city's function as incubator of new businesses.

Cities compete to attract new job opportunities. Business journals periodically rank cities' relative attractiveness, and although the mix of criteria chosen is not "scientific," a high ranking can boost a city's fortunes (see Table 9-1).

Many cities around the world have cultivated the tourist sectors of their economies and attract tourists with historic quarters, fine cuisine, shopping opportunities, and arts (Figure 9-21). Cities can import both performing arts festivals and art works. The City of Memphis, Tennessee, for example, has developed tourism into a key industry. In 1987, it borrowed from Egypt an exhibition titled "Treasures of Ramses the Great" that attracted visitors who injected $83 million into the city economy. This is about $125 per city resident. Tourists coming to visit Elvis Presley's home, Graceland, inject an additional estimated $40 million per year into the city's economy. Tourism, however, often only emphasizes the widening gap between rich and poor, or successful and unsuccessful, in the central cities.

The Shortcomings of Service Economies

Upscale urbanites are only a small fraction of the total inner-city population. The new white-collar jobs being created do not equal the number of blue-collar jobs being lost, and most service jobs pay less. A study in Milwaukee, for example, found that average pay for manufacturing jobs in 1989 was $28,000, but the average pay for service jobs was $19,000. Furthermore, workers who lose their blue-collar jobs often require retraining or education before they can capture one of the new openings. Despite the publicity that the new urbanites and their lifestyle receive, the percentage of jobs in most central cities held by commuters is rising—especially the percentage of the best jobs.

Meanwhile, the labor force participation rates of central-city populations, that is, the percentage of the population that is currently employed or even looking for a job, are falling. In 1992, the overall U.S. participation rate was 66 percent; for Detroit it was 51.7 percent. This suggests not only that a smaller percentage of the city's population was working and paying taxes, but that a higher percentage was receiving public assistance. The gap between average city and suburban incomes widens. In 1960, median family income within New York City was 93 percent of that of suburban families. By 1985, that percentage had fallen to 55 percent. In the same period the incomes of Detroiters fell from 85 percent to 50 percent of suburban incomes; Baltimore declined from 90 percent to 50 percent. In many U.S. metropolitan areas, inner-city incomes are less than one-half of suburban incomes.

TABLE 9-1
Examples of City Rankings

(a) *Money* magazine ranking of the best places to live in the United States	(b) *Fortune* magazine ranking of the best cities for business in the United States
1. Rochester, MN	1. Seattle, WA
2. Madison, WI	2. Houston, TX
3. Minneapolis/St. Paul, MN	3. San Francisco, CA
4. Houston, TX	4. Atlanta, GA
5. Raleigh/Durham, NC	5. New York, NY
6. Goldsboro, NC	6. Raleigh/Durham, NC
7. Sheboygan, WI	7. Denver, CO
8. Grand Forks, ND	8. Chicago, IL
9. Sioux Falls, SD	9. Boston, MA
10. Austin, TX	10. Orlando, FL

Money magazine, September 1993; *Fortune*, November 2, 1992.
Fortune magazine, © 1993 Time Inc. All rights reserved.

FIGURE 9-21

Baltimore's Inner Harbor area offers new attractions and a convention center near the edge of the CBD. The U.S.S. Constellation tied at dockside in the background is the oldest U.S. warship afloat (launched in 1797). The tall building on the waterfront is a World Trade Center, and the modern building at the right is a new aquarium. The low buildings house restaurants and shops.
(Courtesy of Baltimore Area Convention and Visitors Association.)

The Urbanization of African Americans

Chapter 5 recounted how some 1.5 million African Americans migrated from the South to the cities of the North between 1910 and 1945, and an additional 5 million migrated north and west between 1945 and 1970. That second wave arrived in the central cities just as the central cities were losing their ability to provide entry-level job opportunities, and the migrants often faced discrimination in employment and segregation in housing. Deteriorating conditions in new African-American ghettoes eventually triggered civil unrest. A 1965 riot in the black Watts section of Los Angeles left 34 people dead and more than a thousand injured, and it required military occupation of 46 square miles (119 km^2) to halt the violence. The Watts riot was followed by 150 major riots and hundreds of minor ones that summer and the next three summers.

Since the 1960s, the nation has made great strides in civil rights, and a great many of the black urban in-migrants have achieved success. Others, however, have been left behind. If a family's first urban generation failed to find employment, skills,

and upward mobility, the second and third generations may have, also. Many families became trapped in poverty. Numerous aspects of deprivation have concentrated in the central cities: poverty, substandard housing, inadequate education and nutrition, and violent crime, with the deterioration of the nuclear family structure and the formation of violent gangs. Drugs have made everything worse. Murder rates have soared, and life expectancy among black males has tumbled. Los Angeles erupted again in April 1992 when four white police officers were acquitted of beating a black motorist. Over 13,000 police and troops restored order after three days of riots, fires, and looting left over 50 people dead and 4,000 buildings burned. The deteriorating conditions in the central cities has itself become a push factor driving out businesses and middle-class residents.

Scholars who study social conditions in inner-city neighborhoods are formulating a new alternative to the spatial mismatch hypothesis to explain current urban unemployment, especially among second- or third-generation urban blacks. They have learned that the primary qualification employers seek in unskilled workers is reliability, and the best way to find reliable

326 *Chapter 9*

new employees is through recommendations from current employees. Therefore, unskilled urbanites seeking entry-level positions get information about jobs, role models for work, and sponsorship at the workplace through networks of family or friends. The scholars studying such networks argue that many urban black communities, after a generation or two outside the mainstream of labor opportunity, lack these crucial links. The social networks conducive to upward mobility have deteriorated. Professor Philip Kasinitz, a sociologist, has written, "The primary reason for ghetto unemployment is not the lack of nearby jobs but the absence of social networks that provide entry into the job market." The theory that it is the lack of these networks that causes unemployment may be called the **network hypothesis**. Scholars investigating this hypothesis note that significant numbers of inner-city jobs, even those in manufacturing, have often been captured by groups whose residences are distant, while local residents cannot get work. For example, Professor Kasinitz found that Hispanics who commuted from New Jersey were capturing jobs in a manufacturing district of Brooklyn, New York, even though unemployment stayed high in contiguous black neighborhoods. A 1993 study of Indianapolis by geographer Thomas Cooke of Indiana University similarly concluded that "it is not possible to argue that census tract African American male unemployment rates are related to the number of local job opportunities."

These studies do not conclusively replace the spatial mismatch hypothesis, but they introduce a new partial factor of explanation. Assumptions about what *causes* a situation will determine *approaches to altering* that situation. In this case our understanding of the causes of inner-city unemployment will determine what solutions we propose.

Only in pockets of the central cities are conditions at their worst, and even the worst slums of New York City, Chicago, Detroit, and Los Angeles do not compare with the conditions of life for many in Mexico City, Lagos, or Calcutta. The inner-city second or third generation living in deprivation, sometimes called the *underclass*, numbers under 3 million people, less than 1 percent of the national population. Still, the conditions of deprivation and the lack of opportunity contrast starkly with the national self-image.

Racial Segregation in the Inner Cities

The 1990 Census revealed some progress in reducing racial segregation in U.S. cities. Racial isolation is defined as the percentage of African Americans living in census blocks (tracts containing 300 to 400 people) that are at least 90 percent African-American. In the 50 largest metropolitan areas in the United States, that percentage dropped from 43.7 in 1980 to 37 in 1990. Seven of the 50 areas experienced increased segregation, but many metropolises seem integrated, and several others seem to have made substantial steps toward integration. The reduction in black geographic isolation may be due in part to in-migration of Hispanic and Asian minorities rather than to black integration with whites, or it may reflect neighborhoods in transition to minority dominance.

The changing demographic mix in major U.S. cities carries political ramifications, because Hispanics, Asians, and other minorities form new political alliances. When we defined culture in Chapter 2 and ethnicity in Chapter 5, we wrote that both of these may ultimately be matters of self-identification and that outsiders may not be aware of distinctions that are apparent to insiders. Politics among minorities well demonstrates this. For example, in May 1991, Washington, D.C., had an African-American mayor and council president, yet the city suffered riots among its newer disadvantaged minority—its Hispanics. Minorities do not always form one homogeneous "community," nor do Hispanics, nor do Asian Americans. Minority populations include a great and growing diversity of cultures and interest groups. This diversity complicates politics.

Efforts to Redistribute Jobs and Housing

The spatial mismatch hypothesis has dominated thinking about urban unemployment since the 1960s. Therefore, several government programs designed to deal with the problem of inner-city unemployment have addressed the spatial mismatch. Three approaches have been suggested: (1) bring new blue-collar jobs to the cities; (2) move inner-city residents to the suburbs; or (3) transport inner-city workers to suburban jobs.

Some efforts to reindustrialize the central cities rely on **urban enterprise zones**, where manufacturers receive government subsidies. The Federal Budget Act of 1993 called for the designation of 65 urban enterprise communities, and 30 rural zones as well. Several states and cities had already designated zones and granted manufacturers assistance in their poorest communities. Some of these zones have enjoyed success, but the factors that drive industries from the central cities are difficult to overcome.

The second approach is to move poor people from the central city out into new subsidized suburban housing. The federal government actually built suburbs for low-income people in the 1930s: Greendale outside Milwaukee, Greenhills outside Cincinnati, and Greenbelt outside Washington. Today, however, most suburbs zone to exclude subsidized

FOCUS
Changing Employment Patterns in Milwaukee, Wisconsin

Metropolitan Milwaukee exemplifies how an evolving economy can create winners and losers along geographical and racial lines. From the 1940s to the 1970s, Milwaukee won fame for its factories, foundries, and breweries. One in three jobs was in manufacturing. A person with a high school diploma could get a job, buy a house, and live well. Between 1970 and 1990, however, the number of manufacturing jobs shrank from 220,000 to 173,400, while the number of nonmanufacturing jobs grew from 461,200 to 591,000. The metropolis's black population, concentrated in the central city, could not capture those new jobs. Black unemployment rose from 17 percent to 20.1 percent, while white unemployment shrank from 5.3 percent to 3.8 percent. The number of black individuals receiving welfare rose from 64,317 to 71,113, while the number of whites fell from 27,595 to 19,493. By 1991, more than one-half of all Milwaukeeans were on some form of public assistance.

In the late 1980s, Milwaukee was able to recast its manufacturing sector and slightly increase the number of manufacturing jobs. This was partly because the falling dollar increased U.S. exports of manufactured goods and partly because regional manufacturers developed new specialty goods of higher value added. Most of the black manufacturing workers, however, had low seniority, and many lacked the skills necessary to work in the new specialty lines. To operate a machine-tool today, for example, requires computer skills. By 1992, total manufacturing jobs were down to one-quarter of Milwaukee's jobs. Black men stand idle on street corners just blocks from the breweries and factories that used to employ them. Many of the buildings have been converted to offices for tertiary sector employees.

housing or even private apartments with young families. Suburbs prefer expensive single-family homes, because only these pay property taxes sufficient to cover the costs of the services that they and their residents require. Owners of undeveloped property, however, usually want to maximize development on their land, so they battle such *exclusionary zoning*. A clause in the Fifth Amendment to the U.S. Constitution called the *takings clause* protects private property against seizure by the government without just compensation. Property owners argue that exclusionary zoning so severely restrict the use to which owners may put their land that the zoning effectively confiscates the land. This is a fine point in the interpretation of law, but in recent years several state supreme courts have banned exclusionary zoning. Several have actually required local communities to zone to provide all sorts of housing: high-density apartments and low-density single-family homes; as well as housing affordable to people of all income levels.

The Clinton presidential administration has advocated breaking up the concentrations of subsidized housing in the central cities and building more in the suburbs. The government might "bribe" suburban areas to accept a greater share of public housing by offering grants for other community facilities. Another experimental program provides inner-city public housing residents with subsidies or vouchers that can be used as money to rent private apartments in the suburbs.

The third approach to solving the spatial mismatch is to provide inner-city poor people with transportation to suburban job opportunities. Several city governments, private employment agencies, and even suburban employers have instituted dedicated bus services for this purpose.

These three efforts to solve the problem of inner-city unemployment follow from the acceptance of spatial mismatch as the cause of the unemployment. The network hypothesis has only recently been offered as an alternative or complementary explanation. Therefore, few government programs have yet been devised to address the lack of social networks. Perhaps we will see proposals in the 1990s.

Still another potential solution to inner-city unemployment among the unskilled is to educate and train the central-city population to capture the tertiary sector jobs that do open in the central city. The quality of education in the central cities, however, is generally discouraging. Rich suburban schools are equipped with modern gadgetry, whereas inner-city schools often lack basic texts.

Governing Metropolitan Regions

Political geography is the subfield of geography that studies the interaction between political processes and the distributions of all other activities and transformations of the landscape. Political geography can be studied at any scale from local community politics up to international boundary disputes and international law. Conurbations present special problems of interest to political geographers.

The legal boundaries of most U.S. cities were originally drawn to include some surrounding land for future growth, and when the city outgrew its boundaries, it annexed suburban areas. By the 1920s, however, the suburban populations had begun to incorporate themselves to avoid annexation (Figure 9-22). Suburbanites argued that their action upheld the U.S. tradition of local self-government, and many may have felt that by incorporating their own communities they were escaping the problems (and people) of the old city, including high social costs and politics that were often corrupt. Only a few central cities, including Austin, Texas, Charlotte, North Carolina, and

Suburban Incorporation

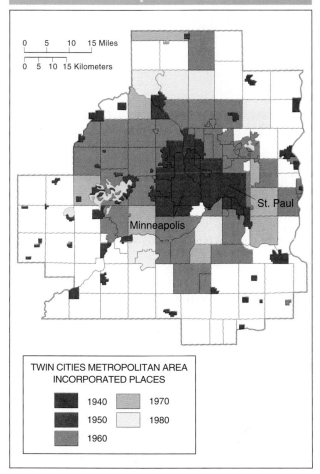

TWIN CITIES METROPOLITAN AREA INCORPORATED PLACES

■	1940	▨	1970
■	1950	□	1980
■	1960		

FIGURE 9-22

Minneapolis and Saint Paul, Minnesota, were surrounded by tiers of independently incorporated municipalities, like the rings of growth of a tree. This pattern of suburban incorporation typifies U.S. metropolitan expansion.

Oklahoma City, can still expand by annexing new suburban areas.

The Proliferation of Governments Today autonomous municipal units form a legal retaining wall around almost every large city in the United States. Metropolitan areas cover a great number of municipalities and also a myriad of *special district governments*, which are incorporated to deal with specific problems. For example, Nassau and Suffolk counties in suburban New York include 2 cities, 13 towns, 95 villages, 127 school districts, and more than 500 special districts, most of which exercise taxing powers for services from garbage collection to hydrant rental. The five-county Los Angeles metropolitan region contains 160 separate governments. Los Angeles county alone has 82, and even entirely within the boundaries of the city of Los Angeles there are 7 independent city governments.

The boundaries among the many jurisdictions are not obvious to everyone, but each government has its own agenda and marks the landscape, and governments jostle for authority and tax dollars. Property taxes soar, but many metropolitan area residents have no idea which government is responsible for which service.

The governing bodies of many of the special districts are not chosen in elections in which each citizen exercises an equal vote. Instead, they are either appointed or else chosen in elections in which votes are weighted in terms of payments for a service, use of a service, or in some other way. As a result, the percentage of public funds spent by officials directly responsible to the voters shrinks. Furthermore, the boundaries of special districts may not conform to those of general-purpose governments, but overlap them. Overlapping or noncoterminous boundaries multiply the difficulties in coordinating the provision of services. All of these factors may discourage voter turnout.

Decisions regarding metropolitan land use and the location of industries, recreation facilities, transport facilities, or new housing cannot be made in the best interests of the entire metropolitan population. They are made on the basis of competition among the local governments, each of which wants to enhance its own property tax base by attracting commercial developments that pay high property taxes but demand little in the way of local services. For example, Santa Clara County, California, has zoned for 250,000 new jobs but only 70,000 new homes. Each community hopes to let surrounding towns cope with the additional costs of schooling, pollution, and congestion.

Many metropolitan areas have created councils of governments (COGs). These are committees of officials representing each of the local governments in the region. COGs, however, exercise limited powers. Each local government can veto any proposed areawide action. Some U.S. metropolitan regions are creating regional governments to address areawide problems. Metropolitan Portland, Oregon, for example, has a "metro government" to manage the region's growth. State governments in Washington, Minnesota, Connecticut, Virginia, and elsewhere are devising new schemes to redistribute or equalize both the revenues and the costs of housing and schooling throughout metropolitan areas. Other state governments are assuming increasing direct responsibility for governing metropolitan areas.

The weakness of metropolitan regional governments in the United States contrasts sharply with models elsewhere (see Figure 9-23). In most European and Latin American countries, either strong metropolitan governments exist or else the national governments themselves oversee land use and the growth of metropolitan regions.

Race and Metropolitan Political Geography

Sometimes the geography of race hinders governmental reorganization of a metropolitan region. In many metropolitan areas, African Americans concentrated in the central city have won political control of the central-city government. A shift of some governmental responsibilities up to a regional government might improve overall governmental administration, but it would dilute the power of the central-city government, and, thus, of the central-city African-American population. The U.S. Department of Justice has occasionally prevented such reorganization. In other cases, the Department of Justice has prevented white-ruled central cities from annexing white-occupied suburban areas because the annexation would dilute the proportional voting strength of African Americans in the central city.

In still other cases, courts have ordered local governments to switch from at-large systems of representation to district systems. In **at-large systems of representation** all members of a governing body are chosen by the entire electorate, whereas in a district system each member represents a specific geographic district. In an at-large system, a slight majority of one group can elect a government that is entirely of its group, and a minority may find it impossible to elect any of its candidates. In a district system, by contrast, a minority neighborhood can elect at least its local representative.

The population of Dade County, Florida, for example, is 50 percent Hispanic, 18 percent non-Hispanic black, and 32 percent non-Hispanic white. In 1992, under a system of at-large representation, non-Hispanic whites held 7 of the 9 seats in the county governing body. In 1993, federal courts required the coun-

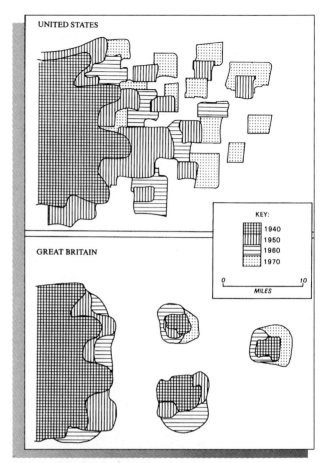

FIGURE 9-23

This schematic drawing contrasts the leapfrogging typical of U.S. suburban growth with the tighter, more controlled suburban growth typical of Great Britain. The British system saves open space and allows for better and cheaper planning and installation of infrastructure.
(Adapted from Marion Clawson and Peter Hall, *Planning and Urban Growth: An Anglo-American Comparison.* Reprinted by permission of *Resources for the Future.*)

ty to create a new governing board with 13 members, each of whom would represent a geographic district. Seven of the districts were mapped with Hispanic majorities, 3 with non-Hispanic black majorities, and 3 with non-Hispanic white majorities. Since this change, the membership of the board has more proportionately reflected the county's population. The new board's first official action is explained in the caption to Figure 7-23.

THE WESTERN TRADITION OF URBAN AND REGIONAL PLANNING

▼▼▼▼▼▼▼▼▼▼▼▼▼▼▼▼▼▼▼▼▼▼▼▼

The process of urban and regional planning applies many of the principles of urban geography to specific situations. The concept of designing an "ideal city" has challenged the best minds for centuries. The founder

330 *Chapter 9*

of Western city planning was probably Hippodamus of Miletus. He laid out that city in today's Turkey according to a grid plan as early as 450 B.C.

The first modern attempt to formulate the needs of the city as a whole was the work of the British visionary Sir Ebenezer Howard. In *Garden Cities of Tomorrow* (1898), he outlined a plan to stop the unbounded growth of the industrial city and to restore it to a human scale. He wanted to relocate population into new medium-sized garden cities in the outlying countryside. These regional cities would be ringed by greenbelts of farmland and parks. All land would be municipally owned, and each town and its surrounding region would be planned as an interlocking whole (Figure 9-24).

Howard built two garden cities just north of London: Letchworth (1904) and Welwyn Garden City (1919). These inspired the Regional Planning Association of America, a private nonprofit organization, to construct two planned communities in the New York City area: Sunnyside Gardens, Queens (1924); and Radburn, New Jersey (1928). Neither is a complete garden city, but both are harmoniously designed and have greatly influenced urban planning in the United States and Europe. Some of the new exurban villages reproduce features of Howard's garden cities.

Probably the most important city planner of the twentieth century was the Swiss architect Charles Édouard Jeanneret-Gris (1887-1965), better known by

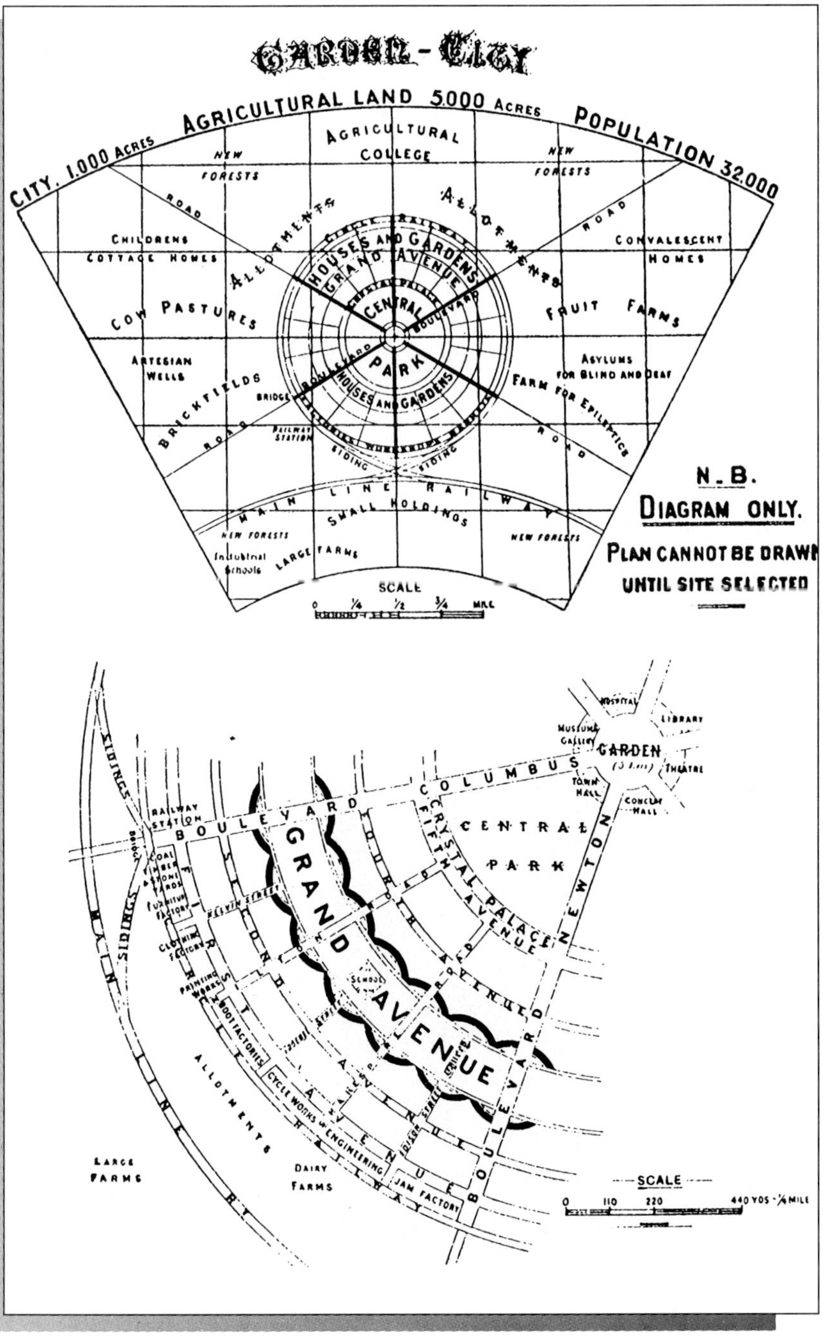

FIGURE 9-24

Sir Ebenezer Howard envisioned planned "Garden Cities" to disperse the concentrations of population in nineteenth-century cities.
(Reprinted with permission from Ebenezer Howard, *Garden Cities of Tomorrow.* London: Faber & Faber, 1946.)

his professional name, Le Corbusier. In a celebrated plan of 1922, he proposed to raze the crowded, run-down historic core of Paris, preserving only the central monuments. In its place he wanted to build a *Radiant City* of tall glass offices and apartments, spaced so far apart that each tower would be surrounded by green space and have a fine and wide view. The concentration of facilities within high-rise slabs would liberate the city from its environment. It could be placed anywhere. Le Corbusier brought together two conceptions: the machine-made environment, standardized, technically perfect to the last degree; and, to offset this, the natural environment, treated as open space, providing sunlight, air, greenery, and views.

Le Corbusier planned Chandigarh, a new capital for the Punjab State in India, in 1950, but the world's supreme Radiant City is Brasília, the capital of Brazil, designed by Lucio Costa in 1957 (Figure 9-25). Unfortunately, the city's gigantic scale demands a completely motorized population. That is the problem with excessive openness—Corbusier's "city in a park" can become a city in a parking lot. The Australian capital at Canberra, planned by Walter Burley Griffin of Chicago, has less openness. It allows walking, and it is generally considered superior (Figure 9-26).

Le Corbusier's 1930 plan for the town of Nemours in Algeria, with a geometric grouping of domino structures, set the international fashion for high-rise slabs for the next 50 years. These ideas were disseminated worldwide by the 1933 Athens Charter of the International Congress of Modern Architecture (CIAM). In the Charter's codification, largely by Le Corbusier, the functions of the city—housing, work, recreation, and transport—provided the city's framework. The Charter called for separation of high-rise development, industrial zones, parks and sports fields, and a hierarchical street system for traffic at different speeds. These ideas diffused to dominate urban planning around the world.

Le Corbusier's theories were enthusiastically adopted by the Communist party governments in Eastern Europe and the former Soviet Union. Cities in these areas constructed massive public housing projects in keeping with Communist doctrine that the state represents the working people and strives to better their standard of living, while emphasizing (and even enforcing) equality. Unfortunately, these *socialist cities* in many countries were so poorly built that they are already decayed and grim.

Many people believe today that the widespread adoption of Le Corbusier's ideas produced a half-century of monotony—not merely of detail and of style, but of insensitivity to place, to the essential difference between one place and another. The high-rise slabs ringing every big city in the world from Mexico City to Singapore look much alike. In 1961, a group of

FIGURE 9-25

Brasília, Brazil's capital since 1960, arose on a largely unpopulated open plateau 603 miles (970 km) northwest of Rio de Janeiro. As this picture reveals, it sometimes lacks water sufficient to keep the grass green or to operate the elaborate fountains. (Courtesy of Professor Barbara Weightman.)

younger architects broke away from CIAM and proclaimed that architecture was more than the art of building. It was the art of transforming people's entire habitat. The School of Architecture at the University of California at Berkeley was reconstituted and renamed the School of Environmental Design.

Most recently, many urban planners worldwide have come to criticize the concept of high-rise living. Low-rise dwellings can achieve the same density of habitation as Le Corbusier's "towers in a park" can, and many people feel more content living in low-rise dwellings. High-rise public housing projects, it turned out, can breed a sense of alienation and helplessness, and many have been razed across the United States (Figure 9-27). Today we still work to design more humane urban environments.

NON-WESTERN MODELS

▼▼▼▼▼▼▼▼▼▼▼▼▼▼▼▼▼▼▼▼▼▼▼▼▼

The concentric zone, sector, and multiple-nuclei city models discussed earlier in this chapter were all devised to describe North American experience, but other models have been sketched to describe characteristics of other cultures.

The three interacting processes identified above—economic factors (including transportation facilities), residential clustering, and government decree—affect all urban settlements, old and new, in all known cultures. These processes sort out the population and the land uses into distinct patterns that can be identified within any city. Figure 9-28, for example, models Latin American cities. There, in contrast to North American models, CBDs thrive. This is partly a result of continuing reliance on public transit and partly because high-income populations choose to live close to the CBD. A commercial spine such as a boulevard extends out from the CBD, and amenities such as opera houses, chic stores, and elegant parks follow this spine. Zones of more modest housing and value surround this elite zone, and the periphery is dominated by squatter slums.

Western Forms Often Overlie Indigenous Forms

Chapter 3 noted that many of the world's great cities sprang into being as a result of trading or political contacts between the native peoples and Europeans. Therefore, most of these great cities were from their

FIGURE 9-26

Canberra became home to Australia's Parliament in 1927, but foreign missions and federal government departments moved here only in the 1950s.
(Courtesy of Australian Overseas Information Service.)

FIGURE 9-27

The Pruitt-Igoe Housing Project in Saint Louis, Missouri, was demolished in 1972 when it was determined that the design of the buildings may have contributed to the social problems among resident families. Other high-rise low-income public housing projects have since been demolished across the country, and low-rise housing is widely preferred for public projects today.
(Courtesy of UPI/Bettman.)

Cities and Urbanization   **333**

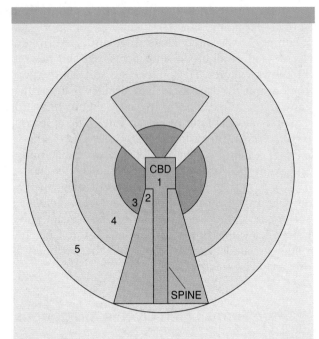

1. Commercial/industrial areas
   CBD = Central Business District, the original colonial city
   SPINE = High quality expansion of the CBD, catering to the wealthy

2. Elite residential sector

3. Zone of maturity
   Gradually improved, upgraded, self-built housing

4. Zone of accretion
   Transitional between zones 3 and 5, modest housing, improvements in progress

5. Zone of peripheral squatter settlements slum housing

FIGURE 9-28

In contrast with the North American models, the Latin American model of urban growth shows an elite residential sector following a spine of high-value land use stretching out from the central business district.
(Adapted with permission from Ernst Griffin and Larry Ford, "A Model of Latin American City Structure," *Geographical Review*, 70 (1980), 406. Reprinted by permission of the American Geographical Society.)

founding planned and built according to European notions of city planning. The internal geography of these cities still shows port zones; enclaves of former European settlements; barracks; colonial government buildings; and racial, religious, and ethnic ghettos based on the role each group played in the city's founding and during the colonial period (Figure 9-29). Western forms continue to be stamped on the world's cities today, as Western architects, engineers, and urban planners, or non-Western individuals educated in Western schools, transform even the older traditional built environments.

Individual cities, however, emerged at different times, for different reasons, and within different cultural contexts. Many great cities still boast historic cores that illustrate indigenous principles of urban planning, and they cannot all be squeezed into three or four simplified models. Chapter 3 noted how settlement patterns can characterize different cultures, but city models based on Western cities alone need to be compared with other traditions.

Islamic Forms

Traditional Islamic cities well illustrate the role of culture in urban form. There are regional differences in cities across the Islamic realm, but most nevertheless show surprising similarity. These characteristics can be identified from Seville, Granada, and Córdoba in

FIGURE 9-29

Most Southeast Asian port cities are the product of Western influence, and thus they reflect Western design forms.
(Adapted from T.G. McGee, *The Southeast Asian City*. Reprinted with permission of Greenwood Publishing Group, Inc. Westport, CT. Copyright © 1967 by Greenwood Publishing Group, Inc.)

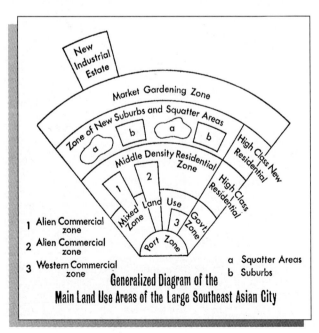

Spain to Lahore in Pakistan, and elements of these principles can be found from Dar es Salaam in East Africa to Davao in the Philippines. These cities may seem chaotic to Westerners at first glance, especially those of us accustomed to grid patterns, but they present an entirely rational structure (Figure 9-30).

The logic of traditional Islamic urban planning is announced in the Koran and has been codified by various schools of Islamic law. Certain basic regulations govern the rights of others and the pursuit of the virtuous life in a densely crowded urban environment. Islamic urban planning recognizes the need to maintain personal privacy; specifies responsibilities in maintaining urban systems on which other people rely, such as keeping thoroughfares or wastewater channels clear; and emphasizes the inner essence of things rather than their outward appearance. This last principle applies as much to the decoration of houses as to purely spiritual issues.

At the heart of the traditional Islamic city stands the Friday mosque, or *jami*, typically the city's largest structure. A number of smaller mosques may be found toward the periphery, but these seldom rival the jami in height. Close to the jami are the main *suqs*, the covered bazaars or street markets. These enclosed shopping arcades prefigured European urban galleries and American enclosed malls. Within the suqs, trades are diffused in relation to the jami. The tradespeople who enjoy the highest prestige, such as book sellers and perfumers, are closest. Farthest away are those who perform the noxious and noisy trades, such as coppersmiths, blacksmiths, and cobblers. The neutral tradespeople, such as clothiers and jewelers, act as buffers.

An immense fortified *kasbat* is attached to the ramparts, on which are located several towers or gates. The kasbat was the place of refuge for the governor or sovereign. It contained not only the govern-

Fez, Morocco

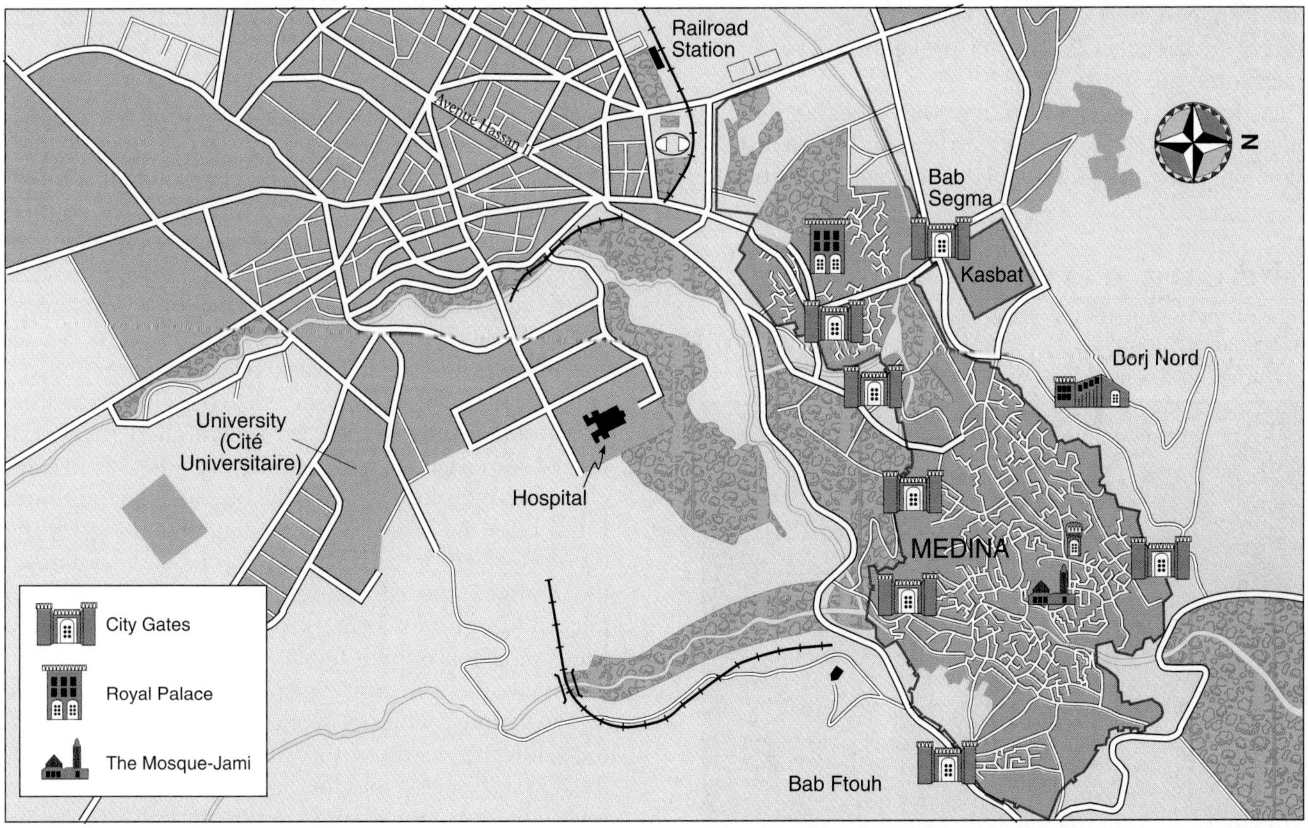

FIGURE 9-30

The Moroccan city of Fez juxtaposes a modern new Western city beside an old Islamic city, the *Medina*. The Medina contains old mosques and narrow constricted streets, and it is surrounded by city walls with great gates. In the dry climate, an historic fountain is a noted spot near the main mosque. It is overlooked by the citadel (*kasbat*). The new city, by contrast, has straight formal avenues, the railroad station, university, and a modern hospital. (Courtesy of the Moroccan National Tourist Office.)

ment buildings and barracks, but had its own small mosques, baths, and shops.

Everywhere else the city is filled in with cellular courtyard houses tied together by winding lanes. Housing is grouped into quarters, or neighborhoods, that are defined acording to occupation, religious sect, or ethnic group. The widest streets usually radiate outward from the core to gates in the city wall. These provide the main axes of movement. Slightly narrower second-order streets serve the major quarters and define their boundaries. Still narrower third-order streets provide access within quarters and are used primarily by people who live in the neighborhood (Figure 9-31). Narrow streets provide vital shade, keep down dust and winds, and use little building land.

Privacy is fundamental to the traditional Islamic city, as clearly demonstrated by the individual courtyard house. Outside walls lining a street are usually left bare and rarely pierced by windows. If windows are necessary, they are placed high above street level (some Islamic texts specify about 5 feet 9 inches— 1.75 m) making it impossible to peer in. Entrances are L-shaped, and doorways opening onto the street rarely face each other, thus preventing any direct views into the house.

Interior courtyards, often with trees and fountains, provide shade in hot climates, but, more important, they provide an interior and private focus for life

FIGURE 9-31

A minaret punctuates the vista down a narrow street of Casablanca, Morocco.
(Courtesy of the Moroccan National Tourist Office.)

sheltered from public gaze. However plain the outside of a house may be, the interior and courtyard may display lavish wealth and decoration. The interior vividness parallels the Koranic emphasis on the richness of the inner self compared to more modest outward appearance.

In the United States, the city closest to exemplifying these values is New Orleans. In the old Spanish sections of the city, misleadingly called today "the French Quarter," houses generally present plain fronts to the street, but they enclose beautiful private courtyards. When Anglo-Saxons began to move to the city after its annexation by the United States in 1803, the Anglo-Saxon rich preferred to build grand homes in a new part of town called the Garden District. Whereas in the French Quarter the courtyard gardens are private inside the houses, in the Garden District the houses sit in the middle of their publicly displayed gardens. This difference is one of cultural preference; it does not result from differences in land prices.

SUMMARY

Every settled society builds cities, because some essential functions of society are most conveniently performed at a central location for a surrounding countryside. The region to which any city provides services and upon which it draws for its needs is its hinterland. Some cities' locations result from site characteristics; others are more the result of convenient situations. Urban environments often suffer from heat, noise, air pollution, and water and waste management problems.

In many traditional societies, the cities' ritualistic and political importance far outweighed their economic importance. As societies develop, however, cities' economic role usually becomes paramount. Cities offer the convenience of agglomeration for the division of labor, and they promote and administer the regional specialization of production. Cities can become important centers of industrial production. Cities also concentrate tertiary sector functions. Some of the workers in any city produce the city's basic sector exports, whereas workers in the nonbasic sector serve the needs of the city's own residents. Jobs in the basic sector multiply jobs in the nonbasic sector. Cities can be classified economically by examining each city's basic and nonbasic sectors and by comparing these sectors among different cities. Central place theory models how cities and hinterlands are distributed across an isotropic plain.

Britain provides the world's first model for urbanization, but urbanization imposed hardship on millions of people. Today urbanization is occurring in many places without concomitant economic development. People migrate from the countryside to the cities because of the balance of push and pull factors, and governments can try to regulate this migration both by reducing the attractiveness of cities and by improving rural life. Urban migrants' industriousness could be an asset for economic growth.

Urban geographers also study the distribution of activities within cities. This distribution may be caused by economic forces, by social factors, or by deliberate actions of the government.

In the United States, growing cities have spilled over their legal boundaries into suburbs. This dispersion has resulted from a complex of government policies, cultural choices, and other economic and social forces. Both housing and jobs have dispersed to the suburbs, and commuter traffic has been redirected. The central cities have suffered economic decline and residential segregation. Efforts are being made to attract jobs back into the central cities and to build networks of information about job opportunity. Attempts have been made to coordinate governmental activities across conurbations.

The concept of designing an "ideal city" has challenged the best minds for centuries. Sir Ebenezer Howard's garden cities plan tried to stop the growth of the industrial city and to restore it to a human scale. Le Corbusier proposed radiant cities of tall office buildings and apartments spaced so far apart that each glass tower is surrounded by green space.

Many of the world's cities were planned and built according to European notions of city planning, and this is still reflected in their internal geography. Western forms continue to be stamped on the world's cities today. Many great cities, however, still boast historic cores that illustrate indigenous principles of urban planning. Islamic cities present one alternative to traditional Western forms.

KEY TERMS

city (p. 303)
hinterland (p. 303)
incorporation (p. 303)
primate city (p. 303)
urbanization (p. 303)
urban geography (p. 303)
agglomeration (p. 305)
primary sector (p. 305)
secondary sector (p. 305)
tertiary sector (p. 305)
service sector (p. 305)
basic sector (p. 305)
nonbasic sector (p. 305)
multiplier effect (p. 305)
microclimate (p. 307)
urban heat island (p. 308)
central place theory (p. 309)
threshold (p. 309)
external economies (p. 310)
internal economies (p. 310)
exurbs (p. 310)
labor-intensive
    activity (p. 311)
capital-intensive
    activity (p. 312)
informal or underground
    sector (p. 314)
central business
    district (CBD) (p. 315)
congregation (p. 316)
segregation (p. 316)

QUESTIONS FOR INVESTIGATION AND DISCUSSION

1. What are the site characteristics of your city? Why was the site selected?

2. What are the situational relationships of your city? What principal transport routes converge on your city?

3. What range of external economies is available in your town's Yellow Pages?

4. What are the biggest employers in your town? Which enterprises do the biggest dollar volume of business? What are your towns' basic economic activities?

5. Are any planned suburbs or exurbs located around your town? What forms of transportation do their residents depend on?

6. How many local general-purpose governments and special-purpose governments are in your metropolitan region, or the metropolitan region nearest to you? Study a map of all the forms of independent government in the area. How are the special-district government ruling boards chosen? What kind of regional planning body or government does the region have?

7. From the study of topographic maps of your local area, where would you recommend new investment in infrastructure? How would you distribute or redistribute park lands, industry, or other facilities?

8. Some people enjoy living in cities; some people flee cities. What are the benefits and the drawbacks of living in cities and the benefits and drawbacks of living in suburbs?

9. What would you do to stop rural-urban migration in any given poor country?

10. In the past, any city's accessibility played a large role in its growth. Today, cities compete for high technology, office- and research-type economic activities that generally employ a highly educated and high-income work force. Therefore, quality of life measurements can affect a city's prosperity. What are the basic measurements of urban or regional quality of life? How would your community score?

ADDITIONAL READING

CRONON, WILLIAM. *Nature's Metropolis: Chicago and the Great West*. New York: W. W. Norton & Co., 1991.

DRAKAKIS-SMITH, DAVID, ed. *Economic Growth and Urbanization in Developing Areas*. New York: Routledge, 1990.

zoning (p. 316)

metropolitan statistical area
(MSA) (p. 318)

consolidated metropolitan
statistical area
(CMSA) (p. 318)

telecommuting (p. 323)

spatial mismatch
hypothesis (p. 324)

gentrification (p. 324)

network hypothesis (p. 327)

urban enterprise
zones (p. 327)

political geography (p. 328)

at-large systems
of representation (p. 330)

FOX, KENNETH. *Metropolitan America: Urban Life and Urban Policy in the United States 1940-1980.* New Brunswick, NJ: Rutgers University Press, 1990.

GARREAU, JOEL. *Edge City: Life on the New Frontier.* New York: Doubleday, 1991.

GUGLER, JOSEPH, ed. *The Urbanization of the Third World.* Oxford: Oxford University Press, 1988.

HART, JOHN FRASER, ed. *Our Changing Cities.* Baltimore. MD: Johns Hopkins University Press, 1991.

KNOX, PAUL, ed. *The Restless Urban Landscape.* Englewood Cliffs, NJ: Prentice Hall, 1993.

LEMANN, NICHOLAS. *The Promised Land: The Great Black Migration and How It Changed America.* New York: Knopf, 1990.

MUMFORD, LEWIS. *The City in History.* New York: Harcourt, Brace & World, 1961.

SORKIN, MICHAEL, ed. *Variations on a Theme Park: The New American City and the End of Public Space.* New York: The Noonday Press, 1992.

10 | A World of Nation-States

Monte Titano is the spine of the Republic of San Marino. This 24 square mile (61 km) independent country lies entirely within Italy, but it has preserved its independence since the 13th century. San Marinans value their separate national identity.

Courtesy of Shaun Egan/Tony Stone Images.

The world political map is probably the most familiar of all maps, because the Earth's division into countries, or states, is the most important territorial organizing principle of human activities. **States** are independent political units that claim exclusive jurisdiction over defined territories and over all of the people and activities within them. The governments are not always able to exercise this jurisdiction completely, but states can encourage or even force patterns of other human activities to conform to the political map. Chapters 7 and 8 noted how the geography of language and religion is affected by the political partitioning of the world. Patriotism, which is a strong emotional attachment to one's country, is itself a powerful cultural attribute. A country is in many ways a fixed culture realm.

The idea that the whole world should be divided up into countries seems natural to us, but it is relatively new in human history (Figure 10-1). This chapter will explain how the idea originated in Europe and how it was diffused worldwide with European conquest. The neatness of the units on today's world political map nevertheless exaggerates the degree to which all peoples accept the current pattern of countries. The map suggests that all the borders are clearly demarcated and that they divide the activities on their two sides, but in fact some activities overlap the borders. The map also suggests that the areas within those borders are politically homogeneous, and this is also false. Governments are only more or less successful in organizing their territory, and no territory can be sealed off. Both civil wars within countries and border wars among countries continue around the world.

This chapter also reviews the political geography of states, analyzing how they demarcate their borders and subdivide their territories. In addition, it tries to measure and map the degree to which various governments either place restrictions on their peoples' freedom or encourage full development of their peoples' potential.

Countries also try to organize their territories economically, that is, to integrate and build national economies. That will be the subject of Chapter 12.

THE DEVELOPMENT OF NATION-STATES IN EUROPE

▼▼▼▼▼▼▼▼▼▼▼▼▼▼▼▼▼▼▼▼▼▼▼▼▼▼

The idea that a state claims exclusive sovereignty over a demarcated space and all of the people and resources within it originated in medieval Europe. Under the Roman Empire, the Roman Catholic church was geographically organized into dioceses, and when the Empire fell, the Church and its system survived. As the Church converted people to Christianity, it also converted them to the idea of territorial political organization. Before conversion, the kings of the peoples in Europe had not ruled over fixed territories, but over groups of followers, wherever they wandered. Government over a group of people rather than a defined territory is called **regnum**. The Church taught the principle of rule over a defined territory, which is called **dominium**. The Merovingian kings (fifth through eighth centuries), for example, called themselves "Kings of the Franks," but the later Capetians (tenth through fourteenth centuries) had settled down and called themselves "Kings of France."

In the nineteenth century, anthropologists insisted that conversion from regnum forms of government to dominium is an evolutionary step in human society. This argument, however, might be interpreted as an excuse for nineteenth-century imperialism. The argument that native forms of government everywhere were "backward" compared to European forms offered a justification for European conquest. At least it served as a rationalization after the conquests. British sociologist Herbert Spencer (1820–1903) extended Darwin's theory of evolution to insist that "Nature's law" called for "the survival of the fittest," even among cultures and whole peoples. This famous phrase was not Darwin's. Spencer's theory is called **social Darwinism**. In the United States it was invoked to buttress the "evolutionary" theories of F. J. Turner that were discussed in Chapter 2.

The Idea of the Nation

A state is a territory, a thing on a map, but a **nation** is a group of people who want to have their own government and rule themselves. The feeling of nationality may be based on a common religion or language, but it does not have to be, as the Swiss and many other nations demonstrate. A group sharing a sense of nationalism may or may not share any other attributes. Nationalism is a cultural concept in its own right.

Nationalism is one expression of **political community**, which is a willingness to join together and form a government to solve common problems. Chapter 9 demonstrated how political community was often lacking across U.S. metropolitan areas. Today the nation is in most places the most powerful level of political community, but in the epilogue we will ask whether worldwide concern over environmental pollution may nurture global political community.

The evolution of nationalism is part of the story of the evolution of legitimacy, the question of who has the right to rule any group. Upon what sort of consent of the governed, if any, is rule based? Even the most totalitarian states today claim to represent the people.

Chapter 8 noted that Christianity explicitly distinguishes religion from secular life. European emperors and kings signified the state; they ruled over earthly matters with the sanction of the Church—by *divine right*—but they required only obedience, not loyalty or personal identification with the state. That is why kings could often assign and reassign thrones among themselves and redraw the political map without significant protest by the people. France's King Louis XIV (reigned 1643–1715) insisted flatly, "I am the state." The Church, however, as protector of people's souls, demanded a greater degree of personal commitment than the sovereign did.

The Protestant Reformation challenged this traditional church-state accommodation. Martin Luther preached that every person was his or her own priest and carried individual responsibility for his or her own soul. This belief sabotaged the divine right of consecrated priests and, by extension, that of consecrated kings. In 1581, the Dutch, who were then subjects of the king of Spain, adopted an Act of Abjuration that renounced (abjured) the theory of divine right and argued that a king had an obligation to rule for the welfare of the people. If he did not, then the people could abjure his rule over them. This turned the notion of divine right upside down. It suggested that the king, or any government, served the people.

Thomas Jefferson drew on the Dutch act when he composed the U.S. Declaration of Independence of 1776, which includes a list of charges against King George III. Prior to that time, even the English themselves had risen up against their King Charles I, defeated his forces in battle, tried him for crimes against "his" people, and executed him in 1649. The subsequent English Bill of Rights (1689) recognized that sovereignty did not lie in the king, but in the people.

The Nation-State

The Swiss philosopher Jean Jacques Rousseau (1712–1778) lay the foundation for allegiance to the state *as* the people. Rousseau believed that in Nature people were merely physical beings, but when they united in a *social contract* they were capable of perfectibility. For Rousseau, politics was a means to moral redemption. Rousseau's ideas swayed France, and the French Revolution gave birth to the French nation. The 1789 *Declaration of the Rights of Man* stated: "The principle of all sovereignty resides essentially in the nation; no body nor individual may exercise any authority that does not proceed directly from the nation."

The nation demanded personal dedication and allegiance from its citizens. Therefore, the perfect state was a **nation-state**, a state ruling over a territory containing all the people of a nation. The theory of the nation-state assumed that nations develop first, and that each nation then achieves a territorial state of its own. Some scholars have argued that several of the nation-states had historic **core areas**, or historic homelands. In many other cases, however, core concentrations of settlement or activity developed only after the state had come into existence.

Within any state, the general set of rules and regulations for governing is called the **regime**, and that is usually formalized in a constitution. The word **government** refers to the people who are actually in power at any time.

Constitutions and laws never fully explain how any government works because each political community has a unique political culture. A **political culture** is the set of unwritten rules—the unwritten ways in which written rules are interpreted and actually enforced. Political communities differ widely, for example, in whether they honor wise elders, rich people, or religious leaders; in their tolerance of bribery; and in the rigidity or laxity with which they enforce laws. Political culture reflects other aspects of a people's culture, such as their religion. This is true everywhere and at every level of government, from city council district up to the U.N. General Assembly.

For example, one scholar examined political life in Italy and found that there are within that one country distinct regional differences in political culture, which he defined as "civic community"—patterns of social cooperation based on tolerance, trust, and widespread norms of active citizen participation (Figure 10-2, p. 344). Furthermore, the distribution of civic community among the regions was already evident as long ago as the thirteenth century. This finding carries extensive ramifications. It suggests, for instance, that long-established patterns of civic community explain both a region's capacity for economic growth and its capacity for democratic self-government. It suggests that political leaders in regions lacking civic community lack the fundamental building blocks upon which stable democracy can be built. Civic community may be almost impossible to create where it does not exist. If this hypothesis is correct, similar studies of whole

The Ages of the World's States Within Predominately Present Borders

FIGURE 10-1

The dates of independence of the world's states emphasize that the current political partitioning of the world is very recent.

countries could indicate which will most probably enjoy economic growth and democracy in the future.

The European Nation-States

The Napoleonic Wars that followed the French Revolution carried the idea of nationalism across Europe. Armies, which formerly had been comprised of hired professionals, became nations in arms. The army was called "the school of the nation," and the entire male population served. Napoleon was defeated in 1815, but the code of laws that he had imposed left a widespread legacy. It swept away aristocratic privileges and strengthened the middle class (Figure 10-3).

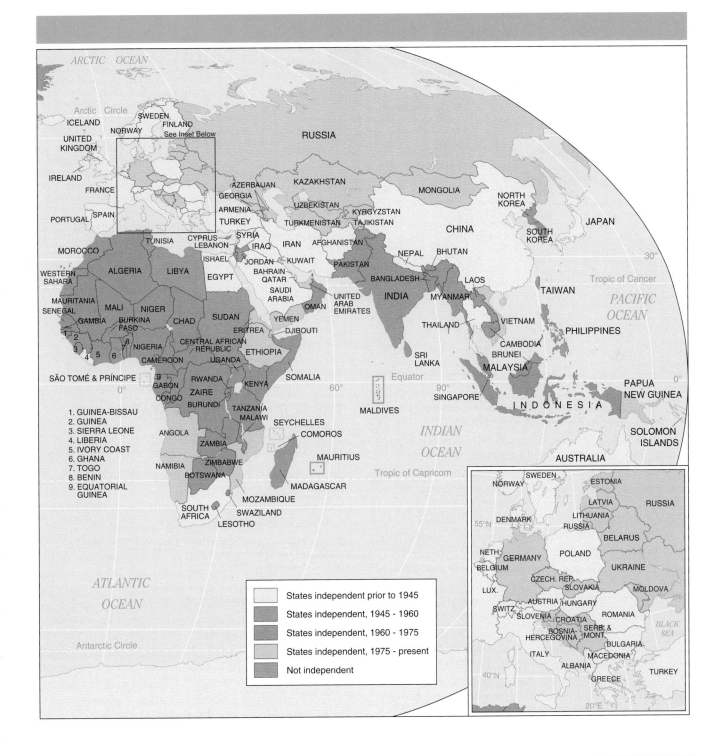

Legend:
- States independent prior to 1945
- States independent, 1945 - 1960
- States independent, 1960 - 1975
- States independent, 1975 - present
- Not independent

1. GUINEA-BISSAU
2. GUINEA
3. SIERRA LEONE
4. LIBERIA
5. IVORY COAST
6. GHANA
7. TOGO
8. BENIN
9. EQUATORIAL GUINEA

The idea of nationalism matured in Europe during the nineteenth century, and new nations struggled to emerge from old empires and feudal states. In some cases, this produced a competitive nationalism, which has caused many wars through the nineteenth and twentieth centuries. For example, several countries claimed the maximum extent of territory over which their people had ever wandered or ruled. This led to overlapping territorial claims.

After World War I, U.S. President Woodrow Wilson advanced the ideal of the nation-state, which he called **national self-determination**. The victors redrew the map of Europe to break up the defeated German, Austro-Hungarian, and Ottoman Turkish

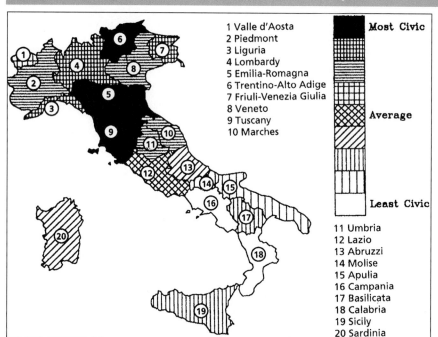

1 Valle d'Aosta
2 Piedmont
3 Liguria
4 Lombardy
5 Emilia-Romagna
6 Trentino-Alto Adige
7 Friuli-Venezia Giulia
8 Veneto
9 Tuscany
10 Marches

Most Civic

Average

Least Civic

11 Umbria
12 Lazio
13 Abruzzi
14 Molise
15 Apulia
16 Campania
17 Basilicata
18 Calabria
19 Sicily
20 Sardinia

FIGURE 10-2

This map measures the relative amount of *civic community* that the scholar Robert Putnam found in each of the regions of Italy. He noted that, as a rule, incomes are high where civic community is high, and incomes are low where civic community is low. He has hypothesized that economic development and welfare can come about only where local political culture is based on strong civic community. (© 1993 The Economist Newspaper Group, Inc. Reprinted with permission.)

empires (but not their own empires) and to grant self-determination to several new European nation-states, including Poland (Figure 10-4). Russia's new Communist ruler, Vladimir Lenin, had criticized the Czarist Russian Empire as "a prison-house of nations." He reorganized the empire under a new totalitarian government disguised as a union of nations. That action will be examined in Chapter 13.

Therefore, it can be argued that several European nations did in fact exist before they achieved their own independent territory and governments, that is, their own states. Even in Europe, however, national governments conscientiously inculcated patriotism in their citizens, and few states have ever achieved a clean match between people and territory. The European map was redrawn again after World War II (Figure 10-5), but several states still claim parts of their neighbors' territory. Territorial claims on a neighbor are called **irredenta**, from the Italian word for unredeemed. Hungary, for example, claims the Romanian province of Transylvania, where many Hungarians live. Hungary was forced to cede Transylvania to Romania in 1920, but it forcibly retook it during World War II and had to surrender it again after that war. In 1989, Hungarian demonstrations in Transylvania triggered the fall of the Romanian Communist regime. Ethnic rioting continued after that, but in 1993, the Romanian government granted cultural concessions to the Hungarian minority, including more schools and government-posted signs in Hungarian.

THE FORMATION OF STATES OUTSIDE EUROPE
▼▼▼▼▼▼▼▼▼▼▼▼▼▼▼▼▼▼▼▼▼▼▼▼▼▼

At the same time as the idea of nationalism was maturing in Europe, the Europeans were actually enlarging their empires. They were not willing to recognize that their colonial subjects had national rights. They argued that non-Europeans were inferior or "not yet ready" for political independence. After World War I the Europeans did not offer national self-determination to their subject peoples outside Europe. The winners just took the losers' colonies while retaining their own colonies. In fact, the individual colonies seldom did represent political or cultural communities. The European imperialists had carved up the world for their own convenience. No matter what level or form of political organization existed among the native peoples, the Europeans drew **superimposed boundaries**. Some colonial boundaries split native political communities, whereas others combined two or more in one colony (Figure 10-6, p. 348).

The Europeans often used native rulers as intermediaries between themselves and the people, especially if a colony included several groups. This form of government was called **indirect rule**. In many cases the imperialist powers even amalgamated groups into new tribes and appointed new kings or, in India, maharajahs. These new groupings soon gained the force of "tradition," leaving today's scholars and citizens a bewildering challenge to define what were genuine precolonial communities. Indirect rule prevented the native peoples from uniting against the imperialists. Later, shared indignation against the colonial power often did provide the only basis for developing nationalisms.

The colonies were administered by bureaucracies made up in some cases of Europeans, in other cases of acculturated natives, and in still others of foreign peoples imported by the Europeans. When the colonies received their political independence, these bureaucracies had a vested interest in maintaining the

FIGURE 10-3

This 1812 painting by Jacques-Louis David was Napoleon's favorite portrait of himself. He is shown neither in imperial robes nor astride a horse in battle, but as a law giver. The pen and scattered documents, the hour on the clock, and the dying candles reveal that the emperor has worked all night on the Law Code. The Code remains today the basis of law in 30 countries in Europe and beyond, including Quebec and the U.S. state of Louisiana.
(Courtesy of National Gallery of Art, Washington, DC: Samuel H. Kress Collection.)

Europe in 1920

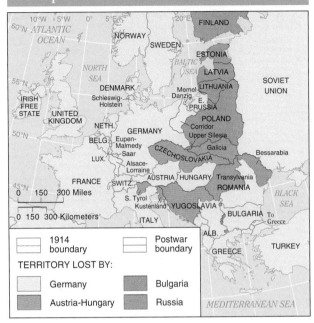

FIGURE 10-4

After World War I the German, Austro-Hungarian, Russian, and Ottoman empires were dismembered, and several new states appeared on the European map. Some of them, such as Yugoslavia and Czechoslovakia, did not represent nations, but were composed of diverse populations. The new Poland was unsatisfied with the eastern border it was originally awarded, and it seized more territory in a war against Russia.

existing units and borders. Therefore, today's world political map is not a map of nations that have achieved statehood. It is a vestige of colonialism.

The Threshold Principle

As late as the 1920s, international diplomats upheld a belief that a nation had to have some minimal population and territory to merit self-determination. This principle, called the **threshold principle**, was responsible for yoking together, for instance, the Czechs and the Slovaks, who had never formed a political community.

The threshold principle was abandoned after World War II, so today the world map reveals many tiny, independent nation-states (Table 10-1, p. 349). Some, such as Singapore, thrive economically, but others rely on foreign subsidies. Few are militarily defensible, and the real independence of some of these microstates is dubious. Four of the smallest, the tiny Caribbean islands of Grenada, Saint Lucia, Saint Vincent, and Dominica, have discussed merging into

Europe in 1946

Acquired by Soviet Union
1. E. Finland
2. Estonia
3. Latvia
4. Lithuania
5. N. East Prussia
6. E. Poland
7. Ruthenia
8. Bukovina
9. Bessarabia

Acquired by Poland from Germany

FIGURE 10-5

After World War II the Soviet Union expanded considerably. It retook the territory that Poland had won in the earlier war and gave Poland German territories, thereby effectively shifting Poland about 150 miles (240 km) to the west. It gave half of German East Prussia to Poland, and it swallowed up the three Baltic states. They had been part of the old Russian Empire but had enjoyed independence since 1919. The USSR also detached Ruthenia from Czechoslovakia and incorporated it into Ukraine, and Bessarabia from defeated Romania. Bessarabia was combined with a part of Ukraine to form a new Republic of Moldavia—today's Moldova. The Soviet Union also seized territory from Finland.

Yugoslavia took territory at the head of the Adriatic Sea from Italy, but the port of Trieste, which Yugoslavia wanted, remained Italian. Compare this map with the front endpaper world political map to see the changes that have occurred since the end of World War II.

one larger state, but even that new state would still have only 500,000 people, and its economy would precariously depend on only two activities: tourism and the export of bananas.

CULTURAL SUBNATIONALISM

When the entire population of a state is not bound by a shared sense of nationalism but is split among several local primary allegiances, then that state is said to suffer **cultural subnationalism**. In some states, many people grant their primary allegiance to traditional groups or nations that are smaller than the population of the whole new state. These traditional identifications may be strong enough to trigger civil war. In other cases, a group's bonds of affinity may extend beyond the state's borders, and this may inspire international disputes. For example, Sri Lanka has long suffered a civil war between the 75 percent of the population that is Sinhalese and Buddhist and the 18 percent that is Tamil and Hindu. The 60 million Tamils in India sympathize with Sri Lanka's Tamils, so India has intervened militarily in Sri Lanka's civil war. Subnationalism is among the forces called **centrifugal forces**, which tend to pull states apart. A strong sense of nationalism shared throughout the whole state is among the competing **centripetal forces**, which bind a state together. Multinational empires have been held together by military force throughout history, but states containing several nations—**multinational states**—can suffer divisive politics.

Civil wars demonstrate that many countries on the world map are not nation-states. At the same time, however, there are many groups that are politically self-conscious and are, therefore, arguably nations, but are politically submerged and do not show up on the map (Figure 10-7, p. 350). Hundreds of African and native American groups might claim national self-determination. In Asia the Kurds were promised a national homeland at the end of World War I, but they remain split and submerged. They continue to fight for independence in Turkey and Iraq.

Submerged nationalities often present threats of local violence, as events in Southeast Asia, the Mideast, Africa, the former Yugoslavia, and throughout the former USSR have continued to demonstrate. The idea of national self-determination retains its powerful attraction, so such local violence may continue.

Subnationalism in Africa

Subnationalism has particularly plagued African states. Somalia, for example, contains only a fraction of the Somali people. They overlap into Ethiopia, Kenya, and Djibouti, so Somalia claims territories from its neighbors. The Ewe people of West Africa are split between Ghana and Togo. Along the southern borders of the Sahara in West Africa, the Taureg people ruled the Mande, Fulani, Songhai, Zarma, and Hausa peoples for centuries before the coming of the European imperialists. Today the Tauregs remain a self-conscious nation, but they are a minority in each of the new countries formed among their former subjects: Algeria, Libya, Mali, and Niger. The Tauregs disrupt all of these countries and even encourage fighting among them.

FOCUS
Geopolitics

Chapter 9 defined political geography as the study of the interaction between political processes and the distributions of all other activities and transformations of the landscape. The term was coined by German geographer Friedrich Ratzel as the title of a book he wrote in 1897. Ratzel's studies focused on the political geography of states, and he suggested that each nation needed *Lebensraum*, room to live. As national populations increased, he wrote, nations might need more Lebensraum.

This idea of Lebensraum was adopted by a school of writers who were at that time drawing analogies between states and living things. They argued that states are a form of biological organism. In order to explain why some states thrive while others perish, these writers combined Ratzel's idea of Lebensraum with Spencer's theory of social Darwinism, and they concluded that competition among states for territory resulted in "the survival of the fittest." In 1916, the Swedish political scientist Rudolf Kjellen published *The State as a Form of Life*, in which he wrote of "the natural and necessary trend towards expansion as a means of self-preservation." Kjellen proposed a new "science" to study states' competitive strategies. He called it **geopolitics**. In 1924, a German professor named Karl Haushofer founded an Institute for Geopolitics in Munich, where he taught that it was natural for strong states to expand at the expense of the weak. These ideas impressed Adolph Hitler, and when Hitler became ruler of Germany, he quoted geopolitical theories to justify Nazi aggression in World War II.

The term geopolitics is still today loosely applied to studies of global military strategies, and it is often confused with political geography. Chapters 11 and 12 will demonstrate, however, that any state's possession of extensive territory—or even of natural resources—does not strictly determine the welfare of its people. Already in 1899, the British scientist Sir William Crookes had refuted geopolitical theories. His book *The Wheat Problem* suggested that technological progress can replace territorial aggression in raising a nation's standard of living. Crookes coined the metaphor "scientific frontiers," but even he would probably be amazed to learn of the technological advances in agriculture described in Chapter 6 and the advances in materials technology discussed in Chapter 11.

Civil Wars in Africa Many countries that have gained independence since World War II have suffered almost endless civil strife. Foreign rule had blanketed and frozen hostilities that were released again with independence. As noted in Chapter 6, civil wars have been partly responsible for recurring famines in several African countries. In some cases, government collapsed in a war of each against all, as in the early 1960s in the country that was then called the Republic of the Congo. (That country is today Zaire, not today's People's Republic of Congo.) In other cases, one or more regions have attempted to break away from the new state. The Igbo people of Nigeria, for example, declared their region to be the independent state of Biafra in 1967, and the central government fought for 3 years to restore its authority over the region. Fatalities were estimated to be over 1 million. In Ethiopia a military junta overthrew the emperor in 1974, but it was in turn overthrown in 1991 by separatist armies representing the two provinces of Eritrea and Tigre. Eritrea received its independence in 1993, and Tigre may achieve independence as well. Meanwhile, Ethiopia absorbed hundreds of thousands of refugees from the Sudanese civil war to Ethiopia's

west and several hundred thousand more from the civil war among clans in Somalia to the south. In Rwanda, the Hutu and the Tutsi have fought periodically, and in 1994 renewed slaughter took hundreds of thousands of lives and exiled additional hundreds of thousands of refugees. Hutu and Tutsi have also fought in Burundi. Angola and Mozambique dissolved in civil wars after gaining independence in 1975. A peace signed in Angola in 1991 lasted only a few months before war broke out again, but it is hoped that the peace signed in Mozambique in 1992 will last. Intertribal fighting broke out in Kenya in 1993 after 30 years of relative peace.

In some cases outsiders have prolonged wars for economic or ideological reasons or just to keep potentially unfriendly neighbors busy at home. Belgium and France, for example, interfered in Zaire and Nigeria. Angola became a battleground for troops from South Africa, Cuba, and even "advisors" from the United States and the Soviet Union, as well as several indigenous competing armies.

Subnationalism helps explain the number of authoritarian governments in Africa. In many countries the armies are the only truly national organizations, and

A World of Nation-States **347**

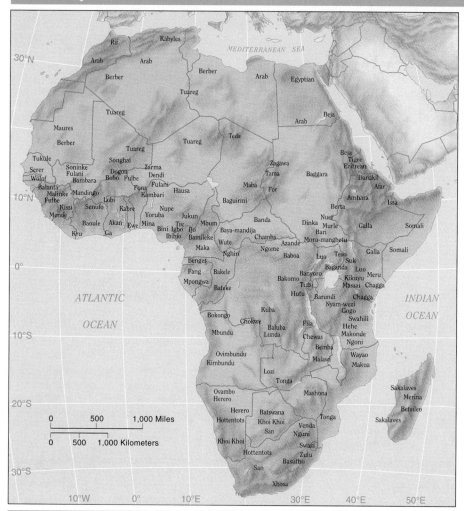

FIGURE 10-6

This map indicates the traditional territories of just a few of Africa's many indigenous peoples. The Europeans superimposed colonial boundaries, and when colonies attained independence as new states, civil wars and international wars broke out because the new states did not represent nations.

they still earn the epithet "the school of the nation." In 1990, 38 of the 43 sub-Saharan African countries had single-party or no-party political systems, and more than half were led by soldiers. The only three that governed themselves by civilian majority rule—Senegal, Gambia, and Botswana—were notably ethnically homogeneous. The authoritarian rulers argued that iron rule was the only alternative to tribalism tearing their countries apart, but this argument was at least partly a rationalization to enable them to hold on to power.

In the early 1990s, however, partly as a result of the end of the Cold War, authoritarian governments in Africa that had been supported by one side or the other began to tumble. In 1991, General Kérékou of Benin became the first African leader ever to leave office peacefully after having been voted out. This was after 18 years of military rule. His example was followed by long-time autocratic rulers of Zambia, Burundi, Madagascar, Niger, Lesotho, Malawi, and Congo. A few authoritarian rulers remain, and the

future will test the argument that authoritarian rule is necessary to hold African countries together.

EFFORTS TO ACHIEVE A WORLD MAP OF NATION-STATES

The complete division of the world into stable and peaceful territorial nation-states may or may not be an ideal goal. At present, however, the nation-state remains the principal political unit, and states use three major strategies to make the map of nations fit the map of states. One is to redraw the international political map. A second is to expel people from any country in which they are not content, or exterminate them; and a third is to forge nations in the countries that exist now. The difficulties and hardships resulting from implementing any of these policies should remind us that the territorial nation-state is only one

TABLE 10-1
A Few of the World's Smallest Independent States

	Area (km^2)	Area (m^2)	Population (in thousands)	GDP (in U.S. dollars)
Bahrain	620	239	551	4000 million
Barbados	430	166	255	1700 million
Grenada	340	131	85	238 million
Maldives	300	116	234	174 million
Monaco	1.9	0.733	30	475 million
Nauru	21	8.1	9	90 million
San Marino	60	23.16	23	400 million
Tuvalu	26	10	9	4.6 million
Vatican City	0.438	0.169	0.802	NA

Source: CIA World Factbook 1993–1994

theory of what might be the best political-geographical division of the Earth. Perhaps territorial nationalism is not a positive ideal. A return to regnum forms of government is improbable, but the universal guarantee of individual rights in any political-territorial framework might be a desirable goal.

 Deciding exactly who is a member of any nation presents problems anyway. Official national definitions, as determined by political leaders, do not always coincide with the self-identification of the people concerned Nor do people necessarily identify themselves with any territorial state that claims to represent them. Nations are generally self-defined, but there are exceptions. For example, sometimes outsiders consider a group of people to be part of a certain nation, but the people concerned reject that identification themselves. During World War II, the British interned anyone who had been born in Germany, including Jews and anti-Fascists. The British felt that anyone must be presumed to owe loyalty to the place of his or her birth. This is not always true.

Sometimes a national majority rejects a group who had considered themselves part of that nation. Throughout the Nazi period the Germans rejected Jews as Germans. During World War II the United States distrusted Japanese Americans and interned them in camps. No Japanese American ever proved traitorous, and many actually fought bravely on the European front. The United States is now paying reparations to families who were interned.

Some nation-states project nationality to citizens of other countries. Israel's Law of Return guarantees citizenship to Jews worldwide. Other people repudiate allegiances that the lands of their ancestors project onto them. Serbia is one of several countries that occasionally draft unlucky tourists who are second- or even third-generation descendants of emigrants. In contrast, other people exploit a presumed nationality. Some worldwide descendants of Irish citizens, for instance, whatever their citizenship, travel on Irish passports. They can win friends by being thought to be Irish because fellow Irish migrated throughout the world, and because Ireland has few enemies in the world today.

Could We Redraw the Map?

Theoretically the world political map could be redrawn until everybody is content being in the state he or she is in. Unfortunately, this would open endless disputes and provoke new wars. Some countries

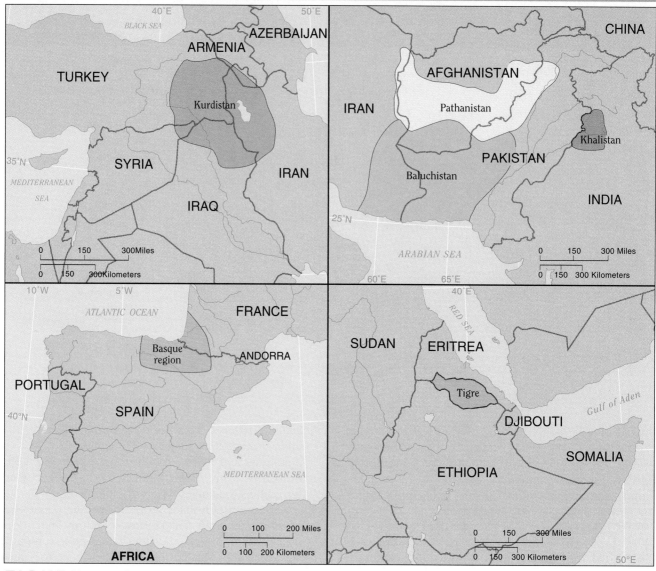

FIGURE 10-7

These are a few of the nations that do not have states of their own on today's world political map but may achieve statehood in the future. The present states, however, may succeed in dissolving these nations.

would split apart, and numerous territories would have to be transferred from one country to another. No satisfactory solution could exist for the states in which different national or ethnic groups intermingle or in which the cities are populated by people of one group and the countryside by people of another group. Disputes would also arise over the distribution of natural resources. Each group would claim the most generously endowed territory as its own, or at least a "fair share" of what had been the entire state's endowment. The Ogoni people of Nigeria, for example, occupy one of the world's richest oil-producing regions, yet the central government claims all mineral rights and distributes profits among all regions of the state. The Ogoni profit little from the oil, but live instead in toxic pollution and poverty. Ogoni independence movements have been squashed quickly.

Despite inevitable problems, political maps continue to be redrawn. In Europe in recent years East and West Germany joined into one country in 1990. Yugoslavia broke up and then fell into bloody war among the remnant republics in 1991, and Czechoslovakia split into two countries—the Czech Republic and Slovakia—on January 1, 1993.

Border Changes Since Independence

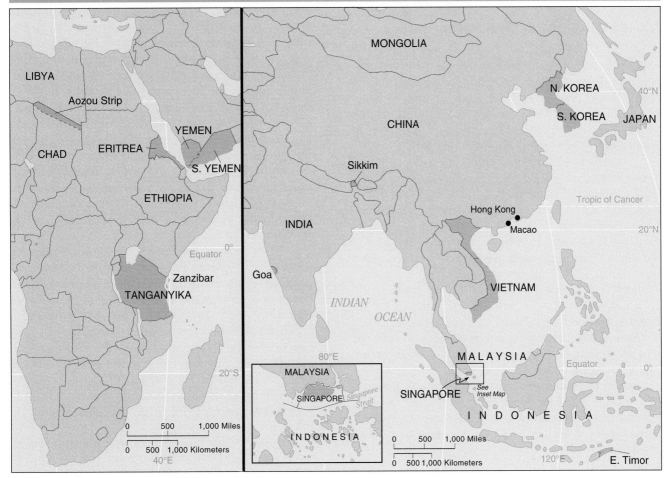

FIGURE 10-8

In Africa, Tanganyika and Zanzibar formed a union as Tanzania in 1964, although the terms of that union may be renegotiated. Zanzibar's historic ties link it with the Arab world; Zanzibar was for many years the capital of Oman. Independent Eritrea appeared in 1993. Chad and Libya have both claimed a stretch of territory across northern Chad about 100 miles (80 km) wide known as the Aozou Strip, but in 1994 the International Court of Justice awarded the land to Chad.

A few colonial borders have been redrawn in Asia. In 1965, Singapore separated from Malaysia. In 1976, North and South Vietnam joined into one Socialist Republic of Vietnam. Korea, which had been annexed by Japan in 1910, was divided by victorious Russian and American troops in 1945, and it remains split into South and North Korea. North Korea tried to conquer the South, but South Korea rebuffed it with United Nations assistance (1950–1953). Yemen, which had received independence from Turkey in 1918, and South Yemen, which received independence from the United Kingdom in 1967, merged in 1990, but war broke out between the two regions in 1994, and they may split again. India annexed the Portuguese colony of Goa in 1961 and the formerly independent country of Sikkim in 1975. In 1975, Indonesia seized the eastern half of Timor from Portugal, but the United Nations refuses to legitimize this action. China will absorb Hong Kong from the United Kingdom in 1997 and Macao from Portugal in 1999.

Few borders have been redrawn in Africa or Asia since decolonization (Figure 10-8). At the 1964 Cairo Conference of the Organization of African Unity, the African states pledged themselves to respect the existing international borders, even though they resent them as a colonial legacy and find it difficult to govern within them. Nevertheless, Eritrea emerged from Ethiopia, and other African states may either break up or else recognize subnational communities by adopting federal forms of government (discussed later in this chapter). At the same time, new frameworks of international cooperation may appear. For example, Angola, Botswana, Mozambique, Namibia, Tanzania, Zambia, Zimbabwe, Lesotho, Malawi, and Swaziland are united in a Southern African Development Community (SADC), and South Africa may join this group.

Mass Expulsions or Genocide

A second strategy used to fit a nation and a state territory is to expel from the state people who are not accepted as members of the nation, or to exterminate them. These policies may be tragic or abominable, but both have actually been implemented when groups have attempted to carry the logic of territorial nationalism to its illogical extreme.

Mass expulsions and genocide occurred in Southern Europe during and after World War I. The Turks massacred Armenians in 1915 and later they expelled about 1.5 million Greeks from Asia Minor. Greece responded by expelling 400,000 Turks. During the 1930s and 1940s, Germany attempted to bring into the country many Germans who lived in other countries and to eliminate Jews within Germany. After World War II Germans were expelled from Poland and Czechoslovakia. Mass expulsions continue. In 1989, Bulgaria expelled about 100,000 ethnic Turks. In 1991, a million Kurds fled or were driven out of Iraq. In 1993, tens of thousands of people of Nepalese origin were driven out of Bhutan. Chapter 8 told how Serbians referred to the slaughter or exile of Muslims in Bosnia-Hercegovina as *ethnic cleansing* (Figure 10-9); this policy extended to *cultural cleansing* of the landscape. The Serbs destroyed Muslim mosques, museums, and even bulldozed cemeteries. In 1994, Hutus may have slaughtered one half of the 1 million Tutsis in Rwanda.

Censuses taken in the Soviet Union in the late 1980s counted some 60 million "displaced Soviets" living outside the regions of their ethnic identity, and many people feared that when the Soviet Union broke apart, the new states would persecute their minorities. Fighting broke out in some areas, and millions of people have migrated back into the homelands of their own ethnic groups. These migrations, however, have not produced tragedies on the scale of the 1947 partition of India and Pakistan (see Chapter 8).

Probably the most ethnically homogeneous states today are Japan and the two Koreas; they are 99 percent homogeneous. China is 92 percent Han. These countries exist more or less within their historical frontiers. Japan may be the closest realization of the concept of a nation-state. The Japanese virtually closed their islands to foreign trade and cultural exchange for hundreds of years and, as noted in Chapter 5, allow very little immigration today.

Forging a Sense of Nationality

Today many existing countries are struggling to weld their populations into nations. This reverses the theoretical order, in which the formation of a nation precedes the achievement of statehood. The significance of this transition for world cultural geography can scarcely be exaggerated. The world political map is struggling to make the maps of other human activities conform to it. Chapters 7 and 8 noted that governments can affect the geography of religion and language. In addition, governments can affect the geography of political community, of law, of land use, and, as will be seen in Chapter 12, of economic activity. Countries can encourage the circulation of people, goods, and ideas within their territories, and they can restrict or discourage circulation across their borders. Countries can never control these activities entirely, but modern governments exercise incomparably greater power over human activities than did governments in the past.

Many units on today's world political map are relatively new. More than half of today's countries, containing about half of the Earth's population and covering a great percentage of the Earth's surface—including virtually all of Africa and South and Southeast Asia—are less than 50 years old. No one can predict how long any world political map will remain fixed, but existing governments are always trying to stabilize it. As noted in Chapter 3, maps of any activity at any time are only temporary. They reflect a balance of the forces for change and the forces for stability.

United States history recounts many struggles to weld one nation out of many diverse groups. In Europe a few nation builders faced their task self-consciously. The Polish hero and later president Joseph Pilsudski said flatly, "It is the state that makes the nation, not the nation the state." Italy was politically unified in the 1860s, and then the statesman Massimo D'Azeglio observed: "We have made Italy; now we must make Italians." Over the past 130 years, the Italian government may not have succeeded. A political movement called the Northern League is growing; it calls for political separation from Italy's South.

European experiences were echoed by Julius Nyerere, the first president of Tanzania. He commented on the African states: "These new countries are artificial units, geographical expressions carved on a map by European imperialists. These are the units we have tried to turn into nations." Many Asian states face the same difficulty, and even many Latin American countries that have been independent for over 150 years still have not welded their populations into nations.

Before many states can forge their populations into nations, they must first undermine the cultural subnationalisms within their borders. This destruction of traditional communities can be profoundly

352 *Chapter 10*

Nations and Ethnic Groups in the Balkans in 1990

FIGURE 10-9

This pattern has defied attempts to delineate nation-states since the nineteenth century. World War I began in 1914 when a Serb assassinated an Austrian archduke in Sarajevo, which was then an Austrian possession but was claimed by Serbs. After World War I Yugoslavia was formed, and that state eventually demarcated six constituent republics. When Yugoslavia dissolved in 1991, the constituent republics of Slovenia, Croatia, Bosnia-Hercegovina, and Macedonia emerged as independent states, while the republics of Serbia and Montenegro stayed together. They claim the name Yugoslavia, but the United States does not recognize any state as the legal successor to Yugoslavia.

Beginning in 1991, Serbian residents of Bosnia-Hercegovina (pink on this map), backed by troops from Serbia, began corralling the Bosnian-Hercegovinin Muslims (green) into ever-shrinking territories or else driving them out of Bosnia-Hercegovina altogether. It is possible that Serbia and Croatia will eventually dispute and divide much of Bosnia-Hercegovina. It is also possible that fighting may break out between Serbs and Albanians in Kosovo, the province of Serbia that has an Albanian population. The Albanian population may be driven out. Alternatively, Albania may attempt to seize the province from Serbia.

destructive psychologically to at least the first generation. Typical is Kenya's campaign to abolish the ancient traditions and distinctiveness of the Masai people, once one of Africa's most powerful nations, and to integrate them into the modern state. The Masai must go to school and conform to a new style of life dictated by the government. They must surrender their age-old lifestyle of nomadic cattle herding, settle down, and take up farming (Figure 10-10).

Some countries have absorbed the leaders of the traditional subnational groups into the new political structure (Figure 10-11). Cameroon, for example, still allows the 17 traditional kings within its borders powers of judgment over minor crimes, and it defers

A World of Nation-States   **353**

FIGURE 10-10

These Masai warriors visiting downtown Nairobi are objects of bewilderment (and, it seems, hostility) to their modern westernized compatriots. Kenya is forcibly acculturating the Masai to modern norms.
(Courtesy of Robert Caputo.)

to their ceremonial importance. In Malaysia, a paramount ruler has been elected every 5 years since independence in 1957 from among the nine traditional Malay sultans. In 1993, these sultans' powers were for the first time curtailed, symbolizing Malaysia's emergence from a traditional monarchic country to a modern democratic country.

India abolished the privileges of its many maharajahs in the 1970s, and it is today the world's largest experiment in bringing diverse groups of people together under democratic political structures. India is afflicted by centrifugal forces based on religion, ethnicity, caste, and language. Several of these distinctions are regional and might encourage separatism, but the Indian civil service, a legacy of British rule, remains a strong centripetal force.

How States Build Nations

Different countries rely on different instruments of nation building, and the choice reflects differences in other aspects of their cultures.

Religion   Chapter 8 noted countries in which a national church was a building block of nationalism: Ireland, Ukraine, and Poland. The Orthodox churches

are traditionally national churches, and some Protestant denominations are also rooted in specific nations: Lutheranism among the Germans, Presbyterianism among the Scots, and Anglicanism in England. The Catholic church is transnational, but the 1965 Vatican Council (called Vatican II) encouraged countries to form national councils of bishops. These councils have begun to reflect cultural differences among countries that are significant enough to surprise and upset the Vatican. The U.S. National Council of Bishops, for example, reflecting U.S. culture, has expressed willingness to allow women a greater role in the Church than the Vatican will accept.

FIGURE 10-11

The son of the Alake (king) of Abeokuta was educated in Britain and returned home to serve as the Chief Justice of the new state of Nigeria. This one family demonstrates the transition from traditional tribal community to new Nigerian nationalism. Many African countries continue to grant symbolic privileges and esteem to traditional rulers such as the Alake.
(Courtesy of National Museum of African Art, Eliot Elisofon Archives, Smithsonian Institution.)

Armed Forces In many countries the armed forces are still "the school of the nation," and a high proportion of the people may be serving (Figure 10-12). The armed forces may consume a high percentage of the national budget, but defense against external enemies is not always their principal function. The army holds the state together, either by force or just by training and socializing young people. Even the U.S. armed forces have been assigned nation-building tasks. In 1947, President Harry Truman ordered their racial integration. This was brought about successfully decades before the society at large was ready to integrate. In many cases the armed forces provide a disciplined labor force for building infrastructure.

Education National school systems are another tool with which states create nations. Rousseau wrote: "Education must give souls a national formation, and direct their opinions and tastes in such a way that they will be patriotic by inclination, by passion, by necessity." Today in France, it is said, the Minister of Education in Paris can look at his watch at any time of day and say precisely what all French schoolchildren are studying.

Most countries have national curricula and textbooks. The telling of national history in those texts is a political issue, and a change in the government can change the interpretation of history. Mexico, for example, introduced in 1992 a new edition of the history book for fourth graders. After 20 years of teaching that Porfirio Díaz, the turn-of-the-century dictator, "was very bad for the life of Mexico, because the people were not given the chance to elect their leaders," the new version emphasizes that Díaz achieved stability, tolerated the Catholic church, built railroads, fostered industrial growth, and attracted foreign investment. These are programs of the current government. The 1992 text also emphasizes the shortcomings of Emiliano Zapata, a revolutionary hero who defended the ejido agricultural system (see Chapter 6).

Schools inculcate youngsters with the society's values and traditions, its political and social culture. This is called the process of **enculturation**, or **socialization**. Schools propagate national culture, literature, music, and artistic traditions. National history and geography lessons leave the rest of the world outside that focus of concern. Classroom maps often promote hostility to neighbors by claiming irredenta.

The United States has traditionally relied strongly on the schools to forge a nation, yet it allows a unique diversity of programs and standards across the country. Education is not mentioned in the Constitution, so responsibility for it remains with the individual states, which have generally delegated the task down to local school districts. International com-

FIGURE 10-12

Why isn't this boy in school? He is a soldier in the army of Laos. Many countries draft high percentages of their people to keep them under control.
(Courtesy of Julia Nicole Paley.)

parisons indicate that U.S. elementary and secondary schools are relatively weak, and this has spurred demands for nationwide standardization of curricula and testing. A 1989 conference of all 50 state governors concluded: "We believe that the time has come…to establish clear national performance goals, goals that will make us internationally competitive." A Congressionally-mandated National Assessment of Educational Progress is working to achieve that goal. Under a standardized system, local districts could still experiment in designing school programs, but standardization could end one of the nation's oldest and most revered traditions of local government.

Manipulating Symbols Every country has a set of national symbols, called an **iconography**, and these

A World of Nation-States   **355**

items are emphasized to schoolchildren. The national flag is one of the most important of these (Figure 10-13). In the United States the flag is probably the most important national icon. States have passed laws forbidding defiling it, but the Supreme Court has overturned these laws and recognized such action as a form of speech protected by the Fifth Amendment. Each nation's national anthem is another element its iconography. In a unique instance of international solidarity, however, the nonpolitical prayer for God's blessing "God Bless Africa" serves Namibia, Tanzania, Zambia, Zimbabwe, and South Africa. It was written by a South African schoolteacher and taken up by the African National Congress as its anthem.

The map is another icon of great power. Children can be taught to respect, even to cherish, the size, shape, topography, resources, and variety of their land. The U.S. weather map has demonstrated appeal on television, and the editors of *USA Today* were surprised to discover that the weather map is one of the paper's most popular features. Few people actually read the map or even care about the weather at the other end of the country, but the map seems to pull all Americans together. The power lies simply in the image. Nevertheless, no country has laws forbidding tearing or burning a national map.

Iconography also includes all the pomp and circumstance of national ceremony—the celebration of holidays, costumes, the designs on national postage stamps, and international sports competitions. A surprising number of these icons were introduced in nineteenth-century Europe in a process British historian E. J. Hobsbawm has called *the invention of tradition*. Many countries ban public displays of the iconography of past regimes, including fascist symbols in Germany, Hungary, and Romania, and Communist symbols in several formerly Communist countries.

Media   National media may reach and sway more of a national population than any other shared experience, even the army or the school system. Two-thirds of India's people, for example, are illiterate, but 80 to 90 percent of them can be reached by state-controlled radio and television.

Government media monopolies may inform the population, educate it, win its allegiance, or command it. The United States is the only country in the world in which commercial television broadcasting preceded public programming, and today only the United States, Sri Lanka, and Norway do not require television stations to donate air time to political candidates.

Many countries protect their national airwaves against foreign programs as a way of safeguarding their national culture. Spain and France, for example,

FIGURE 10-13

In the center of the flag of the Republic of Korea, Yang over Um symbolizes the dualism of the cosmos. The four outer signs read, clockwise from top right, heaven, water, earth, and fire. The flag was designed to suggest the richness of Korean traditional philosophy and culture.

enforce quotas on the amount of radio and television programming and the percentage of movies shown that can be imported. The United States exports popular culture products, so it has protested these quotas. The media, however, are probably the principal definers of U.S. culture. What is an American? A person who knows who Millard Fillmore was? Or a person who watched last year's Super Bowl?

Media that spill over international borders can exert powerful centripetal force. For example, one reason that East Germany never created an East German nation, despite 40 years of trying, was that the government could not prevent East Germans from watching West German television.

Political Parties   Political parties can politicize and mobilize populations, recruit people, and give them a sense of participation. They can broaden the government's base of support, but they can also be divisive.

Political parties play different roles in different countries, and these roles reflect local political cultures. For example, most states have adopted one of two forms of government that grew out of European political experience, but the forms do not actually work the same way everywhere. One model, originated by the United States, has a *strong president* elected independently of the legislature. This model was copied, for the most part, throughout Latin America and later in France's former African colonies. In the alternative model, the *parliamentary form*, the executive is elected from among the members of the legislature. This form was left behind in Britain's former colonies. Both forms assume the existence of several competing political parties. Multiparty democracy may be appropriate in some cultures, but in others one-party government does not necessarily mean totalitarianism. Traditional African political techniques, for instance, are based on consensus building through dialogue. This technique is called *existential democracy*, as distinct from the *adversarial democracy* that Westerners practice. Existential government conforms with traditional African legal systems and religion, which emphasize the maintenance of social harmony rather than retribution.

An example of existential democracy took place in the People's Republic of Congo in 1991, when 1,500 delegates met for 3 months to revise the country's constitution. Revolutionary changes were introduced, and yet few formal votes were taken. The conference reached agreement through debate and consensus, and at its end all delegates dipped their hands into the reflecting pool in the conference hall ceremonially to wash away any ill feelings that might have accumulated. Asian politics are also notably existential, in keeping with traditional Confucian values. The Japanese constitution, for instance, was imposed during the U.S. occupation of Japan, yet the Japanese govern themselves under that constitution in a distinctly Japanese way. Consensus on most issues is reached before any vote is taken, so formal votes are seldom contests or measures of power.

Labor Unions Labor unions may serve as still another building block of nationalism. "Labor" or "workers'" political parties can even govern. In totalitarian states the government may sponsor official labor unions and curtail the formation of independent unions. In Poland under Communist rule, the independent Solidarity labor movement nevertheless overwhelmed the Communist party candidates for parliament and deprived the Communists of any pretense of legitimacy. Independent unions also played a key role in bringing down the Communist government in the former Soviet Union.

Each of these six institutions—a country's religious institutions, armed forces, schools, media, political parties, and labor unions—may be either a centrifugal force or a centripetal force. Geographers study their presence or absence in any country, the mix or balance among them that is unique to each country, and their interaction with other aspects of each country's culture. Furthermore, each institution has a geography within each country. One or more institutions may be equally influential across the entire territory, concentrated in only one region, or spill over the international borders.

Nationalism can be threatened either by regionalism or by internationalism. Whereas regionalism tends to split a country into many parts, internationalism tends to dissolve a country's external borders. Today, many global social and economic developments exert centrifugal force against the centripetal forces of nationalism. Chapters 12 and 13 will examine those global developments.

HOW COUNTRIES ORGANIZE THEIR TERRITORY POLITICALLY

▼▼▼▼▼▼▼▼▼▼▼▼▼▼▼▼▼▼▼▼▼▼▼▼▼

Each state demarcates its territory's borders and subdivides its territory for political representation or administration. This section will examine the methods used and some of the results achieved.

International Borders

Many people think of rivers and mountain ranges as "natural" borders, perhaps because on a map these features often resemble lines. This simple assumption, however, is false, and attempts to enforce it have triggered numerous wars through history. Rivers bind the peoples on their two banks as much as they divide them, and mountain borderlines can be drawn either from one peak to the next or else up and down the mountain valleys that reach between the peaks. Mountain valleys can actually provide foci for political organization. The early Swiss united the valleys on either side of mountain passes. The passes did not divide peoples, but gave them a common interest.

State borders within the United States illustrate various types that can be drawn. Some are rivers (Washington-Oregon and the Mississippi Valley), and some follow mountains (Idaho-Montana, Virginia-West Virginia, North Carolina-South Carolina-Tennessee). Most, however, are geometric lines imposed across the landscape.

Some international borders are defended with minefields and watchtowers. Others are matters of sharp dispute. Guatemala recognized the independence of Belize in 1991 but claims most of its territory. In 1992, El Salvador and Honduras accepted a decision of the World Court about their border, over which they had argued since independence in 1861. India and China fought over their border several times until agreeing, in 1993, to negotiate a mutually satisfactory demarcation. Other international borders, by contrast, are relatively open and free. The U.S.-Canadian border is the world's longest undefended border. The location of an international border can be clear to anyone flying overhead, however, even if there are no border structures. Borders may be evident by contrasting land uses on the two sides, by different types of landholdings, or by discontinuity in the transportation networks (Figure 10-14 and Figure 10-15).

An international border across a sparsely populated area may not be marked or supervised at all, and its exact location may become a matter of dispute only if valuable resources are discovered in the region. This has happened along the inland borders of South American countries. In 1941 Peru attacked Ecuador and annexed 55 percent of Ecuador's territory. The discovery of oil in the former Ecuadorian territory in what is today northern Peru revives the possibility of renewed war between these two countries. The jungle border between Brazil and Venezuela is also disputed. Similar disputes have occurred in Middle Eastern deserts. Under a 1966 agreement, Saudi Arabia and Kuwait share jurisdiction and oil revenues from a 5,700-square-mile (14,763 km) "neutral zone" between them.

Borders may attract complementary economic activities on the two sides. In the 25 counties along the U.S.-Mexican border in the states of California, New Mexico, Arizona, and Texas, the population rose 30 percent between 1980 and 1990, almost 3 times the national figure. Those counties' populations are notably younger, more Hispanic, and poorer than the national norm, yet retail spending per household in those counties is extraordinarily high. Mexicans presumably cross the border to shop. The banks in those 25 counties hold assets far above what might normally be expected—again, presumably, because of Mexicans' deposits. The border zone has also developed complementary manufacturing, as we shall see in Chapter 13.

States often want to control the passage of information as well as of people and of goods over their borders. Today, however, direct people-to-people links via telephone lines or satellites have proliferated and destabilized the distinction between what is inside and what is outside. Telephone connections have been used to break into computers in foreign lands and to read, steal, or tamper with privileged information. This action challenges international law. In which country was the crime committed?

Governments can regulate only from within their own borders, and they cannot regulate broadcasts into a country if those broadcasts originate outside its borders. Electronic broadcasts into a country can be jammed, but efforts to do so are expensive and often unsuccessful. Many countries try to ban or regulate the ownership of satellite dish antennas, but

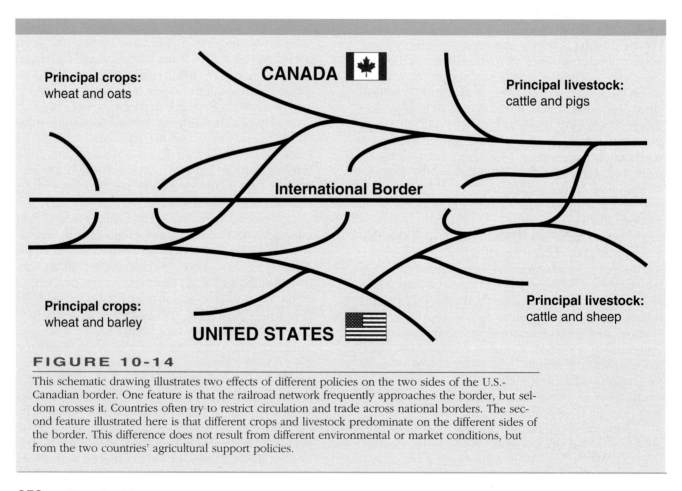

FIGURE 10-14

This schematic drawing illustrates two effects of different policies on the two sides of the U.S.-Canadian border. One feature is that the railroad network frequently approaches the border, but seldom crosses it. Countries often try to restrict circulation and trade across national borders. The second feature illustrated here is that different crops and livestock predominate on the different sides of the border. This difference does not result from different environmental or market conditions, but from the two countries' agricultural support policies.

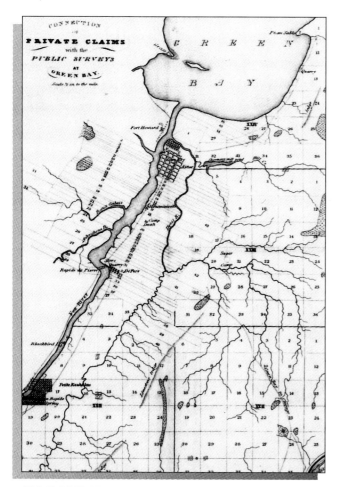

FIGURE 10-15

This early map of the region of Green Bay, Wisconsin, illustrates two different systems of landholding that are usually found in different countries. The first Europeans who settled here were French, and they defined landholdings according to their traditonal *long lot system*, which extends holdings back from the riverbank. Later U.S. surveyors recognized the preexisting long lot claims but subdivided all land not already claimed in a pattern of regular squares, called the *township and range system*. This system was chosen for use in the United States by the Ordinance of 1785.
(Map by S. Morrison, E. Swelle, and J. Hathaway. Courtesy of the State Historical Society of Wisconsin.)

these regulations are often ineffective. Globally, radio is still more important than television is, and U.S. broadcasts of Radio Free Europe, Voice of America, and Radio Free Asia blanket the world.

Some countries welcome international transmissions. In 1990 Pakistan began to allow transmission of U.S. Cable News Network (CNN) into Pakistani homes. Pakistanis have been astonished to see the range of debate allowed in the United States on every issue. Programs always contain messages that are not obvious to the programmers. People in some countries may watch a crime show, for exam-

ple, and remember only that the captured criminals were read their rights. Such direct international dissemination of the U.S. political style might represent a triumph for democracy and sow the seeds for democratic awakenings.

The Shapes of Countries

The shape of a country may affect its ability to consolidate its territory and control circulation across its borders. A circle would be the most efficient shape on an isotropic plain because a circular state would have the shortest possible border in relation to its territory, and that shape would allow all places to be reached from the center with the least travel. States with shapes the closest to this model are sometimes called *compact states*. Poland and Zimbabwe are examples. *Prorupt states* are nearly compact, but they have at least one narrow extension of territory. Thailand is an example. If these extensions reach out to navigable waterways the extensions are called *corridors* (corridors' special importance to landlocked states is discussed in Chapter 13). *Elongated states* are long and thin, such as Chile or Norway, and *fragmented states* consist of several isolated bits of territory. *Archipelago states*, made up of strings of islands, such as Japan or the Philippines, are fragmented states. Still other states, called *perforated states*, are interrupted by the territory of another state enclosed entirely within them. South Africa, for example, is perforated by Lesotho, and Italy is perforated by the Vatican and by San Marino.

The shape of a state's territory may influence the government's ability to organize that territory, but this is not always true. A topographic barrier such as a mountain chain may effectively divide even a compact state. Bolivia and Switzerland, for example, are compact in shape, but mountain chains disrupt their interiors. For some of their regions trade across international borders is easier than trade with other regions of their own country. The people throughout an archipelago state, by contrast, may be successfully linked by shipping.

Before drawing any conclusions about political control from the shape of a state alone, one must consider the distribution of topographic features, of the state's population and resources, and whether any centrifugal forces such as economic or cultural ties straddle the state's borders.

Territorial Subdivision and Systems of Representation

All countries subdivide their territory, and each country has a unique balance of powers between its local

GEOGRAPHIC REASONING

The Drama in China: Can a Country Insulate Its Borders and Isolate Its People?

Questions

1. What methods do governments use to prevent the influx of information and foreign ideas?
2. Can these methods be entirely successful, given modern technology?
3. Do governments have the right to restrict the flow of information? Why or why not?

China still tries to limit the information available to its people, but events in 1989 demonstrated that national borders have become permeable to information. During May and June of 1989, the Chinese government granted U.S. television networks direct satellite transmission facilities in order to report a visit by Soviet Premier Gorbachev. That visit, however, triggered demonstrations in Beijing's Tiananmen Square by students who demanded political liberalization. The Chinese government ordered an end to television transmissions, and the order itself was broadcast live worldwide. This did not stop videofilming. Tapes of continuing demonstrations were smuggled out of China. On June 4, columns of army tanks attacked the protesters in the square, killing untold numbers—perhaps thousands. Tapes of this slaughter were broadcast around the world.

The severity of this repression split the army itself into factions, and factionalism opened the possibility of civil war. This danger made it essential for the Chinese government to monopolize information about what was happening so that various army commmanders could not coordinate their activities against the government. On June 8, however, the Voice of America initiated television satellite broadcasts into China. The broadcasters had no way of knowing who watched the broadcasts or what they did with the information, but they knew that most satellite dishes in China were operated by the army. For the United States to broadcast information directly to the Chinese Army represented outside interference in a volatile domestic political situation. The Voice of America was at the same time broadcasting into China 11 hours each day of Chinese language radio programming to an estimated 60 to 100 million people.

Meanwhile, the 40,000 Chinese students living in the United States faxed daily news summaries and pictures from U.S. news sources to every fax machine in China for which they

governments and its national government. We generally call those countries in which the balance of power lies at the center **unitary governments**, and those in which the balance of power lies with the subunits **federal governments**. Today most national governments are unitary. Few newly independent and poor countries have adopted federalism, perhaps because they fear that centrifugal forces, once recognized, could pull the country apart.

The definitions of unitary states and of federal states represent models, but in fact countries with federal constitutions may be highly centralized through

government by a one-party political system or a military regime. India's federal Constitution allows the prime minister to impose direct rule from New Delhi when law and order have broken down in a state. Even the U.S. Constitution obligates the federal government to "guarantee to every State...a Republican form of government ..." (Article IV, Section 4). All governmental systems actually leave much open to improvization and continual redefinition.

Unitary Government In unitary states the central government theoretically has the power to redraw the

businesses. These senders never knew who was receiving the information or what they were doing with it, but the information did get there, and people in China had no other way of learning what was going on in their country. When the Chinese government later announced a telephone number for informers to call to report the names of "criminal enemies of the people," Chinese students in the United States flooded these lines naming the leaders of the Chinese government themselves. Unfortunately, American press photographs of the demonstrations also helped Chinese authorities identify and capture protest leaders.

The Chinese government succeeded in repressing the student movement, but millions of Chinese for the first time had access to information about events in their own country. This might influence future internal affairs. Since 1989 private television satellite dishes have proliferated. In 1993, the government put new restrictions on their sale, but the Ministry of Electronics and the Chinese Army still compete to manufacture and sell them to pri-

This papier mâché copy of the Statue of Liberty (slightly inexact) was carried through the streets of Beijing in June 1989 by demonstrators demanding political liberty. Many U.S. viewers were deeply stirred by the symbolic sight. The building in the background is the former imperial palace, and it stands in what the Chinese have for thousands of years viewed as the symbolic center of the world, as discussed in Chapter 1. The picture mounted on the palace is of Mao Zedong (1893-1976), the political leader who introduced Communist rule to China in 1949. (Courtesy of Alon Reininger/Contact Press.)

vate households. Millions of Chinese homes are today bombarded by direct satellite transmis-

sions from a great number of international news and commercial enterprises.

boundaries of the subunits. This offers flexibility to accommodate geographic shifts in national population distribution or economic growth. It might actually be a good idea if every country periodically redrew the borders of its internal subdivisions. Interests become vested in any given pattern, however, so unitary governments seldom redraw internal boundaries.

The United States is a federal government, but each of the 50 states is internally unitary. The state governments seldom redraw the boundaries of their counties, however, largely because political parties are organized at the county level, and they resist change.

China is constitutionally "a unitary multinational state," and five regions of minority (non-Han) peoples are granted constitutional status (Figure 10-16). These regions constitute 42 percent of China's territory, but hold only about 8 percent of the Republic's population (Figure 10-17). Other smaller subdivisions provide local autonomy for about 50 national groups.

Federal Government A federal government assumes that diverse regions ought to retain some local autonomy and speak with separate voices in the central government. Federalism also allows each area

A World of Nation-States   **361**

The Peoples of China

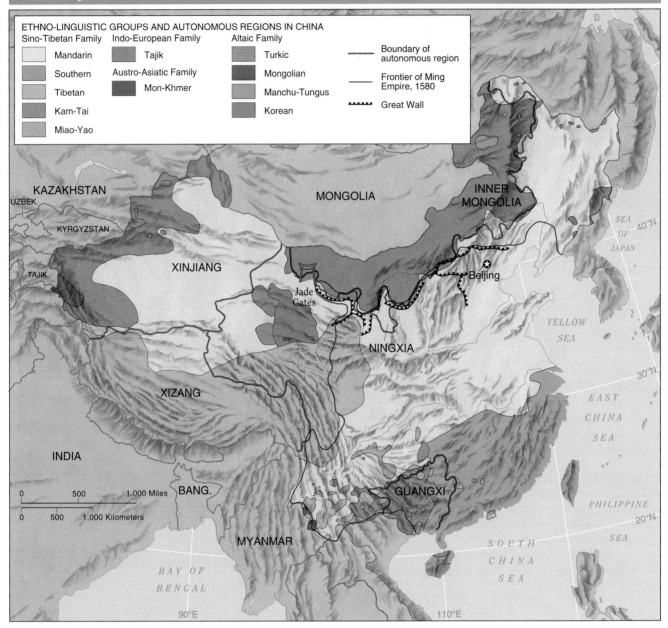

ETHNO-LINGUISTIC GROUPS AND AUTONOMOUS REGIONS IN CHINA

Sino-Tibetan Family: Mandarin, Southern, Tibetan, Kam-Tai, Miao-Yao

Indo-European Family: Tajik

Austro-Asiatic Family: Mon-Khmer

Altaic Family: Turkic, Mongolian, Manchu-Tungus, Korean

Boundary of autonomous region
Frontier of Ming Empire, 1580
Great Wall

FIGURE 10-16

The Jade Gate on the route to Central Asia and the Great Wall against the Mongols mark the traditional limits of Chinese territory. These were the borders of the Ming Empire. Through the centuries, however, Chinese rule has intermittently expanded out beyond these borders to include non-Chinese peoples. China's contemporary constitution guarantees some political and cultural autonomy to five peripheral minority peoples, but the rulers in Beijing must be worried whether the Turkish Muslims of Xinjiang are drawn to their ethnic relations in newly independent Uzbekistan, Kyrgyzstan, Turkmenistan, and Kazakhstan, and whether the Mongols of Inner Mongolia feel affinity with the Mongols in independent Mongolia.

to serve as a laboratory for legislation that can be adopted elsewhere if it works. Wisconsin, for instance, might devise a successful school program that other states might copy, New Mexico a highway program, or Maine environmental legislation. The protection of diversity, however, may perpetuate economic or social inequality.

It is dubious whether the states of the United States are today culture regions or even genuine political communities. Each state controls its own

362 *Chapter 10*

FIGURE 10-17

The Great Wall of China was built to separate ethnic Chinese from the Mongols. It still defines that cultural border fairly well, but the Chinese have extended their political rule beyond the wall to rule about half of the Mongols. (Courtesy of Paul Lau/Gamma-Liaison International.)

education system, which usually inculcates state history and geography, just as education systems do in sovereign nation-states. Each state boasts an extensive iconography: state flags, birds, flowers, and so forth. Texas and Hawaii have histories as independent countries. States have personalities, as revealed by the different types of legislation they pass, and one scholar has even mapped the "dominant political cultures" across the United States (Figure 10-18). If Puerto Rico were to become a state, it would be the most distinct state culture region.

Many observers have suggested redrawing the map of the states. A new pattern could improve the states' conformity with physiographic or economic regions or make each state more compact or equal in population or in economic strength (Figure 10-19). The U.S. Constitution, however, protects the states against restructuring against their will (Article IV, Section 3). California may choose to break up into two or three states. In 1992, 27 of 31 northern counties voted in favor of breakup, and in 1993, the State Assembly voted in favor of a statewide referendum on the issue.

Despite the United States's federal form, it may be merging into one political arena anyway. Two facts provide evidence. First, voter turnout is much higher for presidential elections than it is for House or Senate contests, or even for elections for state and local officials. This suggests a weak sense of identifi-

cation with local representatives. Second, members of Congress are raising increasing percentages of their campaign funds outside their own constituencies. More than half of the U. S. senators, for example, raise more than 50 percent of their funds outside their home states—several over 90 percent. This raises the question of whom, or what constituency, these senators represent. This development arguably corrodes the federal character of U.S. government.

Some countries' federal patterns reflect polyethnicity (for example, Russia, Myanmar, and India). Some countries' internal units are older than the federal framework, and they came together only on the condition that they would retain certain powers (the original 13 United States, Germany, and Canada). Federalism may, on the other hand, serve a country that is relaxing central control. Spain, for example, is devolving toward federalism. Spain's culturally distinctive peripheral provinces have never been welded into one nation, and in recent years they have demanded and received considerable autonomy from Madrid. The Basques enjoy local autonomy, but some demand total independence, and the situation is complicated because their homeland overlaps into France (look back at Figure 10-7). Belgium is decentralizing into three parts: Flanders, Wallonia, and the city of Brussels. In this case, some powers of the national government are devolving down to the regions at the same time as others are being surrendered up to the multinational European Union (discussed in Chapter 13). The national government of Belgium retains few powers.

If loosening the federal ties proves insufficient, a federal state may dissolve in civil war, as happened in the former Yugoslavia.

Canada's Federal System Canada was formed as much to preserve British institutions in North America as in response to any indigenous nationalism. The British North America Act of 1867 created the Confederation of Canada, and other provinces joined the Confederation later (Figure 10-20, p. 366). The 1931 Statute of Westminster removed all legal limitations on Canadian legislative autonomy, and in 1982, the Canadian Constitution was "patriated," meaning that Canadians obtained the right to amend the Constitution in Canada.

Canada's federal government differs from that of the United States. Executive authority is vested in the sovereign (Elizabeth II of Great Britain reigns as Elizabeth I of Canada), who appoints a Governor General, but that office exercises little authority or influence. Legislative authority resides in a bicameral legislature. The 104-member Canadian Senate is appointive, and it too exercises little power. It can-

GEOGRAPHIC REASONING
Corporativism

Questions

1. Should U.S. representatives and senators be restricted to raising funds within the districts they represent?

2. How would that affect government?

3. Would such a plan hinder free political expression?

One alternative to political representation by territorial districts is government by representatives of interest groups. This is called **corporativism**. The representatives may represent labor unions, agricultural cooperatives, chambers of commerce, private or nationalized industries, professionals, and other groups. Corporativism has been tried by Germany, Italy, Ireland, several Latin American countries, France, and several African states that were former French colonies. No countries are governed by corporativist systems today. The U.S. Congress, however, does much of its real work in committees that specialize in various topics (agriculture, labor, and so forth). Each of these committees is lobbied by representatives of groups interested in that topic. Some observers have suggested that this system imitates corporativism.

Political Cultures of the 48 Contiguous States

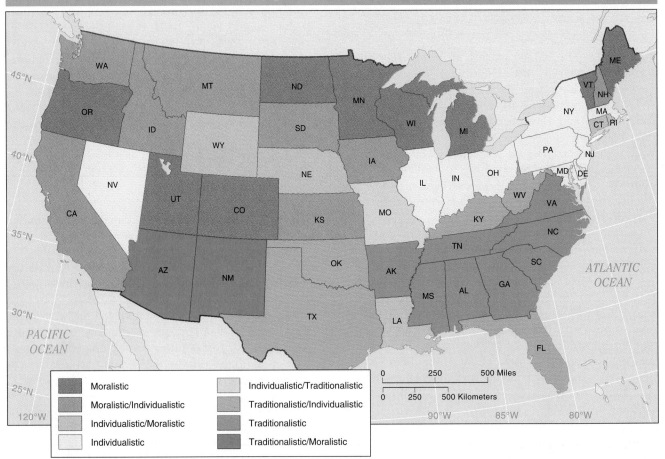

Legend:
- Moralistic
- Moralistic/Individualistic
- Individualistic/Moralistic
- Individualistic
- Individualistic/Traditionalistic
- Traditionalistic/Individualistic
- Traditionalistic
- Traditionalistic/Moralistic

FIGURE 10-18

Political scientist Daniel J. Elazar characterized the dominant political culture in each of the U.S. states in this way. Can you see any coincidences between this map and Figure 8-27, the map of religious affiliation? Would you suggest any hypotheses of explanation? (Map "Dominant Political Cultures by State," adapted from Daniel J. Elazar, *American Federalism: A View From the States,* Third edition. Copyright 1984 by Harper & Row, Publishers, Inc. Reprinted by permission of Harper Collins Publishers, Inc.)

A 38 State U.S.A.

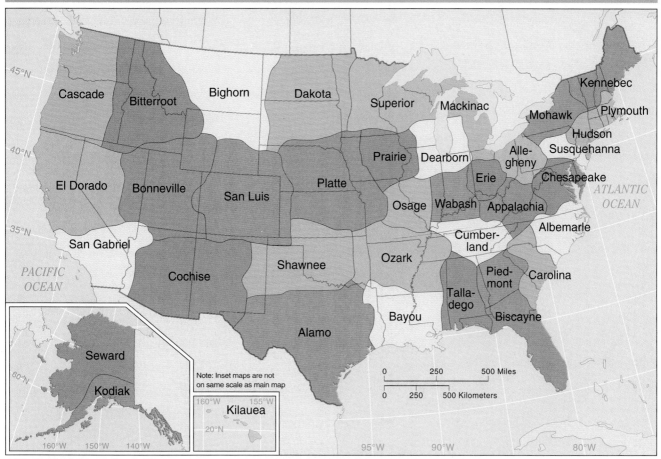

FIGURE 10-19

This redivision of the United States into 38 states was drawn as an intellectual exercise by G. Etzel Pearcy, a distinguished geographer who once served as Geographer to the United States. These territorial units may be more equivalent in population or economic output, or may better reflect true economic or cultural regions than do the present states. What state would you be in? Can you think of any advantages from creating such a state? (Adapted with permission from G. Etzel Pearcy, *A Thirty-Eight State U.S.A.*, Monograph No. 2. Redondo Beach, CA: Plycon Press, 1973.)

not, for example, initiate legislation, although it can delay legislation. Therefore, almost all power is concentrated in the House of Commons, whose 295 members represent districts based on population. Canada's provinces retain more powers than do the individual states of the United States, but they have no significant institutionalized recognition at the level of the national government.

Canada's concentration of power in the Commons has allowed the two most populous provinces, Ontario and Quebec, with over 60 percent of the total national population, to dominate the national agenda. Canadian economic protectionism has benefited manufacturers in these two provinces and disproportionately disadvantaged the peripheral Atlantic and western provinces by denying them

cheaper imports. The federal government's 1980 National Energy Program, for instance, effectively expropriated some $50 billion (Canadian) of windfall profits from Canadian oil producers in the West. As the West's population has increased, however, it has gained political power. Many Canadians are demanding that the members of the Canadian Senate be elected by the various provinces and more clearly represent provincial interests.

Chapter 7 noted that the French-speaking population of Quebec has for years threatened to withdraw from the Canadian Federation. As long ago as 1839, Lord Durham, then Governor General of British North America, wrote in a famous report that Canada was actually made up of "two nations warring in the bosom of a single state." Chapter 13 will examine

A World of Nation-States   **365**

The Canadian Federation

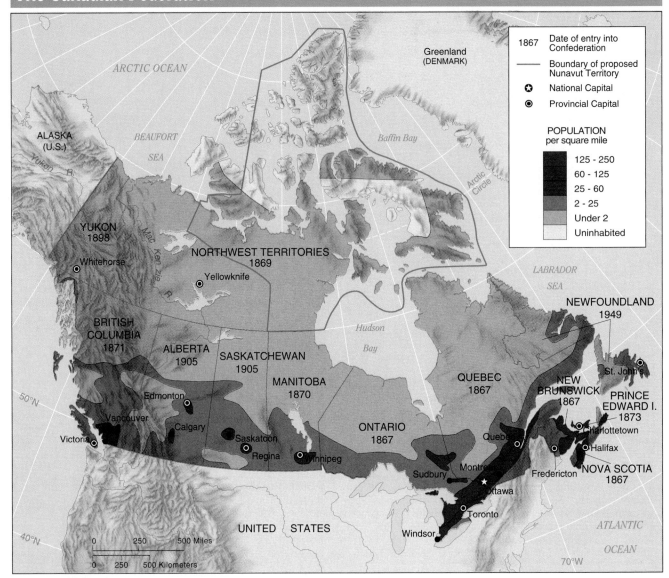

FIGURE 10-20

The units of federal Canada came together only slowly into the country's present configuration. The 1867 British North America Act united the provinces of Upper Canada (Ontario), Nova Scotia, New Brunswick, and Lower Canada (Quebec). In 1869, the Northwest Territories were purchased from the Hudson's Bay Company; Manitoba was carved from this territory and admitted into the confederation in 1870. British Columbia joined in 1871, and Prince Edward Island in 1873. Alberta and Saskatchewan were formed out of previously provisional districts and admitted in 1905. Newfoundland, previously an independent country, joined only in 1949.

how Canada's economic links with the United States affect Canada's internal politics. In 1992, five prominent North American geographers designed a map of a potential political repartitioning of the entire continent by 2050 (Figure 10-21).

Special-Purpose Territorial Subdivisions Several countries divide their territory into economic planning regions, and these can gather almost as much

power as the legal constituent subunits. The former Soviet Union did this, and France has, too. In the United States the Federal Power Commission, the National Labor Relations Board, the Bureau of Reclamation, the Federal Trade Commission, the Federal Communications Commission, and the Federal Reserve System have each subdivided the country into a different pattern of districts. Few of these patterns conform to the pattern of the states, so

366 *Chapter 10*

FOCUS
The Shifting Balance Between State and Federal Powers in U.S. History

United States history exemplifies how the balance of powers between a central government and state governments can shift. The 13 original U.S. states existed before the federal government did. They created a federal government, but they gave that government only specific powers and retained all others. Through the early years of U.S. history, the people identified primarily with their states, and they feared the aggrandizement of federal power. Defenders of states' rights argued that the states protected the people from possible federal tyranny. According to John C. Calhoun of South Carolina, the state governments were "intermediary powers" between the people and the federal government. The doctrine had much to be said for it, despite the fact that it was invoked by the southern states to defend an institution as monstrous as slavery.

The relative balance of power between the states and the central government was ultimately resolved by the Civil War. In conducting the war,

President Abraham Lincoln achieved two ends. First, he upheld the central government's jurisdiction over the entire territory of the United States. States would not be allowed to secede. Second, he exercised federal power directly to each and every citizen. The states could no longer claim a role as "intermediary powers" between their citizens and the federal government. Several federal laws enacted during the war breached the traditional powers of all state governments, North as well as South. These included a federal military draft and direct federal taxes. Therefore, Southerners argued that in fighting for states' rights, they were fighting for Northerners' freedom too.

The states nevertheless retain significant powers, and the states share responsibility with the federal government for programs as diverse as roadbuilding, product liability laws, and welfare. Unfortunately, the prestige of the state governments has continued to sink—as measured, for example, by voter turnout for

state-level elections. In 1988, 51 percent of Americans polled did not even realize that the individual state governments have their own constitutions. Chapter 9 noted that in U.S. metropolitan areas citizen participation and good government is impaired because citizens often do not know what government is responsible for which services. The same is true at the higher levels of government. For example, most people in the United States want restrictions on handgun ownership. It has been argued, however, that the Second Amendment to the U.S. Constitution prevents the federal government from enacting restrictions. Few people realize, however, that the provisions of the Second Amendment do not apply to the states; the Second Amendment has never been, in legal expression, *incorporated*, so states could curtail gun ownership individually. Therefore, if people understood state government powers, they might redirect their lobbying efforts from Washington to state capitals.

when two different districts enforce different policies in two parts of one state, the state government's own powers can be confused or undermined.

The United States is also subdivided by 91 federal district courts and, above them, 11 federal courts of appeal (Figure 10-22). Legal efforts to protect local differences may carry unexpected consequences. Two of today's most controversial topics, abortion and pornography, illustrate this. In 1973, the Supreme Court allowed the states considerable leeway in defining the conditions under which each state would allow abortions (*Roe* v. *Wade*, upheld in *Planned Parenthood* v. *Casey*, 1992). Therefore, many Americans crossed state lines to protest abortions in states other than their own.

Moreover, whereas wealthy women can still cross state borders to get abortions, poor women have to obey the laws of their home state. The question remains on the national agenda, and legislation may yet be standardized throughout the country.

Legislative restriction of pornography and obscenity presents another example of differences in local laws. In 1973, the Supreme Court ruled that courts can regulate expression only "when the average person, applying contemporary community standards, would be likely to find that the material appealed mostly to prurient interest, was patently offensive and was entirely lacking in serious literary, artistic, political or scientific value" (*Miller* v.

North America in the Year 2050

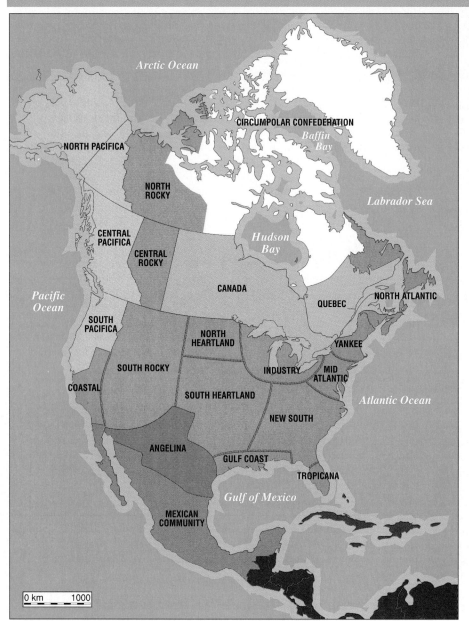

FIGURE 10-21

This map of the potential political partitioning of the continent in the year 2050 was drawn by Dean Louder and Cécyle Trépanier of Université Laval in Quebec City, David Knight of Carleton University in Ottawa, Julian Minghi of the University of South Carolina, and Stanley Brunn of the University of Kentucky. These geographers provoked thoughtful controversy when they splintered the continent into numerous independent states. (Reprinted with permission from *The Gazette*, "Mapping the Future," The Gazette, Montreal, 1992.)

California). This ruling might work in a country of isolated communities, but it is difficult to apply in a country where materials are distributed nationally. The most conservative communities can prosecute what they view as pornography in ways so punitive as to deny the material to other communities. For example, the New York-based Home Dish Satellite Network, which beamed X-rated films to 30,000 subscribers around the country, was driven into bankruptcy after a district attorney in Montgomery County, Alabama (where 30 households received the service) brought criminal charges for violating Alabama's antiobscenity laws. In the same way, the country's most conserva-

tive jurisdictions have prosecuted nationally popular rap recording stars. The Fifth Amendment to the U.S. Constitution prohibits trying the same person twice for the same crime (*double jeopardy*), yet each local jurisdiction can bring charges against nationally distributed materials even if several juries elsewhere have already refused to convict.

Today the federal government has even assumed an active prosecutive role. The Anti-Drug Abuse Act of 1988 grants federal prosecutors the right to bring obscenity charges against materials that cross state lines, even if local authorities will not do so. How can freedom of expression for nationally distrib-

368 *Chapter 10*

The Pattern of U.S. Federal Courts

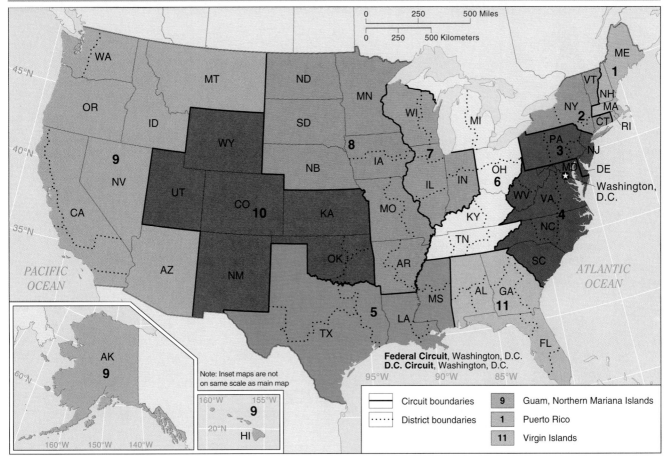

FIGURE 10-22

Any two of these courts frequently interpret federal law differently. Therefore, federal law is not uniform throughout the country until the disagreement is resolved by the Supreme Court.

uted materials be protected against local suits? Is the United States one cultural community? Can it be? Should it be?

Federal Territories Most federal states contain territories that are not included in any of the subunits but that are administered directly from the federal center with, perhaps, some local government. These territories may constitute a significant share of a state's total area, including capital districts, colonies, strategic frontier areas, and federal territories.

Territories usually have only limited local government until some presumed future time when they will be ready for full statehood. If the territories lack resources, they may remain territories indefinitely. Most of the land area of the United States was once federal territory, but the territories were organized, populated, and admitted into the Union as states. By contrast, today some 40 percent of Canada remains in territorial status. Mexico includes two territories that

may become states upon reaching populations of 120,000 and demonstrating "the resources necessary to provide for their political existence."

The U.S. government still owns about one-third of the total national land area (Figure 10-23). In 1993, the state of Alaska sued the federal government for violating the terms of Alaska's statehood by managing federal lands so poorly as to deprive Alaska of its ability to raise money and govern itself. The statehood pact requires that 90 percent of all revenue from oil, mineral, and gas leasing on Alaska's federal lands be given to the state, but the federal government has taken so much land out of potential development that Alaska derives less income than it had hoped. The 1976 Federal Land Policy and Management Act requires the federal government to receive full value for any lands traded away, and Congress still wrestles with the question of how the federal government should exploit or preserve its lands.

A World of Nation-States **369**

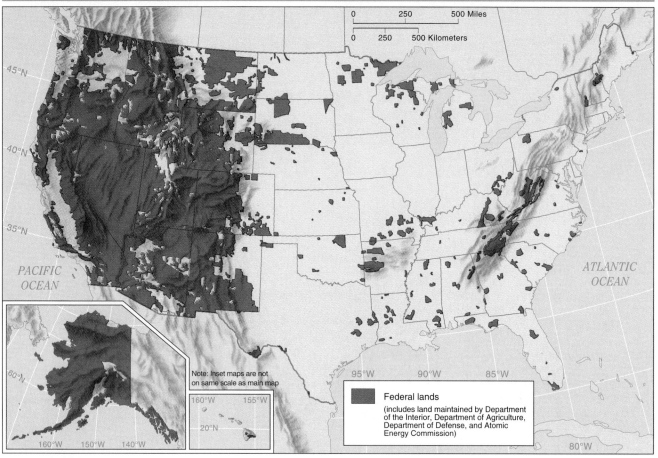

Note: Inset maps are not on same scale as main map

Federal lands (includes land maintained by Department of the Interior, Department of Agriculture, Department of Defense, and Atomic Energy Commission)

FIGURE 10-23

The U.S. government still owns a substantial share of the total national territory.

Capital Cities In some countries the federal capital is also the capital of a component state (Bern is capital of both its own canton and of Switzerland). Austria's federal capital, Vienna, is a federal state. Some federal capitals are governed directly by the federal government (Australia's capital territory, Mexico's Distrito Federal). Residents of Washington, D.C., complain that they have no voice in the federal government; if the city were a separate state it would rank 44th in population. In 1993, for the first time, a measure offering statehood for the District got on the floor of the House of Representatives, but it was defeated.

Several countries have moved their capitals because they believed that their old capitals were international cities grafted onto national life, rather than truly representative of national life. A new capital, it is believed, especially if built inland, will symbolize a rebirth and rededication of a national spirit. In 1918, the Russians moved their capital from Saint Petersburg inland to Moscow; in 1923, the Turks moved theirs from Istanbul to Ankara; Brazilians moved theirs inland to Brasília in 1960; Tanzanians moved theirs inland to Dodoma in 1975; and Nigerians moved theirs to Abuja in 1991. The Japanese, by contrast, moved their capital from inland Kyoto to coastal Tokyo in 1868 to symbolize Japan's opening to the world. Capitals are frequently designed as showplaces of national pride to impress both citizens and foreigners.

HOW TO DESIGN REPRESENTATIVE DISTRICTS

The subfield of political geography that studies voting districts and voting patterns is called **electoral geography**. Chapter 9 explained how minorities can be

FOCUS
Washington, D.C.

No city better illustrates the pride a new nation takes in its capital than Washington, D.C., the world's first city built exclusively as a seat of government.

George Washington selected the spot for "Federal City," and for plans he turned to Pierre L'Enfant (1754–1825). Construction began on the president's house in 1792 and on the Capitol the following year. In 1800 Congress moved from Philadelphia and held its first session in the city, which was named for President Washington just after his death (December 14, 1799).

L'Enfant wrote that the "mode of taking possession of, and improving, the whole district at first must leave to posterity a grand idea of the patriotic interest that promoted it." L'Enfant had been born in Versailles, a city in France in which every avenue and vista radiates from the king's palace to emphasize that power radiates from the king. L'Enfant adopted this metaphor for the capital of the new republic. He radiated all Washington streets from the Capitol to symbolize that in the new republic all power derives from "the people in Congress assembled."

L'Enfant's plan was beautiful, but it was too ambitious. It was designed to accommodate many more residents than the city would house for at least 75 years. Meanwhile, the grandeur of the plan hindered modest beginnings, and the city began to develop haphazardly. In 1889, L'Enfant's plans were reexamined, and in 1901, work began to resurrect his original design. Washington began to assume the gracious aspect it offers today.

Washington, D.C., imitates the design of the French palace and gardens of Versailles, but the symbolism is different. (Both maps reproduced with permission from the Collections of the Library of Congress.)

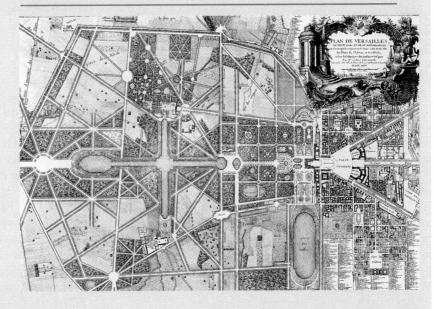

disenfranchised through the use of at-large voting systems. Even in district systems, however, the boundaries of the districts can determine the outcomes of the ensuing elections. This can occur in either one of two ways. First, if the electoral districts are unequal in population, then the ballots cast by some voters outweigh those cast by others. Second, district lines can be drawn in ways that include or exclude specific groups of voters, so that one group gains an unfair advantage. This is called **gerrymandering** (Figure 10-24).

The electoral systems in many countries tolerate inequalities in numerical representation. Chapter 6 noted this in Japan today. In the United States it was true throughout much of history, but in 1962, the Supreme Court ruled that for state and local general-purpose governments the number of inhabitants per legislator in each district must be "substantially equal" (*Baker* v. *Carr*). In 1962, Tennessee was still electing state legislators on the basis of a 1901 apportionment. The population had urbanized, so that by 1962, 1 rural vote was equal to as many as 19 urban votes. Many other states' legislatures contained similarly disproportionate representation, but nationwide redistricting now follows each decennial census. After each census the state legislatures also redistrict for federal representation. The Supreme Court has extended the one-person one-vote rule to all levels of political representation in the United States except the governing boards of the special-district governments in metropolitan areas.

FIGURE 10-24

The original gerrymander was a district created in Massachusetts in 1812 to concentrate the Federalist party vote and thus to restrict the number of Federalists elected to the state senate. The district configuration was at first likened to a salamander, but later it picked up the name of Governor Eldridge Gerry, the Anti-Federalist who signed the districting law. ("It looks like a salamander." "No, by golly; it's a Gerrymander!") (Reproduced from James Parton, *Caricature and Other Comic Art.* New York: Harper Brothers, 1877, p. 316.)

Representation in the U.S. Federal Government

The U.S. federal government itself is exempt from the one-person one-vote principle. Neither the House of Representatives nor the Senate nor the presidency is based on population count. The Constitution says, "Representatives shall be apportioned among the several States according to their respective numbers, counting the whole number of persons in each State, excluding Indians not taxed" (Amendment XIV, Section 2). This means that no district may straddle two states, and courts have ruled that "whole number of persons" means both resident citizens and aliens, even illegal aliens.

The Constitution also specifies that "each State shall have at least one Representative" (Article I; Section 3). Several states have so few residents that without this constitutional protection, they would have to share a representative. There are 435 seats in the House of Representatives. Therefore, assigning one representative to each state leaves 385 seats to be allocated among the states by population. The Constitution's mandates cannot be reconciled to a one-person one-vote ideal. In 1990, the 454,000 people of Wyoming had a representative, but each representative from California represented a district of about 620,000 people, a ratio of 1 to 1.37 (Figure 10-25).

Each state is also represented by two senators, no matter how few residents it has. Therefore, each of the senators from Wyoming represented about 227,000 people, whereas each senator from California represented about 15 million people. This gives the U.S. Senate a 1 to 66 disparity in representation.

The president, in turn, is elected indirectly in the Electoral College, an assembly also based on the states. Each state has as many electors as it has total seats in Congress, and the candidate who wins a majority of a state's popular vote wins all of that state's electoral vote. In 1888, Grover Cleveland won the popular vote over Benjamin Harrison, but Harrison won the electoral vote, and Harrison became president. In the 1992 presidential election

Reapportionment Among the States after the 1990 Census

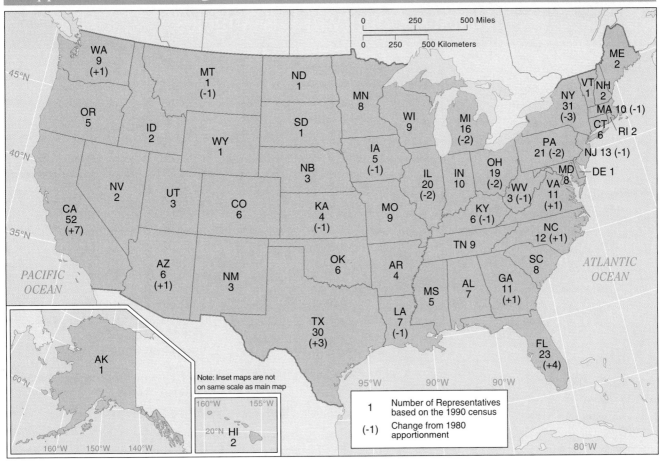

FIGURE 10-25

Seats in the U.S. House of Representatives were reapportioned after the 1990 Census. Fast-growing states such as California gained seats, but those with proportionately declining populations such as New York lost representation.

the Republican incumbent, George Bush, won 38 percent of the national popular vote, but only 31 percent of the electoral vote. Third-party challenger Ross Perot won 19 percent of the popular vote, but no electoral votes, and the winner, Democrat Bill Clinton, won 43 percent of the popular vote and 69 percent of the electoral vote.

Today some states are rapidly increasing in population, and others are holding steady or losing population. Therefore, a population imbalance is widening among the states. Congress is becoming less representative, and a repeat of the experience of 1888 is becoming more possible.

How Gerrymandering Is Done

The following discussion of the techniques of gerrymandering will draw mostly on U.S. examples,

because they will be most familiar to most readers. These techniques, however, apply wherever government is established with representation by territorial districts. Figure 10-26 illustrates the problem of drawing district boundaries in an area of heterogeneous population. It is easier to define the problems of districting than it is to define equitable solutions. Substantial equality of population count in each district is legally required, and contiguity and compactness of districts is desirable because it eases communication within districts. The question still arises, however, of whether homogeneity or heterogeneity of population within a district is preferable. Some people feel that districts should have a common social or economic characteristic, whereas others feel that balanced or integrated districts are preferable.

Computers make it easier to gerrymander, because today any legislator can call up any district on a screen, shift a boundary, and get an instanta-

A World of Nation-States **373**

Questions

1. Does the fact that a particular group is underrepresented or not represented in a governmental body mean that discrimination has occurred?

2. How can discrimination be proved?

3. How much power should the courts or other institutions have to remedy such situations?

In 1990, federal courts ruled that the Los Angeles County Board of Supervisors had used a combination of unconstitutional gerrymandering methods in drawing their voting districts in order to exclude Hispanic people from representation. Hispanics made up 35 percent of the county's total population, but none had ever sat on the five-member Board.

The court recognized that the lack of an Hispanic on the Board did not prove that unconstitutional discrimination had occurred. Alternative hypotheses can partly explain the lack of Hispanic representation. Hispanics may be underrepresented because:

- A high percentage of them are aliens.

- A high percentage fail to register.

- A high percentage are children too young to vote.

- A high percentage simply are not interested in politics.

Each of these hypotheses has some validity, but the court ruled that their combined explanatory power did not explain Hispanics' total exclusion unless intentional discriminatory gerrymandering had also taken place.

Board district lines were redrawn, and a district was created in which Hispanics made up 71 percent of the population, although only 51 percent of registered voters. Voter turnout in the subsequent 1991 special election was only 21 percent, but a Hispanic won the seat on the Board.

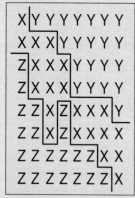

(a) Maximum segregation (b) Maximum integration

FIGURE 10-26

If we want to draw three districts of roughly equal populations, and if we have three population bases (x, y, z) defined by race, income, or any other characteristic, we could draw the boundaries in many different ways: (a) illustrates maximum segregation among the districts; (b) demonstrates maximum integration within each district.

neous readout of what the total population, voting behavior, racial composition, and other characteristics would be of the newly drawn district (Figure 10-27).

Gerrymandering as a Civil Rights Measure

A 1982 amendment to the Federal Voting Rights Act of 1965 banned any redistricting that would have a negative impact on the political representation enjoyed by minority communities. The law insists that when any court is examining a case of redistricting, the court consider "the extent to which members of a protected class have been elected to office." Several states interpreted this to mean that they were mandated to create districts in which the majority of the voters belonged to racial minorities. These districts are called *majority-minority districts*, and they are drawn by practicing the opponent containment technique illustrated as (a) in Figure 10-27. Such majority-minority districts have been drawn at every level of government across the nation, from city

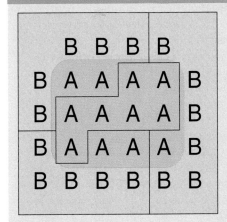
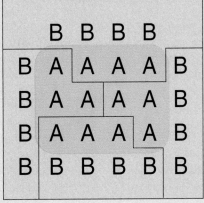
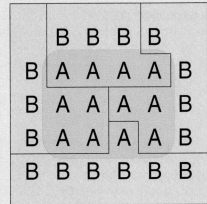

(a) Opponent concentration: A's control one district

(b) Opponent dispersion: A's control no district

(c) Even division: A's and B's each control two districts

FIGURE 10-27

In this figure, the B's have an absolute majority in each case, so they have the power to draw the district lines. Drawing (a) illustrates containment of the opponents (the A's) so that the A candidate wins that one district with an unnecessarily large majority, but A's cannot win anywhere else. Drawing (b) splits up the A's so that they do not form a majority anywhere and cannot elect even one representative.

council seats to congressional districts. Of the 16 African Americans elected to Congress in 1992, 13 represented such districts.

National reapportionment after the 1990 census gave North Carolina 12 seats in the U.S. House of Representatives, and in 1991 the state legislature drew new district lines that created one district with a black majority population, the First District. The U.S. Department of Justice, however, insisted that the state create a second, so that blacks, who represented 22 percent of the state's 1990 population, might control at least 2 of the state's 12 seats. The legislature redrew the state's Twelfth District so that it too had a black majority (Figure 10-28). The Twelfth was a classic gerrymander, winding 160 miles through urban centers across ten counties to pick up high concentrations of black voters. Both North Carolina's First and Twelfth Districts did elect black representatives to Congress in 1992.

In 1992, a group of white voters challenged the Twelfth District, arguing that any law that requires such a gerrymander is unconstitutional because it violates the principle that the Constitution should be "color blind." Some people, however, argue that the United States must be aware of color for awhile as a way of making up for past discrimination and segregation. During arguments before the Supreme Court, Justice Anthony M. Kennedy asked a key question: "Is it the policy of the United States to encourage racial bloc voting?"

In 1993, the Court ruled that the drawing of the Twelfth District may in fact have violated the rights of white voters (*Shaw* v. *Reno*). Creating legislative districts with people of the same race who are otherwise separated "bears an uncomfortable resemblance to political apartheid...and reinforces the perception that members of the same racial group—regardless of their age, education, economic status or the community in which they live—think alike, share the same political interests and will prefer the same candidates at the polls." Such racial stereotyping is legally impermissible. The ruling held that there may be "wholly legitimate purposes" for concentrating members of a

N. C.'s Majority-Minority Districts

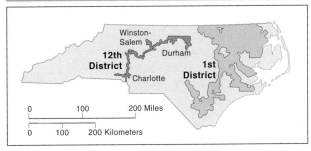

FIGURE 10-28

North Carolina's First and Twelfth congressional districts were designed to elect black representatives to Congress.

racial group. "Traditional districting principles" such as compactness, contiguity, and coterminality with existing political subdivisions could provide such a "race-neutral" rationale. The North Carolina district, however, must be judged by the legal standard of "strict scrutiny" that the Court applies to discrimination on the basis of race. Under this standard, which is difficult to satisfy, North Carolina must show that a "compelling interest" made it necessary to treat people differently on the basis of race. The Court stated that the fact that the legislature drew the district to comply with the Federal Voting Act was not, by itself, a "compelling state interest." The ruling hints that the Supreme Court may overturn deliberately drawn majority-minority districts. The immediate impact of the ruling was to require lower courts to hear the white voters' argument.

Shortly after that Supreme Court ruling, in December 1993, the Federal District Court in Louisiana ruled that the election districts in Louisiana that had been drawn to guarantee the election of two black representatives must be redrawn. Arguments over racial gerrymandering will continue in American life.

ment has won important victories for working people, but in some countries labor unions may protect their members' jobs at the expense of opportunity for other individuals. In the United States, for example, some labor unions have been bastions of racism.

Many countries of diverse populations have *affirmative action* programs designed to lift to national standards of achievement those segments of the population that have suffered historic lack of opportunity. Laws in both Malaysia and Indonesia, for example, favor ethnic Malays over the more entrepreneurially successful Chinese. Nigerian law recognizes efforts of school and job development programs to "reflect the federal character." This is a code for federally mandated ethnic quotas. Affirmative action programs have been crafted in the United States in both the public and the private sectors.

Sovereign states are the units within which laws promote or restrict individual rights, so we can meaningfully analyze and compare the policies of different countries.

The major form of discrimination in the world, however, is sexism, and the most powerful boost to individual achievement is education. Individual freedom is a value sought by all individuals. International variations in these three social characteristics—sexism, educational opportunity, and individual freedom—are examined below.

MEASURING AND MAPPING INDIVIDUAL RIGHTS

▼▼▼▼▼▼▼▼▼▼▼▼▼▼▼▼▼▼▼▼▼▼▼▼▼▼

Countries differ greatly in the restrictions they place on their people's freedom and in the degree to which they encourage and achieve full development of their peoples' potential. Predominant attitudes toward some issues, such as women's rights, may be determined by the underlying culture, which may cross international boundaries. Arab culture, for example, as noted in Chapter 8, seems relatively hostile to women's rights. Nevertheless, sovereign states are the units within which laws promote or restrict individual rights, so we can meaningfully analyze and compare the policies of different countries.

A society in which the most capable people can rise to the top on merit alone is called a **meritocracy**, and most countries claim to be meritocracies. Rigid social stratification can be as unhealthy in a society as arteriosclerosis is in a human body. Hurdles to individual advancement may include stratification of the population by caste, by race, or by employment restrictions. The modern labor move-

Sexism

Almost everywhere women are worse off than men are. They have less power, less autonomy, less money, more work, and more responsibility. National variations in sexism can be approached through a number of quantitative measures. In Chapter 4 we saw that discrimination against women can begin even against children in the womb, as reflected in higher rates of abortions of female fetuses than male fetuses, and that female-to-male ratios in national populations may reveal discrimination against women. Discrimination against female farmers, discussed in Chapter 6, lowers agricultural productivity and the levels of nutrition in many countries.

The United Nations Decade for Women (1975–1985) measured and mapped striking differences in women's welfare around the world. Researchers demonstrated that women in rich countries are generally better off than women in poor countries, but fewer data focus on the contrasts between the welfare of women and that of men in individual countries, and those are the data which must be isolated to compare sexism among countries. One useful statistic might be the differences between the percentages of males and females in school in individual countries (Figures 10-29). A country's failure to educate young girls matters not just in itself, but also because improving female literacy is one of the most effective ways to control population growth and to reduce infant mortality (Figure 10-30).

Some issues are almost exclusively women's issues, such as the female genital mutilation that is common across Sahelian and East Africa and in the Near East, and maternal death rates. Other issues affect both sexes, but principally women: abortion rights, birth rates, equal pay, rape, domestic violence, maternity care or maternity leave, availability of contraception, child welfare and rearing practices, and even marriage and divorce rights.

Some researchers have suggested using the percentage of the labor force that is female as one index of women's rights. The problem with this statistic,

however, is that statistical measurements of the labor force measure only paid occupations, and in many countries women are not paid for their labor—particularly homemaking, farming, or even small-scale domestic manufacturing. Studies show that in 1990, women made up 50 percent of the labor force in the Soviet Union, but everywhere else their participation rate was below that of men. It was about 40 percent in North America, slightly less in Japan, Europe, and Australia and New Zealand, and only 20 percent or even less in South Asia and the Middle East. These figures suggest that the women of Africa are not working. Observation of African life, however, reveals that the women are indeed working; they just are not getting paid. We need better statistical measures of women at work.

Women's participation in politics might serve as another index of sexism. In most democratic countries men got the vote before women, but in 1994, Bhutan and Kuwait were the only countries where

Ratios of Females to Males in Schools

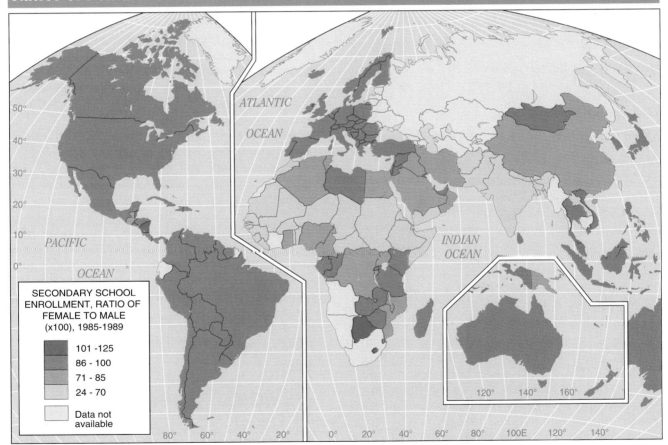

FIGURE 10-29

Varying national ratios of females to males in school may be one comparative measure of discrimination against females. (Data from United Nations, *Handbook of International Trade and Development Statistics*. New York: United Nations Publications. Reprinted by permission.)

FIGURE 10-30

Males and females attend this school in Kenya together, but generally worldwide, females have less chance to attend school than males do.
(Courtesy of Agency for International Development.)

men could vote but women could not. Women have formed political parties in several countries, including Holland, Iceland, Belgium, Spain, and Sweden. Everywhere women are proportionately underrepresented in legislatures, even though several countries reserve legislative seats for women. In Uganda, for example, each of the 34 parliamentary districts reserves an at-large seat for a woman.

The collapse of Communist government in Eastern Europe has, paradoxically, reduced women's rights. The Communist regimes rigorously upheld job security, maternity benefits, rights of divorce, and other social guarantees as a matter of Communist doctrine. Women lost many of these rights in the transition to democracy and free-market economies. Layoffs in state industries struck women hardest because men's jobs were considered of primary importance. Newly elected parliaments in every country have smaller percentages of women than under communism. Abortion rights have been curtailed, and state-financed child care has been ended. Party theory, however, never encouraged husbands to do housework, and therefore working women were often still expected to do the shopping, cooking, and cleaning. Women are not necessarily better off if they are "free" to work outside the home.

Regardless of national constitutions and laws, men's and women's relative welfare can be deter-

mined by their traditional roles in national culture. Many countries provide formally for sexual equality in the law, but few protect specific job, inheritance, property, or marriage rights or protect women from domestic violence. In order to learn more about the world geography of sexism, statistical and analytical tools will have to be improved.

In February 1992, the United Nations sponsored a conference in Geneva, Switzerland, of over 50 wives of heads of state to draw attention to the problems of the world's rural women. This conference may become a recurrent event.

THE WORLD GEOGRAPHY OF EDUCATION

The more advanced a society is, the larger the investment it requires in its human resources, that is, education. The U.S. educator Horace Mann wrote in 1846, "Intelligence is a primary ingredient in the wealth of nations." Ireland is poor in natural resources, yet it advertises in U.S. business magazines that "A Single Investment Has Made Ireland The Wealthiest Nation In Europe: Ireland...invest[s] in education....Ireland is rich in the resource that matters most....skilled people." Ireland has been so successful in educating its people that it must now attract jobs for them or else they emigrate.

A country's wealth and the education levels of its citizenry usually rise together. In the rich countries of Western Europe and Asia and in the United States adult literacy is high. In the rich countries that attract immigrants, significant numbers of people may be literate in a language or languages in addition to or other than the official language of the country. The percentages of literacy drop only slightly in Latin America, but markedly in Africa. Few African countries boast literacy rates, even for the preferred males, of much over 50 percent (Figure 10-31). School enrollment figures tell the same story. The children in the world's rich countries are in school; those in the world's poor countries are not.

THE WORLD GEOGRAPHY OF FREEDOM

Some people in the world live in freedom; others do not, but it is difficult to quantify freedom. Surprises appear when comparing statistics. Many civil libertari-

Literacy Levels

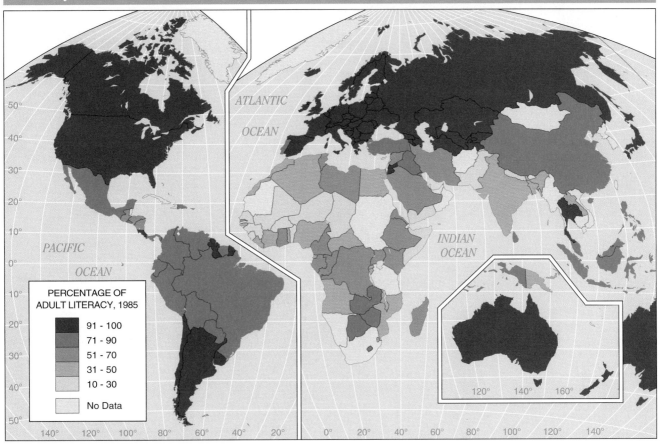

PERCENTAGE OF
ADULT LITERACY, 1985

- 91 - 100
- 71 - 90
- 51 - 70
- 31 - 50
- 10 - 30
- No Data

FIGURE 10-31

Literacy is virtually universal in the rich countries, but this map reveals considerable variations in national literacy rates. (Data from World Resources Institute, *World Resources 1990–91.* New York: Oxford University Press, 1990, pp. 262–63.)

ans, for example, ask why the United States tops all countries in the percentage of its population that is incarcerated. In 1990, 426 Americans per 100,000 of population were incarcerated. South Africa ranked second among countries with 333 out of 100,000 overall, and the Soviet Union, on the verge of collapse, was third overall at 268 out of 100,000. Western European countries range between 35 and 120 per 100,000, and Asian countries count between 21 and 140. Does this mean that the United States is the world's most totalitarian country? Or does it mean only that if you break the laws you are more likely to go to jail in the United States than anywhere else? Many observers might respond that just living in many countries is the equivalent of being in jail. How are such statistics to be interpreted?

Freedom House is a research institute that monitors political and civil freedom around the world. In the measurement of political rights, the institute assigns the highest scores where elections are fair

and free and where the people who were elected rule. The institute defines civil liberties as including freedom of expression, assembly, demonstration, religion, and association. It rates highest those states that protect individuals from political violence and from harm inflicted by courts and security forces. These states have free economic activity and strive for equality of opportunity. Freedom House assigns each country a score of 1 to 7 in each of these two categories and then sums up the scores in an overall rating "Free," "Not Free," or "Partly Free." Freedom House findings as of January 1994 are reproduced in Figure 10-32. These rankings change rapidly as governments rise and fall, but the attempt to rate countries reveals wide disparities in freedom.

In 1991, the United Nations ranked freedom in each country according to a scale of 40 criteria, as reported in 1985. Each criterion was given equal weight, and they ranged from "the right to teach ideas and receive information" to the right to "interracial,

The Geography of World Freedom

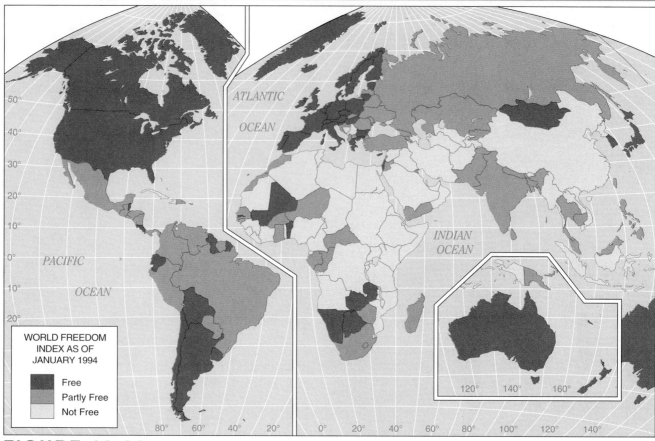

WORLD FREEDOM INDEX AS OF JANUARY 1994

- Free
- Partly Free
- Not Free

FIGURE 10-32

This map results from an effort to quantify the degree of freedom exercised by the population in each country.

interreligious or civil marriage." Sweden and Denmark scored 38; Iraq 0. The United States and Australia tied for sixth place, guaranteeing 33 of the 40 freedoms, although 12 countries ranked above them.

SUMMARY

The Earth's division into sovereign states is the most important territorial organizing principle of human activities. The idea that the whole world should be divided into countries originated in Europe and diffused worldwide with European conquest. The neatness of the units on today's world political map nevertheless exaggerates the degree to which all people accept the current pattern.

A nation is a group of people who want to have their own government and rule themselves. The ideal state is theoretically a nation-state. Several European nations existed before they achieved their own states. The political patterns of the colonies, however, were the conquerors' creations. When independence came, only large units were at first granted independence, but later this threshold principle was abandoned.

Many states suffer from cultural subnationalism. States might employ three strategies to fit the nation to the state territory: redraw the political map; expel people from any country in which they are not content, or exterminate them; or forge nations in the existing states. Redrawing the international map presents many difficulties, expulsion or genocide are abominable, so six national institutions—a country's church or churches, armed forces, schools, media, political parties, and labor unions—become key centrifugal or centripetal forces in any state. Geographers study the presence or absence of these institutions in any country, the unique balance or mix among them in each country, and their interaction with other aspects of each country's culture.

Each state demarcates its territory's borders and subdivides its territory for political representation or administration. Each country has a unique balance of powers between its local governments and its central government. Special districts can undermine the powers of local governments, and most federal states contain territories that are administered directly from the federal center. Capitals are symbolically important. Electoral geographers note that in any system of representative government, voting district boundaries can be gerrymandered to determine the outcomes of the ensuing elections.

Countries vary widely in the degree to which they liberate and train their populations. The major form of discrimination is sexism, and the most powerful boost to individual achievement is education. Individual freedom is a value sought by all individuals. The incidence of each of these three social characteristics varies from country to country.

KEY TERMS

state (p. 340)
regnum (p. 340)
dominium (p. 340)
social Darwinism (p. 340)
nation (p. 340)
political community (p. 340)
nation-state (p. 341)
core area (p. 341)
regime (p. 341)
government (p. 341)
political culture (p. 341)
national
   self-determination (p. 343)
irredenta (p. 344)
superimposed
   boundaries (p. 344)
indirect rule (p. 344)
threshold principle (p. 345)
cultural
   subnationalism (p. 346)
centrifugal forces (p. 346)
centripetal forces (p. 346)
multinational state (p. 346)
geopolitics (p. 347)
enculturation (p. 355)
socialization (p. 355)
iconography (p. 355)
unitary government (p. 360)
federal government (p. 360)
corporativism (p. 364)
electoral geography (p. 370)
gerrymandering (p. 372)
meritocracy (p. 376)

QUESTIONS FOR INVESTIGATION AND DISCUSSION

1. When and how was the pattern of local governments in your state drawn? What responsibilities do the various local governments have? Could you improve the pattern, reassign responsibilities to new districts or to different levels of government?

2. Read the constitutions of a few countries that you believe to be totalitarian and centralized. Do the constitutions reveal that? Does the U.S. Constitution accurately reveal the division of power between state and federal governments today?

3. How much land in your state is owned by the federal government? What is its legal status?

4. Is your state overrepresented or underrepresented in the electoral college?

5. Do you think regions of a country ought to be able to secede? If you favor a vote, do you think that majority approval should be required throughout the entire country or only in the areas seeking to secede?

6. What would be the advantages and disadvantages if your state withdrew from the Union?

7. Make a list of several mountain chains or rivers around the world that serve as international borders. Which act as effective barriers? Which don't? Why?

8. Investigate the symbolism of the flags of a few countries. Look at Australia, the Comoros, Liberia, Mongolia, Namibia, Venezuela, and the new 1994 flag of South Africa.

9. Why might a unitary government be more successsful in Japan than in the United States?

ADDITIONAL READING

Advisory Commission on Intergovernmental Relations, Washington, D.C. various publications.

BARONE, MICHAEL and GRANT UJIFUSA. *The Almanac of American Politics 1994*. Washington, D.C.: The National Journal, 1993. Published biannually.

BLUNDEN, CAROLINE and MARK ELWIN. *Cultural Atlas of China*. New York: Facts on File, 1983.

Country Reports on Human Rights Practices for 1993. Washington, D.C.:U.S. Department of State, 1994.

DAVIDSON, BASIL. *The Black Man's Burden: Africa and the Curse of the Nation-State*. New York: Times Books/Random House, 1992.

Freedom in the World: Political Rights and Civil Liberties. New York: Freedom House, 1994.

FREEMAN-GRENVILLE, G.S.P. *The New Atlas of African History*. New York: Simon and Schuster, 1991.

GREENFIELD, LIAH. *Nationalism: Five Roads to Modernity*. Cambridge, MA: Harvard University Press, 1992.

Historical Atlas of Canada. In three volumes: Vol. 1 *From the Beginning to 1800*, edited by R. Cole Harris. Vol. 2 *The Land Transformed 1800-1891*, edited by R. Louis Gentilcore. Vol. 3 *Addressing the Twentieth Century*, edited by Donald Kerr and Deryck W. Holdsworth. University of Toronto Press.

HOBSBAWM, ERIC and TERENCE RANGER, eds. *The Invention of Tradition*. Cambridge: Cambridge University Press, 1983.

MORRILL, RICHARD. "Gerrymandering." An issue of *Focus*. New York: American Geographical Society, 1991.

MURRAY, JOCELYN, ed. *Cultural Atlas of Africa*. New York: Facts on File, 1981.

NUSSBAUM, M. and J. GLOVER, eds. *Human Capabilities: Women, Men, and Equality*. Oxford: Oxford University Press, 1993.

Political Handbook of the World. Binghamton, New York: CSA Publications; The State University of New York, annual.

Statesman's Year-Book. New York: Macmillan Press and Saint Martin's Press, annual since 1864.

WARING, MARILYN. *If Women Counted*. New York: Harper & Row, 1989.

11 | World Mineral, Energy, and Water Resources

This sulfur is removed from Canadian natural gas because burning the gas with the sulfur would cause acid rain. The sulfur itself, however, is an indispensable raw material for a great variety of industrial processes, so it is exported around the world.

Courtesy of Avcom International Ltd.

The world population continues to rise, and each one of the Earth's billions of people aspires to superior material welfare. These aspirations impose tremendous pressures on the Earth's supplies of minerals, energy, and water. This chapter examines world supplies and distributions of these resources. This chapter will also investigate whether there are enough resources to satisfy the Earth's population and to provide rising standards of living for millions of people.

In general, inventories of resources depend on the definition of a resource. A **resource** may be defined as anything which can be consumed or put to use by humankind. Using this definition, humans are almost continuously expanding both the number of things that we define as resources and also our supplies of many individual resources. Humans are constantly devising ways of using individual minerals or other items that we previously thought were useless, and we are also learning to produce greater quantities of many of these individual commodities. Petroleum, for example, was thought to be practically worthless 150 years ago, yet today we regard it as a key natural resource. Furthermore, so much of it is being discovered regularly that per capita supplies of oil are rising. The basic perspective of this chapter is optimistic, while at the same time we must recognize that many people live in conditions of deprivation. That is humankind's tragedy.

WORLD MINERAL RESOURCES: FUEL AND NONFUEL

▼▼▼▼▼▼▼▼▼▼▼▼▼▼▼▼▼▼▼▼▼▼▼▼▼▼▼▼

There are many types of resources—climatic, biotic, and cultural—but most people think of mineral resources first when the word is mentioned. *Organic minerals*, including coal, petroleum, and natural gas, contain carbon compounds, and these are useful fuels. Most minerals, however, are inorganic. About 3,000 minerals are known, but fewer than 100 are common, and even fewer are of major importance to humankind. Useful nonmetallic minerals include the organic minerals plus clay, salt, building stone, and gemstones. *Metals* are elements that are usually heavy, reflect light, can be hammered and drawn, and are good conductors of heat and electricity. Roughly 75 percent of the chemical elements are metals. Most metals occur naturally as minerals in combination with nonmetallic elements. These naturally occurring mixtures from which metals are extracted are called *ores*.

Mineral resources are not evenly distributed throughout the world. Some areas contain substantial quantities and varieties, whereas other areas have almost none of any value. Few clear conclusions can be drawn about how this distribution relates to the distribution of human population or of national wealth. The key to wealth is not simply the possession of raw material resources, but the ability to transform them into items valuable for use or in international trade. Processing ores and manufacturing are not necessarily concentrated where the minerals are mined. Great quantities of mineral raw materials are constantly being transported from mines to factories around the world. Finally, the distribution of the final consumption of minerals is different from the distribution of manufacturing. Geographers investigate the principles that determine the distributions of each of the activities from the mine to the ultimate consumer.

The citizens of the world's rich countries consume or enjoy the use of vastly greater quantities of most minerals than do the citizens of the world's poor countries. This has caused some observers to worry about future supplies, or even to assert that all people on Earth cannot be raised to the level of material welfare enjoyed in today's rich countries. Each person could not possibly have so many *things*. Technology, however, is astonishingly flexible, and that fact offers reason for optimism that higher standards of living can be achieved for all.

Resources and Reserves

The **crustal limit** of a mineral is the total amount of that mineral in the Earth's crust. This amount is determined by geology, and it is fixed. The amount of that crustal limit that is currently or potentially extractable is called a **resource**. Enormous quantities of minerals are known to lie on the ocean floor, but the expense of harvesting these is still so great that few of these deposits are currently developed. Off-shore platforms, however, tap oil and natural gas deposits from deep beneath the sea floors.

The **reserves** of any mineral are the amounts that can economically be recovered for use. Thus, the word "reserves" is not a geographical or geological term, but an economic term. This should be remembered when we speak of the "limits" of the quantities of any item or commodity. The "limits" to the amounts of most commodities that can be found or produced are usually economic—they are "limits" at a given price; they are seldom "absolute" limits. More can be found or produced for a higher price. Prices ration demand. If the price of any commodity rises, demand for that commodity shrinks, but, at the same time, new supplies can be brought to market. For example, at a price of $10 per ton, world reserves of a certain mineral might be 1 billion tons. At a price of $20 per ton, however, world reserves of that mineral might be 5 billion tons. Many factors affect the price of any mineral, and, therefore, the profitability of extracting it and the amount of it considered reserves. The price of a mineral fluctuates according to supply, demand, and the perception of future supplies and demand.

For several years the prices of most minerals have been falling on world markets. Many circumstances can contribute to the fall of the price of a mineral. In one case, the price falls if the cost of extracting the mineral falls. This often happens because of improved technology. For example, drillers can get much more oil out of each well than they used to. The greater availability of oil puts downward pressure on world oil prices. Miners are everywhere exploiting mineral deposits that would have been considered waste rock just a few years ago. Nevada mines, for instance, can afford to dig up a ton of rock to get back as little as 0.025 ounces of gold, worth today about $9. Coal mining in the United States mechanized between 1975 and 1990, and this increased efficiency so greatly that the price of coal dropped 44 percent in real terms (Figure 11-1). What are only resources of a mineral in one country could actually be counted among reserves in another country that is more technologically advanced. Thus, measures of holdings of mineral reserves in different countries can be deceptive. An increase in the capital invested in mining could multiply reserves.

Mineral Substitutes

Some minerals can be replaced in many uses by cheaper substitutes, and the introduction of a cheaper substitute reduces demand for the mineral. Some metals, such as antimony, tin, cadmium, selenium, and tellurium, are easily replaced. Tin has been losing out to glass, steel, aluminum, and plastics.

Materials can be divided into five *materials families:* natural petroids, or minerals; synthetic petroids, such as ceramics and glass; metals; natural polymers, derived from plants and animals; and plastics, which are synthetic polymers. Plastic is the world's first new material since the Bronze Age. John Wesley Hyatt (1837-1920) developed the first synthetic polymer when he combined a form of cellulose, the main constituent of plant tissue, with camphor. The result was a chemical chain, called *celluloid,* made by repeating simple molecules known as monomers. Celluloid was first used as a substitute for ivory in billiard balls. Rayon was synthesized late in the nineteenth century, and the plastic resin Bakelite

FIGURE 11-1

In the technique of longwall coal mining shown here, a rotating cutting drum moves back and forth along the face of the coal seam. This efficient method of mining greatly lowers the price of coal.
(Courtesy of the U.S. Bureau of Mines.)

followed in 1907. A whole family of synthetic fibers, beginning with nylon in the 1930s, replaced silk. Since then polystyrenes, polyvinyls, polyacrylics, and polyurethanes have appeared. Hydrocarbons are their principal constituents, and today they are all products of the petroleum industry.

Plastics have replaced metals, wood, and fibers in many products. In 1993, the United States produced about 36 million tons of plastic resins. This is less than the 100 million tons of crude steel consumed in that year, but it is significantly more than the 4.5 million tons of aluminum or the 2.2 million tons of copper consumed. Stronger plastics are being introduced; some are already stronger than steel (Figure 11-2). In the future, more plastics may be made of agricultural raw materials.

The life-spans of synthetic polymers are generally reckoned in terms of a few decades or less. They can be sensitive to light, air, and heat, and they degrade. The materials in the space suits worn by the astronauts in the 1960s and 1970s were well-chosen for the needs immediately at hand, but they are already oxidizing and degrading in the collections of the Smithsonian Institution. This degrading can be a problem, but it suggests that many plastics can be recycled, as discussed below.

Sometimes different metals replace others as technology advances. In the United States, for example, many industrial products substitute aluminum and

World Mineral, Energy, and Water Resources **385**

FIGURE 11-2

These plumbing fixtures are not made out of metal, but of DuPont Delrin acetal resin. They are easier to manufacture than parts made of metal, and they last longer. (Courtesy of DuPont.)

other metals for steel. Manufacturers are also finding more uses for magnesium. Magnesium is currently used to make aluminum harder and to help in the production of various other metals. Magnesium has a density two-thirds that of aluminum, so it gives more strength and stiffness per pound than any common metal. Magnesium is especially runny when molten, so it is easy to cast into complex shapes. It can be extracted from seawater in a clean process without mines or slag heaps. It is easy to reclaim and recycle. For years it has been used in the manufacture of Volkswagens in Brazil, but, in response to demands that cars be made lighter, General Motors, Chrysler, and Ford announced in 1993 a joint research program to increase the use of magnesium in U.S. cars.

Strong but lightweight ceramics are replacing metals in many uses. Ceramics are basically com-posed of sand. In the future, manufacturers will make more and more products out of sand, a raw material that is inexpensive and abundant throughout the world. Even concrete, an old invention, is being made so much better and stronger that it is replacing both metals and plastics in a wide variety of uses. In 1993, a French laboratory succeeded in embedding fine steel wires, or microfibers, in concrete, making the use of conventional reinforcement bars unneces-sary. This reduces the amount of concrete needed to build any structure by two-thirds to three-quarters. Concrete may be one of the surprise "miracle prod-ucts" of the future.

Biomimetics, the study of the structure and function of biological materials as models for artifi-cially created ones, is making astonishing progress. Some scientists are duplicating the multilayered struc-ture of seashells to make ceramics that are tough but not brittle, while others are unlocking the keys to the strength of spider webs, in the hope of producing threads stronger than Kevlar, the substance now used to make bulletproof vests. Scientists are studying deer antlers to learn why this tough, light material absorbs shocks so well. The strength, suppleness, and dura-bility of new laboratory-created biomimetic materials exceed those of many of our current natural raw materials.

Platinum is today one of the world's most valu-able minerals. It is used chiefly in jewelry and in cat-alytic converters, devices that reduce air pollution from automobiles. The price of platinum has risen with demand for catalytic converters, but carmakers say that they are close to developing a catalytic con-verter that does not use platinum. When they do, the price of platinum will collapse. Platinum jewelry, therefore, although beautiful, is perhaps not a good investment. Gold also makes beautiful jewelry, but its primary role is as a standard of monetary exchange. If gold ever loses this role, its price could collapse.

Scientists often find entirely new uses for min-erals—such as platinum in catalytic converters. Petroleum has been recognized and used widely as a fuel for only about 130 years. The manufacture of aluminum metal is a twentieth-century development. The use of some minerals, on the other hand, such as asbestos, declines because of technological or ecological obstacles to their use.

Dematerialization

New and more efficient industrial processes squeeze more output from each unit of raw material input. This is called the **dematerialization** of manufacturing. Cellular telephones, for instance, are being introduced

386 *Chapter 11*

in poor countries that lack telephone wiring. They save raw materials by leap-frogging the technology of wiring. Even where wiring is still necessary, copper is being replaced by hair-thin optical fibers that are made, basically, of sand. As technology increases supplies and cuts demands for raw materials, the world uses less raw material input per unit of total economic output. In 1990, the world used about 25 percent less copper per unit of gross product than in 1970; 40 percent less iron ore, and 50 percent less tin.

Dematerialization has its limits. Some products need a minimum mass to function properly. Heavy cars, generally speaking, are safer than light ones. Dematerialization is also not necessarily environmentally beneficial. Lightweight products may be thrown away, rather than repaired.

Renewable and Recyclable Resources

Chapter 6 noted that animal or vegetable natural resources such as fish, livestock, and trees naturally renew themselves, so they can be harvested at the rate at which they reproduce themselves through time. Supplies diminish only if they are harvested at a rate above their natural reproduction rate.

Other resources can be reused, and these are called **recyclable resources**. *Recycling* is a term used to describe two distinct processes. One is the recovery and reuse of industrial wastes. Industrial wastes can pollute, but recycling them or rerouting them to serve as the raw material for another industrial process both reduces pollution and increases the supply of raw materials. For example, the banks of the Houston ship canal in Texas are lined with petroleum refineries and other factories that use petrochemicals as raw materials. Pipelines exchange raw materials and waste products among these factories in such complicated ways that the region is sometimes referred to as the "spaghetti bowl." Industrialized societies today worry less about using up the world's raw materials than about cleaning up the emissions and discharges that result from industrial processes.

Another form of recycling is the reuse of obsolete or discarded goods for their material content (Figure 11-3). Six factors interact to determine whether or not this type of recycling is feasible. One factor is the cost of virgin materials. A rise in the price of virgin materials encourages recycling, but a fall in their price discourages recycling. World prices for lead, for instance, fluctuate up and down, so the percentage of lead in U.S. car batteries that was recycled in various years through the 1980s fluctuated between 90 percent and 75 percent. Generally, virgin materials are cheaper than recycled materials, and that fact discourages recycling. A second economic

FIGURE 11-3

In 1992, the U.S. aluminum industry recycled beverage cans at the rate of about 114,000 per minute. (Courtesy of Reynolds Metal Company.)

factor is the cost of collecting and sorting the materials to be recycled. This is usually very high, and this cost is another factor that discourages recycling. In the United States, for instance, the costs of collecting and sorting municipal waste are so high that only 10 to 12 percent of it is being recycled, compared to the 80 to 85 percent that theoretically could be. Processing costs can also be high.

If discarded materials are not recycled, they must be disposed of somewhere. Today about 80 percent of solid waste in the United States goes into sanitary landfills. Therefore, the cost of landfill space is a third factor in the economics of recycling. The rising price of landfill space encourages recycling. Many communities across the United States have instituted recycling programs, and many people incorrectly assume that the purpose of these efforts is to conserve scarce raw materials. Most of these recycling programs, however, have actually been instituted to slow the rate at which landfill space is being filled (Figure 11-4).

The life of some landfills can be extended by incinerating some of the material they contain to provide power and by sifting the rest of the material for raw materials that can be reclaimed. Lancaster, Pennsylvania, for example, has been experimenting with a program that today supplies electricity for 15,000 homes and provides the municipality with a generous profit. Many people, however, protest the construction of waste incinerators. They fear that the emissions from the incinerators may contain toxic materials.

A fourth factor to be measured in the feasibility of recycling is the cost of energy used to recycle a

World Mineral, Energy, and Water Resources   **387**

FIGURE 11-4

New York City's Fresh Kills landfill is the world's largest, but it will soon be filled to capacity. As the cost of acquiring new landfill sites rises, recycling of refuse becomes more cost-competitive.
(Courtesy of City of New York Department of Sanitation.)

product. Energy can usually be saved by recycling. For example, recycling aluminum scrap saves the United States 95 percent of the energy that would have been needed to make new aluminum from ore. Recycled iron and steel result in energy savings of 74 percent; recycled copper, 85 percent; and recycled paper, 65 percent. A rise in the price of energy, therefore, encourages recycling.

A fifth factor is whether any nation has its own supplies of a commodity or relies on imported supplies. A nation may prefer recycling to importing because of the effect on the nation's trade balance.

A sixth factor affecting recycling is noneconomic; it is peoples' attitudes. Many people favor recycling because they think that recycling is environmentally beneficial and thrifty. They hold these opinions even though the processes involved in recycling some chemicals are among the most polluting chemical processes known, and recycling may not in fact be economical. Nevertheless, public demand for recycling is growing.

In balancing these six factors, the United States is making great efforts to recycle raw materials. In 1992, scrap supplied 40 percent of the coun-

try's total consumption of aluminum; 42 percent of copper; 46 percent of iron and steel; 71 percent of lead; 24 percent of zinc; 43 percent of stainless steel; and 33 percent of paper. In 1992, about 18 percent of plastics were also recycled. Every plastic container made in the United States today shows a number inside the three-arrow recycling symbol; this number indicates which of seven major groups of resins was used in the manufacture of that container (Table 11-1). These seven groups must be sorted carefully, but once the containers are sorted, the plastic can be recycled.

In 1992, Germany became the first nation to require business and industry to collect and recycle all packaging—the bottles, cans, cardboard, paper, and plastic that contained everything from toothpaste to refrigerators. Those regulations call for an average of 80 percent of packaging materials to be recycled by mid-1995. France and Austria have adopted modified versions of this program, and several other countries have shown interest. Germany has also considered instituting mandatory recycling—forcing manufacturers to take back and recycle their products once their useful life has ended, to make the economy a closed loop. A circulatory economy is not 100 percent possible, but it may be a good theoretical goal. The German government claims that the policy makes companies rethink the ways that they manufacture, prompting them to avoid nonessential components and materials that cannot be recycled.

These considerations of new mining techniques, mineral substitutes, dematerialization, and recycling help us understand how many factors affect the supplies and prices of raw materials. Their prices on international commodity markets fluctuate from moment to moment. These prices determine, in turn, the amount of reserves of any material. A labor stoppage at a mine in South America can drive up the price of a mineral on the New York and London markets within minutes. Rumors of a discovery of new supplies of another mineral in Africa can lower the price of that one. The announcement of the invention of a successful substitute at a scientific laboratory in Asia can collapse the price of any mineral. Political upheaval can bring about sudden changes in a supplier country's marketing policies. For example, Russia is one of the world's largest storehouses of raw materials. Under communism, it supplied raw materials to the other regions of the USSR and to its Eastern European Communist allies at prices far below world market prices. Russia continued this policy for awhile after the fall of communism, at enormous cost to itself. At the same time, however, Russia sold large quantities of raw materials on world mar-

388   *Chapter 11*

TABLE 11-1
The Seven Categories of Labeled Resins and Their Uses

Resin name	Typical uses and characteristics	Recycled products
Polyethylene terephthalate (1 PET)	Soft drink bottles, peanut butter-type jars; used in 20-30% of all bottles. The most expensive resin; keeps oxygen out and bubbles in.	Carpets, soda bottles, fiberfill for parkas, skis, tennis ball fuzz.
High density polyethylene (2 HDPE)	Milk and water jugs, liquid detergents; 50-60% of all bottles. Cheap, strong, forms good handles; dyed many colors.	Base cups for soft drink bottles, stadium seats, trash cans.
Vinyl or polyvinyl chloride (3 V)	Blister packs, cooking oils, food wrap, shampoo; 5-10% of bottles. Very clear, resists degradation by oils.	Floor mats, pipes, hose, mud flaps.
Low density polyethylene (4 LDPE)	Lids, squeeze bottles, bread bags; 5-10% of bottles. Flexible.	Garbage can liners, grocery bags.
Polypropylene (5 PP)	Syrup, ketchup, yogurt, margarine, bottle caps; 5-10% of bottles. Moisture-resistant, flexible, doesn't deform when filled with hot liquid.	Miscellaneous items like paint buckets, manhole steps.
Polystyrene (6 PS)	Coffee cups, meat trays, packing "peanuts," plastic utensils, videocassette boxes. Light but brittle, can be rigid or air-foamed.	Trays, flower pots, trash cans, pipe.
Other (7)	Various resins. Metals, glues, or other contaminants may be mixed in.	Plastic lumber.

Source: Copyright © 1992 by *The New York Times* Company. Reprinted by permission.

kets. These sales lowered world prices of many materials and harmed the economies of several African and Latin American countries that also export these materials. Russian political-economic policy will continue to affect world supplies and prices of many key materials.

For all of these reasons, tables or maps of measurable reserves of major minerals should be dated to the very hour of their composition. Furthermore, for all of these reasons, it is dubious to say that humankind could be "running out of" any "essential" resource. When any resource becomes scarce, its price rises. When a mineral's price rises, (1) people look for more of it, and they often find more; (2) economically recoverable reserves rise; (3) people try to recycle it; (4) people develop ways of using it more efficiently; and (5) substitutes become price-competitive. Major raw material supplies and prices can be followed in the financial section of any good general newspaper, or in the *Wall Street Journal.*

DO MINERAL RESERVES GUARANTEE WEALTH OR POWER?

Few minerals are so vital that possessing them gives the producing country real leverage in world affairs or guarantees wealth. Some minerals, however, are so important to industrial processes that they are referred to as **strategic minerals**. Chromium, for instance, is necessary in the manufacture of stainless steel. The leading producers of chromium are South Africa and Russia. The United States generally relies on supplies from South Africa and Turkey. The United States is the world's largest importer of strategic minerals, but the cost of switching suppliers if supplies from specific countries or regions were interrupted, however, would actually be small. The United States also stockpiles strategic and vital minerals.

One often hears casual statements that one country "depends" on another for its supply of a certain raw material, but this remark has little meaning. Global supplies of most materials are plentiful, and buyers can usually choose among several potential suppliers. Therefore, whenever we hear that one country "depends" on another for a commodity, what is usually meant is that the one country is buying the commodity cheapest from that source. Alternative sources are almost always available, although at, perhaps, slightly higher prices.

We can say that one country's economy "depends" on imports of a particular commodity from a particular source only when the importer is receiving the commodity at a subsidized price, as in the case of Russia's allies. For example, Hungarian, Polish, and Ukrainian industries depend on receiving raw materials from Russia at prices below world market prices. If Russia insists that those countries pay world market prices for their raw materials, the industries could collapse. Therefore, several countries' economies are dependent on Russia. Russia uses this economic dependency to wring political concessions from those countries—military cooperation, for example, or their vote in international councils.

Possessing most minerals does not guarantee wealth. On the contrary, if any country relies on the export of a mineral for a high percentage of its foreign exchange earnings, then that country's economy is vulnerable to disruption by a fall in the price of that mineral. To avoid a decline in prices for minerals, producing countries have occasionally joined together in *cartels*, organizations which agree to control and limit production. Cartels have seldom succeeded, however, because members fail to cooperate fully. Any nation can gain a short-term advantage by dumping supplies on world markets, thus collapsing prices. An example of an unsuccessful cartel is the Tin Council, with 22 member countries, which went broke in 1985. The Organization of Petroleum Exporting Countries (OPEC) is an example of a successful cartel. We will examine its success below.

Manufacturing Adds Value to Raw Materials

Many people assume that the countries that possess natural resources should be rich. One of the great paradoxes of world geography, however, is that there is little coincidence between the countries that possess raw materials and those that are rich. If the possession of raw materials were the key to wealth, then Zaire and Mexico, for example, would count among the richest countries in the world, and Japan and Switzerland would count among the poorest. In fact, Zaire and Mexico are poor, and Japan and Switzerland are rich.

Today, world markets exist for natural resource raw materials, and almost all natural resources are abundant. Therefore, selling raw materials on world markets does not make a country rich. When a raw material is transported, processed, or manufactured into something, however, value is added to it, and the transporter, processor, or manufacturer can keep a portion of that value added as profit. The key to wealth is not raw materials, but adding value to raw materials. In Chapter 6 we saw this principle at work in the marketing of agricultural products. Wisconsin and Switzerland add value to milk by converting it to butter and cheese, and Wisconsin and Switzerland prosper by selling these higher-value products.

Figure 11-5 reveals that the location of the processing and ultimate consumption of many minerals (the third and fourth columns of pie graphs) is not the same as the location of the production (the second column of graphs). Many poor countries export their raw materials in an unprocessed state. They do not develop manufacturing to add value to these raw materials. Other countries import raw materials, process them, manufacture items out of them, and enjoy their ultimate use. Specific rich countries that import and process great quantities of minerals include Germany, Japan, the United Kingdom, and

FIGURE 11-5

These 16 pie graphs illustrate a tremendous amount of information about world distribution and trade in four major minerals in 1990. For each mineral, a row of four pie graphs shows, first, where the reserves of that mineral are found; second, where the greatest production of that mineral is; third, where that mineral is processed; and fourth, where that mineral is ultimately consumed. If you look across the graphs for any individual mineral, you can see that the countries that have the greatest reserves are not necessarily the ones producing the most, and that the countries producing the most are not necessarily processing it or enjoying its ultimate consumption.

The colors chosen for the countries convey additional information. The darker colors represent the world's rich countries, and the pale colors represent the world's poor countries. A glance at the page reveals that the column of pie graphs on the left side of the page—the column of pie graphs showing where the reserves are—is lighter in color than the column of pie graphs at the right edge—the column of pie graphs showing where the consumption is. This shows that many minerals are found in the poor countries, but the processing and consumption of those minerals occur in the rich countries. Looking across the top row illustrating bauxite, for example, we can see that Guinea holds 26 percent of the world's bauxite but consumes a negligible percentage of the world's aluminum, whereas the United States has a negligible share of the world's bauxite reserves but consumes 26 percent of the world's aluminum.

BAUXITE

Reserves	Production	Aluminum production	Aluminum consumption

Guinea 26%
Australia 20%
Brazil 13%
Jamaica 9%
India 5%
Guyana 3%
Greece 3%
Suriname 3%

Australia 36%
Guinea 16%
Brazil 9%
Jamaica 8%
USSR 5%
Suriname 3%
China 3%
Yugoslavia 3%
India 3%
Hungary 3%

Australia 29%
U.S. 13%
USSR 10%
Suriname 4%
Jamaica 4%
China 4%
Venezuela 4%
Brazil 4%
India 3%
W. Germ. 3%

U.S. 26%
Japan 12%
USSR 10%
W. Germ. 7%
France 4%
China 3%
Italy 3%
U.K. 2%
Canada 2%
India 2%

COPPER

Reserves	Production	Refinery capacity	Consumption

Chile 25%
U.S. 17%
USSR 11%
Zaire 8%
Australia 5%
Zambia 5%
Canada 4%
Philippines 3%
Poland 3%
Peru 2%

Chile 17%
U.S. 17%
Canada 8%
USSR 8%
China 5%
Poland 5%
Zambia 5%
Zaire 4%
Peru 4%
Australia 3%

U.S. 20%
Chile 10%
Japan 10%
USSR 9%
Zambia 5%
Canada 5%
Belgium 4%
China 4%
Poland 3%
W. Germ. 3%

U.S. 21%
Japan 12%
USSR 12%
W. Germ. 7%
China 4%
Italy 4%
France 4%
U.K. 3%
Belgium 3%
So. Korea 2%

IRON ORE

Reserves	Production	Crude steel production	Crude steel consumption

USSR 36%
Australia 16%
Brazil 10%
Canada 7%
U.S. 6%
China 5%
India 5%
So. Africa 4%
Sweden 2%

USSR 26%
Brazil 17%
Australia 12%
China 11%
U.S. 6%
India 5%
Canada 4%
So. Africa 4%
Sweden 2%

USSR 21%
Japan 14%
U.S. 12%
China 8%
W. Germ. 5%
Brazil 3%
Italy 3%
France 2%
So. Korea 2%
Poland 2%

USSR 22%
U.S. 14%
China 11%
Japan 10%
W. Germ. 4%
Italy 3%
Poland 2%
India 2%
So. Korea 2%
U.K. 2%

TIN

Reserves	Production	Smelting capacity	Consumption

China 25%
Brazil 20%
Malaysia 19%
Indonesia 11%
USSR 5%
Thailand 5%
Australia 3%
Bolivia 2%

Brazil 24%
Malaysia 14%
China 13%
Indonesia 13%
USSR 8%
Bolivia 7%
Thailand 6%
Australia 4%

Malaysia 31%
Brazil 13%
Thailand 11%
China 9%
Indonesia 8%
USSR 5%
Bolivia 4%
Spain 4%
U.S. 2%
Mexico 2%

U.S. 16%
Japan 14%
USSR 13%
W. Germ. 8%
China 6%
U.K. 4%
France 3%
So. Korea 3%
Brazil 3%
Italy 3%

 Australia
 Belgium
Bolivia
Brazil
Canada
Chile
China
France
Greece

 Guinea
Guyana
Hungary
India
Indonesia
Italy
Jamaica
Japan
Malaysia

 Mexico
 Peru
Philippines
Poland
South Africa
South Korea
Spain
Suriname
Sweden

 Thailand
United Kingdom
United States
USSR
Venezuela
West Germany
Yugoslavia
Zambia
Zaire

Belgium. These are world manufacturing centers. Figure 11-5 and the concept of value added should help you begin to understand *why* the poor countries are poor and the rich countries are rich. The reasons for the current distribution of world manufacturing will be examined in greater detail in Chapter 12.

WORLD RESOURCES, PRODUCTION, AND CONSUMPTION OF MINERALS
▼▼▼▼▼▼▼▼▼▼▼▼▼▼▼▼▼▼▼▼▼▼▼▼▼

Mineral deposits are unevenly distributed over the Earth, and this has favorably endowed some countries in relation to their sizes, their populations, or both. The effort that has been expended on exploration varies greatly, and a lot more of any given mineral might be found if geologists only looked harder. Some countries, such as the United Kingdom and Japan, have been explored extensively for minerals. Others, such as Brazil and Zaire, remain largely unexplored. Only guesses can be made about the quantities of any minerals yet to be discovered or about what new uses might be found for minerals for which little or no use is now known. Scientists do know that the distribution of minerals is related to processes in the formation of the Earth's crust. As geologists learn more about these processes and their distribution, they can make better guesses about where new mineral resources might be found.

Nonfuel Minerals

The richest sources of nonfuel minerals have proved to be geological shields of old, worn-down rock. The geography of these shields has located great concentrations of the world's nonfuel mineral resources in the United States, Canada, Australia, South Africa, Brazil, and Russia (Figure 11-6).

Several populous countries have few or virtually no reserves of nonfuel minerals. These include some rich countries, such as France or Japan, that can import nonfuel minerals, but also countries that have limited industrial development, such as Bangladesh and Nigeria. Nigeria has struggled to build a multi-billion-dollar steel mill at Ajaokuta despite the country's lack of coal and iron ore, but economists have criticized this effort as an example of a developing country misdirecting investment to win national prestige.

Some countries boast huge deposits of just one or two minerals. Chile, for example, has about 25 percent of the world's copper reserves, but it has few other valuable minerals. Copper exports represent one-third to one-half of Chile's annual exports. The Chilean economy exemplifies one that is tied to world prices of one ore. This makes it vulnerable to world price shifts. If the price of copper dips, Chileans suffer.

The countries with the largest reserves of any mineral are not necessarily the ones that produce the most of it. Some countries have substantial reserves of a mineral, but they lack the technology or the capital to exploit their known deposits. They may be crippled by local civil unrest or by administrative incapacity, they may lack the necessary transportation infrastructure, or they may be the victims of international boycotts. In other cases, some countries reduce their output to conserve the resource.

Energy and Fuels

The world distribution of energy supplies differs from the distribution of nonfuel minerals. There are several main fuels, and each has its own preferred uses and problems associated with its use. Overall, so many new reserves are being discovered that total world fuel reserves per capita are rising.

About 85 percent of total human energy use is commercial energy, that is, energy produced for sale rather than for direct use. Most of the noncommercial energy used is the traditional fuels wood and dung, and deforestation has triggered serious local energy crises among the poorest people in some of the world's poorest countries. Hydroelectric power is an important source of renewable energy, but the other renewable forms of energy production—solar, tidal, and wind power—contribute today only negligible total amounts to total world energy use.

 The maps of the world geography of fuel energy supplies reveal the same characteristics as do the maps of nonfuel minerals: The countries that have the greatest reserves of any resource are not necessarily those that produce the most, and the countries that produce the most do not necessarily consume the most (Figure 11-7). Furthermore, those countries that consume the most on a per capita basis are not necessarily significant producers (Figure 11-8, p. 395).

Not all countries use energy in the same ways; each has a distinct *energy budget*. For example, in 1989 the United States reported that 30 percent of its commercial energy was used by industry, 35 percent for transportation, and 18 percent for residential purposes. In Japan, by contrast, 46 percent of all commercial energy was used by industry, 24 percent for transportation, and 12 percent for residential purposes. In most poor countries, high percentages of their commercial energy is used in industry, and low percentages are used in transport or for residential purposes. China, for example, has reported that 64 per-

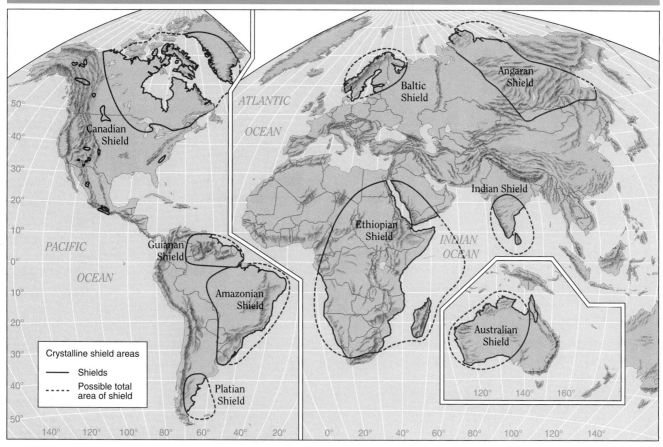

FIGURE 11-6

The cores of continental landmasses, ancient rock complexes called shields, are often rich in valuable minerals.

cent of its commercial energy is used for industry, only 5 percent for transport, and 19 percent for residential purposes. We may hypothesize that in most poor countries traditional noncommercial fuels are used in homes, and that people walk or ride an animal where they want to go. Generally they do not travel as much as people do in rich countries.

Petroleum

Petroleum is the world's principal source of commercial energy, supplying 37 percent of world needs. Global oil reserves in January 1993 were just over 1 trillion barrels (Figure 11-9, p. 396). A barrel is 42 gallons (159 L). This figure includes only barrels of liquid petroleum. Additional trillions of barrels of oil are chemically locked into geological formations known as **oil shales** and **tar sands**. The cost of releasing liquid petroleum from these formations is higher than that of extracting oil from wells, so esti-

mates of quantities are inexact, and this petroleum is not counted among world reserves. The United States, Canada, and Venezuela each contain over 1 trillion barrels of oil in shales or sand formations. Each time during the past 100 years that scientists have predicted the imminent commercialization of oil from shales or sands, however, new discoveries of liquid petroleum have dropped the price from that source. Therefore, oil shales and sands have never been significantly developed commercially. Liquid oil remains today the number one commodity in world trade (Figure 11-10, p. 397).

The Organization of Petroleum Exporting Countries

The Organization of Petroleum Exporting Countries (OPEC), founded in 1960, is a cartel that has enjoyed some success in limiting world supplies in order to maintain high prices. Its 12 members in 1994 were

Major Importers and Exporters of Commercial Energy Supplies

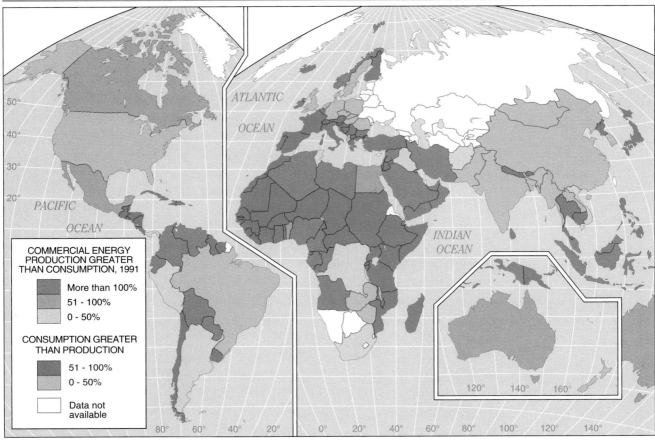

FIGURE 11-7

This map illustrates the major importers and exporters of commercial energy supplies. Some of the greatest energy-deficit nations spend a considerable share of their national earnings on imported energy. (Data from the World Resources Institute, *World Resources 1994–95*. New York: Oxford University Press, 1994, pp. 334–35.)

Iran, Iraq, Kuwait, Libya, Saudi Arabia, Venezuela, Qatar, Indonesia, the United Arab Emirates, Algeria, Nigeria, and Gabon. These countries together held about 77 percent of world petroleum reserves, and the cartel has exercised substantial but not total power to manipulate supplies.

The OPEC countries triggered a global energy crisis in the 1970s. In 1973, the Arab countries that dominate OPEC tried to use their control of oil as a political weapon in world affairs. They raised the price of oil from about $3 per barrel to $13 per barrel and imposed an embargo on the United States and other international supporters of Israel. A second increase in 1979 raised the price to $35 per barrel by 1981. These price increases were not "economic," that is, they were not the result of scarcity of oil or of rising production costs. In fact, at that time oil from the Arab states could be produced for less than $1 per barrel.

The non-Arab OPEC countries raised prices along with the Arabs because many of them (Indonesia, for example) needed oil export revenues. The price of oil is internationally denominated in dollars, so the countries whose currencies were rising against the dollar, most notably Japan and what was then West Germany, were cushioned against the price increases. Many of the world's poorest countries, however, suffered a crippling blow to their development hopes.

OPEC and the Industrial World   As oil prices rose, the industrial countries became frightened that they had become too dependent on OPEC oil. They began to develop alternative energies, such as solar energy and shale oils, and to promote nuclear power. They also developed new energy-efficient manufacturing techniques, so the consumption of energy per dollar of real economic output in them shrank. Japanese oil

Energy Consumption Per Capita

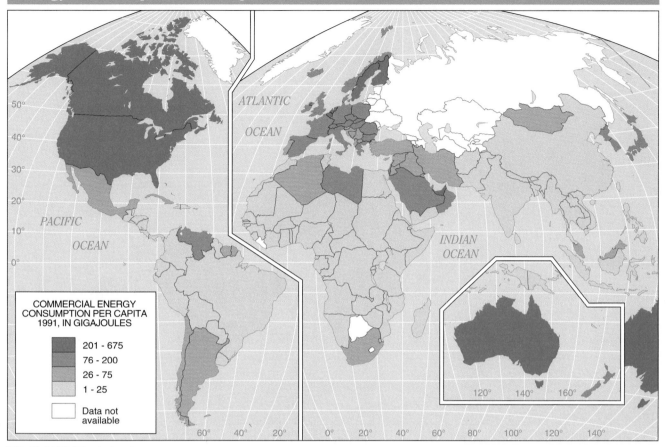

FIGURE 11-8

Energy consumption per capita is one of the most revealing statistical measures of people's standard of living. The nations where people enjoy the highest per capita consumption of energy are not, however, necessarily the same as those having the greatest energy resources. Comparison of this figure with Figure 11-7 reveals that several countries in West Africa and South America produce significant energy surpluses, yet their own per capita consumption is very low. What could explain this? Why don't the people of these countries consume more of their energy supplies to raise their own standards of living instead of exporting energy? What hypotheses of political or economic conditions in these countries can you formulate? (Data from the World Resources Institute, *World Resources 1994–95*. New York: Oxford University Press, 1994, pp. 334–35.)

imports actually dropped between 1973 and 1979, while the economy grew by 20 percent. In the United States, the 1975 Energy Policy Conservation Act mandated the adoption of energy-saving technologies and launched aggressive conservation and technology programs to achieve energy independence.

High prices stimulated new worldwide exploration. This resulted in discoveries of large reserves and in new production in Alaska, the North Sea, Mexico, Brazil, India, and Yemen. As explained above, when prices rose, reserves rose, and supplies were brought to market from new locations. The combined result of conservation in the industrial markets and production from new sources was wild price fluctuations, shrinking demand for oil worldwide, and, eventually, falling revenues for the OPEC coun-

tries. By 1986, oil prices had collapsed below $10 per barrel. OPEC had no choice but to lower its own prices. This restimulated worldwide demand.

Between 1973 and 1986, OPEC learned three things about the worldwide oil market: the best way to keep the world dependent on oil is to keep oil affordable; it is essential to avoid price wars; and in order to lessen competition from alternative fuels, consumers in the industrial countries must be assured that their oil needs will always be met.

After OPEC lowered its prices, the oil-producing areas that had been developed in the late 1970s decreased their production. They could not compete with Middle Eastern oil on the basis of price alone. The consumer countries also slackened their efforts to find alternative sources of energy.

World Petroleum Production 1992 and Reserves at Year End

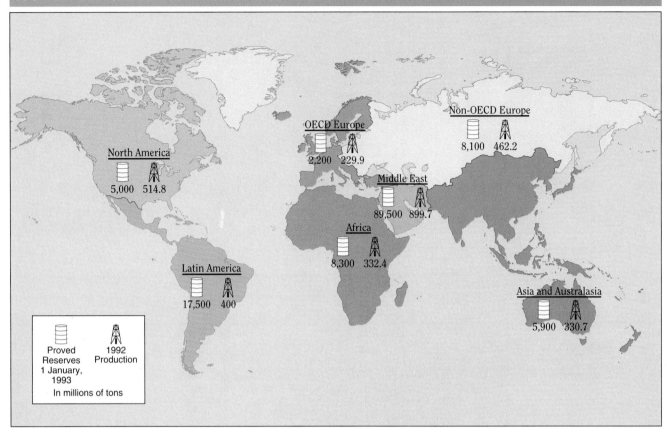

FIGURE 11-9

(Data courtesy of The British Petroleum Company p.l.c.)

Today OPEC Members Add Value to Their Product Today each OPEC country's national oil company is no longer content merely to sell crude petroleum on world markets. Rather, each company is trying to control supplies from the ground to the ultimate consumer and to capture the profit from value added at each step. Thus, producers buy their own fleets of ships to transport their oil, their own refineries to refine it, their own chemical companies to manufacture products from it, and their own international distribution networks to retail the products. These national oil companies are replacing the largest private companies in Europe and the United States. Between 1980 and 1990, Saudi Arabia almost quadrupled its domestic refining capacity (Figure 11-11, p. 399). It also purchased half of Texaco's refining and marketing system in the eastern United States, while other Saudi interests purchased the huge Fina Corporation of Dallas. The National Petroleum Corporation of Nigeria is a partner in a major oil refinery in Kansas, and the state-owned Petroleos de Venezuela owns Citgo, with its Louisiana and Illinois refineries and its network of over 7,000 service stations across the United States. Other state-owned petroleum corporations active in U.S. markets included those of Mexico, Colombia, and Norway. By 1992, less than 20 percent of U.S. refinery capacity was owned by U.S. corporations. The Venezuelan and Kuwaiti petroleum corporations owned refining and distribution networks throughout Europe, and Libya owned a refinery in Germany and a distribution network in Italy.

The process is spreading throughout the industrializing countries of Asia, where petroleum consumption is growing rapidly. East Asian economies are among the world's fastest-growing, and their thirst for oil is increasing. China and Indonesia are the region's two biggest producers, but China became a net importer in 1993, and Indonesia will soon. Despite new discoveries in Thailand, Malaysia, Papua New Guinea, Myanmar, and Vietnam, Asia may import more Middle Eastern oil by the year 2000, and OPEC members have begun to buy up Asian refining and distributing capacity. In 1992, Saudi Arabia contracted to build a refinery in Japan. The Middle East

Major Routes of World Trade in Petroleum 1992

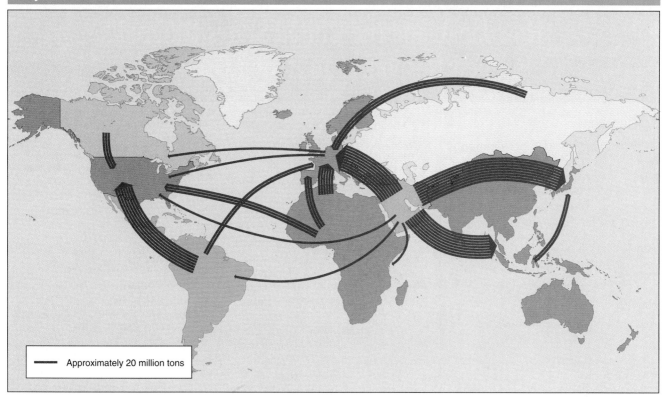

Approximately 20 million tons

FIGURE 11-10

(Data courtesy of The British Petroleum Company p.l.c.)

has enough oil to satisfy Asian needs, but new Asian demands may put upward pressure on world prices.

Asian thirst for oil may trigger two political repercussions: One will be increasing Asian interest in Middle Eastern politics. Already in 1992, China was known to sell arms to Iran in exchange for oil. The second result will be intensifying disputes over territorial claims to the South China Sea, where several nations believe new oil deposits may be found.

The expansion of oil companies from OPEC nations into the economies of their customer nations might seem to contain potential dangers for the oil-importing nations, but the presence of the oil producers directly in their markets also carries with it an interest to help promote economic stability. The economic interdependence of nations is escalating.

OPEC Production Quotas The biggest problem within OPEC is that of assigning production quotas to member nations. The cartel parcels out production according to a formula of percentages of current production, but OPEC's members differ greatly

in their wish to use oil supplies as a weapon in world affairs, as well as in their oil reserves and their needs for income. In 1991, per capita GNPs of OPEC countries ranged from $340 in Nigeria and $610 in Indonesia to $20,140 in the United Arab Emirates. Iraq's claim to the territory of Kuwait predated the discovery of Kuwait's great oil wealth, but certainly Kuwait's wealth intensified Iraq's interest in pressing its claim. In 1990, per capita income in Iraq was about $1,950, compared to about $16,000 in Kuwait. Had Iraq been successful in its attempt to annex Kuwait, Iraq probably would have increased oil production, thereby lowering world oil prices. Algeria and Indonesia demand big quotas because they have immediate financial needs and limited reserves.

OPEC members may be less likely to obey OPEC's quota system as they accumulate refining and international marketing capacity. They will be more likely to produce the amount of crude petroleum best suited to meet the needs of their own shipping interests, refineries, and marketing outlets. This may ulti-

GEOGRAPHIC REASONING
How Japan and the United States Differed in Their Responses to Rising Oil Prices

Questions

1. Could the U.S. government plan and institute energy-saving programs as the Japanese government did?
2. Do you think that the U.S. federal government should have power to regulate Americans' lifestyles, as well as American industry, in order to achieve greater energy independence?

Through the 1980s, oil prices on world markets fell. United States imports rose, domestic production declined, and the country abandoned the energy-conservation initiatives of the 1970s. By August 1990, when Iraq's invasion of Kuwait precipitated fears of a disruption of global supplies, U.S. domestic gasoline prices were the lowest they had been in real dollars (adjusted for inflation) since 1950. In 1989 and 1990, imports supplied about one-half of U.S. oil consumption, up from only 28 percent in 1982.

Japan's policies after 1973 contrast sharply with those of the United States. In 1973, oil accounted for 90 percent of Japan's energy imports and nearly the same proportion of total energy consumption. Japan responded to the escalating price of oil in ways that reduced the country's future vulnerability. Government bureaucrats devised plans to shift the economy away from energy-intensive industries such as aluminum smelting. They imposed energy conservation requirements on all aspects of Japanese life, raised the price of gasoline, and sped up the construction of nuclear power plants. Between 1973 and 1990, Japan's total economic output doubled without any increase in national energy consumption. Japan's industries are so energy-efficient that whenever the price of oil rises, Japanese manufacturers actually become stronger relative to their competitors in North America or Europe.

---

mately destroy OPEC. Ecuador, a founding member of OPEC, withdrew in 1993 because it wished to produce more oil than its quota.

Some of the greatest potential producers have not increased production because they do not presently need the additional revenues. Kuwait needs money to rebuild after the devastation of the 1990-1991 occupation and war, but other Arab states still have huge reserves and little immediate needs. The richest Arab oil countries have hundreds of billions of dollars invested overseas, and that sum is growing merely by earning interest. In the mid-1980s, Kuwait began earning considerably more each year from international investments than from exports of oil. Several Arab nations could live forever from their pools of accumulated capital without ever again tapping their pools of oil.

Saudi Arabia has traditionally commanded 25 to 30 percent of OPEC's total production quota, and it has the capability to triple its exports overnight. If Saudi Arabia did that, world oil prices would plum-

met. Thus, Saudi Arabia holds the power to dictate world oil prices.

At the time of the Iraqi invasion of Kuwait in 1990, the cost to produce Middle Eastern oil was about $2 per barrel. Saudi Arabia and the United Arab Emirates alone had proven reserves sufficient to supply the entire world's current consumption for about 30 years. Allowing the principal producing countries a profit well above their production costs and also considering that they are selling nonrenewable resources, market economic forces would have set the world price of oil at about $10 per barrel. The world market price at that time, however, was over $20 per barrel (then OPEC's target price), and it quickly rose to $40. This price defied the logic of supply and demand, but it reflected fears concerning the security of supplies. The principal oil-exporting countries reaped immense profits, and the world's poor oil-importing countries suffered another devastating blow. In 1994, several OPEC members produced more than their quotas to build up financial

FIGURE 11-11

Refineries such as this one in Saudi Arabia add considerable value—and thus profit—to the production of crude oil.
(Courtesy of Aramco World Magazine.)

reserves, and at the same time consumption in the industrial world fell because of a recession. The price of oil fell to $15 per barrel. Rising world demand and OPEC efforts to limit production continue to exert upward pressure on prices.

Petroleum Production in the Former Soviet States

Through the 1980s, the Soviet Union was the world's leading producer of petroleum. Ninety-two percent of production was in Russia, 4 percent in Kazakhstan, and 2 percent in Azerbaijan. At the end of 1992, Russia held 5 percent of world oil reserves—compared to 3 percent held by the United States. Huge new resources are being discovered regularly, and there are still many unexplored but promising areas. Reserves in Russia and Kazakhstan may prove to total more than those of Saudi Arabia. The former Soviet states might increase their oil production enough to flood world markets and collapse world oil prices. This has happened before. In the 1880s and again in 1928 the Russian Empire suddenly increased its oil exports to raise money. In the early 1990s, however, Russia was still selling oil to formerly-communist states at a price below the world market price. Selling that oil on world markets would have raised several hundred million dollars per day for Russia.

In the early 1990s, oil production in Russia and the other Republics fell, due to outdated equipment, lack of capital to buy spare parts, and political unrest.

The governments of several Republics invited Western oil companies to share in exploration and production of oil, and several major contracts were signed. The companies supply the capital and the expertise to increase production. Delays occurred in negotiating contracts with each of the newly independent Republics and in getting projects launched. Also, the countries that lie between landlocked Kazakhstan and Azerbaijan and seaports (whether by way of Russia, Turkey, or Iran) want a share of profits from new exports. Production and marketing might nevertheless be increasing by the late 1990s.

The Future of Petroleum Supplies

The world is not going to run out of oil. New supplies will be brought to world markets if the price rises high enough, as it has before. If OPEC continues trying to keep prices high by limiting production, then world oil prices toward the year 2000 will probably be determined by six factors:

1. The steady increase in demand, especially among the Asian industrializing countries.
2. The price at which new supplies from former Soviet states can reach world markets.
3. The ability of OPEC members that need oil export income to sustain their production levels.
4. Possible discoveries of previously unknown supplies.
5. New technologies to exploit oil shales and tar sands, which could bring massive quantities of oil onto world markets at low prices.
6. Improving technologies of energy alternatives, either fuel or renewable sources, such as solar and wind resources, could reduce the demand for oil.

Coal

Coal is the world's second most important commercial fuel source. It supplies about 30 percent of total global energy. China, which relies on coal for more than three-fourths of its total energy needs, is the world's largest consumer.

At the end of 1992, the United States held 23 percent of world coal reserves; the countries of the former Soviet Union, 23 percent (in Ukraine, Russia, and Kazakhstan); and China, 11 percent. In general, however, coal reserves are widespread and abundant, and the use of coal is expected to rise (Figure 11-12). Furthermore, new technologies are increasing the efficiency with which the energy in coal can be convert-

World Coal Reserves at the End of 1992

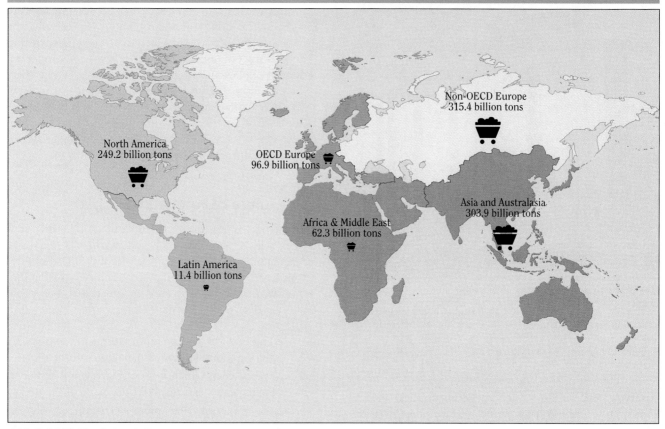

North America
249.2 billion tons

OECD Europe
96.9 billion tons

Non-OECD Europe
315.4 billion tons

Asia and Australasia
303.9 billion tons

Africa & Middle East
62.3 billion tons

Latin America
11.4 billion tons

FIGURE 11-12

World coal reserves are more widely distributed than world oil reserves. (Data courtesy of The British Petroleum Company p.l.c.)

ed into electricity. The leading exporters have been the United States, Australia, South Africa, and Canada, but new suppliers to world markets may join these.

The Use of Coal in the United States In 1973, the U.S. Federal Power Commission predicted that coal's share of U.S. electricity generation would decline from 46 percent to 30 percent by 1990. By 1993, however, it had risen to about 55 percent. Coal offers the virtues of being widely available in the United States and cheap to mine.

Mining and burning coal, however, does entail a number of externalities, and some ecologists and economists argue that when these externalities are figured into the cost of the coal, coal's price advantage erodes. For one thing, mining coal is dangerous. For about every 5 million person-hours worked in coal mines, one miner dies. That meant 54 deaths in the United States in 1992, as well as about 9,000 serious injuries.

In addition, using coal presents considerable environmental costs. Coal mines produce large quantities of mineral wastes. Some of these are acidic or

even toxic, and these wastes often drain into local streams or percolate acids into soil and soil water supplies. It is difficult to place an exact value on the fish thus killed, whether their value is measured as food, as a potential recreational opportunity, or simply as an amenity.

New techniques of surface mining, where the material on top of the coal is scraped away, are cheaper than mining coal from deep shafts, but the land is scarified unless restored to its original contour, topsoil is replaced, and vegetation is replanted. The federal government has recently begun to require that coal companies repair the damage, thus internalizing these costs and raising the price of coal (Figure 11-13).

Burning coal releases pollutants that contribute to the air's content of particulate matter and also to the ingredients of acid rain. These pollutants can be reduced by adding scrubbers to smokestacks. Scrubbers are expensive, but technology is reducing their cost and improving their effectiveness.

The use of oil imposes many externalities on society as well, including pollution, disease, and, arguably, armed forces to protect sources and foreign

FIGURE 11-13

The dragline in photograph (a) first removes the topsoil and then scoops the coal out of these deposits in Illinois. Photograph (b) illustrates that stripmined lands can be returned to productive agriculture.
(Photographs courtesy of CONSOL Coal Group.)

aid to win friends. If the United States were to add all negative externalities to the cost of oil and coal, analysis might reveal that the alternative energy sources discussed in following sections are actually cheaper than either of those fuels.

Natural Gas

Finding natural gas has always been a disappointment to oil drillers, because bringing natural gas to market requires expensive pipelines or other equipment that is not available everywhere. Therefore, the world uses about five times as much oil as natural gas. Enormous volumes of natural gas have been found but left untapped, even in the world's poorest countries. Gas is nevertheless more efficient than oil in many applications, and burning it produces virtually no sulfur dioxide or nitrogen oxides and less than half the carbon dioxide of oil or coal.

At the end of 1992, world proved reserves of natural gas were 140 trillion cubic meters, which is the energy equivalent of about 900 billion barrels of oil. The distribution of natural gas reserves is not the same as that of oil (Figure 11-14). Russia had 34 percent of world reserves of natural gas. Iran, with the world's second highest reserves, had only 14 percent, and no other country held over a 5 percent share. Many countries that import oil actually have their own substantial resources of gas (Figure 11-15). A 1989 World Bank study identified eight less-devel-

oped countries that rely on imported petroleum and yet could deliver domestic natural gas to their own population centers for less than one-quarter of the cost of oil: Bangladesh, Pakistan, Tanzania, Egypt, Tunisia, Thailand, India, and Morocco. If these countries had the capital to tap their own natural gas, they could lower their oil import bills, enjoy cheaper and cleaner domestic power, and profit from gas exports.

Liquefying gas makes it easier to transport. The world's first commercial liquefaction plants were built in Algeria in the 1960s. They purify gas and cool it to −256°F (−160°C). Then the gas can be sold to the highest bidder, rather than to the closest one. The global demand for liquefied natural gas (LNG) has grown at 17 percent per year for 20 years. Today, Japan consumes two-thirds of the world's LNG, which satisfies 11 percent of Japan's total energy demand. South Korea, Taiwan, and other Asian countries are turning to Middle Eastern LNG. Abu Dhabi and other Arab states are ordering construction of huge new LNG tankers and planning to capture the value added by shipping the gas to their customers. By 1997, Abu Dhabi will be supplying LNG to Japan and Korea (Figure 11-16, p. 404). Nevertheless, in 1992 only 15 percent of the total natural gas produced crossed international borders (4 percent as LNG), compared to half of the oil produced.

The United States might profit from greater reliance on natural gas. Demand is growing steadily, and in 1989 it first passed that for industrial oil. Newly discovered reserves have kept pace with pro-

World Mineral, Energy, and Water Resources   **401**

Proved Natural Gas Reserves at the End of 1992

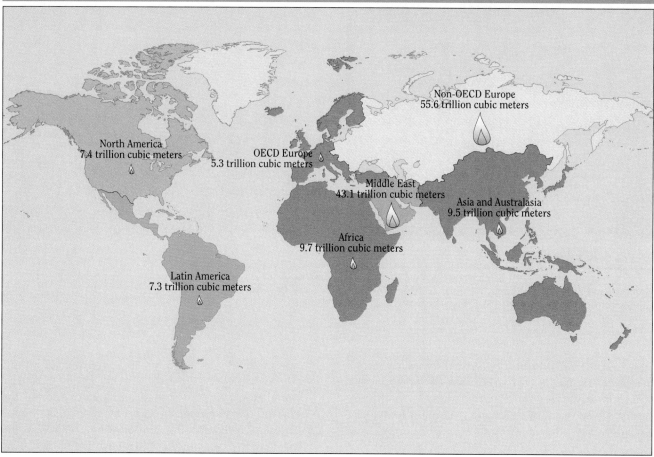

North America
7.4 trillion cubic meters

OECD Europe
5.3 trillion cubic meters

Non-OECD Europe
55.6 trillion cubic meters

Middle East
43.1 trillion cubic meters

Asia and Australasia
9.5 trillion cubic meters

Africa
9.7 trillion cubic meters

Latin America
7.3 trillion cubic meters

FIGURE 11-14

(Data courtesy of The British Petroleum Company p.l.c.)

duction, and about 90 percent of U.S. consumption comes from domestic production. Huge reserves remain in the Arkoma Basin in Arkansas and Oklahoma, in the Tide Formation in New Mexico, and in pockets throughout coal seams in the Rocky Mountains. If the nation were magically to transform all of its electricity production and all of its car engines to natural gas and methanol, the economy could run more cheaply, pollution problems would decline, independence in international affairs could be more easily achieved, emissions of greenhouse gases could be cut, and the trade deficit could fall.

Nuclear Power

By 1994, about 430 nuclear reactors had been constructed in 26 countries. They provided about 17 percent of the world's electricity, or 5 percent of total world energy use. Several energy-deficit countries have turned to nuclear power, particularly for generating electricity. In 1993, France relied on nuclear power for over 70 percent of its electricity; Belgium, 66 percent; South Korea, 53 percent; and Hungary and Taiwan, 49 percent each. In the United States nuclear generators provided only 19 percent of electricity. China, Argentina, Brazil, India, and Pakistan were the only developing countries operating commercial reactors.

The number of new nuclear plants being built is falling sharply. The United States has about a quarter of the world's plants, but no new one has been ordered and not cancelled since 1974. Every country in Europe that now has nuclear reactors has a moratorium on new construction except France. Several poor countries have built plants that are not yet running, or else they have canceled projects halfway through. Only Japan and South Korea are today forging ahead with nuclear power.

There are several reasons why the world is turning away from nuclear energy. One is that the rich countries are finding that they need less electricity than had been forecast just a few years ago. Another

402 *Chapter 11*

Major Trade Movements of Natural Gas in 1992

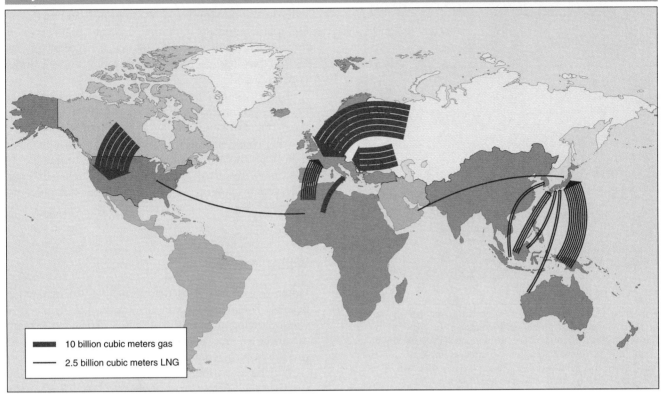

10 billion cubic meters gas

2.5 billion cubic meters LNG

FIGURE 11-15

International trade in natural gas is rising rapidly. The gas usually moves through pipelines, and liquid natural gas (LNG) moves in specially-built ships. (Data courtesy of The British Petroleum Company p.l.c.)

reason is public fear of pollution and of explosions. Accidents at Three Mile Island in the United States, at Chernobyl in Ukraine, and, in 1993, in Tomsk, Russia, reinforced these fears. Many of the nuclear reactors in operation in Eastern Europe, Ukraine, Belarus, and Russia were being closed down in the early 1990s because they presented unacceptable levels of pollution or danger. The world is just now learning the extent of nuclear contamination of Russia, Ukraine, and Kazakhstan by explosions and inadequate disposal of nuclear wastes under the Soviet government. Worries about safety also make it more expensive to find new sites or to store waste.

The costs of the rival energy sources oil, coal, and gas are relatively low, and it is hard to calculate the actual cost of nuclear power. Four factors are relevant: capital costs (which are very high), costs of operation and maintenance (low), costs of fuel (low), and the costs of decommissioning a plant at the end of its life, which are unknown. The oldest nuclear plants are starting to age. In the United States, 49 of the country's 112 plants will be at least 20 years old by 1995. They were licensed to run 40 years, and it is

unknown whether they can be, should be, or will be refitted. Some companies are already closing plants prematurely in order to save the cost of refurbishing them. When a plant is decommissioned, its highly radioactive core must be disposed of, and so must a lot of radioactive waste. Nobody knows what to do with any of this material.

In the foreseeable future, the main growth in demand for power will come in the poor countries, and they probably will not develop nuclear energy. Nuclear plants need complex, highly trained work forces, which make them inappropriate for most poor countries. Most poor countries might be better off spending their money on natural gas or on improving the cleanliness of burning coal.

Despite its dangers, nuclear power does not impose some of the environmental problems associated with the use of other fuels: mining practices that scarify and pollute the countryside; polluting spills during transportation; the release of gases that contribute to global warming; and acid rain. Nuclear power may still turn out to be the environmentally preferred source of energy.

World Mineral, Energy, and Water Resources **403**

FIGURE 11-16

This is an artist's rendering of one of four new LNG tankers being constructed at Finnish shipyards for the Abu Dhabi National Oil Company by the Norwegian Kvaerner Moss Corporation. Norwegian spherical tank technology won the bidding on this billion-dollar contract. Each ship will be 265 yards (289 m) long and 44 yards (48 m) wide, with a carrying capacity of 135,000 cubic meters of gas. By 1997, these ships will be transporting LNG from the Arabian Gulf to Japan.
(Courtesy of Kvaerner Moss Corporation.)

The nuclear reactors currently in operation rely on nuclear fission, in which the nuclei of certain heavy atoms decay, giving off energy. The most common element used in reactors is uranium, which is abundant in the world, particularly in Australia and South Africa. Nuclear energy could become far more attractive if scientists could harness nuclear fusion, the sun's energy source, as an alternative to fission reactors. Fusing hydrogen atoms into helium releases huge bursts of energy. The principal problem of fusion is that the process occurs at extremely high temperatures. Low-temperature fusion has been achieved (in Europe in 1991 and, using a different process, in the United States in 1993), but the amount of energy required to control the fusion was greater than the amount of energy produced. Research continues on controlled low-temperature fusion.

Japan has invested heavily in nuclear energy to reduce its reliance on imported oil. Nuclear plants today provide about one-third of Japan's electricity, and Japan plans to have nuclear power generating 40 percent of its electricity by 2000. As the only country ever to have suffered a nuclear bombing, however, Japan has understandable hesitations about nuclear

power. Nevertheless, the technological sophistication of Japanese plants is so high that the level of radioactivity in the cooling water in them is 1/500 of that in U.S. plants, and the radiation doses to plant workers are only 7 to 10 percent of those in U.S. plants. In 1993, Japan had 38 plants in operation and 15 under construction.

The United States obtains a smaller share of its energy from nuclear power than do some other countries, but its overall energy consumption is so great that it produces more nuclear power than any other nation. In 1992, the United States accounted for 32 percent of the total global output of nuclear energy, followed by France (17 percent), Japan (11 percent), Germany (8 percent), and Russia (5 percent).

Water Power

Water power is a particularly cheap form of energy, because it is relatively easy to tap the gravity force of water flowing naturally to the sea. Turbines convert the force of the falling water into electricity. Thus, several tropical countries that have great amounts of rainfall have enormous water power potential. The short stretch of the Congo River in Zaire, for example, where the river tumbles off the high African plateau, offers enormous amounts of potential energy.

Electricity, however, is technically difficult to transport. Few countries successfully export it, and then only to immediate neighbors. Norway, for example, receives orographic precipitation where the westerlies rise against Norway's mountains (look back at Figures 1-17 and 1-18). Norwegian hydroelectric facilities tap the energy of the water running down the moutainsides to the sea, and Norway exports electricity to its neighbors. Generally, however, hydropower has to be used wherever it is tapped.

Technology has increased the efficiency of turbines and thereby multiplied the percentage of the world's potential hydropower that can be realized. The water does not have to fall far to produce electricity in what are called *low-head dams*. Hoover Dam on the Colorado River, built between 1931 and 1936, has a drop of over 1,000 feet (305 m), but today efficient dams can be built with drops of less than 20 feet (6 m).

The world distribution of installed hydroelectric capacity differs greatly from the distribution of potential. Canada consumed 14 percent of the world total in 1992; the United States 12 percent; Brazil 10 percent; Russia 7 percent; China 6 percent; and Norway 4 percent. Zaire and Indonesia, however, have estimated potential as great as either Canada or the United States.

Future Alternative Energy Supplies

The current major sources of energy will not necessarily be those of the future, so a word must be said about what alternatives are on the horizon. Each of these may turn out to be either a turning point or just a mirage. Each is today too costly to compete with the current sources of power, and it cannot be foretold when or where any may reach the price level at which it could replace power from petroleum, coal, or natural gas, but scientific breakthroughs could make any of the following renewable sources feasible.

The price of photovoltaic cells, which make electricity directly from sunlight, is falling rapidly. Photovoltaic cells are now commonly found on pocket calculators, but larger-scale applications of the technology are being explored, and the efficiency with which the cells convert solar energy into electricity is rising steadily. Worldwide production of photovoltaic energy has grown from 3.3 megawatts in 1980 to some 56 megawatts in 1992. Because photovoltaic cells can be built to individual site specifications, no matter how small the local need may be, their installation eliminates the need for expensive infrastructure in networks of powerlines, called grids. Already in the United States, for example, if a new home is one-third of a mile or farther from existing powerlines, it may be cheaper to install photovoltaic cells than to run lines to the house. Other remote applications—lighting for bus stops, parking lots, and billboards—are proliferating in the rich countries, and this source of energy might turn out to be the cheapest to install in the developing countries.

At several locations around the world concentrations of giant windmills, called *wind farms*, have been built (Figure 11-17). The wind must be steady, not too strong or too weak (Figure 11-18). Today the cost of wind power may be almost competitive with fossil-fuel-fired electricity generation in some areas. Small local windmills also eliminate the need for investment in power grids. With technological improvement in the efficiency of the blades and of the turbines that turn the wind energy into electrical energy, wind may supply more of our energy.

Burning trees or crops grown especially for that purpose is another potential renewable source of energy. In 1993, some 0.5 percent of U.S. electricity was produced in that way, and that percentage is expected to double by 2010. Georgia Power Company in Atlanta, for example, burns as much as 2,000 tons of peanut shells and scrap wood each month in place of coal. Using this material is 30 percent cheaper than coal, and it reduces harmful emissions.

France has built facilities to tap the power of the Atlantic Ocean tides, but most scientists regard harnessing tidal power as prohibitively expensive.

FIGURE 11-17

These experimental windmills in California incorporate new designs in blades, increasing efficiency of the conversion of wind power to electricity.
(Wind plant slide courtesy of U.S. Windpower, Inc.)

Potential Wind Energy

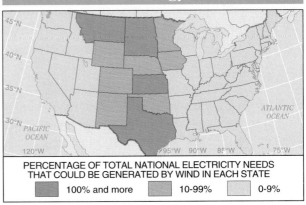

PERCENTAGE OF TOTAL NATIONAL ELECTRICITY NEEDS THAT COULD BE GENERATED BY WIND IN EACH STATE

100% and more	10-99%	0-9%

FIGURE 11-18

In the United States, Kansas, Nebraska, and the Dakotas are "the Saudi Arabia of windpower"—having enough to supply all U.S. electricity needs. (Data courtesy of U.S. Windpower, Inc. Copyright © 1992 by *The New York Times* Company. Reprinted by permission.)

Conservation of energy would greatly extend the use of the energy that we produce. Thus, energy conservation has often been called a potential "new source" of energy. Economists have calculated that the United States, for example, could achieve enormous overall energy savings if more Americans used mass transit—or even if they chose smaller, more efficient cars—and if they insulated their homes better. By one estimate, 85 percent of the energy spent heating a typical U.S. freestanding house is wasted.

WORLD WATER RESOURCES

▼▼▼▼▼▼▼▼▼▼▼▼▼▼▼▼▼▼▼▼▼▼▼▼▼▼

Water is essential to life. It makes up 50 percent to 97 percent of the weight of all plants and animals and about 70 percent of the human body. Water helps maintain the Earth's climate and dilute environmental pollutants. In addition, it is a vital resource for agriculture, manufacturing, transportation, and countless other human activities. About 17 percent of the Earth's cultivated area is irrigated, but since much irrigated land is harvested more than once a year, its contribution to total world harvests may be as high as 35 percent. Cheap supplies of water are necessary for many industries. Water is used as a coolant, a solvent, in washing applications, and for dilution of pollutants. Nevertheless, humankind has not shown care in managing Earth's water resources.

The Earth's surface is 71 percent water, but little of that is directly usable. Over 97 percent of the water on Earth is salty, and under 3 percent is fresh. Over two-thirds of that fresh water supply is frozen in ice caps and glaciers. Only 0.3 percent of all water occurs in the rivers, lakes, and streams, and the rest is held in underground reservoirs. Surface fresh water is continually replenished by precipitation, but much of that is polluted even as it falls, or it is polluted or wasted as it runs off the land to the sea.

The continuous interchange of moisture between the Earth and atmosphere is known as the **hydrologic cycle** (Figure 11-19). Liquid water (primarily from the oceans) evaporates into the gaseous air, subsequently is reconverted to a liquid or solid state, and returns to the Earth as precipitation.

The fresh water that we use comes from two sources: groundwater and surface water runoff. Precipitation that does not seep down into the ground or return to the atmosphere by evaporation or transpiration is called surface water; it becomes runoff—fresh water that flows into streams, rivers, and lakes. Surface water can be withdrawn from streams, rivers, lakes, and reservoirs for human uses, but only part of the total annual runoff is available for use. Some of it flows in rivers to the sea too rapidly to be captured, and some must be left in streams for wildlife and to supply downstream areas.

Some precipitation seeps into the ground. Most of this water is eventually lost to the atmosphere by evaporation from the upper layers of the soil and by transpiration from leaves. Some of this water, however, slowly percolates down deeper into the Earth. There it fills pores and fractures in spongelike, or permeable, layers of sand, gravel, and porous rock. This area, where all available pore spaces are filled, is called the zone of saturation. The porous, water-bearing layers of underground rock are called **aquifers**, and the water in them is **groundwater**. Aquifers are replenished, or *recharged*, naturally by precipitation, but this process is slow—decades to hundreds of years—compared to the more rapid replenishment of surface water supplies. If the withdrawal rate exceeds an aquifer's recharge rate, the aquifer will eventually be exhausted.

Landscape transformation shifts the proportion of rainfall that follows each possible path. Reduction of the vegetative cover, for example, increases the amount of runoff and decreases seepage into the soil and aquifer recharge. As a result, less soil moisture and groundwater are available during the dry season, and during the rainy season the increased runoff intensifies flooding and soil erosion.

In global terms, there is enough usable fresh water to meet the domestic, industrial, and irrigation needs of at least twice the current world population. Neither the water nor the population, however, is evenly distributed (compare world precipitation, Figure 1-18, with world population distribution, Figure 4-1). Some countries receive generous supplies of water per capita. Iceland, for example, has been estimated to have enough excess precipitation to provide 878,000 cubic yards (672,000 cubic meters) of water per person per year. Other countries are water-poor. At the other extreme, Bahrain, Kuwait, and other Arab Gulf states have no rivers and limited groundwater. They rely increasingly on desalinization of seawater by oil-powered stations. Furthermore, citizens of different countries use different amounts of water. The rich countries use ever-increasing amounts. The average U.S. resident uses 62 times as much each year as the average resident of Ghana.

Sixty-nine percent of the water withdrawn from the Earth is used for agriculture, about 23 percent for industry, and 8 percent for domestic use. This allocation also differs greatly from one country to another. In the United States, for example, industry claims 46 percent, agriculture 42 percent, and domestic uses 12 percent. In China, by contrast, industry takes only 7 percent, agriculture 87 percent, and domestic needs 6 percent.

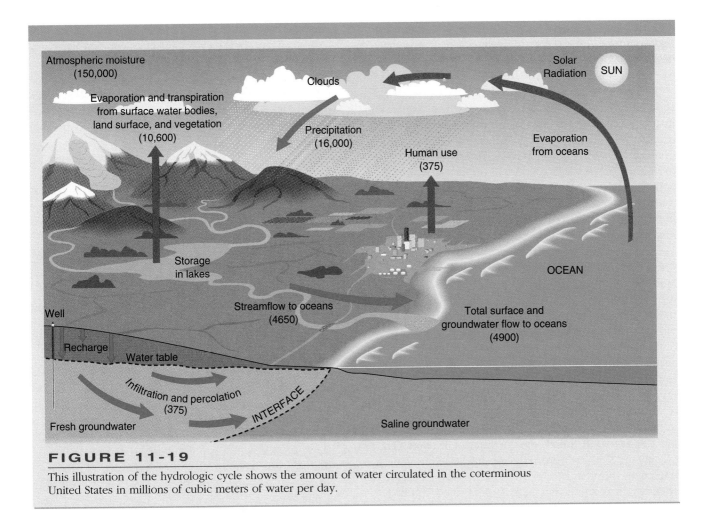

Atmospheric moisture
(150,000)

Clouds

Solar Radiation SUN

Evaporation and transpiration
from surface water bodies,
land surface, and vegetation
(10,600)

Precipitation
(16,000)

Human use
(375)

Evaporation
from oceans

Storage
in lakes

OCEAN

Well

Streamflow to oceans
(4650)

Total surface and
groundwater flow to oceans
(4900)

Recharge

Water table

Infiltration and percolation
(375)

INTERFACE

Fresh groundwater

Saline groundwater

FIGURE 11-19

This illustration of the hydrologic cycle shows the amount of water circulated in the coterminous United States in millions of cubic meters of water per day.

Problems of Water Scarcity

Many scholars consider the availability of enough fresh water to meet human needs one of the most serious long-range problems facing the United States and many other parts of the world. Tapping of groundwater to increase supplies should be avoided unless the groundwater supplies will be replenished, but groundwater depletion is already common in parts of India, China, Mexico, Thailand, North Africa, the Middle East, and the United States.

China has about 22 percent of the Earth's population but only 8 percent of its fresh water, and acute problems are arising in the drier regions in the north. Urbanization and industrialization demand new water supplies. In the 1950s, Beijing wells drew water from 16 feet (4.9 m) below the surface, but today the city's wells reach down an average of 160 feet (49 m). The prospect of shortages has prompted consideration of moving China's government or building an aquaduct from the south. Over 100 cities, mostly in the northern and coastal regions, have suffered shortages in recent years. In addition, several cities' water supplies are suffering from industrial pollution or contamination by pesticides.

Droughts can affect the ability of many areas to raise enough food to support their populations, and occasional droughts still afflict millions of people and cause thousands of deaths per year. In the summer of 1988, about 43 percent of the United States suffered severe drought. The grain harvest dropped by 31 percent and crop losses amounted to billions of dollars. This was one of the most expensive natural disasters in U.S. history. In many less-developed countries, people have no choice but to try to survive on drought-prone land. They often strip the land of trees, cultivate poor soils, and let their livestock overgraze grasslands. This land degradation increases the severity of long-term drought by reducing the amount of rainfall absorbed and released slowly by vegetation and soil. Each new drought can lead to a downward spiral of further land degradation and desertification, lower food and fuelwood productivity, and more poverty.

Many of the world's major river systems are shared by two or more countries. Thus, international clashes over water rights may increase. In Central

World Mineral, Energy, and Water Resources **407**

Asia, for example, five newly independent countries that splintered off from the former Soviet Union now share two overused rivers, the Amu Darya and the Syr Darya (Figure 11-20).

Actions in Nepal and in India partly explain why Bangladesh suffers flooding in the wet season and insufficient water supplies in the dry season. Runoff caused by upriver deforestation submerged about one-third of Bangladesh during floods in 1993, and thousands of people drowned. In the dry season India controls the flow of the Ganges River into Bangladesh through a dam that India built in 1975. A 1977 agreement guaranteed Bangladesh enough water until 1982, and this agreement was amended until 1988. As of 1994, however, the agreement had not been prolonged. India has refused to give Bangladesh the amount of water that Bangladesh received in 1977, insisting that Indians need more water themselves for industrialization, expanded irrigation, and to keep up with the other needs of India's growing population.

In many cases water shortages can be dealt with either by increasing local supplies (capturing more runoff in dams or tapping groundwater) or by conserving local supplies, especially through more efficient water use. Only about 37 percent of all irrigation water in the world, for instance, is actually taken up by crops; the rest is lost in runoff, evaporation, or leakage. The percentage lost could be reduced. In the 1960s, Israeli engineers pioneered drip irrigation, which conserves water by piping only as much as crops need and delivering it directly to their roots. This method reduces the average amount of water needed per irrigated acre, boosts crop yields, removes the threat of parasitic dis-

FIGURE 11-20

The Aral Sea in Central Asia was once the world's fourth largest lake, but the rivers that flowed into it were diverted to irrigate Central Asian cotton fields. It has shrunk to 60 percent of its former size.
(Courtesy of Novosti/Lehtikuva/SABA.)

ease spread by irrigation canals, and reduces soil salination. Israel and the arid Muslim Central Asian countries have put religious differences aside and shared irrigation technology. If Israel provided its own immediate neighbors with technical assistance, this would benefit all countries involved, reduce competitive pressures for limited regional water supplies, and perhaps lessen regional political tensions as well.

U.S. Water Supplies

Overall, the United States has plenty of fresh water. Much of the country's runoff, however, is not in the right place, occurs at the wrong time, or is contaminated by industrial or agricultural activities. Most of the eastern half of the United States receives ample precipitation, whereas most of the West receives too little.

In parts of the eastern United States the major water supply problems are not overall shortages, but problems of flooding, inability to supply the water needs of large urban areas, and water pollution. For example, the 3 million residents of New York's Long Island draw all of their water from an aquifer that is being contaminated by industrial wastes and leaking septic tanks and landfills. Also, ocean salt water is being drawn into the aquifer as fresh water is withdrawn faster than it is naturally recharged.

In the western states, by contrast, shortage of runoff is the major problem. Most of the western United States receives too little precipitation to grow crops without irrigation, so 85 percent of western water is used for agriculture. Today water tables are dropping rapidly throughout this area as farmers and cities deplete groundwater aquifers. Federally subsidized dams, reservoirs, and water transfer projects provide water to farms and cities at such low prices that there is little incentive for conservation.

Many major U.S. urban areas are located in areas projected to suffer water shortages by 2000 (Figure 11-21). Several of the fastest growing arid or semiarid states in the West, especially Arizona and California, already undergo periodic shortages.

In the relatively well-watered East, the laws regulating water access and use are based on the doctrine of **riparian rights**. This system basically gives to anyone whose land adjoins a flowing stream the right to use water from the stream, as long as some water is left for downstream owners. Most of the states of the West, by contrast, rely on the system of **prior appropriation**. Under this system, the first user of water from a stream establishes a legal right for continued use of the amount originally withdrawn. This means that later water users have little access to the resource, especially during droughts. The system causes unnecessary use and waste because to hold on to their

408 *Chapter 11*

Predicted Water Shortages in Metropolitan Areas

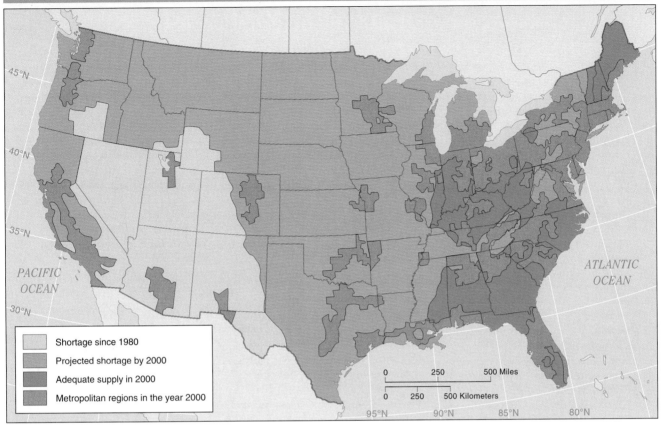

Shortage since 1980
Projected shortage by 2000
Adequate supply in 2000
Metropolitan regions in the year 2000

FIGURE 11-21

Several major metropolitan areas in the contiguous 48 states face water shortages in coming years. (Adapted from United States Geological Survey.)

rights, users must keep withdrawing and using a certain amount of water, even if they do not need it. This penalizes not only the populations of the West's growing cities, but even those farmers who use water-conserving irrigation methods by causing them to surrender future water rights.

In the United States overall, about half of the drinking water (96 percent in rural areas and 20 percent in urban areas), 40 percent of the irrigation water, and 23 percent of all fresh water used is withdrawn from underground aquifers. Overuse of groundwater can cause or intensify depletion, intrusion of salt water into aquifers, and groundwater contamination. Subsidence (the dropping or sinking of ground levels) is another problem. Some areas of the San Joaquin Valley of California have sunk almost 30 feet (9 m), damaging houses, factories, pipelines, highways, and railroads. Currently about one-quarter of the groundwater withdrawn in the United States is not replenished. In the arid and semiarid Texas Gulf region this figure rises to almost 80 percent (Figure 11-22, p. 413).

Water Pollution

Worldwide, pollution of water supplies by both organic and industrial wastes is increasing. Organic wastes are biodegradable—that is, they can be broken down by bacteria and other organisms—but waters can be overloaded and their oxygen supplies depleted. In addition, organic wastes also contain pathogens that can be transmitted to people who use the water. Industrial wastes may not be easily degraded, and they too may be toxic.

Pollution of rivers and lakes is potentially reversible with current knowledge and technology, but less is known about the quality of the Earth's groundwater reserves. Groundwater is cut off from the oxygen in the atmosphere, and therefore its capacity for self-purification is low. In the United States and Europe, where groundwater is a significant source of fresh water, between 5 percent and 10 percent of all wells have already been contaminated. This contamination is less visible than the pollution of popular beaches, so it

FOCUS
Water in the Middle East

"The next war in the Middle East will be fought over water, not politics," warned Egypt's Minister of Foreign Affairs, Boutros Boutros-Ghali, in 1985. Today Boutros-Ghali is the UN Secretary General, and localized water conflicts in the Middle East continue to raise tensions. The area has one of the highest population growth rates in the world, and agricultural productivity relies heavily on irrigation. At the same time, fast-growing cities demand more water.

Water shortages in Israel are growing acute. Already by 1992, Israel's water use was exceeding its renewable supply by 15 percent, and its yearly water deficit was worsening. The Jordan River basin is shared by Israel, Jordan, the occupied West Bank, and part of Syria. The Jordanians use less than half as much water as the Israelis, per capita, but Jordan's needs are rising, and in 1990, Jordan's King Hussein declared that water was the only issue that could take him to war with Israel. Israel hesitates to surrender any land it occupies in the West Bank, because 25–40 percent of Israel's water supply comes from an aquifer that is recharged there. Israel has severely restricted the amount of water that West Bank Arabs

can draw from this aquifer, even as Israel continues overdrawing the aquifer for its own uses. This inequity has greatly angered the Arabs. Furthermore, Israel fears surrendering the Golan Heights, seized from Syria, because the Heights forms part of the catchment area for the Sea of Galilee, Israel's largest reservoir and the source for the National Canal that transports water to Israel's drier south. Israel blocks joint Jordanian and Syrian efforts to build a dam on the Yarmuk River, a tributary of the Jordan, because that could reduce flows to the Jordan. Meanwhile, Israel's coastal aquifers are overdrawn and currently suffering from the invasion of seawater.

The Tigris and Euphrates Rivers still have excess water after regional needs are met, but the three countries that share the rivers—Turkey, Syria, and Iraq—argue over new demands. Both rivers rise in the mountains of eastern Turkey. The Tigris runs directly through Iraq to the Persian Gulf, and the Euphrates crosses both Syria and Iraq before reaching the Gulf. Turkey has undertaken a massive water development scheme to increase hydroelectric power generation and to expand Turkey's area under irrigation. Turkey's Ataturk Dam already

holds back the Euphrates River, filling a reservoir that is expected to hold more than ten times the volume of the Sea of Galilee. The Ataturk Dam anchors a plan that includes 22 dams and 19 power plants. Syria and Iraq fear that this project could foil their own development plans and leave them short of water. Both started complaining of shortages of water and power already in 1990, when Turkey began holding back the river to fill the Ataturk reservoir.

provokes less public outcry, but it is just as serious. Some of the advanced countries have recently begun to battle contaminating processes, but for much of the world's population, clean water remains precious and rare. Waste treatment is practically nonexistent in most poor countries.

The United Nations proclaimed the 1980s the International Drinking Water Supply and Sanitation Decade. It targeted the provision of safe drinking water and appropriate sanitation for everyone on Earth by 1990. Those goals were not met, but about 1.3 billion

people did receive a clean water supply, and 750 million received sanitation facilities for the first time (Figure 11-23, p. 415). The decade also brought about some improvements in drilling, water treatment, and sanitation technologies that were cheaper and more efficient than those regularly installed when the decade began.

Throughout the developing world, water-related diseases are the biggest single killer of infants and the principal cause of illness in adults. Therefore, money invested in preventive medicine in the form of clean water supplies buys better health care than money

410 *Chapter 11*

Water Supplies in the Mideast

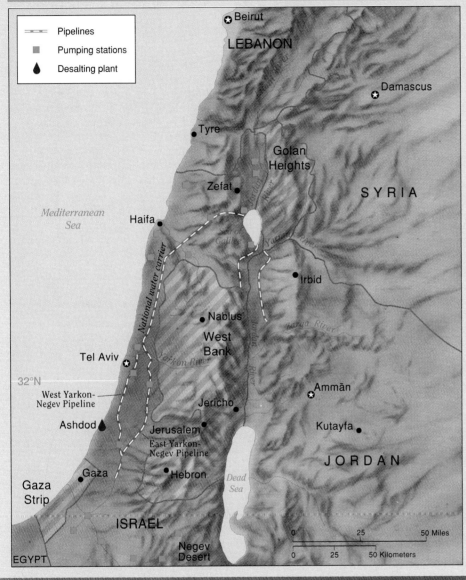

Israel hesitates to surrender the West Bank or the Golan Heights for reasons of water security, as well as military security. (Adapted from United States Geological Survey.)

invested in expensive new hospitals. In many areas urbanization has occurred so fast that authorities have not been able to provide adequate water or sanitation to millions of people. In sub-Saharan Africa, for example, the number of people enjoying access to clean water doubled during the decade, but the number of urban residents without access to clean water rose 29 percent. Countries throughout Latin America, Africa, and the countries of the former Soviet Union are reporting new outbreaks of diseases suppressed by modern medicine, including gastrointestinal diseases

such as giardiasis, typhoid, and cholera. The major rivers that flow through cities are choked with sewage and industrial waste. In the rural areas, many populations in poor countries lack the fuel to boil water to purify it. Many developing nations, in a rush to industrialize, continue to poison their primary water supplies with the unregulated wastes of factories and the runoff of agricultural pesticides and chemicals.

In September 1990, representatives of 115 nations met in New Delhi, India, to bring the decade's efforts to a formal conclusion. The conference work-

FOCUS
Competition for Water in California

The world's largest water reservoir and distribution system is the California Water Plan. It transports water from the mountains of northern California, where two-thirds of the state's precipitation falls, to mostly arid and semiarid central and southern California, which uses 80 percent of the water in the state.

From Mount Shasta in the north to Grapevine Grade some 500 miles (800 km) to the south, California's Central Valley is one of the most productive agricultural regions in the world. It supplies one-quarter of all the crops grown in the United States. About 85 percent of all water used in California goes to farmers, while competition for this water intensifies from California's growing cities and suburbs. In 1990, California alfalfa fields alone took more water than the people of the San Francisco Bay area and Los Angeles combined. Most farmers have paid about $10 per acre-foot (the amount of water that would cover one acre to a depth of one foot; 270,000 gallons) for water from the Central Valley Project, a system of 20 dams and 500 miles of canals that supplies 3 million acres of farmland and some urban users in the San Francisco Bay area. Most cities, by contrast, pay up to $250 per acre-foot. The result has been one of the world's great agricultural absurdities: irrigated green fields of alfalfa, cotton, and rice—crops more suited to wet lands than to a desert, and more than 50 percent of federally irrigated land in California is devoted to crops that are already in surplus.

California farmers may be encouraged to turn from rice and alfalfa to higher value crops that consume less water. The state already produces about $900 million worth of flowers each year.

The 1992 federal Omnibus Water Act lets farmers sell water rights to cities. Letting farmers enjoy large and riskless profits from such sales might encourage the transfer of 10 percent of the farm water, enough to supply the cities and abrogate needs for new dams. The farmers are unhappy, however, that 800,000 acre-feet of water will be diverted for the environment. The federal Endangered Species Act requires water regulators to maintain high water levels in the Sacramento River to preserve the environment of the Chinook salmon. This will require taking land out of agricultural production, forcing some 2,000 farmers to leave the land and costing as much as $1 billion a year in lost output.

Meanwhile, urbanization is eroding the farmers' political base. California is the biggest supplier of farm produce in the United States, but agriculture accounts for only 3.3 percent of California's jobs, compared with 5.2 percent nationally. The allocation of water among farmers, cities, and salmon will remain hotly contested.

The California Water Plan

These massive public projects transfer large quantities of water to arid and semiarid regions where population continues to grow. Much of the water, however, is dedicated to agriculture.

Areas of Water Supply Problems

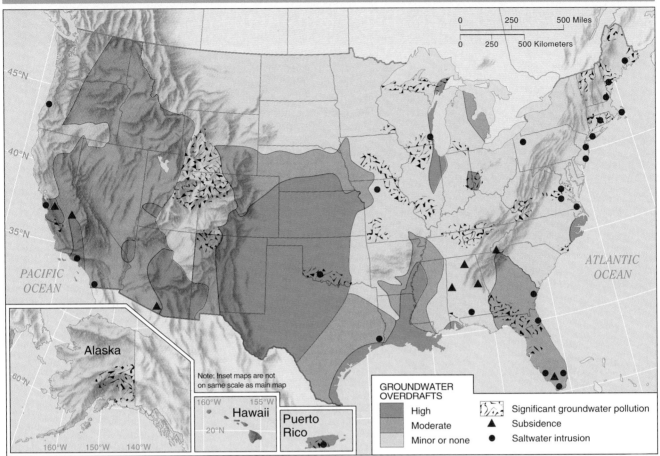

FIGURE 11-22

(Adapted from United States Geological Survey.)

ing paper warned that "some 80 countries, supporting 40 percent of the world's population, already suffer from serious water shortage." The World Health Organization estimated that 1.2 billion people in developing countries were still without safe water, 243 million in urban areas and 989 million in rural areas.

The United Nations has set as its next goal "Health and Sanitation for All" by the year 2000.

SUMMARY

A resource is anything which can be consumed or put to use by humankind. Humans are almost continuously expanding both the number of things that we define as resources and also our supplies of many individual resources. Mineral resources are not evenly distributed throughout the world, and the geography of world mineral production does not directly reflect the distribution of supplies. Furthermore, the places

where minerals are manufactured into useful and valuable items are not necessarily the same places where the raw materials were produced. Finally, the distribution of the final consumption of minerals is different from the distribution of manufacturing. Trade in minerals in all stages of their processing and manufacture into finished items makes up a great share of all international trade. There is no clear coincidence between the countries that possess raw materials and those that are rich. The key to wealth is not raw materials, but adding value to raw materials. The citizens of the world's rich countries consume or enjoy the use of vastly greater quantities of most minerals than do the citizens of the world's poor countries.

The amount of the crustal limit of a mineral that is currently or potentially extractable is called a resource. Reserves are the amounts that can economically be recovered for use. Thus, "reserves" is an economic term. Many factors affect the price of any mineral, and, therefore, the amount of it considered

World Mineral, Energy, and Water Resources   **413**

Questions

1. Do you think that water conservation measures should be enforced? By what level of government?

2. Should farmers be told how much water they can withdraw, or what crops they can grow?

3. How should water be reserved for the growing demands of the region's urban areas?

A large and important aquifer system in the High Plains that contains the Ogallala formation has been undergoing depletion for several decades. Stretching from southern South Dakota to northwest Texas, the High Plains Aquifer supplies about 30 percent of the groundwater used for irrigation in the United States. This water irrigates about one-fifth of all U.S. cropland, which yields wheat, sorghum, cotton, corn, and 40 percent of the country's grainfed beef cattle.

The High Plains Aquifer contains a massive amount of water, but it has an extremely low recharge rate because it underlies a region of low annual rainfall. Today the overall rate of withdrawal is about 8 times the natural recharge rate, but withdrawal rates as high as 100 times the recharge rate are taking place in parts of Texas, New Mexico, Oklahoma, and Colorado. At the present rate of withdrawal, much of the aquifer will be dry by 2020. Even before this happens, however, the high cost of pumping water from greater depths forces farmers to turn from thirsty crops such as corn and sugar beets to crops that demand less water, such as

wheat. The amount of irrigated land already is declining in five of the eight states drawing on the reservoir, because of the cost of pumping water from depths as great as 6,000 feet (1830 m).

The most severe depletion has occurred in northwest Texas, where heavy pumping for irrigation began to expand rapidly in the 1940s. As of 1990, 24 percent of the Texas portion of the High Plains Aquifer had been depleted, a loss of 164 billion cubic meters—equal to nearly 6 years of the entire state's water use for all purposes. As pumping costs have risen and some irrigation has become uneconomical, the irrigated area in northwest Texas has shrunk rapidly, falling from a peak of 2.4 million hectares in 1974 to 1.6 million hectares in 1989, a drop of one-third.

Depletion of the aquifer could be delayed if farmers switched to crops with low water needs and began using water conservation measures, but depletion could not be prevented in the long run. Most farmers are likely to continue to draw as much water as possible from this shared resource in order to increase their short-term profits.

The High Plains Aquifer Region*

*As defined by the 1982 Six State High Plains Ogallala Aquifer Regional Study

The High Plains Aquifer underlies parts of 6 states.

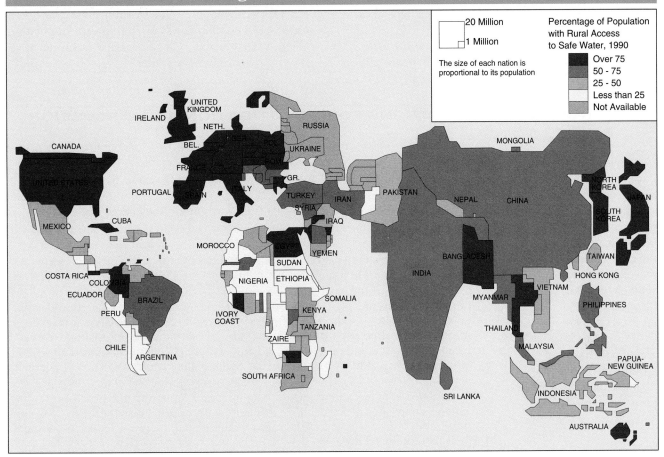

FIGURE 11-23

This cartogram reveals that many rural people still suffer limited or unsafe supplies of drinking water. The percentages of urban residents with access to safe drinking water are usually slightly higher. (Data from the World Resources Institute, *World Resources 1994–95*. New York: Oxford University Press, 1994, pp. 274–75.)

reserves. These include the technology of extraction, substitutes, and the possibility of recycling.

Few minerals are so vital that possessing them guarantees the producing country leverage in world affairs or wealth. If any country depends on the export of a mineral for a high percentage of its foreign exchange earnings, then that country's economy is vulnerable to disruption by a fall in the price of that mineral. Producing countries have occasionally joined together in cartels.

The richest sources of nonfuel minerals have proved to be shields. Rich nonfuel mineral resources lie in the United States, Canada, Australia, South Africa, Brazil, and Russia.

Petroleum is the world's principal source of commercial energy. OPEC members hold about 77 percent of world petroleum reserves and 40 percent of world natural gas reserves, but additional trillions of barrels of oil are locked into oil shales and tar sands. OPEC

members are trying to add value to their product from the ground to the ultimate consumer and to capture the profit from value added at each step. The biggest problem within OPEC is that of assigning production quotas to member nations. Factors affecting future oil supplies are the steady increase in demand, developments in the Commonwealth of Independent States, the ability of OPEC members to sustain production levels, new technologies to exploit oil shales and tar sands, and technologies of energy alternatives.

Coal is the world's second most important commercial fuel source, but its use entails externalities. Natural gas is efficient and clean. Reserves are widespread, and liquefaction makes it easy to transport. Few new nuclear plants are being built because the rich countries need less electricity than had been forecast just a few years ago, and the public fears pollution and explosions. Nuclear fusion could bring advantages over current fission techniques. Several tropical countries have enormous hydroelectric energy potential, but hydropower generally has to be used locally. The price of photovoltaic cells is falling, and we might soon exploit the tides, wind, and burning biomass.

The hydrologic cycle provides plentiful water, but humankind has not shown care in managing Earth's water resources. Most of the water is used for agriculture, but the share varies from country to country. Pollution is increasing. Local shortages can be dealt with either by increasing or by conserving local supplies. For much of the world's population, clean water remains precious, and waste water treatment is practically nonexistent.

KEY TERMS

a resource
  (in general) (p. 384)
crustal limit (p. 384)
resources of
  a mineral (p. 384)
reserves (p. 384)
biomimetics (p. 386)
dematerialization (p. 386)
recyclable resources (p. 387)
strategic minerals (p. 389)
oil shale (p. 393)
tar sand (p. 393)
hydrologic cycle (p. 406)
aquifer (p. 406)
groundwater (p. 406)
riparian rights (p. 408)
prior allocation (p. 408)

QUESTIONS FOR INVESTIGATION AND DISCUSSION

1. Is recycling practiced in your community? Does any local industry rely on recycled raw materials?

2. What is the source of your city's water supply? Does the water have to be treated? How is it treated?

3. What fuels do local power plants use? Where do they come from?

4. Seek explanations for cases in which a given country has substantial mineral reserves but does not utilize them locally.

5. What hampers the exploitation of potential hydroelectric power in some of the countries listed in the text? What political factors? What economic factors?

6. The OPEC cartel may crumble for any of several reasons. Among the most probable reasons are that its members may quarrel over the allocation of production quotas; new liquid resources may be discovered and brought to market from non-members; or technology may reduce the price of oil from sands or shales to a level competitive with liquid oil. How would any of these developments affect the world distribution of wealth? world trade? world political affairs? the production of energy in the United States or its import? the mix of energy supplies used in the United States?

7. Almost any day's newspaper contains stories assuming that possession of raw materials translates into wealth, or that some international dispute revolves around possession of raw materials. Find a few of these and analyze whether the newspaper's assumption is accurate.

8. Explain the importance to any nation of reliable energy and raw material sources.

9. In economic terms, there is no such thing as a "shortage." Any commodity has a price. That price may be determined either by politics or by "the market," that is, self-interested buyers and sellers. Explain this.

ADDITIONAL READING

ADAMS, W. M. *Wasting the Rain*. Minneapolis: University of Minnesota Press, 1992.

AGNEW, CLIVE and EWAN ANDERSON. *Water Resources in the Arid Realm*. London: Routledge, 1992.

DORIAN, JAMES, ed. *CIS Energy and Minerals Development*. Geojournal Library Series, vol. 25. New York, 1993.

EARNEY, FILLMORE. *Marine Mineral Resources*. New York: Routledge, 1990.

REISNER, MARC. *Cadillac Desert: The American West and Its Disappearing Water*. New York: Penguin Books, revised and updated, 1993.

U.S. Department of the Interior. Bureau of Mines. *Mineral Facts and Problems*. Washington, D.C., annually.

U.S. Department of the Interior, Bureau of Mines. *Minerals Yearbook*. Washington, D.C., annually.

Water: The Power, Promise, and Turmoil of North America's Fresh Water. National Geographic Special Edition. Washington, D.C.: National Geographic, 1993.

Water Environment Federation. *A National Water Agenda for the 21st Century*. Alexandria, VA, 1993.

World Energy Council. *Energy for Tomorrow's World*. New York: St. Martin's Press, 1993.

YERGIN, DANIEL. *The Prize: The Epic Quest for Oil, Money and Power*. New York: Simon & Schuster, 1991.

12 | National Paths to Economic Growth

The contrast between the traditional architecture of Nepal in the foreground and the modern factory in the background emphasizes the rapidity with which many nations are striving to develop.

Courtesy of Cary Wolinsky/Stock Boston.

*M*ost people agree that one of the principal tasks of any government is to promote the welfare of its people, including their economic welfare. Each country devises a program to build its national economy and also to ensure that economic well-being is distributed across the national territory. Therefore, the world political map is also a map of local economies and economic policies. Many states need economic growth not only to raise standards of living but also because economic growth may convince the population that the state works for them, that it is at least a success economically.

Some countries have achieved high incomes and standards of living for their people. Others have not. Throughout this book we have generally referred to the *rich countries* and the *poor countries*. The terms "rich" and "poor" are fairly straightforward and descriptive. Other terms that are often used include references to the *developed countries* and the *undeveloped countries* or *underdeveloped countries*. These words, however, have sometimes been used to suggest that development is a process, a direction in which all countries are headed, that some have gone farther in this direction than others, and even that there may be something "wrong" with the countries "lagging behind." This chapter emphasizes that countries' levels of wealth and standards of living are actually difficult to measure and compare. There are many ways to measure economic status and welfare, and these measures do not always rank countries in the same order.

This chapter is divided into four sections. The first section examines several ways of measuring countries' wealth, and then it maps the rich and the poor countries. The second section of the chapter examines the reasons why wealth has been created in some places and not in others. We will investigate the meaning of the popular term "The Third World." The third section of this chapter investigates national economic-geographic policies. Each country has a unique way of organizing its economy, managing the distribution of economic activities within its territory, and regulating its participation in world trade. The final section looks at how the scale of international trade today is creating a global division of labor. The organization of any activity throughout the entire globe—treating the whole globe as one place—is called **globalization**. For example, huge corporations, such as General Electric and Nestlé, own and coordinate production and marketing facilities worldwide. Economic globalization has far outpaced cultural or political integration, and tensions among these three sorts of activities generate much of today's international news.

ANALYZING AND COMPARING COUNTRIES' ECONOMIES

When the study of statistics first developed in the nineteenth century, British Prime Minister Benjamin Disraeli doubted its value. He knew that countries count things differently, that countries might lie in order to exaggerate their power or achievement or to conceal relative backwardness, and that even with the best of intentions, numbers often are not reliable and are open to misinterpretation. Disraeli concluded, "There are three kinds of lies: lies, damn lies, and statistics."

 Consider the following: In 1989, the leading Japanese economic newspaper, *Nihon Keizai Shimbun*, trumpeted that as of 1987 Japan had become the world's richest country. It had $43.7 trillion worth of land, factories, stocks, and other wealth, whereas the United States in that year had, according to the U.S. Federal Reserve Bank, $36.2 trillion worth of assets. Japan's population was a little less than one-half that of the United States, so Japan's wealth per capita was more than double that of the United States. This wealth was due partly to all-time highs on the Japanese stock market, and also to very high Japanese land prices. The city of Tokyo was "worth more" than the entire United States of America.

This statistic may not seem to make sense. Rising land prices do not make a nation richer. They simply represent a transfer of wealth to land owners from those who wish to own land, in this case, overwhelmingly, other Japanese. Also, within a few months of this announcement, the Japanese stock market lost 60 percent of its value. Trillions of dollars of Japanese assets evaporated. By the time you read this, that wealth may have reappeared and disappeared several times again.

Most people think that statistics describe facts, but actually statistics are matters of choice and value. The vocabulary used to measure national wealth is based on theories devised by the British economist John Maynard Keynes (1883-1946). The next few pages examine two statistics commonly used to measure a country's wealth: gross domestic product and gross national product.

MEASURES OF GROSS PRODUCT AND THEIR LIMITATIONS

The two most commonly used measures of a country's wealth are its gross domestic product and its gross national product. **Gross domestic product**

(GDP) is the total value of all goods and services produced within a country. **Gross national product (GNP)** is the total income of a country's residents, no matter where it comes from, minus any amount paid out to foreigners abroad. Profits that a U.S. company receives from overseas, for instance, are included in the GNP of the United States, but they are not included in the nation's GDP. Most countries' net income from abroad is small compared to their domestic economy, so there is little difference between the two statistics. Some countries, however, either receive large payments from abroad or send money abroad, so these countries will record a significant difference between their GNP and their GDP. In 1990, for example, Kuwait enjoyed income from many investments abroad, so the country's GNP was 135 percent of its GDP. The U.S. GNP is usually about 101 percent of its GDP. If profits are taken out of a country by owners in other countries, then that country's GNP can be smaller than its GDP. Canada's GNP, for example, is usually about 5 percent smaller than its GDP, Ireland's 13 percent smaller, and Brazil's about 14 percent smaller. These statistics suggest that significant shares of these three countries' economies are owned by foreigners. When you want to compare standards of living in different countries, GNP is generally more useful than GDP, but GDP provides a better guide when you want to analyze a country's internal economy.

Both GNP and GDP are deceptive. Both begin with the idea that most goods and services are of no value or of little value if they cannot be exchanged. They have no value in their simply being used. Therefore, these measuring techniques underestimate the activities of hundreds of millions of people who provide entirely for their own needs or who exchange very little except through barter. For example, many farmers buy little food, but eat food they raise themselves. Therefore, farmers' total output and income are probably both undercounted. A peasant farmer in China whose total annual income is recorded as $250 may eat food that would cost a resident of Chicago $1,000 in the local supermarket. Farmers usually make up a greater percentage of the population in the poor countries than in the rich countries, so undercounting farmers' production distorts the statistics of the poor countries more than the statistics of the rich countries. Measures of gross product underestimate the material welfare of many people in the poor countries. By noting this, however, we do not mean to exaggerate the quality of life those people may have or to suggest that they live well.

Because statistics undercount subsistence areas, they exaggerate the degree to which modern areas, especially cities, dominate national economies. In extreme cases, a single city can provide most of a country's measurable output. Abidjan, with 15 percent of the Ivory Coast's population, accounts for 70 percent of all economic and commercial transactions in the country. São Paulo, with 10 percent of Brazil's population, contributes one-quarter of that country's measurable economic activity. Statistics particularly overlook the work of women (Figure 12-1). Unpaid homemaking has no counted value, so if homemakers did each others' cooking instead of their own, measured national product would soar.

Statistics cannot measure activity that is illegal and therefore clandestine. Colombia's profitable drug exports do not seem to exist. Neither do illegal drugs in the United States, yet the federal government

FIGURE 12-1

The grain carried in the bucket on this woman's head is not counted in measures of Niger's national product, nor is her labor. If the grain were carried in a truck, however, it would increase Niger's gross product.
(Courtesy of Agency for International Development, Carl Purcell.)

National Paths to Economic Growth **419**

admits that the drug trade is one of the nation's largest industries. Nor does gross product include the activities in the informal sector (discussed in Chapter 9).

The Geography of Exchange Rates

Attempts to compare economies are further complicated by the fact that there are no common measures of value. Gross products are measured in local currencies, and the exchange rates among currencies are changeable and may be manipulated. This is of special concern to geographers, not only because it hinders comparison of one place with another but also because variations in exchange rates affect the flows of goods, investment, and people that geographers study. Between 1985 and 1990, for example, the U.S. government deliberately lowered the value of the U.S. dollar by 50 percent relative to the currencies of Japan and the Western European countries. This had extensive geographic results. American goods were cheaper for foreigners to buy, so exports of U.S. manufactured goods rose 80 percent from 1985 to 1990; new exports boosted the output of the nation's manufacturing belt (the industrial region from Pittsburgh to Green Bay; see Figure 9-3). Foreigners invested billions of dollars in the United States, and millions of foreign tourists came to visit. In 1989, for the first time ever, foreign visitors spent more money in the United States than U.S. citizens spent abroad. As the dollar fell, goods flowed out, but tourists and investment flowed in.

These new flows of people and goods diffused culture. Tourists took away impressions of American life, and Americans received new impressions of foreigners. Many more Europeans and Japanese could afford U.S. clothes, cassettes, compact disks, videos, and other cultural exports when those goods' prices fell by one-half. Millions more saw U.S. movies and television shows.

Reciprocally, the exports to the United States from foreign trading partners sagged, their economies suffered, and prosperity shifted across their national landscapes. Billions of dollars that U.S. manufacturers had planned to invest in factories abroad were invested in the United States instead, and communities from the Philippines to Brazil went without new vocational training programs, roads, and schools. Places dependent on U.S. tourists suffered recessions. Geographers watch shifting exchange rates to understand changing patterns of local prosperity.

Furthermore, exchange rates are very different from what economists call *purchasing power parity*. Parity attempts to measure what goods or services a certain amount of money will buy at different places. Therefore, when you read that annual incomes in a

certain country are a certain amount of money, do not assume that the people there live at the same level of material welfare that that amount of money would buy where you live. For example, some 40 million Indians live in households with annual incomes of over 900,000 rupees. The exchange rate value of that is about $30,000. The purchasing power of that amount, however, approximates an income of $600,000 in the United States.

Gross Product and the Environment

Measures of gross product fail to assess environmental damage. In standard techniques of bookkeeping, machines and buildings are counted as capital assets. As they age, their declining value is deducted from income. When a country's natural resources are exploited, however, annual depreciation is not deducted. Many countries sell off their timber and minerals, destroy their fisheries, mine their soils, and deplete their aquifers, and national accounting treats the proceeds as current income. The loss of the natural resources is not deducted as a depreciation of national assets.

If natural resources were treated as assets, then statistics might demonstrate that protecting the environment is sensible economic policy. Many economists are working to improve the calculation of a new statistic, **gross sustainable product (GSP)**, that subtracts from gross product the value of natural resources that are destroyed or depleted. Costa Rica's GNP, for example, grew on an average by 4.6 percent per year between 1970 and 1989. More than a quarter of this apparent growth disappeared, however, when a calculation of GSP adjusted for depreciation of Costa Rica's natural resources.

In 1994, the U.S. government first attempted to calculate at least part of a national GSP. The report analyzed use of oil, gas, coal, and other minerals. It estimated that proven oil and mineral reserves would add 3 percent to 7 percent to the nation's business assets. Over the past three decades, the value of new reserves discovered has roughly offset the depletion of reserves. "As a result," concluded the report, "the treatment of mineral resources as productive capital would have little effect" on calculations of domestic product. Future reports will add renewable commodities such as lumber, food, and fish, and the government hopes eventually to quantify the cost of air and water pollution. These costs would lower national product.

Many economists doubt the value of a GSP. They point out that the countries that are rich today exploited their environments in the past at unsustainable rates, but that after these countries had achieved wealth, they were able to repair much environmental damage. Perhaps GSP is irrelevant to a poor population struggling to raise its standard of living to some minimum level, but once the population attains that level of welfare, then it can afford to maintain or even repair the environment. We do not know for certain.

The Gross National Product and the Quality of Life

Figure 12-2 is a world cartogram with each country drawn in proportion to its GNP, and it is strikingly different from a territorial map. Some of the smallest countries, Belgium, for instance, have the biggest

GNPs. Figure 12-2 also differs from the population cartogram pictured in Figure 4-2. Some countries with small populations have enormous economies. Figure 12-3 establishes categories of GNP per capita on a regular world map, and Figure 12-4 then transfers that information onto the population cartogram. A comparison of these figures reveals that neither population nor territorial size explains wealth.

Noneconomic Measures of National Welfare

Per capita gross product is only one measure of national welfare. A picture of world standards of living and welfare can be improved by combining it with other statistics. We have already examined world death rates (Figure 4-8), food supplies (Figure 6-22), literacy (Figure 10-31), freedom (Figure 10-32), availability of safe water supplies (Figure 11-23), and per

The Relative Sizes of the Nations' Economies

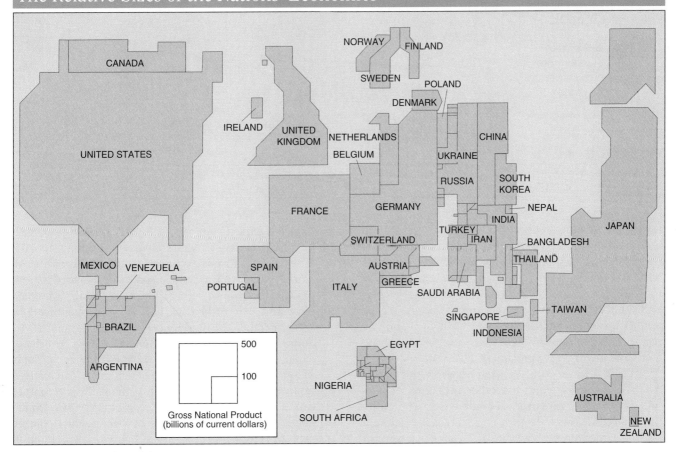

FIGURE 12-2

The size of the countries on this cartogram corresponds to their gross national products. The United States, Europe, and Japan grow considerably beyond their territorial relative size. In contrast, Africa shrinks. This cartogram contrasts sharply with the world population cartogram (Figure 4-2). The world's most populous countries are not necessarily those with the greatest economic output. (Data from the World Resources Institute, *World Resources 1994–95*. New York: Oxford University Press, 1994, pp. 256–57.)

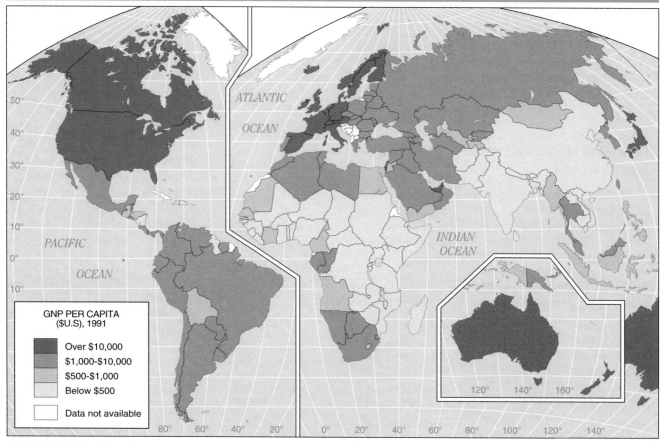

Per Capita Gross National Product

FIGURE 12-3

(Data from the World Resources Institute, *World Resources 1994–95.* New York: Oxford University Press, 1994, pp. 256–57.)

GNP PER CAPITA ($U.S), 1991
- Over $10,000
- $1,000-$10,000
- $500-$1,000
- Below $500
- Data not available

capita energy consumption (Figure 11-8). You should not be surprised to discover that there is a great deal of coincidence among these statistics. The countries with the lowest death rates generally have the best food supplies, education, and safe drinking water, and their citizens consume the most energy. These are all measures of the standard of living, and the highest ranking countries are much the same as those that have the highest per capita GNPs.

 These coincidences between per capita product and the other measures of welfare, however, are not exact. Therefore, the United Nations tries each year to balance different statistical measures of welfare to compile a **human development index (HDI)**. This index compares three statistics among countries: purchasing power, life expectancy, and adult literacy (Figure 12-5). HDI rankings differ from those based on any one criterion alone. In the report for 1993, for example, Uruguayans reported a GNP of $2,620 per capita, but they enjoyed purchasing power of $5,916 per capita;

their life expectancy was 72.2 years, and 96.2 percent of adults were literate. That gave Uruguay an HDI rank of 30. Citizens of the Republic of Korea (South Korea) reported a per capita GNP of $5,450, but enjoyed per capita purchasing power of $6,733, could expect to live 70.1 years, and 96.3 percent of adults were literate. The Republic had an HDI rank of 33. Saudi Arabia (per capita GNP $7,070) scored per capita purchasing power of $10,989, life expectancy of 64.5 years, and 62.4 percent literacy, for an HDI rank of 84. The HDI ranking of these three countries reversed their ranking by per capita GNP. The United States in 1993 ranked tenth in per capita GNP but sixth in HDI.

Rankings that balance several criteria are problematic because the results depend on the variables chosen and the weight assigned to each variable. Several statistical yearbooks published by the United Nations and the World Bank allow you to browse for yourself among the quantified descriptions of life in each country. The characteristics generally attributed

Categories of Per Capita Gross National Product

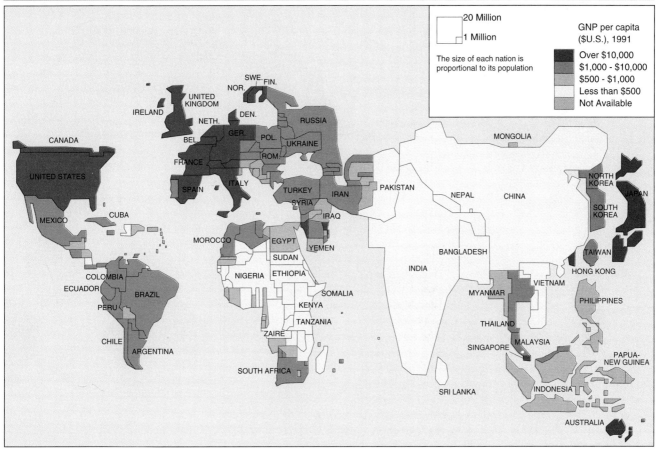

FIGURE 12-4

This figure maps the categories of per capita GNP on the cartogram on which the size of each country reflects its population. Thus, this cartogram presents visually what share of the Earth's population enjoys each category of GNP per capita. Countries that occupy over half of the space on this cartogram are colored yellow. That illustrates that over half of the world's population lives in countries in which per capita GNP is less than $500. (Data from the World Resources Institute, *World Resources 1994-95*. New York: Oxford University Press, 1994, pp. 256-57.)

to wealth and "the good life" are found in some countries but not in others.

Variations within countries can be even greater than variations among countries. Internal variations may reveal uneven national economies, inequity, or social injustice. The U.S. 1990 infant mortality rate was 9.2 infant deaths per 1,000 live births. The figures for the various states ranged from 6.2 in Maine (the same as in Denmark) to 12.4 in Georgia (the same as in Bulgaria). The rate in Washington, D.C. was 20.7, the same as on the Indian Ocean island-nation of Mauritius.

Statistics examined per capita through time reveal a relationship between population growth and changes in the statistic measured. For example, a country's GNP can grow, but if the population grows faster than the GNP, GNP *per capita* will fall. For the countries of tropical Africa, per capita incomes rose in the 1960s, flattened out in the 1970s, and fell in the 1980s. This was due in part to the general deterioration of African economies during the period, and in part to rapid population growth.

PRE-INDUSTRIAL, INDUSTRIAL, AND POST-INDUSTRIAL SOCIETIES

▼▼▼▼▼▼▼▼▼▼▼▼▼▼▼▼▼▼▼▼▼▼▼▼▼▼

Geographers are interested in more than the size of each country's economy. They also want to know what the people are doing for a living and which activities are producing the wealth in each country. Chapter 9 noted that over a long period most coun-

National Paths to Economic Growth   **423**

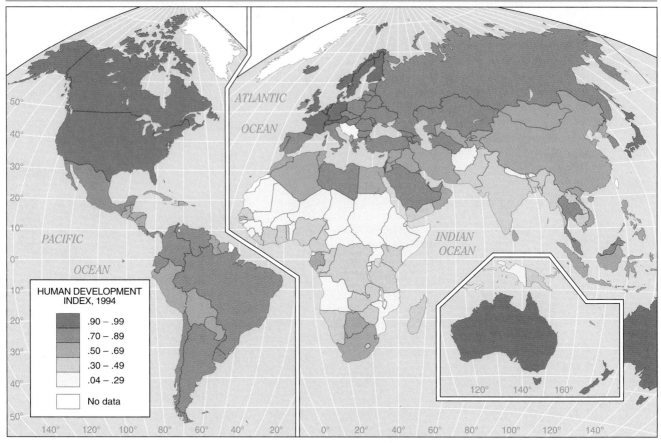

HUMAN DEVELOPMENT
INDEX, 1994

.90 – .99
.70 – .89
.50 – .69
.30 – .49
.04 – .29
No data

FIGURE 12-5

This map indicates where life is relatively good, as measured by the UN's Human Development Index.

tries that are rich today experienced shifts in the distribution of jobs among their economies' primary, secondary, and tertiary sectors. This shift was defined as **sectoral evolution**. The best way to examine any country's economy is to study the relative importance of each sector.

Sectoral Evolution Changes National Employment

Before industrialization, most of a country's labor force is occupied in the primary sector (Figure 12-6). Societies with the bulk of their employment in this sector are called **pre-industrial societies**. Many societies are still today pre-industrial, and even with high percentages of their workers in agriculture, many can barely feed themselves (Figure 12-7).

As some countries industrialized, many workers found employment in factories, and the *proportion* of

the labor force employed in the primary sector declined. This is true despite the fact that both the *total number* of people in the primary sector as well as the value of the primary sector's output may have risen. The workers in the sector usually became more productive. The primary, extractive activities remain crucial to many economies, but they provide a diminishing share of jobs.

Many jobs lost in agriculture were initially replaced by new opportunities in industry, and the proportion of workers in the secondary sector increased until, at no precisely defined point, certain societies came to be called **industrial societies**. In the United States, for example, manufacturing employment overtook farm employment during World War I, and by 1925 blue-collar workers in manufacturing industries had become the largest single occupational group. Many of the richest countries today are industrial societies.

Continuing evolution of some countries' economies has drawn a higher percentage of workers

424 *Chapter 12*

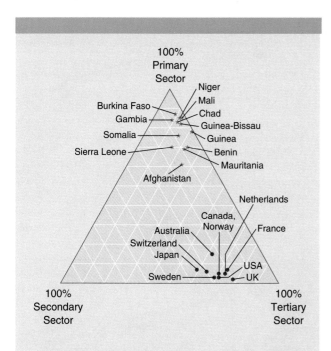

FIGURE 12-6

This figure shows the proportion of the labor force in each of 20 countries that works in each of the three sectors of that country's economy. If the symbol for the country is near the top, its labor force is concentrated in the primary sector. A position at the lower left reveals a concentration in the secondary sector, and a position at the lower right indicates a concentration in the tertiary sector. The 20 countries shown are the top 10 and the bottom 10 ranked by HDI, except that Mauritania has been substituted for Djibouti, for lack of statistics. Note that the workers in the poor countries are concentrated in the primary sector, whereas higher percentages of those in the rich countries work in the secondary and tertiary sectors. (Data from UNCTAD.)

into the tertiary sector, producing services instead of goods. Services accounted for almost 50 percent of all jobs in the United States by 1929. Sometime in the 1940s, the proportion first exceeded 50 percent, and the United States became the world's first **post-industrial society**. By 1990, tertiary employment represented about 75 percent of the nation's jobs. The trend toward increasing percentages of all jobs in the tertiary sector has been observed in most other advanced nations. It has also been noted in a few poor nations whose tertiary sectors are well integrated into the international economy—tourist destinations, for example.

At first both the secondary and the tertiary sectors increased their shares of total employment as the primary sector's share dropped, but in recent years in the United States the proportion of workers in the

secondary sector has begun to fall as well. The nation began losing manufacturing jobs as a percentage of all employment about 1960, and even the absolute number of manufacturing jobs has been declining since at least 1970. This has not happened in all rich countries, and some economists fear that it signals a decline in the competitiveness of U.S. manufacturing.

Measuring How Each Sector Contributes to Gross Product

The share that each sector of an economy contributes to gross product is not necessarily the same as the share of the national labor force that sector employs (Figure 12-8). Some countries have high proportions of their workers in the primary sector, but those workers produce relatively little of the measurable national output.

As a rule, the value added per worker in manufacturing is usually highest, so the secondary sector contributes a larger share of total product than it employs of the labor force (Figure 12-9, p. 428). For example, even though both the share of the U.S. labor force in manufacturing and the total number of jobs in manufacturing are falling, that sector's contribution to GNP has held fairly steady since World War II—between 20 and 23 percent annually.

The ability to add value to a raw material and to capture some of that value-added as profit helps explain why the miller usually gets richer than the farmer, the manufacturer richer than the miner—both the miller and the manufacturer are in the secondary sector. This helps explain why some *places* get rich while others stay poor.

WHY SOME COUNTRIES ARE RICH AND SOME COUNTRIES ARE POOR

If you compare the map of per capita GNP (Figure 12-3) and the map of HDI rankings (Figure 12-5) with the figures of resources in Chapter 11 (Figures 11-5 and 11-7) you can see that the countries most richly endowed with raw materials do not necessarily have the highest per capita products or standards of living. Conversely, some of the world's richest peoples enjoying the highest standards of living live in environments with meager endowments. The geography of resources alone does not explain the geography of wealth.

The theory of environmental determinism discussed in Chapter 2 would find this inexplicable, but the study of geography teaches that the key to wealth

National Paths to Economic Growth   **425**

Percentage of the Labor Force Working in Agriculture

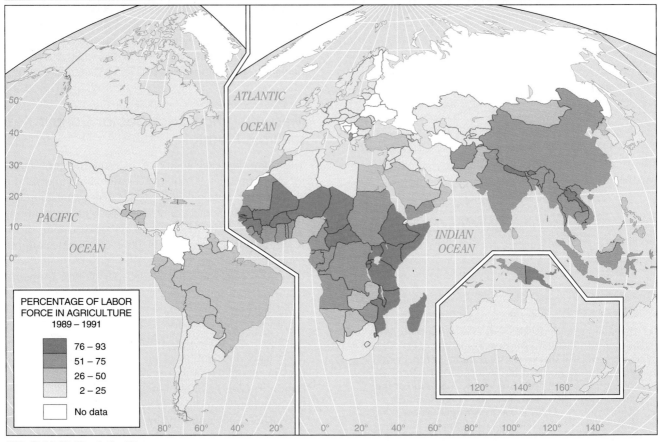

FIGURE 12-7

If you compare this map with Figure 6-22, per capita food supplies, you will see that many countries with high percentages of their labor forces in agriculture are barely able to feed themselves, whereas several countries that have only small percentages of their labor forces working in agriculture enjoy food surpluses. What can explain this paradox? The explanation is that economic and technological development have allowed some countries to invest in agriculture, raise agricultural yields, and at the same time release workers for manufacturing and service activities. (Data from World Resources Institute, *World Resources 1992–93*. New York: Oxford University Press, 1992, pp. 264–65.)

is not having raw materials, but adding value to raw materials. This principle of adding value was first introduced in the discussion of marketing dairy products in Chapter 6, and we met it again in Chapter 11, where we saw that many poor countries do not manufacture to add value to their raw materials. Other countries import raw materials, process them, manufacture items out of them, and enjoy their ultimate use.

In economic discussions it is common to use the image that all products "flow" from their sources as raw materials to their ultimate consumers; as a result, the economic activities closer to the consumers are called **downstream activities**. Downstream activities include processing raw materials, manufacturing and transporting goods, and even advertising, insuring, marketing, and financing the flow of goods. Downstream activities add value, and therefore they are profitable.

The value added to raw materials downstream in the secondary and tertiary sectors surpasses the value of the original raw materials, whether the raw materials are mineral or agricultural. The value added by refining copper ore and manufacturing things out of it, for example, quickly surpasses the value of copper ore. The value added by grinding wheat into flour is greater than the value of the wheat. This principle holds true no matter how valuable the original raw material may be. Diamonds are valuable, but the value added to them by diamond cutters and polishers is greater than the value of the original uncut diamonds. The greater the value added, the greater the potential for profit.

The places that capture the greatest value added prosper, while the places that export raw materials do not enjoy the same growth. They may even have

426 *Chapter 12*

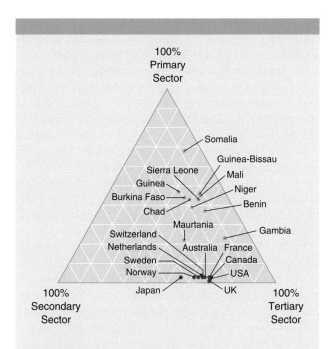

FIGURE 12-8

This figure shows the origin of the GDP, by sector, for the same top 10 and bottom 10 countries ranked by HDI that were shown on Figure 12-6. Value added per worker is usually highest in the secondary sector, so the percentage of a country's GDP that is produced in the secondary sector almost always exceeds the percentage of the country's labor force in that sector. In poor countries, the large percentage of the labor force that works in the primary sector produces a relatively small share of GDP. (Data from UNCTAD.)

to buy goods manufactured out of raw materials that they originally exported themselves. As long as Chile exports copper to the United States, for example, and buys back goods manufactured out of copper, the United States will grow richer than Chile. As long as Botswana exports uncut diamonds to Israel and buys back engagement rings, the Israelis will grow richer than the Botswanans.

The term value added has even been extended to *conceptual value added*, as in advertising. If a $10 million advertising campaign convinces people to spend $100 million more for Brand X than for Brand Y, when those brands are, in fact, nearly identical, then $90 million of value has been added to Brand X "conceptually." That $90 million can provide a lot of jobs. The conceptual value added to products by the advertising industry in New York City today is far higher than the value added by manufacturing in many of the world's greatest manufacturing cities. If consumers will buy a particular brand that they have come to trust consistently, they will try different goods sold under that brand name ("I like Cobb's

brand plum jam, so I will try Cobb's brand strawberry jam."). The right to use that brand name, or *trademark*, is itself a valuable commodity—sometimes more valuable than any good to which it may be attached. Selling brand name cookies is more profitable than growing wheat or grinding the wheat into flour. The geography of adding value to goods downstream helps explain what people do for their living and why some places prosper.

The Japanese Economy

Japan is the most astounding and paradoxical success story in economic geography today. The Japanese islands have few natural resources, but on this meager natural geographical endowment the Japanese have developed a great economy. How?

The answer lies in Japan's cultural resources and in its trading patterns. The key cultural resources include the people's education and labor skills, technology, and the country's style and degree of cooperative organization. All of these are carefully managed by the government. Japan's participation in international exchange is also carefully regulated. The Japanese import raw materials, transform them into manufactured goods, and export these finished goods, plus an array of valuable services, around the world (Figure 12-10). The Japanese also invest their profits and savings around the world. Japan's economic success demonstrates how a place can prosper by importing raw materials and adding value in manufacturing, or by providing other downstream services.

A country can achieve economic growth by steadily increasing the value that it adds to raw materials. Table 12-1 gives the approximate value added in manufacturing certain products. In the 1950s, Japan was a relatively poor country, but it began to manufacture the products at the bottom of the scale. These items are of relatively low value-added, and their manufacture does not require the highest technology. Japan has built its prosperity by steadily progressing up the scale in its exports. As it has done this, it has surrendered the manufacture of the items that are low in the value-added scale. Japan invests in factories to manufacture these products in less-developed countries that are dominated economically by Japan: Malaysia, Thailand, and Indonesia. The process continues: In the early 1990s, Japan's production of lower-technology consumer electronic equipment began to fall, whereas its production of the highest-technology wide screen televisions, minidisk and digital compact cassettes, electronic components and devices, and industrial electronic equipment continued to rise.

National Paths to Economic Growth **427**

Alfred Weber's Theories of the Location of Manufacturing

Most countries strive to industrialize to capture the high value added by manufacturing. One of the earliest scholars to identify the locational determinants of manufacturing was Alfred Weber (1868–1958). Weber's analysis focused on the role of transport costs, and he devised models that differentiate material-oriented manufacturing from market-oriented manufacturing.

Material-oriented manufacturing locates close to the source of the raw material for one of two reasons. In the first case, the raw material is heavy or bulky, and the manufacturing reduces that weight or bulk. The steel industry is an example. The value added in manufacturing steel is low relative to the cost and difficulty of transporting the raw materials (iron ore, coal, and water), so steel mills generally locate where the raw materials can be found or cheaply assembled. The second type of material-oriented manufacturing locates near the raw material

because the raw material is perishable and needs immediate processing. Canning, freezing, and the manufacture of cheese from milk are examples.

Market-oriented manufacturing, Weber's second category, locates close to the market, either because the processing increases the perishability of the product (baking bread, for instance), or because the processing adds bulk or weight to the product. A can of a soft drink, for example, consists of a tiny amount of syrup, plus water, in a can. The syrup can be transported easily and cheaply, but canned soda is heavy and expensive to move. Therefore, the water should be added and the beverage canned close to the market.

Weber elaborated his models of the location of manufacturing by adding additional considerations, such as the availability of a labor force. In that three-factor model, manufacturers locate factories to minimize the cost distance from three points: the location of the raw materials, the labor force, and the market. The optimum location for any spe-

Output of the Secondary Sector as a Percentage of GDP

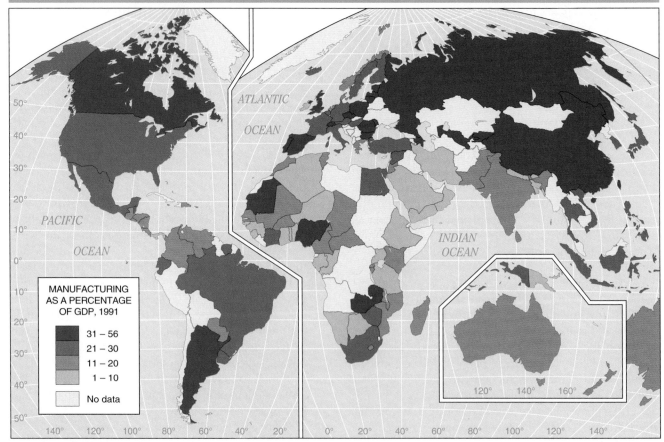

FIGURE 12-9

The term *industrial society* has no exact formal definition, but this map reveals the countries in which industry contributes the highest share of GDP. (Data from UNCTAD.)

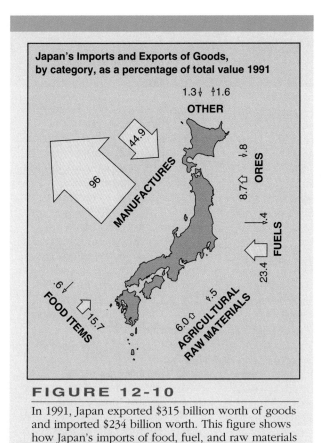

Japan's Imports and Exports of Goods, by category, as a percentage of total value 1991

OTHER 1.3↓ ↑1.6

MANUFACTURES 44.9 / 96

ORES 8.7↑ ↓.8

FUELS ↓.4 / 23.4

FOOD ITEMS .6↓ / 15.7↑

AGRICULTURAL RAW MATERIALS 6.0↑ ↓.5

FIGURE 12-10

In 1991, Japan exported $315 billion worth of goods and imported $234 billion worth. This figure shows how Japan's imports of food, fuel, and raw materials are converted to manufactured goods for export. Japan imports eight tons of fuel, food, wood, and other raw materials for each ton of goods it exports.

TABLE 12-1
Value Added in Manufacturing Products

Product	Added value ($/pound) rough 1990 estimates
Satellite	20,000
Jet fighter	2,500
Supercomputer	1,700
Airplane engine	900
Jumbo jet	350
Video camera	280
Mainframe computer	160
Semiconductor	100
Submarine	45
Color television	16
Numerically-controlled machine tool	11
Luxury car	10
Standard car	5
Cargo ship	1

cific manufacturer depends on the balance of these costs in his or her business.

The Global Distribution of Manufacturing Today

Several things have changed since Weber's original studies: Transportation costs have steadily fallen (look back to Table 3-1), and the value added in manufacturing has increased as manufactured products have become more and more sophisticated. Another factor that has changed is the extent of world trade. Weber's studies focused on the role of locational determinants for manufacturing within countries. Today, as trade binds world regions, the question of locational determinants may be addressed on a global scale.

Transport costs and the value of raw materials are steadily falling as a percentage of the final value of many manufactured goods. The value of the iron, copper, and other raw materials that go into a television-video unit is only a minuscule percentage of the value of the completed unit. Because transport costs

are falling and the value added in manufacturing is rising, high-value-added manufacturing is increasingly footloose (see Chapter 3).

It is still necessary to investigate why manufacturing does not always locate in the countries that have raw materials. Local processing would add value and generate local employment and wealth. Many resource-rich countries, however, have never been able to develop local industries. We must identify other locational determinants for manufacturing that are more important than the location of the raw materials. What are they? What locational determinants have made it possible for some countries to grow rich from industrialization while others, equally or even better endowed with raw materials, remain poor?

Alfred Weber's models considered the locations of the raw materials, the labor force, and the market, and the transportation costs among these

GEOGRAPHIC REASONING
The United Fruit Company

Questions

1. Why is it that oil producers can capture downstream activities but food producers cannot?

2. How many alternative sources of oil are there? How many alternative sources of food products?

The story of the United Fruit Company exemplifies the geography of value added. In the nineteenth century the company established banana plantations in Central America. Several countries came to depend economically on the export of bananas, and many aspects of their local cultures were consequently transformed. Their politics, for example, were manipulated internally by banana interests and externally by relations with the United States; their social structures reflected peoples' positions in the dominant export-oriented cash crop economy; the peoples' diet came to depend on imported staples. The United Fruit Company owned the banana plantations and the ships that transported the bananas, and it marketed the bananas. Eventually many

Central Americans began to resent the foreign control of the plantations, and they accused the company of "Yankee imperialism." The company responded by abruptly surrendering many of its plantations to local governments.

This might sound like a financial sacrifice, but what really happened? The company was no longer tied to one supply (its own plantations), so it could bargain with several potential suppliers. The producers increased production in order to raise their incomes, but they increased production faster than demand could be raised. This only lowered prices. Competition among the suppliers reduced the price of bananas still further. As a result, the new plantation owners got poorer. Transporting bananas and marketing them—

places. At least five other considerations, however, must be added to these (Table 12-2). Each of these considerations is an *input*, or ingredient, for manufacturing. One is *capital*, because manufacturing is increasingly capital-intensive. A second is *technology*. Third, footloose industries seek places with *hospitable governmental regulations*; that usually means low taxes and little environmental regulation. *Political stability* is a fourth consideration. Manufacturers hesitate to invest in volatile political environments. *Inertia* is still another consideration in the geography of manufacturing. Factories are major investments, and networks of suppliers and trained labor forces develop around them; therefore, they are not quickly abandoned. Since Weber's time, the balance among locational determinants for many industrial processes has tipped away from those Weber considered. The balance most relevant to each manufacturing process is different, but the inputs most important to today's highest-value-added manufacturing processes are skilled labor, capital, and technology. An environment of political stability is essential.

Europe today has much of the world's manufacturing, and the reasons for this stem from historical consideration of the locational determinants (Figure 12-11). Europeans brought about the Industrial Revolution first, giving Europe a technological lead, and then Europe politically and economically overwhelmed most

TABLE 12-2
Locational Determinants for Manufacturing

Factors considered by Alfred Weber	Additional factors that are important today
Raw materials	Capital
Labor force	Technology
Market	Governmental Regulations
Transport Costs	Political Stability
	Inertia

430 *Chapter 12*

adding conceptual value through name-brand advertising—is considerably more profitable than growing them, and it supposedly causes far fewer headaches for the company's management.

Eventually the company even changed its name to Chiquita Brands International, and it is now applying its valuable trademark to marketing other fruits from a great variety of sources worldwide. By focusing on downstream activities, the company transformed itself from a raw material supplier into a marketing company. Its profits fattened, while the countries that supplied the raw material got poorer.

A similar tale could be told about almost any raw material or food supplier in the world today and its relationship with multinational corporations that control

every aspect of handling and processing goods from the raw material to the final consumers. Compare the story of the United Fruit Company with the story of

OPEC's petroleum marketing networks told in Chapter 11. In that case, the suppliers of raw materials are themselves seizing control of downstream activities.

Chiquita Brands International's modern refrigerator ship Edythe L. carries Central American fruits to Northern Europe in containers that can be loaded directly from the ship on to railroad cars or trucks. This containerized shipping offers great convenience. (Courtesy of Chiquita Brands International.)

of the rest of the world. Europeans generally located the manufacturing in their own countries in order to retain the value added. It was not their intention to enrich their colonies. In some cases they actually intended to retard the development of manufacturing in the colonies. For example, from 1651 until 1776, at the very beginning of the Industrial Revolution, Britain's Navigation Acts required that American raw materials be shipped to Britain so that Britain could capture the value added by manufacturing from them. Iron from New Jersey had to be sent to Britain, where nails were manufactured to be shipped back to New Jersey. Colonists' resentment of this exploitation helped trigger the U.S. War of Independence. Many colonies in Africa and Asia later suffered the same deliberate restraint on their industrialization.

When manufacturers located factories in countries other than their own, they generally chose other rich countries, where they could find markets and labor forces. Today, the rich countries still have the technological lead and the best-trained labor forces. They usually are politically stable, and they provide the eventual market for most goods. This geographical system has

prevailed for so long that it has a tremendous inertia. Raw materials continue to flow from the poor countries to the rich countries, where they are transformed into more valuable goods. This pattern has been labeled *economic colonialism* or *economic imperialism*. It is noteworthy, however, that neither Japan nor Switzerland, two countries with wealth but without natural resources, ever had a major *political* empire.

The Problems of Today's Poor Countries

Today's poor countries have not been able to capture the value added by downstream activities. Most of them export only unprocessed raw materials (Table 12-3), and they may be dependent on the export of just one or two raw materials (Figure 12-12, p. 434). The welfare of their people rises and falls with world prices over which they have no control, and world prices of commodities have slowly declined. In 1992, the world market prices of the 33 leading nonfuel commodities in world trade were only 54 percent of what they had been in 1975.

Value Added by Manufacturing, 1990

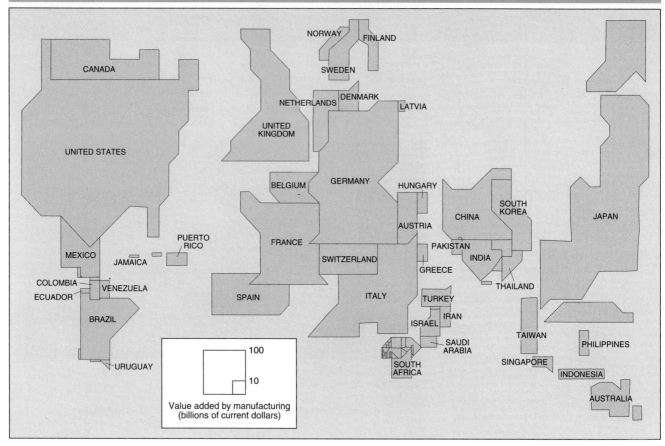

FIGURE 12-11

This cartogram indicates the value added by manufacturing in each country. These statistics are difficult to gather, so several countries do not have reliable figures, but the major manufacturing countries are included here. (Data courtesy of the World Bank.)

Exporters of some raw materials have tried to form cartels, such as OPEC, but few have succeeded. If production of a commodity can easily be increased, then any nation can gain a short-term advantage by dumping stockpiles on world markets, causing prices to collapse. The 74-member International Coffee Organization, which had kept coffee prices high, broke down in 1989, and world coffee prices halved in 4 months. In the countries dependent on coffee exports, suddenly there was no money for children's schoolbooks; national roads went unpaved; hospitals ran out of medicines. In 1993, the 28 major coffee exporters tried to reform the cartel as the Association of Coffee Producing Countries. Erratic swings in commodity prices can be as harmful to national economies as steadily falling prices.

In their efforts to industrialize, today's poor countries suffer from a number of problems that reinforce one another. Poor countries are not as important as markets as the rich countries are. In addition, cheap manufactured goods from the rich countries can flood their domestic markets. Their poverty increases their political instability. They need capital, and their people need education. Poor countries can, however, offer three locational determinants: raw materials, hospitable regulatory environments, and inexpensive labor. Workers in the poor countries receive relatively low wages, and they will often work in conditions worse than workers in the rich countries will. Poor countries' governments seldom complain about pollution coming from a new factory that provides jobs. Later in this chapter we will see how many corporations from the rich countries are today industrializing the poor countries to take advantage of these factors.

Locational Determinants Migrate

The geography of locational determinants is not fixed forever. The locational considerations for manufactur-

TABLE 12-3
A Few Countries' Commodities Exports

Country	Export	Percentage of total exports
South and Central America		
Bolivia	Natural gas	54
	Nonferrous metals	16
	Tin	15
Chile	Copper	32
	Fruit and nuts	11
	Concentrated metal ore	10
	Animal feed	8
El Salvador	Coffee	57
Venezuela	Crude petroleum	38
	Petroleum products	27
Africa		
Liberia	Iron ore	62
	Crude rubber	18
Nigeria	Crude petroleum	95
Tanzania	Coffee	50
	Spices	11
	Cotton	10
	Copper	5
Zaire	Copper	50
	Coffee	24
	Crude petroleum	16
	Diamonds	8
Asia		
Bangladesh	Clothing	21
	Textiles	30
	Fresh and salted fish	13
Fiji	Sugar	70
Indonesia	Crude petroleum	38
	Natural gas	19
	Plywood	6
Iran	Crude petroleum	87
The United States (For comparison)		
Road motor vehicles		9
Office machines		7
Aircraft		7
Misc. electric machinery		6
Misc. nonelectric machines		5
Non-electric-powered machinery		4
Organic chemicals		3

Source: UN *Handbook of International Trade and Development Statistics.*
(Data Courtesy UNCTAD, 1990. Reprinted with permission.)

ing are increasingly footloose, so manufacturing can migrate around the world if the mix of attributes of any place changes and attracts it.

There are, in fact, reasons why the geography of determinants will *continually* change. Chapter 3 already suggested relevant factors: new products, new technologies, new raw materials, new sources of traditional raw materials, new technologies of manufacture, new governments with new policies, growing and shrinking labor supplies, and the opening and development of new markets. At each place the local balance among the relevant factors continuously evolves. For example, if any place attracts manufacturing because it offers a low-wage labor force, its local standard of living and wages will rise. As local wages rise, the industries that developed there to take advantage of the low wages will be driven away. They will migrate to still poorer places where wages are lower. Thus, each place hopes to enjoy a continuous evolution of industries and an upward spiral of wages and prosperity as the skills of local workers increase.

The manufacture of men's dress shirts for the U.S. market offers an example. Shirt making is labor intensive, so manufacturers are continuously searching for cheaper labor. In the 1950s, U.S. shirt manufacturers located in Japan to use the low-paid workers there. When the costs of labor and real estate rose in Japan, the companies moved to Hong Kong. As Hong Kong's factories gave way to offices, the shirt makers moved again, first to Taiwan and Korea and then, in the 1970s and 1980s, to China, Thailand, Singapore, Indonesia, Malaysia, and Bangladesh. Toward the end of the 1980s and into the early 1990s, Costa Rica, the Dominican Republic, Guatemala, Honduras, and Puerto Rico have become shirt-manufacturing centers.

These Caribbean basin countries offer cheap labor, and U.S. manufacturers who locate in them enjoy U.S. tax advantages granted under a U.S. government program, the Caribbean Basin Initiative, that was designed to help these countries industrialize. These countries also offer proximity to the United States. Relative proximity means that oversight is easier and that shipments can be faster. Speed is important for three reasons. First, fashion demands faster introduction of new styles—a consideration once restricted to women's clothing but now relevant to men's clothing as well. Second, renting a ship for an additional week is expensive. Third, capital tied up as shirts in the hold of a ship in transit is not earning any profit. One shipload of shirts may be worth $10 million. The interest on that money each day that the ship is on its way to the United States adds to the retail cost of the shirts.

Interest Rates and Economic Growth

Interest rates, or what economists call "the cost of money," concern students of diffusion, exchange, and

Export Concentration Indices

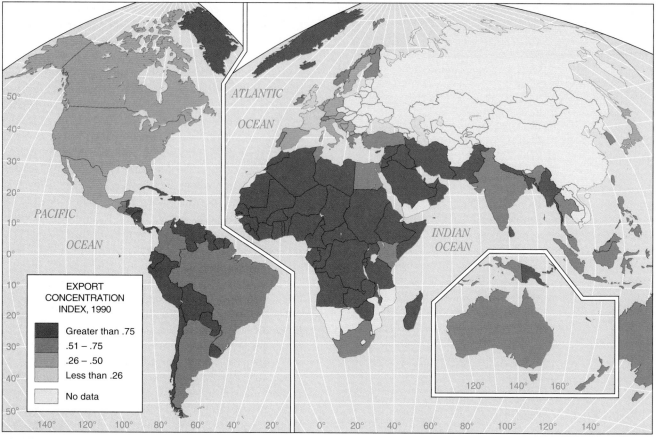

Export Concentration Index, 1990

	Greater than .75
	.51 – .75
	.26 – .50
	Less than .26
	No data

FIGURE 12-12

The export concentration index mapped here ranges from 0 to 1, with 1 representing extreme concentration. As a rule, poor countries have the highest indices. Production of any one product may fall or the market for it may be lost, so the economies of countries that rely on few exports are generally vulnerable to production falls or world market shifts. If, on the other hand, a country is able to export a variety of products, it is more economically secure. The rich cartelized oil exporters deviate from the rule. (Data from UNCTAD.)

trade for three reasons. One reason is that people generally invest their money wherever interest rates are highest. That tendency, and all of the following discussion, follows from what we learned about *economic man* in Chapter 2. A place that offers high returns can attract capital investment, and capital investment may bring development.

Second, whenever anything is in transit, the value that it represents is idle. The owner is therefore losing an opportunity to invest that value in some other activity that might pay a return every minute. A financial return sacrificed by leaving capital invested in one form or activity rather than another is called an **opportunity cost**. Goods in transit cost their owner not only transportation charges, but also opportunity costs, and that is why people want things to move *fast*. If an entire country has poor transportation, much of its capital is

tied up in raw materials or in materials in transit. Fast and efficient transportation releases capital for productive investment and thus boosts economic growth.

A third reason why interest rates concern geographers is that transportation infrastructure is one of any society's greatest investments—roads, bridges, railroads, airports, pipelines, and so forth. Money is usually borrowed to finance this construction, and the interest rates on that debt can determine what gets built and how facilities are operated. For example, the builders of the English Channel tunnel borrowed billions of dollars to construct it. The interest rates on that debt will determine the fees the owners must charge users of the tunnel. The fees will affect who uses the tunnel, and who uses the tunnel will influence the rate of redistribution of activities across northern Europe.

434 *Chapter 12*

Technology and the Future Geography of Manufacturing

The balance among the inputs to manufacturing is continuously changing. For example, the value of the labor input is generally a shrinking percentage of the value of manufactured goods. In the consumer electronics industry it is no more than 5 percent to 10 percent.

The importance of technology, by contrast, is increasing. When Japan was industrializing, it bought technology. By one estimate, the Japanese paid foreign manufacturers a total of only $10 billion for patents and licenses between 1950 and 1980. That must rank as the shrewdest investment any nation ever made. Today, Japan has joined the United States, Germany, Great Britain, France, the Netherlands, Switzerland, and the other most advanced countries among recognized holders of key patents. The world geography of manufacturing *tomorrow* may be guessed by mapping investment in research. It is difficult to study relative investment worldwide, but some analysts feel that U.S. investment in research may be falling behind that in Japan and other international economic competitors, measured both as a percentage of GNP and in actual money terms.

Technology and capital contribute the most rapidly increasing shares of the final value of most manufactured goods, and it is difficult for poor countries to develop these on their own. The combination of those two facts would seem to be bad news for today's poor countries. However, technology and capital are exceptionally footloose. In the manufacture of computer disk drives, for example, the raw materials are an insignificant fraction of the drives' final value, so computer disk drives can be easily manufactured and distributed worldwide from almost anywhere. Why then are more than one-half of all of them manufactured in Singapore? They were not invented there. The decisive locational determinants for this high-value product are the availability of inexpensive skilled labor, technology, capital, and political stability in Singapore. Each of these factors, however, is footloose. How long will Singapore continue to dominate world manufacture of disk drives?

The development of computer software is one of the world's fastest growing industries, and it is even more footloose than the manufacture of computer hardware. All it requires is a personal computer, an electric plug, and a skilled programmer. This multi-billion-dollar industry can migrate on an airplane. In 1990, the United States commanded 57 percent of the $110 billion global market for software and related services. This value is greater than the total production of almost any of the world's raw materials. In the same year, Japan held 13 percent; France, 8 percent; Germany, 7 percent; Britain, 6 percent; Canada, 3 percent, and all other countries, 6 percent of the software industry. How long will this distribution last?

Air freight costs are falling, and because more manufactured products are becoming smaller and of higher value, they can be moved by air. Fort Worth, Texas, has opened the world's first airport at which planes can taxi directly from the runway into an industrial park. Sponsors hope that it will attract manufacturers that rely on far-flung foreign and domestic suppliers and that sell their finished goods to buyers around the globe. Could such an airport-industrial park just as well be in Alaska or on an island in the South Pacific?

In 1993, the Union Bank of Switzerland measured the future economic competitiveness of nations based on such important variables as current investment in physical capital, education, and research and development. The top five countries that emerged from this study were China, Israel, Japan, Korea, and Singapore. The United States tied with Germany for 19th place. The Bank's economists suggested that if current trends continue, per capita income in Korea will equal that in the United States by 2008. These calculations are projections, subject to the limitations of projections that were noted in Chapter 4. The very publication of such projections might stimulate policy changes that can change the future. As the report noted, the growth rate in the United States should increase "as it adopts the innovations that give the other countries their advantage."

More than ever before in history, maps of the world geography of wealth, of manufacturing, and of economic development are exceptionally kaleidoscopic.

Where Is the Third World?

The term "Third World" was introduced by the French demographer Alfred Sauvy in 1952. It reflected attitudes during the Cold War that followed World War II. An ideological line was drawn between the capitalist countries led by the United States and the Communist-ruled countries headed by the Soviet Union. At the height of the diplomatic pulling and tugging, it was expected that all countries should line up on one side or the other. Many did so, but a few, lead by presidents Tito of Yugoslavia, Nasser of Egypt, Nehru of India, Nkrumah of Ghana, Makarios of Cyprus, Sukarno of Indonesia, and Emperor Haile Selassie of Ethiopia, clung to a precarious neutrality. These maverick states came to be known, collectively,

FOCUS
The Economy of Singapore

Singapore's exports equal roughly 175 percent of its GDP. How can a country export one and three quarters the amount that it produces? The answer is that Singapore does two things: First, it imports raw materials and semifinished products and re-exports them after adding value by further processing; second, it serves as a transshipment point for other countries' exports. A transshipment point is called an *entrepôt*.

During its first 25 years of independence (1965–1990), Singapore lured world manufacturers with an inexpensive but highly trained and disciplined labor force and government-supported factories. Singapore achieved a per capita income comparable to those of Ireland and Spain. Today Singapore wants to increase the value it adds by evolving from dependence on factory workers to researchers, software engineers, and biotechnology specialists. New research on computer screens, for example, the advancing edge of technology, is locating in Singapore. By 1994, Singapore maintained its 40 percent share of the world market in computer disk drives by shipping some assembly work to Malaysia, where labor costs are lower.

Singapore is also trying to capture air traffic as it has captured shipping by building Asia's largest airport, called

Airtropolis. Airtropolis's director has said, "Since its founding Singapore has been an entrepôt for goods and trade. As with cargo, so with air passengers—we import and export, and we hope to provide a bit of value-added on the way." Singapore profits handsomely if each traveler passing through Airtropolis

buys just a cup of coffee and a magazine.

Singapore's transport facilities plus its political stability enhance its role as an Asian service center. Today financial services provide one-third of Singapore's gross product, and the city-state hopes to become a regional medical center.

Singapore

Singapore, at the narrow Straits of Malacca, was not founded by the peoples native to the region. The situation's strategic value was most immediately recognizable to a globally seafaring people, so it was the British who founded the city as a naval base early in the nineteenth century.

as the **Third World**, to distinguish them from the **First World** of the Western bloc and the **Second World** of the Soviet bloc.

Over the years the term Third World gained economic and sociological connotations. The First World was interpreted as a haven of science and rational decision making: progressive, technological,

efficient, democratic, and free. The First World included as allies in the Cold War all reasonably well-to-do, non-Communist countries, even though some of them were politically repressive, such as several Latin American states, South Africa, South Korea, and Taiwan. The Second World was also defined as modern, powerful, and technologically sophisticated, but it was dominated by totalitarian Communist governments. In fact the countries of the Second World never achieved economic growth equivalent to that of the First World, and the collapse of Communist regimes in 1989 exposed their true poverty.

Nevertheless, in contrast with the First and Second Worlds, the Third World was interpreted as a world of underdevelopment, overpopulation, irrationality, religion, and political chaos. The term also carried a racial, or even racist, aspect: Most Third World countries were nonwhite. Much of the Third World had still been organized as colonies in 1939, but it "emerged" after World War II. According to the theory, it was poor but struggling to develop. The Third World societies were interpreted as a zone of competition between the First and Second World models of modernity (Figure 12-13).

Today the terminology of First, Second, and Third Worlds is clearly out of date. A region, by definition, must exhibit meaningful homogeneity. Historically, the unifying element in the Second World was the existence of Communist regimes.

With the collapse of these regimes in Eastern Europe and the former Soviet states, however, the Second World as a political bloc ceased to exist. Similarly, the term Third World lumped together peoples with extremely diverse histories and cultures who inhabit a broad range of physical environments that possess a great variety of natural and cultural resources. If we consider just the countries whose leaders made up the original political Third World—Yugoslavia, Egypt, India, Ghana, Cyprus, Indonesia, and Ethiopia—they cannot be homogenized economically, culturally, or sociologically. It is condescending and ultimately insulting to refuse to recognize and to appreciate their differences. Indiscriminate categorizing not only prevents understanding, it prevents helping. If we want to understand and to help the world's poor countries, we must recognize and appreciate their differences.

In the 1990s the term Third World will disappear. It will be remembered as a relic of a political Cold War mentality that regarded political alignment in a bygone struggle as the most significant distinction among all the world's nations and cultures.

The Third and Fourth Worlds

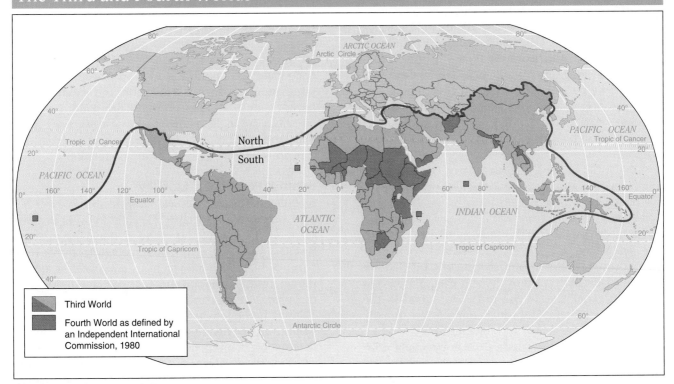

FIGURE 12-13

The term *Third World* confused Cold War political alliance with economic status, so efforts were made to "refine" it. Some scholars subdivided the Third World into a poorest "Fourth World." Others tried to differentiate a rich "North" of the world from a poor "South."

THE VARIETY OF NATIONAL ECONOMIC-GEOGRAPHIC POLICIES

▼▼▼▼▼▼▼▼▼▼▼▼▼▼▼▼▼▼▼▼▼▼▼▼▼▼

Each country organizes its domestic economy, manages the distribution of economic activities within its territory, and regulates its participation in world trade and investment.

Political Economy

Each country defines a set of principles to organize its economic life. The study of these principles is **political economy**. Direct government participation ranges theoretically from a *Communist system*, in which the government owns both the natural resources and the productive enterprises in the name of the workers, to a *capitalist system*, in which the state defers to private enterprise. The economy of each country lies between these two extremes. Most countries have mixed economies.

Even among those countries generally considered capitalist, there is a great range of government involvement in the economy. Each country's political economy usually reflects other characteristics of that country's culture. The United States, for example, generally favors a system that minimizes the government's role. This is called **laissez-faire capitalism**, from the French for "leave us alone," a phrase supposedly said long ago by a French businessman to a government bureaucrat. The U.S. government has never developed an explicit industrial policy, but its regulations, research funding, and defense budget have nevertheless greatly influenced the national economy.

Most Latin American countries, by contrast, have traditions of considerable government ownership of the nation's assets and industries. In the late 1980s however, these governments began privatizing the assets (see Chapter 6). The same is true in Western Europe, where, as of 1991, nationalized companies still accounted for about 25 percent of GNP in France, 11 percent in Germany, and even 3.5 percent in Britain, which began the privatizing trend in the early 1980s. Eastern European countries are rapidly privatizing as they abandon communism.

In several Asian countries the government does not own many companies outright, but it plans and regulates the economy. This is called **state-directed capitalism**, and it conforms with these countries' Confucian bureaucratic cultural traditions (discussed in Chapter 8). South Korea, for example, is considered a capitalist, free-market economy, yet the gov-

ernment proposes national economic plans and subsidizes new industries. Industry is dominated by combinations of corporations that are affiliated by interlocking directorships and ownership. The economies of Japan and Taiwan are similarly organized, but in the United States such combinations would be broken up as illegal *trusts*.

The economies of the Communist countries were often referred to as *planned economies*. This term, however, does not recognize the significant amount of government planning in the state-directed capitalist economies, whose recent success has provoked the suggestion that the United States adopt an explicit industrial policy. Many Americans, however, distrust government planning as both an intrusion on liberty and as economically inefficient.

Each country also manipulates its national budget and expenditures, interest and exchange rates, and money supply to promote growth and high employment. The tools for manipulating these are imperfect, but national governments' economic planning and management, the sheer weight of their revenue and expenditure, and their role in redistributing income have made national economic policies more central to most peoples' lives than ever before.

Domestic Economic Geography

Countries try to boost their economic growth by organizing their populations, territory, and resources for production. A strong sense of nationalism can encourage the population to work together, but the population also needs education and training. The territory must be at peace, and the country should constitute a single market for raw materials, goods, and labor. Another factor that promotes economic growth is internal mobility of the population.

Internal Population Mobility   Internal population mobility assists each individual in realizing his or her own potential, and it also allows employers to draw on the potential of the entire national population. Workers would normally be expected to move from areas of high unemployment to areas of low unemployment, so variations in regional unemployment can provide a measure of labor mobility and national cultural homogeneity. Great variations suggest that something is discouraging people from moving. Factors preventing mobility may include peoples' attachment to place or family; racial or ethnic animosity among different localized groups; lack of nationwide availability of certain cultural products and services (religious services, foods, or other consumer goods); regional variations in unionization and difficulty in joining unions; variations in wage scales;

variations in workers' benefits and whether workers lose them by relocating; and nationwide acceptance of degrees or certification for professionals.

Italy exemplifies a country in which unemployment rates in one region, the south, are usually two or more times as high as in the north. This reflects continuing cultural differences between the regions. Unemployment rates vary across the United States, but on the whole Americans are a mobile population. Almost 20 percent move to a different house each year; almost 3 percent to a different state. This illustrates the relative homogeneity of the national culture.

States With Frontiers

Not all states occupy and utilize their full territory. The world population map (Figure 4-1) reveals that many countries have core areas of dense population and development but also **frontier** areas, undeveloped regions that may offer potential for settlement. Environmental limitations may hinder exploitation of frontiers. Brazil, for example, contains vast sparsely settled areas in which the government has encouraged settlement, but the results have often been economically disappointing and ecologically disastrous. Similarly, the island of Java is only 7 percent of Indonesia's national territory, but it is home to 60 percent of the population. The Indonesian government has tried to relocate settlers to other islands, but many new settlements have had difficulty supporting themselves.

Some African states seem overpopulated relative to their current ability to feed themselves, yet they are only sparsely populated. With an end to civil wars and with greater investment in agriculture, some of these territories could support greater populations. Canada's northern regions are poorly suited to agriculture (compare Figure 1-23 with Figure 10-20), but they are exploited for mineral, timber, and hydropower resources.

Expanses of the Siberian region of Russia have never been densely settled or even fully explored for resources, yet most of Russia's population remains west of the Urals (Figure 12-14). Salaries in Siberia are higher than those paid in European Russia, but settlement is still slow. New resources are discovered regularly, but enormous investment will be required to develop them at the site or else to bring them out for manufacturing elsewhere.

How Do Governments Distribute Economic Activities?

Most countries try to maintain a fairly equal standard of living throughout their territory. Great disparities of wealth from region to region are a centrifugal force that may pull the state apart. Economic competition among regions may dominate national politics as politicians weigh the domestic regional impact of every program and devise new ones to reduce disruptive imbalances.

The geography of any country's resource endowment will favor some regions over others. The relative fortunes of regions may change, depending on the discovery of new resources, shifting trade patterns, or patterns of innovation. In the area of today's Belgium, for example, the lowland coastal Flemings grew rich from trade and commerce during the medieval and Renaissance periods, and they dominated the highland Walloons. Later the discovery of coal and the industrialization of Wallonia brought Walloons prosperity. Today the decline of coal and steel manufacturing and the development of a trade and service economy has swung the pendulum of prosperity back to the Flemings.

The national distribution of prosperity can also change as the sectors of the national economy evolve. As employment shifts from sector to sector, job opportunity shifts from place to place. The expansion of the secondary and tertiary sectors, for example, brings urbanization. These shifts might occur faster than workers can be retrained or relocated, so certain regions of a country may suffer while others thrive.

Countries often devise special development programs for poor regions, just as cities or states designate urban enterprise zones (discussed in Chapter 9). For example, the United States created the Tennessee Valley Authority (TVA) in 1933, an Appalachian Regional Commission in 1964, and a Lower Mississippi Delta Development Commission in 1988. Income and welfare were significantly below national levels in these three regions, and federal programs attempted to boost their economies. Table 12-4 shows that regional incomes in the U.S. South, which historically have lagged behind those in the rest of the country, are catching up.

A government may locate factories or other enterprises that it owns in poor regions, or it may distribute offices, research institutes, and military installations. Governments may lure private enterprises by offering subsidies, tax waivers, free development sites, or loans. The Spanish government, for example, subsidizes international businesses that invest in Spain's poorer provinces, and Italy subsidizes investment in its south. India encourages firms to locate in officially designated "backward areas" by offering low-interest loans, tax reductions, and low freight rates on government railroads for products from the designated areas.

National Paths to Economic Growth　　**439**

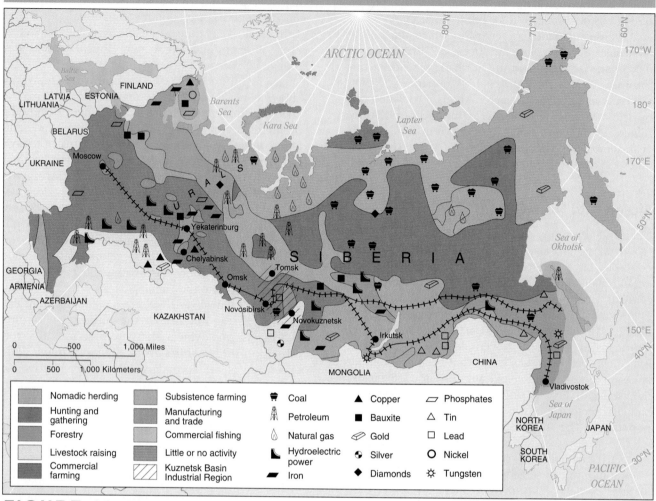

FIGURE 12-14

Siberia remains almost uninhabited, and enormous amounts of capital will be needed to exploit its raw materials.

TABLE 12-4
U.S. Regional Incomes as a Percentage of National Income

	1840	1860	1880	1900	1920	1930	1940	1950	1960	1970	1980	1989
United States	100	100	100	100	100	100	100	100	100	100	100	100
Northeast	135	139	141	137	132	138	124	115	114	106	105	115
North Central[a]	68	68	98	103	100	101	103	106	101	100	100	96
South	76	72	51	51	62	55	65	72	80	85	86	90
West	—	—	190	163	122	115	125	114	105	101	103	104
Mean deviation (%)				37			28			12	9	

[a]North Central is equivalent to what was the 'West' before the Civil War.
Source: U.S. Bureau of the Census.

No matter what economic policies a government devises, some regions may dominate a national economy while others suffer population losses. The central region of the United States has been losing population since World War II. Kansas alone has over 2,000 officially registered ghost towns (Figure 12-15). The Census Bureau estimates that Iowa, North Dakota, and Nebraska will suffer net population declines between 1990 and 2000, while the South and Southwest, known as the Sunbelt, will continue to absorb most U.S. population growth.

National Transportation Infrastructures

Transportation and communication allow different territories to specialize production and to trade. This territorial division of labor is comparable to the division of labor among workers. The design and regulation of a country's transport infrastructure are two important influences on internal geography. In some countries transportation is slow and difficult. Some regions are practically inaccessible (Figure 12-16). In other countries, virtually every place is easily and cheaply accessible from every other place (Figure 12-17).

In much of Southeast Asia and in parts of South America and Africa, some regions are in open rebellion, and the central governments invest in roads specifically to "occupy" the territory. Colombia, for example, has recently discovered enormous oil deposits beneath its eastern plains, but exploiting these deposits will require the central government to seize control of this region from drug dealers and independence movements.

Economic growth, however, is usually the explanation for investment in transportation. Fast and efficient transportation releases capital for productive investment, allows the exploitation of natural resources, regional specialization of production, and domestic trade among regions. India exemplifies a country whose economic growth is held back by an inadequate road system. Traffic in passengers and goods on India's roads has increased thirtyfold since

Year of Maximum Populations

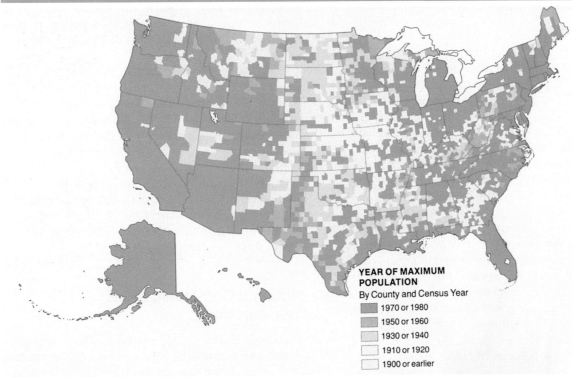

YEAR OF MAXIMUM POPULATION
By County and Census Year
- 1970 or 1980
- 1950 or 1960
- 1930 or 1940
- 1910 or 1920
- 1900 or earlier

FIGURE 12-15

Some counties in the central U.S. farming region and some old cotton farming centers in the South counted their maximum populations before 1900. The populations of many other areas across the Great Plains and in Appalachia peaked by 1940, when the prosperity that had come from farming and mining was past. The national population today seems to be concentrating at the margins of the country. (Reprinted with permission from National Geographic Society, *Historical Atlas of the United States*. Washington, DC: National Geographic Society, 1988, p. 77.)

FIGURE 12-16

Malaysia's new roads cut through forests and open frontier regions to development.
(Courtesy of Caterpillar, Inc.)

independence, but the total road mileage in India has increased only fivefold. Roads throughout India are overcrowded, and thousands of villages lack all-weather roads. The government hopes to connect every village of more than 500 people to an all-weather road by the year 2000. Accessibility would allow regional specialization of production and cash cropping, especially of high value crops such as fruits, that spoil quickly. This could improve the Indian diet.

In many countries the pattern of the transport network is not integrated, but consists of lines penetrating into the country from the ports. These roads or railroads are called **tap routes**. Just as some trees have tap roots that reach deep down into the soil to draw up water, so these tap routes stretch into the interiors from the seaports to bring out the exploitable wealth. Tap routes are often the legacy of colonialism or the product of neo-colonialism, and they facilitate getting into or out of a country, but they do not make it easy to get around inside it (Figure 12-18, p. 446).

Regulating Transportation Most countries blend public and private transportation services, but mapping a network is only a first step toward understand-

ing what moves, where, and why. Lines on a map are static, but conditions and charges that are "hidden" behind the map can manipulate movement. Many countries manage freight rates on government-owned railroads, as India does, for example, to distribute industry or other activities.

In all countries trucks, railroads, and airlines compete for traffic. They may also compete for government subsidies, and these subsidies can in turn affect domestic economic geography. In Kenya, for example, the trucking industry has successfully prevented national investment in railroads. In most European countries, in contrast, government-owned railroads receive generous subsidies. In the United States the trucking industry, the oil industry, the highway-construction industry, and the car-driving public make up a lobby powerful enough to overwhelm the railroad interests. As a result, trucks pay only a fraction of their real highway-use costs, and they do not pay for the many negative externalities that they impose, such as air pollution, traffic accidents, and costs due to congestion. This subsidy to the trucking industry enables it to capture much long-haul traffic that is actually best suited to railroads. Railroads still handle much U.S. bulk traffic (Figure 12-19, p. 447), but freight is shifting to trucks, and railroad trackage is being abandoned. Thus, in the United States as in Kenya, political competition distorts economic competition and locks inefficiency into the transportation system.

Countries may also manipulate domestic freight rates to direct traffic to favored seaports, and the port facilities themselves may be subsidized to attract business and create jobs. The Netherlands, for example, organizes its national transport network and fixes prices on it to capture an international hinterland for its port of Rotterdam.

These examples provide only a hint of the manipulation of charges built into each country's transportation infrastructure. Geographers often learn that the answer to their question Why there? lies hidden in these policies.

National Communications Infrastructures

The dissemination of information can be as important as the dissemination of goods and materials. A wealth of sources of information arguably both reveals and also supports both economic development and political liberty. There are several ways of distributing information, and the availability of each type can be measured (Figure 12-20, p. 447). Home mail delivery, for example, is guaranteed only in the world's richest countries. The availability of telephones is another

World Roads and Highways

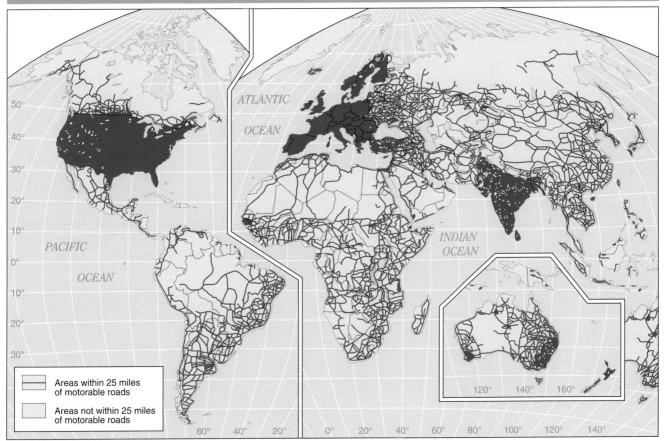

FIGURE 12-17

Compare this map with the world population distribution shown in Figure 4-1. Many densely settled regions, such as Eastern China, would benefit from improved road networks. (Adapted from Rand McNally, *Goode's World Atlas*, 18th edition. 1990 by Rand McNally R.L. 92-5-76.)

indicator of connectivity. The United States and Canada have almost 40 percent of the world's telephones; Africa about 1 percent. The fax machine, which uses telephone wires, is rapidly replacing postal service in places where telephone service is available.

Chapter 9 pointed out that millions of Americans telecommute, and millions of U.S. homes are already plugged into two electronic highways—television cables that bring information and entertainment, and telephone wires that bring voice communication and a great range of additional services to homes equipped with personal computers. Either of these competing technologies might eventually replace the other, and improvements in communication through airwaves (cellular telephones, for example) offer additional possibilities. What the outcome of this three-way technological competition will be, and what roles the government and private industry will play, are matters of national debate. Other rich countries are also experimenting with national electronic highways.

NATIONAL TRADE POLICIES

Countries differ in the degree to which they participate in world trade and circulation. Each country wants to develop its own economy, to ensure its national security, and, to some degree, to protect its national culture or identity. Chapter 6 noted how some countries subsidize domestic food production as a national security measure. Countries also protect national security industries—armaments, for example.

The Import-Substitution Method of Growth

A government may protect a national industry from imported competition as only a temporary measure. This is done in the belief that newly developing industries, called **infant industries**, cannot com-

National Paths to Economic Growth **443**

FOCUS
The U.S. Transport Infrastructure

Regional specialization of land uses reflects environmental endowments, but specialization cannot occur without transportation and exchange. The United States has one of the world's finest transportation infrastructures. Its development allowed the exploitation of the nation's raw materials, the creation of a single mass market for consumer goods, and regional specialization of production.

Today, however, complaints are rising about the deterioration of the U.S. transport infrastructure. The Federal Highway Administration reported that in 1990, 57.3 percent of U.S. primary and secondary roads were in need of repair, and 39.1 percent of the nation's bridges were structurally deficient or obsolete. In that year Americans spent over 2 billion person-hours tied up in traffic

jams, and airport congestion caused periodic gridlock from coast to coast. Late deliveries of goods cost the national economy an estimated \$35 billion. Estimates of what the United States must spend to repair what it has and to build what more it needs are as high as \$3.3 trillion. How to raise and allocate investment in infrastructure might soon be a major national issue.

pete with imports. Infant industries manufacture only small quantities of the product, so manufacturing costs per unit are high. Eventually, however, the industry might build a national market for its prod-

uct and produce larger quantities of it. Then the cost per unit falls. This is called achieving **economies of scale**. When economies of scale have been achieved in protected national markets,

U.S. Land Resources

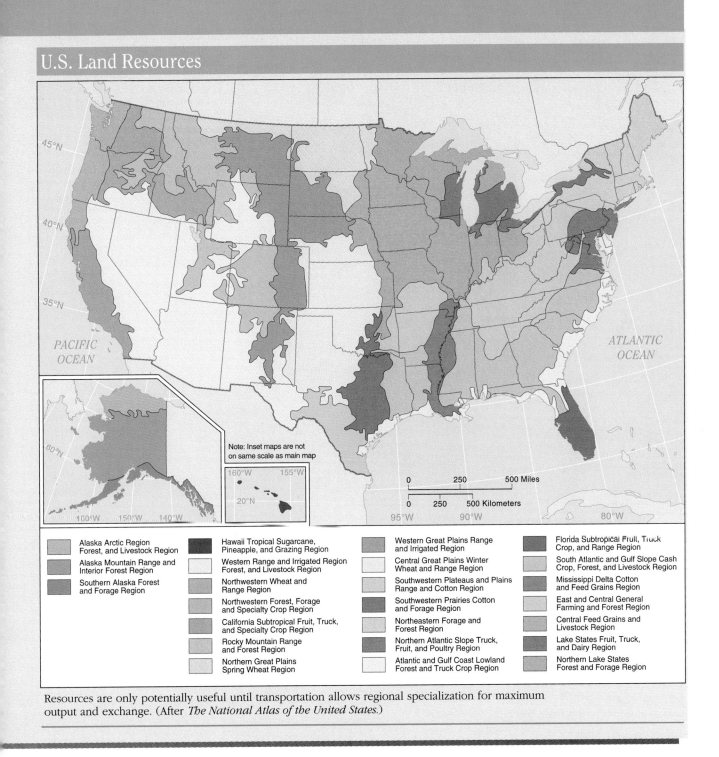

Note: Inset maps are not on same scale as main map

- Alaska Arctic Region Forest, and Livestock Region
- Alaska Mountain Range and Interior Forest Region
- Southern Alaska Forest and Forage Region
- Hawaii Tropical Sugarcane, Pineapple, and Grazing Region
- Western Range and Irrigated Region Forest, and Livestock Region
- Northwestern Wheat and Range Region
- Northwestern Forest, Forage and Specialty Crop Region
- California Subtropical Fruit, Truck, and Specialty Crop Region
- Rocky Mountain Range and Forest Region
- Northern Great Plains Spring Wheat Region
- Western Great Plains Range and Irrigated Region
- Central Great Plains Winter Wheat and Range Region
- Southwestern Plateaus and Plains Range and Cotton Region
- Southwestern Prairies Cotton and Forage Region
- Northeastern Forage and Forest Region
- Northern Atlantic Slope Truck, Fruit, and Poultry Region
- Atlantic and Gulf Coast Lowland Forest and Truck Crop Region
- Florida Subtropical Fruit, Truck Crop, and Range Region
- South Atlantic and Gulf Slope Cash Crop, Forest, and Livestock Region
- Mississippi Delta Cotton and Feed Grains Region
- East and Central General Farming and Forest Region
- Central Feed Grains and Livestock Region
- Lake States Fruit, Truck, and Dairy Region
- Northern Lake States Forest and Forage Region

Resources are only potentially useful until transportation allows regional specialization for maximum output and exchange. (After *The National Atlas of the United States.*)

then imports will be allowed to compete. This policy of protecting domestic infant industries is called the **import-substitution** method of economic growth.

This form of economic protectionism helped some countries industrialize in the past, including, to some degree, the United States, but it involves economic risks. If a country raises tariffs (taxes on

Tap-Routes in Africa

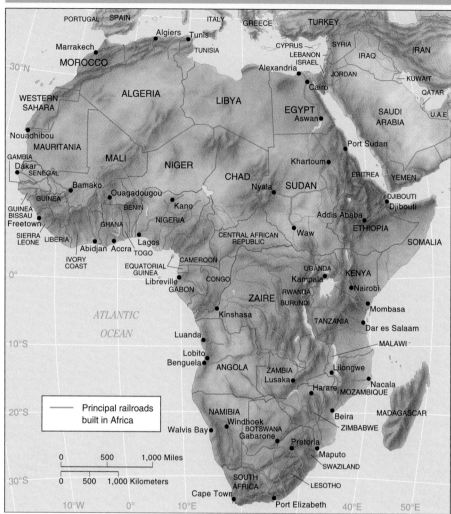

FIGURE 12-18

Africa's coastline offers few natural harbors, and the continent is relatively impenetrable by water because many of its rivers drop off the continental interior through rapids and waterfalls only a few miles inland from the coast. The Europeans built railroads to penetrate and to haul out agricultural products and minerals. Today, railway lines provide the only freight service in roadless areas, but deterioration of the rail network hikes transport costs and lowers income for many African states. Many lines shown on this map are disrupted by civil strife or else they are inadequately maintained, so they offer only intermittent service.

imports) to protect a particular industry, all national consumers are subsidizing that industry. Some protected industries may *never* be able to survive competition with imported goods, so the national public may go on subsidizing them. This may be an inefficient investment of national resources. For example, the United States imposes an average 48 percent tariff on imported apparel; this protection costs U.S. consumers $145,000 per job saved in the U.S. apparel industry. The manufacturers and workers, however, lobby for continuing protection from foreign competition.

Export-Led Economic Growth

The alternative to the import-substitution method of economic growth is the **export-led method**. Countries that adopt this program welcome foreign investment to build factories that will manufacture goods for international markets. In that way they can achieve economies of scale immediately. Export-led policies rely on global capital markets to facilitate international investment and global marketing networks to distribute the products.

The countries which have grown the fastest in recent decades have generally followed export-led programs. The success of these policies demonstrates again how, increasingly, what happens *at* places is the result of what happens *among* places.

Trade Policies Affect the National Distribution of Wealth

International trade and investment sometimes multiply regional economic inequalities within a country. If a country adopts protectionism, all regions of the country may not profit equally, and that can intensify regionalism. If a government adopts the alternative export-led growth policies, outside investment might concentrate in favored regions. China best illustrates how disruptive this can be.

A Case Study: The Political-Economic Geography of China

China exemplifies on a colossal scale how foreign investment can intensify regional imbalances in economic growth and how the centrifugal force of these imbalances can threaten national stability. China has always had an economic and cultural dichotomy between an outwardly-focused southeast coastal region and an inwardly-focused central region. Chinese capitals were in the north or center, and the government limited the trading activities of the merchants in the southern coastal cities. In the nineteenth century, however, European powers forcibly "opened" China's southeastern ports to trade, and industrial development, inspired by international trade, clustered along the coast. The Chinese government keenly felt this humiliation.

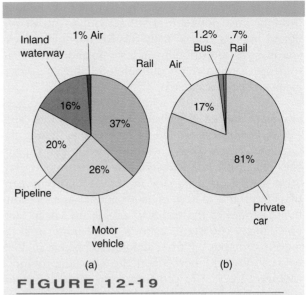

(a)   (b)

FIGURE 12-19

These pie graphs illustrate the percentage shares of movement of ton-miles of freight (a) and passenger miles (b) in the United States in 1991. In the United States today, freight moves by rail and truck; people move by private car and by air. Rail passenger services, which once dominated U.S. intercity traffic and still do in some European countries, have shrunk.

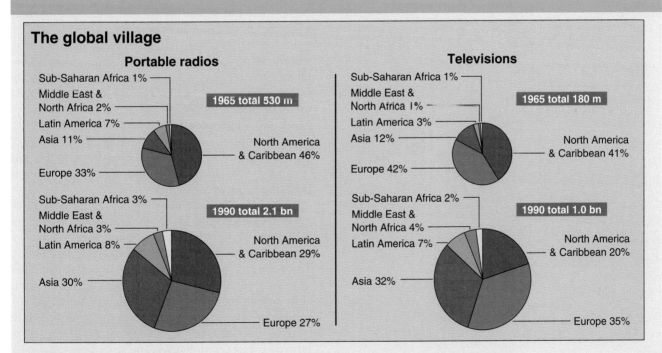

FIGURE 12-20

The numbers of radios and televisions has increased everywhere in the world, but the rise of the percentage share found in Asia is dramatic. (Data courtesy of The Economist Newspaper Group, Inc. © 1992. Reprinted with permission.)

National Paths to Economic Growth   **447**

The Communist government that Mao Zedong brought into power in 1949 set out to boost the inland regions out of their relative economic backwardness by redistributing industry inland. From 1949 to 1979, foreign trade plummeted; China's coastal areas and cities stagnated.

Mao died in 1976, and in 1979 the Chinese government reversed its policies. It designated five "special economic zones" and 14 "open cities" along the coastline (Figure 12-21), where relatively free capitalistic practices were allowed. The designated coastal areas leaped to create new wealth. Hong Kong manufacturers and a host of international corporations built factories in the zones to exploit the cheap land and labor. The state sector's percentage of total industrial output fell from almost 100 percent in 1978 to as low as 14 percent in the coastal provinces (Figure 12-22). The provincial governments themselves invest in entrepreneurial undertakings. China had engaged in very little foreign trade in 1979, but between 1978 and 1988 both exports and imports quintupled. By 1988 foreign trade represented one-third of the national income, an extraordinarily high proportion for such a big country.

FIGURE 12-22

Traditional Chinese fish, rice, and tea have yielded to hamburgers, fries, and soft drinks at this McDonald's in Guangdong, China, part of a special economic zone in which foreign investment has created a thriving industrial concentration. The sign on the wall sets prices in Hong Kong dollars, reflecting the integration of the local economy with that of nearby Hong Kong.
(Courtesy of Francis Li/Gamma Liaison.)

Economic Zones in China

In 1979, China designated special zones and cities where capitalist-style economic growth has boomed.

FIGURE 12-21

As China's seaboard provinces grew in wealth, the inland provinces slipped behind noticeably, and the historic contrast between the interior and the periphery reemerged (Figure 12-23). The government hopes that the inland areas will eventually benefit from the more capitalistic policies. Large investments continue to pour into coastal China, and by 1991, exports had soared to $73 billion and imports to $64 billion. In that year, for the first time since 1949, the percentage of total industrial output that came from the state sector fell below 50 percent. The country's aged and debt-ridden state-owned industries contribute progressively less to national output each year. Many survive only on central bank loans, and mounting debt threatens bankruptcy. China hopes to privatize many of these industries.

As a result of China's slowing rate of population growth (see Chapter 4), its surge in food production (see Chapter 6), and foreign investment, China has achieved a rate of economic growth estimated at 6 or 7 percent per year. This has come about with repressive government, and any number of factors might slow or halt this development, including political disruption, environmental disruption, an energy supply crisis, or inadequacies in China's infrastructure. If the growth continues, however, the population of China, a full 22 percent of the human species long living in poverty, might achieve considerable material comfort within 50 years. This is one of the most important "news items" of the late twentieth century.

448 *Chapter 12*

Variations in Per Capita Income in China 1989

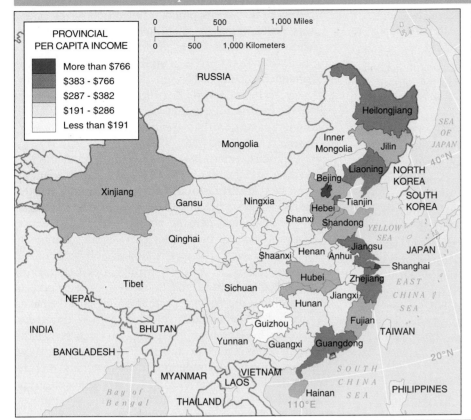

PROVINCIAL PER CAPITA INCOME

- More than $766
- $383 - $766
- $287 - $382
- $191 - $286
- Less than $191

FIGURE 12-23

In 1989 overall per capita income in China was $383, but regional income gaps were widening. (Source: Asian Development Bank.)

Nevertheless, China's leaders delicately balance trade against self-sufficiency, and the interests of inland regions against coastal regions. If leaders fear that control is slipping out of their hands, they may still be willing to repress the population and shock and offend world opinion, as they did in 1989 (see Chapter 10). The residents of Hong Kong have been promised that they will be allowed to continue living under a capitalist system after Hong Kong is merged back with China in 1997, but the Chinese government may choose to suppress capitalism again.

Taiwan Taiwan is a principal source of investment funds for China. The 20 million people of this island-nation enjoy a per capita income of almost $8,000 per year, contrasted to under $500 in China. This contrast underlines a bitter political division.

In the 1930s, a civil war raged in China between the Communists, led by Mao, and the Kuomintang, a political party led by General Chiang Kai-shek (1887–1975). In 1949, the Communists drove the Kuomintang offshore to Taiwan, where it established a separate government. The ethnic Chinese form a minority among the native Taiwanese. In Taiwan state-directed capitalism has achieved phenomenal economic growth, and recently Taiwanese have invested and transferred low-wage, low-technology manufacturing such as shoes and plastic items to China.

Both the government in Beijing and that on Taiwan claim that there is only one China and that they represent it, but at present China and Taiwan are in fact two separate countries. The United Nations General Assembly ousted the Taiwan government and seated the People's Republic in its place in 1971, and the United States recognized the People's Republic as the sole government of China in 1978. Some political accommodation will probably be reached in the 1990s between the two governments. The two might agree on joint economic ventures and joint representation in foreign affairs, for instance, without completely merging their domestic governments. Alternatively, Taiwan might negotiate virtual independence. The economic integration of Hong Kong, Taiwan, China, and the considerable economic vitality of the Overseas Chinese is proceeding rapidly.

THE FORMATION OF THE GLOBAL ECONOMY

▼▼▼▼▼▼▼▼▼▼▼▼▼▼▼▼▼▼▼▼▼▼▼▼▼

International trade has been carried on for millenia, but since new means of transportation and communication began to bind the world more closely in the late nineteenth century, trade has multiplied faster than total world production almost without interruption. In 1991, it totaled almost $7 trillion (Figure 12-24). This reflects increasing regional specialization and exchange.

Trade has accelerated so fast and multiplied to such a degree that international management consultant Peter Drucker has written, "The world economy has become a reality, and one largely separate from national economies. The world economy strongly affects national economies; in extreme circumstances it controls them." Chapter 6 noted the increasing share of food production that is entering world trade, but national protectionism has stymied the creation of a free global market for foods. A global market for manufactured goods is evolving, and today states are negotiating to create a global market for services.

Capital and Commodities

Two developments in economic globalization accelerated in the late nineteenth century. One was the evo-

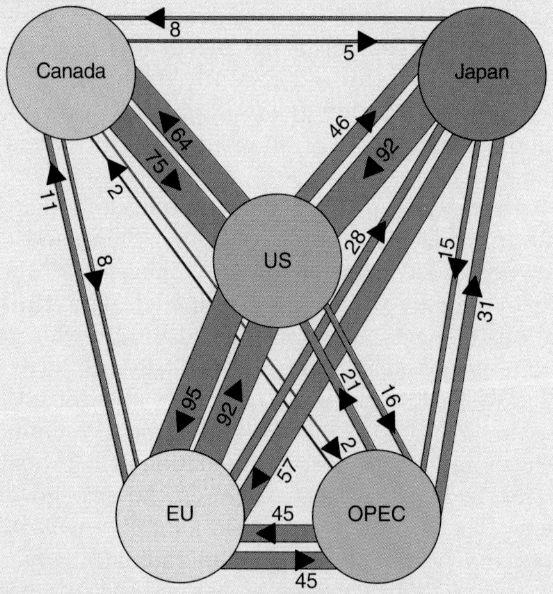

FIGURE 12-24

International trade among selected countries or blocs, in billions of dollars. (Data from *UNCTAD Handbook of International Trade and Development Statistics, 1992.*)

lution of a world market for certain primary products, foods, and minerals. Chapters 6 and 11 traced the results of that development. The second development was accelerating international flows of investment. These investments were at first limited to shares of stock and bonds representing minority holdings in foreign companies. Foreign direct investment, however, soon followed. **Foreign direct investment (FDI)** means investment by foreigners in wholly-owned factories that are operated by the foreign owner.

Today, enterprises that own and coordinate production and marketing facilities in several countries are called **multinational** or **transnational corporations**. Their evolution generally followed four stages. In the first stage, demand abroad was met by export of the item from the corporation's home country. In the second stage, the corporation established production facilities abroad to supply specific markets abroad; exports from the home country dropped. In stage three, the foreign production facilities began to supply foreign markets other than their own local markets. In stage four, the foreign production facilities began to export back into the home country. These stages appeared in trade in primary materials—copper and oil, for example; they were later repeated in the evolution of international manufacturing, and they are now being repeated again in the tertiary sector.

Multinational Manufacturing

Modern transnational manufacturing was launched in 1865, when the German chemical magnate Alfred Bayer built a factory in Albany, New York. Today technology allows the integration of manufacturing processes worldwide. For example, in the spring of 1990, a model paraded a prototype of a new fashion on a runway in Paris. Observers working for The Limited, America's largest retail garment chain, sketched the garment and faxed the sketch from Paris to a factory in Foshan, China. There an inexpensive copy of the garment was mass-produced, and this was then distributed to The Limited stores throughout the United States within 1,000 hours. This was weeks before the original designer had actually produced the garment.

The ownership and range of activities of transnational corporations are so widespread that it is difficult to tell which corporations are corporate citizens of which countries. Among the largest exporters from Taiwan, for example, are AT&T, RCA, and Texas Instruments. These are generally considered U.S. corporations, yet exports of these firms from Taiwan to the United States account for a good share of the U.S. trade deficit with Taiwan. The United

The Manufacture of Automobiles for the U.S. Market

The manufacture of automobiles in the United States demonstrates FDI to capture the U.S. market. Automobile manufacturing companies that are headquartered in the United States, Japan, Germany, South Korea, and a few other countries operate manufacturing, assembling, and marketing facilities around the world. For foreign firms, manufacturing in the United States substitutes for exporting cars to the United States. The amount a manufacturer invests is determined by the relative costs of manufacturing in the United States compared to other international locations, and also by the U.S. tariffs on imported automobiles. In 1982, the Japanese Honda corporation opened an assembly plant in Marysville, Ohio, and by 1990, that plant assembled more cars than any other in the United States.

In 1989, the Honda Accord became the first foreign car to become the number-one selling model in the United States. That car was a hybrid built by workers in both Japan and the United States, and it was exported around the world from both countries. In 1990, cars made in U.S. factories by Japanese manufacturers held about 30 percent of the total U.S. market, and slightly more than one-half of all Japanese nameplate cars sold in the United States had been made in the United States.

Many transnational car manufacturers either own substantial percentages of other transnational car companies or have partnerships for the manufacture or marketing of particular products in particular countries. Therefore, today a car's nameplate does not reveal which flag the corporation flies, where the corporation's stockholders live, or where the car was either designed or manufactured.

As long as U.S. residents buy millions of cars each year, thousands of U.S. workers will probably work in auto-manufacturing plants in the United States. We cannot predict, however, what the names of the manufacturing companies will be or who will own them. It is possible that by the year 2000 General Motors, Ford, and Chrysler will not be manufacturing cars in the United States, but that they will have moved all of their production to foreign countries, and that all domestic production will be performed by "foreign" firms.

This Honda factory in Marysville, Ohio, has brought new management techniques and technologies to the United States. Here steel parts are stamped for assembly. (Courtesy of Andy Snow/SABA.)

States takes the position that any company that employs and trains U.S. workers and that adds value in the United States is a U.S. corporation, no matter which flag it flies at its international headquarters or where those headquarters are.

The Geography of Foreign Direct Investment

Foreign direct investment is today many times the amount of foreign aid being given, and it has been much more successful in triggering economic growth. In the 1980s, global FDI grew three times faster than world trade and four times faster than total world output, to a total of about $2 trillion. This flow of FDI greatly increased and significantly redistributed the world's productive capacity.

Companies from an increasing number of countries have become active investors in other countries. Transnational corporations from Brazil, Malaysia, Indonesia, Mexico, and several other countries reached around the world. In 1989, however, the United States, the combined European Union countries, and Japan still accounted for 81 percent of FDI. The United Nations refers to these three sources as *the Triad*.

In 1989, U.S. firms accounted for 30 percent of global FDI, and the United States was also the world's largest recipient of FDI from other countries (Figure 12-25). The list of foreign firms in the United States lengthens daily. They account for 11 percent of U.S. manufacturing jobs and are creating more new ones than U.S. manufacturers are, and they are among the leading exporters of goods from the United States (about 23 percent of total exports in 1993). They are

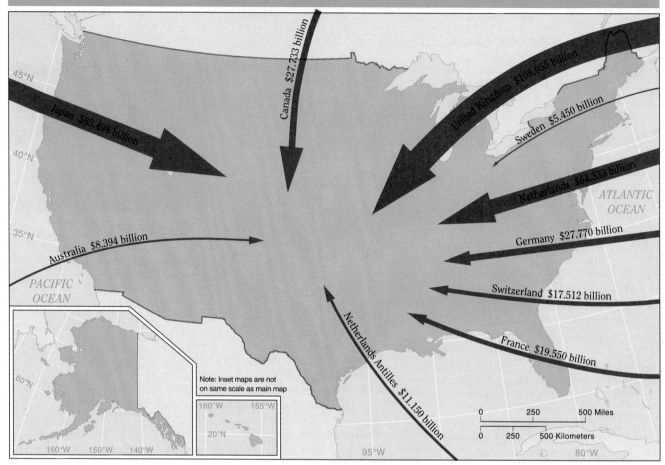

FIGURE 12-25

These were the countries with the largest direct investments in the United States at the end of 1990. (Data from U.S. Department of Commerce.)

carrying on research, introducing new technology and management techniques, and changing Americans' daily lives in the workplace as well as Americans' consumption habits.

FDI in the Developing Countries Three trends characterize FDI in the developing countries. First, the share of FDI that the Triad is allocating to the poor countries is declining. The rich countries are increasingly investing in each other.

Second, FDI has become increasingly geographically selective. During the 1980s, the countries that attracted most investment were countries that chose the export-led method of economic growth. Their economies boomed. Four Asian nations attracted FDI and grew so fast during the 1980s that they earned the nickname "The Four Tigers": Singapore, South Korea, Taiwan, and Hong Kong (technically a colony

of Great Britain, but considered a country in international trade statistics). These areas reached European levels of prosperity within a decade, and they have already become sources of capital and technology for the next tier of Asian developing lands: Thailand, China, Malaysia, and Indonesia. Indonesia, for example, long relied on oil as its principal export, but between 1983 and 1993 FDI in Indonesian manufacturing allowed the country to boost the share of its exports from nonenergy goods from 20 percent to over 60 percent. Vietnam and other nations now court investment. Even India, which has been committed to import-substitution growth since 1947, has begun to seek foreign investment.

A third characteristic of the pattern is that each Triad member has a majority share of the FDI in a cluster of countries that have become its economic satellites. These clusters help explain world patterns of trade. For example, in recent years the United

States has had a trade deficit in electronic consumer goods with Thailand. Table 12-5 shows that Thailand falls within the Japanese cluster; Thai electronics factories are mostly owned by Japanese corporations. Thus, the consumer electronics that the United States imported from Thailand profited the Japanese corporations as well as the Thai economy. In a similar fashion, many of Europe's imports from Latin America are the products of U.S. corporations.

Developing countries compete to attract investment, and even rich countries that fail to maintain roads, education, and other national infrastructure, may fail to attract new investment and lose their manufacturing base. Geographers watch capital flows carefully, because the pace of development within individual countries is increasingly determined by their ability to attract foreign capital. The world map of wealth continues to change kaleidoscopically.

Some observers have hypothesized additional ramifications of the rise in global FDI. In order to attract private investment, in contrast to government aid or even bank loans, governments must openly reveal national economic and social conditions. Furthermore, consistency in the rule of law is obligatory. Therefore, perhaps, the substitution of private investment for foreign aid and bank loans may have a substantial liberalizing and democratizing influence worldwide.

The International Tertiary Sector

When most people hear the words "foreign trade," they think of oil and iron, bananas and coffee, cars and clothes being shipped around the world. It is easy to imagine such "things." The international economy, however, has transcended trade in primary products and manufactured goods. Trade in the tertiary sector, in services, approached $1 trillion in 1991, and this is growing faster than trade in manufactured goods. Services are also taking a growing share of international investment.

In many cases corporations that provide services to manufacturing corporations in their homelands have followed these out into the world arena (Figure 12-26). This partly explains the internationalization of legal counsel, business consulting, accounting, and advertising (Figure 12-27). Many professionals, such as architects and physicians, market their skills around the world. For many nations, welcoming tourists has become a major element of their economies. So has welcoming foreign students to a nation's schools. Financial services, entertainment, and still other services are produced for global markets.

TABLE 12-5
Triad Foreign Investment Clusters

United States	European Union	Japan
Argentina	Brazil	Hong Kong
Bolivia	CIS states	South Korea
Chile	Croatia	Thailand
Colombia	Czechoslovakia	
Mexico	Hungary	
Panama	Poland	
Philippines	Slovenia	
Saudi Arabia	Yugoslavia	
Venezuela		

Note: Clusters identified by the UN Centre for the Study of Transnational Corporations, 1991. Czechoslovakia and Yugoslavia have each dissolved into several new states, but the new states remain within the designated spheres of economic influence.

FIGURE 12-26

Electronic Data Systems Corporation coordinates worldwide computer consulting services from this main operations room in Texas.
(Courtesy of Electronic Data Systems.)

One Firm's Accounting Offices Throughout the World

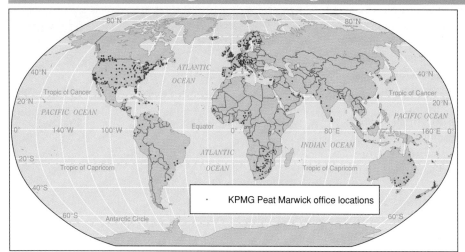

FIGURE 12-27

KPMG Peat Marwick, a joint Dutch and U.S. firm, is the world's largest accounting firm, with offices in all the major economic capitals. (Courtesy of KPMG Peat Marwick.)

International services demonstrate the same trends as trade in goods. Countries are losing their autonomy to an increasingly global market. Transnational corporations treat the world as "one place" and exploit resources, locate facilities, and market their products accordingly. The United States is the world's leading exporter of services, and the nation enjoys regular trade surpluses in them. In education, for example, the United States earned a net $4.3 billion in 1990. Other U.S. tertiary exports include real estate development and management, accounting, medical care, business consulting, computer software development, legal services, advertising, security and commodity brokerage, and architectural design (Figure 12-28).

United States entertainment productions dominate the world's airwaves and movie screens. Movies, music, television programs, and home videos together account for a significant annual trade surplus.

The Globalization of Finance

The FDI that is redrawing the world map of manufacturing is only a fraction of the vast amounts of capital moving around the world. A single global capital market has formed. Multinational banking has developed, and telecommunications make it possible to monitor and trade in national currencies, stocks, and bonds

FIGURE 12-28

This stunning glass pyramid is the new entrance to the Louvre museum in Paris. It was designed by the architect I.M. Pei, a U.S. citizen who was born in China but lives in New York. The same year this building was opened, Mr. Pei opened a new symphony hall in Dallas and a 70-story skyscraper for the Bank of China in Hong Kong. Architectural design has become an important export from the United States.
(Courtesy of Pei Cobb Freed & Partners/Diede von Schawen.)

GEOGRAPHIC REASONING
Selling U.S. Culture

Questions

1. Is it bad, good, or a matter of no importance that U.S. film studios grow more sensitive to foreign sensibilities and tastes? Could this make U.S. moviegoers more cosmopolitan?

2. Should the U.S. government protect culture producers and forbid their sale to foreign interests?

Many U.S. citizens are flattered and enriched if U.S. culture products are sold abroad, but some observers are troubled by two new considerations about this trade. One is the question of whether increasing dependence on foreign markets will affect what is produced. This would affect what Americans have an opportunity to see themselves. The second question is whether U.S. citizens should sell the corporations that actually produce the culture. Is the United States selling its cultural birthright?

Studies have found that if a movie studio, for example, is owned by a foreign investor, or if a significant share of the studio's profits come from showing its work abroad, then the treatment of foreign countries and their cultures is notably more sympathetic than when the ownership and intended audience was restricted to the United States. Furthermore, it has been noted that the production of films with sophisticated dialogue has generally given way to comedies with easily accessible visual humor and to simple action pictures. It is difficult to measure exactly to what degree this may be happening to increase foreign sales, but it is changing the nature of the film entertainment available to U.S. audiences.

By 1990, 50 percent of the revenues of America's seven largest movie studios came from abroad. That percentage was rising rapidly, and foreign countries present the greatest opportunity for increasing sales. By 1991 the four largest movie studios in the United States were foreign-owned: MGM/UA, owned by an Italian corporation; Twentieth Century Fox, Australian; Columbia Pictures, Japanese; and Universal Studios, Japanese. New films made by the Disney Corporation are also financed by Japanese investors. Will Matsushita Corporation, the Japanese owner of Universal Studios, let the studio make a movie about World War II? When the chairman of the board of Matsushita was asked, he said he could not answer.

---

listed anywhere in the world instantaneously. By 1992, over $600 billion was traded daily on the world's foreign exchange markets. This is over 100 times the average daily merchandise trade, and during that year an additional $5 trillion worth of stock changed hands on the world's major stock exchanges. The International Organization of Securities Commissions (IOSC) monitors this activity, but no international agency records ownership, even by national source, of these assets.

Nations are expected to regulate their own stock exchanges, and the home countries of global banks are responsible for supervising the banks' operations, but countries do not agree on how to do these things. Many giant banks have located their international headquarters in places renowned for lenient banking regulations. The 100 square mile (260 km^2) British Caribbean colony of the Cayman Islands (population 25,000) is one of the world's greatest financial centers, holding assets of over $400 billion in 1992. It has, however, almost no vaults, tellers, security guards, or even bank buildings. Assets are held electronically in computers. Many banks consist of one-room offices, and local laws protect transactions' confidentiality. For example, international agencies cannot trace the profits of the international drug trade.

Today Japan is one of the world's principal sources of capital, so the regulations governing the Tokyo stock market affect investments and jobs throughout the world. The Arab oil states are another source of capital, so a costly Mideast war raises the interest rate on U.S. college tuition loans. Bank policies in Frankfurt, Germany, and Johannesburg, South Africa, affect the checking account charges of U.S.

banks. Financial manipulations in Rio de Janeiro and Bombay affect the security of many U.S. elderly people's pensions, which have been invested abroad, and the interest rates set at the Federal Reserve Bank in Minneapolis affect business taxes in Indonesia. The explanations of the "where" and "why there" of local economic affairs may lie in global finance.

Tourism

Tourism is, by some accounts, the world's largest industry. The World Travel and Tourism Council estimates 1991 travel and tourism activity to have generated $3.4 trillion in gross output, to have employed 183 million people, and to have produced 10.2 percent of world GDP. Tourism is the top earner of foreign exchange for many individual countries, and tourism advances many countries' sectoral evolution. In 1991, tourism became the largest export industry of even the United States, earning a $10.6 billion surplus. Tourism is expected to grow everywhere by at least 5 percent per year, with even higher rates in the poor countries.

Citizens of the rich countries account for the largest tourist expenditures, and the rich countries are also the most popular destinations (Figure 12-29). When income from tourism is measured as a share of total national exports, however, the importance of tourist income to the poor countries becomes clear. Turkey, for example, quadrupled the value of its

exports of goods between 1980 and 1989, but the value of its tourist revenues increased twice as quickly. By 1989, Turkey's revenues from tourism were 22 percent of the value of Turkey's exports of goods. Thailand was the world's fastest growing economy from 1988 to 1990, and despite Thailand's rapid industrialization in those 3 years, tourism grew faster than trade. By 1990, it represented 6 percent of Thailand's GDP, and it was the country's leading foreign exchange earner.

The tourist potential of any spot depends upon its "Three A's:" accessibility, accommodations, and attractions. Many tourists find sunshine and a beach sufficient attraction (Figure 12-30). Recreational opportunities draw millions more to mountains. Tropical regions or areas spectacularly endowed with wildlife may feature **ecotourism**, which is travel to see distinctive examples of scenery, unusual natural environments, or wildlife (Figure 12-31). The rich bird life on the islands of Lake Nicaragua, the lions of Kenya, and the gorillas of Rwanda are important tourist destinations. Ecotourism today provides the money needed to save many countries' natural environments as well as many species of rare animals.

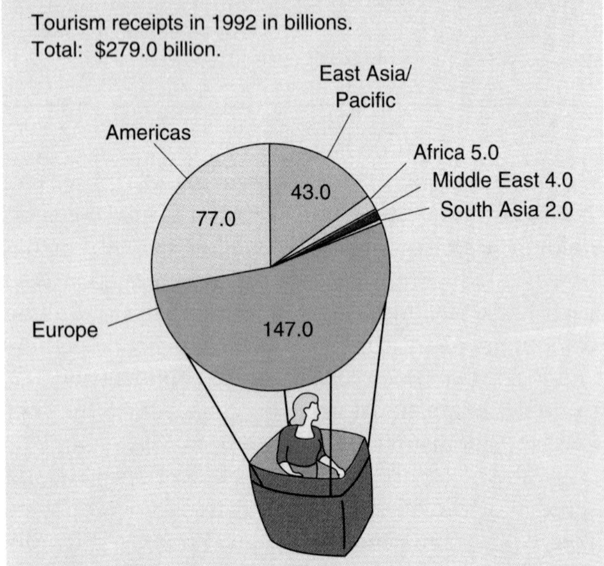

FIGURE 12-30

In the late 1940s, the Mexican government chose Acapulco to be the first focus of government-promoted tourism. The city enjoys year-round sunshine and temperatures of around 80°F (27°C), and sandy beaches stretch for 10 miles (16 km) along its coastline. Luxury hotels are set against a majestic backdrop of mountains and evergreen tropical vegetation. Mexico's foreign exchange earnings from tourism tripled from 1945 to 1950 and are still among the world's highest today. Acapulco was an old port city. From 1565 until 1815, galleons sailed from here annually to Manila in the Philippines, exchanging silver for porcelain, spices, and silks.
(Courtesy of the Mexican Tourist Office.)

Tourism receipts in 1992 in billions.
Total: $279.0 billion.

East Asia/Pacific 43.0
Americas 77.0
Africa 5.0
Middle East 4.0
South Asia 2.0
Europe 147.0

FIGURE 12-29

The rich countries enjoy the world's highest tourism receipts, but tourist revenues substantially boost per capita incomes in many poor countries.

FIGURE 12-31

The government of Belize depends on income from ecotourism to protect this rare jaguar. (Courtesy of Jeanne White/Photo Researchers.)

Overall, the countries best endowed for tourism have both natural and cultural attractions, pleasant climates, good beaches, and reasonably well-educated populations. Political stability is a necessity.

Benefits and Drawbacks of Tourism Tourism creates many jobs and offers wide multiplier effects (see Chapter 9). For example, a new hotel creates demand for a host of related goods and services. Tourism may also stimulate the local agricultural economy, and it is generally a high-value-added and labor-intensive industry that trains local labor. Some things that tourists need must be imported, but as a general rule, the more developed and complex the rest of the economy is, the greater the economic gains from tourism.

Tourism does have drawbacks for the host country. Chapter 3 noted that tourism may corrupt local culture. It may also encourage prostitution and provoke inflation. Too many tourists may overwhelm the natural environment.

The Global Office

Many developing countries export clerical work. They have educated young people willing to work for wages lower than those in the rich countries, but their domestic economies cannot employ them. Large corporations need clerical assistants and computer operators, and telecommunications allows the corporations to tap these pools of labor wherever they are. United States companies have been moving clerical operations out of urban centers for decades. Now they can relocate those jobs around the world. New Yorkers' bank credit card charges are processed in South Dakota.

Why not in Paraguay? The Boeing and Bechtel multinational corporations have located software development facilities that provide many jobs in Ireland. Why not in Costa Rica, the Sudan, or Madagascar?

The relocation of these jobs can boost national economies and can also spread cultural values and practices to other countries. By the year 2000 your auto insurance claim may be processed in India or Egypt; these countries are importing computer hardware and diligently training software professionals. The time and cost distance to these places is, for electronic transmissions, negligible.

Governing International Trade

The rules regulating business vary so widely from nation to nation that the terms of trade between countries require political regulation. The United Nations first proposed an International Trade Organization in 1947, and a preliminary General Agreement on Tariffs and Trade (GATT) was signed in that year by 23 countries. GATT provided an outline of rules, but the organization to elaborate and enforce the rules was never created. Therefore, GATT, which was intended as an interim arrangement, long remained as the only instrument laying down internationally accepted trade rules. By 1994, 125 countries had signed the agreement, and most others followed GATT rules. Under GATT procedures, trade disputes lead to the convening of panels of neutral specialists at GATT headquarters in Geneva.

GATT was created to cover world trade in raw materials and manufactured goods, and it was never extended to cover trade in farm products and in services, even as trade in them grew to represent about one-third of all world trade by 1990. GATT also failed to provide protection for intellectual property, such as patents and copyrights. United States firms claim to lose over $60 billion per year because of foreign copying of their intellectual property such as books, videos, and pharmeceuticals without paying fees. Many nations felt that GATT had to be expanded to include these items of trade.

From 1986 through 1993, a round of GATT negotiations (the eighth, known as the "Uruguay Round" after the country where the talks were first convened) met to define terms of international trade in these items. The final 22,000 page agreement lowered many remaining trade barriers and defined a new organization, the World Trade Organization (WTO), to replace GATT on January 1, 1995. The new agreement tightened rules to protect intellectual property and liberalized trade in banking, insurance, and tourism, but fell short of the goal of negotiating new agreements on foods, intellectual property (including cultural products), civil aircraft, and financial services. Countries cannot agree on the degree to which a country's labor standards—the rules it enacts and enforces to protect the rights and advance the interests of working people—should be open to international scrutiny or negotiation. Some nations fear that the WTO will encroach on national sovereignty by defining manufacturing standards, environmental and labor regulations, and other *domestic* conditions in its efforts to "harmonize" conditions of trade. Negotiations will continue on these issues.

Other international organizations oversee specific aspects of international commerce, such as the IOSC mentioned above. The London-based International Accounting Standards Committee tries to harmonize corporate accounting procedures. The International Labor Organization (ILO) works to improve international labor conditions, and a few international trade union confederations exist, but none exercises significant power or influence.

SUMMARY

Every country devises a program to build and to regulate its national economy and to distribute economic well-being across the national territory. Gross national product, gross domestic product, and gross sustainable product are three ways of measuring wealth, although each undervalues subsistence production

and unpaid labor. Maps of these values reveal great differences in global incomes. The Human Development Index reveals standards of living. Welfare may vary greatly within countries, and statistics examined per capita and through time may reveal a relationship between population growth and economic growth.

Sectoral evolution reveals that some countries are pre-industrial, some industrial, and some post-industrial. Various sectors' contribution to national product is not the same as their share of employment.

The world distribution of wealth is not the same as the distribution of raw material resources. Downstream secondary and tertiary sector activities add value and produce wealth. Today's map of global manufacturing is a result of historical developments modified by new balances among relevant locational determinants, including capital and technology, as well as labor and raw materials. Several locational determinants are themselves footloose, so manufacturing will probably be redistributed continually.

The term Third World combines economic measures with cold war political values. Political events have reawakened our attention to differences among poor countries.

Each nation's political economy usually reflects other aspects of local culture. Nations try to knit and develop the national territory and to maintain an equal standard of living throughout it. Transportation networks allow regional specialization and trade, and their improvement releases national resources for production. Communication infrastructures both reflect and boost economic development as well as, potentially, political liberty.

Import-substitution policies protect national growth behind tariff walls, whereas export-led policies encourage participation in international trade as a source of capital and as markets. Both methods entail both economic risks and political risks.

International trade has multiplied faster than total world production. Capital has been invested internationally, and world markets for primary products, foods, and minerals evolved. Today, manufacturing industries are achieving globalization, and flows of FDI are redistributing world productive capacity. The United States is both the largest source of FDI and also the largest recipient. Each Triad member dominates FDI in a cluster of countries. A global capital market has formed, and international trade in tertiary sector services is growing. Tourism, including ecotourism, is today arguably the world's largest industry. GATT's role in defining international trade rules is to be replaced by a new World Trade Organization in 1995.

KEY TERMS

globalization (p. 418)

gross domestic
product (GDP) (p. 418)

gross national
product (GNP) (p. 419)

gross sustainable
product (GSP) (p. 420)

human development
index (HDI) (p. 422)

sectoral evolution (p. 424)

pre-industrial society (p. 424)

industrial society (p. 424)

post-industrial society (p. 425)

downstream activities (p. 426)

material-oriented
manufacturing (p. 428)

market-oriented
manufacturing (p. 428)

opportunity cost (p. 434)

First World (p. 436)

Second World (p. 436)

Third World (p. 436)

political economy (p. 438)

laissez-faire
capitalism (p. 438)

state-directed
capitalism (p. 438)

frontier (p. 439)

tap route (p. 442)

infant industries (p. 443)

economies of scale (p. 444)

import-substitution
economic growth (p. 445)

export-led economic
growth (p. 446)

foreign direct
investment (FDI) (p. 450)

multinational
or transnational
corporations (p. 450)

ecotourism (p. 456)

QUESTIONS FOR INVESTIGATION AND DISCUSSION

1.   At the opening of international currency markets on January 7, 1994, the following values were quoted against the U.S. dollar:

Australian dollar	0.6849
Brazilian cruzeiro	0.0031
British pound	1.48
German mark	0.5732
Indian rupee	0.0319
Japanese yen	0.008869
South African rand	0.2946

Check these values in today's newspaper. Calculate the percentage each has risen or fallen against the dollar or against one another. What flows of goods, of investment, or of tourists should these fluctuations have triggered?

2.   From the *United Nations Demographic Yearbook*, pick your own list of three or four variables to compare the quality of life in various countries.

3.   Can you hypothesize an explanation for the variations in infant mortality recorded across the United States?

4.   Did the United States develop in a sustainable way or at a sustainable rate between 1870 and 1910? Does it today?

5.   Could foreign ownership of businesses hurt the economy of a developing country?

6.   What countries have frontiers? What barriers exist to development of those areas? How might these barriers be overcome? Investigate and compare the real development potential of the frontiers of Canada, Russia, and a South American or African country.

7.   What physical and cultural attributes are most conducive to economic development?

8.   Compare Figures 10-31 (world literacy) and 10-32 (world freedom) with Figure 12-3 (GNP per capita). Can you hypothesize any cause-and-effect relationships?

9.   In today's small world, international developments carry widespread ramifications. What effect did the fall of Eastern European Communist governments have on Africa? On Latin America? Why did Nigeria's Foreign Minister say, "The Cold War is over, and Africa lost"?

ADDITIONAL READING

AHMAD, YOSUF and SALAH EL SERAFY, ERNST LUTZ, eds. *Environmental Accounting For Sustainable Development.* Washington, D.C.: The World Bank, 1992.

CRUICKSHANK, J. "The Rise and Fall of the Third World: A Concept Whose Time Has Passed." *World Review*, February 1991.

GINSBURG, NORTON. "Natural Resources and Economic Development." *Annals of the Association of American Geographers* 47 (3): 197-212.

HARRISON, LAWRENCE. *Who Prospers? How Cultural Values Shape Economic and Political Success.* New York: Basic Books, 1992.

The International Bank for Reconstruction and Development. *World Development Report.* New York: Oxford University Press, annually for the World Bank.

The International Bank for Reconstruction and Development. *The East Asian Miracle: Economic Growth and Public Policy.* New York: Oxford University Press for the World Bank, 1993.

MABOGUNJE, AKIN. *The Development Process: A Spatial Perspective.* London: Unwin Hyman, second edition 1989.

OMARA-OJUNGU, and PETER HASTINGS. *Resource Management in Developing Countries*. London: Longman, 1992.

OWEN, WILFRED. *Transportation and World Development*. Baltimore, MD: Johns Hopkins University Press, 1987.

PEET, RICHARD and NIGEL THRIFT, eds. *New Models in Geography: The Political Economy Perspective*. London: Unwin Hyman, 1989, 2 vols.

REICH, ROBERT. *The Work of Nations*. New York: Alfred Knopf, 1991. The 1992 Vintage Paperback has an addendum.

United Nations Conference on Trade and Development. *Handbook of International Trade and Development Statistics*. New York: United Nations, annually.

The World Bank. *The World Bank Atlas*. Washington, D.C., annually.

13 International Organization: Regional Organizations and Global Coordination

The opening ceremonies for the 1994 winter Olympics at Lillehammer, Norway, included a colorful parade of national flags and contestants in national costumes. The Games' purpose is to foster international understanding through athletics, although this often seems overwhelmed by nationalistic displays. The International Olympic Committee is an independent international body.

*I*n 1522, the Spanish explorer Sebastian de El Cano completed the circumnavigation of the globe that had begun under the command of Ferdinand Magellan, who had been killed in the Philippines. This voyage proved that the world was one spherical stage for human activity. The many local dramas being played on that stage, however, were widely separated, little related to one another, or even totally oblivious of one another. Political and economic events remained focused at the local level.

The European voyages of exploration connected all peoples, and accelerating improvements in transportation and communication seemed to shrink the world. The British geographer Sir Halford Mackinder identified the period from 1492 until his time (1904) as "the Columbian Era," during which the world became connected. During the "post-Columbian era," he wrote, the Earth would become "a closed political system...of worldwide scope. Every explosion of social forces...will be sharply re-echoed from the far side of the globe."

This post-Columbian era has so far witnessed two contradictory trends. One has been the shattering of European empires into a great number of new sovereign states (Figure 13-1). At the same time, however, the globalization of activities and the increasing interdependence of societies and economies have restricted countries' freedom of action. We have already seen international ramifications of any nation's environmental policies, continuing migrations, world markets for food, minerals, manufactured goods, and capital. Frameworks of international collaboration must be designed to regulate these activities.

Individual countries remain the highest level of absolute territorial sovereignty. Organizations that coordinate activities among two or more countries are called **international organizations**, whereas organizations that actually exercise power over countries, by contrast, are called **supranational organizations**. Countries surrender some powers to supranational organizations, but each country can escape that control—by making war, if necessary. Therefore, the allocation of powers in either international or supranational organizations is actually flexible and open to continual renegotiation. A *commonwealth* is another kind of international organization that exercises limited powers over its members.

International and supranational organizations are not new in history. Alliances between independent governments have always been made, including efforts to coordinate governmental policies. Today some international organizations coordinate military defense against common enemies. Today's most active international and supranational organizations, however, work for economic or broader cultural goals. They

are trying to achieve across several countries many of the same things that sovereign states try to achieve within their individual borders. These goals include creating unified markets for materials, goods, and services; knitting areawide transport networks; ensuring free travel throughout the territory; and even guaranteeing common standards of political or civil rights.

A group of countries that presents a common policy when dealing with other countries is a **bloc**. Most blocs are regional, and they form an intermediate level of territorial organization between the individual states and global organizations. The new World Trade Organization discussed in Chapter 12 includes most of the world's countries, but its goals are restricted to regulating trade. The United Nations organization, by contrast, attempts to coordinate global action on a wide variety of issues.

This chapter starts by examining three regional blocs. The three examined are particularly significant because the participants have great economic and military power and because in each case the members have surrendered considerable autonomy to a central body or at least to a common policy. Two blocs—the European Union and the North American Free Trade Area—are coordinating or harmonizing increasing numbers of activities across regions. New markets are opened to manufacturers, and new opportunities are opened for travel, education, and the diffusion of goods, ideas, and people. The third regional bloc examined—the Commonwealth of Independent States (CIS)—is evolving to replace the Union of Soviet Socialist Republics. That Russian-dominated multinational empire centralized political and economic organization across a full one-sixth of the Earth's land surface. Commonwealth government and policies will emerge through the 1990s, but the CIS framework probably will not centralize control across this vast territory as strictly as the government of the former Soviet Union did.

After reviewing regional international blocs, this chapter investigates a few problems of coordination and cooperation that confront the entire global community and describes the organizations that exist to resolve international issues.

THE EUROPEAN UNION: INTERNATIONAL TERRITORY BEING KNIT TOGETHER

▼▼▼▼▼▼▼▼▼▼▼▼▼▼▼▼▼▼▼▼▼▼▼▼▼

World War II ended in Europe in 1945 with the surrender of the Axis powers: Germany, Italy, Romania, Bulgaria, and Hungary. Japan had been an Axis Asian ally. The principal victorious Allied powers were the

The World in 1900

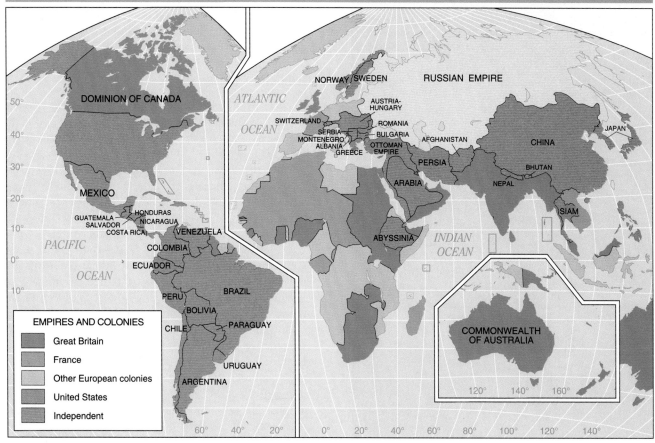

FIGURE 13-1

In 1900 there were far fewer independent actors in international affairs than there are today. Compare this map with today's political map on this book's front endpaper.

United States, Great Britain, Canada, France, the Union of Soviet Socialist Republics, and China. In Europe the Allies had squeezed the Axis powers from West and East, and their victorious armies met in the middle of the continent.

The Formation of Blocs in Europe

The Allies themselves, however, were already divided by mistrust, and Europe soon split into two competitive blocs. British Prime Minister Winston Churchill said, "From Stettin in the Baltic to Trieste in the Adriatic, an Iron Curtain has descended across the continent." To the east, Soviet armies installed satellite regimes in Poland, Hungary, Romania, Czechoslovakia, and Bulgaria. Local Communist leaders seized power in Yugoslavia and Albania without Russian help. The constitutions of all these countries were amended to guarantee Communist party dominance in national politics, and their economies were largely collectivized.

Germany itself was split into four zones of military occupation. Already by 1949, however, the U.S., French, and British zones were united functionally, and a new Federal Republic of Germany (West Germany), with Bonn as its capital, gained sovereignty in 1955. The Soviet Union reacted by granting its occupation zone independence as the German Democratic Republic (East Germany). Germany's capital city of Berlin had also been jointly occupied, so it too was split between West and East Germany, even though it lay about 125 miles (201 km) inside East Germany. Its western sector was recognized as a part of West Germany, and East Berlin became the capital of East Germany (Figure 13-2). Austria, which had been annexed by Germany, was also split into four occupation zones, and they were reunited in a newly independent state in 1955.

The Iron Curtain across Europe followed fairly closely the far older cultural division that Chapters 7 and 8 demarcated in orthography and religion: The West was defined by Roman Catholicism or

FIGURE 13-2

West Berlin showcased Western prosperity in the middle of East Germany. As a result, so many East Berliners fled into West Berlin that in August 1961, East Germany built a wall across the city to seal in its citizens. This infamous Berlin Wall fell on November 9, 1989.
(Courtesy of R. Bossu/Sygma.)

Protestantism and the Roman script, and the East by Orthodox Christianity and the use of the Cyrillic script. The Iron Curtain deviated from this line, however, in that Estonia, Latvia, Lithuania, Poland, Eastern Germany, Czechoslovakia, Hungary, Croatia, and Slovenia had never been "Eastern European" culturally, yet they were east of the Iron Curtain.

The division of Europe into two competing blocs was furthered by Soviet seizure of territory during and after World War II (look back at Figure 10-5). In 1975, 35 countries—mostly European, but including the United States and other World War II participants—met in Helsinki, Finland, and agreed to accept European borders as inviolable. The conference did not, however, rule out revision by peaceful agreement, and European borders have been redrawn since then. In 1990, the Federal Republic of Germany

and the German Democratic Republic united and recognized Berlin as the capital of the united country. In 1991, Yugoslavia split apart into warring new states, and in 1993, Czechoslovakia split into two countries.

Military Pacts

The postwar political division of Europe was accompanied by its division into two military alliances. In the West, the **North Atlantic Treaty Organization (NATO)**, established in 1949, united Belgium, Denmark, France, Great Britain, Iceland, Italy, Luxembourg, the Netherlands, Norway, and Portugal with Canada and the United States. Greece, Turkey, West Germany, and Spain joined later. In 1955, the *Warsaw Pact* linked the USSR, Albania (which withdrew in 1968), Bulgaria, Czechoslovakia, East Germany, Hungary, Poland, and Romania. The USSR used this treaty to justify the continued occupation of Eastern European countries by hundreds of thousands of Soviet troops, at the expense of the occupied countries. These troops put down occasional anti-Communist uprisings. The Western and Eastern blocs faced one another in a protracted *cold war*. Although the two blocs never engaged directly in combat, they competed for economic growth and for influence among the new countries that gained their independence after World War II.

The pattern of alliances in Europe dissolved in 1989. The USSR directed its attention to its own internal problems and released its grip on the countries of Eastern Europe. These countries, in turn, repudiated the privileged role of the Communist party in their own countries, and within a few months they scheduled free elections. The Warsaw Pact formally dissolved in 1991.

NATO recognized the end of the Cold War in 1990. A statement declared, "The Atlantic Community must reach out to the countries of the East that were our adversaries in the Cold War, and extend to them the hand of friendship." NATO has nevertheless survived as a mutual defense and political structure, and in 1993 several members of the former Warsaw Pact actually petitioned to join NATO. A Partnership for Peace program has scheduled joint military exercises between NATO and former Warsaw Pact members—including Russia.

Economic Blocs

The Marshall Plan In 1947, the United States announced a plan of financial aid to war-shattered Europe. This plan was eventually known as the **Marshall Plan** in honor of Secretary of State George

C. Marshall (Figure 13-3). Between 1948 and 1952, the United States helped the European countries rebuild their economies. Aid was offered to the Soviet Union and to the countries of Eastern Europe, but Soviet dictator Joseph Stalin (1879–1953) rejected it. The Marshall Plan became a program for the economic rehabilitation of Western Europe.

Comecon The Soviet Union and its Eastern European satellites formed a competing economic pact, the Council for Mutual Economic Assistance (COMECON), that achieved a partial integration of their national economies. COMECON was the only case in history in which the dominant power supplied raw materials, including energy supplies, to its satellites. In 1991, COMECON formally disbanded, and the East European countries began to negotiate new economic links with the more prosperous countries of Western Europe.

Some of the East European economies grew under their Communist regimes, enough to surpass

FIGURE 13-3

George C. Marshall (1880–1959) is shown here (at right) with Mrs. Marshall and President Eisenhower in 1957, at ceremonies commemorating the tenth anniversary of the Marshall Plan. During World War II, Marshall served as Army Chief of Staff and rose to five-star rank, as in the painting behind himself. As Secretary of State (1947–1949) he devised the Marshall Plan for the recovery of Europe, for which he won the Nobel Peace prize in 1953. He also served as Secretary of Defense (1950–1951). During World War II Dwight D. Eisenhower (1890–1969) was Supreme Commander of the Allied Expeditionary Force in Europe, and then Eisenhower served as U.S. president (1953–1961). (Courtesy of the George C. Marshall Foundation.)

the Soviet Union itself in per capita measures. They did not, however, keep up with Western Europe. Czechoslovakia, for example, had been a rich country before World War II, the equal of Germany in technology and productivity per worker, and ahead of France. During World War II its industries were not damaged, yet by 1990 productivity per worker in Czechoslovakia, although higher than that in the Soviet Union, was less than one-half that of West Germany or France.

Today most of Eastern Europe's socialized industries are being returned to private ownership. This requires the formerly Communist countries to develop new stock and bond markets, private banks, and accounting systems. Under communism, Eastern European industries were inefficient and hugely overstaffed, and the new governments have struggled to make them competitive. This has triggered economic dislocations, including high unemployment. East Germany, for example, was estimated during the 1980s to be the world's eighth most powerful industrial economy. When it unified with West Germany, however, analysis of its industries revealed that most were so dilapidated, polluting, and lacking in modern infrastructure that their privatization realized just a fraction of the original estimate of their value. Collectivization boosted agricultural output in Hungary, Bulgaria, and Czechoslovakia, and these areas may retain agricultural cooperatives.

Reprivatization will also require that the new governments and the new private sectors agree on sharing responsibility for cleaning up polluted Eastern European environments (Figure 13-4). The Communist governments' industrial drives had compelled managers to meet production targets regardless of the cost or hazards to people's health or to the environment. Unbridled industrial spewing and spilling carried disastrous consequences. It will take years and enormous sums of money to clean up Eastern Europe's contaminated environment and to restore its people to health.

THE EUROPEAN UNION TODAY

The Marshall Plan encouraged economic cooperation among the European countries, and in 1952, six of the Western European nations united their industrial economies to form the European Coal and Steel Community: Belgium, the Netherlands, Luxembourg, West Germany, France, and Italy. The success of this union encouraged further cooperation, and the European Economic Community (EEC) and the European Atomic Energy Community (Euratom) came

Pollution in Eastern Europe, 1990

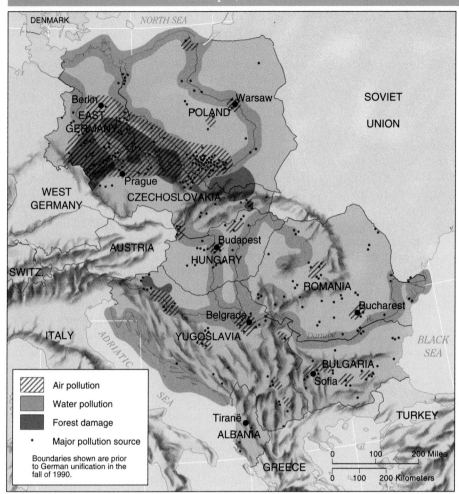

FIGURE 13-4

When Communist governments in Eastern Europe collapsed and Western observers were allowed to examine industrial facilities, they found a degree of pollution that they could scarcely believe.

into existence in 1957. All three merged into one European Community (EC) in 1967. Denmark, Ireland, and the United Kingdom joined in 1973; Greece in 1981; and Spain and Portugal in 1986 (Figure 13-5). In 1993, these 12 countries signed a new treaty in Maastricht, the Netherlands, pledging further coordination and changing the formal name of their organization to the **European Union (EU)**. The combined population of the twelve countries is almost 350 million, but four additional countries with 26 million people—Sweden, Norway, Finland, and Austria—are scheduled to join on January 1, 1995, if the people of each of the four agree in individual referenda.

There are three kinds of economic groupings. A **free trade area** has no internal tariffs, but its members are free to set their own tariffs on trade with the rest of the world. A **customs union** enforces a common external tariff. A **common market** is a still closer association of states; it is a customs union, but, in addition, common laws create similar conditions of production within all members. The EU is forming a full common market, and it even declares political merger to be an eventual goal.

The members of the European Union share many broad cultural patterns, and these cultural foundations provide a basis for their successful cooperation. Common traditions include democratic ideals and parliamentary institutions; civil rights and legal codes; Judeo-Christian ethics; respect for scientific inquiry; artistic traditions, including Classicism, Realism, and Romanticism; and humanism and individualism. In fact, cooperation within the Union has bolstered democracy in certain countries—Portugal, Spain, and Greece—that previously were only dubiously committed to it. It may be difficult to extend Union membership to some states, such as Turkey, that are pressing for entry but do not share these European traditions.

The European states have several goals. One is to end the wars that have plagued the continent. Another is economic growth, and a third reason for European unification is the urge to play a role in world affairs. Economic strength has largely replaced

The European Union and the European Free Trade Association

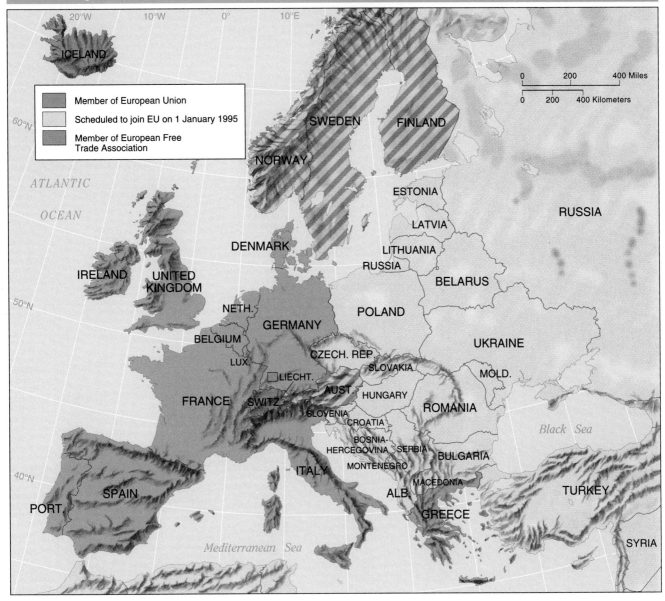

FIGURE 13-5

colonialism or military might as a measure of prestige, and no European nation alone can command prominence on the world stage. United Europe, however, can assume a leading position.

European Union Government

The Union is governed by four institutions: the European Commission, the Council of Ministers, the Parliament, and the European Court of Justice. Figure 13-6 illustrates the path of Union legislation. The European Commission proposes legislation and is responsible for administration. It sits in Brussels,

Belgium, which has enjoyed considerable growth and prosperity as de facto capital of the Union. The Council of Ministers, comprised of one minister per member nation, who represents his or her government, is the final legislative body. The 1985 Single Europe Act defined a wide range of issues on which the Council decides by majority vote. This considerably reduced individual countries' sovereignty. The European Parliament sits in Strasbourg, France. The apportionment of its 567 seats is a compromise between numerical democracy and equality among nations, but its members have been elected directly by the citizens of the member countries since 1979. In the Parliament the members who represent affiliat-

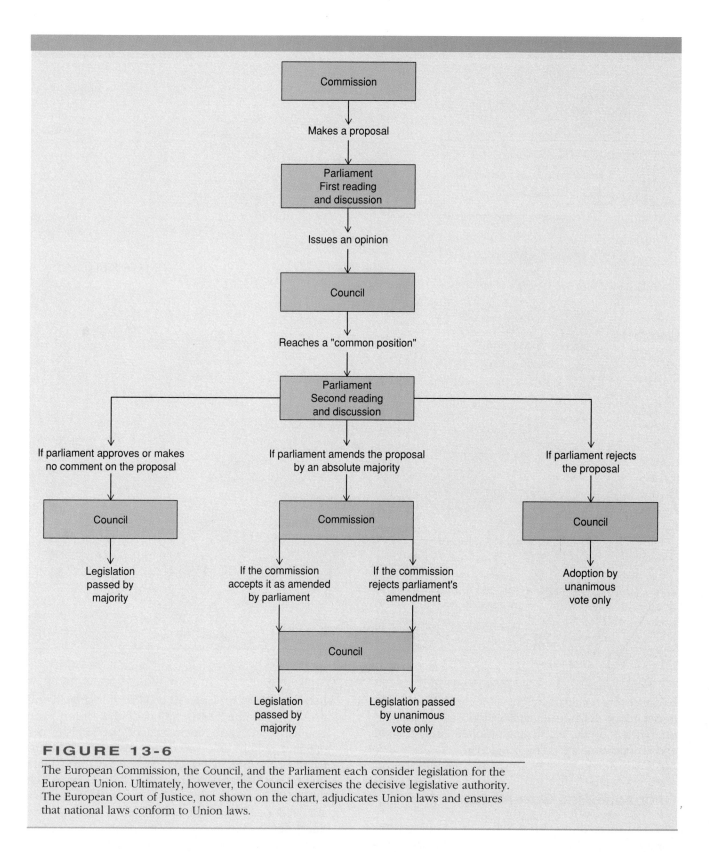

FIGURE 13-6

The European Commission, the Council, and the Parliament each consider legislation for the European Union. Ultimately, however, the Council exercises the decisive legislative authority. The European Court of Justice, not shown on the chart, adjudicates Union laws and ensures that national laws conform to Union laws.

ed political parties from different countries join together as international voting blocks. Therefore, the Parliament and to a degree the Commission are genuinely supranational.

The European Parliament is not, however, equivalent to a national legislature. It can scrutinize proposed Union laws, offer amendments, vote against directives, veto the actions of any branch of the

Union, or dismiss the European Commission, but it cannot initiate legislation, select the executive, or levy taxes. Parliament is the most democratic of Union institutions, and it has been gaining power.

European law is adjudicated by the 13-judge European Court of Justice. In cases of conflict between Union law and national laws, the supremacy of Union law is being steadily affirmed. Union institutions have become so important to the member states that, for example, the German government estimated that 50 percent of all German legislation passed in recent years was based on Union directives.

European Union Integration

The Single Europe Act of 1985 called for the creation of one economic community by the end of 1992, and this goal was largely achieved. Integration of the Union territory as "one place," economically at least, can be seen in five aspects of its evolution: (1) creation of one market for goods, (2) creation of one market for capital, (3) transportation policies, (4) social policies, and (5) regional policies.

The Creation of One Market for Goods One of the Union's first acts was to establish a common agricultural policy. Overall, however, agriculture accounts for less than 4 percent of total Union gross product. European manufacturers would like to achieve the economies of scale in one European market that U.S. manufacturers have long enjoyed in the huge U.S. domestic market. Many European firms have merged across national borders, and Union bureaucrats have standardized business regulations. Standards cover aspects of manufacturing as various as the fat content of specific food products and the mechanical tolerance allowed in gearboxes. The United States and other countries that wish to export to Europe are already adopting these European standards—even for domestic production.

The evolution of one continental market should boost economic growth and lower prices to consumers. The sacrifice required, however, is the individuality of the national cultures. Both U.S. tourists and Europeans have already observed that the various European national cultures are less distinct than they were just twenty or thirty years ago. Styles of dress, cuisine, architecture, and other artifacts are merging.

Creation of One Market for Capital In 1978, the European Monetary System introduced its own monetary unit, the European Currency Unit (ECU). This theoretical unit of account did not exist in the form of actual currency, but in 1991 the 12 governments committed themselves to the creation of a single currency and a

regional bank by the year 1999. A single currency will end the confusion that arises every time a traveler crosses a border; beyond that, the absence of exchange rate fluctuations and the simplification of accounting procedures will promote trade. A nation's control over its own currency, however, affects control over its national budget, expenditures, and domestic interest rates. Therefore, acceptance of a single currency and central bank is such an important surrender of national sovereignty that EU members may delay this step.

Transport System The members of the European Union are redesigning and rebuilding Europe's transport system from 12 national webs into a single coordinated web to cover their whole territory. Much of the transportation infrastructure will require modification. All railroads, for example, must use the same gauge (the width between the tracks) and electricity systems, and all highways, bridges, and tunnels must have the same construction specifications. Union transport policy encourages the movement of freight by railways rather than highways to conserve energy and reduce pollution. An integrated network of high-speed trains is to be superimposed on the existing national networks (Figure 13-7). This network is to be completed by 2025. European air transport also will need coordination by a new supranational equivalent to the U.S. Federal Aviation Administration (Figure 13-8).

Regional Policy The Single Europe Act commits the countries "to strengthen economic and cultural cohesion." Living standards and quality of life ought to be equalized throughout the Union area, just as nations strive to equalize incomes throughout their territories (Figure 13-9). The Union furnishes funds for basic infrastructure, such as transport and telephone service.

Union planning can designate regions that straddle international borders. For example, the Commission is planning to build roads and clean polluted rivers along the Spain-Portugal border. Private companies will then be encouraged to increase industrial investment in the north of the region and to develop tourist facilities in the south.

Social Policy The Union members have tried to strengthen cultural ties. International committees are rewriting school history books, for example, to soften nationalistic antagonisms. University admissions policies, academic standards, and definitions of degrees are being harmonized. Union institutions are also encouraging and subsidizing international scientific research projects. Numerous pan-European cultural festivals are celebrated, including the naming each year of a "European Cultural Capital," a city chosen for its pan-European cultural-historical significance.

Europe's Planned New Rail Network

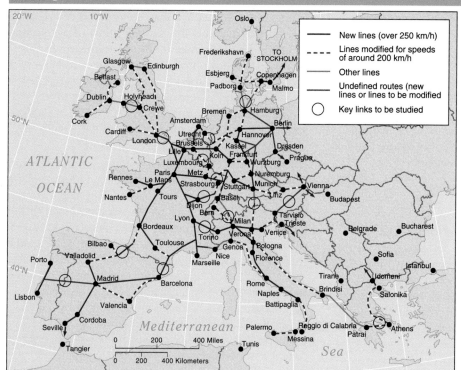

FIGURE 13-7

In an area as compact and densely populated as Europe, city-to-city train travel is a sensible alternative to air travel at little or no additional time cost. The new system will require laying 4,600 miles (7400 km) of new track and improving 12,000 (19,300 km) more. (Data from the European Commission, 1990.)

FIGURE 13-8

This map shows Europe's air control systems. Western Europe has only one-half the density of flights of the northeastern United States, but the inefficient system of 22 national air control systems operating from 42 air control centers causes flight delays. This system will be reconfigured. (Data from Association of European Airlines.)

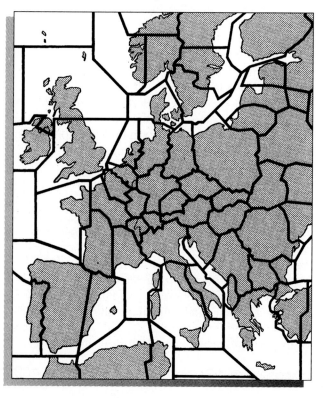

Most specific Union social actions, however, deal with questions of worker health and safety standards, the international transfer of pension rights, the guarantee of rights to migrant workers, and mutual recognition of technical qualifications. The Workers' Charter regulates the workweek, overtime pay, holidays, and other terms of employment. In addition, the European Court of Justice has enforced standard civil rights and antidiscrimination legislation. Union citizens may travel, settle, and work anywhere within the Union, whereas non-Union citizens face common immigration and visa requirements.

Through the years, statistical descriptions of life in the EU member countries have converged. These include statistics as diverse as median educational attainment of the population, degree of urbanization, percentage of females in the labor force, crude birth rates and total fertility rates. This convergence demonstrates that life in the member countries is growing measurably more alike.

Expansion of the Union

When the EU first formed, a few Western European countries chose not to join. Austria, Liechtenstein, Finland, Iceland, Norway, Sweden, and Switzerland united in a *European Free Trade Association (EFTA)*.

Variations in Income Across Europe

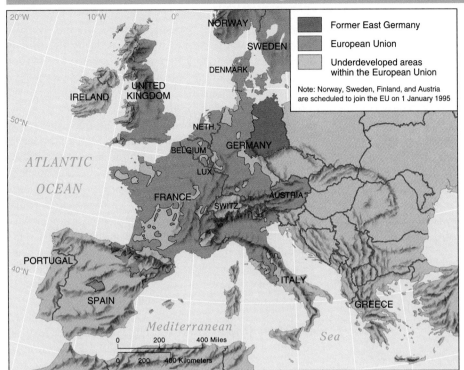

Former East Germany

European Union

Underdeveloped areas within the European Union

Note: Norway, Sweden, Finland, and Austria are scheduled to join the EU on 1 January 1995

FIGURE 13-9

Incomes are not everywhere equal throughout the Union territory, and the Union devises programs to assist the less-well-off areas.

On January 1, 1994, five EFTA nations (all but Switzerland and Liechtenstein) joined with the EU to create one 17-nation free trade region known as the **European Economic Area (EEA)**. Not all EEA members participate in all of the types of integration that are occurring across the EU territory, but the EEA does extend free movement of goods, capital, services, and people across five more countries (four of which are, in fact, scheduled to be full EU members by 1995. Thus, only Iceland, Switzerland, and Liechtenstein will be in the EEA and EFTA but outside the EU).

In 1993, the Union announced that Poland, Hungary, the Czech Republic, Slovakia, Romania, and Bulgaria may join when they have demonstrated stable democracy, respect for human rights, and functioning market economies. Estonia, Latvia, and Lithuania will join the EEA on January 1, 1995. Turkey, Cyprus, and Malta are "EU Associates," enjoying privileged access to EU markets, and Russia signed a trade agreement in June 1994.

Union members also extend privileged access to Union markets to 66 of their former colonies. Ninety percent of the population in the 66 countries is African, but several countries are in the Caribbean and Pacific Basins. This favoritism for former colonies, however, may not last. The beneficiaries' agricultural products compete with those from Southern Europe, and Italy, Greece, and Portugal want to protect European markets for their own produce. The favoritism also prompt protests from the United States and Latin American countries, which oppose discrimination against their products.

Common Foreign Policy

The EU members consult among themselves in foreign affairs, but they lack a common foreign policy. Their inability to speak with one voice reveals that they remain far from united. The Union members could not agree on a common foreign policy either during the 1991 war against Iraq or during the long civil war that accompanied the breakup of Yugoslavia. Negotiations on developing a common foreign policy will continue.

The collapse of Communist party governments in Eastern Europe presents the EU countries with tremendous opportunities, but also tremendous challenges. If the EU countries can consolidate the gains they have achieved and also reach out to help the countries of Eastern Europe, a new age of European affluence and prestige may open. United Europe could be a global superpower commanding wealth and influence comparable to or even greater than that of the United States.

THE UNRAVELING OF THE FORMER UNION OF SOVIET SOCIALIST REPUBLICS

▼▼▼▼▼▼▼▼▼▼▼▼▼▼▼▼▼▼▼▼▼▼▼▼▼

Throughout the late 1980s, the Union of Soviet Socialist Republics, then the largest country on Earth, unraveled. A new, looser association, called the Commonwealth of Independent States (CIS), reformed among 12 of its 15 constituent units, but we cannot know how much political or economic coordination the CIS will maintain across the enormous territory. The CIS powers will be defined through the 1990s. In the early 1990s, civil wars broke out around the margins of the former Union, and several years after the end of the USSR, the political situation remains unstable.

The Russian Empire, Revolution, and Reorganization

At the same time that the Western European countries were building overseas empires, Russia built an empire across Asia, conquering the peoples along its path as did other imperial powers (Figure 13-10). During World War I, the empire suffered internal rebellion as well as external attack. The Communists, led by Vladimir Lenin (1870–1924), allied themselves with rebellious nationalist minorities and pledged to reorganize and reform not only Russia but also Russia's relationship with the subject nationalities.

The Communists seized control in 1917, and, despite their promises, they struggled to hold all of the empire's peoples and territory. They replaced the anti-Russian propaganda that had helped them gain power with a new propagandistic ideal of federation. Theoretically, each subject nation of the old empire would experience an individual revolution of liberation from czarist and capitalist oppression, and then the nationalities would reunite in a new supranational union. This union, the Union of Soviet Socialist Republics, came into being in 1922. Not all of the subject nations were willing to reunite. Several fought for real independence and had to be reconquered. Others actually gained freedom. Poland, Finland, Estonia, Latvia, and Lithuania all emerged as independent countries after World War I, and the province of Bessarabia broke away to join Romania. The Soviet Union regained much of this lost territory during and after World War II. In 1990, the Soviet Union covered 8.6 million square miles (22.4 million km²), and its population of about 290 million made it the Earth's third most populous country, after China and India.

The Political Geography of the USSR and the Commonwealth

The Soviet Union was subdivided into 15 *union republics*, each of which was theoretically the homeland of a national group (Figure 13-11). Several Republics contained within them the homelands of less numerous ethnic groups, and these groups enjoyed lesser territorial autonomy. Russia was the largest republic, occupying fully 76 percent of the total territory of the USSR, and Moscow was the capital of both the Russian Republic and also of the USSR.

The Communist rulers redrew internal borders and even transferred whole populations. Between 1921 and 1991, there were 90 changes of borders among the republics. The Crimea, for instance, was taken from Russia and given to Ukraine in 1954. The dissolution of the Union rekindled a few historic border disputes, and although the founding document of the CIS recognizes all current borders, these borders may be redrawn in coming years.

Russian Imperialism The individual Soviet republics ostensibly enjoyed considerable autonomy, but in fact all power was centralized in Moscow. Russian Communists dominated in the organs of government, in the Communist party, and in the leading positions in society. Russians even held controlling positions in the non-Russian republics.

Russians attempted to acculturate the other peoples to a Russian norm—a process called **Russification**. The Russian language, for example, was preferred for central government affairs, and it was one of two official languages in all non-Russian republics. Furthermore, historic national churches were persecuted. The position of the 55 million Muslims in the Soviet Union was especially volatile. Kazakhstan, Turkmenistan, Uzbekistan, Tajikistan, and Kyrgyzstan in Central Asia plus Azerbaijan in the Caucasus region gave the Soviet Union one of the world's largest Islamic populations, and high birth rates among Muslims were increasing their percentage of the total USSR population. Many Muslims, however, were politically dissatisfied in the Union, and their dissatisfaction threatened the political stability of the USSR. Migration of Russians into the other homelands was still another form of Russian imperialism. In 1990, some 25 million Russians lived in the 14 other republics, more than half of them in Ukraine or Kazakhstan.

Developments from 1985 to the Dissolution of the USSR in 1991

In 1985, when Mikhail Gorbachev assumed power as Union president, the Soviet economy was stagnant. To

The Historic Expansion of Russia

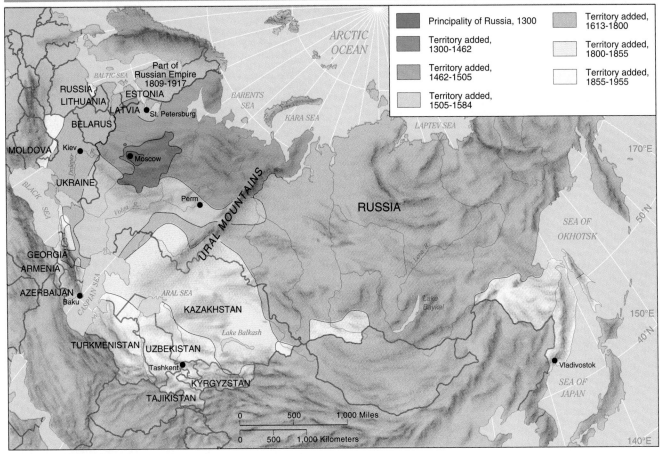

FIGURE 13-10

Russians' expansion from their homeland around Moscow began in the mid-sixteenth century, and by 1647, Russian cavalrymen had reached the Pacific. During the seventeenth and eighteenth centuries, Russians pushed south to the Black Sea, and Russian explorers crossed over the Bering Strait to incorporate Alaska (which Russia sold to the United States in 1867) and to locate settlements as far south as California. In the nineteenth and early twentieth centuries, Czarist power advanced still further into Eastern Europe and Central Asia.

revive the economy, Gorbachev cut loose the satellites of Eastern Europe and launched two liberalizing initiatives in the Soviet Union itself. One was *glasnost*, a loosening of restraints against freedom of speech and the press. The second was *perestroika*, a restructuring of the economy and of politics. In 1990 Gorbachev repealed the Communist party's constitutional monopoly on political power. The president of Russia itself, Boris Yeltsin, renounced communism and then won reelection to his office. This gave Yeltsin a unique political legitimacy within the dominant republic.

Gorbachev gambled that his policies could unleash individual initiative and productivity while retaining the centralized control of the Party and of the central Union government. Gorbachev lost. The tensions unleashed by *glasnost* and *perestroika*

brought down the old governmental system before Gorbachev could bring a new one into place.

Ethnic frustrations rose to the surface. Nationalist sentiments had been intensified by the spread of mass education, echoes of decolonization abroad, and resentment of Russian hegemony. The central government proposed a new treaty to redefine the relationship between the central government and the constituent republics, but the day before the treaty was to be signed, August 19, 1991, reactionary elements of the Communist party, the army, and the KGB (state police) arrested Gorbachev and attempted to take over the central government. By that time, however, the governments of the individual republics had already gained too much power, and the coup failed. The government of Russia itself held out against the

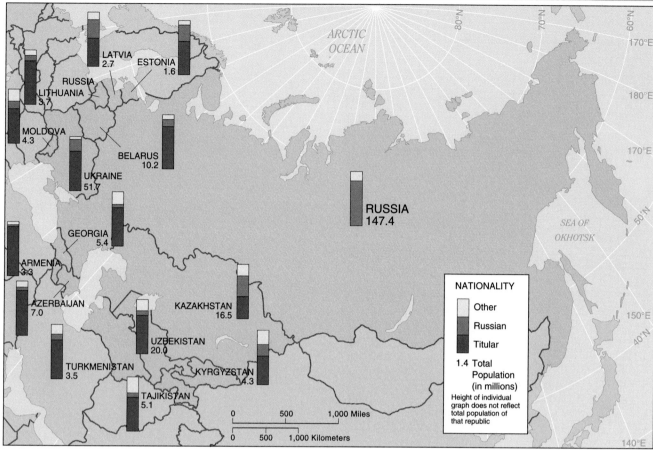

The Political and Population Geography of the USSR in 1990

FIGURE 13-11

coup. Within weeks, most of the republics had declared their independence from the USSR and outlawed the Communist party.

The Formation of the Commonwealth of Independent States

On December 9, 1991, representatives of Belarus, Russia, and Ukraine met in the Belarussian town of Brest and declared "the USSR, as a subject of international law and geopolitical reality, is ceasing its existence." The three leaders created the new Commonwealth that was declared to be "neither a state nor a superstate structure," but only a set of "coordinating bodies" with its "seat" (not capital) in Minsk, the capital of Belarus. Two weeks later, 11 more republics joined as "equal co-founders" of the Commonwealth. Georgia held out but joined in 1993. On December 26, 1991, the Soviet Parliament formally voted the USSR out of existence.

The allocation of powers between the newly independent republics and the new Commonwealth structure is yet to be defined. Joint agreements are being negotiated on foreign affairs, transportation and communication, environmental matters, and economic affairs. The three Slavic republics (Russia, Ukraine, and Belarus), however, forged ahead and formed a "single economic space," with one currency and central bank, in 1993.

Continuing Political Fragmentation within the Republics The successor states of the USSR have had to renegotiate their relationships with the former autonomous regions within them. Because the threshold principle for the size of states has been abandoned worldwide, several minority regions have claimed the right to national self-determination and full independence. The government in Moscow has granted varying degrees of autonomy to 88 constituent units, 20 of which were ethnically based republics (Figure 13-12, p. 478). The Parliament creat-

474 Chapter 13

FOCUS
The Baltic Republics

The three Baltic republics of Estonia, Latvia, and Lithuania seceded from the Soviet Union and were internationally recognized as independent countries in 1991. Their departure reduced the Soviet Union's territory only 0.8 percent and its population only 2.9 percent, but these three republics enjoyed high productivity and incomes.

The Baltics were independent democracies after World War I, but the Soviet Union had forcibly reincorporated them in 1939. From 1939 until 1991, the United States never formally recognized that reincorporation. When the republics held plebiscites on independence in 1990 and 1991, even the ethnic Russians in them supported independence, swayed, perhaps, by the relative wealth and freedom in their adopted homelands.

The Baltic republics are of strategic importance to Russia because they lie between it and the Baltic Sea. They even isolate bits of Russian territory. The capitals of Estonia (Tallinn) and of Latvia (Riga) are seaports. The capital of Lithuania, however, Vilnius, is inland. Lithuania's seaport is Klaipéda, which was historically the German city of Memel. Russia took this city from Germany at the end of World War II, and Russia insists that it remains Russian property. Russian access to the port of Kaliningrad (historic German Königsberg) is even more impor-

tant. That port, also taken from Germany, is Russia's only ice-free port, so it was home to the Soviet Baltic fleet. Kaliningrad has been declared a *free economic zone* where international investors receive economic privileges.

The Baltic republics were the three republics that could most probably quickly shift their trading patterns outside the Union, and they are the only three that have continued to decline to join the CIS. They are trying to attract foreign investment to develop economies

dedicated to agriculture and light industry, rather like the Finns, to whom the Estonians are ethnically related. By early 1994, all three countries' currencies had stabilized and were freely convertible on world markets. The three states will probably continue to depend on Russia for raw materials and fuel. Russia is withdrawing its soldiers, but it is also threatening that allegations of discrimination against ethnic Russians gives Russia a right to interfere in these three countries.

The Baltic republics regained their independence in 1991, and today they are struggling to rebuild their cultural and economic links with Western Europe.

ed under the new Russian Constitution of 1993 will redefine these regions' rights.

Through the early 1990s, civil wars broke out in several of the newly independent republics, including Moldova and Georgia. Various combatants appealed for help to the Russian army, and this made other fac-

tions fear reimposition of Russian rule. Western observers could seldom be certain of the exact allegiance of the warring factions, and many even hypothesized that the government in Moscow was not itself always certain of the allegiance of fighting units.

FOCUS
Ukraine

Ukraine was the second most powerful of the Union republics. It accounted for about 21 percent of total Union population and gross product, and an even higher percentage of agricultural products. Ukrainian nationalism has always been strong; the Ukrainians fought the Russians for their independence after World War I and again after World War II, and anti-Russian animosity remains a powerful force.

The industrial might of the Ukraine, like that of East Germany, looked greater than it has turned out to be. The eastern Ukraine contains a great concentration of industry, but most of these factories are obsolete, and they may represent more of a liability than an asset to a new economy. The basin of the Donets River (called the *Donbass*) has been the Russian

Empire's main coal producer for more than a century, but costs of mining are rising.

When the Commonwealth was created, Ukraine disputed several points with Russia. For one, Ukraine demanded the right to have its own army. Ukraine held enough of the former Soviet nuclear arsenal to make it the world's third leading nuclear military power. Early in 1994, Ukraine agreed to surrender its missiles in exchange for U.S. financial aid. Possession of Crimea was a second point of contention. The population of Crimea is about 67 percent Russian, and local leadership defied Ukrainian sovereignty. Ukrainian leaders feared that if Ukraine loses Crimea, then Donbass (about 40 percent Russian) might want to break away too. Ukraine and Russia also disputed ownership of the former Soviet Black Sea

fleet. The fleet is actually rusting and worthless, but its port facilities could be valuable to Russia.

Ukrainian leaders have bargained with Russia from a position of weakness, because the economy of the independent Ukraine collapsed. Some 30 percent of Ukrainian GDP had been defense-related, and that sector evaporated. Also, the government was slow to reform the economy. It was reluctant to privatize, for example, so inefficient state-owned industries piled up losses, and Ukraine remained dependent upon Russia for fuel and raw materials. Already by 1993, in accepting the single economic space, Ukraine surrendered considerable control over its economy. In 1994, Ukraine, the "breadbasket" of the former Union and potentially one of the richest agricultural regions on Earth, imported food.

The Economic Geography of the USSR and of the Commonwealth

One of the reasons for the political dissolution of the USSR was its economic failure. In turn, the political dissolution threatens worse economic circumstances for the peoples of the 15 new countries. This is because of the way the territory was organized economically under the Soviet regime. Great economic variations existed among the republics (Table 13-1), but the two principle characteristics of Soviet economic development were *autarky* for the Union as a whole, and *centralized state control*. The successor

states will struggle against the legacy of these two principles for many years.

Communist Economic Organization The economic organization imposed on the USSR was dictated by Communist theory. This did not allow anyone to own "means of production" in such a way as to profit from the labor of others. Only a little private property and a few private-sector services were allowed. Natural resources were the property of the state.

Agriculture was brutally collectivized. New lands were opened to agriculture, notably the steppes

LAND USE
- Grazing (cattle, pigs)
- Crop cultivation (flax, potatoes) and some grazing
- Intensive crop cultivation (beets, sunflowers, wheat, barley)
- Suburban cultivation

PRINCIPAL CASH CROPS
- Cotton
- Fruit
- Tobacco

MINING PRODUCTS
- Coal
- Iron Ore

Ukraine is by any measure—population, size, productivity, or potential strength—a major new country in the world community.

of northern Kazakhstan, but Soviet agricultural output was slow to increase. Despite the immense territory of the former Union, conditions across much of that territory—particularly northern Russia and the desert regions of central Asia—are unfavorable for agriculture (Figure 13-13, p. 480). Agriculture also suffered from mismanagement, lack of worker incentive, lack of investment, and government policies that kept farm prices low. These factors kept productivity low, and the poor internal transportation system compounded the difficulties. Each of the 15 new countries can improve its individual agricultural output by resolving these problems, and there is no reason why

each could not become much more productive. Much of Russia is virtually useless for agriculture, but many areas could be agriculturally productive. In the early 1990s, Russia was breaking up the agricultural collectives and redistributing land holdings into smaller cooperatives and private farms. Moldova, Ukraine, the Baltics, Kazakhstan, and the countries in the Caucasus (Georgia, Armenia, and Azerbaijan) could become significant food exporters.

Under communism, industry was owned and organized by the state, guided by planning bureaucracies. Heavy industry was emphasized, and it made immense strides. By the 1980s, the Soviet

Russia's 20 Autonomous Republics

FIGURE 13-12

These 20 ethnically based "autonomous republics" within Russia are renegotiating their legal powers with the government in Moscow.

Union had become the world's largest producer of many heavy industrial goods, but it lagged behind the West in light industry, in the production of consumer goods, and, despite superlative institutes of higher education and scientific research, in high-technology products.

Distribution of Industrial Production We do not have enough space to provide a detailed economic geography of the Soviet Union or the successor Commonwealth. It is necessary, however, to understand a few attributes of the former Soviet economy in order to understand the immensity of the economic problem facing the newly independent states.

Russia was overwhelmingly the dominant republic in the Union. In 1990, with just over half the Soviet population, Russia produced 91 percent of the USSR's oil, 75 percent of its natural gas, 55 percent of its coal, 58 percent of its steel, 81 percent of its timber, 50 percent of its meat, 48 percent of its wheat,

and 60 percent of its cement. Russia's treasury subsidized inefficient industries in other republics, and Russia supplied raw materials to other republics at prices below world market prices.

Both production and internal trading were highly centralized. Factories received their instructions from the central government. They sent goods to other factories and shops according to a series of Five-Year Plans. There were few direct contacts between one enterprise and another, and fewer still between any enterprise and consumers. Enterprises did not have any alternatives to the suppliers and customers assigned by the state planning bureaucracy. Enterprises specialized in a few products, and there was little or no competition. Enterprises could not change from one product to another, from one supplier to another, or from one customer to another, as demand changed. This system allowed no flexibility. Some consumer goods, for example, piled up unwanted in warehouses while the managers of the factories that had made them received medals for

TABLE 13-1
The Economies of the Soviet Republics at the Time of the Dissolution of the Soviet Union

Republic	Per capita GNP* (dollars)	Economic profile**
Latvia	6740	Advanced industry producing high-quality electronics and consumer goods; imports energy and raw materials
Estonia	6240	Most industrialized Baltic republic; largely self-sufficient in agriculture
Belarus	5960	Diverse highly productive industry, although dependent upon imported raw materials; important producer of fertilizers and synthetic fibers; net exporter of mixed agricultural products
Lithuania	5880	Diverse manufacturing base; imports raw materials, energy, and grain
Russia	5810	Produced more than 60 percent of total Union income; diverse exports; imports include specialty crops
Armenia	4710	Important machine-building industry; agricultural exports include specialty crops, fruits, and preserves
Ukraine	4700	Second largest economy in the Union; major heavy industry and also the Union's "breadbasket," producing about 25 percent of total Soviet agricultural output
Georgia	4410	Highly diverse economy, but exports principally specialty crops, tea, citrus fruits, wine
Moldova	3830	Intensive agriculture and food processing, with a small agricultural machinery industry
Azerbaijan	3750	Industry largely based on Caspian oil; agricultural sector mostly fruits and vegetables
Kazakhstan	3720	Major supplier of power, fuel, chemicals, metals and agricultural products
Turkmenistan	3370	Natural gas and irrigated land; important fishing industry in the Caspian Sea
Uzbekistan	2750	Largest central Asian economy; indigenous chemical industry based on natural gas; produced two-thirds of Union's cotton
Tajikistan	2340	Exports hydroelectric power and minerals; also fruits, vegetables, and processed foods
USSR	5000	

*Courtesy of PlanEcon, Inc., Washington, D.C.
**Adapted from "The Soviet Republics: A Political and Economic Overview," Office of the Geographer, U.S. Dept. of State, February 1991.

having exceeded production targets. Meanwhile, other goods in short supply had to be rationed.

Another important geographic characteristic of industrialization during the Soviet regime was that industries were scattered all over the Union's huge territory, but the production of any given item was highly concentrated in individual factories. Many of the most basic goods in any economy, such as

The Environment of Russia and the States of the Former U.S.S.R.

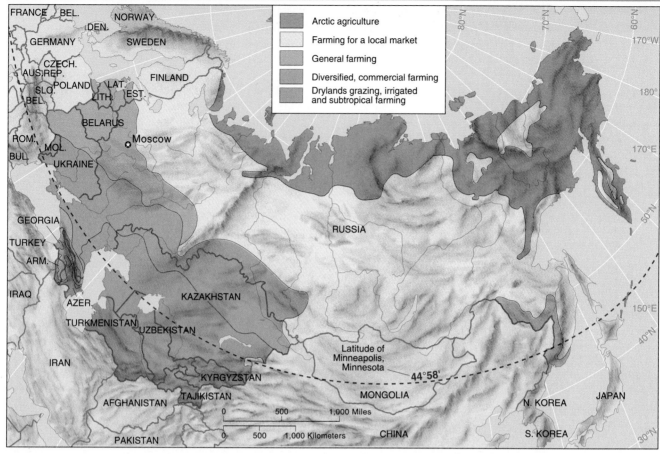

Legend:
- Arctic agriculture
- Farming for a local market
- General farming
- Diversified, commercial farming
- Drylands grazing, irrigated and subtropical farming

FIGURE 13-13

Despite the immense size of the former Soviet Union, much of this territory was not agriculturally productive—partly for environmental reasons, and partly for political-economic reasons.

polypropylene, stainless-steel pipes, and concrete mixers, were produced entirely or almost entirely by a single factory. Between 30 and 40 percent of the total value of Soviet manufactured goods was produced on single sites. The centralization, rigidity, and geographic particularization of production in the Soviet Union bound the republics in trade.

The legacy of this highly centralized command economy is that none of the successor states has a balanced economy. Most are dependent on other former republics for essential supplies, and most rely on enterprises in other former republics as their sole markets. Furthermore, many of their manufactured goods—particularly the consumer items—are of such poor quality that they cannot find new export markets. It will be difficult to unravel the countries' economic interdependence, and the process of unraveling will impose considerable hardship.

Economic Collapse and the New Economies

The Soviet economy had begun unraveling already in the late 1980s. All across the Union, local political authorities had defied Moscow and sought to achieve individual self-sufficiency by (1) seizing control of local resources and dictating terms and prices for their sale; (2) imposing direct restraints on trade; and (3) raising prices for goods produced locally. This triggered the collapse of the centrally planned economic system. Newly independent successor states have tried to reach trade agreements among themselves, but complications have caused delays. Delays that cut off established flows of raw materials, foods, or industrial parts led to a decline in trade, and a decline in vital trade has caused standards of living to fall. Some individual countries are introducing eco-

480 *Chapter 13*

nomic reforms and seeking new international trade and investment partners, but it will be difficult to replace the economic interconnections established under the USSR.

Most of the new countries have begun privatizing their economies, but they are doing so at different rates. These privatizations have usually taken the form of either employee stock ownership plans, in which the workers assume ownership (which Communist propaganda had insisted was theirs all along), or of private sales to domestic or even foreign investors. In other cases, privatization has meant seizure of the assets by the members of the old Communist government bureaucracies. It is said that in many cases *privatizatsia* (privatization) turned into *prikhvatizatsia* (to grab). Nevertheless, by midsummer 1994, 15 thousand factories (60 percent of the industrial workforce) and 80 thousand shops had been privatized in Russia.

Meanwhile, the complexity of international trade with the new countries can be illustrated with one 1992 transaction. In 1992, Marc Rich & Co. of Switzerland devised a $100 million deal involving enterprises and governments in five newly independent countries. The parties themselves, hampered by political tension and lack of business skills, could not have devised it. Rich & Co. purchased 70,000 tons of raw sugar from Brazil on the open market. The sugar was shipped, via a Baltic seaport, to Ukraine, where it was processed at a local refinery. After paying the refinery with part of the sugar, Rich & Co. sent 30,000 tons of refined sugar 6,000 miles east to several Siberian oil refineries, which needed sugar to supply their work forces. The refineries did not have hard currency, so they offered to pay for the sugar with oil products. Much of what was available, however, was low-grade gasoline, which has few export markets. One of these, however, is Mongolia. Rich & Co. has long traded with that country, so the company moved 130,000 tons of oil products there. In payment, the Mongolians turned over 35,000 tons of copper concentrate. The company sent most of that back across the border to Kazakhstan, where it was refined into copper metal. Then the metal was shipped westward to a Baltic port and out to the world market where, several months after the deal began, Rich & Co. earned a profit in hard currency.

Several of the 15 new countries may turn to some form of capitalism. State-directed capitalism, such as characterizes South Korea, Taiwan, and Japan today, worked in Russia in the decades before 1914. Russia had the fastest economic growth rate in Europe. Today, Russia alone is still one of the world's largest economies, and its schools produce one-third of the world's doctorates in science and engineering.

In the future, Russia, even if standing alone, will almost certainly still be the largest and one of the most powerful nations on Earth. At present, however, demographic figures from the early 1990s reveal a society in crisis. The birth rate and the total fertility rate are dropping, while death rates are rising rapidly. Life expectancies are actually shortening. If these trends continue, Russia's population will shrink sharply in coming years.

Political ramifications of economic hardship appeared in several of the new states already within one or two years of independence. Many people had been overly optimistic about the improvements in their lives that political independence could bring. When independence brought a decline in standards of living, former Communist party officials running under new party labels won popular elections by demanding that "the good old days" be brought back.

Environmental Pollution

The Constitution of the USSR said that "citizens of the USSR are obliged to protect nature and conserve its riches." They failed to do so. The Soviet regime passed environmental protection measures, but these measures were ignored during the drive for industrialization. Few environmental laws incorporated penalties, and most industries were attached to ministries of the central government, so they were not subject to any local legislation. Agricultural chemicals were often used without regard to their poisonous consequences, and industries were technologically backward and grossly inefficient in their use of raw materials.

In 1990, Alexei Yablokov, director of the Institute of Biology at the Academy of Sciences, estimated that 20 percent of Soviet citizens lived in "ecological disaster zones," and 35 to 40 percent more in "ecologically unfavorable conditions." These conditions were spread among the 15 republics, and the new countries face enormous cleanup tasks.

New Centripetal and Centrifugal Forces in the CIS

The Soviet Union was held together by raw power exercised by the Communist party. Today we must ask what centripetal forces bind the CIS. Some new states may seek Russian military protection, but others feel that Russia is the greatest threat to their new independence. Russia has threatened intervention in several states to protect the rights of ethnic Russian minorities. Several CIS states might be held together by shared economic interests. Russia offers cheap raw materials to those states that cooperate with it in other political or military pacts. These centripetal forces, however, will be balanced by powerful centrifugal forces.

Centrifugal forces tending to pull the Commonwealth apart include the animosity that many minority nationalities feel toward their Russian former rulers. Furthermore, outside countries are making friendly overtures to many CIS members. In the western part of the Commonwealth, for example, many Moldovans want to rejoin Romania. The European Union may eventually extend membership not only to the former Eastern European satellites, but even to the Baltic states, Ukraine, Belarus, and perhaps Russia itself.

To the south, the newly independent Central Asian Islamic countries may retain economic links to Russia, but religion and other cultural ties pull them away. Several Islamic countries compete for influence. Turkey, which is democratic and secular, a NATO member and an EU associate member, may be a model for these peoples, but the Tajiks are related to the Iranians in language and culture, although they are Sunni Muslims, whereas the Iranians are Shiite. The Azerbaijani of the Caucasus region are also Shia Muslims. Iran has paid for new mosques and schools in Azerbaijan, and it is initiating joint economic ventures. Saudi Arabia has also opened banks, schools, and mosques throughout several republics, and it has extended offers of liberal financial assistance. China is investigating trade opportunities, and even Japan and South Korea have invested in Kazakhstan and Kyrgyzstan.

The dissolution of the Union of Soviet Socialist Republics has created new units on the world political map and released powerful new political, religious, cultural, and economic forces. These forces will cause global redistributions of cultural, political, and economic activities.

THE FORMATION OF A NORTH AMERICAN TRADE BLOC

▼▼▼▼▼▼▼▼▼▼▼▼▼▼▼▼▼▼▼▼▼▼▼▼▼▼▼▼

On January 1, 1994, the North American Free Trade Agreement (NAFTA) came into effect, linking the United States, Canada, and Mexico. One impetus for this pact is fear of being closed out of the European market; NAFTA came into effect on the same day as the EEA. The action is also partly an effort to extend economic aid to Mexico and to develop the potential of that country, where per capita incomes and standards of living are considerably below those in the other two North American partners.

The agreement does not propose a full customs union or common market. There are no plans for any supranational governing body, nor do the North American partners look toward political merger. The three partners, however, outstrip the sixteen-member European Union in total area (8.2 million square miles to 1.4 million) and in total gross product ($6 trillion to $5.4 trillion), but the total EU population will be slightly larger (365 million to 376 million).

The Canada-United States Free Trade Agreement

Canada and the United States signed a Free Trade Act in 1988, by the terms of which all barriers to trade in goods and services will be eliminated between the partners over a 10-year period that started on January 1, 1989. The treaty also removes or reduces the hurdles to cross-border investment, government procurement, agricultural sales, and movements of employees.

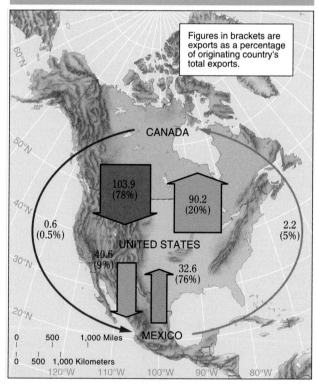

North American Trade Flows 1992

FIGURE 13-14

This figure shows trade among NAFTA partners in billions of U.S. dollars. Canada takes about 20 percent of U.S. exports each year, more than twice Japan's share, and the United States buys three-quarters of Canada's exports. Trade between Canada and Mexico, however, is limited.

Economic Links In 1988, Canada and the United States were already by far the world's two leading trading partners (Figure 13-14), as well as major investors in each other's economies. Some 72 percent of foreign investment in Canada came from the United States, representing one-third of all U.S. investment abroad. Canada was the fourth largest investor in the United States, and since 1988 many more Canadian businesses have established new U.S. operations or bought U.S. businesses. United States corporations were generally strong supporters of the pact, because Canadian tariffs were two to three times higher than U.S. tariffs.

Canada's labor unions fought the pact. They pointed out that government benefits in Canada are generally much more generous than they are in the United States, and the unions feared a drop to U.S. standards. In Canada labor unions represent 40 percent of the labor force, whereas in the United States unions represent only 19 percent, and in 1990 hourly compensation for production workers in manufacturing in Canada was 108 percent of that in the United States.

Canada's farmers also fought the pact. Canadian agriculture has been dominated by relatively high-cost small farms, and they feared competition. Canadian agriculture is at a competitive disadvantage because of the harsh climate, vast distances, and small scale of regional markets. Canadian small producers also suffer competition from big food-processing firms that can buy more cheaply in the United States.

Cultural and Nationalistic Considerations

Canadians have long feared the economic and cultural attraction of the United States as a centrifugal force disrupting Canadian nationalism, and they have tried to define and to defend their nationalism as distinctively different from that of the United States. If you look back on Figure 10-20, you will notice a thin line of Canadian settlement along the border and an almost empty empire stretching away to the north. Some 75 percent of Canadians live within 150 miles (241 km) of the U.S. border, where they absorb U.S. news and culture. In contrast, only 12 percent of Americans live within 150 miles of Canada. One Canadian leader warned that the pact would "throw us into the north-south pull of the United States. And that will reduce us, I'm sure, to an economic colony of the United States, because when the economic levers go, the political independence is sure to follow." Canadian means of transportation and communication have always been designed to foster east-west linkages, as illustrated in Figure 10-14. Nevertheless, the international border to the south is punctured by more than 75 million border crossings per year, by common telephone lines, pipelines, computer links, power lines, contracts, and cross-ownership, and even by common professional sports allegiances.

Economic union with the United States affects Canada's political cohesion in still another way. The people of Quebec continue to consider secession from Canada, but nobody is sure whether secession would jeopardize Quebec's economic ties with the United States or allow Quebec to negotiate more favorable ties. Thus, U.S. attitudes may play a role in encouraging or discouraging Quebecois separatism. In a 1991 poll, 18 percent of Quebecois said that they would actually like their province to join the United States. Politicians in several other Canadian provinces have said that if Quebec secedes, the Canadian Federation might dissolve, and several provinces might request to join the United States as states.

Canadian cultural nationalists argue that economic union with the United States threatens the existence of distinctly Canadian broadcasting, publishing, and related cultural industries, but the pact provides for protection of these industries. It may be difficult even today, however, to distinguish Canadian cultural products from those of the United States. The majority of Canadians neither watch Canadian television, read Canadian books, listen to Canadian music, nor go to Canadian movies and plays, whereas many U.S. citizens are probably unaware of the Canadian nationality of authors Robertson Davies and Margaret Atwood, reporters Peter Jennings and Morley Safer, and entertainers Rich Little, K. D. Lang, Jason Priestley, and Michael J. Fox. About 10 percent of the chief executives of the largest U.S. corporations are Canadian-born.

Early Results Five years after the U.S.-Canadian free trade agreement took effect, economists were uncertain of its impact. Some Canadian industries seem to have been unable to compete, but a recession in the United States, an overvalued Canadian dollar, and high interest rates all may have caused job losses in Canada. The number of manufacturing jobs in Canada fell from 2.1 million in 1988 to about 1.8 million in late 1993, but Canada's trade surplus with the United States grew from $10.6 billion in 1989 to $25 billion in 1993, thereby creating, economists believed, as many as 50,000 more jobs. Increasing exports may begin to increase employment.

Mexico

On January 1, 1994, a new trade agreement between the United States and Mexico went into effect. Tariffs and other barriers to trade among the United States, Canada, and Mexico will be brought down in four 5-

year steps. Mexico is the third largest market for U.S. goods, after Canada and Japan, and about three quarters of Mexico's exports are sent to the United States.

The integration of Mexico into a North American trade pact might be more difficult than the integration of the U.S. and Canadian economies. The United States and Canada are both rich, post-industrial societies with populations that are growing slowly. Mexico is a less-developed nation with a population still growing rapidly. Its population grew from 25 million in 1950 to 85 million in 1985, and it is expected to mushroom to over 100 million by 2000. To provide work for this rapidly growing labor force, Mexico needs economic growth and hopes to attract foreign investment. Economic growth in Mexico might reduce the push for Mexicans to migrate, legally or illegally, to the United States. Mexico's Hispanic culture and language contrast with the British traditions and English language held in common in the two countries to the north (except in Quebec), but that cultural contrast may not be as troublesome in building a free trade pact as it would be if the three countries were building a full common market.

The continental free trade pact might remap

FIGURE 13-15

United States Vice President Al Gore (left) debated businessman Ross Perot in 1993 during national consideration of the free trade pact with Mexico. Perot, who had been a contender for the presidency in the 1992 elections, was strongly against the pact, and he still attracted a nationwide following. If the United States signed the pact, he warned, we would hear "a giant sucking sound" of U.S. jobs going south to Mexico.
(Courtesy of AP/Wide World Photos.)

continental economic geography. In the geography of agriculture, for example, the United States and Canada may provide all of North American cereals. This can harm the subsistence corn farming population of Mexico, but it is hoped that new opportunities will open for Mexican farmers to specialize in fruits and vegetables. Farming interests in California, Texas, and Florida insisted that free trade in many agricultural specialties of these states be delayed. The greatest political battle in the United States, however, was not over agricultural goods, but was sparked by fears of potential job losses in U. S. manufacturing (Figure 13-15).

Mexican Industrial Policies Mexico's economy has been state-directed and centralized through most of the twentieth century. In 1982, some 70 percent of the nation's economy was in the hands of the state. Mexico took up a new direction, however, in 1986. Mexico changed its national economic policies from import-substitute to export-led. The government privatized much of its state sector, selling off factories, shopping centers, copper mines, and public utilities. Only PEMEX, the state oil monopoly created when Mexico's oil industry was nationalized in 1938, was viewed as politically untouchable. In 1990, it accounted for 5 percent of Mexico's GDP and 22 percent of export revenue. Mexico also liberalized rules for foreign investment and succeeded in attracting investment in new factories.

Maquiladoras Many U.S. companies were active in the Mexican economy even before the trade pact was signed. They had moved labor-intensive operations to Mexican factories called *maquiladoras*. Mexico imported components duty-free; these were assembled and the products re-exported to the United States. The United States charged tariffs only on the value added, which was low because Mexican wages are low (Figure 13-16). The Ciudad Juarez/ El Paso area has the largest number of such jobs (Figure 13-17). Between just 1985 and 1990, the value of Mexico's manufactured exports doubled, and those exports are of increasing value added.

The Free Trade Debate and Final Pact

Many arguments were raised both for and against free trade with Mexico. Many U.S. workers feared that their jobs would migrate to Mexico. In 1990, hourly compensation for production workers in manufacturing there were just 12 percent of those in the United States. Compared to U.S. factories, Mexican factories have long working hours, more flexible work rules, lax safety standards, and poor pollution controls.

FIGURE 13-16

This young worker is inspecting a circuit board for a television tuner at the Zenith assembly plant in Matamoros, Mexico. These parts are manufactured in the United States, but assembled in Mexico.
(Courtesy of Zenith.)

NAFTA establishes a three-nation Labor Council, but already U.S. labor unions are working to improve the conditions of labor in Mexico.

United States manufacturers, however, responded that Mexican wages are lower than U.S. wages because Mexican productivity is lower than U.S. productivity; that is, each Mexican worker's output is much lower than the output of each U.S. worker. Productivity is usually determined by capital investment per worker. Some economists have argued that much low-wage manufacturing is going to leave the United States anyway, and it is better for it to migrate to Mexico than, for instance, to Southeast Asia. This is because when the Mexican economy grows, an estimated 15 percent of each additional dollar of income is spent on U.S. goods and services. This produces

new U.S. exports. Growth in Southeast Asia does not have the same reciprocal effect on U.S. exports.

The result of the pact with Mexico will also be determined by the ultimate destination of the products manufactured in Mexico by U.S. corporations. If these products stay in Mexico or are exported to the rest of the world, that will be good for both Mexico and the United States. If these products are imported into the United States, however, that will lower costs for U.S. consumers, but it will also increase the U.S. trade deficit and eliminate manufacturing jobs in the United States.

European and Asian corporations have invested in Mexico, perhaps intending to reexport goods from Mexico into the United States. This is likely to cause trade disputes. Negotiators of trade pacts must agree what percentage of the total value of a good entering one country must have been added in the second country for that product to qualify as a product of the second country. This is called a **local content requirement**. Arguments over local content requirements typically arise whenever two countries negotiate free trade pacts that are not customs unions. Disagreements have arisen between the United States and Canada about whether goods assembled in Canada out of components made in other countries should enjoy free access to the U.S. market.

While U.S. corporations have been investing in labor-intensive industries in Mexico, Mexico's own global corporations have shifted capital-intensive investments north. The Mexican cement company Cemex, for example, North America's largest, purchased several large U.S. producers in the 1980s.

Mexican Politics The political situation in Mexico may stall Mexican growth and integration with the U.S. economy. Mexico has been ruled by one political party since 1929, and its political apparatus makes little distinction between the party and the state. Accusations of corruption and repression abound. Furthermore, new investment in Mexico might intensify the split between Mexico's modernizing economic sector and the great share of the population that remains in subsistence poverty.

On New Year's Eve, 1994, just as the U.S.-Mexican pact took effect, a native American guerrilla organization calling itself the Zapatista National Liberation Army, in honor of Emiliano Zapata (discussed in Chapter 10), seized four towns in Mexico's southern state of Chiapas (see Figure 13-18). The rebels were motivated in part by fears that imports of U.S. corn and other market pressures would wipe out their primitive agricultural economy. "The free trade agreement," said the leader of the group, "is a death certificate for the Indian peoples of Mexico." Chiapas is one of the poorest states in Mexico, and largely

MAQUILA SCOREBOARD

FIGURE 13-17

Most maquiladoras are concentrated along the U.S.-Mexican border, but this map shows that already by October 1993, U.S. corporations had spread assembly facilities throughout Mexico. (Courtesy of *Twin Plant News*, Vol. 9, No. 3, October 1993.)

The following numbers are supplied by Secretariat of Commerce and Industrial Development (SECOFI) offices, maquiladora associations and economic development organizations in Mexico and the United States. Not every city is listed, but totals for maquiladoras and employees are provided for all of Mexico.

MAQUILAS

2 0 2 9

EMPLOYEES

4 6 7 2 4 1

	NO. OF MAQUILAS	NO. OF EMPLOYEES
BAJA CALIFORNIA		
Tijuana	513	71,490
Mexicali	138	20,223
Tecate	82	5,024
Ensenada	39	2,730
BAJA CALIFORNIA SUR		
La Paz	8	662
SONORA		
San Luis Río Colorado	15	3,000
* Nogales	73	20,280
Naco	4	1,200
Agua Prieta	26	7,500
Hermosillo	15	5,000
Guaymas	2	900
Cd. Obregón	3	1,726
Other cities	39	7,506

* Total includes surrounding areas.

	NO. OF MAQUILAS	NO. OF EMPLOYEES
CHIHUAHUA		
Palomas	5	137
Cd. Juárez	274	109,713
Cd. Chihuahua	66	29,010
Ojinaga	1	300
Other Cities	17	7,357
COAHUILA		
Cd. Acuña	42	18,451
Piedras Negras	35	8,032
Torreón	40	5,031
Saltillo	6	2,077
Monclova	3	1,226
Other cities	28	1,441
NUEVO LEON		
* Monterrey	95	18,821
TAMAULIPAS		
Nuevo Laredo	58	10,000
Reynosa	61	30,000
Matamoros	76	38,268
Valle Hermoso	13	2,500

	NO. OF MAQUILAS	NO. OF EMPLOYEES
DURANGO		
Gómez Palacio	50	8118
JALISCO		
Guadalajara	45	19,500
YUCATAN		
* Mérida	31	5354
MORELOS	15	1,189
MEXICO, D.F.	27	3,475
SINALOA	5	
ZACATECAS	4	
SAN LUIS POTOSI	9	
AGUASCALIENTES	10	Number of
GUANAJUATO	36	Employees
MICHOACAN	2	Not Available
HIDALGO	13	
VERACRUZ	2	
CAMPECHE	3	

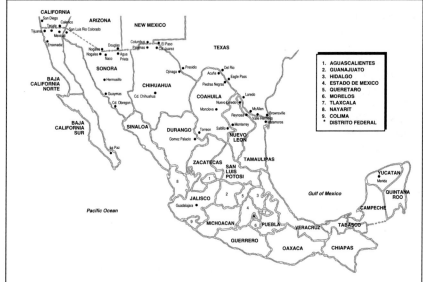

1. AGUASCALIENTES
2. GUANAJUATO
3. HIDALGO
4. ESTADO DE MEXICO
5. QUERETARO
6. MORELOS
7. TLAXCALA
8. NAYARIT
9. COLIMA
* DISTRITO FEDERAL

Indian in population. The Mexican government suggested that the insurrection was supported by Roman Catholic priests, followers of liberation theology discussed in Chapter 8. The government has negotiated land tenure and other issues with the rebels, but the uprising served as a reminder of how far Mexico has to go to achieve the goals of economic welfare and political equity for all of its peoples.

Creating a Western Hemisphere Free Trade Region

United States President George Bush proclaimed a goal of forging a free trade region of the entire Western Hemisphere, "Enterprise for the Americas," and President Clinton agreed. Chile has been suggested as a likely early partner to NAFTA. Many international trade agreements were signed in the Western Hemisphere in the early 1990s. The United States renewed the Caribbean Basin Initiative, discussed in Chapter 12. Venezuela, Colombia, Peru, Bolivia, and Ecuador agreed to achieve free trade among themselves (the Andean Pact). Mexico signed free trade agreements with Chile, Colombia, and Venezuela. El Salvador, Honduras, Guatemala, Nicaragua, Costa Rica, and Panama agreed to work toward economic integration. In addition, the English-speaking countries of the Caribbean united in a Caribbean Union and Common

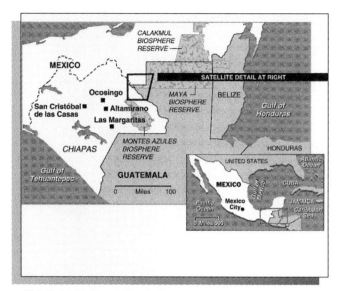

FIGURE 13-18

The satellite photograph of the border between Guatemala and Mexico's state of Chiapas reveals the devastation of the rainforest on the Mexican side. This devastation has been caused by chopping down the forest and then trying to farm the poor soils too intensively, and by extensive commercial logging. Food production has not kept up with population growth.
(Map courtesy of *The New York Times*, Jan. 9, 1994; photograph by NASA/John C. Stennis Space Center.)

Market (CARICOM). Colombia, Mexico, and Venezuela committed themselves to free trade by 2003, and Chile signed trade agreements with Argentina, Bolivia, Colombia, Venezuela, and Mexico. Argentina, Brazil, Paraguay, and Uruguay united in a pact called MERCOSUR. All of these agreements are *free investment* pacts as much as *free trade* pacts.

In Latin America as in Europe, the unification of transport infrastructures reflects the expanding territory being organized. For example, a new railroad tunnel and new gas and oil pipelines are piercing the high Andes mountains between Chile and Argentina. Brazil and Argentina reconciled the gauges of their railroad systems, and a new highway is under construction from São Paulo to Buenos Aires. By the year 2000, Mercosur nations plan to complete a five-nation inland waterway that will allow barges to travel 2,140 miles (3424 km) down the Paraguay and Paraná Rivers from eastern Bolivia to the sea. By October 1995, Brazil hopes to inaugurate a separate system of river locks that will allow barges to travel 1,500 miles (2400 km) from Paraguay to São Paulo.

Only 15 percent of total Latin American trade in 1990 was with other Latin American countries. This is a small amount compared with the level of trade typical between industrialized neighbors, and it suggested that the various Latin American states did not produce items they could exchange to mutual profit. The signing of the new trade agreements, however, might encourage specialization of production, increases in productivity, and intraregional trade and growth.

Intraregional trade is soaring. The United States is the principal supplier of imported goods to most Latin American countries, so growth in Latin America usually boosts the U.S. economy.

Other Regional International Groups

The European Union, the Commonwealth of Independent States, and the new Western Hemisphere trade area are the three most important regional international organizations today, but many others exist (Figure 13-19). Some of these are primarily military-defensive. With the end of the Cold War, however, those may turn into general economic and cultural associations. For example, the Association of Southeast Asian Nations (ASEAN) began as a military alliance, but in 1992, it became a free trade organization. Other organizations, such as the Southern African Development Community (SDAC), are economic. The purpose of each is simply to provide a framework for consultation to reach agreement whenever possible in areas of mutual concern. In some cases, however, the members of these organizations can agree on very little and have even gone to war with one another. The Organization of African Unity, for example, and the Arab League, serve few functions.

The fluidity of world economic and political groupings in the 1990s has stimulated many theories of what new groupings might form in the future. The distinguished political scientist Samuel Huntington has predicted that "the fundamental source of conflict [in

A Few World Regional Organizations

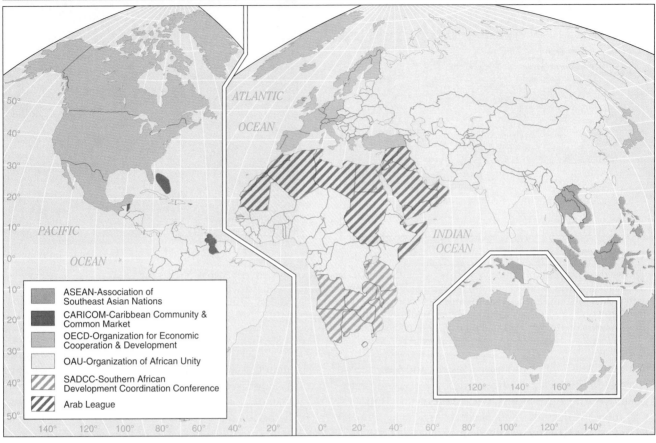

FIGURE 13-19

These are just a few of the many international organizations that provide frameworks for international cooperation.

coming decades] will not be primarily ideological or primarily economic. The great divisions among humankind and the dominating source of conflict will be cultural." He has warned of a coming "clash of civilizations." He defines a civilization as "the highest cultural grouping of people and the broadest level of cultural identity people have short of that which distinguishes humans from other species," and his list of major civilizations today includes "Western, Confucian, Japanese, Islamic, Hindu, Slavic-Orthodox, Latin American and possibly African civilization." With few exceptions, these are the major religious groupings and regions discussed in Chapter 8. Professor Huntington suggests that these realms are the cultural equivalents of the world's geological tectonic plates discussed in Chapter 1. If that is true, then the areas where they meet might experience virtually unending strife. Geographers have traditionally referred to such areas as *shatterbelts*. This book has already mentioned a few, including Bosnia (Muslims caught between Eastern and Western Christians), Sudan (an Arab

Muslim north battling the Black Christian and animist South), and Kashmir (between Pakistan and India).

It remains unclear, however, whether the peoples within each of Professor Huntington's "civilizations" share any real affinity, or why those affinities might trigger hostilities against other groups. Culture matters, but Huntington's hypothesis of global culture wars simplifies at least the impact of migration discussed in Chapter 5 and the economic forces discussed in Chapter 12. Some observers have predicted that the Earth may be organizing into three major blocs—Europe, North America, and East Asia. If this division becomes rigid, it may splinter the global economy and impede global trade and political agreements.

THE COLLAPSE OF EMPIRES

Chapter 12 noted that the international economy can work smoothly only if all states agree on the rules of

488 *Chapter 13*

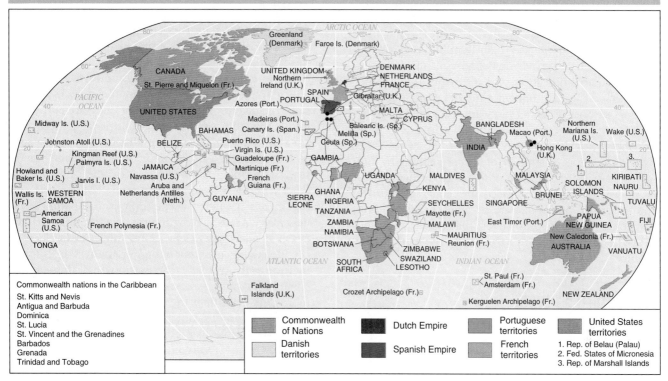

FIGURE 13-20

A few empires linger, whereas the breakup of others left behind cooperative organizations.

trade. Today, however, it is especially difficult to reach international agreements on any issue because the number of states has risen dramatically.

The following section discusses the breakups of a few individual empires that gave birth to large numbers of new states. In some cases frameworks of international cooperation linger; in others, quarrels over jurisdiction still haunt the world political map. The few remnants of empires can be seen in Figure 13-20. Since the breakup of the USSR, China and India are the greatest remaining multinational states.

British Empire to Commonwealth

At its peak in 1900, the British Empire covered over one-quarter of the Earth's land surface. This empire has gradually been transformed into a loose association of independent states, the Commonwealth of Nations. Most of the United Kingdom's former colonies have chosen to join, and today the Commonwealth has 51 members. Most have become republics, but Queen Elizabeth remains head of state in 16 countries in addition to the United Kingdom, including Canada, Australia, and Jamaica. The British legal system, language and education system, the Anglican church, and other cultural tra-

ditions linger in each of the countries. The Commonwealth cannot form a common economic or foreign policy, but it offers a framework for consultation and cooperation for the achievement of common ends, where they exist. As the United Kingdom turns toward its European Union partners, however, the future role of the Commonwealth is uncertain.

Britain retains a few small colonies around the world, and some of them present special problems. Hong Kong is a 409-square-mile (1060 km^2) colony at the mouth of China's Canton River. Since Hong Kong Island was acquired in 1841, it has developed great wealth as a transshipment point, a manufacturing center for apparel and consumer electronics, and as a financial service center. Hong Kong's 5.7 million people have enjoyed civil liberties. In 1985, Britain agreed to return Hong Kong to Chinese rule in 1997, but tens of thousands of people are leaving now and taking billions of dollars.

Britain retains the Falkland Islands ("Malvinas" in Spanish) against claims by Argentina, which even invaded the islands in 1982, and Britain retains Gibralter against claims to it by Spain. Residents of both colonies have repeatedly voted to continue their current status.

International Organization: Regional Organizations and Global Coordination **489**

Northern Ireland's status is open to dispute. The United Kingdom gave independence to 26 of Ireland's 32 counties in 1921, but the 6 counties of Northern Ireland (Ulster) elected to remain within the United Kingdom. The Irish Republic regards the 1921 partition as provisional and remains formally committed to unification. Great Britain and Ireland agree that Ulster can remain a province of Britain for as long as most of its people want it to, but since the 1960s, it has been wracked by violence. The province's Roman Catholics, about 40 percent of the total population, complain of discrimination by the majority of Protestants, and both Catholic and Protestant paramilitary groups are guilty of murder and terrorism. Birth rates are so much higher among the Roman Catholics, however, that Ulster might be overwhelmingly Roman Catholic early in the 21st century.

The French Empire

France built up one empire in the seventeenth and eighteenth centuries but lost most of its colonies during an eighteenth-century rivalry with Britain. During the nineteenth century France built up a second empire, beginning with the occupation of Algeria in 1830 and culminating in 1919 with the assumption of rule in Syria, Togoland, and Cameroon. The French believed that their subject peoples would mature not to independence but to full representation in the government in Paris. The four French colonies of Martinique, Guadaloupe, Réunion, and Guiana were organized as Overseas Departments of France, and each was granted representation in the French National Assembly in 1946.

France granted independence to most of its other colonies in the 1950s, but it offered various forms of continuing alliance. Still today France is the largest donor of aid to African countries and maintains the largest non-African military forces on the continent. France intervened unilaterally in Rwanda in 1994 to stop a civil war there. Thirteen former colonies use an international currency, the CFA Franc, whose exchange rate is stabilized against the French franc, but a 50 percent devaluation of the CFA franc in January 1994 signaled France's decreasing willingness to support its former colonies.

The present French Republic encompasses, in addition to mainland France and the four Overseas Departments, two Territorial Collectivities and four Overseas Territories. France also possesses a number of islands. New Caledonia holds important nickel reserves, but the other territories are generally poor and sparsely populated.

The Successor States of the Ottoman Empire

When the Ottoman Turkish Empire collapsed and a new Turkish nation was born in the 1920s, most of the Turkish Empire in the Middle East was divided between France and Britain. New states were eventually granted independence, but the map of today's states results from the interests of the colonial powers rather than the sentiments of the local people, and several borders have been disputed for 75 years. Israel was carved out to provide a homeland for the Jewish people, and bitter confrontation over this land—historic Palestine—continues. Furthermore, the region has remained a cauldron of competing loyalties. Many Arabs bestow their loyalty on units that are much smaller than states (clans, tribes, or families) or on ideas that are much bigger than states. These include pan-Arabism, the notion that there exists some bond among all Arabic-speaking peoples that is more legitimate than the modern Arabic states, and a universal bond of Islam.

The Middle East is plagued by border disputes, rulers whose power is based on foreign interests rather than popular support, religious animosities, and political and economic intervention by international oil companies and banks. Several wars have troubled the region since World War II. Lebanon has suffered civil war, Iran has fought Iraq, and in 1990 Iraq attempted to annex Kuwait. Israel has been able to defend its independence, and it also has made its desert land bloom and built a strong economy. To accomplish this, Israel has relied on generous financial assistance from the U.S. government (in recent years, about $500 per Israeli per year) plus billions of additional dollars from U.S. private citizens. The 1993 agreement between Israel and the Palestine Liberation Organization may open wider Arab diplomatic recognition of Israel, peace, and coordinated regional economic development.

The Empire of the United States

The United States organized most of its territories and admitted them to the Union fairly quickly. With the granting of statehood to Alaska and Hawaii in 1959, the United States incorporated even overseas territories, as France has.

Former Colonies The United States took Cuba from Spain in 1898 but granted it full independence in 1934 and today retains only a naval base in the island's Guantanamo Bay. The United States also took the Philippines from Spain in 1898, granted them

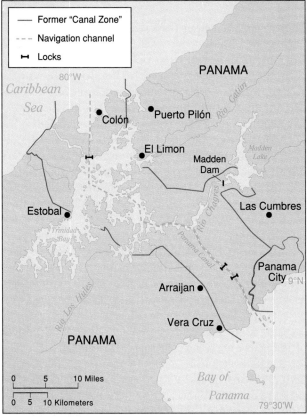

The Panama Canal

Former "Canal Zone"
Navigation channel
Locks

FIGURE 13-21

The Panama Canal is too small to accommodate many of today's ships, so an international group representing both governments and private interests in Panama, the United States, and Japan is considering reconstructing the Canal.

independence in 1946, but retained military bases until 1990.

Panama and Liberia have been politically tied to the United States so closely for so long that their genuine independence may be questioned. The United States provoked Panama's uprising for independence from Colombia in 1903, and then, by prearrangement, leased the Canal Zone from the new country. A 1979 treaty allows Panama to take responsibility for the Canal in 2000, but until then the United States operates it through an independent federal agency supervised by five Americans and four Panamanians (Figure 13-21). All nine are appointed by the president of the United States.

In 1989, the United States invaded Panama to interdict drug traffic to the United States and to remove from power and bring to the United States for trial a military dictator, Manuel Noriega, as an international drug trafficker. The United States continues to occupy the country, but, according to the U.S. Department of State, drug traffic has increased.

Some 14 percent of the world's merchant ships are registered in Panama. This would seem to be an astonishingly high percentage for such a small, poor country. Many of these ships, however, are owned by U.S. corporations, which register the ships in Panama to evade U.S. taxes and regulation. The Panamanian flag is known as a *flag of convenience.*

The United States maintains a "special relationship" (an official State Department term) with Liberia, which was founded as a haven for freed U.S. slaves and which received independence in 1847. Inequities in wealth and political power fuel antagonism between the dominant Americo-Liberians (in 1990, only 5 percent of the total population of 2.5 million) and the indigenous Africans, and the indigenous population is itself split among ethnic groups. A civil war that opened in the 1980s continued into the 1990s, despite the presence of an international peace-keeping force representing five West African states.

The United States maintains military facilities in Liberia and gives Liberia several hundred million dollars in aid each year. The U.S. dollar is Liberia's official currency. The Liberian flag is another flag of convenience. Liberia registers about 6 percent of the world's merchant marine fleet, representing 13 percent of total world gross tonnage. The fleets of both Panama and Liberia may actually be owned by corporations from many countries, but in both cases U.S. ownership is probably most significant.

U.S. Colonies Today The United States still retains a modest empire of islands. Their total population is about 4 million, most of whom are citizens of the United States. The five largest colonies—Puerto Rico, the Virgin Islands, Guam, Samoa, and the Northern Marianas—have locally elected governors and legislatures. They are subject to U.S. laws, although they have no direct voice in political processes. All use U.S. currency and depend on the United States for their economic well-being.

Puerto Rico The United States granted Puerto Rico territorial status in 1917 and elevated it in 1952 to the unique status of a free commonwealth. Puerto Rico enjoys internal autonomy, but Puerto Ricans are also under the jurisdiction of the federal government. They do not vote in presidential elections, and their resident commissioner in Congress can vote only in committee meetings. The United States and Puerto Rico share a common market and monetary system, but Puerto Ricans pay no federal taxes, and some federal customs and excise duties are paid back into the island's treasury. In November 1993, Puerto Rico held a nonbinding plebiscite on the island's future. Some 73 percent of Puerto Rican voters turned out,

48 percent of whom opted to continue the commonwealth status. Forty-six percent voted for statehood, and 4 percent voted for independence.

Most arguments being forwarded in the debate over independence are economic. The densely populated, resource-poor island now enjoys a standard of living far above that of any other Caribbean or Latin American nation, so it may be considered to be either a rich Caribbean country or, conversely, a poor part of the United States. Puerto Rico's 3.4 million people would rank it twenty-eighth among the states in population, but its per capita income ($6,600 in 1990) was only two-thirds that of Mississippi, currently ranked last.

Statehood would end federal tax benefits for U.S. companies that build facilities on the island. These laws have encouraged industrialization, but mainland states complain that they drain jobs from other parts of the United States. Overall, manufacturing accounts for 40 percent of Puerto Rico's economic output and for about 18 percent of employment. Agriculture accounts for only 1.5 percent of output and 4 percent of employment, far below the levels in tropical developing countries.

Puerto Ricans are Spanish-speaking, and many wish to retain their linguistic and cultural traditions. This might make it difficult for Puerto Rico to assimilate fully into U.S. life. Mainlanders have never been asked whether they are willing to accept Puerto Ricans into full Union status, nor probably will they be. If the Puerto Ricans ever choose independence, however, terms of continuing financial assistance plus leases on U.S. military bases (a full 13 percent of the territory) will have to be renegotiated.

The Monroe Doctrine The United States extends its military power over the entire Western Hemisphere by the terms of the Monroe Doctrine, which was promulgated in 1823. The doctrine declares that the United States considers dangerous to its peace and safety any attempt on the part of European powers to extend their systems of government to any point in the Western Hemisphere. The 1904 Roosevelt Corollary adds, "Chronic wrongdoing, or an impotence that results in a general loosening of the ties of civilized society, may...require intervention...and in the Western Hemisphere the...Monroe Doctrine may force the United States...to the exercise of an international police power." In the first nine decades of the twentieth century, the United States invaded and occupied an average of five Latin American countries per decade. United States military assistance has supported many undemocratic Latin American governments.

GLOBAL GOVERNMENT

The Dominican Father Francisco de Vitoria (1486-1546) earned recognition as "The Father of International Law" by arguing that each people has the right to its own ruler, its "natural lord." Therefore, de Vitoria argued, relations among nations would have to recognize mutual rights. Subsequent centuries, however, have shown only token respect for this principle.

Discussions of world order are generally based on four principles: (1) state sovereignty, (2) self-determination, (3) democratic governments within states, and (4) the universal protection of human rights. These four principles, however, may be imperfect, and they often conflict with one another. As the number of states has multiplied, the number of actual and possible conflicts among states has grown. The causes of these conflicts include aggressive ambitions, as with Iraq in the early 1990s; border disputes or rival claims on territory, as between Chad and Libya; and domestic crises or policies that have effects abroad, causing other states to threaten to intervene, as with Yugoslavia. World flows of refugees are an international problem. The necessity for common action in areas of common concern, such as the oceans or the global environment, is another pressure for global government. International society as a whole, however, has neither the centralized government, judicial system, and police that characterize a state; nor does it even have a consensus on what constitutes an international crime.

The first lasting international political structure was the League of Nations. The League was founded by the Allies after World War I, but several nations, including the United States, refused to join. The League provided a forum for discussion of world problems, and it carried out many humanitarian projects. It exercised little real power, however, and was unable to resolve the disputes that eventually led to World War II.

The United Nations

In 1942, 26 states that were allied against the Axis powers joined in a Declaration by the United Nations, pledging to continue their united war effort. A meeting of Allies in San Francisco in 1945 drew up a Charter for a United Nations organization to continue in existence after the war, and by the end of 1945 the new organization had 51 members. That number had risen to 184 member states by 1994.

The United Nations General Assembly serves as a rudimentary legislature of a world government, but it has no power of enforcement (Figure 13-22). The

FIGURE 13-22

The United Nations found a permanent home on the east side of Manhattan Island, where its headquarters buildings were designed by an international committee of architects, including Le Corbusier. The high slab of the Secretariat dominates the group, with the domed General Assembly to the north. The land legally is not part of the United States of America.
(Courtesy of Rafael Macia/Photo Researchers.)

power that the United Nations does exercise is vested in the 15-member Security Council. Five members of the Council (the United States, China, France, the United Kingdom, and Russia—originally the USSR) hold their seats permanently, and the remaining 10 members are elected by the General Assembly for two-year terms. This allocation of seats reflects the allocation of world power in 1945. In the 1990s, Germany and Japan, defeated in World War II but rich countries today, may win permanent membership status. The UN Secretariat, with the Secretary General at its head, handles all administrative functions.

The Security Council has occasionally voted for the UN to field armies. United Nations forces fought in Korea from 1950 to 1953, and in 1990 a Security Council resolution gave the United States and its allies sanction to force Iraq to retreat from Kuwait. In a greater number of cases, the UN has sent observer troops to patrol world trouble spots, and these incidents have been increasing. Examples include Kashmir; the Golan Heights; Lebanon; Cambodia; Yugoslavia; the Mediterranean island of Cyprus, split by feuding Greeks and Turks; and Haiti in 1993. In 1992, the Security Council voted to send a U.S.-led force to Somalia, which was wracked by civil war, to guarantee the distribution of international food relief supplies. The purpose of that mission expanded to restoring central civil government to Somalia.

In many cases the UN has assumed responsibility for monitoring the freedom of national elections. The idea that international governments can be legitimized by international standards represents a reversal of international law.

The International Court of Justice in the Hague is another branch of the United Nations that, on the surface, works like a branch of government. The litigant states, however, must choose to come before it, and the court has no power to enforce its decisions.

UN Special Agencies

The United Nations has more than 40 specialized agencies. Some facilitate communications among member states: the Universal Postal Union, the International Civil Aviation Organization, the World Meteorological Organization, and the Intergovernmental Maritime Consultative Organization. The UN International Telecommunications Union, for example, has allocated frequencies on the radio spectrum, thus allowing international expansion of cellular telephones, satellite computer transmissions, and digital radio broadcasting. Other UN agencies encourage international cooperation toward specific humanitarian goals: the Food and Agriculture Organization; the World Tourist Organization; the United Nations Educational, Scientific, and Cultural Organization; the International Atomic Energy Agency; the International Bank for Reconstruction and Development; and the International Monetary Fund. The UN Conference on Trade and Development (UNCTAD) works to assist the economic development and trade of the poor countries. In 1992, for example, UNCTAD developed a standardized computer customs collection system that was soon in use in over 50 countries.

Is National Sovereignty Inviolable?

The idea of national sovereignty means that international borders are inviolable. No matter how monstrous any regime may be, no matter how much it persecutes its own people or oppresses minorities, no outside state or international agency has the right to interfere. This is true even when internal persecution triggers international flows of refugees. There have been cases of widespread famine tolerated by the starving country's government (Ethiopia in the 1980s) or caused by a civil war (Somalia). In some cases a government's resistance to outside assistance, or chaos in the country, prevented international rescue efforts from succeeding or even taking place.

Events surrounding the 1991 United Nations war against Iraq, however, signaled the beginnings of a change in policy. The internationally agreed-on purpose of the war was only to force Iraq to retreat from Kuwait, not to bring down the government of Iraq. After Iraq's expulsion from Kuwait, however, international outrage forced the United Nations to establish "secure zones"—territories within Iraq within which the United Nations would not allow the Iraqi government to bomb its own citizens (Figure 13-23). Shiites remain relatively "secure" south of the 32nd parallel, and north of the 36th parallel the Kurds hope to establish the state they have long been promised.

In September 1991, U.S. President Bush, speaking before the UN General Assembly, insisted that international sanctions against Iraq remain in force as long as Saddam Hussein "remains in power." Referring to "nationalist passions," Bush said that no one can "promise that today's borders will remain fixed." "Despots ignore," he said, "the heartening fact that the rest of the world is embarked on a new age of liberty," and he called upon the United Nations to take up "the important business of promoting values... ." This could be interpreted as a challenge for the United Nations to interfere in the internal affairs of member states. Many representatives of despotic regimes must have felt challenged. In 1993, the UN created a War Crimes Tribunal to try Serbian leaders for "crimes against humanity."

JURISDICTION OVER THE EARTH'S OPEN SPACES

▼▼▼▼▼▼▼▼▼▼▼▼▼▼▼▼▼▼▼▼▼▼▼▼▼

One of the most important issues requiring international agreement is the adjudication of the rights that individual states have in the Earth's open spaces. Open spaces include the Arctic, Antarctica, and the world's seas.

The Arctic and Antarctica

Seven countries have territory north of the Arctic Circle: the United States, Canada, Denmark, Finland, Iceland, Norway, Sweden, and Russia. In 1991, these countries pledged cooperation in monitoring Arctic pollution and protecting the region's plant and animal life. In 1992, however, Russia admitted that for years the Soviet navy had dumped radioactive wastes in Arctic waters. Scientists are investigating the results of this action.

Canada claims dominion over the uninhabited Arctic Islands north of Canada all the way to the North Pole, but the United States has challenged Canada's claims and attempted to realize a passage for shipping from Alaska's North Slope across the Arctic to the U.S. East coast. The issue remains in dispute.

Antarctica has never had any permanent inhabitants, but seven states claim overlapping sovereignty. Australia, New Zealand, Chile, and Argentina, the world's most southerly states, claim sovereignty on the basis of proximity. The claims of Britain, Norway, and France are based on explorations (Figure 13-24). Neither the United States nor Russia, which support

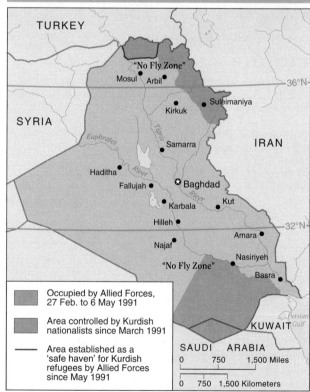

Protected Zones in Iraq

FIGURE 13-23

The United Nations war against Iraq did not topple Iraq's government, but several years after the war's end the United Nations still protects certain Iraqis from their own government.

494 Chapter 13

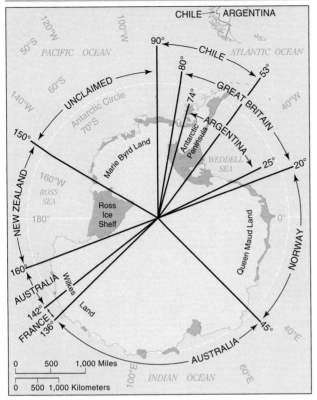

Claims to Antarctica

FIGURE 13-24

The United States does not recognize any of these overlapping claims to Antarctica.

most of the scientific research there, make any territorial claims, nor do they recognize other nations' claims. Forty states are parties to a 1959 Antarctic Treaty which resolves that "Antarctica shall continue forever to be used exclusively for peaceful purposes and shall not become the scene or object of international discord." The treaty prohibits military installations, but it opens the continent to all nations for scientific research and study. The treaty also prohibits signatories from harming the Antarctic environment.

Despite the terms of this treaty, however, some countries, led by Malaysia, demand that exploration should be undertaken for Antarctic resources and that any resources found should be exploited for the good of all nations. Other nations, led by France and Australia, insist that Antarctica should be set aside as a wilderness preserve.

Territorial Waters

The nation-states have divided up all the land on Earth (except Antarctica), but international agreement negotiates the question of how far out to sea the territorial claim of a country can reach.

The Dutch jurist Hugo Grotius (1583–1645) first argued that the world's seas are open space, *mare liberum*, which no political unit can claim. Each coastal state, however, has claimed coastal waters for defensive purposes. A later Dutch jurist, Cornelius van Bynkershoek, accepted this in his book *De Domina Maris* (1702). The limit of sovereignty was set at 3 nautical miles from shore, and this was generally accepted for over 200 years. All distances at sea are measured in nautical miles, an international standard defined as one minute of latitude—6,076 feet (1,852 meters). Territorial sea is measured from the low tide mark, and disputes between states arising from irregular coastlines or islands are resolved by individual negotiations. A 1958 International Conference on the Law of the Sea set rules for ascertaining the seaward extension of land borders.

The United States accepted a 3-mile limit in 1793, but in 1945, President Truman broke that international covenant and claimed sole right to the riches of the continental shelf up to 200 nautical miles out. A **continental shelf** is an area of relatively shallow water that surrounds most continents before the continental slope drops more sharply to the deep sea floor. The 1958 Sea Law Conference agreed on a water depth of 656 feet (200 meters) as the definition of a continental shelf (Figure 13-25).

The Truman Proclamation claimed shelf mineral rights, but it did not claim control over fishing or shipping in the seas over the shelf beyond the 3-mile territorial limit. It nevertheless triggered extended claims to fishing rights by other nations. In 1976, the United States extended its own claims to exclusive fishing rights up to 200 nautical miles out. As noted in Chapter 6, about 90 percent of the world's marine fish harvest is caught within 200 miles of the coasts.

The 1982 United Nations Convention on the Law of the Sea In 1982, the United Nations proposed a new law of the sea treaty. The procedure for adoption of international treaties is as follows: Any number of countries can sign a treaty, but the treaty is not in force until a specified number of those signers *ratify* the treaty. Alternatively, a number of countries which did not sign a treaty may *accede* to it. At the end of a specified period after the minimum number of countries has ratified a treaty, it comes into force among the countries that ratified or acceded to it. They are called *parties* to the treaty. For example, on November 17, 1993, Guyana became the sixtieth country to ratify the UN Law of the Sea Treaty; therefore, the treaty came into force among its parties on November 16, 1994.

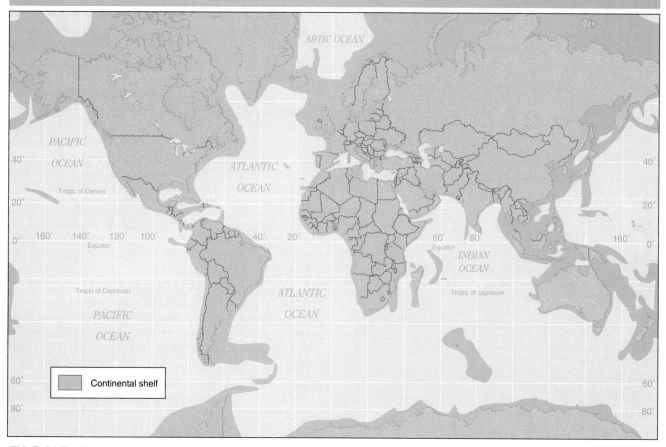

FIGURE 13-25

The various continental shelves extend out as little as a few hundred yards or as far as over 600 miles.

Once a treaty becomes effective, revision becomes very difficult.

The Law of the Sea Treaty authorizes each coastal state to claim a 200-mile exclusive economic zone (EEZ), in which it controls both mining and fishing rights. Therefore, possession of even a small island in the ocean grants a zone of 126,000 square miles (326,000 km^2) of sea around that island. This might help explain why France retains many small island colonies. France claims a full 10.4 percent of the world total EEZ. The United States claims 8.4 percent, New Zealand 5.8 percent, and Indonesia 4.7 percent. No other state claims over 4 percent.

Another clause of the 1982 treaty guarantees ships **innocent passage** through the waters of one state on the way to another. The extension of countries' territorial waters, however, means that countries now claim many of the world's narrow waterways. They remain open in times of peace, but they can be closed in time of war (Figure 13-26).

The United States refused to sign the treaty in protest against provisions calling for international-ization of seabed mineral resources and creation of an International Seabed Authority to control mining. The United States insisted on rights for free enterprise. A compromise on the disputed provisions was reached in 1994 and the United States intends to become a party. Several of the treaty's other provisions did serve U.S. interests. For example, one provision allows nations to declare a 12-mile territorial limit. Therefore, in 1983, U.S. President Reagan announced that the United States would regard all but the seabed provisions as law, even though the country had not signed the treaty. In accordance with the treaty, on December 28, 1988, the United States extended its territorial limit to 12 miles. Given the length of the U.S. coastline, this seaward extension enlarged the territory of the United States by some 185,000 square miles.

Landlocked States

Some 42 states are landlocked and without sea coasts. Several other states are not totally landlocked,

496 *Chapter 13*

Landlocked States and Strategic Straits

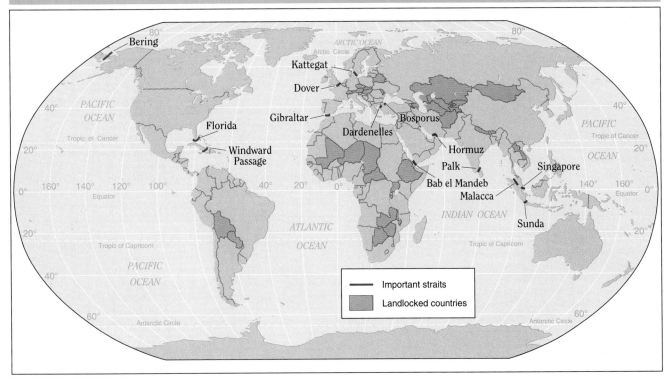

FIGURE 13-26

This map shows the world's landlocked states and some of the world's most important narrow sea lanes. As states have extended their seaward claims, many of these narrow waterways have been claimed by the adjacent states.

but their own coasts are unsuitable for port development, so they rely on neighbors' ports. Landlocked states must secure the right to use the high seas, the right of innocent passage through the territorial waters of coastal states, port facilities along suitable coasts, and transit facilities from the port to their own territory.

Landlocked or partially landlocked states may gain access to the sea in one of three ways. First, any navigable river that reaches the sea may be declared open to the navigation of all states. Freedom of navigation on rivers that flow through several countries was first proclaimed by France in 1792. The revolutionary government proclaimed that the freedom of rivers was a "natural law." International commissions regulate navigation on many international rivers, and often these same commissions guard against pollution and regulate the drawing of irrigation waters from the river (Figure 13-27).

Second, a landlocked state may obtain a corridor of land reaching either to the sea or to a navigable river. Several countries have long, thin extensions out to seaports (Figure 13-28). Some of these, such as

Zaire's corridor to the Atlantic Ocean, are important transport routes, but others, such as Namibia's Caprivi Strip to the international Zambezi River, serve no significant traffic function.

The third way that a landlocked state can gain access to the sea is to obtain facilities at a specific port plus freedom of transit along a route to that port. Coastal states have signed international conventions promising to assist the movement of goods across their territories from landlocked states without levying discriminatory tolls, taxes, or freight charges. Chile, for example, helped build a railroad connecting La Paz, Bolivia, to the Chilean port of Arica, and Chile guarantees free transit. Bolivia claims the two miles of seacoast north from Arica to the Peruvian border, but Peru also disputes that territory with Chile. Argentina grants Bolivia a free zone at the Argentine city of Rosario on the Parana River, and in 1993, Peru granted Bolivia a free trade zone in the port of Ilo and agreed to help pave the roads from Ilo to La Paz. In 1993, Ethiopia joined the ranks of the landlocked states when its coastal province of Eritrea gained independence. Eritrea

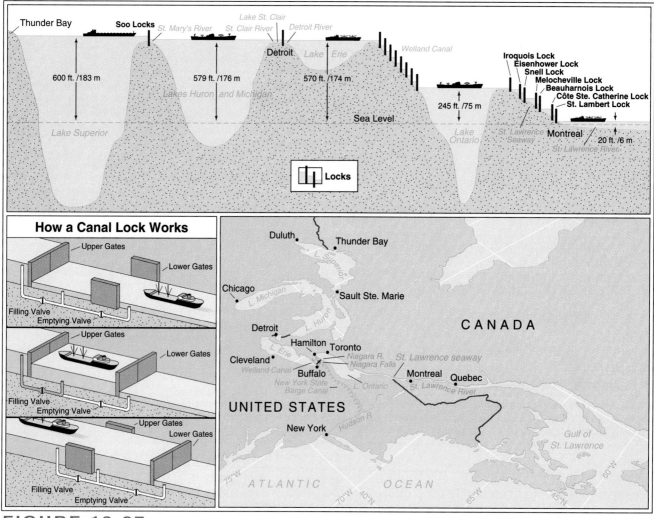

FIGURE 13-27

The Saint Lawrence Seaway, completed jointly by the United States and Canada in 1959, is an example of international cooperation on an international waterway. It allows ocean vessels to reach from the Atlantic Ocean into the Great Lakes; the joint project included the construction of hydroelectric power plants.

promised to assist Ethiopian import and export trade, but in fact most of Ethiopia has long relied on transit via the port of Djibouti (look back at Figure 12-18).

Coastal states can profit by granting transit rights. The coastal state's railroads have a captive customer. Coastal states' ports gain extra business and opportunity to serve as break-of-bulk points for processing imported or exported raw materials (see Chapter 9). The fact that a landlocked state misses these opportunities can hinder its economic development. Landlocked states also lose out in the apportionment of the resources of the continental shelves, exclusive economic zones, and fishing opportunities.

The High Seas

All nations agree that beyond territorial waters are high seas, where all nations should enjoy equal rights. Several problems, however, threaten any area where rights are not specifically assigned. These include depletion of resources (such as over-fishing) and pollution. International conventions regulate each of these potential abuses of the high seas. The 1972 London Dumping Convention, for example, signed by 70 countries, originally banned the dumping of radioactive and other "highly dangerous" wastes at sea, but it has been extended to ban dumping of all forms of industrial wastes by 1995. The

A Few Corridors

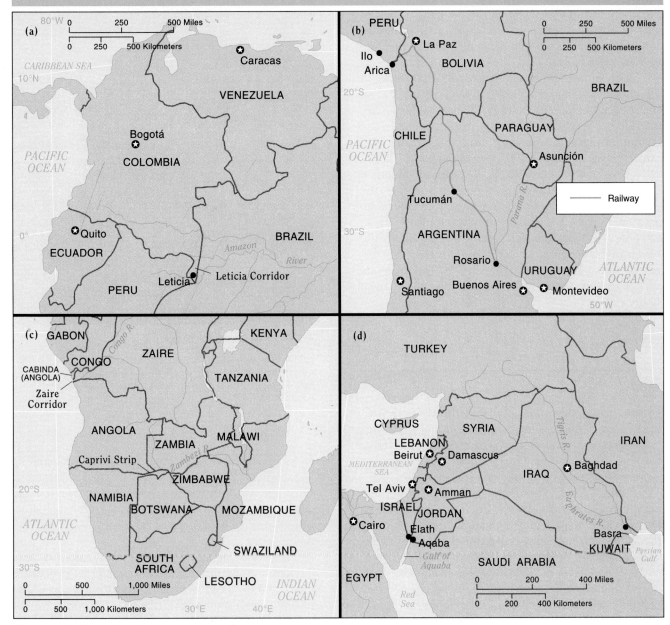

FIGURE 13-28

Several countries that would otherwise be landlocked have narrow corridors of territory reaching to provide access to the sea or to a navigable river. (Note: Sucre is technically the capital of Bolivia, but La Paz is the "seat of government.")

enforcement of international agreements such as these is uncertain, and their effect is limited.

Airspace

The question of how far up a nation's sovereignty extends is as complex as the question of how far out to sea it extends. Ships enjoy innocent passage through territorial waters, but airplanes have never been granted innocent passage to fly over countries. Airlines must negotiate that right, and states commonly prescribe narrow air corridors, the altitudes at which aircraft must fly, and even the hours of the day the passages are open. United States President Dwight Eisenhower proposed an "Open Skies" treaty in 1955, which would allow the planes of one country to fly over others. An Open Skies Treaty was signed in March 1992, in Helsinki, Finland, and in

1993, the U.S. Senate voted to ratify. The treaty establishes conditions for unarmed observation flights over the entire territory of participants. In order to prevent secret military spying, the treaty stipulates that all sensing equipment used in these flights must be commercially available.

Diplomats cannot agree on the altitude at which national airspace ends. One possible definition of national airspace would limit it to the lowest altitude at which artificial unpowered satellites can be put into orbit at least once around the Earth. That ranges between 70 and 100 miles (113 and 161 km).

No country wants to be the first to militarize space, but suggestions have been made that the eventual parceling-out of space is almost inevitable, as a measure of national security. Countries have so far refrained from extending territorial claims to the moon or planets, but several have insisted that they reserve the right to do so (Figure 13-29). Our study of geography, therefore, is so far restricted to the Earth.

FIGURE 13-29

When U.S. astronaut Neil A. Armstrong first walked on the moon on July 20, 1969, he planted a U.S. flag. The United States denies that the gesture was intended as a claim. (Courtesy of NASA.)

SUMMARY

International organizations coordinate activities among countries, whereas supranational organizations exercise powers over countries. Most active international or supranational organizations today are working for broad economic or social goals.

After World War II, the Allies split into two competitive blocs divided by an Iron Curtain across Europe. To the East, the Communist nations formed the Warsaw Pact military organization and the COMECON economic organization. Both of these collapsed in 1991, and most of their members have since petitioned to join the Western organizations. The Western nations formed the NATO military organization and the economic community known today as the European Union. Its 16 members (scheduled for January 1995) are committed to forming one economy and eventually one government. The total territory is merging into one market for goods and capital, one transport network, and one region of common social and regional development policies. The European Free Trade Association (EFTA) and the EU are one free trade region which may expand to include Eastern European countries.

The Union of Soviet Socialist Republics was a Russian-dominated multinational empire. It collapsed in 1991 and has been replaced by the Commonwealth of Independent States (CIS). Shared economic interests is a major centripetal force among its members.

The United States, Canada, and Mexico are merging their economies. NAFTA will probably slowly redistribute economic activities throughout North America.

The British Empire has been transformed into the Commonwealth of Nations, but Britain retains a few colonies, as France does. The United States also retains a modest empire of islands.

As the number of states has grown, the number of actual and possible conflicts among them has. The UN General Assembly serves as a rudimentary world legislature, but power is vested in the Security Council. The United Nations has many specialized agencies.

The rights of different countries concerning the Earth's open spaces, including the Arctic, Antarctica, and the seas, are negotiated. Landlocked states must secure access to the seas and a port facility. Beyond territorial waters are high seas. National airspace might be limited to the lowest altitude at which artificial satellites can be put into orbit at least once around the Earth.

KEY TERMS

international
   organization (p. 462)
supranational
   organization (p. 462)
bloc (p. 462)
North Atlantic Treaty
   Organization
   (NATO) (p. 464)
Marshall Plan (p. 464)
European Union (EU)
   (p. 466)
free trade area (p. 466)
customs union (p. 466)
common market (p. 466)
European Economic
   Area (EEA) (p. 471)
Russification (p. 472)
local content
   requirement (p. 485)
continental shelf (p. 495)
innocent passage (p. 496)

QUESTIONS FOR INVESTIGATION AND DISCUSSION

1.   Investigate and analyze the assignment of seats in the European Parliament. Does it represent population? Economic power?

2.   List factors of European cultural unity and disunity. How will various factors boost or retard cooperation among the countries and economic development?

3.   Most U.S. citizens know very little about Canada. Why?

4.   Could you recommend a weighted system of voting in the UN General assembly—by population, wealth, or territory?

5.   Investigate a few of the projects of the specialized agencies of the United Nations, such as UNICEF, the FAO, or WHO.

6.   French television commentator Christine Ockrent has said, "The only true pan-European culture is American culture." What did she mean by that? Can we say that the only true global culture is American culture?

7.   How has physical geography encouraged or discouraged the formation of multinational bonds in Europe, in the CIS states, or in North America? Are there obvious complementarities in products? Do transportation or settlement patterns act as centripetal forces?

8.   What impact will the EU have on the role of the United States in world affairs?

9.   The international community sometimes enforces economic sanctions against a country to force its government to change its policies or actions. Are such sanctions always successful? Whom do they harm?

10.   What might happen to Antarctica if a practical way is found of extracting its resources?

ADDITIONAL READING

ADAMS, WILLIAM JAMES, ed. *Singular Europe: Economy and Polity of the European Community After 1992*. Ann Arbor: University of Michigan Press, 1992.

ASLUND, ANDERS and RICHARD LAYARD, EDS. *Changing the Economic System in Russia*. New York: Saint Martin's Press, 1992.

Canadian Journal of Regional Science (Vol 13. No. 2/3). Special double issue devoted to the spatial impact of the U.S.-Canadian free trade agreement. Québec: Université du Québec.

DAWSON, ANDREW H. *A Geography of European Integration*. London: Belhaven Press, 1993.

EICKELMAN, DALE F., ed. *Russia's Muslim Frontiers*. Bloomington: Indiana University Press, 1993.

HARDT, JOHN P. and RICHARD F. KAUFMAN, eds. *The Former Soviet Union in Transition*. Published for the Joint Economic Committee, U.S. Congress. Washington, D.C: Government Printing Office, 1993.

HOMER-DIXON, THOMAS FRASER. "On the Threshold: Environmental Changes as Causes of Acute Conflict." *International Security*, Fall, 1991.

MCKNIGHT, TOM. *Regional Geography of the United States and Canada*. Englewood Cliffs, NJ: Prentice Hall, 1992.

MILNER-GULLAND, ROBIN and NIKOLAI DEJEVSKY. *Cultural Atlas of Russia and the Soviet Union*. New York: Facts on File, 1991.

PETERSON, D. J. *Troubled Lands: The Legacy of Soviet Environmental Destruction*. Boulder, CO: Westview Press, 1993.

PRESCOTT, JOHN ROBERT VICTOR. *Maritime Political Boundaries of the World*. New York: Methuen, 1986.

United Nations. *The Law of the Sea*. New York, 1983.

WHITING, VAN R. *The Political Economy of Foreign Investment in Mexico*. Baltimore, MD: Johns Hopkins University Press, 1992.

WILLIAMS, ALLAN. *The European Community: The Contradictions of Integration*. Cambridge, MA: Blackwell Publishers, 1991.

Epilogue: Protecting the

Political community was defined in Chapter 10 as a willingness to join together and form a government to solve common problems. Chapters 12 and 13 listed a number of the problems and challenges facing all humankind that are increasing global political community and cooperation. These include conduct among states, jurisdiction and regulation of use of the Earth's open spaces, the high seas, Antarctica, and the regulation of the evolving global economy. Protection of the global environment is another challenge to all humankind. Unlike the competitive national goals of building economic or military might, this goal is positive and inclusive, and it cannot be achieved by any one country acting alone. Environmental matters threaten the internal political stability of some countries, and they are a source of friction among countries.

International bilateral cooperation was pioneered by the United States and Canada. Those countries were the first two to allow a citizen of one country who is affected by environmentally harmful activity in another country to sue in the second country to stop or modify that action and to collect damages. Such bilateral cooperation has been continued. The United States and Canada have, for example, acted jointly to clean up pollution in the Great Lakes. In 1992, the United States and Mexico signed an agreement to cooperate in cleaning up pollution along that border as well.

Global cooperation to protect the environment was first proposed by U.S. President John Kennedy in an address to the UN General Assembly in September 1963. He urged international action "to protect the forest and wild game preserves now in danger of extinction; to improve the marine harvests of food from our oceans; and to prevent the contamination of air and water by industrial as well as nuclear pollution." The UN sponsored a Conference on the Human Environment in Stockholm, Sweden, in 1972, which was attended by more than 100 nations representing over 90 percent of the world's population. The Conference issued a 26-point *Declaration on the Human Environment*. That document recognized humankind's "solemn responsibility to protect and improve the environment for present and future generations." The near-universality of acceptance of this document marked it as a milestone in global cooperation.

In 1973, 108 countries signed a Convention on International Trade in Endangered Species of Wild Flora and Fauna; this led to a 1990 world ban on trade in ivory. In 1994, for the first time, the United States imposed trade sanctions on another country (Taiwan) for refusing to halt the sale of products made from endangered species—tiger bones and rhinocerous horns. Earlier, in 1987, the world's richest countries had agreed to ban the importation of products made with ozone-destroying chlorofluorocarbons.

The UN called a second conference on global environmental matters in Oslo, Norway, in 1988. That conference, titled "Sustainable Development," emphasized that economic growth must go hand-in-hand with preservation of natural resources.

Leaders of 172 countries attended the Third United Nations Conference on the Global Environment in Rio de Janeiro, Brazil, in June 1992. They fashioned an environmentally-sound approach to growth that hinges on relieving poverty, conserving

FIGURE 1

These photographs taken (a) ten years ago and (b) recently illustrate the considerable die back in Germany's Black Forest caused by acid rain.
(Courtesy of Régis/Bossu/Sygma.)

Global Environment

resources, and cutting pollution before the 21st century. This is called *Agenda 21*. In addition, most representatives signed two landmark treaties: one to combat global warming (the Climatic Change Convention), and one to protect the diversity of plant and animal species (the Biodiversity Treaty). Both were ratified by enough countries to come into effect by January 1994. The Biodiversity Treaty commits countries to conserve species within their borders, protect endangered species, expand protected areas, and promote public awareness of the need for conservation. The treaty's most controversial clause establishes rules for compensating developing countries for the commercial use of their genetic materials. As noted in Chapter 1, the rich biological diversity of many of the world's poor tropical countries has often been exploited by the rich countries. The new treaty gives the poor countries a strong economic incentive to conserve their biodiversity.

Many participants at the Rio conference expressed dismay that the conference avoided discussing two topics of great and growing significance for most of humankind: population growth and urbanization. Both of these topics affect the environment in general and the human environment in particular, but other international conferences have achieved collaborative progress on defining goals in these areas.

The principal international division on many issues of environmental protection is between the rich countries and the poor countries. The rich countries advise the poor countries to develop their economies only in sustainable ways and at sustainable rates. These words of advice, however, seem to contradict both the past experience and even the contemporary behavior of the rich countries. Many of the rich countries polluted their own environments in the process of growing rich, and as their wealth has increased, so has their ability to repair environmental damage. Nevertheless, they are still responsible for much of the Earth-threatening pollution. Their burning of fossil fuels, for example, makes the greatest contribution to greenhouse gases (Figure 1).

Many of the poor countries are skeptical of the advice from the rich countries. Many of them desperately need to increase their per capita product and are willing to do this in any way they can. Perhaps only after having reached some minimum level of per capita product will they be able to consider the sustainability of what they have achieved.

Are the rich countries willing to help the poor countries reach this level of per capita development? What is the responsibility of the rich countries to reduce their own pollution before they admonish or advise the poor countries on their policies? (See Figure 2.) If environmental pressures result directly from population increases, is family planning, perhaps, the most important of all potential environmental protection measures? These are among the most important questions facing humankind. Watch for the evolution of a global political community to confront these issues, or, better, *participate* as an educated citizen in forming such a community.

FIGURE 2

The smoke from an oil refinery obscures the setting sun in Torrance, California.
(Courtesy of Kathleen Campbell/Gamma-Liaison.)

Appendix:
Map Projections

A map projection is a method of portraying the rounded surface of the Earth on a flat surface. No matter how this is done, something will be wrong with the shapes and relative sizes of regions and the distances and directions among places.

The basic principle of projecting a map is direct and simple. Imagine a transparent globe on which are drawn meridians, parallels, and continental boundaries and that has a light bulb in its center. A piece of paper in the form of a geometric figure (such as a cylinder or a cone) is placed over the globe (see Figure A-1). When the bulb is lighted, the lines drawn on the globe are "projected" outward onto the paper cylinder or cone. Those lines can then be sketched on the paper, and the paper laid out flat, producing a map projection.

In actuality, very few map projections have ever been constructed by the direct projection of data from a globe onto a piece of paper. Nearly all projections have been derived by mathematical computation. Their common feature is that each projection shows the correct location of latitude and longitude on the Earth's surface. In other words, each projection consists of an orderly rearrangement of the geo-

Theory of Map Projection

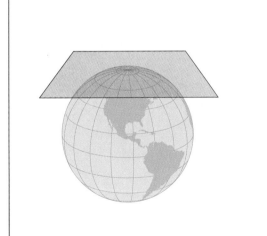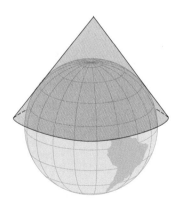

The three common geometric figures used in projections are a plane, a cylinder, and a cone.

FIGURE A-1

graphic grid transposed from the globe to the map. The difference among projections is the difference among the grid arrangements.

There is no way to avoid distortion completely, so no map projection is perfect. Each of the many hundreds of map projections has been designed as a compromise from reality to achieve some purpose. Each projection has some advantage over the others, but it also has its own particular limitations.

More than a thousand map projections have been devised. Most of them can be grouped into just a few families, based on their derivation. Projections in the same family generally have similar properties and relative distortion characteristics. Figure A-2 provides examples of four major families: cylindrical, elliptical, azimuthal, and conic projections.

A cylindrical projection is derived from the concept of projection onto a paper cylinder that is tangential to, or intersecting with, a globe. Most cylindrical projections are designed so that the cylinder is tangent to the globe (that is, just touches the globe) at the equator. This produces a right-angled grid net-

work (which means that meridians and parallels meet at right angles) on a rectangular map. There is no distortion at the circle of tangency, but distortion increases progressively away from this circle. Thus, most cylindrical projections display little distortion in low latitudes but enormous distortion in the polar regions. Cylindrical projections are generally used for maps of the whole world. The Mercator projection is an example of a cylindrical projection.

Elliptical projections are oval shaped and display the whole world, although their central sections sometimes are used for maps of smaller areas. In most elliptical projections a central parallel (usually the equator) and a central meridian (generally the prime meridian) cross at right angles in the middle of the map, which is a point of no distortion. Distortion normally increases progressively toward the outer margins of the map. Parallels are usually arranged parallel to one another, but meridians (apart from the central meridian) are shown as curved lines.

An azimuthal projection is derived by the perspective extension of the geographic grid from a

Four Sample Projections

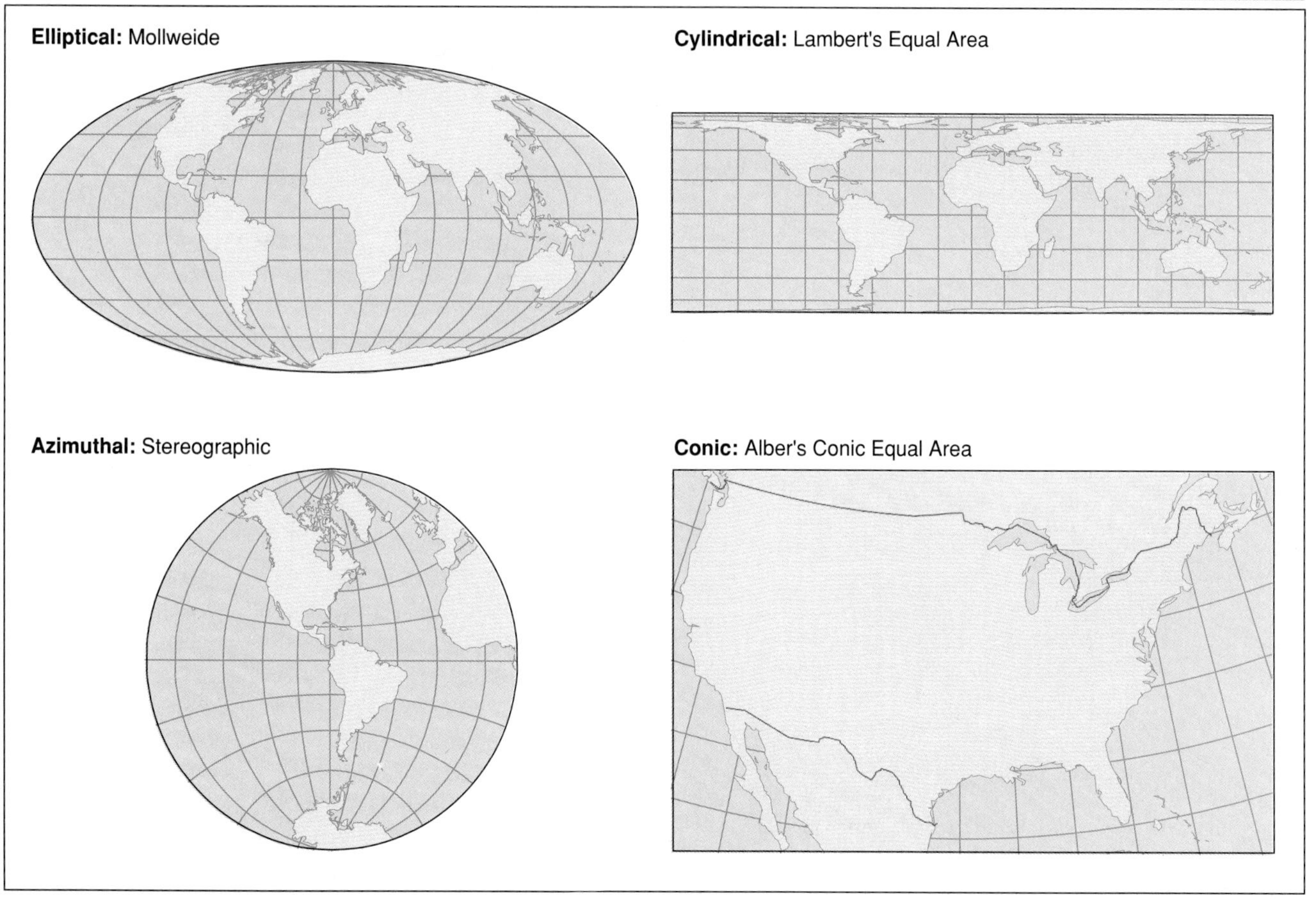

Elliptical: Mollweide

Cylindrical: Lambert's Equal Area

Azimuthal: Stereographic

Conic: Alber's Conic Equal Area

FIGURE A-2

globe to a plane that is tangent to the globe at some point. There is no distortion at the point of tangency, but distortion increases progressively away from that point. No more than half the Earth (a hemisphere) can be displayed with any success on an azimuthal projection. Thus, azimuthal projections have a logical "look" because they portray the Earth as you might view a globe, or an astronaut's perspective from space. They have the same drawbacks as globes: Only half the Earth can be seen at one time, although they can be useful for focusing attention on a specific region of the world.

A conic projection is conceived as having one or more cones set tangent to, or intersecting, a portion of the globe, and the geographic grid is project-ed onto the cone(s). Normally the apex of the cone is considered to be above a pole, which means that the circle of tangency coincides with a parallel, which becomes the standard parallel of the map. Distortion increases progressively away from this parallel. Consequently, these projections are best suited for regions of east-west orientation in the middle latitudes, which makes them particularly useful for maps of the United States, Europe, and China. It is impractical to use conic projections for more than one-fourth of the Earth's surface (a semihemisphere). They are particularly adapted for mapping smaller areas.

There are dozens of other projections that do not fit into any of the four families discussed above. Each has its specific uses and limitations.

Glossary

A

Aborigine: The name given to any indigenous people, but, in particular, those of Australia.

absolute location: The location of a place as pinpointed in terms of the global geographic grid.

acculturation: The process of adopting some aspect of another culture.

acid rain: The deposition of acidic materials from the atmosphere on to the Earth's surface.

African diaspora: The migration of black peoples out of Africa.

agglomeration: The bringing of people and activities together in one place for greater convenience.

agricultural revolution: The application of science and technology to agriculture, resulting in greatly increased yields and releasing workers.

alphabet: A system of writing in which letters represent sounds.

animism: A belief in the ubiquity of spirits or spiritual forces.

anthropogenic: Human-caused.

apartheid: A policy of racial segregation enforced in South Africa 1948–1993.

aquaculture: Herding or domesticating aquatic animals and farming aquatic plants.

aquifer: A porous, water-bearing layer of underground rock.

arithmetic density: The number of people per unit of area.

artifact: A material object of culture; literally, "a thing made by skill."

at-large system of representation: A representative political system in which all members of a governing body are chosen by the entire electorate, not from districts.

autarky: Economic independence.

B

Bantustan: One of ten segregated areas into which the former white-dominated government of South Africa intended to move all black Africans.

basic sector: The part of a city's economy that is producing exports.

behavioral geography: The study of how we perceive our environment and of how our thoughts and perception influence our behavior.

biome: A large recognizable assemblage of plants and animals in functional interaction with its environment.

biomimetics: The study of biological materials as models for creating artificial ones.

biophilia hypothesis: The theory that the years of evolution during which humans constantly and intimately interacted with nature imbued us with a deep, genetically-based emotional need to affiliate with the rest of the living world.

bloc: A group of countries that presents a common policy when dealing with other countries.

brain drain: The emigration of a country's best-educated people and most skilled workers.

built environment: Houses and other relatively durable cultural landscape artifacts.

C

capital-intensive activity: An activity in which a large amount of capital is invested per worker.

carrying capacity: The ability of agricultural land to support population.

cartogram: A map-like image designed to convey the magnitude of something rather than exact spatial locations.

caste: A group in the rigid social hierarchy of Hinduism.

census: An official enumeration of the population; national censuses usually include additional information relating to social and economic conditions.

central business district (CBD): The traditional core of the city, concentrating office buildings and retail shops.

central place theory: A model of the distribution of cities across an isotropic plain.

centrifugal forces: Forces that tend to pull states apart.

centripetal forces: Forces that bind a state together.

challenge-response theory: The theory that people need the challenge of a difficult environment to put forth their best effort and to build a civilization.

city: A concentrated nonagricultural human settlement.

climatic climax vegetation: A stable association of vegetation in equilibrium with local soil and climate.

cognate: A word that clearly looks or sounds like one in another language to which it is related historically.

cognitive behavioralism: The theory that people react to their environment as they perceive it.

coincidental: Occurring together.

commercial agriculture: Raising food to sell.

commercial revolution: A tremendous expansion of global trade between about 1650 and 1750.

common market: A customs union within which common laws create similar conditions of production.

507

Confucianism: A philosophical system based on the teachings of K'ung Fu-tzu (c. 551–479 B.C.).

congregation: Residential clustering by choice.

consolidated metropolitan statistical area (CMSA): Two or more contiguous MSAs.

continental shelf: An area of relatively shallow water that surrounds most continents before the continental slope drops more sharply to the deep sea floor.

core area: The historic homeland of a nation.

corporativism: Government by representatives of interest groups.

creole: A pidgin that has survived long enough to become a mother tongue.

crude birth rate: The annual number of live births per 1,000 people.

crude death rate: The annual number of deaths per 1,000 people.

crustal limit: The total amount of a mineral in the Earth's crust.

cultural diffusion: The spreading of cultural attributes.

cultural ecology: The study of the ways societies adapt to environments.

cultural geography: The study of the geography of human cultures.

cultural hearth: The place where a distinctive culture originated.

cultural imperialism: The substitution of one set of cultural traditions for another, either by force or by degrading those who fail to acculturate and rewarding those who do.

cultural landscape: A landscape that reveals the many ways people modify their local environment.

cultural mosaic: A phrase describing Canada and suggesting that various cultural groups retain individuality there.

cultural realm: The region throughout which a culture prevails.

cultural subnationalism: The splitting of a state population among several local primary allegiances.

culture: A bundle of attributes of shared behavior or belief. These may include virtually anything about the way a people live.

customs union: A free trade area that enforces a common external tariff.

D

deductive geography: The approach to understanding geography that starts from general assumptions about reality, creates models and then compares real-world examples with those models.

dematerialization: The reduction of the amount of raw material necessary to manufacture goods.

demographic transition model: A model that describes the historical experience of population growth in the countries that are today rich.

demography: The analysis of a population in terms of specific characteristics, such as age or income levels.

dependency ratio: The ratio of the combined population less than 15 years old and adult population over 65 years old to the population of those between 15 and 64 years of age.

desertification: The expansion of desert conditions into areas that previously were not deserts, usually as a product of human-induced environmental degradation superimposed on a natural drought situation.

dialect: A minor variation within a language.

dialectology: The systematic study of dialects.

dialect geography (linguistic geography): The study of dialects across space.

diffusionism: The theory that aspects of civilization were developed in very few places and then diffused from those places.

diminishing returns: Diminishing returns exist when, in adding equal amounts of one factor of production, such as fertilizer or labor, each successive application yields a smaller increase in production than the application just preceding.

distance decay: The fact that the presence or impact of any cultural attribute usually diminishes away from its hearth area.

diversity index: The number of different plant or animal species per unit of area.

domestication: The process of adapting plants and animals to intimate association with humankind, to the advantage of humankind.

dominium: The principle of rule over a defined territory.

doubling time: The number of years it would take any country's population to double at present rates of increase.

downstream activities: Economic activities away from raw materials and closer to goods' ultimate consumers.

dump: To sell an item for less than it costs to produce.

E

ecological niche: A particular combination of physical, chemical, and biological factors that a species needs in order to thrive.

economic geography: The study of how various peoples make their living and what they trade.

economic man: A theoretical actor who, in any situation, has all the information necessary to make decisions, is completely rational, and tries to minimize the effort he expends to achieve any goal and to maximize the profitability of any action.

economies of scale: The economic factors determining that as the number of units of a good produced increases, the production cost per unit generally falls.

ecosystem: All the organisms in a given area and the totality of interactions among the organisms and between the organisms and the nonliving portion of the environment.

ecotourism: Travel to see distinctive examples of scenery, unusual natural environments, or wildlife.

electoral geography: The study of voting districts and voting patterns.

ejido: A Mexican form of land tenure in which a peasant community collectively owns a piece of land and the natural resources and houses on it.

electronic highway: Electronic methods of communication, such as e-mail, facsimile machines, and computer modems.

emigration: Movement away from a place.

enculturation (socialization): Teaching youngsters a society's values and traditions, its political and social culture.

endemic: A state of equilibrium between humans and a microorganism whereby the latter is widespread in a community without causing extensive death.

environmental determinism: The simplistic belief that human events can be explained entirely as the result of the effects of the physical environment.

environmentalist theories: Theories that emphasize the role of the environment in human life.

epidemic: A sudden and virulent outbreak of a disease.

epidemiological transition: A shift in the most common causes of death in a society as that society accumulates wealth.

epidemiology: The study of the incidence, distribution, and control of disease.

equator: The Earth's imaginary midline everywhere equidistant between the poles.

ethnic group: A dubious term suggesting that a particular cultural group is in the minority or "not normal" for a particular place and time.

ethnocentrism: A tendency to judge foreign cultures by the standards and practices of one's own, and usually to judge them unfavorably.

etymology: The study of the origin and history of words.

European Economic Area (EEA): A 17-nation free trade area.

European Union (EU): A bloc of 12 European countries (scheduled to be 16 in 1995) enjoying free trade and committed to eventual full political union.

Evangelical Protestantism: Protestant groups that most actively spread faith in individuals' ability to change their lives (with God's help).

evolutionism: The theory that a culture's sources of change were embedded in the culture from the beginning, so the course of development was internally determined.

exclusive economic zone (EEZ): A 200 mile zone within which a coastal state controls both mining and fishing rights.

export-led economic growth: A national economic policy of welcoming foreign investment to build factories that will manufacture goods for international markets.

external economies: The range of goods and services in a city that can be rented or hired temporarily.

externalities: Any ramifications of an activity that do not rebound directly to the responsible party and are not, therefore, accounted for in that party's costs or profit.

external refugees: People who seek refuge in foreign countries.

exurbs: The settlements that make up the outermost ring of expanding metropolitan areas.

F

federal government: A form of government in which a central government shares power with subunits.

feng-shui: The placement and design of temples, gravesites, homes, business establishments, and even whole cities to guarantee good luck or happiness.

fertility rate: The number of children born per year per 1,000 females.

First World: During the cold war, all reasonably well-to-do, non-Communist countries.

fishery: A concentration of aquatic species suitable for commercial harvesting.

folk culture: Culture that preserves traditions.

food chain: A series of linkages in which one organism eats another; the second organism is consumed by a third, and so on.

footloose activities: Activities that can move or relocate freely.

foreign direct investment (FDI): Investment by foreigners in wholly-owned enterprises that are operated by the foreigner.

free trade area: An international territory having no internal tariffs, but its members are free to set their own tariffs on trade with the rest of the world.

friction of distance: The effort, time, or cost necessary to move or transport items.

frontier: An undeveloped region that may offer potential for settlement.

fundamentalism: The strictest adherence to traditional religious beliefs.

G

garden or park: A landscape that humans have molded according to aesthetic notions.

gentrification: The occupation and restoration of select urban residential neighborhoods by successful urban white collar workers.

Geographic Information System (GIS): The use of computer systems to organize, store, update, and map information, and to investigate those data.

geographical dialect continuum: A chain of dialects or languages spoken across an area; speakers at any point in the chain can understand people who live in adjacent areas, but they find it difficult to understand people who live farther along the chain.

geomancy: The attempt to foretell the future by means of studying configurations of landmarks.

geopolitics: A pseudo-science studying "the natural and necessary trend towards [national] expansion as a means of self-preservation;" today often loosely applied to studies of military strategies.

gerrymandering: The drawing of voting district lines in ways that include or exclude specific groups of voters, so that one group gains an unfair advantage.

ghetto: A segregated residential community.

globalization: The organization of any activity treating the entire globe as one place.

government: The people actually in power at any time.

gravity model: A mathematical formula to express the gravitational force that one object exerts on another; social scientists seek analogous formulas in their study of human activities.

green revolution: The development of higher-yielding and hardier strains of crops through advances in botanical science.

greenhouse effect: The atmosphere's effect of allowing short-wave radiation from the sun to pass through to the Earth's surface but inhibiting the escape of long-wave radiation into space. This heats the lower atmosphere.

Grimm's Law: The rules describing the regular shifts in sounds that occurred when the Indo-European languages diverged from one another.

gross domestic product (GDP): The total value of all goods and services produced within a country.

gross national product (GNP): The total income of a country's residents, no matter where it comes from, minus any amount paid out to foreigners abroad.

gross sustainable product (GSP): An experimental statistic that subtracts from gross product the value of natural resources that are destroyed or depleted.

groundwater: The water in an aquifer.

growth poles: Cities a government favors for economic growth.

H

hajj: A Muslim pilgrimage to Mecca.

hinterland: The region to which any city provides services and upon which it draws for its needs.

historical consciousness: A people's consciousness of past events insofar as that consciousness influences their present behavior.

historical geography: The study of the geography of the past and how geographic distributions have changed.

historical materialism: The belief that technology has historically increased humankind's control over the environment and improved material welfare, and that this improvement prompts other historical events and movements.

human development index (HDI): A statistic that combines purchasing power, life expectancy, and adult literacy to compare the quality of life around the world.

human geography: The study of the geography of human groups and activities.

hunter-gatherers: Primitive people who live on what they can hunt or harvest from the Earth.

hydrologic cycle: The continuous interchange of moisture between the Earth and atmosphere.

hypothesis: A proposition tentatively assumed in order to draw out all of its consequences and so test its accord with all facts an investigator can gather.

I

iconography: A state's set of symbols, including a flag and anthem.

ideogram: In writing, a character that represents a word or concept.

immigration: Movement into a place.

import-substitution economic growth: A national economic policy of protecting domestic infant industries.

incorporation: The process of defining a city territory and establishing a government.

indigenous peoples: As defined by the UN, "descendants of the original inhabitants of a land who were subjugated by another people coming after them."

indirect rule: The imperialist use of native rulers as intermediaries between themselves and the people.

inductive geography: The approach to understanding geography that begins with specific cases and leads up to general conclusions.

industrial revolution: The evolution first in Europe between about 1750 and 1850 from an agricultural and commercial society to an industrial society relying on inanimate power and complex machinery.

industrial society: A society with a significant share of its output from the secondary sector.

inertia: The force that keeps things stable or fixed in place.

infant industries: Newly developing industries that probably cannot compete with imports.

infant mortality rate: The number of infants per thousand who die before reaching 1 year of age.

informal or underground sector: The economic activities that do not appear in official accounts.

infrastructure: Fixed assets in place, such as buildings, dams, and roads.

innocent passage: The internationally-guaranteed right of the ships of one state to pass through the territorial waters of another on the way to a third.

interculturation: Blending two cultures, as in efforts in India to incorporate pre-Christian cultural and Hindu concepts into Roman Catholic practice.

internal economies: Those goods and services that a large company can provide for itself.

internal refugees: People who are forced to flee from one to another region of a country.

international organization: An organization that coordinates activities among two or more countries.

irredenta: Territory that one state claims from another.

isogloss: A line around places where speakers use a linguistic feature in the same way.

isotropic plain: In deductive geographic studies, a theoretical perfectly flat surface with absolutely no variations across it.

J

jus sanguinis: The definition of one's citizenship based on the citizenship of his or her ancestors.

jus soli: The definition of one's citizenship based on his or her place of birth.

K

kinesics: The study of nonverbal visual communication.

L

labor-intensive activity: An activity that employs a high ratio of workers to invested capital.

laissez-faire capitalism: A capitalist system that minimizes the government's role in the economy.

language: A set of words, plus their pronunciation and methods of combining them, that is used and understood to communicate within a group of people.

language family: The languages which are related by descent from a common protolanguage.

latitude: The location of a place measured as angular distance north and south of the equator.

law: A dubious term often applied to a theory that has withstood testing over a long time.

liberation theology: The belief in putting the problems of overcoming poverty at the heart of Christian theology.

life expectancy: The average number of years that a newborn baby within a given population can expect to live.

lingua franca: A second language held in common for international discourse.

liquidity: The ability of an asset to be easily converted from one form to another; cash, for example, is more liquid than a house.

living language: A language actually used today.

local content requirement: A definition of the percentage of the total value of a good entering one country that must have been added in the second country for that product to qualify as a product of the second country.

locational determinants: The specific factors considered when people choose where to locate an activity.

longitude: The location of a place measured as angular distance east and west from the prime meridian.

M

Maori: The native people of New Zealand.

man the satisficer: A theoretical actor who does not always seek to maximize profit or return, but seeks only to gain satisfaction or a satisfactory return.

map projection: A method of portraying the Earth or any portion of it on a flat map.

market-oriented manufacturing: Manufacturing that locates close to the market either because the processing increases the perishability of the product or because the processing adds bulk or weight to the product.

Marshall Plan: A plan of financial aid for the economic rehabilitation of Europe after World War II, named for U.S. Secretary of State George C. Marshall.

material-oriented manufacturing: Manufacturing that locates close to the source of the raw material either because the raw material is heavy or bulky or because the raw material is perishable.

melting pot: Name given to the United States in a 1914 novel of that title, suggesting that ethnic and racial differences among immigrants melt together to form one culture.

mental map: The ideas that we have about places, whether our ideas are true or false.

mentifact: The mental aspects of culture, such as a group's poetry, music, language, and religion.

meritocracy: A society in which the most capable people can rise to the top on merit alone.

mestizo: A person of either mixed white-native American or white-native Philippino ancestry.

metropolitan statistical area (MSA): According to the U.S. Census Bureau, "an integrated economic and social unit with a recognized large population nucleus."

microclimate: Local climatic conditions, as in cities, that differ from those of surrounding areas.

migration chain: A network of social and communication linkages that attracts migrants to follow others.

model: An idealized, simplified representation of reality.

monotheism: Belief in the existence of one God.

monsoon: A seasonal reversal of winds, accompanied by distinct precipitation patterns.

mulatto: A person of mixed white-black ancestry.

multinational or transnational corporation: An enterprise that produces and markets goods in several countries.

multinational state: A state containing several nations.

multiplier effect: The fact that jobs in a city's basic sector multiply jobs in the nonbasic sector.

N

nation: A group of people who want to have their own government and rule themselves.

national self-determination: The idea of the nation-state as defended by U.S. President Woodrow Wilson after World War I.

nation-state: A state ruling over a territory containing all the people of a nation and no others.

nativism: An attitude or policy favoring the native inhabitants of a country as against immigrants.

natural landscape: A landscape without evidence of human activity.

natural population increase (decrease): The difference between the number of births and the number of deaths.

network hypothesis: The theory that central city unemployment is caused by a lack of social networks.

nonbasic sector: The part of a city's economy serving the needs of the city itself.

North Atlantic Treaty Organization (NATO): A military bloc founded in 1949 and today including Belgium, Denmark, France, Great Britain, Iceland, Italy, Luxembourg, the Netherlands, Norway, Portugal, Canada, the United States, Greece, Turkey, Germany, and Spain.

O

official language: The language in which government business is conducted and official records are kept.

oil shales and tar sands: Geological formations that contain oil, but which must be treated to release the oil.

opportunity cost: A return sacrificed by leaving capital invested in one form or activity rather than another.

orthography: The study of writing, or a system of writing.

Overseas Chinese: Chinese migrants who often remain linked across international borders by bonds of family, clan, and common home province.

P

pandemic: A worldwide epidemic.

pastoral nomadism: A style of life that does not have fixed residences, but drives flocks from place to place to find grazing lands and water.

pathetic fallacy: A figure of speech that either personalizes nature by giving it the ability to think or to act, or else that suggests that nature encourages specific feelings in people.

pathogen: A disease-causing organism.

physical geography: The study of the characteristics of the physical environment.

physiological density: The density of population per unit of arable land.

pidgin: A system of communication which has grown up among people who do not share a common language, but who want to talk with each other.

plate tectonics: The theory that the upper portion of the Earth's crust consists of a mosaic of rigid plates embedded in an underlying, somewhat plastic layer.

political community: A willingness to join together and form a government to solve common problems.

political culture: The set of unwritten ways in which written rules are interpreted and actually enforced.

political economy: The study of individual countries' organization and regulation of their economies.

political geography: The study of the interaction between political processes and the distributions of all other activities and transformations of the landscape.

polyglot state: A country that grants legal equality to two or more languages.

polytheism: The worship of many gods.

popular culture: The culture of people who embrace innovation and conform to changing norms.

population geography: The study of the distribution of humankind across the Earth.

population projection: A forecast of the future population, assuming that current trends remain the same or else change in defined ways.

population pyramid: A graphic device that shows the shares of a nation's population by age groups.

positive checks: According to Thomas Malthus, premature deaths of all types, such as those caused by war, famine, and disease.

possibilism: The theory that the physical environment itself will neither suggest nor determine what people will attempt, but it may limit what people can profitably achieve.

post-industrial society: A society with the bulk of its economic activity in the tertiary sector.

pre-industrial society: A society with the bulk of its economic activity in the primary sector.

preventive checks: According to Thomas Malthus, human actions designed to limit population growth.

prime meridian: The meridian passing through the Royal Observatory in Greenwich, England, from which longitude is measured.

primate city: A large city concentrating a national population or national political, intellectual, or economic life.

primary sector: That part of the economy that extracts resources directly from the Earth, including agriculture, fishing, forestry, and mining.

prior allocation: A legal system under which the first user of water from a stream establishes a legal right for continued use of the amount originally withdrawn.

privatize: Give or sell government assets to private individuals or investors.

proselytize: Try to convert others to your religious beliefs.

proto-Indo-European: The common ancestor of all Indo-European languages.

protolanguage: The ancestor that is common to any group of several of today's languages.

proxemics: The study of how people perceive and use space.

pull factors: Considerations that attract people to new destinations.

push factors: Anything that makes a person want to leave a place and seek a better life elsewhere.

R

race: A relatively minor biological difference within one species.

racism: The false belief in the inherent superiority of one race over another and the linking of human ability, potential, and behavior to racial inheritance.

recyclable resources: Resources that can be reused.

refugee: As defined by the 1951 Geneva Convention, someone with "a well-founded fear of being persecuted in his country of origin for reasons of race, religion, nationality, membership of a particular social group or political opinion."

regime: A general set of rules and regulations for governing, usually formalized in a constitution.

region: A territory that exhibits a certain uniformity.

regional geography: An inventory analysis of all characteristics of any individual place.

regnum: Government over a group of people rather than a defined territory.

relative location: The location of a place relative to other places.

renewable resources: Animal or vegetable resources, such as fish or forests, that naturally renew themselves and can be harvested at that rate.

replacement rate: A total fertility rate of 2.1, which stabilizes a population.

reserve or preserve: An area that has been set aside to save a natural environment, including, usually, wildlife.

reserves: The amounts of a resource that can economically be recovered for use.

resource (in general): Anything which can be consumed or put to use by humankind.

resources of a mineral: The amount of the crustal limit that is currently or potentially extractable.

riparian rights: A legal system giving to anyone whose land adjoins a flowing stream the right to use water from the stream, as long as some water is left for downstream owners.

Russification: Russian attempts to acculturate the other peoples in the former USSR.

S

sacerdotalism: The belief that a church or priest intercedes between God and humankind.

scientific method: A step-by-step technique of investigating phenomena.

scientific revolution in agriculture: A complex of factors including new pesticides, fertilizers and even genetic engineering of crops to improve agricultural productivity.

Second World: During the cold war, the Soviet bloc dominated by totalitarian Communist governments.

secondary sector: That part of the economy that transforms raw materials into manufactured goods.

sectoral evolution: A shift in the concentration of activity from an economy's primary sector to its secondary and tertiary sectors.

secularism: A lifestyle or policy that deliberately ignores or excludes religious considerations.

segregation: Residential clustering as the result of discrimination.

service sector: That part of the economy that services the primary and secondary sectors.

shamanism: A belief in the power of mediums (shamans) who characteristically go into autohypnotic trances, during which they are thought to be in communion with the spirit world.

Shiite Muslims: Muslims who believe that Muhammad's son-in-law Ali should have been the first caliph.

Shinto: The ancient native religion of Japan.

site: The characteristics of the absolute location of a place.

situation: The characteristics of the relative location of a place.

social Darwinism: The theory of British sociologist Herbert Spencer that "Nature's law" calls for "the survival of the fittest," even among cultures and whole peoples.

sociofact: The ways culture groups do things, such as interpersonal arrangements, the family, and education.

soil: An infinitely complex and varied mixture of disintegrating mineral particles, decaying organic matter, living organisms, gases, and liquid solutions.

sojourner: A migrant who intends to stay in a new location only long enough to save capital to return home to a higher standard of living.

spatial mismatch hypothesis: The hypothesis that central city unemployment is caused by the removal of job opportunity to the suburbs and the concentration of the poor in central city low-income housing.

speech community: A group of people who speak together.

standard language: The way any language is spoken and written according to formal rules of diction and grammar.

state: An independent political unit that claims exclusive jurisdiction over a defined territory and over all of the people and activities within it.

state-directed capitalism: A capitalist system in which the government does not own many enterprises outright, but it plans and regulates the economy.

strategic minerals: Minerals that are particularly rare and important to industrial processes.

subculture: A smaller bundle of attributes shared among a smaller group within a larger, more generalized culture group.

subsistence agriculture: Raising food only for oneself, not to sell.

Sunni Muslims: Muslims who approve the historic order of Muhammad's first four successors; today about 85 percent of all Muslims.

superimposed boundaries: Boundaries imperialists drew to establish colonies over natives' territorial demarcations.

supranational organization: An organization that exercises power over countries.

syncretic religions: Religions that combine Christianity with traditional practices.

T

Taoism: A philosophy derived from the book *Tao-te Ching* attributed to a legendary Chinese philosopher Laotze; it advocates a contemplative life in accord with nature.

tap route: A transport network that consists of lines penetrating into the country from its ports.

telecommuting: Working at home at a computer terminal connected to an office.

temperature inversion: An abnormal condition during which temperature increases at higher altitudes.

territoriality: A drive—possibly biological—to lay claim to territory and defend it against members of one's own species.

tertiary sector: That part of the economy that services the primary and secondary sector.

theocracy: A form of government where a church rules directly.

theory: A hypothesis that has been given probability by experimental evidence.

Third World: During the cold war, the nonwhite countries seen as characterized by underdevelopment, overpopulation, irrationality, religion, and political chaos.

threshold: The minimum number of potential customers that is needed for a product or service to be offered.

threshold principle: The belief that a nation must have some minimal population and territory to merit self-determination.

topical or systematic geography: The study of universal laws or principles that apply to all places; topics may be as diverse as the geography of soils (pedology), of life forms (biogeography), of politics (political geography), of economic activities (economic geography), and of cities (urban geography).

topography: Surface relief.

toponymy: The study of place names.

topophilia: "Love of place;" the study of how people evaluate landscapes according to personal emotional or sentimental criteria.

total fertility rate: The number of children an average woman in a given society would have over her lifetime.

U

undocumented, irregular, or illegal immigrants: People who cross borders without completing legal papers.

unitary government: A form of government in which the balance of power lies with the central government.

urban enterprise zones: Areas within which governments create generous conditions for enterprises to encourage the creation of jobs.

urban geography: The geographic study of cities.

urban heat island: A dome of heat over a city created by urban activities and conditions.

urbanization: The process of concentrating people in cities.

V

value added by manufacturing: The difference between the value of a raw material and the value of a product manufactured from that raw material.

vector: An intermediate host of pathogens traveling from one human or animal to another.

virtual reality: A mental state—a theoretical "place"—created by intense involvement either with interactive electronic devices or else with distant people through electronic devices.

W

workers' remittances: Money that migrant workers send home from abroad.

Z

zambo: A person of mixed black-native American ancestry.

zero population growth: A stabilized total population.

Zionism: The belief that the Jews should have a homeland of their own.

zoning: Restricting or prescribing the use to which parcels of land may be put.

Index

515

516 *Index*

World: Physical

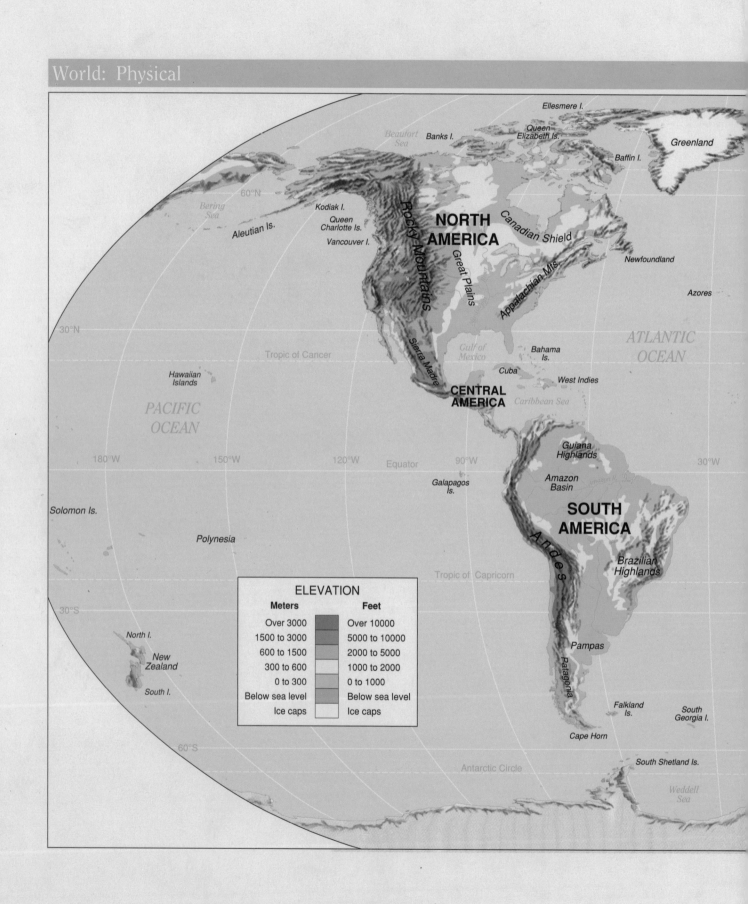

ELEVATION

Meters		Feet
Over 3000		Over 10000
1500 to 3000		5000 to 10000
600 to 1500		2000 to 5000
300 to 600		1000 to 2000
0 to 300		0 to 1000
Below sea level		Below sea level
Ice caps		Ice caps